Perspectives in Nucleoside and Nucleic Acid Chemistry

WILEY-
VCH

Perspectives in Nucleoside and Nucleic Acid Chemistry

M. Volkan Kisakürek
Helmut Rosemeyer (Eds.)

Verlag Helvetica Chimica Acta · Zürich

Weinheim · New York · Chichester
Brisbane · Singapore · Toronto

Dr. Helmut Rosemeyer
Laboratorium für Organische
und Bioorganische Chemie
Universität Osnabrück
Barbarastrasse 7
D-49069 Osnabrück

This book was carefully produced. Nevertheless, editor and publishers do not warrant the information contained therein to be free of errors. Readers are advised to keep in mind that statements, data, illustrations, procedural details, or other items may inadvertently be inaccurate.

Published jointly by
VHCA, Verlag Helvetica Chimica Acta, Zürich (Switzerland)
WILEY-VCH, Weinheim (Federal Republic of Germany)

Editorial Directors: Dr. M. Volkan Kisakürek, Pekka Jäckli
Production Manager: Birgit Grosse

Cover Design: Bettina Bank

Library of Congress Card No. applied for.

A CIP catalogue record for this book is available from the British Library.

Die Deutsche Bibliothek – CIP-Cataloguing-in-Publication-Data

A catalogue record for this publication is available from Die Deutsche Bibliothek

ISBN 3-906390-21-7

Printing: Konrad Triltsch, Print und Digitale Medien, D-97199 Ochsenfurt-Hohestadt
Printed in Germany

Preface

The chemistry of nucleosides and nucleic acids continues to be a rapidly developing field of study. Many of the most important recent advances in medicinal chemistry, such as the application of the antisense and antigene concept to drug discovery, have occurred in this field with the development of novel nucleoside- and nucleotide-based antiviral and antitumor drugs. The realization of atypical base-pairing interactions involving natural as well as synthetic nucleosides has extended the genetic alphabet and expanded horizons in molecular-recognition applications. New synthesis strategies, novel protecting groups and solid supports, and combinatorial chemistry have helped spur progress in drug-development and genome-sequencing research.

This volume, comprised of contributions written by internationally recognized experts, covers cutting-edge developments in recent nucleoside and nucleic acid research. The majority of the manuscripts were originally dedicated to Prof. Dr. *Frank Seela*, University of Osnabrück, on the occasion of his 60th birthday and have appeared in *Helvetica Chimica Acta*. The topics covered by the contributions consist of the most-recent synthesis innovations, including novel protecting-group and solid-support strategies and combinatorial approaches, nucleic acid labeling for the development of reporter molecules, spectroscopy and structural studies, investigations of enzymatic reaction mechanisms, thermodynamic and computational investigations, stability assessments, and medicinal applications. The comprehensive scope of this collection will be valued by *synthetic*, *physical*, *organic*, *bioorganic*, and *medicinal* chemists working on all aspects of nucleoside and nucleic acid research. This book will also prove an invaluable introduction to the state-of-the-art in nucleoside and nucleic acid chemistry to aid researchers new to the field.

October 2000

M. Volkan Kısakürek
Helmut Rosemeyer

Contents

Part I. Nucleosides and Nucleotides

Part II. Oligonucleotides

The Synthesis of Bicyclic N^4-Amino-2'-deoxycytidine Derivatives

by **David Loakes**, **Rita Bazzanini**[1]), and **Daniel M. Brown***

Medical Research Council, Laboratory of Molecular Biology, Hills Road, Cambridge, CB2 2QH, UK
(Tel: (+ 44) 1223 248011; Fax: (+ 44) 1223 412178; e-mail: dmb@mrc-lmb.cam.ac.uk)

Nucleosides which have ambivalent tautomeric properties have value in a variety of nucleic acid hybridization applications, and as mutagenic agents. We describe here synthetic studies directed to stable derivatives of this kind of nucleoside based on N^4-aminocytosine. Treatment of the 4-(1H-1,2,4-triazol-1-yl)-5-(chloroethyl)pyrimidinone nucleoside derivative **5** with hydrazine leads to formation of the 6,6-bicyclic pyrimido-pyridazin-7-one **3**, and with methylhydrazine to the corresponding fixed tautomeric 1-methyl derivative **7** (*Scheme 1*). If these cyclization reactions are carried out in the presence of a base, the 6-ring bicyclic derivatives undergo rearrangement to their corresponding 5-ring pyrrolo-pyrimidin-2-one analogues **8** (*Scheme 2*). In the reaction of the triazolyl derivative **5** with 1-[(benzyloxy)carbonyl]-1-methylhydrazine, spontaneous cyclization gives the 5-ring derivative **13** related to **8** rather than the open-chain product **12** (*Scheme 4*). Reaction of an acetylated analogue of triazolyl derivative **5** with 1,1-dimethylhydrazine gives rise to some of the open-chain product **9**, but it too cyclizes to a product that we have assigned the structure of the 6,6-ring quaternary ammonium salt **11** (*Scheme 3*).

Introduction. – Nucleosides which are capable of base-pairing with more than one of the natural DNA/RNA bases are mutagenic. Such analogues are of use not only to explore aspects of chemical mutagenesis [1][2], but also as tools in molecular biology. We have extensively examined the mutagenic nucleoside **1**, which behaves as both thymidine and deoxycytidine. Its 5'-triphosphate has been used for random mutagenesis [3][4], whilst in oligonucleotides, it has been used in primers for PCR [5] and to study H-bonding patterns in DNA duplexes [6][7]. The analogue, however, does not behave indiscriminately as either T or C, but has a bias towards T. For this reason, we have for some time been examining alternative analogues that may shift this balance. The 5,6-ring (ribo) analogue **2** was prepared [8] in the expectation that the smaller ring size might have an effect on the tautomeric ratio. This, whilst appearing to behave more as a cytidine analogue in its ^{1}H-NMR spectrum, proved to be too unstable to investigate further.

N^4-Hydroxycytosine derivatives have tautomeric constants (K_T) of the order of 10 with the imino (thymine-like) form predominating [9]. N^4-Aminocytosine derivatives exist predominately in the amino form, with K_T of around 30 in H_2O. Whilst these compounds are highly mutagenic *in vivo*, N^4-(alkylamino)cytosine derivatives are significantly less mutagenic [10]. This suggests that bicyclic analogues should show ambivalent base-pairing, but be less potent chemical mutagens. We, therefore, chose to

---

[1])   Present address: Consiglio Nazionale delle Recerche, I.Co.C.E.A. Area della Ricerca, Via P. Gobetti, 101, I-40129, Bologna.

investigate N^4-amino bicyclic analogues, such as **3**. The parent bicyclic compound **3** proved unstable; presumably it is susceptible to aerial oxidation. Therefore, we have attempted to prepare alkylated derivatives in the expectation that they would be stable, and that such compounds could then be used to investigate their ambivalent H-bonding behaviour in oligonucleotides.

Results and Discussion. – The 5-(2-chloroethyl)-2′-deoxyuridine derivative **4** [11] was converted to its C^4-triazolyl derivative **5**, which was then subjected to a series of reactions with various hydrazines. Thus, reaction with anhydrous hydrazine led to the rapid displacement of the C^4-triazolyl group, followed by a slower cyclization with the 5-(2-chloroethyl) group to give the bicyclic product **3** (*Scheme 1*). The product is rather unstable, readily degrading to a number of products, even if stored at 4°. Treatment of the C^4-triazolyl derivative **5** with methylhydrazine under similar conditions led to an essentially single product, though in a slower reaction. Although in equivalent displacement reactions the N-atom carrying the Me group is the more nucleophilic, we hoped to obtain some of the desired **6** [10]. The structure of the 3′,5′-di-O-acetyl rather than 3′,5′-di-O-toluoyl derivative was elucidated by NOE experiments (no NOE on irradiation of MeN (3.14 ppm), irradiation of CH$_2$N (t, 2.97 ppm) → enhancement of the exchangeable signal and NOE at CH$_2$(4) (2.51 ppm)). From this data, we deduced that the structure was not that of **6**, but of the regioisomer **7**.

It was observed that if the above cyclization reactions were carried out in the presence of a base, a second minor product was also formed (this product was also slowly formed on standing, particularly in solution). These minor products were rather difficult to separate from the first-formed 6,6-bicyclic products **3** and **7**. Therefore, **3** and **7** were treated with Et$_3$N or pyridine whereupon each was converted into this second product **8a** and **8b**, respectively, the methylated derivative **7** rearranging much slower than **3** (*Scheme 2*). Thus, the transformation of **3** in pyridine at 50° was complete after 16 h, whereas, under the same conditions, **7** had only reacted to *ca.* 50%. The rearrangement also occurred in 2,6-lutidine. The structures of **8a,b** were established as the 5-membered ring isomers from their ¹H-NMR spectra. The isomer **8a** was further characterized by conversion to its crystalline hydrazone with benzaldehyde, thus confirming the presence of the free NH$_2$ group.

The product **8a**, derived from **3**, showed a *s* (2 H) at 4.82 ppm for an unsubstituted NH$_2$ group. For the Me-substituted product **8b**, MeN group was a *d* (3.34 ppm), whilst the NH proton was a *q* (5.30 ppm). The MeN *d* collapsed to a *s* in the presence of D$_2$O.

Scheme 1

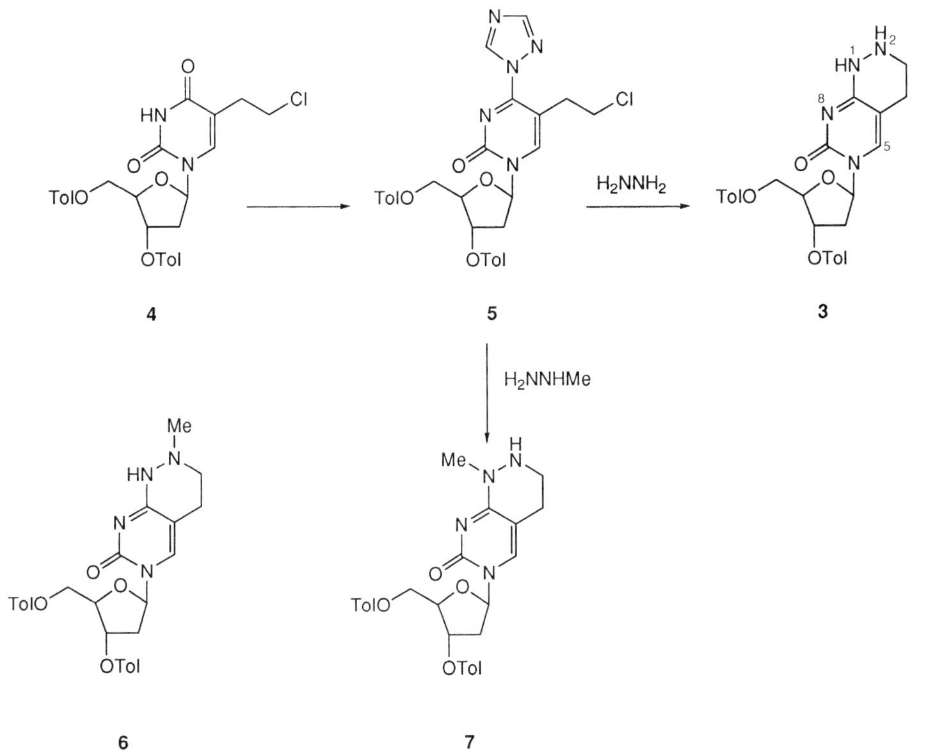

Scheme 2

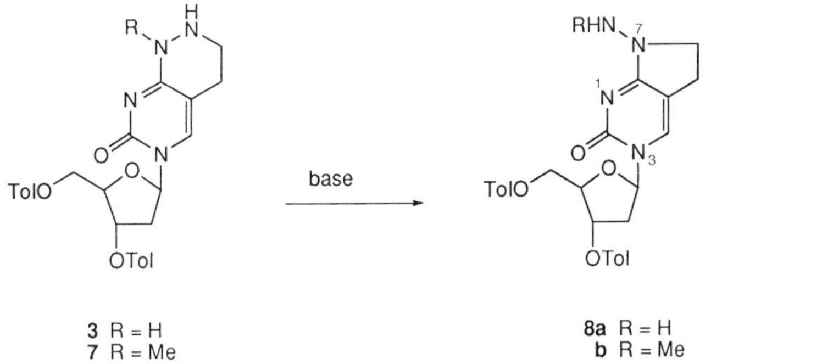

To investigate the formation of the 5-membered ring bicyclic analogues further, the 3′,5′-di-*O*-acetyl-substituted analogue of **5** was reacted with 1,1-dimethylhydrazine (*Scheme 3*). This was expected to give **9**, which we then planned to use to examine whether cyclization would occur leading to the 5-ring **10**. Displacement of the triazolyl group of the di-*O*-acetyl analogue of **5** by 1,1-dimethylhydrazine in tetrachloroethane at 100° overnight gave the expected product **9** in low yield, besides a major product which was polar, water-soluble, and very difficult to isolate in pure form. In an earlier

Scheme 3

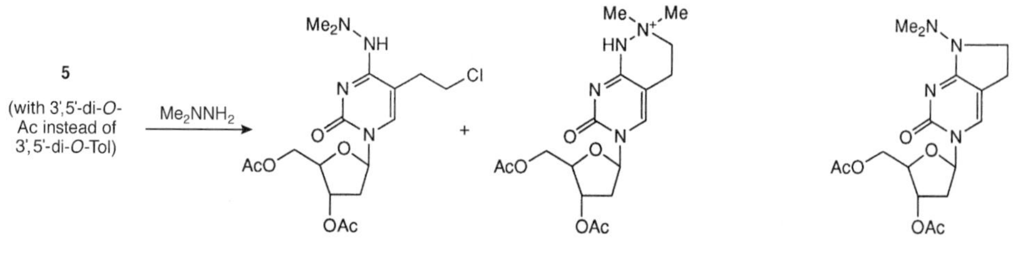

preliminary report [12], we suggested that this product was the quaternary ammonium salt **11**. This assignment was based on the fact that the product is water-soluble and polar: it has a mass spectrum with M^+ at m/z 381 and a ^{1}H-NMR spectrum corresponding to the proposed structure **11**. Isolation of the pure product has been attempted by a variety of methods including ion-exchange and reversed-phase HPLC, but has so far eluded us and is the subject of further work. It is anticipated that we may be able to demethylate the purified product to give the desired bicyclic di-*O*-acetyl analogue of **6**. Interestingly, we have no evidence, despite attempting the reaction many times under a variety of conditions, that ring contraction to the 5-ring analogue **10** occurred.

The ^{1}H-NMR spectrum of the salt **11** shows a *s* (6 H) at 3.15 ppm for Me$_2$N$^+$, whereas for the dimethylhydrazine product **9**, there is a *s* (6 H) at 2.50 ppm. This is consistent with the change in chemical shift for Me$_2$N to Me$_2$N$^+$. Thus, the data are entirely consistent with the structure of **11** being the 6-ring quaternary ammonium derivative.

As the methylamino residue of methylhydrazine is evidently the more nucleophilic group [13], it was decided to use a protected methylhydrazine derivative as an alternative route to produce the analogue **6**. Once deprotected, the methylamino residue could undergo cyclization with the 5-(chloroethyl) side chain. Thus, reaction of the triazolyl derivative **5** with 1-[(benzyloxy)carbonyl]-1-methylhydrazine [14] gave a product that we initially believed to be the expected compound **12** in a slow reaction (*Scheme 4*). On this assumption, we reductively removed the (benzyloxy)carbonyl (Z) group using either Pt or Pd in MeOH. However, the product that we obtained was not the desired product but again the 5-membered-ring derivative **8b**, as established by its ^{1}H-NMR spectrum, and comparison with the product **8b** obtained by ring contraction from **7** (see above). We therefore assumed that the product had preferentially cyclized or rearranged with N^4 as the (benzyloxy)carbonyl protecting group was being removed. Thus, the reduction was carried out in acid (AcOH or 0.1M HCl), as we hoped that protonation of the amino group would prevent ring closure after removal of the Z protecting group. This would then enable cyclization to occur to give the desired regiospecific product following neutralization. However, the product **14** obtained was again a 5-ring derivative related to **8**; moreover, under these conditions, the methylamino group had been reductively cleaved (*Scheme 4*).

Subsequently, we found that the intermediate (Z-protected) reaction product, initially assigned the structure **12**, was in fact the cyclized reaction product **13** (see

Scheme 4

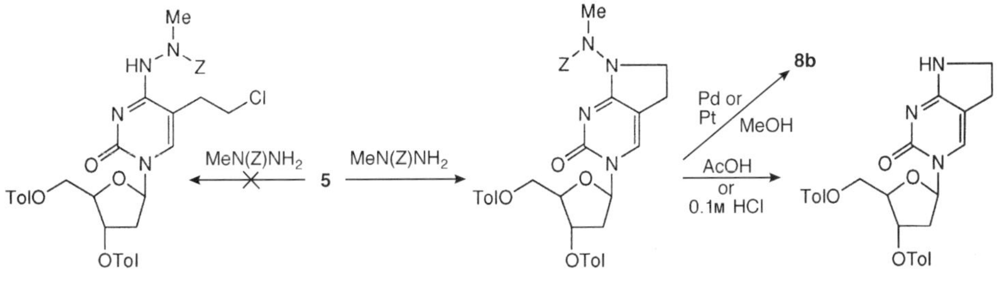

12   13   14

Scheme 4). This was confirmed by the MS of **13**, which showed an ion at m/z 675 ($[M +$ Na]$^+$), corresponding to the loss of HCl from **12**. Thus cyclization must have occurred after displacement of the triazole moiety and prior to reduction. The open-chain intermediate **12** was never observed in this reaction despite using a variety of alternative reaction conditions. The formation of the 5-membered ring bicyclic product is surprising. In this connection, when the triazole **5** was treated with ammonia, the cytidine derivative **15** was formed which did not undergo cyclization (*Scheme 5*). Neither did the *N*-hydroxycytidine derivative **16** [15] cyclize to give the corresponding bicyclic product. The presence of the Z group at the methylhydrazine renders it particularly unreactive, both in terms of the initial nucleophilic displacement of the C^4-triazolyl group and towards further cyclization.

Scheme 5

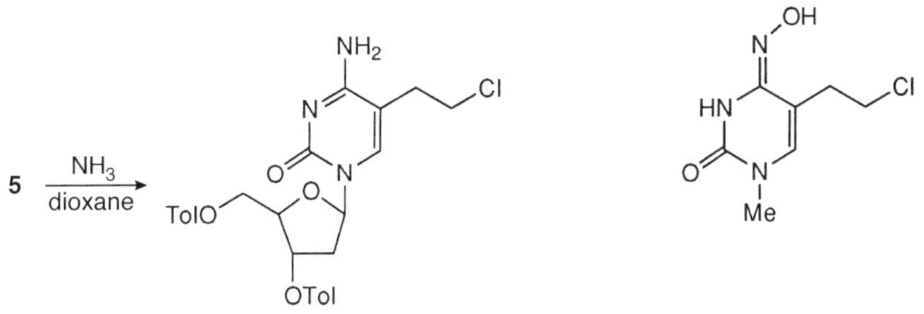

15   16

It is of interest to speculate on the nature of the formation of the bicyclic compound **13**. We have previously experienced the fact that the chloroethyl side chain is particularly unreactive towards nucleophilic displacement reactions, except for intramolecular cyclization. Therefore, the formation of **13** probably does not arise by first displacing chloride, followed by cyclization. However, the cyclization of the chloroethyl group to form a bicyclic compound is, nevertheless, a slow reaction, compared to the nucleophilic displacement of the C^4-triazolyl group. Although the Z-protected hydrazine is particularly unreactive, it appears that, once the initial reaction to displace the triazole moiety has occurred, cyclization with the chloroethyl group is spontaneous.

Lastly, we turn to the question of the mechanism whereby the 6,6-ring hydrazine derivatives of type **7** rearrange to 5,6-ring isomers, carrying an N^4-amino function in the cytosine moiety (see *Scheme 2*). The evidence is strong that the first-formed products are the 6,6-ring bicyclic compounds of type **7** and result from a displacement first of the triazolyl residue of **5** followed by ring closure. Rearrangement is base-catalysed and occurs in the cases where the N^4-amino N-atom carries a proton. Two mechanisms suggest themselves (*Scheme 6*). In the first, an intramolecular displacement at C(4) by the N^4-amino group occurs, *via* a strained transition entity **17**. Alternatively, if the base is acting as a nucleophile, an intermediate **18** may be postulated which would lead in a second displacement reaction to the thermodynamically and kinetically favoured 5-membered ring product. The fact that the rearrangement occurs in 2,6-lutidine strongly suggests that the first mechanism is the more likely route.

Scheme 6

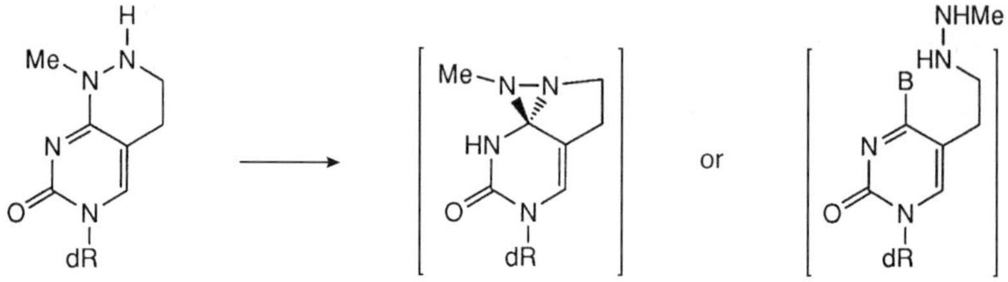

7   17   or   18

Experimental Part

General. Unless otherwise stated, reactions were worked up as follows: After removal of the solvent, the product was dissolved in CHCl₃ and washed with aq. sodium hydrogen carbonate soln. The combined org. fractions were dried (Na₂SO₄) and evaporated. TLC: pre-coated F_{254} silica gel plates. Column chromatography (CC): *Merck* silica gel *60* or reversed-phase column *LiChroprep RP-18* (*Merck*). M.p.: *Gallenkamp* melting point apparatus; uncorrected. UV Spectra: *Perkin-Elmer-Lambda-2* spectrophotometer; in 10% MeOH/H₂O unless otherwise stated; λ_{max} (ε) in nm. ¹H-NMR Spectra: *Bruker DRX 300*; in (D₆)DMSO, unless otherwise stated; δ in ppm, *J* values in Hz. NOE Experiments: *Bruker-AMX-500* spectrometer. Mass spectra: *Kratos MS890*; in *m/z* (rel. %).

5-(2-Chloroethyl) 1-[2-deoxy-(3,5-di-O-(p-toluoyl)-β-D-ribofuranosyl]-4-(1H-1,2,4-triazol-1-yl)pyrimidin-1(1H)-one (**5**). To a suspension of 1*H*-1,2,4-triazole (4.72 g, 68.3 mmol) in dry MeCN (100 ml) at 0°, phosphoric trichloride (1.27 ml, 13.7 mmol) was added dropwise, and the mixture was stirred at 0° for 15 min. After this time, dry Et₃N (11.5 ml, 82 mmol) was added and the mixture stirred at 0° for a further 20 min, then at r.t. for 10 min. A soln. of 5-(2-chloroethyl)-1-[2-deoxy-3,5-di-*O*-(*p*-toluoyl)-β-D-ribofuranosyl]pyrimidine-2,4-(1*H*,3*H*)-dione [11] (2.4 g, 4.5 mmol) in dry MeCN (10 ml) and DMF (10 ml) was then added dropwise with vigorous stirring. The mixture was kept under Ar at r.t. overnight and then evaporated. The residue was dissolved in CHCl₃, the soln. washed with aq. NaHCO₃ soln., dried, and evaporated, and the orange syrup purified by CC (silica gel, AcOEt/hexane 1:1): 2.27 g (86%) of **5**. White solid. UV: 323 (5200), 242 (33200), min. 216 (14200). M.p. 155–157°. ¹H-NMR: 2.28 (*s*, *Me*C₆H₄); 2.39 (*s*, *Me*C₆H₄); 2.59–2.68 (*m*, 1 H–C(2′)); 2.87–3.17 (*m*, 1 H–C(2′), CH₂*CH₂*Cl); 3.57 (*t*, *J* = 7.1, CH₂*CH₂*Cl₂); 4.56–4.74 (*m*, H–C(4′), 2 H–C(5′)); 5.62–5.65 (*m*, H–C(3′)); 6.28 (*t*, *J* = 6.5, H–C(1′)); 7.22, 7.36, 7.76, 7.93 (4 *d*, 8 arom. H (Tol)); 8.37 (*s*, H–C(6)); 8.40 (*s*, CH(triazole)); 9.36 (*s*, CH (triazole)). FAB-MS: 578.8 ([*M* + H]⁺). HR-MS: 578.18268 ([*M* + H]⁺, C₂₉H₂₉N₅O₆³⁵Cl⁺; calc. 578.18066; deviation –3.50 ppm).

6-[2-Deoxy-3,5-di-O-(p-toluoyl)-β-D-ribofuranosyl]-2,3,4,6-tetrahydropyrimido[4,5-c][1,2]pyridazin-7(1H)-one (**3**). To a soln. of **5** (0.25 g, 0.43 mmol) in dry MeCN (10 ml), anh. hydrazine (20 μl, 0.64 mmol) was added, and the soln. was stirred at r.t. for 2 h. The soln. was concentrated and chromatographed (5% MeOH/CHCl$_3$): 0.19 g (87%) of **3**. White solid. UV: 278 (10900), 244 (33300), min. 269 and 218; pH 1: 274 (10600), 240 (31850), min. 269 and 233; pH 12: 274 (10500), 240 (31750), min. 268 and 233. ^{1}H-NMR: 2.37 (*s*, *Me*C$_6$H$_4$); 2.38 (*s*, *Me*C$_6$H$_4$); 2.43–2.63 (*m*, CH$_2$, 2 H–C(2′)); 3.32 (br. *s*, NH); 3.49–3.69 (*m*, CH$_2$N); 4.44–4.48 (*m*, H–C(4′)); 4.50–4.63 (*m*, 2 H–C(5′)); 5.56 (br. *s*, H–C(3′)); 6.28–6.37 (*m*, H–C(1′)); 7.10 (*s*, H–C(5)); 7.31–7.36 (*m*, 4 arom. H (Tol)); 7.86–8.31 (*m*, 4 arom. H (Tol)); 9.64 (br. *s*, NH). FAB-MS: 505.9 ([*M* + H]$^+$). HR-MS: 505.20813 ([*M* + H]$^+$, C$_{27}$H$_{29}$N$_4$O$_6{}^+$; calc. 505.20871; deviation 1.10 ppm).

6-[2-Deoxy-3,5-di-O-(p-toluoyl)-β-D-ribofuranosyl]-2,3,4,6-tetrahydro-1-methylpyrimido[4,5-c][1,2]pyridazin-7(1H)-one (**7**). To a soln. of **5** (0.25 g, 0.43 mmol) in MeCN (10 ml), methylhydrazine (35 μl, 0.66 mmol) was added and the soln. stirred at r.t. for 8 h. The soln. was evaporated and the residue chromatographed (5% MeOH/CHCl$_3$): 0.19 g (85%) of **7**. Off-white foam. UV 285 (12100), 243 (38700), min. 218 and 273; pH 1: 303 (14950), 244 (41700), min. 216 and 269; pH 12: 285 (13200), 239 (37600), min 229 and 266. ^{1}H-NMR: 2.22–2.32 (*m*, 2 H–C(4)); 2.37 (*s*, *Me*C$_6$H$_4$); 2.38 (*s*, *Me*C$_6$H$_4$); 2.37–2.54 (*m*, 2 H–C(2′)); 2.88 (*m*, CH$_2$N); 3.12 (*s*, MeN); 4.46–4.48 (*m*, H–C(4′)); 4.51–4.65 (*m*, 2 H–C(5′)); 5.52 (*t*, *J* = 6.9, NH); 5.56–5.59 (*m*, H–C(3′)); 6.33 (*t*, *J* = 6.3, H–C(1′)); 7.31–7.37 (*m*, 4 arom. H (Tol)); 7.34 (*s*, H–C(5)); 7.86-7.92 (*m*, 4 arom. H (Tol)). FAB-MS: 519.8 ([*M* + H]$^+$). HR-MS: 519.22802 ([*M* + H]$^+$, C$_{28}$H$_{31}$N$_4$O$_6{}^+$; calc. 519.22437; deviation − 7.0 ppm).

In the same manner the 3′,5′-di-*O*-acetyl instead of 3′,5′-di-*O*-toluoyl derivative was prepared. White foam. UV: 292, 229, min. 250; pH 1: 305, min. 252. ^{1}H-NMR: 2.06 (*s*, Ac); 2.07 (*s*, Ac); 2.24–2.30 (*m*, 2 H–C(2′)); 2.51 (*m*, 2 H–C(4)); 2.97 (*t*, *J* = 5.7, CH$_2$N); 3.14 (*s*, MeN); 4.12–4.17 (*m*, H–C(4′)); 4.22–4.25 (*m*, 2 H–C(5′)); 5.15–5.19 (*m*, H–C(3′)); 5.59 (*t*, *J* = 5.7, NH); 6.24 (*t*, *J* = 6.7, H–C(1′)); 7.38 (*s*, H–C(5)). EI-MS: 366 ([*M* + H]$^+$). HR-MS: 366.1572 ([*M* + H]$^+$, C$_{16}$H$_{22}$N$_4$O$_6{}^+$; deviation 3.2 ppm).

7-Amino-3-[2-deoxy-3,5-di-O-(p-toluoyl)-β-D-ribofuranosyl]-3,5,6,7-tetrahydro-2H-pyrrolo[2,3-d]pyrimidin-2-one (**8a**). A soln. of **3** (200 mg, 0.6 mmol) in MeCN (10 ml) and Et$_3$N (1 ml) was heated under reflux overnight. Alternatively, a soln. of **3** was stirred in pyridine at r.t. overnight. After evaporation, the product was chromatographed (5% MeOH/CHCl$_3$): 146 mg (73%) of **8a**. White foam. UV: 283 (10400), 245 (32000), min. 270 and 215; pH 1: 286 (11100), 244 (29800), min. 270 and 215. ^{1}H-NMR: 2.37 (*s*, *Me*C$_6$H$_4$); 2.38 (*s*, *Me*C$_6$H$_4$); 2.44–2.58 (*m*, 2 H–C(2′), CH$_2$); 3.58 (*t*, *J* = 7.5, CH$_2$N); 4.22–4.23 (*m*, H–C(4′)); 4.50–4.63 (*m*, 2 H–C(5′)); 4.82 (*s*, NH$_2$); 5.54–5.56 (*m*, H–C(3′)); 6.36 (*t*, *J* = 7.8, H–C(1′)); 7.27 (*s*, H–C(4)); 7.32–7.36 (*m*, 4 arom. H (Tol)); 7.86–7.91 (*m*, 4 arom. H (Tol)). FAB-MS: 505.9 ([*M* + H]$^+$), 527.9 ([*M* + Na]$^+$). HR-MS: 527.18983 ([*M* + Na]$^+$, C$_{27}$H$_{28}$N$_4$O$_6$Na$^+$; calc. 527.19067; deviation 1.60 ppm).

Benzaldehyde Hydrazone of **8a**: After treatment of **8a** with benzaldehyde in CH$_2$Cl$_2$, the soln. was evaporated and the residue chromatographed (2% MeOH/CHCl$_3$): pale yellow solid which recrystallized from EtOH. M.p. 200–202°. UV: 327 (23900), 240 (21200), min. 272 and 222; pH 1: 334 (17800), 247 (19600), min. 296 and 226; pH 12: 328 (24200), 240 (21000), min. 272. ^{1}H-NMR: 2.35 (*s*, *Me*C$_6$H$_4$); 2.39 (*s*, *Me*C$_6$H$_4$); 2.43–2.63 (*m*, 2 H–C(2′)); 2.71–2.88 (*m*, CH$_2$); 3.94–4.00 (*m*, CH$_2$(6)); 4.51–4.58 (*m*, H–C(4′)); 4.59–4.67 (*m*, 2 H–C(5′)); 5.57–5.60 (*m*, H–C(3′)); 6.36 (*t*, *J* = 7.5, H–C(1′)); 7.31–7.37 (*m*, 4 arom. H (Tol)); 7.40–7.47 (3 H, *m*, 3 arom. H (Ph)); 7.63 (*s*, H–C(4)); 7.74 (1 H, *s*, 1 arom. H (Ph)); 7.77 (*s*, N=CH); 7.90 (*s*, 1 arom. H(Ph)); 7.86–7.95 (*m*, 4 arom. H (Tol)). FAB-MS: 593.24 ([*M* + H]$^+$). HR-MS: 593.23926 ([*M* + H]$^+$, C$_{34}$H$_{33}$N$_4$O$_6{}^+$; calc. 593.23999; deviation 1.20 ppm).

3-[2-Deoxy-3,5-di-O-(p-toluoyl)-β-D-ribofuranosyl]-3,5,6,7-tetrahydro-7-(methylamino)-2H-pyrrolo[2,3-d]pyrimidin-2-one (**8b**) (*Method A*). A soln. of **7** (150 mg, 0.29 mmol) in MeCN (10 ml) and Et$_3$N (0.5 ml) was heated under reflux overnight. After evaporation, the product was chromatographed (5% MeOH/CHCl$_3$) to give a white solid which was recrystallized from EtOH: 124 mg (83%) of **8b**. UV: 283 (10800), 244 (31900), min. 269 and 217; pH 1: 293 (12800), 244 (32900), min. 268 and 216; pH 12: 283 (11900), 240 (31700), min. 265 and 229. M.p. 175–176°. ^{1}H-NMR: 2.37 (*s*, *Me*C$_6$H$_4$); 2.38 (*s*, *Me*C$_6$H$_4$); 2.49–2.65 (*m*, 2 H–C(2′), CH$_2$(5)); 3.34 (*d*, MeN, *s* after D$_2$O wash); 3.59 (*t*, *J* = 7.5, CH$_2$(6)); 4.43–4.63 (*m*, H–C(4′), 2 H–C(5′)); 5.30 (*q*, *J* = 5.7, exchangeable NH); 5.50–5.60 (*m*, H–C(3′)); 6.34 (*t*, *J* = 7.6, H–C(1′)); 7.31–7.36 (*m*, H–C(4), 4 arom. H (Tol)); 7.85–7.91 (*m*, 4 arom. H (Tol)). FAB-MS: 519.3 ([*M* + H]$^+$). HR-MS: 519.22380 ([*M* + H]$^+$, C$_{28}$H$_{31}$N$_4$O$_6{}^+$; calc. 519.22437; deviation 1.10 ppm). Anal. calc. for C$_{28}$H$_{30}$N$_4$O$_6$: C 64.9, H 5.8, N 10.8; found: C 64.93, H 5.87, N 10.80.

6-(3,5-Di-O-acetyl-2-deoxy-β-D-ribofuranosyl)-1,2,3,4,6,7-hexahydro-2,2-dimethyl-7-oxopyrimido[4,5-c][1,2]pyridazinium (**11**). To a soln. of the 3′,5′-di-*O*-acetyl-substituted analog of **5** [16] (1 g, 1.7 mmol) in tetrachloroethane (25 ml), 1,1-dimethylhydrazine (0.4 ml, 5.3 mmol) was added and the soln. heated at 100° overnight (TLC: two main products, one with *R*$_f$ 0). The product was extracted (H$_2$O/CHCl$_3$), and each of the two layers was evaporated. The org. layer was chromatographed (5% MeOH/CHCl$_3$) to give an off-white foam, which was characterized as *5-(2-chloroethyl)-1-(3,5-di-O-acetyl-2-deoxy-β-D-ribofuranosyl)-N^4-(dimethylamino)cytosine* (**9**; 0.32 g, 33%). Off-white foam. ^{1}H-NMR: 2.06 (*s*, Ac); 2.07 (*s*, Ac), 2.22–2.36 (*m*, 2 H–C(2')); 2.50 (*s*, Me$_2$N); 2.95–3.05 (*m*, CH$_2$); 3.14 (*s*, CH$_2$Cl); 4.12–4.19 (*m*, H–C(4')); 4.23–4.25 (*m*, 2 H–C(5')); 5.14–5.19 (*m*, H–C(3')); 5.59 (br. *s*, NH); 6.24 (*t*, *J* = 6.7, H–C(1')); 7.38 (*s*, H–C(6)). EI-MS: 417 (*M*$^+$).

The aq. layer was evaporated and purified by CC (reversed-phase *C-18* silica gel, H$_2$O → 25% MeOH/H$_2$O): 0.28 g (31%) of **11**. Brown foam. ^{1}H-NMR: 2.06 (*s*, Ac); 2.07 (*s*, Ac); 2.12–2.32 (*m*, 2 H–C(2')); 2.79 (*t*, *J* = 5.3, CH$_2$); 3.15 (*s*, Me$_2$N$^+$); 3.44 (*t*, *J* = 5.6, CH$_2$N); 4.06–4.10 (*m*, H–C(4')); 4.20–4.27 (*m*, 2 H–C(5')); 5.14–5.16 (*m*, H–C(3')); 6.24–6.30 (*m*, H–C(1')); 7.23 (*s*, H–C(6)); 8.24 (*s*, NH). EI-MS: 381 (*M*$^+$).

7-[[(Benzyloxy)carbonyl]methylamino]-3-[2-deoxy-3,5-di-O-(p-toluoyl)-β-D-ribofuranosyl]-3,5,6,7-tetrahydro-2H-pyrrolo[2,3-d]pyrimidin-2-one (**13**). To a soln. of **5** (0.5 g, 0.865 mmol) in tetrachloroethane (10 ml), 1-[(benzyloxy)carbonyl]-1-methylhydrazine [14] (0.38 g, 2.14 mmol) was added and the soln. heated at 85° overnight. After dilution with CHCl$_3$ and workup as described, the product was chromatographed (AcOEt/hexane/MeOH 1:1:0 → 7:3:0.1): 0.52 g (97%) of **13**. Off-white solid. UV: 283 (10400), 241 (34800), min. 267 and 219; pH 1: 300 (9900), 245 (24200), min. 269. ^{1}H-NMR: 2.37 (*s*, *Me*C$_6$H$_4$); 2.39 (*s*, *Me*C$_6$H$_4$); 2.53–2.72 (*m*, 2 H–C(2'), CH$_2$(5)); 3.08 (*s*, MeN); 3.60–3.82 (*m*, CH$_2$(6)); 4.48–4.63 (*m*, H–C(4'); 2 H–C(5')); 5.09–5.14 (*m*, CH$_2$); 5.56–5.58 (*m*, H–C(3')); 6.33 (*t*, *J* = 7.3, H–C(1')); 7.22–7.39 (*m*, 8 arom. H); 7.56 (*s*, H–C(4)); 7.84–7.92 (*m*, 5 arom. H). FAB-MS: 675.2448 ([*M* + Na]$^+$). HR-MS: 675.2448 ([*M* + Na]$^+$, C$_{36}$H$_{36}$N$_4$O$_8$Na$^+$; calc. 652.2431; deviation 2.50 ppm).

Pyrrolopyrimidinone **8b** (*Method B*). To a soln. of **13** (0.4 g, 0.6 mmol) in dry MeOH (20 ml), 10% Pd/C (or PtO$_2$) catalyst (50 mg) was added and the soln. stirred under H$_2$ for 2 h. The suspension was filtered through *Celite* and the cake washed with MeOH. The solvent was evaporated and the residue chromatographed (5% MeOH/CHCl$_3$): 0.25 g (79%) of **8b**. White solid. For data, see above (*Method A*).

3-[2-Deoxy-3,5-di-O-(p-toluoyl)-β-D-2-ribofuranosyl]-3,5,6,7-tetrahydro-2H-pyrrolo[2,3-d]pyrimidin-2-one (**14**). To a soln. of **13** (0.4 g, 0.6 mmol) in AcOH (20 ml), *Adam*'s catalyst (50 mg) was added and the soln. stirred under H$_2$ for 4 h. Workup and FC as described for **8b** (*Method B*) gave **16** (0.13 g, 43%). White solid. UV: 277 (8400), 244 (32800), min. 265 and 216; pH 1: 285 (9500), 244 (31600), min. 269 and 220. ^{1}H-NMR: 2.37 (*s*, *Me*C$_6$H$_4$); 2.38 (*s*, *Me*C$_6$H$_4$); 2.42–2.49 (*m*, 2 H–C(2')); 2.61–2.69 (*m*, CH$_2$); 3.49 (*t*, *J* = 7.8, CH$_2$(6)); 4.43–4.44 (*m*, H–C(4')); 4.50–4.63 (*m*, 2 H–C(5')); 5.55–5.57 (*m*, H–C(3')); 6.34 (*t*, *J* = 7.9, H–C(1')); 7.32–7.37 (*m*, H–C(6), 4 arom. H (Tol)); 7.86–7.92 (*m*, 4 arom. H (Tol)); 7.99 (*s*, NH). FAB-MS: 490.7 ([*M* + H]$^+$). HR-MS: 490.19792 ([*M* + H]$^+$, C$_{27}$H$_{28}$N$_3$O$_6$$^+$; calc. 490.19781; deviation − 0.20 ppm).

5-(2-Chloroethyl)-1-[2-deoxy-3,5-di-O-(p-toluoyl)-β-D-ribofuranosyl]cytosine (**15**). A soln. of **5** (0.25 g, 0.43 mmol) in ammonia-saturated dioxane (10 ml) was stirred at r.t. overnight. The soln. was evaporated and the product chromatographed (5% MeOH/CHCl$_3$) to give a white solid. Attempts to recrystallize resulted in gel formation: 0.22 g (97%) of **15**. UV: 248 (23000), 272 (sh), min. 232; pH 1: 285 (11000), 245 (29300), min. 220 and 268. ^{1}H-NMR: 2.37 (*s*, *Me*C$_6$H$_4$); 2.39 (*s*, *Me*C$_6$H$_4$); 2.44–2.57 (*m*, 2 H–C(2'), CH$_2$); 3.53 (*t*, *J* = 7, CH$_2$Cl); 4.47–4.64 (*m*, H–C(4'), 2 H–C(5')); 5.57–5.59 (*m*, H–C(3')); 6.31 (*t*, *J* = 7.6, H–C(1')); 7.09 (br. *s*, NH$_2$); 7.30–7.36 (*m*, 4 arom. H (Tol)); 7.47 (*s*, H–C(6)); 7.85–7.92 (*m*, 4 arom. H (Tol)). FAB-MS: 526.9 ([*M* + H]$^+$), 548.9 ([*M* + Na]$^+$). HR-MS: 548.15620 ([*M* + Na]$^+$, C$_{27}$H$_{28}$N$_3$O$_6$^{35}ClNa$^+$; calc. 548.15643; deviation 0.40 ppm).

We thank Dr. *David Neuhaus* for the NOE experiments, Dr. *Mao Jun Guo* for helpful discussions, and *Nycomed Amersham plc* for financial assistance.

REFERENCES

[1] M. J. Stone, A. N. R. Nedderman, D. H. Williams, P. Kong Thoo Lin, D. M. Brown, *J. Mol. Biol.* **1991**, *222*, 711.
[2] K. Negishi, D. M. Williams, Y. Inoue, K. Moriyama, D. M. Brown, H. Hayatsu, *Nucleic Acids Res.* **1997**, *25*, 1548.
[3] M. Zaccolo, D. M. Williams, D. M. Brown, E. Gherardi, *J. Mol. Biol.* **1996**, *255*, 589.
[4] M. Zaccolo, E. Gherardi, *J. Mol. Biol.* **1999**, *285*, 775.
[5] F. Hill, D. Loakes, D. M. Brown, *Proc. Natl. Acad. Sci. U.S.A.* **1998**, *95*, 4258.
[6] M. H. Moore, L. Van Meervelt, S. A. Salisbury, P. Kong Thoo Lin, D. M. Brown, *J. Mol. Biol.* **1995**, *251*, 665.
[7] G. S. Schuerman, L. Van Meervelt, D. Loakes, D. M. Brown, P. K. T. Lin, M. H. Moore, S. A. Salisbury, *J. Mol. Biol.* **1998**, *282*, 1005.

[8] D. Loakes, D. M. Brown, *Nucleosides Nucleotides*, **1994**, *13*, 679.

[9] D. M. Brown, M. J. E. Hewlins, P. Schell, *J. Chem. Soc. (C)* **1968**, 1925.

[10] A. Nomura, K. Negishi, H. Hayatsu, *Nucleic Acids Res.* **1985**, *13*, 8893.

[11] H. Griengl, M. Bodenteich, W. Hayden, E. Wanek, W. Streicher, P. Stutz, H. Bachmayer, I. Ghazzouli, B. Rosenwirth, *J. Med. Chem.* **1985**, *28*, 1679.

[12] D. Loakes, D. M. Brown, *Nucleosides Nucleotides* **1995**, *14*, 291.

[13] D. E. Butler, S. M. Alexander, J. W. McLean, L. B. Strand, *J. Med. Chem.* **1971**, 1052.

[14] S. A. Dutta, J. S. Morley, *J. Chem. Soc., Perkin Trans. 1*, **1975**, 1712.

[15] P. Kong Thoo Lin, D. M. Brown, *Heterocycles* **1989**, *29*, 1735.

[16] P. Kong Thoo Lin, D. M. Brown, *Nucleic Acids Res.* **1989**, *17*, 10373.

Synthesis and Biological Activity of 2'-Fluoro-D-arabinofuranosylpyrazolo[3,4-*d*]pyrimidine Nucleosides

by **Anita T. Shortnacy-Fowler**, **Kamal N. Tiwari**, **John A. Montgomery**, **Robert W. Buckheit, Jr.**, and **John A. Secrist III***

Southern Research Institute, P.O. Box 55305, Birmingham, AL 35255-5305 USA

and **Frank Seela**

Laboratorium für Organische und Bioorganische Chemie, Institut für Chemie, Universität Osnabrück, Barbarastrasse 7, D-49069 Osnabrück

Coupling of 2-fluoro-3,5-di-*O*-benzoyl-α-D-arabinofuranosyl bromide with 4-methoxypyrazolo[3,4-*d*]pyrimidine gave an α-D/β-D mixture of N^1- and N^2-coupled products. All the anomers were separated and deblocked to yield the corresponding nucleosides. The β-D-anomer **7** was converted to the 4-amino derivative **11**, which was deaminated by adenosine deaminase to give the 4-oxo compound **12**. Compound **7** showed significant activity against human cytomegalovirus and hepatitis B virus, and compound **11** showed activity against human herpes virus 8. All the compounds were noncytotoxic in several human tumor-cell lines in culture.

1. Introduction. – Since the approval of 6-mercaptopurine by the FDA in 1953 for the treatment of human cancer, eight other nucleobases or nucleosides thereof have been approved for cancer treatment. All of the nucleosides are 2'-deoxyribonucleosides or biochemical analogs (*i.e.*, nucleosides that behave metabolically like 2'-deoxyribonucleosides) [1]. In the ongoing program to develop antiviral and anticancer agents in our laboratory, we have developed fludarabine [2][3], identified the anticancer activity of cladribine [4], and are developing clofarabine [5], another rationally designed drug which is undergoing clinical trials at the present time. This latter drug, which contains an F substituent at C(2) of the furanose ring in the *arabino*-configuration, behaves metabolically like a 2'-deoxyribonucleoside[1]) [6].

To explore other nucleosides containing the fluoro-*arabino*-sugar, we turned to purine ring analogs. *Seela* et al. have prepared the 2'-deoxyribonucleosides [7] and the arabinonucleosides [8] of the pyrazolo[3,4-*d*]pyrimidine ring system. Prior work had shown that 4-aminopyrazolo[3,4-*d*]pyrimidine was active against experimental animal tumors [9] but proved too toxic for human use [10].

We report herein the synthesis and biological activity of some pyrazolo[3,4-*d*]-pyrimidine nucleosides with an F substituent in the 2'-*arabino*-position.

2. Results and Discussion. – Since the introduction of phase-transfer glycosylation [11] and the sodium-salt procedure [12] for nucleoside synthesis, various modified

---

[1]) In the *ribo*-configuration, the nucleosides behave biologically like ribonucleosides [6].

purines have been coupled with 1-α-halo sugars. In theory, these coupling reactions proceed mainly by an S_N^2 mechanism that should provide the desired β-D-anomer as the major product. In practice, the outcome depends on various factors, including the particular heterocycle and the halo sugar used and the reaction solvent and temperature. In our present work (*Scheme 1*), we used the sodium-salt method [12] to couple 4-methoxypyrazolo[3,4-*d*]pyrimidine (**1**) [13] with α-bromo sugar **2** [14]. However, these reactions gave poor yields and an unfavorable distribution of products, caused in part by the low solubility of the sodium salt of **1** (*Table 1, Entries a and b*).

Scheme 1

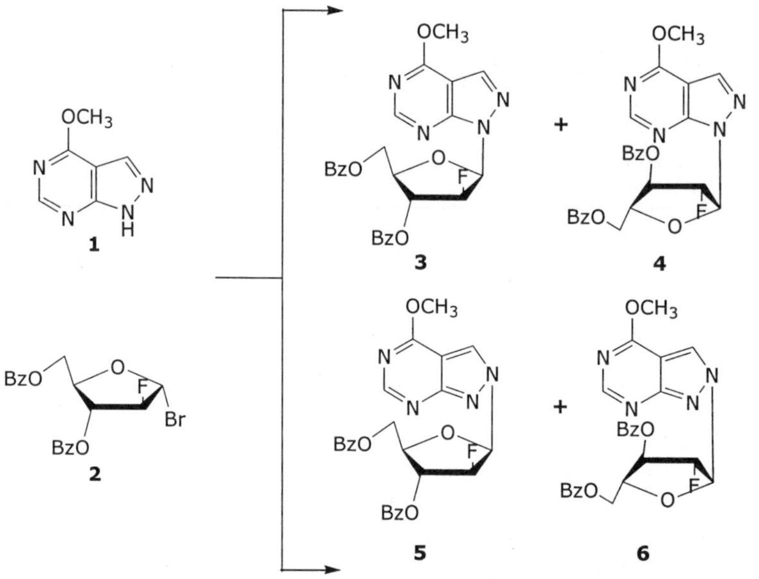

Table 1. *Coupling Studies*

Entry	Base [equiv.]	Conditions	Nucleoside distribution	Total yield
a [12]	NaH (1.6)	DME, 45°, 3 h	**3** (6%), **4** (4%), **5** (33%), **6** (–)	43%
b [12]	NaH (1.5)	MeCN, r.t., 23 h	**3** (24%), **4** (17%), **5** (14%), **6** (8%)	63%
c [15]	DBU[a]) (1.0)	MeCN, r.t., 4 h	**3** (45%), **4** (17%), **5** (26%), **6** (6%)	94%

[a]) 1,8-Diazabicyclo[5.4.0]undec-7-ene.

Employing a new method reported by *Jungmann* and *Pfleiderer* [15], we treated an MeCN suspension of **1** with an equimolar amount of 1,8-diazabicyclo[5.4.0]undec-7-ene (DBU) to provide a clear solution. This easily soluble DBU salt of **1** was reacted with **2** to give a 94% total nucleoside yield (*Table 1, Entry c*). Debenzoylation of **3**–**6** with MeONa/MeOH produced **7**–**10** (*Scheme 2*) [7]. As depicted in *Scheme 3*, nucleoside **7** was aminated in conc. aqueous NH₃, giving **11**, which was readily deaminated by adenosine deaminase (HPLC experiment) [7][8].

Scheme 2

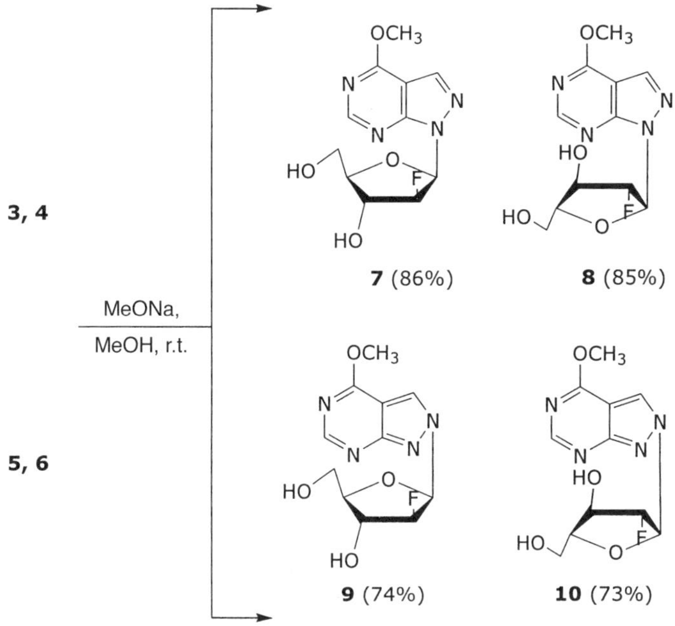

3, 4

MeONa,
MeOH, r.t.

5, 6

7 (86%)   8 (85%)

9 (74%)   10 (73%)

Scheme 3

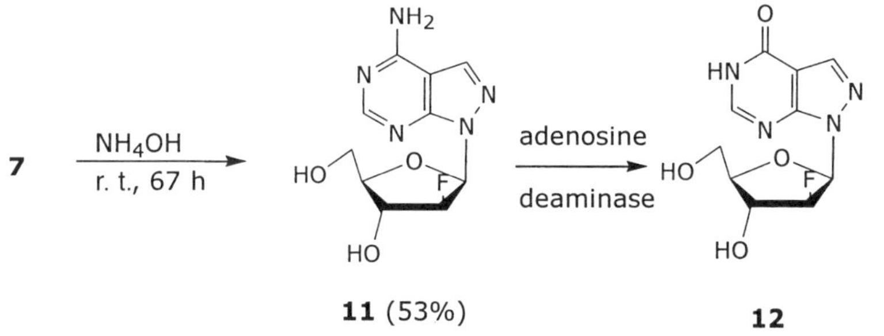

7   NH₄OH
      r. t., 67 h

adenosine

deaminase

11 (53%)   12

Compounds **3–11** were characterized by UV, NMR, and mass spectra. The anomeric configuration and point of attachment of the sugar to the heterocyclic moiety was confirmed by ¹H-NMR spectra with decoupling experiments and coupling constants. Correlation with published literature on 2′-deoxynucleosides in this series further verified assignments [7]. ¹⁹F-NMR Spectra of β-D-anomers **7** and **9** showed signals upfield from CFCl₃ at − 204 and − 202 ppm, respectively. In contrast, signals from the α-anomers **8** and **10** appeared at − 194 and − 189 ppm, respectively.

The cell-culture cytotoxicities of the target compounds **7–11** were determined against eight human cancer-cell lines, namely CCRF-CEM (leukemia), CAKI-I (renal), DLD-1 (colon), LOXIMVI (colon), NCI-H-23 (lung), PC-3 (prostate), SNB-7 (CNS), and ZR-75-1 (mammary) [16]. All compounds were found to be noncytotoxic at the highest level tested (40 µg/ml). In addition, **7**, **9**, and **11** were screened for activity against a variety of viruses (*i.e.*, CMV, influenza A, HBV, HHV-8, HSV-1, and respiratory syncytial). Those compounds with significant antiviral activity are identified in *Table 2*. Compound **7** showed good activity against CMV and in the HBV screen [17]. Its isomer **9** exhibited moderate activity only against CMV. Compound **11**, which is the 6-amino analog of **7**, was active against HHV8 [18].

Table 2. *Virus Screening Summary*

Virus Cell line Compound	IC_{50} [µM][a])	TC_{50} [µM])[b])
Cytomegalovirus (CMV AD169) in MRC-5 cells		
Ganciclovir[c])	8.6	> 1000
7	0.42	> 10
9	3.8	> 10
Hepatitis B virus (HBV EC50) in HepG2 2.2.15 cells		
3TC[c])	0.14	> 1000
7	0.98	> 100
Herpes virus 8 (HHV8) in TPA-induced BCBL-1 cells		
Cidofovir[c])	0.63	> 350
11	0.59	> 10

[a]) IC_{50}: Inhibitory concentration at 50% cell viability.   [b]) TC_{50}: Toxic concentration at 50% cell viability. [c]) Control drug.

This investigation was supported by the *National Cancer Institute, National Institutes of Health* (P01-CA34200). The authors are indebted to Dr. *J. M. Riordan* for the recording and interpretation of NMR data, to Mr. *M. Richardson* for mass-spectral data, to Ms. *J. Bearden* for UV and elemental analyses, and to Ms. *S. Campbell* for HPLC analyses. We gratefully acknowledge Dr. *W. Parker*, Dr. *W. Waud*, and Ms. *D. Adamson* for cell-culture data.

Experimental Part

General. 2-Fluoro-1,3,5-tri-*O*-benzoyl-α-D-arabinofuranose was purchased from *Pfanstiehl Laboratories, Inc.* TLC: *Analtech* precoated (250 µm) silica gel (*GF*) plates. Flash chromatography (FC): 230–400-mesh silica gel from *E. Merck.* HPLC: *Hewlett-Packard 1100 Series* liquid chromatograph with a *Phenomenex SphereClone 5 µ ODS (1)* column (250 × 4.6 mm), UV monitoring (254 nm). M.p.: *Mel-Temp* apparatus; uncorrected. UV Spectra: *Perkin-Elmer lambda 9* spectrometer in MeOH; extinction coefficients ($\varepsilon \times 10^{-3}$) in parentheses. ^{1}H-NMR Spectra: *Nicolet NT-300 NB* spectrometer, at 300.635 MHz; chemical shifts (δ) in ppm downfield from Me$_4$Si. ^{19}F-NMR Spectra: *Bruker CXP-200* spectrometer, at 188.2 MHz; chemical shifts (δ) in ppm upfield relative to CFCl$_3$ (δ 0.0). MS: *Varian/MAT 311A* double-focusing mass spectrometer in the fast-atom-bombardment (FAB) mode. Microanalyses were performed by the *Molecular Spectroscopy Section* of *Southern Research Institute.* Where solvents are noted as part of the elemental analyses, they were seen in the ^{1}H-NMR spectra in the proper amounts.

1- and 2-(2-Deoxy-2-fluoro-3,5-di-O-benzoyl-β-D- and -α-D-arabinofuranosyl)-4-methoxy-1H- and -2H-pyrazolo[3,4-d]pyrimidine (**3–6**). To a suspension of *4-methoxy-1H-pyrazolo[3,4-d]pyrimidine* (**1**; 329 mg, 2.20 mmol) in anh. MeCN (24 ml) under Ar was added 1,8-diazabicyclo[5.4.0]undec-7-ene (DBU; 329 µl, 2.20 mmol), and the mixture was stirred 15 min. The resulting clear soln. was treated dropwise over 5 min with a soln. of **2** (1.02 g, 2.42 mmol) in MeCN (6 ml). The soln. was stirred at r.t. for 4 h and then evaporated to a yellow foam. This material was dissolved in a minimum of CH$_2$Cl$_2$ and applied to a flash column containing 50 g silica gel previously equilibrated with hexane/AcOEt 3:1. Elution with a gradient from 3:1 to 1:1 hexane/

AcOEt provided N^1-isomers followed by N^2-isomers. After further chromatography of mixed bands, four pure isomers were obtained for a 94% total nucleoside yield. A small sample of each was crystallized from i-PrOH for analysis.

Data of N^1-*β*-D-*Isomer* (**3**; 486 mg, 45%): m.p. 148–150°. TLC (hexane/AcOEt 2:1): R_f 0.42. HPLC: MeCN/H$_2$O 65:35. UV (MeOH): 231 (32.2), 266 (sh). ^{1}H-NMR (CDCl$_3$): 8.60 (*s*, H−C(6)); 8.15 (*s*, H−C(3)); 8.06 (*m*, 4 H$_o$ of Ph); 7.62 (*m*, 2 H$_p$ of Ph); 7.42 (*m*, 4 H$_m$ of Ph); 6.94 (*br. d*, *J* = 8, H−C(1′)); 6.64 (*m*, *J*(3′,4′) = 6, *J*(3′,F) = 24, H−C(3′)); 5.73 (*ddd*, *J*(2′,3′) = 4, *J*(2′,F) = 62, H−C(2′)); 4.93 (*J*(5′a,5′b) = 14, *J*(4′,5′a) = 4, H−C(5′)); 4.80 (*J*(4′,5′b) = 6, H−C(5′)); 4.16 (*s*, MeO). MS 493 ([M + H]$^+$). Anal. calc. for C$_{25}$H$_{21}$FN$_4$O$_6$·0.10 H$_2$O (494.27): C 60.75, H 4.32, N 11.34; found: C 60.50, H 4.48, N 11.32.

Data of N^1-*α*-D-*Isomer* (**4**; 186 mg, 17%): m.p. 92–93°. TLC (hexane/AcOEt 2:1): R_f 0.54. HPLC: MeCN/H$_2$O 65:35. UV (MeOH): 231 (31.4), 266 (sh). ^{1}H-NMR (CDCl$_3$): 8.62 (*s*, H−C(6)); 8.12 (*s*, H−C(3)); 8.02 – 8.10 (*m*, 4 H$_o$ of Ph); 7.40 – 7.65 (*m*, 6 arom. H); 6.90 (*dd*, *J*(1′,2′) = 2, *J*(1′,F) = 20, H−C(1′)); 6.14 (*ddd*, *J*(2′,3′) = 2, *J*(2′,F) = 54, H−C(2′)); 5.90 (*m*, *J*(3′,4′) = 2, *J*(3′,F) = 18, H−C(3′)); 4.93 (*dt*, H−C(4′)); 4.74 (*J*(5′a,5′b) = 14, *J*(4′,5′a) = 4, H−C(5′)); 4.64 (*J*(4′,5′b) = 5, H−C(5′)); 4.18 (*s*, MeO). MS 493 ([M+H]$^+$). Anal. calc. for C$_{25}$H$_{21}$FN$_4$O$_6$·0.20 H$_2$O (496.07): C 60.53, H 4.35, N 11.29; found: C 60.42, H 4.30, N 11.30.

Data of N^2-*β*-D-*Isomer* (**5**; 282 mg, 26%): m.p. 175–176°; TLC: (hexane/AcOEt 1:1), R_f 0.30. HPLC: MeCN/H$_2$O 65:35. UV (MeOH): 224 (32.0), 260 (12.4). ^{1}H-NMR (CDCl$_3$): 8.68 (*s*, H−C(6)); 8.39 (*br. d*, H−C(3)); 8.10 (*m*, 4 H$_o$ of Ph); 7.56 (*m*, 6 arom. H); 6.58 (*dd*, *J*(1′,2′) = 2, *J*(1′,F) = 20, H−C(1′)); 5.82 (*m*, H−C(2′)); 5.48 (*dd*, *J*(3′,4′) = 2, *J*(3′,F) = 50, H−C(3′)); 4.86 (*m*, H−C(4′), H−C(5′)); 4.72 (*m*, H−C(5′)); 4.15 (*s*, MeO). ^{19}F-NMR (CDCl$_3$ + CFCl$_3$): − 201.32. (*J*(2) = 50, *J*(3) = 16.4, *J*(3) = 32.7). MS 493 ([M + H]$^+$). Anal. calc. for C$_{25}$H$_{21}$FN$_4$O$_6$·0.20 H$_2$O (496.07): C 60.53, H 4.35, N 11.29; found: C 60.60, H 4.18, N 11.32.

Data of N^2-*α*-D-*Isomer* (**6**; 67 mg, 6%): m.p. 128–130°; TLC (hexane/AcOEt 1:1): R_f 0.40; HPLC: MeCN/H$_2$O 65:35. UV (MeOH): 223 (28.7), 261 (10.1). ^{1}H-NMR (CDCl$_3$): 8.68 (*s*, H−C(6)); 8.36 (*s*, H−C(3)); 8.12 (m, 4 H$_o$ of Ph); 7.54 (*m*, 6 arom. H); 6.52 (*br. dd*, *J* = 14, H−C(1′)); 6.18 (*br. dd* *J* = 50, H−C(2′)); 5.75 (*br. dd*, *J*(3′,4′) = 5, *J*(3′,F) = 20, H−C(3′)); 4.96 (*dt*, H−C(4′)); 4.75 (*m*, 2 H−C(5′)); 4.14 (*s*, MeO). MS 493 ([M + H]$^+$). Anal. calc. for C$_{25}$H$_{21}$FN$_4$O$_6$·0.20 H$_2$O (496.07): C 60.53, H 4.35, N 11.29; found: C 60.62, H 4.20, N 11.18.

1-(2-Deoxy-2-fluoro-β-D-arabinofuranosyl)-4-methoxy-1H-pyrazolo[3,4-d]pyrimidine (**7**). A suspension of **3** (486 mg, 0.99 mmol) in MeOH (12 ml) was treated in one portion with 0.50N MeONa in MeOH (5.9 ml). The mixture became a clear soln. after 10 min and was stirred an additional 2 h. The mixture was neutralized to pH 6 with glacial AcOH and evaporated. This residue was purified by prep. TLC (*Analtech GF*, 20 × 20 cm, 2,000 μ) with development in CHCl$_3$/MeOH 95:5. The product band was extracted with MeOH, and the extract was evaporated. The isolated material was crystallized from hot H$_2$O (10 ml) to give pure **7** (242 mg, 85%). M.p. 135–136°. TLC (CHCl$_3$/MeOH 9:1): R_f 0.55. HPLC: H$_2$O/MeCN 80:20. UV (MeOH): 246 (8.44), 265 (sh, 4.83). ^{1}H-NMR ((D$_6$)DMSO): 8.66 (*s*, H−C(6)); 8.42 (*s*, H−C(3)); 6.76 (*br. d*, *J* = 8, H−C(1′)); 5.90 (*br. d*, *J* = 6, HO−C(3′)); 5.54 (*dt*, *J*(2′,3′) = 6, *J*(2′,F) = 52, H−C(2′)); 4.83 (*t*, HO−C(5′)); 4.74 (*m*, H−C(3′)); 4.14 (*s*, MeO); 3.84 (*m*, H−C(4′)); 3.66 (*m*, 2 H−C(5′)). ^{19}F-NMR (DMSO + CFCl$_3$): − 204.11 (*J*(2) = 53.41, *J*(3) = 18.53, *J*(3) = 0). MS 285 ([M + H]$^+$). Anal. calc. for C$_{11}$H$_{13}$FN$_4$O$_4$·0.30 H$_2$O (289.65): C 45.61, H 4.73, N 19.34; found: C 45.52, H 4.70, N 19.24.

1-(2-Deoxy-2-fluoro-α-D-arabinofuranosyl)-4-methoxy-1H-pyrazolo[3,4-d]pyrimidine (**8**). Compound **4** (139 mg, 0.28 mmol) was treated with 0.50N MeONa as described for **3**. Similar workup and crystallization from H$_2$O provided pure **8** (68 mg, 85%). M.p. 163–164°. TLC (CHCl$_3$/MeOH 9:1): R_f 0.57. HPLC: H$_2$O/MeCN 80:20. UV (MeOH): 246 (8.44), 265 (sh, 5.02). ^{1}H-NMR ((D$_6$)DMSO): 8.92 (*s*, H−C(6)); 8.42 (*s*, H−C(3)); 6.48 (*dd*, *J*(1′,2′) = 4, *J*(1′,F) = 18, H−C(1′)); 6.02 (*d*, *J* = 6, OH−C(3′)); 5.84 (*ddd*, *J*(2′,3′) = 4, *J*(2′,F) = 54, H−C (2′)); 4.96 (*br. t*, HO−C(5′)); 4.40 (*m*, H−C(3′)); 4.12 (*s*, MeO); 4.10 (*m*, H−C(4′)); 3.60 (*m*, 2 H−C(5′)). ^{19}F-NMR (DMSO + CFCl$_3$): − 193.70 (*J*(2) = 54.5, *J*(3) = 16.35, *J*(3) = 23.98). MS: 285 ([M + H]$^+$). Anal. calc. for C$_{11}$H$_{13}$FN$_4$O$_4$ (284.25): C 46.48, H 4.61, N 19.71; found: C 46.40, H 4.62, N 19.51.

2-(2-Deoxy-2-fluoro-β-D-arabinofuranosyl)-4-methoxy-2H-pyrazolo[3,4-d]pyrimidine (**9**). A suspension of **5** (250 mg, 0.51 mmol) in MeOH (6 ml) was treated with 0.50N MeONa in MeOH (3 ml). The mixture became a clear soln. after 5 min and was neutralized with glacial AcOH after 4 h. The soln. was concentrated to a small volume and applied to a prep. TLC plate (*Analtech GF*, 20 × 20 cm, 2,000 μ) that was developed four times in CHCl$_3$/MeOH 9:1. The residue from the MeOH plate extract solidified when triturated with acetone to yield **9** (107 mg, 73%). M.p.: softens at 167°, decomp. at 210°. TLC (CHCl$_3$/MeOH 9:1): R_f 0.38. HPLC: 0.01M NH$_4$H$_2$PO$_4$/MeOH 70:30 (pH 5.1). UV (MeOH): 260 (9.64), 285 (sh). ^{1}H-NMR ((D$_6$)DMSO): 8.92 (*s*, H−C(3)); 8.58 (*s*, H−C(6)); 6.46 (*br. t*, *J* = 6, H−C(1′)); 6.04 (*s*, HO−C(3′)); 5.36 (*ddd*, *J*(2′,3′) = 6, *J*(2′,F) = 54, H−C(2′)); 5.20 (*br. m*, HO−C(5′)); 4.55 (*br. dd*, H−C(3′)); 4.10 (*s*, MeO); 3.96 (*m*, H−C(4′)); 3.76 (*m*,

2 H−C(5′)). ^{19}F-NMR (DMSO + CFCl$_3$): − 202.08 ($J(2) = 52.3, J(3) = 6.54, J(3) = 18.0$). MS: 285 ($[M + H]^+$).
Anal. calc. for C$_{11}$H$_{13}$FN$_4$O$_4$ · 0.25H$_2$O (288.75): C 45.76, H 4.71, N 19.40; found: C 45.96, H 4.60, N 19.02.

2-(2-Deoxy-2-fluoro-α-D-arabinofuranosyl)-4-methoxy-2H-pyrazolo[3,4-d]pyrimidine (**10**). Compound **6**
(50 mg, 0.10 mmol) was reacted with 0.50N MeONa as described for **5**. Purification by prep. TLC (*Analtech GF*,
10 × 20 cm, 1000 μ) was accomplished with one development in CHCl$_3$/MeOH 9 : 1. The MeOH plate extract
was solidified by Et$_2$O trituration to provide **10** (21 mg, 67%). M.p.: softens at 130°, decomp. at 200°; TLC
(CHCl$_3$/MeOH 9 : 1): R_f 0.49. HPLC: 0.01M NH$_4$H$_2$PO$_4$/MeOH 70 : 30 (pH 5.1). UV (MeOH): 260 (9.49), 285
(sh). ^{1}H-NMR ((D$_6$)DMSO): 8.87 (*s*, H−C(3)); 8.60 (*s*, H−C(6)); 6.44 (*dd*, $J(1′,2′) = 2$, $J(1′,F) = 16$,
H−C(1′)); 6.04 (*br. s*, HO−C(3′)); 5.62 (*ddd*, $J(2′,3′) = 2$, $J(2′,F) = 52$, H−C(2′)); 5.10 (*br. s*, HO−C(5′)); 4.35
(*m*, H−C(3′), H−C(4′)); 4.10 (*s*, MeO); 3.60 (*m*, 2 H−C(5′)). ^{19}F-NMR (DMSO + CFCl$_3$): − 188.66 ($J(2) =$
52.54, $J(3) = 15.3, J(3) = 20.71$). MS: 285 ($[M + H]^+$). Anal. calc. for C$_{11}H_{13}FN_4O_4$ · 0.30H$_2$O · 0.30 CH$_3$CO$_2$Na
(314.26): C 44.33, H 4.65, N 17.83; found: C 44.42; H 4.50, N 17.55.

4-Amino-1-(2-deoxy-2-fluoro-β-D-arabinofuranosyl)-4-methoxy-1H-pyrazolo[3,4-d]pyrimidine (**11**). A
soln. of **7** (38 mg, 0.13 mmol) in conc. NH$_4$OH (10 ml) was stirred at r.t. for 67 h. The solvent was evaporated,
and the residue was purified by prep. TLC (*Analtech GF*, 10 × 20 cm, 1000 μ) with two developments in CHCl$_3$/
MeOH 5 : 1 containing 1% conc. NH$_4$OH. The product isolated from the MeOH extract of the plate band was
crystallized from hot MeCN to yield pure **11** (19 mg, 53%). M.p. 208−210°; TLC (CHCl$_3$/MeOH 5 : 1 + 1%
NH$_4$OH): R_f 0.40. HPLC: 0.01M NH$_4$H$_2$PO$_4$/MeOH 70 : 30 (pH 5.1). UV (MeOH): 260 (9.59), 275 (11.1).
^{1}H-NMR ((D$_6$)DMSO): 8.20 (*s*, H−C(3), H−C(6)); 7.82 (*br. dd*, NH$_2$); 6.60 (*dd*, $J = 6$, H−C(1′)); 5.86 (*br. d*,
HO−C(3′)); 5.40 (*ddd*, $J(2′,3′) = 8$, $J(2′,$ F$) = 52$, H−C(2′)); 4.84 (*br. t*, HO−C(5′)); 4.74 (*m*, H−C(3′)); 3.80
(*m*, H−C(4′)); 3.65 (*m*, 2 H−C(5′)). MS: 270 ($[M + H]^+$). Anal. calc. for C$_{10}$H$_{12}$FN$_5$O$_3$ (269.24): C 44.61,
H 4.49, N 26.01; found: C 44.68, H 4.59, N 26.33.

REFERENCES

[1] J. A. Montgomery, Antimetabolites in 'Cancer Chemotherapeutic Agents,' Ed. W. Foye, American
Chemical Society, Washington, DC, 1995, pp. 47−109.
[2] R. W. Brockman, F. M. Schabel, Jr., J. A. Montgomery, *Biochem. Pharmacol.* **1977**, *26*, 2193.
[3] S. J. Wright, L. E. Robertson, S. O'Brien, W. Plunkett, M. J. Keating, *Blood Rev.* **1994**, *8*, 125.
[4] D. A. Carson, D. B. Wasson, J. Kaye, B. Ullman, D. W. Martin, Jr., R. K. Robins, J. A. Montgomery, *Proc. Natl. Acad. Sci. U.S.A.* **1980**, *77*, 6065.
[5] J. A. Montgomery, A. Shortnacy-Fowler, S. D. Clayton, J. M. Riordan, J. A. Secrist, III, *J. Med. Chem.* **1992**, *35*, 397.
[6] W. B. Parker, S. C. Shaddix, L. M. Rose, D. S. Shewach, L. W. Hertel, J. A. Secrist, III, J. A. Montgomery, L. L. Bennett, Jr., *Mol. Pharmacol.* **1999**, *55*, 515.
[7] F. Seela, H. Winter, M. Möller, *Helv. Chim. Acta* **1993**, *76*, 1450; F. Seela, H. Steker, *ibid* **1985**, *68*, 563.
[8] F. Seela, W. Bourgeois, H. Winter, *Nucleosides Nucleotides* **1991**, *10*, 713.
[9] H. E. Skipper, R. K. Robins, J. R. Thomson, C. C. Cheng, R. W. Brockman, F. M. Schabel, Jr., *Cancer Res.* **1957**, *17*, 579.
[10] R. K. Shaw, R. N. Shulman, J. D. Davidson, D. P. Ball, E. J. Frei, III, *Cancer* **1960**, *13*, 482.
[11] F. Seela, H.-D. Winkeler, *J. Org. Chem.* **1983**, *48*, 3119.
[12] Z. Kazimierczuk, H. B. Cottam, G. R. Revankar, R. K. Robins, *J. Am. Chem. Soc.* **1984**, *106*, 6379.
[13] R. K. Robins, *J. Am. Chem. Soc.* **1956**, *78*, 784.
[14] C. H. Tann, P. R. Brodfuehrer, S. P. Brundidge, C. Sapino, Jr., H. G. Howell, *J. Org. Chem.* **1985**, *50*, 3644.
[15] O. Jungmann, W. Pfleiderer, *Tetrahedron Lett.* **1996**, *37*, 8355.
[16] J. A. Secrist, III, K. N. Tiwari, A. T. Shortnacy-Fowler, L. Messini, J. M. Riordan, J. A. Montgomery, S. C. Meyers, S. E. Ealick, *J. Med. Chem.* **1998**, *41*, 3865.
[17] W. A. Tatarowicz, N. S. Lurain, K. D. Thompson, *J. Virol. Methods* **1991**, *35*, 207; B. F. Korba, J. L. Gerin, *Antiviral Res.* **1992**, *19*, 55.
[18] R. H. Shoemaker, S. M. Halliday, T. M. Fletcher, R. W. Buckheit, Jr., E. A. Sausville, submitted.

A Novel Route for the Synthesis of Deoxy Fluoro Sugars and Nucleosides

by Igor A. Mikhailopulo* and Grigorii G. Sivets

Institute of Bioorganic Chemistry, National Academy of Sciences, 220141 Minsk, Acad. Kuprevicha 5, Belarus
(phone/fax: +375/172/648324; e-mail: igormikh@ns.igs.ac.by)

The reaction of (diethylamino)sulfur trifluoride (DAST) with methyl 5-*O*-benzoyl-β-D-xylofuranoside (**1**) followed by column chromatography afforded the riboside **2** (62%) and the *ribo*-epoxide **3** (18%) (*Scheme 1*). Under similar reaction conditions, the α-D-anomer **4** gave the riboside **5** and the difluoride **6** in 60 and 9% yield, respectively. Treatment of the β-D-xyloside **10** with DAST gave, after chromatographic purification, the riboside **11** as the principal product (48%; *Scheme 2*). These results suggest that the C(3)−O−SF$_2$NEt$_2$ derivatives were initially formed in the case of the xylosides studied. The distinctive feature of the reaction of DAST with the β-D-arabinoside **12** consists in the formation of a 3- or 5-benzylideneoxoniumyl-substituted intermediate on one of the consecutive transformations, which finally give rise to the inversion of the configuration at C(3) affording the xylosides **17** (18%) and **18** (55%); the lyxoside **14** was also isolated from the reaction mixture in a yield of 25% (*Scheme 3*). In the presence of the non-participating 5-*O*-trityl group, *i.e.*, from the reaction products of **21** with DAST, the compounds **23** and **24** were isolated in 16 and 52% yield, respectively (*Scheme 4*). It may be thus reasonable to conclude that, in the case of the β-D-arabinosides **12** and **21**, the principal route of the reaction is the formation of the intermediate C(2)−O−SF$_2$NEt$_2$ derivative. Unlike **21**, the α-D-arabinoside **26** was converted to the *lyxo*-epoxide **25** (53%) and the lyxoside **27** (14%), which implies the intermediate formation of the C(3)−O−SF$_2$NEt$_2$ derivative (*Scheme 5*).

Introduction. – The synthesis of diverse deoxy fluoro pentofuranosides as key intermediates in a convergent approach to the corresponding nucleosides has been reported over the last decade from our laboratory [1]. In turn, a number of deoxyfluoro nucleosides displayed significant biological activity (for a review, see, *e.g.*, [2]) and, at the 5′-triphosphate level, are used as versatile probes for the DNA and RNA polymerases [1a][3]. Moreover, the incorporation of deoxyfluoro nucleosides into oligonucleotides imparts extraordinary biophysical and biochemical properties to fluorinated oligomers as compared to their unmodified counterparts [1h][4–8].

Despite the widespread interest in the chemistry of deoxyfluoro nucleosides, no generally applicable chemical methods have been available for the introduction of an F-atom [2][9]. The approaches to the synthesis of deoxyfluoro nucleosides may be divided into two main groups: *i*) glycosylation of heterocyclic bases with universal carbohydrate precursors, and *ii*) pentofuranose-ring fluorination of nucleosides. Utilizing the first approach (see, *e.g.*, [1b–d,g,i–l]), which has inherent advantages [10], we focused our attention on the development of practical methods for the preparation of deoxy fluoro sugars, which may be transformed in the universal glycosylating agents.

Recently, we have investigated the ring fluorination of methyl pentofuranosides with non-protected *trans*-arranged 2′- and 3′-OH groups under the action of (diethylamino)sulfur trifluoride (DAST), which results in the formation of 2,3-*cis* arranged deoxy fluoro pentofuranosides [11]. Although the regioselectivity of this transformation was rather low and the yields of the desired deoxy fluoro derivatives were moderate, it became evident that this method is of practical utility due to its simplicity and the mildness of the reaction conditions. This study was continued and expanded, and was especially focused on the influence of the configuration at the anomeric center and the nature of the 5-*O*-blocking group on the course of the transformation.

Results and Discussion. – *Chemical Transformations.* Both starting methyl xylosides **1** and **4** were prepared in three steps from D-xylose, *viz.*, D-xylose was transformed to 1,2-*O*-isopropylidene-α-D-xylofuranose [12] in 80% yield, careful benzoylation of which in CH_2Cl_2 in the presence of Et_3N [13] gave the 5-*O*-benzoyl derivative (95%), which was finally treated with I_2 in MeOH [14] under reflux for 4 h to afford a mixture of the desired β-D- and α-D-xylosides in a ratio of *ca.* 1:1 in 69% yield (*Scheme 1*). Following chromatographic purification, the pure homogeneous **1** and **4** were isolated. Alternatively, the reaction of methyl 2,3-anhydro-5-*O*-benzoyl-β-D-lyxofuranoside with potassium benzoate in DMSO gave, after workup and subsequent chromatography as described previously [11], the xyloside **1** (23%) and the arabinoside **12** (22%). Treatment of methyl 5-*O*-benzoyl-β-D-xylofuranoside (**1**) with DAST in a molar ratio of 1:6 in anhydrous CH_2Cl_2 at room temperature for 19 h, followed by column chromatography (silica gel), afforded the β-D-riboside **2** and the *ribo*-epoxide **3** in 62 and 18% isolated yield, respectively. Under similar reaction conditions, the corresponding α-D-anomer **4** was completely transformed within 4 h at room temperature, and after column chromatography, the α-D-riboside **5** and the difluoride **6** were isolated in 60 and 9% yield, respectively [11]. These results tend to suggest that the $C(3)-O-SF_2NEt_2$ derivatives were initially formed in the case of both xylosides **1** and **4**. Interestingly, the $C(3)-O-SF_2NEt_2$ derivative of the β-D-anomer **1** mainly underwent intermolecular attack by an F-anion along with an intramolecular nucleophilic attack by the O-atom at C(2). On the contrary, we did not observe the

Scheme 1

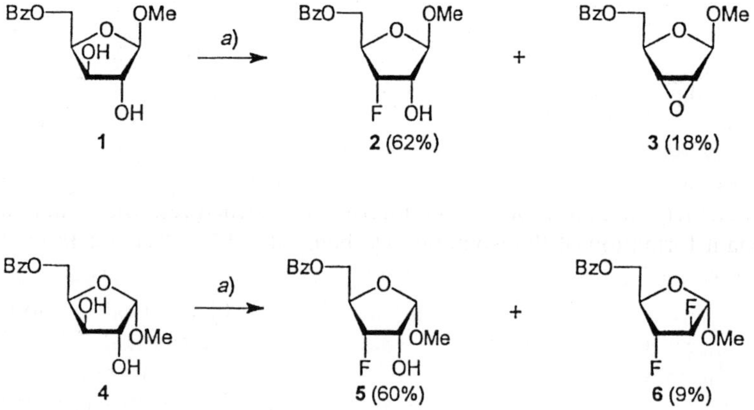

(*a*) DAST (**1** or **4**/DAST 1:6 (molar ratio)), anh. CH_2Cl_2, 20°, 19 (**1**) or 4 h (**4**).

formation of the corresponding epoxide in the case of the α-D-anomer **4**, but the fluoride **5** reacted with an excess of DAST to afford the difluoride **6** instead (*Scheme 1*). The most likely explanation of this observation may be the different conformations of the $C(3)-O-SF_2NEt_2$ derivatives of the β- and α-D-anomers **1** and **4**. It is noteworthy that the described preparation of 3-fluoro-3-deoxyribosides **2** and **5** from D-xylose offers a useful alternative to methods previously published [1b][11].

In extension of this work, we studied the reaction of DAST with 9-(5-*O*-benzyl-β-D-xylofuranosyl)adenine (**10**). The latter was prepared in three steps from methyl 5-*O*-benzyl-β-D-xylofuranoside (**7**) [11] (*Scheme 2*). Benzoylation of xyloside **7** quantitatively gave benzoate **8**, which was coupled with persilylated N^6-benzoyladenine under standard conditions [1b][15] to give the blocked nucleoside **9** in 63% isolated yield. Debenzoylation of the latter afforded the β-D-nucleoside **10**. The reaction of **10** with DAST in CH_2Cl_2 in the presence of pyridine (CH_2Cl_2/pyridine 13:1, (v/v)) at room temperature for 5 h gave a complex mixture from which the riboside **11** was isolated as the principal product (48% based on consumed **10**) besides the starting nucleoside **10** (24%). Once again, this result points to the interaction of DAST predominantly with the OH group at C(3'). Addition of pyridine was necessary to dissolve the starting nucleoside **10** before the addition of DAST (*Scheme 2*).

Scheme 2

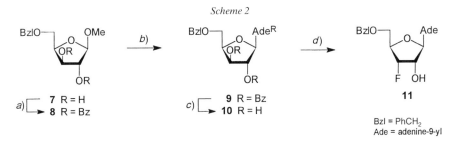

a) BzCl, pyridine, 20°, 18 h; 92%. b) 8-persilylated N^6-benzoyladenine/SnCl$_4$ 1.0:1.5:2.94 (molar ratio), MeCN, reflux for 15 min, 20° for 30 min; 63%. c) Saturated (at 0°) NH$_3$/MeOH soln., 20°, 24 h; 70%. d) DAST (**10**/DAST 1:6 (molar ratio)), anh. CH$_2$Cl$_2$/pyridine 13:1 (v/v), 20°, 5 h; chromatography (SiO$_2$); 48% based on **10** consumed.

The course of the reaction was further studied on the example of methyl arabinosides. In this case, all reactions were performed under identical conditions (molar ratio sugar/DAST 1:6, CH_2Cl_2 as solvent, stirring at room temperature for 5 h, chromatographic isolation of the products as described previously [11]). At variance with the *xylo*-benzoates **1** and **4**, the stereochemical course of the reaction of *arabino*-benzoate **12** (for its preparation, *vide supra*) was dependent upon the participation of the 5-*O*-benzoyl function in one of the intermediate structures. The rather surprising predominant formation of the isomeric *xylo*-benzoates **17** (18%) and **18** (55%) most likely involved a 3- or 5-benzylideneoxoniumyl-substituted transient intermediate **16** (*Scheme 3*). One can speculate that the interaction of DAST with the arabinoside **12** results in the formation of the isomeric $C(3)-O-SF_2NEt_2$ derivative **13** and the C(2) counterpart (not shown) in a ratio of *ca.* 1:3. The former undergoes an intermolecular attack by an F-anion furnishing the lyxoside **14**, while the latter gives, in a similar manner, the methyl 5-*O*-benzoyl-2-deoxy-2-fluoro-β-D-ribofuranoside (not shown), which reacts with an excess of DAST to afford **15**. Our data cannot exclude the reverse sequence of transformations, *viz.*, initial formation of a 3- or 5-benzylideneoxoniumyl-substituted intermediate, followed by activation of $OH-C(2)$, followed by nucleophilic displacement of the $C(2)-O-SF_2NEt_2$ function by an F-anion leading to the common intermediate **16**. Conformational disposition of the 5-*O*-benzoyl group facilitates an intramolecular attack by the carbonyl O-atom at the C(3)-atom, giving rise to a 3- or 5-benzylideneoxoniumyl-substituted intermediate, which, upon workup, is transformed into **17** and **18** (*Scheme 3*).

Scheme 3

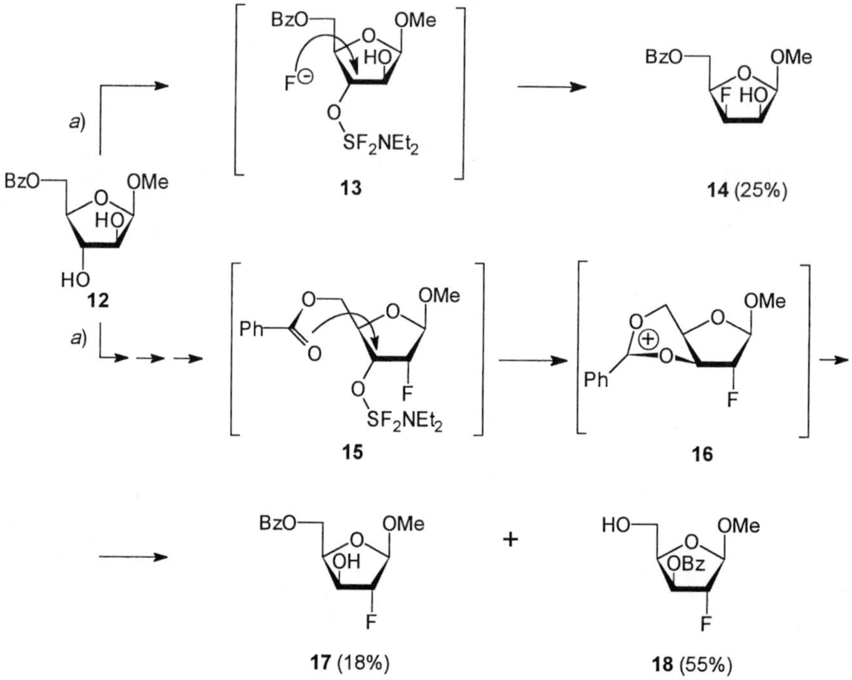

a) DAST(**12**/DAST 1:6 (molar ratio)), anh. CH_2Cl_2, 20°, 5 h; chromatography (SiO_2 *Woelm* (20% water)).

Further support for the above considerations on the initial step of the DAST reaction with arabinoside **12** was given by the reaction with the 5-*O*-tritylated arabinoside **21**. Indeed, in the presence of a non-participating trityl group, the lyxoside **23** and the riboside **24** were formed in a ratio of *ca.* 1:3 (*Scheme 4*). Note that the reaction of methyl 5-*O*-benzyl-β-D-arabinofuranoside with DAST yielded the corresponding lyxoside and riboside in the ratio of *ca.* 1:2 [11]. The arabinoside **21** was prepared from D-arabinose by the modification of a literature procedure [16], in which D-arabinose was treated with *ca.* 0.3N HCl/MeOH at room temperature for 3–4 h leading to the predominant formation of methyl α-D-arabinofuranoside. In our experiments, treatment of D-arabinose with 0.18N HCl/MeOH at room temperature for 5.5 h furnished a mixture of the β-D- and α-D-anomeric methyl arabinosides **19** and **20**, respectively (β-D/α-D *ca.* 2:3 according to ^{13}C-NMR data; combined yield 79%).

Scheme 4

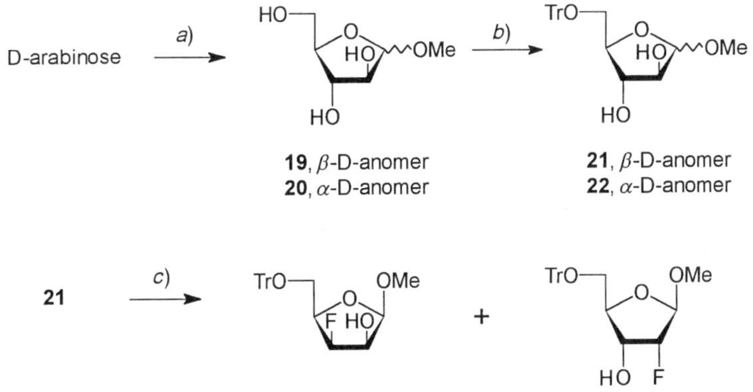

19, β-D-anomer
20, α-D-anomer

21, β-D-anomer
22, α-D-anomer

23   **24**

a) 0.18N HCl/MeOH, 20°, 5.5 h; **19/20** 79%; β-D/α-D *ca.* 2:3. *b*) TrCl, pyridine, 4-(dimethylamino)pyridine, 20°, 18 h; 60–70°, 4 h; 25% of **21**, 46% of **22**. *c*) DAST (**21**/DAST 1:6 (molar ratio)), anh. CH$_2$Cl$_2$, 20°, 18 h, chromatography (SiO$_2$), 16% of **23**, 52% of **24**.

Tritylation of this anomeric mixture followed by column chromatography afforded the 5-*O*-trityl derivatives **21** and **22** in 25 and 46% isolated yield, respectively.

In contrast to the β-D-arabinoside **21** and the methyl 5-*O*-benzyl-β-D-arabinofuranoside [11] as well, the reaction of methyl 5-*O*-benzyl-α-D-arabinoside **26** with DAST as described for its β-D-counterpart [11] furnished, after standard workup and subsequent chromatography, the *lyxo*-epoxide **25** as the main product, along with the lyxoside **27** in 53 and 14% isolated yield, respectively (*Scheme 5*). From this result, it may be reasonable to conclude that the principal route of the reaction is the formation of the C(3)−O−SF$_2$NEt$_2$ intermediate, which mainly undergoes an intramolecular nucleophilic attack at C(3) by the neighboring O-atom at C(2), to furnish the *lyxo*-epoxide **25**. It is noteworthy that the rate of conversion of **26** to the products was much slower than the analogous reaction with the β-D-anomer [11]. This reactivity displays a good resemblance with the reactivities of the pair of β-D- and α-D-xylosides **1** and **4**, respectively.

NMR Spectroscopic Studies. The assignment of the structures of the furanosides described here was based on NMR data (*Tables 1–3*). Confirmation of the F-atom position resulted from large one-bond coupling constants ^1J(C,F) of *ca.* 180–190 Hz,

Scheme 5

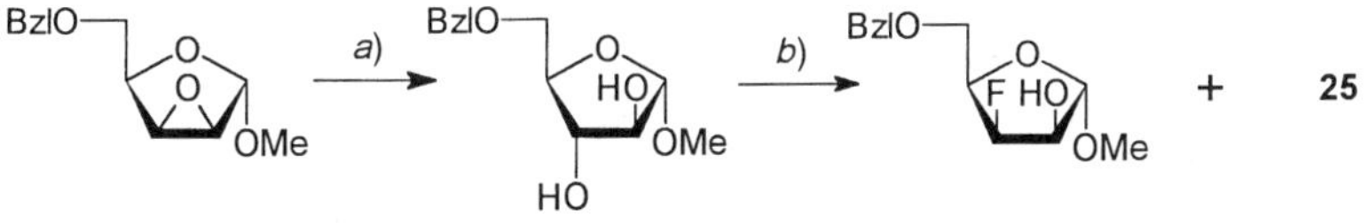

25   **26**   **27**

a) KOBz, DMSO, reflux, 1 h; sat. (at 0°) NH$_3$/MeOH, 20°, 18 h; 76%. *b*) DAST (**26**/DAST 1:6 (molar ratio)), anh. CH$_2$Cl$_2$/pyridine 18:1 (*v/v*), 20°, 5 h, chromatography (SiO$_2$); 14% of **27**; 53% (based on the consumed **26**) of 53%

Table 1. *¹H-NMR Chemical Shifts* (CDCl$_3$) *of* **1–5, 10–12, 14, 17, 18, 21,** *and* **23–26.** δ(H) in ppm.

	H–C(1) (H–C(1'))	H–C(2) (H–C(2'))	H–C(3) (H–C(3'))	H–C(4) (H–C(4'))	H–C(5) (H–C(5')) (H'–C(5'))	H'–C(5) (H'–C(5'))	Others[a]
1	4.90 (s)	4.26 (br. d)	4.16 (br. dd)		4.48–4.76 (m)		2.97 (d, OH–C(3)); 2.07 (d, OH–C(2))
2	4.93 (br. s)	4.27 (br. s)	5.20 (dt)		4.38–4.70 (m)		2.66 (br. s, OH–C(2))
3	5.02 (s)	3.86 (d)	3.75 (d)		4.38–4.54 (m)		
4	5.0 (d)	4.18 (t)	4.31 (t)		4.36–4.74 (m)		
5	4.98 (d)	4.22 (m)	4.91	4.60 (dt)	4.38–4.58 (m)		2.88 (dd, OH–C(2))
10[b]	5.90 (d)	4.37–4.28 (m)	4.06 (br. t)	4.37–4.28 (m)	3.83 (dd)	3.70 (dd)	8.23 (s, H–C(8)); 8.16 (s, H–C(2)); 6.04 (d, OH–C(3')); 5.96 (d, OH–C(2')); 7.34 (br. s, NH$_2$); 7.20 (br. s, Ph); 4.52 (s, PhCH$_2$)
11[c]	6.12 (d)	4.68 (ddd)	5.12 (dd)	4.52 (dm)	3.76 (dm)	3.69 (dd)	8.24 (s, H–C(8)); 8.10 (s, H–C(2)); 7.36 (br. s, NH$_2$, Ph); 4.55 (s, PhCH$_2$)
12	4.80 (d)	4.06–4.18 (m)	4.25 (t)	4.06–4.18 (m)	4.53 (dd)	4.37 (dd)	
14	ca. 4.90	4.20 (m)	5.04 (dm)		4.36–4.70 (m)		2.96 (d, OH–C(2))
17	5.08 (d)	4.93 (d)	4.39 (m)	4.62 (m)	4.72 (dd)	4.51 (dd)	2.98 (d, OH–C(3))
18	5.14 (d)	5.19 (dd)	5.63 (ddd)	4.68 (dt)	4.25 (br. d)		2.54 (br. s, OH–C(5))
21	4.82 (d)	4.02–4.16 (m)		3.95 (dt)	3.25 (d)		7.20–7.60 (m, Ph); 2.64 (br. s, OH–C(2), OH–C(3))
23	4.88 (d)	4.02–4.40 (m)	4.90 (dt)	4.02–4.40 (m)	3.32–3.48 (m)		7.20–7.60 (m, Ph); 2.80 (d, OH–C(2))
24	5.23 (d)	4.76 (dd)	4.35 (dm)	4.07 (m)	3.36 (dd)	3.23 (dd)	7.20–7.60 (m, Ph); 1.98 (d, OH–C(3))
25	4.94 (br. s)	3.74 (br. d)	3.62–3.66 (m)	4.18 (br. t)	3.62–3.66 (m)	3.66 (m)	7.26–7.40 (m, Ph); 4.54, 4.60 (2d, PhCH$_2$)
26	4.90 (s)	3.96 (s)	4.0 (br. s)	4.20 (br. m)	3.64 (dd)	3.72 (dd)	7.26–7.40 (m, Ph); 4.64, 4.54 (2d, PhCH$_2$)

[a] δ(H) of MeO: 3.36–3.50 ppm; δ(H) of BzO: ca. 7.20–7.70 (m) and ca. 8.08 (d).   [b] In (D$_6$)DMSO.   [c] In CDCl$_3$/CD$_3$OD.

Table 2. *Coupling Constants* $^3J(H,H)$ *and* J(H,F) *of* **1–5**, **10–12**, **14**, **17**, **18**, **21**, *and* **23–26**[a]). *J in Hz.*

	J(1,2) (J(1',2'))	J(2,3) (J(2',3'))	J(3,4) (J(3',4'))	J(4,5) (J(4',5'))	J(4,5') (J(4',5"))	J(1,F) (J(1',F))	J(2,F) (J(2',F))	J(3,F) (J(3',F))	J(4,F) (J(4',F))	Others
1	<1.0	<1.0	3.0	n.d.	n.d.	–	–	–	–	J(3,OH) = 10.5, J(2,OH) = 4.35
2	1.5	4.5	4.5	n.d.	n.d.	1.5	n.d.	54.0	n.d.	
3	<1.0	2.9	<1.0	n.d.	n.d.	–	–	–	–	
4	3.75	3.75	3.75	n.d.	n.d.	–	–	–	–	
5	5.0	5.70	1.45	4.0	4.0	<1.0	24.0	56.0	≈22	J(5,5') = J(2, OH) = 12.0, J(F,OH) = 1.5
10	1.75	1.0	5.0	3.5	7.0	–	–	–	–	J(5',5") = 10.5, J(3,OH) = 5.25, J(2,OH) = 3.5
11	7.15	4.5	0.7	3.0	2.6	<1.0	23.4	54.0	26.0	J(5',5") = 9.75, J(5',F) = 1.3
12	4.3	7.4	7.4	3.2	5.2	–	–	–	–	J(5,5') = 12.0
14	4.5	5.0	n.d.	n.d.	n.d.	<1.0	25.0	50.0	n.d.	J(2,OH) = 11.5
17	<1.0	0.9	4.0	4.5	6.0	10.0	48.5	11.5	<1.0	J(5,5') = 10.0, J(3,OH) = 10.0
18	<1.0	2.5	6.2	4.5	4.5	14.0	50.5	20.5	<1.0	
21	4.2	n.d.	6.3	4.5	4.5	–	–	–	–	
23	5.4	4.2	4.2	n.d.	n.d.	<1.0	≈22	54.0	28.8	J(2,OH) = 10.8
24	<1.0	3.9	8.0	3.5	7.5	10.5	53.5	25.0	<1.0	J(5,5') = 10.5, J(3,OH) = 9.0
25	<1.0	2.5	n.d.	6.0	6.0	–	–	–	–	
26	<1.0	<1.0	<1.0	2.5	2.5	–	–	–	–	J(5,5') = 10.5

[a]) n.d.: not determined.

exhibited in the ^{13}C-NMR spectra by the F-substituted C-atoms. Large geminal constants $^2J(H,F)$ of *ca.* 48–55 Hz, displayed in the ^{1}H-NMR spectra were of the same diagnostic value. Thus, compounds **2**, **5**, **6**, **11**, **14**, **23**, and **27** clearly show fluorination at C(3), whereas **17**, **18**, and **24** are 2-deoxy-2-fluoro derivatives. The assignments of configuration for most of the compounds synthesized were based primarily on ^{13}C-NMR data (*Table 3*), taking into account previous empirical correlations of the effect of configuration of vicinal substituents in the furanose ring on the $\delta(C)$ values of the atoms bearing these groups [1i][18].

The conformational analysis of the furanose rings of the compounds described above was performed by the PSEUROT (Version 6.2) program, which calculates the best fits of three experimental $^3J(H,H)$ coupling constants ($^3J(H-C(1),H-C(2))$), $^3J(H-C(2),H-C(3))$, and $^3J(H-C(3),H-C(4))$) to the five conformational parameters (P and ψ_m for both N- and S-type conformers and corresponding mol fractions) [19]. In the PSEUROT program, a minimization of the differences between the experimental and calculated couplings is accomplished by a nonlinear *Newton-Raphson* minimization. This procedure is enhanced if the ratio of the number of data points *vs.* the number of optimized parameters increases. Three $^3J(H,H)$ values are of limited value for conformational analysis of a pentofuranose ring, especially if the equilibrium under consideration represents a mixture of conformations present in comparable proportions. Moreover, it is not axiomatic that a two-state $N \leftrightarrow S$ model

Table 3. *¹³C-NMR Chemical Shifts (CDCl₃) and Coupling Constants J(C,H) and J (C,F) of* **1**–**6**, **12**, **14**, **17**, **18**, **21**, *and* **23**–**26**. *δ*(C) *in ppm, J in Hz.*

	C(1)	C(2)	C(3)	C(4)	C(5)	MeO	J(C(1),F)	J(C(2),F)	J(C(3),F)	J(C(4),F)	J(C(5))F
1	108.6 (J = 171.1)	79.6 (J ≈ 156)	76.4 (J ≈ 160)	80.9 (J ≈ 151.0)	64.3 (J ≈ 148.4)	55.3 (J ≈ 143.4)	–	–	–	–	–
2	108.0 (J = 173.6)	74.5 (J = 155.5)	92.4 (J = 160.4)	78.5 (J = 151.0)	64.2ᶜ (J = 148.7)	55.6 (J = 141.5)	4.0	14.9	187.4	25.3	4.5
3	102.5 (J = 169.8)	56.4ᵈ (J = 194.4)	55.1ᵈ (J = 190.6)	76.1 (J = 150.9)	64.2 (J = 149.1)	55.4 (J = 142.8)	–	–	–	–	–
4	101.9 (J = 172.5)	76.7ᵈ (J = 151.0)	76.3ᵈ (J = 151.0)	78.0 (J = 147.2)	63.3 (J = 149.1)	55.9 (J = 142.7)	–	–	–	–	–
5	102.4 (J = 174.1)	72.5 (J = 147.6)	90.5 (J = 165.3)	80.5 (J = 149.8)	63.8 (J = 149.8)	55.8 (J = 143.2)	<2.0	16.5	185.5	25.3	10.3
6	106.1 (J = 174.5)	96.0ᵈ (J = 162.5)	98.4ᵈ (J = 162.5)	80.2 (J = 150.4)	63.6 (J = 150.4)	55.0 (J = 142.4)	1,F3 <2.0; 1,F2: 36.5	186.5ᵈ; 2,F3: 30.3	180.0ᵈ; 3,F2: 28.1	4,F2 <2.0; 4,F3: 28.7	<2.0
12ᶜ	102.6 (J = 174.9)	76.7	80.3	78.5	65.8 (J = 149.4)	55.9 (J = 143.6)	–	–	–	–	–
14	101.8 (J = 171.7)	73.1 (J = 146.6)	89.7 (J = 147.8)	77.7 (J = 144.7)	63.5 (J = 151.0)	55.8 (J = 143.4)	<2.0	16.7	190.2	17.8	14.8
17	106.0 (J = 176.1)	96.8 (J = 145.3)	73.6 (J = 166.1)	80.9 (J ≈ 150)	63.7 (J = 149.7)	55.5 (J = 143.4)	32.6	183.0	25.6	<2.0	<2.0
18	106.6 (J = 172.0)	98.3 (J = 163.6)	75.7 (J = 153.5)	81.4 (J = 150.3)	61.4 (J = 144.7)	55.9 (J = 141.5)	35.3	182.7	29.1	<2.0	<2.0
21	101.8 (J = 173.6)	77.0ᵈ (J ≈ 150)	78.0ᵈ (J ≈ 150)	80.8 (J = 143.4)	64.9 (J = 142.5)	55.4 (J = 143.4)	–	–	–	–	–
23	101.6 (J = 174.2)	73.1 (J = 151.1)	89.9 (J = 164.3)	78.9 (J = 145.2)	63.1 (J = 142.7)	55.2 (J = 143.6)	<2.0	16.7	190.1	18.4	12.6
24	105.0 (J = 173.0)	93.9 (J = 168.3)	71.6 (J = 164.3)	81.7 (J = 149.4)	71.4 (J = 141.5)	55.3 (J = 141.5)	29.5	179.5	15.9	<2.0	<2.0
25	102.1 (J = 173.6)	56.1ᵈ (J = 192.8)	54.2ᵈ (J = 191.8)	74.9 (J = 147.4)	68.4 (J = 142.5)	55.4 (J = 141.5)	–	–	–	–	–
26ᶠ	109.5 (J = 173.6)	78.8ᵈ (J = 154.7)	78.2ᵈ (J = 152)	86.2 (J = 147.2)	69.4 (J = 143.4)	54.8 (J = 142.7)	–	–	–	–	–

ᵃ) δ(C) of BzO: 165.8–167.3 (s, C=O); 128.4–129.0 (dd, ¹J(C,H) ≈ 158, ³J(C,H) ≈ 7, Cₘ); 133.1–133.8 (dt, ¹J(C,H) ≈ 160, ³J(C,H) ≈ 8, Cₚ); 129.3–130.4 (dt, ¹J(C,H) ≈ 162, ³J(C,H) ≈ 6, Cₒ; t, ³J(C,H) ≈ 8, Cᵢₚₛₒ).   ᵇ) The CH₃O signal in the ¹H-coupled ¹³C-NMR shows an additional splitting into a d (³J(CH₃O, H–C(1)) = 2.8–4.7 Hz; not observed (<2.0 Hz) for **1** and **6**; the C(1) signal in the ¹H-coupled ¹³C-NMR is a dm: ³J(C(1), CH₃O) ≈ 3.0 and ³J(C(1), H–C(4)) ≈ 3.0 for **2**, **3**, and **5**. ³J(C(1), CH₃O) = 4.0 and ³J(C(1), H–C(4)) = 4.0 Hz for **12**, **14**, **17**, **18**, and **22**.   ᶜ) The C(5) signal displayed an additional J(C,H) of 4.5 (**2**), 2.2 (**5**), 4.0 (**12**), 6.0 (**14**), and 3.5 Hz (**17**) (tentatively assigned to ³J(C(5), H–C(3)).   ᵈ) Data (δ and related J's) may be interchanged.   ᵉ) The C(3) and C(4) signals may be interchanged.   ᶠ) The C(2) signal displays additional couplings: ³J(C(3), H–C(1)) = 3.0 and ³J(C(4), H–C(2)) = 2.2 Hz, resp. ᶠ) The C(2) signal displays an additional coupling, tentatively assigned to ³J(C(2), H–C(4)) (3.7 Hz).

may accurately describe pentofuranose rings with other than β-D-*ribo*-configuration. Thus, two approaches are of interest to define more accurately the pseudorotational parameters P and ψ_m for two N- and S-conformers. *Serianni* and co-workers have shown that some of the $J(C,H)$ coupling constants are equally valuable conformational probes for defining a rather narrow N- or S-domain of the pseudorotational wheel of the pentofuranose ring [20]. The main problem associated with this approach is that the $J(C,H)$ coupling constants can be correctly measured only in ^{13}C-enriched molecules. With reference to deoxy fluoro nucleosides, *Chattopadhyaya* and co-workers have very recently developed a new *Karplus*-type relation between vicinal $^3J(H,F)$ coupling constants and the corresponding $H-C-C-F$ torsion angles [21]. The use of temperature-dependent $^3J(H,F)$ coupling constants in combination with $^3J(H,H)$ greatly facilitates the conformational analysis of pentofuranose rings because of the overwhelming increase of the number of experimental data points over the puckering parameters P and ψ_m [21].

We have qualitatively examined the $^3J(C,F)$ spin-couplings as an additional conformational probe of furanose rings in solution. The conformational behavior was evaluated by the PSEUROT analysis of the $^3J(H,H)$ values only essentially as it was described previously [22]. The resulting optimized geometries of N- and S-pseudo-rotamers are presented in *Table 4*.

Table 4. *Pseudorotational Parameters of Some Selected Compounds*

	P_N	$\psi'_{m(N)}$	P_S	$\psi_{m(S)}$	r.m.s.	$\|\Delta J_{max}\|$	% S
1	18.6	29.9	108[a])	34[a])	0.000	0.00	15
2	9.3	41[a])	220.0	38[a])	0.060	0.08	53
4	19.0	46[a])	121.1	46[a])	0.001	0.00	61
5	− 9[a])	44[a])	147.5	27.9	0.169	0.25	100
11	10[a])	39[a])	189.9	35.4	0.006	0.00	98
12	− 13.0	42[a])	137.4	38[a])	0.081	0.13	6
17	− 15.1	28.7	34[a])	108[a])	0.000	0.00	14
23	− 30.5	37.0	108[a])	46[a])	0.545	0.75	0
24	27.9	45.6	198[a])	44[a])	0.000	0.00	16
26	9[a])	30[a])	156.0	31.5	0.000	0.00	99

[a]) The values indicated were fixed during the final calculations.

The factors affecting the conformation of the pentofuranose rings of nucleosides in solution have been exensively investigated during last years by *Chattopadhyaya* and coworkers (see [21] and ref. cit. therein). The sugar moieties of nucleosides are involved in a two-state $N \leftrightarrow S$ pseudorotational equilibrium, which is driven by the relative strength of various *gauche* and anomeric stereoelectronic effects. It was shown that the stronger *gauche* effect of an F-substituent, due to its high electronegativity, governs the overall conformation of the pentofuranose rings [5][21][23]. Less is known regarding the contribution of a MeO group replacing a heterocyclic base to the conformational behavior of pentofuranose rings [24].

The dominating population of the S-conformer $(^2T_1 \leftrightarrow {}^2E)$ of the α-D-riboside **5** is apparently caused by the *gauche* effects of the $F-C(3)-C(4)-O(4)$, $F-C(3)-C(2)-OH$, and $HO-C(2)-C(1)-OMe$ fragments. In such a conformation, the $F-C(3)-$

C(4)−C(5) and F−C(3)−C(2)−C(1) fragments are in a *anti*-periplanar (*ca.* 170°) and *gauche* (*ca.* 90°) arrangement, respectively, which is consistent with the corresponding $^3J(C,F)$ values of 10.3 and <2.0 Hz (*Table 3*). Changing the anomeric configuration from α-D to β-D gives rise to an equal population of the *N*- and *S*-type puckered conformers of the β-D-riboside **2**, which result mainly from the competing *gauche* interactions (F−C(3)−C(4)−O(4), F−C(3)−C(2)−OH, and HO−C(2)−C(1)−OMe fragments for *S*-type, and F−C(3)−C(2)−OH and HO−C(2)−C(1)−O(4) fragments for *N*-type). Note that, on the basis of the aforementioned *gauche* effects alone, the *N*- and *S*-conformers would not be equally populated in the β-D-riboside **2**. However, the coupling constants $^3J(C(1),F)$ (4.0 Hz) and $J(F,C(5))$ (4.5 Hz) of **2** take the intermediate values and are consistent with the presence of comparable proportions of the *N*- and *S*-conformers.

It is noteworthy that the conformational behavior of the β-D-riboside **2** substantially differs from that of 3′-deoxy-3′-fluoroadenosine (P_s 168, $\psi_{m(S)}$ 40.0; *S* 97%) [5] and of its 5′-*O*-benzyl derivative **11** as well. This may be due to the different anomeric effects of the MeO group and the adenine base. On the contrary, the conformational properties of the β-D-riboside **24** are closely related to those of 2′-deoxy-2′-fluoroadenosine [4]. In the case of the β-D-riboside **24**, the values of $^3J(H−C(1),F)$ (10.5 Hz) and $^3J(H−C(3),F)$ (25.0 Hz) are in good qualitative agreement with the predominant *N*-type ($^3E \leftrightarrow {}^3T_4$) conformation. Another interesting finding consists in that the population of the *S*-conformer increases by *ca.* 45% by going from the β-D-anomers **1** and **2** to the respective α-D-counterparts **4** and **5**. A more dramatic conformational change occurs on going from β-D-arabinoside **12** to its α-D-anomer **26**.

The $^3J(C(4),F)$ (< 2.0 Hz) of xyloside **17** is in accord with predominant population of the *N*-conformer ($_2E$), for which the F−C(2)−C(3)−C(4) fragment is in the *gauche* orientation (*ca.* 90°). The migration of the benzoyl group from the 5-*O* to the 3-*O* position is accompanied by remarkable conformational changes, which are clearly reflected in the $^3J(H−C(1),F)$ and $^3J(H−C(3),F)$ values of **18** (*Table 2*). Unexpectedly, attempts to perform the PSEUROT analysis of the 3-benzoate **18** failed. Although we found a number of pseudorotational parameters with good (< 0.2) root mean square (r.m.s.) deviations of the fit, the most populated conformations of **18** are not consistent with the $^3J(C(4),F)$ <2.0 Hz. In a similar way, the PSEUROT analysis of lyxosides **23** and **27** led to the pseudorotational parameters with rather large r.m.s. values (see, *e.g.*, the data for **23**; *Table 4*) which are, however, compatible with the $^3J(C(1),F)$ values of <2.0 Hz. These preliminary data tend to suggest that conformational behavior of lyxosides and, to some extent, xylosides cannot be adequately described by the two-state $N \leftrightarrow S$ pseudorotational equilibrium. More definitive conclusions can, however, be drawn after detailed conformational analysis with both $^3J(H,H)$ and $^3J(H,F)$ coupling constants [23].

In conclusion, the scope and limitations of ring fluorination of pentofuranosides containing free secondary OH groups under the action of DAST were established. We demonstrated that the synthesis of some fluorinated carbohydrates may be achieved in good overall yield starting from commercially available sugars. This approach does provide a useful alternative to the previously described methods.

I.A.M. is deeply grateful to the *Alexander von Humboldt Foundation*, Bonn/Bad-Godesberg, Germany, for partial financial support of this work. The authors are thankful to Dr. *Natalja B. Khripach*; Institute of Bioorganic Chemistry, Minsk, for the measurement of the NMR spectra.

Experimental Part

1. *General.* The solns. of compounds in org. solvents were dried (Na_2SO_4) for 4 h. Column chromatography (CC): silica gel *60* (70–230 mesh ASTM; *Merck*, Darmstadt, Germany), except where otherwise indicated. TLC: *Silufol UV$_{254}$* (Czech Republic); eluents: hexane/AcOEt 1:2 (*A*), hexane/AcOEt 4:1 (*B*), $CHCl_3$/MeOH 15:1 (*C*), and hexane/AcOEt 1:1 (*D*). M.p.: *Boetius* apparatus (Germany); not corrected. UV Spectra: *Specord M-400* spectrometer (*Carl Zeiss*, Germany). CD Spectra and $[\alpha]_D^{25}$: *J-20* spectropolarimeter (*JASCO*, Japan). ¹H- and ¹³C-NMR Spectra: *AC-200* spectrometer equipped with an *Aspect 3000* data system (*Bruker*, Germany) at 23° and 200.13 (¹H) and 50.325 MHz (¹³C); $CDCl_3$ soln., unless otherwise stated; δ values in ppm downfield from internal $SiMe_4$; assignments of $\delta(H)$, when possible, by selective homonuclear decoupling experiments.

2. *Methyl 5-O-Benzoyl-β-D-xylofuranoside* (**1**) *and Methyl 5-O-Benzoyl-α-D-xylofuranoside* (**4**). To a soln. of syrupy 5-*O*-benzoyl-1,2-*O*-isopropylidene-α-D-xylofuranose [12][13] (4.95 g, 16.82 mmol) in anh. MeOH (95 ml), crystalline I_2 (0.95 g) was added, and the mixture was heated under reflux for 4 h. After cooling, the mixture was poured into sat. aq. $Na_2S_2O_3$ soln. (150 ml) and extracted with $CHCl_3$ (3 × 200 ml), the combined org. extract washed with sat. aq. NaCl soln. (100 ml), dried, and evaporated, and the oily residue (4.92 g) submitted to CC (silica gel; 200 ml), linear gradient of hexane/AcOEt 1:2 (1.5 l) in hexane/AcOEt 7:1 (1.5 l): 1.6 g (35%) of **1**, and 1.52 g (34%) of **4**.

Data of **1**: M.p. 107–108° (from Et_2O/hexane). $[\alpha]_D^{25} = -46.0$ ($c = 1.0$, $CHCl_3$). TLC (*A*): R_f 0.49. Anal. calc. for $C_{13}H_{16}O_6$ (268.26): C 58.20, H 6.02; found: C 57.93, H 5.82.

Data of **4**: TLC (*A*): R_f 0.43.

3. *Reaction of DAST with* **1** *and* **4**. To a soln. of **1** (0.24 g, 0.89 mmol) in anh. CH_2Cl_2 (5 ml), DAST (0.71 ml, 5.36 mmol) was added and the mixture stirred at r.t. for 19 h. After cooling to 0°, the mixture was poured into sat. cold aq. $NaHCO_3$ soln. (60 ml), the aq. phase extracted with CH_2Cl_2 (3 × 80 ml), the combined org. extract dried and evaporated, and the residue chromatographed (silica gel *L* (*Chemapol*, Czech Republic, 40/100 μm; 80 ml), linear gradient of hexane/AcOEt 2:1 (350 ml) in hexane/AcOEt 6:1 (350 ml): 41 mg (18%) of **3** and 150 mg (62%) of **2**.

Methyl 5-O-Benzoyl-3-deoxy-3-fluoro-β-D-ribofuranoside (**2**). M.p. 68–70° (from Et_2O/hexane) $[\alpha]_D^{25} = -82.0$ ($c = 1.0$, $CHCl_3$). TLC (*B*): R_f 0.17. Anal. calc. for $C_{13}H_{15}FO_5$ (270.28): C 57.77, H 5.59; found: C 57.64, H 5.76.

Methyl 2,3-Anhydro-5-O-benzoyl-β-D-ribofuranoside (**3**). Syrup. TLC (*B*): R_f 0.40.

In a similar way, **4** (0.23 g, 0.86 mmol) was treated with DAST (0.68 ml, 5.14 mmol) in CH_2Cl_2 (5 ml) at r.t. for 4.5 h: 20 mg (9%) of **6** and 140 mg (60%) of **5**.

Methyl 5-O-Benzoyl-2,3-dideoxy-2,3-difluoro-α-D-arabinofuranoside (**6**): Syrup. TLC (*B*): R_f 0.62.

Methyl 5-O-Benzoyl-3-deoxy-3-fluoro-α-D-ribofuranoside (**5**): Syryp. TLC (*B*): R_f 0.29.

4. *5'-O-Benzyl-3'-deoxy-3'-fluoroadenosine* (**11**). To a soln. of methyl 5-*O*-benzyl-β-D-xylofuranoside [1b] (**7**; 0.3 g, 1.18 mmol) in anh. pyridine (3.5 ml), benzoyl chloride (0.33 ml, 2.84 mmol) was added, and the mixture was stirred at r.t. for 18 h. Standard workup followed by CC (silica gel (100 ml), linear gradient of hexane/AcOEt 10:1 (0.5 l) in hexane (0.5 l) gave 0.50 g (92%) of syrupy *2,3-di-O-benzoyl-5-O-benzyl-β-D-xylofuranoside* (**8**). TLC (*B*): R_f 0.67. ¹H-NMR ($CDCl_3$): 3.48 (*s*, MeO); 3.79 (*d*, $J = 6.0$, 2 H–C(5)); 4.46, 4.54 (2*d*, $J = 12$, PhCH_2); 4.82 (*dt*, $J = 6.0$, 5.5, H–C(4)); 5.10 (*s*, H–C(1)); 5.46 (*d*, $J = 1.8$, H–C(2)); 5.76 (*dd*, $J = 1.8$, 5.5, H–C(3)); 7.38–7.62, 8.0–8.08 (2*m*, 3 Ph).

A mixture of **8** (0.50 g, 1.08 mmol), $SnCl_4$ (0.37 ml, 3.17 mmol) and the bis(trimethylsilyl) derivative of N^6-benzoyladenine (obtained from 0.39 g (1.62 mmol) of N^6-benzoyladenine) in anh. MeCN (10 ml) was refluxed for 15 min and then allowed to cool to r.t. under stirring for an additional 30 min. After standard workup, the residue was purified by CC (silica gel; 100 ml), linear gradient of heptane/AcOEt 1:1 (0.5 l) in heptane (0.5 l), then heptane/AcOEt 1:2 (0.4 l): 0.46 g (63%) of *N^6-benzoyl-9-(2,3-di-O-benzoyl-5-O-benzyl-β-D-xylofuranosyl)adenine* (**9**). Foam. TLC (*C*): R_f 0.90. ¹H-NMR ($CDCl_3$): 3.84 (*dd*, $J = 5.1$, 11.0, H–C(5')); 3.92 (*dd*, $J = 4.5$, 11.0, H'–C(5')); 4.52, 4.60 (2*d*, $J = 12$, PhCH_2); 4.86 (*m*, $J = 4.2$, 4.5, 5.1, H–C(4')); 5.92 (*dd*, $J = 4.2$, 2.5, H–C(3')); 6.25 (*t*, $J = 2.5$, H–C(2')); 6.50 (*d*, $J = 2.5$, H–C(1')); 5.46 (*d*, $J = 1.8$, H–C(2)); 5.76 (*dd*, $J = 1.8$, 5.5, H–C(3)); 7.40–7.66, 7.88–8.12 (2*m*, 4 Ph); 8.46 (*s*, H–C(2)); 8.68 (*s*, H–C(8)).

Standard debenzoylation of **9** followed by CC (silica gel (130 ml), linear gradient of $CHCl_3$/EtOH 8:1 (0.6 l) in $CHCl_3$ (0.6 l)) afforded 0.16 g (70%) of 9-(5-O-benzyl-β-D-xylofuranosyl)adenine (**10**). M.p. 83–85° (from CH_2Cl_2/MeOH). $[\alpha]_D^{25} = -41.0$ ($c = 0.54$, MeOH). TLC (*C*): R_f 0.17. UV (EtOH): 207 (23375), 260 (13880). CD (EtOH; $[\Theta] \cdot 10^{-3}$ (λ in nm)): -22.3 (220), -5.4 (250). Anal. calc. for $C_{17}H_{19}N_5O_4$ (357.40): C 57.13, H 5.36, N 19.60; found: C 57.00, H 5.52, N 19.41.

To a soln. of **10** (49 mg, 0.14 mmol) in CH_2Cl_2 (2 ml) and anh. pyridine (0.15 ml), a soln. of DAST (0.11 ml, 0.83 mmol) in the same solvent mixture (1.6 ml) was added, and the mixture was stirred at r.t. for 5 h. After dilution with CH_2Cl_2 (30 ml), the mixture was washed with sat. aq. $NaHCO_3$ soln. (30 ml), the aq. phase extracted with CH_2Cl_2 (5 × 35 ml), the combined org. extract dried and evaporated, and the residue submitted to CC (silica gel (50 ml), linear gradient of $CHCl_3$/MeOH 11:1 (0.35 l) in $CHCl_3$ (0.35 l)): 18 mg (48% based on the amount of consumed **10**) of **11** and 12 mg (24%) of recovered **10**.

Data of **11**: M.p. 180–181° (from EtOH). $[\alpha]_D^{25} = -71.0$ ($c = 0.65$, MeOH). TLC (*C*): R_f 0.24. UV(EtOH): 207 (28880), 260 (14600). CD (EtOH; $[\Theta] \cdot 10^{-3}$ (λ in nm)): -8.2 (217), -11.7 (260). Anal. calc. for $C_{17}H_{18}FN_5O_3$ (359.40): C 56.82, H 5.05, N 19.49; found: C 57.01, H 4.83, N 19.20.

5. Reaction of DAST with Methyl 5-O-Benzoyl-β-D-arabinofuranoside (**12**). The arabinoside **12** was prepared in two steps from methyl 2,3-anhydro-β-D-lyxofuranoside [17]. Standard benzoylation of the latter followed by the treatment of the syrupy 5-*O*-benzoate with KOBz in DMSO as described previously [11] gave, after CC separation, xyloside **1** (yield 23%; TLC (*A*): R_f 0.49) and syrupy arabinoside **12** (yield 22%; TLC (*A*): R_f 0.31). TLC of the mixture before CC showed the presence of two compounds with higher mobility than **1** and **12**, which were presumably the di-*O*-benzoyl derivatives of the latter and were not investigated.

To a soln. of **12** (0.32 g, 1.19 mmol) in CH_2Cl_2 (7 ml), DAST (0.95 ml, 7.18 mmol) was added and the mixture stirred at r.t. for 5 h and worked up as described for the reaction of **1**. CC (silica gel containing 20% of H_2O (*Woelm*, Germany, 80 ml), hexane/AcOEt 11:1 (400 ml), then hexane/AcOEt 1:2 (350 ml)) afforded the following syrupy-like compounds, in order of elution: *methyl 5-O-benzoyl-2-deoxy-2-fluoro-β-D-xylofuranoside* (**17**; 33 mg, 18%), *methyl 3-O-benzoyl-2-deoxy-2-fluoro-β-D-xylofuranoside* (**18**; 100 mg, 55%), *methyl 5-O-benzoyl-3-deoxy-3-fluoro-β-D-lyxofuranoside* (**14**, 45 mg, 25%), and the starting **12** (140 mg), TLC (*D*): R_f 0.69, 0.55, 0.46, and 0.19, resp.

6. Reaction of DAST with Methyl 5-O-Trityl-β-D-arabinofuranoside (**21**). Arabinoside **21** was prepared from D-arabinose in two steps by the following modification of a known procedure [16]: To a stirred suspension of D-arabinose (1.0 g, 6.66 mmol) in anh. MeOH (25 ml), a freshly prepared HCl soln. in MeOH (resulting from the addition of acetyl chloride (0.4 ml) to MeOH (6 ml) at 0°) was added, and stirring was continued at r.t. The mixture was homogeneous after *ca.* 5 h. After additional 0.5 h stirring, the mixture was neutralized by powdered $(NH_4)HCO_3$ to pH 7.0–7.5. Insoluble $(NH_4)HCO_3$ was filtered off and washed with MeOH (10 ml), silica gel (30 ml) was added to the combined MeOH solns., and the mixture was evaporated. The residue was put on top of a column packed with silica gel (50 ml) and submitted to CC ($CHCl_3$ (150 ml), then acetone (250 ml)): 0.86 g (79%) of oily *methyl β-D/α-D-arabinofuranoside* (**19/20**). ^{13}C-NMR ((D_6)DMSO): **19/20** 2:3; **19** 104.9 (C(1)); 78.2 (C(2)); 75.8 (C(3)); 83.9 (C(4)); 64.5 (C(5)); 55.0 (MeO); **20**: 109.6 (C(1)); 82.6 (C(2)); 77.6 (C(3)); 84.4 (C(4)); 62.0 (C(5)); 55.5 (MeO).

To a soln. of **19/20** (0.86 g, 5.24 mmol) in anh. pyridine (17 ml), 4-(dimethylamino)pyridine (0.71 g, 5.81 mmol) and trityl chloride (1.77 g, 6.35 mmol) were added, and the mixture was stirred first at r.t. for 18 h and then at 60–70° for 4 h. The mixture was allowed to cool to r.t. and poured into ice/water (80 ml), the org. phase separated after the ice was melted, the aq. phase washed with AcOEt (3 × 100 ml), the combined org. soln. washed with 5% aq. $NaHCO_3$ soln. (80 ml), dried, and evaporated, and the residue chromatographed (silica gel (150 ml), linear gradient (0 → 50%) of AcOEt (1.0 l) in hexane (1.0 l)): 0.97 g (46%) of **22** and 0.65 g (30%) of **21**.

Methyl 5-O-Trityl-α-D-arabinofuranoside (**22**): TLC (*D*): R_f 0.42. ^{1}H-NMR ($CDCl_3$): 3.42 (*dd*, $J = 2.0$, 10.5, H–C(5)); 3.66 (*s*, MeO); 3.89 (*dd*, $J = 2.5$, 10.5, H'–C(5)); 3.89 (br. *s*, H–C(3)); 3.98 (*s*, H–C(2)); 4.12 (*m*, H–C(4)); 5.00 (*s*, H–C(1)); 7.18–7.50 (*m*, 3 Ph).

β-D-Anomer **21**: M.p. 56–58° (from Et_2O/hexane). $[\alpha]_D^{25} = -52.0$ ($c = 1.0$, $CHCl_3$). TLC (*D*): R_f 0.27. Anal. calc. for $C_{25}H_{26}O_5$ (406.52): C 73.87, H 6.45; found: C 73.75, H 6.76.

As described for the reaction of **12**, **21** (0.25 g, 0.61 mmol) in CH_2Cl_2 (6 ml) was treated with DAST (0.52 ml, 3.93 mmol) at r.t. for 18 h. CC (silica gel (70 ml), linear gradient of hexane/AcOEt 3:1 (0.5 l) in hexane) gave 130 mg (52%) of **24** and 40 mg (16%) of **23**.

Methyl 2-Deoxy-2-fluoro-5-O-trityl-β-D-ribofuranoside (**24**): Syrup. TLC (*B*): R_f 0.41.

Methyl 3-Deoxy-3-fluoro-5-O-trityl-β-D-lyxofuranoside (**23**): TLC (*B*): R_f 0.27.

7. *Reaction of DAST with Methyl 5-O-Benzyl-α-D-arabinofuranoside* (**26**). Compound **26** was prepared by treatment of methyl 2,3-anhydro-5-O-benzyl-α-D-lyxofuranoside (**25**) [16][1b] (0.7 g, 2.96 mmol) with KOBz (1.4 g, 8.74 mmol) in DMSO (12 ml) under reflux for 1 h. Similarly to the synthesis of **12** (see above), TLC (*A*, R_f 0.31) of the residue after workup showed two main products, **26** (*D*; R_f 0.25) and a faster moving compound (*D*; R_f 0.51), probably the benzoyl derivative of **26**. The oily residue was dissolved in MeOH (40 ml), the soln. saturated at 0° with ammonia, stored at r.t. for 18 h, and evaporated, and the residue submitted to CC (silica gel (50 ml), linear gradient of hexane/AcOEt 1:1 (0.5 l) in hexane/AcOEt 1:8 (0.5 l)): syrupy **26** (0.57 g, 76%).

The reaction of **26** with DAST was performed as described previously for its β-D-anomer [11]. In contrast to the latter, **26** (0.14 g, 0.55 mmol) reacted very slowly and afforded, after stirring at r.t. for 5 h followed by standard workup and chromatography, *methyl 2,3-anhydro-5-O-benzyl-α-D-lyxofuranoside* (**25**; 37 mg, 53%; TLC (*D*): R_f 0.87) and *methyl 5-O-benzyl-3-deoxy-3-fluoro-α-D-lyxofuranoside* (**27**; 10 mg, 14% based on the amount of consumed **26**; TLC (*D*): R_f 0.64), and recovered **26** (65 mg). ^{1}H- and ^{13}C-NMR for **27**: in fair agreement with those previously reported for the same compound obtained by an alternative method [1i].

REFERENCES

[1] a) I. A. Mikhailopulo, T. I. Pricota, N. E. Poopeiko, G. G. Sivets, E. I. Kvasyuk, T. V. Sviryaeva, L. P. Savochkina, R. Sh. Beabealashvilli, *FEBS Lett.* **1989**, *250*, 139; b) I. A. Mikhailopulo, N. E. Poopeiko, T. I. Pricota, G. G. Sivets, E. I. Kvasyuk, J. Balzarini, E. De Clercq, *J. Med. Chem.* **1991**, *34*, 2195; c) I. A. Mikhailopulo, G. G. Sivets, T. I. Pricota, N. E. Poopeiko, J. Balzarini, E. De Clercq, *Nucleosides Nucleotides* **1991**, *10*, 1743; d) I. A. Mikhailopulo, T. I. Pricota, N. E. Poopeiko, T. V. Klenitskaya, N. B. Khripach, *Synthesis* **1993**, 700; e) I. A. Mikhailopulo, G. G. Sivets, N. E. Poopeiko, N. B. Khripach, *Nucleosides Nucleotides* **1995**, *14*, 383; f) I. A. Mikhailopulo, G. G. Sivets, N. E. Poopeiko, N. B. Khripach, *ibid.* **1995**, *14*, 381; g) N. E. Poopeiko, J. Poznanski, I. A. Mikhailopulo, D. Shugar, T. Kulikowski, *ibid.* **1995**, *14*, 435; h) E. N. Kalinichenko, T. L. Podkopaeva, N. E. Poopeiko, M. Kelve, M. Saarma, I. A. Mikhailopulo, J. E. van den Boogaart, C. Altona, *Recl. Trav. Chim. Pays-Bas* **1995**, *114*, 43; i) I. A. Mikhailopulo, G. G. Sivets, N. E. Poopeiko, N. B. Khripach, *Carbohydr. Res.* **1995**, *278*, 71; j) G. V. Zaitseva, G. G. Sivets, Z. Kazimierczuk, J. Vilpo, I. A. Mikhailopulo, *Bioorg. Med. Chem. Lett.* **1995**, *5*, 2999; k) N. E. Poopeiko, N. B. Khripach, Z. Kazimierczuk, J. Balzarini, E. De Clercq, I. A. Mikhailopulo, *Nucleosides Nucleotides* **1997**, *16*, 1083; l) G. V. Zaitseva, A. I. Zinchenko, V. N. Barai, N. I. Pavlova, E. I. Boreko, I. A. Mikhailopulo, *ibid.* **1999**, *18*, 687; m) I. A. Mikhailopulo, G. G. Sivets, N. B. Khripach, *ibid.* **1999**, *18*, 689.

[2] F. Viani, in 'Enantiocontrolled Synthesis of Fluoro-Organic Compounds: Stereochemical Challenges and Biomedical Targets', Ed. V. A. Soloshonok, John Wiley & Sons Ltd., 1999, Chapt. 13, pp. 419–449.

[3] Z. G. Chidgeavadze, A. V. Scamrov, R. Sh. Beabealashvilli, E. I. Kvasyuk, G. V. Zaitseva, I. A. Mikhailopulo, G. Kowollik, P. Langen, *FEBS Lett.* **1985**, *183*, 275; G. V. Zaitseva, E. I. Kvasyuk, N. E. Poopeiko, T. I. Kulak, V. E. Pashinnik, V. I. Tovstenko, L. N. Markovskii, I. A. Mikhailopulo, *Bioorg. Khim.* **1988**, *14*, 1275; E. I. Kvasyuk, G. V. Zaitseva, L. P. Savochkina, I. A. Mikhailopulo, Z. G. Chidgeavadze, R. Sh. Beabealashvilli, P. Langen, *ibid.* **1989**, *15*, 781; G. E. Wright, N. C. Brown, *Pharmacol. Ther.* **1990**, *47*, 447; A. A. Krayevsky, K. A. Watanabe, 'Modified Nucleosides as anti-AIDS Drugs: Current Status and Perspectives', Bioinform, Moscow, 1993, p. 211.

[4] R. H. Griffey, E. Lesnik, S. Freier, Y. S. Sanghvi, K. Teng, A. Kawasaki, C. J. Guinosso, P. D. Wheeler, V. Mohan, P. Dan Cook, in 'American Chemical Society Symposium Series 580. Carbohydrate Modifications in Antisense Research', Eds. Y. S. Sanghvi and P. Dan Cook, American Chemical Society, Washington, DC, 1994, Chapt. 14, p. 212.

[5] J. E. van den Boogaart, E. N. Kalinichenko, T. L. Podkopaeva, I. A. Mikhailopulo, C. Altona, *Eur. J. Biochem.* **1994**, *221*, 759.

[6] M. R. Player, P. F. Torrence, *Pharmacol. Ther.* **1998**, *78*, 55.

[7] H. Ikeda, R. Fernandez, A. Wilk, J. J. Barchi, Jr., X. Huang, V. E. Marquez, *Nucleic Acids Res.* **1998**, *26*, 2237.

[8] B. Reif, V. Wittmann, H. Schwalbe, C. Griesinger, K. Woerner, K. Jahn-Hofmann, J. W. Engels, *Helv. Chim. Acta* **1997**, *80*, 1952.

[9] P. Herdewijn, A. Van Aerschot, L. Kerremans, *Nucleosides Nucleotides* **1989**, *8*, 65.

[10] H. Vorbrueggen, in 'Nucleoside Analogues. Chemistry, Biology, and Medical Applications', Eds. R. T. Walker, E. De Clercq, and F. Eckstein, Plenum Press, New York, 1979, Vol. 26, Ser. A, NATO Advanced Study Institute, pp. 35–69.

[11] I. A. Mikhailopulo, G. G. Sivets, *Synlett* **1996**, 173.

[12] J. Moravcova, J. Capkova, J. Stanek, *Carbohydr. Res.* **1994**, *263*, 61.

[13] C. R. Johnson, D. R. Bhumralkar, *Nucleosides Nucleotides* **1995**, *14*, 185.

[14] W. A. Szarek, A. Zamojski, H. N. Tiwari, E. R. Ison, *Tetrahedron Lett.* **1986**, *27*, 3827.

[15] A. A. Akhrem, E. K. Adarich, L. N. Kulinkovich, I. A. Mikhailopulo, E. B. Poschastieva, V. A. Timoshchuk, *Dokl. Akad. Nauk SSSR* **1974**, *219*, 99; N. E. Poopeiko, E. I. Kvasyuk, I. A. Mikhailopulo, M. J. Lidaks, *Synthesis* **1985**, 605; J. A. Maurins, R. A. Paegle, M. J. Lidaks, E. I. Kvasyuk, I. A. Mikhailopulo, *Bioorg. Khim.* **1986**, *12*, 1514.

[16] R. K. Ness, H. G. Fletcher, Jr., *J. Am. Chem. Soc.* **1958**, *80*, 2007.

[17] B. R. Baker, R. E. Schaub, J. H. Williams, *J. Am. Chem. Soc.* **1955**, *77*, 7.

[18] A. A. Akhrem, I. A. Mikhailopulo, A. F. Abramov, *Org. Magn. Res.* **1979**, *12*, 247.

[19] J. van Wijk, C. Altona, 'PSEUROT 6.2. A Program for the Conformational Analysis of the Five-Membered Rings', University Leiden, July, 1993; C. A. G. Haasnot, F. A. A. M. de Leeuw, C. Altona, *Tetrahedron* **1980**, *86*, 2783, L. J. Rinkel, C. Altona, *J. Biomol. Struct. Dyns.* **1987**, *4*, 621.

[20] A. S. Serianni, in 'NMR of Biological Macromolecules', 'NATO ASI Series H, Cell Biology', Vol. 87, Ed. C. I. Stassinopoulou, Springer-Verlag, Berlin, 1994, p. 293; C. A. Podlasek, J. Wu, W. A. Stripe, P. B. Bondo, A. S. Serianni, *J. Am. Chem. Soc.* **1995**, *117*, 8635; C. A. Podlasek, W. A. Stripe, I. Carmichael, M. Shang, B. Basu, A. S. Serianni, *J. Am. Chem. Soc.* **1996**, *118*, 1413.

[21] C. Thibaudeau, J. Plavec, J. Chattopadhyaya, *J. Org. Chem.* **1998**, *63*, 4967.

[22] F. Seela, H. Debelak, H. Reuter, G. Kastner, I. A. Mikhailopulo, *Tetrahedron* **1999**, *55*, 1295.

[23] W. Guschlbauer, K. Jankowski, *Nucleic Acids Res.* **1980**, *8*, 1421; D. M. Cheng, L.-S. Kan, P. O. P. Ts'o, Y. Takatsuka, M. Ikehara, *Biopolymers* **1983**, *22*, 1427.

[24] J. Raap, J. H. van Boom, H. C. van Lieshout, C. A. G. Haasnot, *J. Am. Chem. Soc.* **1988**, *110*, 2736.

A Practical Synthesis of 3'-Thioguanosine and of Its 3'-Phosphoramidothioite (a Thiophosphoramidite)

by **Jasenka Matulic-Adamic** and **Leonid Beigelman***

Department of Chemistry & Biochemistry, *Ribozyme Pharmaceuticals, Inc.*, 2950 Wilderness Place, Boulder, Colorado 80301, USA

Starting from guanosine, an efficient method for the synthesis of 3'-thioguanosine (see **13**) and of its 3'-phosphoramidothioite (see **23**), suitable for automated incorporation into oligonucleotides, was developed. Reaction of 5'-N^2-protected guanosine with 2-acetoxyisobutyryl bromide afforded stereoselectively the 2'-*O*-acetyl-3'-bromo-β-D-xylofuranosyl derivative **3**, which was converted to a 7:3 mixture of the *S*-acyl ribofuranosyl intermediates **5** or **6** and the 3',4'-unsaturated by-product **4**. The *S*-acylated nucleosides **5** and **6** were then converted in three steps to 5'-*O*-(4,4'-dimethoxytrityl)-3'-*S*-(pyridin-2-ylthio)-3'-thioguanosine (**11**), which served as a common intermediate for the preparation of free 3'-thionucleoside **13** and 3'-thionucleoside 3'-phosphoramidothioite **23**.

Introduction. – Oligonucleotides containing 3'-*S*-phosphorothioate linkages have attracted increasing interest as probes for studying the interaction of nucleic acids and their processing enzymes. In particular, these analogs have been used as probes in the elucidation of the roles of metal ions in phosphoester transfer reactions catalyzed by RNA [1][2] and ribonucleoprotein enzymes [3].

As part of our studies of chemically modified hammerhead ribozymes, we recently demonstrated [4] that the previously developed '5-ribo' nuclease-stabilized hammerhead motif can be further refined by systematic incorporation of 1-(β-D-xylofuranosyl)adenine (xA) and 1-(β-D-xylofuranosyl)guanine (xG) in place of the conserved ribopurine residues of the catalytic core. Modified ribozymes substituted with xA at positions A15.1 and A6 demonstrated catalytic activity close to the activity of the parent stabilized ribozyme and an improved nuclease stability, effectively reducing the number of unstabilized residues from 5 to 3. Unfortunately, analogous guanosine substitutions at positions G5, G8, and G12 substantially lowered catalytic rates. Based on these results, we wanted to incorporate 3'-deoxy-3'-thioguanosine in place of these highly conserved [5][6] guanosine residues in the stabilized catalytic domain. Replacing the sugar 3'-O-atom by the larger, more electropositive S-atom should favor the 3'-*endo* sugar pucker, making these analogs very good mimics of RNA [7–9]. At the same time, this modification is expected to increase ribozyme resistance to nuclease degradation [2][10]. In addition, 3'-thio analogs, when incorporated into the ribozyme substrate cleavage site, can also serve for the study of the mechanism of cleavage in the presence of divalent metal ions [1][3].

The synthesis of 3'-*S*-phosphorothioate-linked deoxyribodinucleotides by solution chemistry [3][11–13] and solid-phase chemistry was reported [1][14].

The synthesis of ribonucleotide 3'-*S*-phosporothiolate analogs has been limited to the preparation of UspU [10] and IspU [2] dimers by solution chemistry. Recently, *Sun et al.* [15] described the synthesis of U, C, G, and I 3'-*S*-phosphoramidothioites and their incorporation into RNA by standard phosphoramidite solid-phase synthesis. This work enabled direct incorporation of 3'-thioribonucleosides into oligonucleotides.

Two general approaches were investigated in the reported syntheses of 3'-thioribonucleosides. In the first approach, the target 3'-thioribonucleosides were prepared by condensation of an appropriately protected 3-thioribose derivative with the desired nucleoside base [16–18]. While glycosylation reactions using pyrimidines proceed in high yields and with N(1) regioselectivity [19], purine bases generally give more complex mixtures because both N(7) and N(9) of the purine base are reactive towards glycosylation [20]. *Sun et al.* [15] reported the first synthesis of 3'-thioguanosine derivatives by the above-described approach. Noticeably, peracylated 3-thioribose reacted with persilylated N^2-acetylguanine to provide the condensation product in *ca.* 40% yield. Subsequent synthetic steps proceeded with moderate yields resulting in a low overall yield of 3'-phosphoramidothioite.

The second general approach to the preparation of 3'-thioribonucleosides is based on the S_N2 displacement of an appropriate leaving group at C(3') of the protected xylonucleoside. Synthesis of 3'-thioadenosine [21], 3'-thiouridine [10], and 3'-thioinosine [2][22] starting from preformed nucleosides has been reported. So far, this approach has not been applied to the synthesis of 3'-thioguanosine. Here we describe a novel and efficient synthesis of 3'-thioguanosine (**13**) and of its protected 3'-phosphoramidothioite **23** from guanosine as a starting material.

Results and Discussion. – We envisaged that (3'-bromo-3'-deoxy-β-D-xylofurano-syl)guanosine, which can be prepared in one step from guanosine and 2-acetoxyiso-butyryl bromide (= 2-acetoxy-2-methylpropanoyl bromide; AcOibuBr; *Mattocks-Moffatt* reagent) [23], will provide an attractive starting material for the introduction of the 3'-thioribo functionality by nucleophilic displacement. Unfortunately, the reaction of guanosine with AcOibuBr results in a low yield of the mixture of bromo acetates of *xylo-* and *arabino*-configuration [24]. In general, reactions of base-unprotected purine nucleosides with this reagent result in the mixtures of *trans*-bromo acetates [23][24]. It was noted [25] that the reaction of N^2,5'-*O*-dibenzoylguanosine with AcOibuBr leads predominantly to the 2'-*O*-acetyl-3'-bromo-3'-deoxy-β-D-xylo-furanosyl derivative. On the other side, it was recently reported [26] that the reaction of N^2-[(dimethylamino)methylene]guanosine (**1**), with AcOibuBr proceeded stereo-selectively, yielding exclusively the 3'-bromo-3'-deoxy-β-D-xylofuranosyl derivative.

Encouraged by this last report [26], we decided to apply the same reaction conditions to the suitably 5'-protected N^2-[(dimethylamino)methylene]guanosine derivative **2** (*Scheme 1*). The 5'-protection in **2** should reduce the complexity of the product mixture by eliminating the possible formation of the mixture of 5'-OH, 5'-(2,5,5-trimethyl-4-oxo-1,3-dioxolanyl), and/or 5'-*O*-acylated derivatives in the reaction with AcOibuBr [24][26]. In this manner, identification of the reaction products becomes straightforward. We chose the 5'-*O*-(*tert*-butyl)diphenylsilyl ('BuPh$_2$Si) protection because of its relatively high stability towards acidic conditions generated during the reaction with AcOibuBr in moist MeCN. This group is also expected to undergo selective cleavage in the presence of *S*-acyl groups. Thus, reaction of **1** [27] with 'BuPh$_2$SiCl proceeded quantitatively to afford the 5'-*O*-silyl derivative **2**, which

Scheme 1

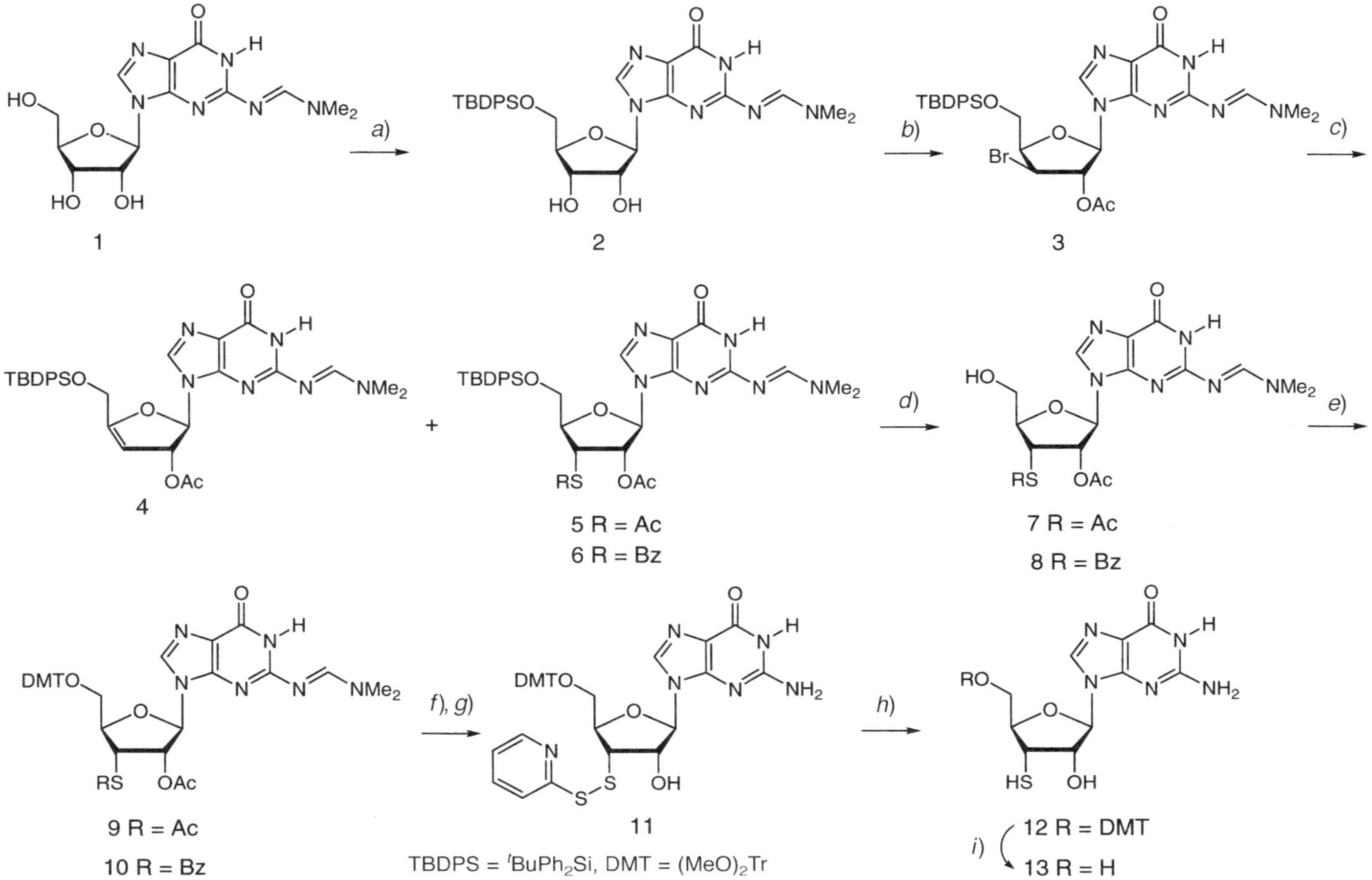

TBDPS = ^{t}BuPh$_2$Si, DMT = (MeO)$_2$Tr

a) ^{t}BuPh$_2$SiCl, Py, r.t., 16 h.   *b*) Me$_2$C(OAc)COBr, MeCN, H$_2$O, r.t., 3 h.   *c*) KSAc or KSBz, DMF, 60°, 10 h.   *d*) Bu$_4$NF·3H$_2$O, AcOH, THF, r.t., 5 h.   *e*) (MeO)$_2$TrCl, Py, r.t., 4 h.   *f*) 40% aq. MeNH$_2$ soln., r.t., 16 h.   *g*) 2,2'-Dithiobis[pyridine], DMF, 60°, 10 h.   *h*) Dithiothreitol (DTT), CHCl$_3$, r.t., 3 h.   *i*) 1N HCl in MeOH, DTT, r.t., 3 h.

reacted smoothly with AcOibuBr to yield the desired 3′-bromo-3′-deoxy-β-D-xylofur-anosyl derivative **3** in high yield. Only the *xylo*-isomer was obtained as judged by ^{1}H-NMR analysis. This result indicates that the nature of the guanosine protection at the NH_2 group directs the stereochemical outcome of this reaction. Reaction of **3** with potassium thioacetate or potassium thiobenzoate in DMF yielded the 3′-*S*-acetyl or 3′-*S*-benzoyl derivatives **5** and **6**, respectively, along with 3′,4′-unsaturated derivative **4**. The 3′,4′-unsaturated derivative **4** was easily identified by the characteristic absence of H−C(4′) and a down-field shift of H−C(3′) in the ^{1}H-NMR spectra. Compound **4** is formed by a competing elimination reaction analogous to the one observed with the carbohydrate analog [18]. The ratio was 7:3 in favor of the substitution products **5** or **6**, which could not be separated from the elimination product **4** at this stage. The mixture **4/5** or **4/6** was then treated with tetrabutylammonium fluoride (Bu_4NF) buffered with an excess of AcOH, and the product mixture was separated by chromatography to give the desired 5′-deprotected derivatives **7** and **8**, respectively, in good yield. The unsaturated derivative **4** was unstable under the acidic reaction conditions, and no attempt was made to isolate the products of its degradation. When triethylamine trihydrofluoride ($Et_3N \cdot 3\,HF$) was used for the deprotection, desilylation was incomplete.

According to our experience, it is highly desirable to keep guanosine derivatives protected with lipophylic groups during synthetic transformations because of the solubility problems and, therefore, low isolated yields when working with unprotected guanosines. Thus, **7** and **8** were 5′-reprotected in high yields with the 4,4′-dimethoxytrityl (($MeO)_2$Tr) group to give the fully protected derivatives **9** and **10**, respectively (*Scheme 1*). The $(MeO)_2$Tr group provided a hydrophobic tag that simplified workup and purification of subsequent synthetic intermediates. Next, **9** and **10** were converted to 3′-(pyridin-2-yldithio) derivative **11** in 85 and 80% yield, respectively, in aqueous $MeNH_2$ solution for *S*-deacylation followed by the *in situ* SH protection by disulfide exchange with 2,2′-dithiobis[pyridine] in DMF. It was reported that the removal of a 2′-*O*-acyl protection in ribofuranosyl derivatives similar to **9** and **10** proceeds with difficulty [15][22]. We found that 40% aqueous $MeNH_2$ solution easily removed acyl and *N*-protecting groups from **9** and **10** and was the base of choice because, contrary to aqueous NH_3 solution, it completely solubilized the fully protected substrates. To synthesize the free nucleoside **13**, **11** was treated with dithiothreitol (DTT) in $CHCl_3$. When Et_3N was added to the reaction mixture, the reaction was faster than in its absence, but at the same time, **12** was converted to its Et_3NH^+ salt with the SH group. Final deprotection of the $(MeO)_2$Tr group of **12** to give **13** was achieved with 1N HCl in MeOH in the presence of DTT which quenched the released $(MeO)_2$Tr cation. In the absence of DTT, quantitative *S*-alkylation took place. To the best of our knowledge, this is the first report on the synthesis of 3′-thioguanosine (**13**).

It is worth noting that when unbuffered Bu_4NF in THF was used to desilylate the mixture **4/5**, the *S*-acetyl protecting group was also cleaved, leading to the formation of the disulfide **14** and of the 5′-desilylated 3′,4′-unsaturated derivative **15** (*Scheme 2*), which were separated by column chromatography (silica gel), **14** being, however, invariably contaminated with Bu_4NF. Attempted rechromatography of **15** led to its decomposition. Disulfide **14** was then 5′-$(MeO)_2$Tr-protected to afford **16**.

It appeared to us that, at this stage, the selective removal of the 2′-*O*-acetyl group of **16**, followed by the introduction of the 2′-*O*-[(*tert*-butyl)dimethylsilyl] ($^tBuMe_2$Si)

Scheme 2

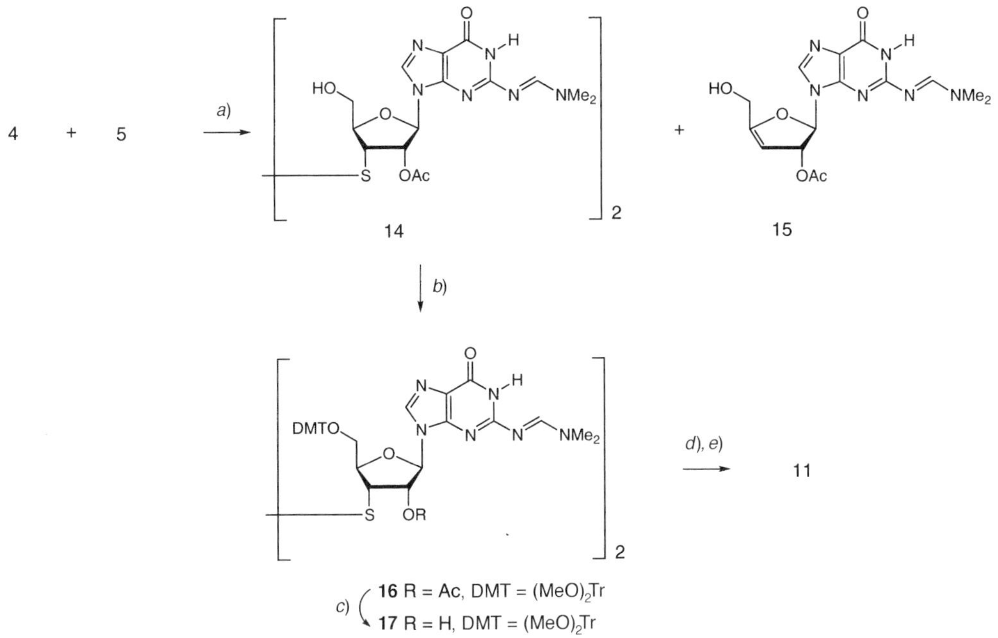

a) 1M Bu$_4$NF in THF, r.t., 3 h.   b) (MeO)$_2$TrCl, Py, r.t., 4 h.   c) Ion exchange with *AGIX8* (OH$^-$) or *Amberlyst A-26* (CN$^-$), MeOH, 55°, 16 h.   d) 40% aq. MeNH$_2$ soln., r.t., 16 h.   e) 2,2'-dithiobis[pyridine], DMF, 60°, 10 h.

protection, subsequent reduction of the 3'-disulfide and 3'-*S*-phosphitylation would be the shortest way to prepare the desired 3'-phosphoramidothioite building block. Reactions of **16** with mild deacylating agents like basic ion exchangers in OH$^-$ or CN$^-$ form [28] selectively removed the 2'-*O*-acetyl protection (→**17**), but at the same time, the nucleoside was strongly adsorbed by the resin, resulting in low recoveries. We, therefore, abandoned this approach and instead applied the same strategy, *i.e.* base-catalyzed *S*-deacetylation followed by *S*-protection with the pyridylinylthio group, as used for the preparation of 3'-(pyridin-2-yldithio) derivative **11** from *S*-acylated derivatives **9** or **10**. In this manner, **11** was obtained from the disulfide **16** in 67% yield.

Our approach to the 3'-phosphoramidothioite synthesis is shown in *Scheme 3*. Reaction of **11** with dimethylformamide dimethyl acetal yielded the desired *N*-protected derivative **18** in 23% yield. Unfortunately, this reagent also effected the cleavage of the *S*-(pyridinylthio) protecting group, leading to formation of disulfide **17** in 33% yield. The 2'-OH function in the alcohol **18** was smoothly protected with the ^{t}BuMe$_2$Si group using (*tert*-butyl)dimethylsilyl trifluoromethanesulfonate (^{t}BuMe$_2$SiTf)(→**19**). Alternatively, 5'-*O*-(MeO)$_2$Tr derivative **11** was first 2'-*O*-silylated with ^{t}BuMe$_2$SiCl^1) to afford **20** and then *N*-protected using isobutyric anhydride (ibu$_2$O) in the presence of 4-(dimethylamino)pyridine (DMAP), yielding

---

^{1})   The ^{t}BuMe$_2$SiTf reagent caused partial *N*-silylation of *N*-unprotected guanosine, making it unsuitable for the silylation of **11** (unpublished results).

Scheme 3

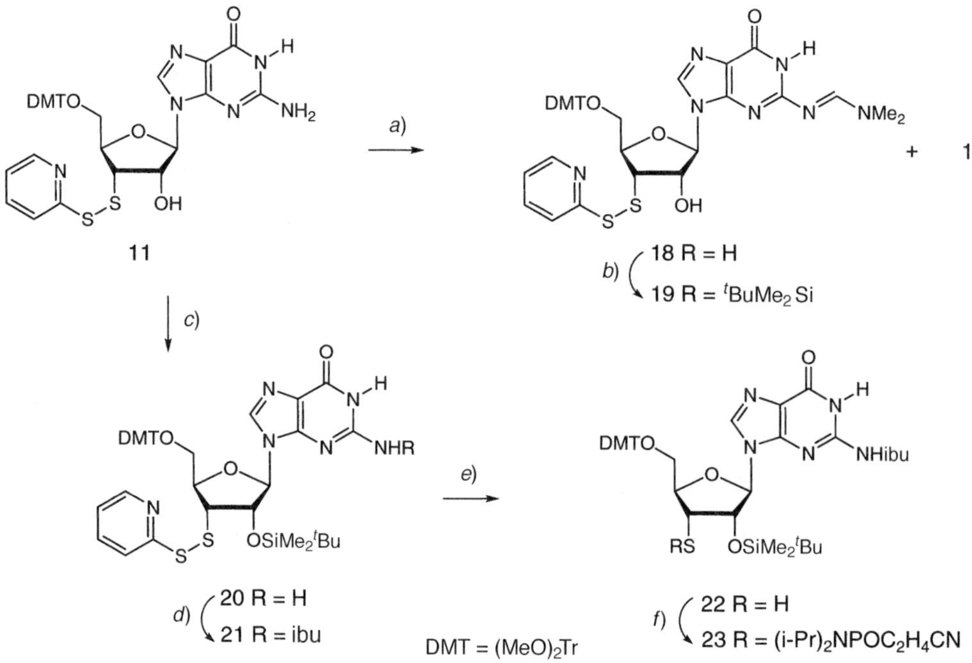

a) Me₂NCH(OMe)₂, Py, r.t., 16 h. b) 'BuMe₂SiTf, Py, r.t. 5 h. c) 'BuMe₂SiCl, Py, 1H-imidazole, r.t., 16 h.
d) ibu₂O, Py, DMAP, r.t., 16 h, then 50°, 5 h. e) DTT, CHCl₃, Et₃N. f) ('Pr)₂NP(Cl)OC₂H₄CN, 'Pr₂EtN,
1-methyl-1H-imidazole, r.t., 2 h.

the fully protected **21**. In the absence of DMAP, no reaction occurred. On the other hand, reaction of **20** with isobutyryl chloride led to *N*-bis-acylation. Reduction of **21** with DTT afforded the 3′-SH derivative **22**, which appeared as a mixture of two rotamers in the ¹H-NMR spectrum. Resonances of the major rotamer were in accordance with those reported by *Sun et al.* [15]. Phosphitylation of **22** under standard conditions afforded 3′-phosphoramidothioite **23**.

Conclusion. – An efficient synthesis of 3′-thioguanosine and its 3′-*S*-phosphorami-dothioite from guanosine was devised. Transformation of the easily accessible key intermediate **3** to 3′-(pyridin-2-yldithio) synthon **11**, followed by protection of the 2-NH₂ and 2′-OH group, reduction, and phosphitylation provided the target 3′-*S*-phosphoramidothioite **23** in 13% overall yield from guanosine. Keeping all synthetic intermediates protected with lipophylic groups enabled their chromatographic purification and, consequently, a good recovery of the products. Incorporation of 3′-*S*-phosphoramidothioite **23** into ribozymes and subsequent mechanistic studies are in progress and will be published in due course.

Experimental Part

General. All reactions were carried out under a positive pressure of Ar in anh. solvents. Commercially available reagents and anh. solvents were used without further purification. Anal. TLC: *Merck* silica gel *60 F₂₅₄* plates (Art. 5554). Flash column chromatography (FC): *Merck* 0.040-0.063 mm silica gel *60*. ^{1}H- and ^{31}P-NMR Spectra: at 400.075 and 161.947 Hz, resp. in CDCl$_3$, unless stated otherwise; chemical shifts δ in ppm rel. to SiMe$_4$ and H$_3$PO$_4$, resp. Mass spectra: fast-atom bombardment (FAB) method.

5'-O-[(tert-Butyl)diphenylsilyl]-N^2-[(dimethylamino)methylene]guanosine (**2**). To a stirred soln. of N^2-[(dimethylamino)methylene]guanosine (**1**) [24] (5.5 g, 16.3 mmol) in pyridine (100 ml), (*tert*-butyl)diphenyl-silyl chloride (6.2 ml, 23.8 mmol) was added under Ar. The mixture was stirred at r.t. for 16 h, then quenched with MeOH (20 ml), and evaporated to a syrup. The residue was precipitated from EtOH/Et$_2$O: **2** (9 g, 96%). ^{1}H-NMR ((D$_6$)DMSO + D$_2$O): 8.46 (*s*, CH=N); 7.89 (*s*, H$-$C(8)); 7.58$-$7.31 (*m*, 2 Ph); 5.81 (*d*, *J*(1',2') = 4.8, H$-$C(1')); 4.46 ('*t*', *J*(2',1') = 4.8, H$-$C(2')); 4.23 ('*t*', *J*(3',2') = 5.0, H$-$C(3')); 3.97 (*m*, H$-$C(4')); 3.84 (*dd*, *J*(5',4') = 2.8, *J*(5',5'') = 12.0, H$-$C(5')); 3.74 (*dd*, *J*(5'',4') = 4.4, *J*(5'',5') = 12.0, H'$-$C(5'')); 3.05 (*s*, H'$-$C(5')Me); 2.97 (*s*, Me); 0.94 (*s*, *t*-Bu). HR-MS (FAB$^+$): 577.26095 (C$_{29}$H$_{36}$N$_6$O$_5$Si$^+$, [*M* + H]$^+$; calc. 577.2595).

1-[2-O-Acetyl-3-bromo-5-O-[(tert-butyl)diphenylsilyl]-3-deoxy-β-D-xylofuranosyl]-N^2-[(dimethylamino)-methylene]guanine (**3**). To a soln. of **2** (5.8 g, 10 mmol) and H$_2$O (0.12 ml) in MeCN (130 ml) at 0°, 2-ace-toxyisobutyryl bromide (5.56 ml, 38 mmol) was added, and the mixture was stirred at r.t. for 3 h. The soln. was poured into sat. aq. NaHCO$_3$ soln. (100 ml) and extracted with CH$_2$Cl$_2$ (3 × 200 ml). The combined org. layers were dried (Na$_2$SO$_4$) and evaporated: chromatographically pure (6 g, 87%). White foam **3**. ^{1}H-NMR: 8.97 (br. *s*, NH); 8.62 (*s*, CH=N); 7.82 (*s*, H$-$C(8)); 7.73$-$7.31 (*m*, 2 Ph); 6.09 (*d*, *J*(1',2') = 1.6, H$-$C(1')); 5.92 (*d*, *J*(2',1') = 1.6, H$-$C(2')); 4.36 (*m*, H$-$C(4')); 4.42 (*m*, H$-$C(3')); 4.06 (*dd*, *J*(5',4') = 5.6, *J*(5',5'') = 10.4, H$-$C(5')); 3.97 (*dd*, *J*(5'',4') = 6.4, *J*(5'',5') = 10.4, H'$-$C(5')); 3.17 (*s*, 1 MeN); 3.07 (*s*, 1 MeN); 2.19 (*s*, Ac); 1.07 (*s*, *t*Bu). HR-MS (FAB$^+$): 681.1850 (C$_{31}$H$_{37}$BrN$_6$O$_5$Si$^+$, [*M* + H]$^+$; calc. 681.1856).

1-[2-O-Acetyl-5-O-[(tert-butyl)diphenylsilyl]-3-deoxy-β-D-glycero-pent-3-enofuranosyl]-N^2-[(dimethylami-no)methylene]guanine (**4**) *and* *2'-O,3'-S-Diacetyl-5'-O-[(tert-butyl)diphenylsilyl]-N^2-[(dimethylamino)methy-lene]-3'-thioguanosine* (**5**). To a soln. of **3** (5.4 g, 7.9 mmol) in dry DMF (50 ml), potassium thioacetate (2.7 g, 23.6 mmol) was added, and the mixture was stirred at 60° for 16 h and then evaporated to a syrup. The residue was partitioned between an aq. NaHCO$_3$ soln./brine 1:1 and CH$_2$Cl$_2$, the org. layer dried (Na$_2$SO$_4$) and evaporated, and the residue chromatographed (silica gel, gradient 2$-$10% MeOH/CH$_2$Cl$_2$): **4/5** (4.8 g). Yellowish foam. ^{1}H-NMR: **4/5** 3 : 7; 8.65 (*s*, CH=N, **5**); 8.61 (*s*, CH=N, **4**); 7.73 (*s*, H$-$C(8), **5**); 7.52 (*s*, H$-$C(8), **4**); 7.66$-$7.27 (*m*, Ph); 6.39 (*d*, *J*(1',2') = 1.4, H$-$C(1'), **4**); 6.03 (*s*, H$-$C(1'), **5**); 6.02 (*d*, *J*(1',2') = 1.4, H$-$C(2'), **4**); 5.83 (*s*, H$-$C(2'), **5**); 5.44 (*s*, H$-$C(3'), **4**); 4.75 (*dd*, *J*(3',2') = 5.8, *J*(3',4') = 10.2, H$-$C(3'), **5**); 4.28 (*m*, 2 H$-$C(5'), **4**); 4.18 (*ddd*, *J*(4',5') = 2.2, *J*(4',5'') = 4.8, *J*(4',3') = 10.2, H$-$C(4'), **5**); 3.90 (*dd*, *J*(5',5'') = 11.5, *J*(5',4') = 2.2, H$-$C(5'), **5**); 3.82 (*dd*, *J*(5'',5') = 11.5, *J*(5'',4') = 4.8, H'$-$C(5''), **5**); 3.14 (*s*, 1 MeN, **4**); 3.09 (*s*, 1 MeN, **5**); 3.08 (*s*, 1 MeN, **4**); 3.04 (*s*, 1 MeN, **5**); 2.31 (*s*, AcS, **5**); 2.16 (*s*, AcO, **5**); 2.13 (*s*, AcO, **4**); 1.07 (*s*, *t*Bu); 0.99 (*s*, *t*Bu).

When potassium thiobenzoate was used instead of potassium thioacetate, an inseparable mixture of **4** and *2'-O-acetyl-3'-S-benzoyl-5'-O-[(tert-butyl)diphenylsilyl]-N^2-[(dimethylamino)methylene]-3'-thioguanosine* (**6**) was obtained in a similar yield and ratio to the unsaturated derivative **4** as above.

2'-O,3'-S-Diacetyl-N^2-[(dimethylamino)methylene]-3'-thioguanosine (**7**). To the above mixture **4/5** (0.9 g) in THF (15 ml), AcOH (0.37 ml, 6.5 mmol) was added, followed by Bu$_4$NF · 3 H$_2$O (0.82 g, 2.6 mmol). The mixture was stirred at r.t. for 5 h, then diluted with CH$_2$Cl$_2$, and washed with H$_2$O and 10% aq. NaHCO$_3$ soln. The aq. layers were back-washed with CH$_2$Cl$_2$ and the combined org. layers dried (Na$_2$SO$_4$) and evaporated. FC (silica gel, gradient 2$-$10% MeOH/CH$_2$Cl$_2$) afforded **7** (300 mg, 46% from **3**). Yellowish foam. ^{1}H-NMR: 8.87 (br. *s*, NH); 8.78 (*s*, CH=N); 7.73 (*s*, H$-$C(8)); 5.80 (*d*, *J*(1',2') = 2.0, H$-$C(1')); 5.76 (*dd*, *J*(2',1') = 2.0, *J*(2',3') = 6.4, H$-$C(2')); 4.94 (*dd*, *J*(3',2') = 6.4, *J*(3',4') = 9.6, H$-$C(3')); 4.22 (*d*, *J*(4',3') = 9.6, H$-$C(4')); 4.04 (br. *s*, OH$-$C(5')); 3.99 (*d*, *J*(5',5'') = 12.0, H$-$C(5')); 3.71 (*d*, *J*(5'',5') = 12.0, H'$-$C(5')); 3.19 (*s*, 1 MeN); 3.04 (*s*, 1 MeN); 2.34 (*s*, AcS); 2.13 (*s*, AcO). HR-MS (FAB$^+$): 439.1405 (C$_{17}$H$_{22}$N$_6$O$_6$S$^+$, [*M* + H]$^+$); calc. 439.1355).

2'-O-Acetyl-3'-S-benzoyl-N^2-[(dimethylamino)methylene]-3'-thioguanosine (**8**). According to the same pro-cedure as above, **8** was synthesized from **4/6** in *ca*. 70% yield. ^{1}H-NMR: 8.90 (br. *s*, NH); 8.57 (br. *s*, NH); 7.69 (*s*, H$-$C(8)); 5.88 (*m*, H$-$C(1'), H$-$C(2')); 5.30 (*m*, H$-$C(3')); 4.34 (*d*, *J*(4',3') = 9.2, H$-$C(4')); 4.04 (*d*, *J*(5',5'') = 12.8, H$-$C(5')); 3.80 (*dd*, *J*(5'',OH) = 9.6, *J*(5',5'') = 12.8, H'$-$C(5')); 3.69 (br. *s*, OH$-$C(5')); 3.25 (*s*, MeN); 3.10 (*s*, 1 MeN); 2.16 (*s*, AcO). HR-MS (FAB$^+$): 501.1561 (C$_{17}$H$_{22}$N$_6$O$_6$S$^+$, [*M*+H]$^+$; calc. 501.1556).

2'-O,3'-S-Diacetyl-3'-deoxy-5'-O-(4,4'-dimethoxytrityl)-N^2-[(dimethylamino)-methylene-3'-thioguanosine (**9**). To a soln. of **7** (720 mg, 1.64 mmol) in dry pyridine (15 ml), (MeO)$_2$TrCl (1.1 g, 3.3 mmol) was added. The mixture was stirred at r.t. for 4 h, quenched with MeOH, and evaporated to a syrup which was partitioned between 5% aq. NaHCO$_3$ soln. and CH$_2$Cl$_2$. The org. layer was washed with brine, dried (Na$_2$SO$_4$), and evaporated, and the residue purified by FC (silica gel, gradient 1–5% MeOH/CH$_2$Cl$_2$ **9** (0.85 g, 70%). Colorless foam. ^{1}H-NMR: 8.69 (s, CH=N); 8.58 (br. s, NH); 7.69 (s, H–C(8)); 7.38–6.74 (m, H–C(8), 12 arom. H); 6.06 (dd, J(2′,3′) = 6.4, J(2′,1′) = 1.2, H–C(2′)); 5.82 (d, J(1′,2′) = 1.2, H–C(1′)); 4.73 (dd, J(3′,4′) = 10.6, J(3′,2′) = 6.4, H–C(3′)); 4.21 (dq, J(4′,3′) = 10.6, J(4′,5′) = 3.0, J(4′,5″) = 4.4, H–C(4′)); 3.78 (s, 2 MeO); 3.36 (m, 2 H–C(5′)); 3.07 (s, 1 MeN); 3.05 (s, 1 MeN); 2.26 (s, AcS); 2.15 (s, AcO). HR-MS (FAB$^+$): 741.2692 (C$_{38}$H$_{40}$N$_6$O$_8$S$^+$, [M + H]$^+$; calc. 741.2707).

2′-O-Acetyl-3′-benzoyl-3′-deoxy-5′-O-(4,4′-dimethoxytrityl)-N^2-[(dimethylamino)methylene]-3′-thioguanosine (**10**). As described for **9**, **8** was converted to **10** in 69% yield. ^{1}H-NMR: 8.80 (s, CH=N); 8.65 (br. s, NH); 7.70 (s, H–C(8)); 7.88–6.66 (m, 19 arom. H); 6.17 (d, J(2′,3′) = 5.8, H–C(2′)); 5.86 (d, J(1′,2′) = 1.2, H–C(1′)); 5.08 (dd, J(3′,4′) = 10.4, J(3′,2′) = 5.8, H–C(3′)); 4.31 (m, H–C(4′)); 3.67 (s, 2 MeO); 3.45 (m, 2 H–C(5′)); 3.06 (s, 2 MeN); 2.15 (s, AcO). HR-MS (FAB$^+$): 803.2855 (C$_{43}$H$_{42}$N$_6$O$_8$S$^+$, [M + H]$^+$; calc. 803.2863).

5′-O-(4,4′-Dimethoxytrityl)-3′-(pyridin-2-ylthio)-3′-thioguanosine (**11**). *a*) A soln. of **9** (530 mg, 0.38 mmol) in 40% aq. MeNH$_2$ (50 ml) was kept at r.t. for 16 h. The mixture was evaporated and the residual syrup dissolved in Ar-purged DMF (30 ml) containing 2,2′-thiobis[pyridine] (340 mg, 1.54 mmol). The mixture was heated at 60° for 10 h and then evaporated to a syrup. FC (silica gel, gradient 1–12% MeOH/CH$_2$Cl$_2$) afforded **11** (460 mg, 85%). Colorless solid. ^{1}H-NMR: 10.64 (br. s, NH); 8.39 (m, 1 H, Py); 7.83 (s, H–C(8)); 7.73–6.72 (m, 16 arom. H); 6.50 (d, J(OH,2′) = 4.80, OH–C(2′)); 6.45 (br. s, NH$_2$); 5.81 (d, J(1′,2′) = 2.4, H–C(1′)); 4.83 (m, H–C(2′)); 4.34 (m, H–C(4′)); 4.09 (dd, J(3′,2′) = 6.00, J(3′,4′) = 7.8, H–C(3′)); 3.70 (s, 2 MeO); 3.11 (dd, J(5′,5″) = 11.2, J(5″,4′) = 4.8, H′–C(5′)). HR-MS (FAB$^+$): 711.2076 (C$_{36}$H$_{34}$N$_6$O$_6$S$_2^+$, [M + H]$^+$; calc. 711.2060).

b) By the same procedure as above, but starting from *S*-benzoyl derivative **10**, target **11** was prepared in 80% yield.

c) Starting from **16** (830 mg, 1.12 mmol) and under the above conditions, derivative **11** (570 mg, 67%) was obtained.

5′-O-(4,4′-Dimethoxytrityl)-3′-thioguanosine (**12**). To the soln. of **11** (240 mg, 0.34 mmol) in CHCl$_3$ (14 ml), dithiothreitol (DTT; 125 mg, 0.81 mmol) was added, and the mixture was stirred at r.t. for 3 h. It was then evaporated to a syrup, and the product was precipitated by addition of peroxide-free Et$_2$O. The precipitate was filtered off, washed with Et$_2$O, and dried: 230 mg of crude **12**. ^{1}H-NMR ((D$_6$)DMSO): 10.63 (br. s, NH); 7.86 (s, H–C(8)); 7.32–6.80 (m, 13 arom. H); 6.49 (br. s, NH$_2$); 5.81 (s, H–C(1′)); 4.43 (d, J(2′,3′) = 4.8, H–C(2′)); 3.93 (m, H–C(4′)); 3.79 (dd, J(3′,2′) = 4.8, J(3′,4′) = 9.6, H–C(3′)); 3.71 (s, 2 MeO); 3.16 (dd, J(5″,5′) = 10.4, J(5″,4′) = 4.8, H′–C(5′)).

3′-Thioguanosine (**13**). The mixture of crude **12** (230 mg, 0.33 mmol) and DTT (150 mg) was dissolved in 1N HCl/MeOH (12 ml), and the mixture was kept at r.t. for 3 h. It was then evaporated and the residue co-evaporated with toluene two times. Addition of AcOEt afforded a precipitate which was filtered off, washed well with AcOEt and dried: **13** (90 mg, 79%). The product was reprecipitated from H$_2$O. ^{1}H-NMR (CD$_3$OD): 8.10 (s, H–C(8)); 5.91 (s, H–C(1′)); 4.37 (d, J(2′,3′) = 5.2, H–C(2′)); 3.97 (m, H–C(4′), H–C(5′)); 3.82 (dd, J(5″,5′) = 13.0, J(5″,4′) = 3.4, H′–C(5′)); 3.64 (dd, J(3′,2′) = 5.2, J(3′,4′) = 9.6, H–C(3′)). HR-MS (FAB$^+$): 300.0767 (C$_{10}$H$_{13}$N$_5$O$_4$S$^+$; [M + H]$^+$); calc. 300.0767.

3′,3′′′-Dithiobis[2-O-acetyl-3′-deoxy-N^2-[(dimethylamino)methylene]guanosine] (**14**) *and 1-(2-O-Acetyl-3-deoxy-β-D-glycero-pent-3-enofuranosyl)-N^2-[(dimethylamino)methylene]guanine* (**15**). To **4/5** (4.8 g) in THF (100 ml), 1M Bu$_4$NF in THF (10 ml) was added. The mixture was stirred for 3 h at r.t. and then evaporated to a syrup. FC (silica gel, gradient 2–10% MeOH/CH$_2$Cl$_2$) yielded the faster eluting **15** (1 g, 35% from **3**). Colorless foam. ^{1}H-NMR: 8.96 (br. s, NH); 8.57 (s, CH=N); 7.65 (s, H–C(8)); 6.41 (s, H–C(2′)); 6.04 (s, H–C(1′)); 5.42 (m, H–C(3′)); 4.32 (m, 2 H–C(5′)); 3.19 (s, MeN); 3.06 (s, MeN); 2.11 (s, Ac).

The slower eluting **14** was obtained as a yellowish solid (0.9 g, 29% from **3**). ^{1}H-NMR ((D$_6$)DMSO): 11.34 (br. s, NH); 8.53 (s, CH=N); 8.00 (s, H–C(8)); 5.97 (d, J(1′,2′) = 2.4, H–C(1′)); 5.89 (dd, J(2′,1′) = 2.4, J(2′,3′) = 6.0, H–C(2′)); 5.23 (t, J(OH,5′) = 5.6, OH–C(5′)); 4.11 (m, H–C(4′)); 4.02 (dd, J(3′,2′) = 6.0, J(3′,4′) = 8.4, H–C(3′)); 3.78 (m, H–C(5′)); 3.60 (m, H′–C(5′)); 3.10 (s, MeN); 3.00 (s, MeN); 2.06 (s, AcO). HR-MS (FAB$^+$): 791.2355 (C$_{30}$H$_{38}$N$_{12}$O$_{10}$S$_2^+$, [M + H]$^+$; calc. 791.2354).

3′,3′′′-Dithiobis[2′-O-acetyl-3′-deoxy-5′-O-(4,4′dimethoxytrityl)-N^2-[(dimethylamino)methylene]guanosine] (**16**). To the soln. of **14** (400 mg, 0.5 mmol) in dry pyridine (10 ml), (MeO)$_2$TrCl (508 mg, 1.5 mmol) was added and the mixture stirred 4 h at r.t. MeOH (10 ml) was added and the soln. evaporated to dryness. The residue was partitioned between sat. NaHCO$_3$ soln. and CH$_2$Cl$_2$, and the org. layer washed with brine, dried (Na$_2$SO$_4$), and evaporated to a syrup. FC (silica gel, gradient 2–10% MeOH/CH$_2$Cl$_2$) yielded **16** (620 mg, 71% yield). Yellowish foam. ^{1}H-NMR: 8.72 (br. s, NH); 8.01 (s, CH=N); 7.48–7.21 (m, H–C(8), 13 arom. H); 6.25

($d, J(2',3') = 4.8$, H$-$C(2')); 5.78 (s, H$-$C(1')); 4.00 (m, H$-$C(3'), H$-$C(4')); 3.78 (s, 2 MeO); 3.50 (br. s, 2 H$-$C(5')); 3.14 (s, 1 MeN); 3.13 (s, 1 MeN); 1.82 (s, AcO). HR-MS (FAB$^+$): 1395.4943 ($C_{72}H_{74}N_{12}O_{14}S_2^+$, $[M + H]^+$; calc. 1395.4967).

3',3'''-Dithiobis[3'-deoxy-5'-O-(4,4'dimethoxytrityl)-N^2-[(dimethylamino)methylene]guanosine] (**17**). *a*) To a soln. of **16** (60 mg, 0.04 mmol) in dry MeOH, ion exchange resin *AG1X8* (OH$^-$) (1 g) was added. The mixture was stirred at 55° for 16 h, the resin filtered off and washed well with hot MeOH, and the filtrate evaporated: pure **17** (16 mg, 28%). Colorless solid. ^{1}H-NMR ((D_6)DMSO): 11.34 (br. s, NH); 8.48 (s, CH$=$N); 7.92 (s, H$-$C(8)); 7.31$-$6.73 (m, 13 arom. H); 6.27 (d, J(OH,2') = 5.2, OH$-$C(2')); 5.88 (d, J(1',2') = 1.2, H$-$C(1')); 4.60 (m, H$-$C(2')); 4.16 (m, H$-$C(4')); 4.08 (m, H$-$C(3')); 3.65 (s, 2 MeO); 3.65 (m, 2 H$-$C(5')); 3.02 (s, 1 MeN); 2.97 (s, 1 MeN). HR-MS (FAB$^+$): 1311.4746 ($C_{68}H_{70}N_{12}O_{12}S_2^+$, $[M + H]^+$; calc. 1311.4756).

b) With *Amberlyst A-26* (CN$^-$) under the above conditions, **17** was obtained from **16** in 21% yield.

5'-O-(4,4'-Dimethoxytrityl)-N^2-[(dimethylamino)methylene]3'-S-(pyridin-2-ylthio)-3'-thioguanosine (**18**) *and* **17**. To the soln. of **11** (400 mg, 0.56 mmol) in dry pyridine (5 ml), dimethylformamide dimethyl acetal (1.2 ml, 9 mmol) was added and the mixture stirred at r.t. for 16 h. Solvents were removed *in vacuo* and the residue chromatographed (silica gel, gradient 1 $-$ 50% MeOH/CH$_2$Cl$_2$). Fractions containing the faster-running material gave **18** (110 mg, 23%). ^{1}H-NMR: 8.73 (br. s, NH); 8.53 (s, CH$=$N); 8.49 (m, 1 H, Py); 7.71 (s, H$-$C(8)); 7.62 (m, 1 H, Py); 7.44$-$6.79 (m, 15 arom. H); 6.09 (s, H$-$C(1')); 4.55 ($d, J(2',3') = 4.8$, H$-$C(2')); 4.23 ($dq, J(4',3') = 10.5$, $J(4',5') = 2.8$, $J(4',5') = 3.4$, H$-$C(4')); 4.14 ($dd, J(3',2') = 4.8$, $J(3',4') = 10.5$, H$-$C(3')); 3.78 (s, 2 MeO); 3.57 ($dd, J(5'',5') = 10.6$, $J(5',4') = 2.8$, H$-$C(5')); 3.41 ($dd, J(5',5'') = 10.6$, $J(5'',4') = 3.4$, H$-$C(5')); 3.08 (s, 1 MeN); 3.05 (s, 1 MeN).

Fractions containing the slower running compound gave **17** (120 mg, 33%). Colorless foam. ^{1}H-NMR: identical to that of **17** obtained by the above procedures.

2'-O-[(tert-Butyl)dimethylsilyl]-5'-O-(4,4'-dimethoxytrityl)-N^2-[(dimethylamino)methylene]-3'-S-(pyridin-3-ylthio)-3'-thioguanosine (**19**). To a soln. of **18** (110 mg, 0.14 mmol) in dry pyridine (1 ml), ^{t}BuMe$_2$SiTf (0.103 ml, 0.45 mmol) was added. The mixture was stirred at r.t. for 5 h, then quenched with MeOH, and evaporated. The residue was dissolved in CH$_2$Cl$_2$, the org. phase washed with 5% aq. NaHCO$_3$ soln. and brine, dried (Na$_2$SO$_4$), and evaporated, and the syrup submitted to FC (silica gel, gradient 1 $-$ 10% MeOH/AcOEt): **19** (90 mg, 71%). Colorless solid. ^{1}H-NMR: (br. s, NH); 8.50 (s, CH$=$N); 8.37 (m, 1 H, Py); 7.81 (s, H$-$C(8)); 7.50$-$6.74 (m, 16 arom. H); 5.98 ($d, J(1',2') = 2.4$, H$-$C(1')); 4.75 ($dd, J(2',3') = 5.0$, $J(2',1') = 2.4$, H$-$C(2')); 4.50 (m, H$-$C(4')); 3.99 ($dd, J(3',2') = 5.0$, $J(3',4') = 8.4$, H$-$C(3')); 3.76 (s, 2 MeO); 3.62 ($dd, J(5'',5'') = 10.9$. $J(5',4') = 2.2$, H$-$C(5')); 3.38 ($dd, J(5',5'') = 10.9$, $J(5'',4') = 4.4$, H$-$C(5')); 3.06 (s, 1 MeN); 3.04 (s, 1 MeN); 0.93 (s, tBu); 0.17 (s, Me); 0.10 (s, Me). HR-MS (FAB$^+$): 880.3357 ($C_{45}H_{53}N_7O_6S_2Si^+$, $[M + H]^+$; calc. 880.3346).

2'-O-[(tert-Butyl)dimethylsilyl]-5'-O-(4,4'-dimethoxytrityl)-3'-S-(pyridin-2-ylthio)-3'-thioguanosine (**20**). To a soln. of **11** (410 mg, 0.58 mmol) in dry pyridine (36 ml), 1H-imidazole (2.36 g, 35 mmol) and ^{t}BuMe$_2$SiCl (4.29 g, 28 mmol) were added. The mixture was stirred at r.t. for 16 h and then evaporated to a syrup. The residue was partitioned between CH$_2$Cl$_2$ and sat. aq. NaHCO$_3$ soln., the org. layer washed with H$_2$O, dried (Na$_2$SO$_4$), and evaporated, and the syrup submitted to FC (silica gel, gradient 1 $-$ 10% MeOH/CH$_2$Cl$_2$): **20** (430 mg, 85%). White foam. ^{1}H-NMR ((D_6)DMSO): 11.99 (br. s, NH); 8.39 (m, 1 H, Py); 7.84 (s, H$-$C(8)); 7.71 (m, 1 H, Py); 7.62$-$6.72 (m, 15 arom. H); 6.41 (br. s, NH$_2$); 5.80 ($d, J(1',2') = 4.4$, H$-$C(1')); 5.04 ('t', $J(2',3') = 4.4$, H$-$C(2')); 4.39 (m, H$-$C(4')); 4.01 ('t', $J(3',4') = 6.4$, H$-$C(3')); 3.69 (s, 2 MeO); 3.13 ($dd, J(5'',5') = 11.0$, $J(5'',4') = 4.6$, H'$-$C(5')); 0.82 (s, tBu); 0.08 (s, 1 Me); 0.06 (s, 1 Me). HR-MS (FAB$^+$): 825.2977 ($C_{42}H_{48}N_6O_6S_2Si^+$, $[M + H]^+$; calc. 825.2924).

2'-O-[(tert-Butyl)dimethylsilyl]-5'-O-(4,4'-dimethoxytrityl)-N^2-isobutyryl-3'-S-(pyridin-2-ylthio)-3'-thioguanosine (**21**). To the soln. of **20** (310 mg, 0.38 mmol) in dry pyridine (5 ml), isobutyric anhydride (0.19 ml, 1.14 mmol) and 4-(dimethylamino)pyridine (46 mg, 0.38 mm) were added. The mixture was stirred at r.t. for 16 h and at 50° for 5 h, then quenched with MeOH (2 ml), and evaporated. The syrup was partitioned between CH$_2$Cl$_2$ and 5% aq. NaHCO$_3$ soln., the org. layer washed with brine, dried (Na$_2$SO$_4$), and evaporated and the syrup submitted to FC (silica gel, gradient 1 $-$ 5% MeOH/CH$_2$Cl$_2$): **21** (320 mg, 95%). Colorless foam. ^{1}H-NMR: 11.94 (br. s, NH); 8.36 (m, 1 H, Py); 7.85 (m, 1 H, Py); 7.80 (s, H$-$C(8)); 7.58$-$6.71 (m, NH, 15 arom. H); 5.83 ($d, J(1',2') = 5.2$, H$-$C(1')); 5.22 ('t', $J(1',2') = 5.2$, H$-$C(2')); 4.50 (m, H$-$C(4')); 4.27 ($t, J(3',4') = 6.4$, H$-$C(3')); 3.76 (s, MeO); 3.75 (s, MeO); 3.56 ($dd, J(5',5'') = 11.0$, $J(5',4') = 1.8$, H$-$C(5')); 2.98 ($dd, J(5'',5') = 11.0$, $J(5'',4') = 3.0$, H'$-$C(5')); 1.69 (m, Me$_2$CH); 0.94 ($d, J = 7.2$, Me); 0.76 ($d, J = 7.2$, Me); 0.88 (s, tBu); 0.11 (s, Me); 0.06 (s, Me). HR-MS (FAB$^+$): 895.3380 ($C_{46}H_{54}N_6O_7S_2Si^+$, $[M + H]^+$, calc. 895.3343).

2'-O-[(tert-Butyl)dimethylsilyl]-5'-O-(4,4'-dimethoxytrityl)-N^2-isobutyryl-3'-thioguanosine (**22**). To the soln. of **21** (340 mg, 0.38 mmol) in CHCl$_3$ (20 ml), Et$_3$N (0.4 ml) and dithiotreitol DTT (140 mg, 0.91 mmol) were added, and the mixture was stirred for 1 h at r.t. The mixture was then washed with sat. aq. NaHCO$_3$ soln.

and H_2O, dried (Na_2SO_4) and evaporated. FC (silica gel, gradient 0.5–2% MeOH/CH_2Cl_2) afforded **22** (270 mg, 90%). ^{1}H-NMR of the major rotamer: 11.92 (br. s, NH); 7.93 (s, H$-$C(8)); 7.63–6.80 (m, NH, 15 arom. H); 5.83 (d, $J(1',2') = 2.8$, H$-$C(1')); 4.74 (dd, $J(2',1') = 2.8$, $J(2',3') = 5.4$, H$-$C(2')); 4.13 (br. d, $J(4',3') = 7.6$, H$-$C(4')); 3.78 (s, MeO); 3.73 (s, MeO); 3.74 (m, H$-$C(3')); 3.63 (dd, $J(5',5'') = 11.0$, $J(5',4') = 1.2$, H$-$C(5')); 3.27 (dd, $J(5'',5') = 11.0$, $J(5'',4') = 3.0$, H'$-$C(5')); 1.63 (d, J(SH,3') $= 8.4$, SH); 2.08 (m, Me$_2$CH); 1.09 (d, $J = 6.8$, Me); 0.98 (d, $J = 6.8$, Me); 0.91 (s, tBu); 0.14 (s, Me); 0.08 (s, Me). HR-MS (FAB$^+$): 786.3354 ($C_{41}H_{51}N_5O_7SSi^+$, $[M+H]^+$; calc. 786.3357).

2'-O-[(tert-Butyl)dimethylsilyl]-5'-O-(4,4'-dimethoxytrityl)-N^2-isobutyryl-3'-thioguanosine 3'-(2-Cyanoethyl Diisopropylphoramidothioite) (**23**). Phosphitylation of **22** as described by *Sun et al.* [15] afforded a product which was purified by FC (0.5% EtOH/CH_2Cl_2 containing 1% Et$_3$N). The final product was obtained as a white powder by precipitation from toluene/pentane at 0°: 76% yield. ^{31}P-NMR: 163.5 (s); 159.6 (s). HR-MS (FAB$^+$): 986.4406 ($C_{50}H_{68}N_7O_8PSSi^+$, $[M+H]^+$; calc. 986.4435).

REFERENCES

[1] J. A. Piccirilli, J. S. Vyle, M. H. Caruthers, T. R. Cech, *Nature (London)* **1987**, *361*, 85.

[2] L. B. Weinstein, D. I. Earnshaw, R. Cosstick, T. R. Cech, *J. Am. Chem. Soc.* **1996**, *118*, 10341.

[3] E. J. Sontheimer, S. Sun, J. A. Piccirili, *Nature (London)* **1997**, *308*, 801.

[4] J. Matulic-Adamic, A. T. Daniher, A. Gonzalez, L. Beigelman, *Bioorg. Med. Chem. Lett.* **1998**, *9*, 157.

[5] L. Beigelman, J. A. McSwiggen, K. G. Draper, C. Gonzalez, K. Jensen, A. M. Karpeisky, A. S. Modak, J. Matulic-Adamic, A. B. DiRenzo, P. Haeberli, D. Sweedler, D. Tracz, S. Grimm, F. E. Wincott, V. G. Thackray, N. Usman, *J. Biol. Chem.* **1995**, *270*, 25702.

[6] D. B. McKay, *RNA* **1996**, *2*, 395.

[7] W. Guschlbauer, K. Jankowski, *Nucleic Acids Res.* **1980**, *8*, 1420.

[8] C. Thibaudeau, I. Plavec, N. Garg, A. Papchikhin, J. Chattopadhyaya, *J. Am. Chem. Soc.* **1994**, *116*, 4038.

[9] I. Plavec, C. Thibaudeau, J. Chattopadhyaya, *J. Am. Chem. Soc.* **1994**, *116*, 6558.

[10] X. Liu, C. B. Reese, *Tetrahedron Lett.* **1996**, *37*, 925.

[11] R. Cosstick, J. S. Vyle, *Nucleic Acids. Res.* **1990**, *18*, 829.

[12] X. Li, D. M. Andrews, R. Cosstick, *Tetrahedron* **1992**, *48*, 2729.

[13] X. Li, G. K. Scott, A. D. Baxter, R. J. Taylor, J. S. Vyle, R. Cosstick, *J. Chem. Soc., Perkin Trans. 1* **1994**, 2123.

[14] J. S. Vyle, B. A. Connolly, D. Kemp, R. Cosstick, *Biochemistry* **1992**, *31*, 3012.

[15] S. Sun, A. Yoshida, J. A. Piccirilli, *RNA* **1997**, *3*, 1352.

[16] K. J. Ryan, E. M. Acton, L. Goodman, *J. Org. Chem.* **1968**, *33*, 1783.

[17] X. Cao, M. D. Matteucci, *Bioorg. Med. Chem. Lett.* **1994**, *4*, 807.

[18] M. Elzagheid, A. Azhayev, M. Oivanen, H. Lonnberg, *Collect. Czech. Chem. Commun., Special Issue* **1996**, *61*, S42.

[19] H. Vorbruggen, B. Bennua, *Chem. Ber.* **1981**, *114*, 1279.

[20] M. J. Robins, R. Zou, Z. Guo, S. F. Wnuk, *J. Org. Chem.* **1996**, *61*, 9207.

[21] R. Mengel, H. Griesser, *Tetrahedron Lett.* **1977**, 1177.

[22] A. P. Higson, G. K. Scott, D. J. Earnshaw, A. D. Baxter, R. A. Taylor, R. Cosstick, *Tetrahedron* **1996**, *52*, 1027.

[23] A. F. Russell, S. Greenberg, J. G. Moffatt, *J. Am. Chem. Soc.* **1973**, *95*, 4025.

[24] M. J. Robins, J. S. Wilson, D. Madej, N. H. Low, F. Hansske, S. F. Wnuk, *J. Org. Chem.* **1995**, *60*, 7902.

[25] T. C. Jain, J. G. Moffat, 'Abstracts, 165 National Meeting of the American Chemical Society', Dallas, Texas, April, 1973, CARB 15.

[26] G.-X. He, N. Bishofberger, *Tetrahedron Lett.* **1995**, *39*, 6991.

[27] T. L. Sheppard, A. T. Rosenblatt, R. Breslow, *J. Org. Chem.* **1994**, *59*, 7243.

[28] J. Herzig, A. Nudelman, H. E. Gottlieb, B. Fisher, *J. Org. Chem.* **1986**, *51*, 727.

Novel Nucleoside Analogues with Fluorophores Replacing the DNA Base

by Christoph Strässler, Newton E. Davis[1]), and Eric T. Kool*[1])

Department of Chemistry, University of Rochester, Rochester, NY 14627, USA

We describe the preparation and fluorescence properties of a set of new nucleosides in which a known hydrocarbon or oligothiophene fluorophore replaces the DNA base at C(1) of the deoxyribose moiety (see **3a – f**). These compounds are potentially useful as probes in the study of the structure and dynamics of nucleic acids and their complexes with proteins. In addition, they may find use as fluorescent labels for nucleic-acid-based biomedical diagnostics methods. The fluorophores conjugated to deoxyribose at C(1) in the α-D-form include terphenyl, stilbene, terthiophene, benzoterthiophene, and pyrene. Also included is a non-fluorescent spacer in which cyclohexene replaces the DNA base. The nucleosides are derived from brominated fluorophore precursors and *Hoffer*'s 2-deoxy-3,5-di-*O*-(*p*-toluoyl)-D-ribofuranosyl chloride. The emission maxima of the free nucleosides range from 345 to 536 nm. Also described are the 5'-(dimethoxytrityl) 3'-*O*-phosphoramidite derivatives **5a – f**, suitable for incorporation into oligonucleotides by automated synthesizers.

Introduction. – Fluorescence methods are extremely widespread in chemistry and biology. They give useful information on structure, distance, orientation, complexation, and location for biomolecules [1]. In addition, time-resolved methods are increasingly used in measurements of dynamics and kinetics [2]. As a result, many strategies for fluorescence labeling of biomolecules, such as nucleic acids, have been developed [3]. In the case of DNA, one of the most convenient and useful methods for fluorescence labeling is to add a fluorescent moiety during the DNA synthesis itself; this avoids the extra steps required for post-synthesis labeling and purification. The majority of labels commonly used during DNA synthesis are attached to the DNA by tethers that are often 5 to 11 atoms long; these flexible tethers can at times be problematic, since they allow the dye to tumble independently of the DNA and make the location of the dye difficult to determine precisely [4]. There are very few examples of dye conjugates that hold the dye close to the DNA, thus avoiding these problems. Among the known dyes of this class are ethenodeoxyadenosine [5] and 9*H*-purin-2-amine deoxyriboside [6]. These latter two compounds have modified DNA bases that are themselves fluorescent, and they have found much use as probes of enzymatic activities such as DNA synthesis, editing, and repair [7–9].

We have taken a related approach to developing new fluorescent labels for DNA [10]. Rather than modifying an existing DNA base, however, we have simply replaced it by another flat aromatic structure, *i.e.*, by a hydrocarbon rather than by a heterocyclic N-containing base. This substitution is particularly attractive because it creates only a

[1])   Current address: Dept. of Chemistry, Stanford University, Stanford, CA 94305-5080, USA (fax: (650) 725-0259; e-mail: kool@chem.stanford.edu)

small perturbation to the natural DNA structure and allows for close interaction, including possible stacking, with the neighboring DNA helix. There are many known hydrocarbon-based fluorophores, many with high quantum yields and with varied excitation and emission characteristics. Moreover, their lack of functional groups makes them relatively simple to work with in preparing conjugates.

We previously reported early studies on the incorporation of 4-methyl-1*H*-indole, naphthalene, phenanthrene, and pyrene fluorophores at the C(1)-position of 2-deoxy-D-ribose [10][11]. In a similar strategy, *Coleman* and co-workers recently reported the substitution of a coumarin dye at C(1) [12]. Our 4-methyl-1*H*-indole derivative recently found use as a fluorescent reporter of DNA-repair activities [13]. In addition, the derivative with a pyrene moiety at C(1) in the *α*-D-form has been shown to be useful in DNA diagnostics strategies, where it efficiently forms excimers with neighboring pyrene labels [14]. We demonstrated further that the corresponding pyrene derivative in the *β*-D-form stabilizes DNA helices markedly (due to its low polarity) [15][16], and that it can be enzymatically incorporated into the DNA helix [17]. Thus, this new nucleic acid labeling strategy is beginning to find useful applications.

In our ongoing studies, we wished to generate a series of new nucleosides with improved fluorescence characteristics, increasing the range of emission wavelengths over those we previously studied. Such compounds might be more generally useful in biophysical and diagnostics experiments. Here we report an expanded set of new fluorescent hydrocarbons and oligothiophenes conjugated to 2-deoxy-D-ribose. These new compounds significantly broaden the range of fluorescence properties available for automated incorporation into DNA.

Results and Discussion. – We have previously described the preparation of *C*-nucleosides of type **3** by cadmium- or zinc-mediated reaction of *Grignard* derivatives of aromatic compounds of type **1** with *Hoffer*'s chlorosugar (= 2-deoxy-3,5-di-*O*-(*p*-toluoyl)-D-ribofuranosyl chloride) (*Scheme 1*) [18][19]. The primary product in this coupling reaction is the 1-coupled product of type **2** in the *α*-D-form generated with retention of configuration. Although the *α*-D-orientation is not the same as for natural *β*-D-nucleosides, *α*-D-nucleosides are also known to form DNA-like helices [20], and models suggest that they can still interact well with natural bases in neighboring positions. Our approach in this study was to choose a set of known fluorophores for which a bromo derivative **1** was readily available and which had varied absorption and emission characteristics.

Besides this set of five fluorophores (see **3a** – **e**), we also prepared a *C*-nucleoside **3f** with cyclohexene at the 1-position as a nonfluorescent spacer. Fluorophores are usually quenched by neighboring DNA bases [21], and we observed this to be the case for the pyrene derivative **3d** as well [10]. Thus, we designed the cyclohexene compound to be inserted, if desired, between fluorophores and natural DNA bases to possibly limit any quenching that might occur. Cyclohexene was chosen rather than saturated cyclohexane because the former has sp^2 geometry at the point of attachment, the same as the other analogues and the natural bases.

Thus, we purchased 1-bromopyrene (**1d**) and prepared bromobenzoterthiophene **1b** [22], 4-bromo-1,1′:4′,1″ terphenyl (**1c**) [23], 4-bromostilbene (= 4-bromo-1,1′-(ethene-1,2-diyl)bis[benzene]; **1e**) [24], and 1-bromocyclohexene (**1f**) [25] according

to literature procedures. The 5-bromo-2,2′: 4′,2″-terthiophene (**1a**) was prepared from
2-bromothiophene *via* **6** and **7** as shown in *Scheme 2*. Subsequent coupling reactions
with *Hoffer*'s chlorosugar [26] were performed generally as described previously, with
moderate to modest success, giving yields of 23–55% for the coupled products **2**. The
pyrene nucleoside **3d** was also prepared by this approach, as previously described [10].
The toluoyl protecting groups on all the nucleosides **2a–f** were then removed to
generate the free nucleosides **3a–f**. These were examined for their fluorescence
characteristics (see below). For future studies, in which these fluorophores are
incorporated into DNA, we carried on with the nucleosides **3a–f** to prepare the 5′-
(dimethoxytrityl)-protected nucleosides **4a–f** and then the 3′-phosphoramidites **5a–f**.

The syntheses proceeded as expected, except in the benzoterthiophene case. During
the synthesis of the nucleosides **2b** and **3b**, significant decomposition was observed. To
determine whether it was a thermal or a photodecomposition, an NMR sample of the
di-*O*-toluoylnucleoside **2b** in CDCl$_3$ was kept exposed to fluorescent room light, and a
similar sample was kept wrapped in aluminium foil. ^{1}H-NMR Spectra were measured
immediately after preparation of the two samples, and then after 1 day and after 7 days.
We found that the spectrum of the aluminium-foil-protected sample remained
unchanged (not shown), whereas decomposition of the nucleoside **2b** could be
observed in the light-exposed sample after 7 days. Subsequently, we carried out all
experiments with the benzoterthiophene derivatives and limited light exposure by
covering glassware with aluminium foil. The best success in coupling **1b** with *Hoffer*'s
chlorosugar was observed with the debrominated form of **1b**, *i.e.*, direct deprotonation

Scheme 1

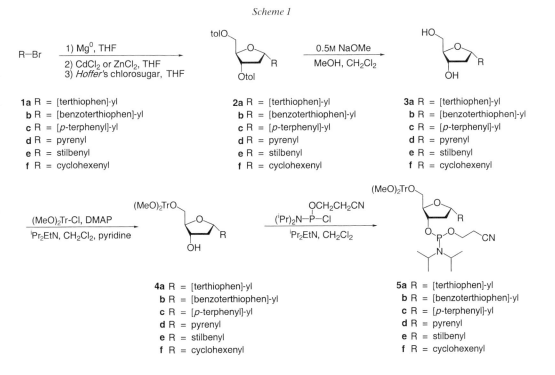

1a R = [terthiophen]-yl
b R = [benzoterthiophen]-yl
c R = [*p*-terphenyl]-yl
d R = pyrenyl
e R = stilbenyl
f R = cyclohexenyl

2a R = [terthiophen]-yl
b R = [benzoterthiophen]-yl
c R = [*p*-terphenyl]-yl
d R = pyrenyl
e R = stilbenyl
f R = cyclohexenyl

3a R = [terthiophen]-yl
b R = [benzoterthiophen]-yl
c R = [*p*-terphenyl]-yl
d R = pyrenyl
e R = stilbenyl
f R = cyclohexenyl

4a R = [terthiophen]-yl
b R = [benzoterthiophen]-yl
c R = [*p*-terphenyl]-yl
d R = pyrenyl
e R = stilbenyl
f R = cyclohexenyl

5a R = [terthiophen]-yl
b R = [benzoterthiophen]-yl
c R = [*p*-terphenyl]-yl
d R = pyrenyl
e R = stilbenyl
f R = cyclohexenyl

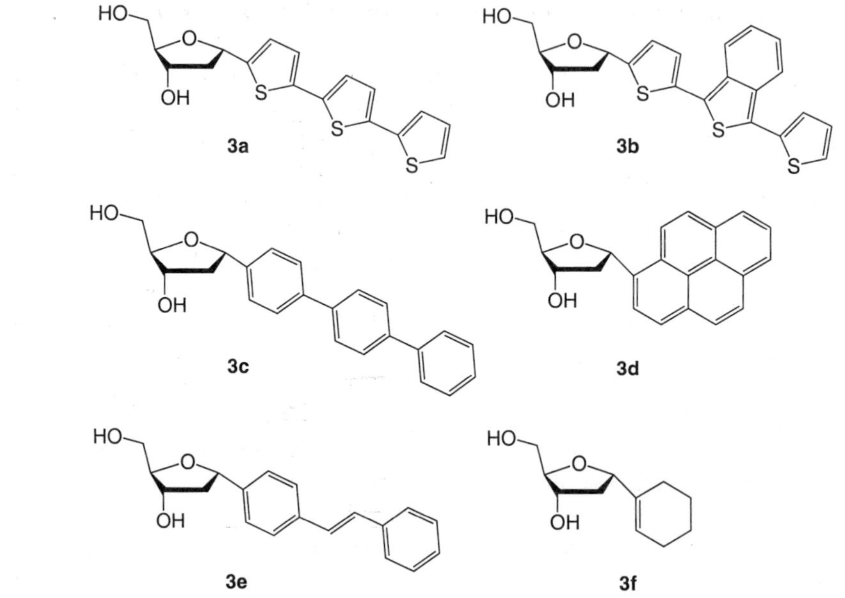

3a

3b

3c

3d

3e

3f

Scheme 2

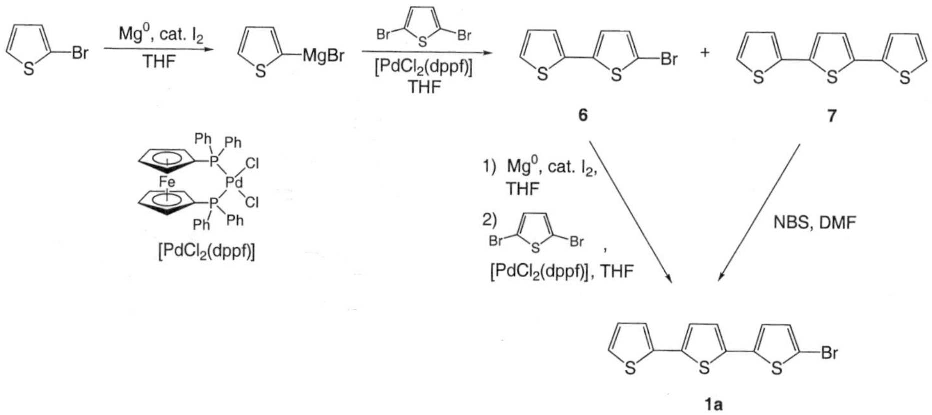

6

7

1a

formed the organolithium species which was then exchanged with CdCl$_2$ to give the analogous organocadmium-mediated reaction.

Absorption and emission spectra of 10 µM solutions of the deprotected nucleosides **3a–c,e** and of the earlier described pyrene nucleoside **3d** [10] were measured in deoxygenated MeOH at room temperature. Excitation spectra were also measured (data not shown) at the emission maxima, and the spectra were identical to the absorption curves shown in the *Figure*. The quantum yields for the five compounds (*Table*) were determined with quinine sulfate and fluorescein as standards. The results show absorption maxima ranging from 285 nm for terphenyl derivative **3c** to 437 nm for

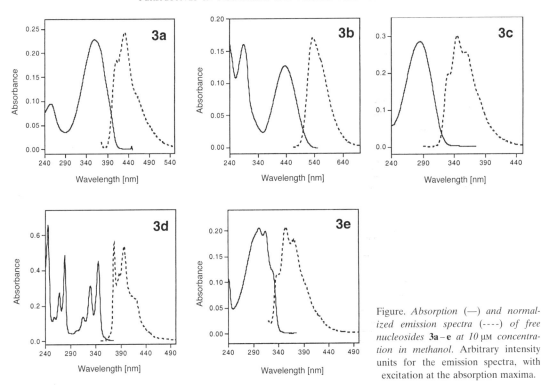

Figure. *Absorption* (—) *and normalized emission spectra* (----) *of free nucleosides* **3a – e** *at 10 μM concentration in methanol.* Arbitrary intensity units for the emission spectra, with excitation at the absorption maxima.

benzoterthiophene derivative **3b**, which appears yellow-orange in solution under incandescent light. Emission maxima range from 345 nm for terphenyl derivative **3c** (a violet-blue fluorophore) to 536 nm for the benzoterthiophene derivative **3b** (which fluoresces bright yellow).

Not surprisingly, there is little difference in the absorbance and fluorescence spectra of the nucleosides **3a – e** and the corresponding free fluorophores. The quantum yields of the terthiophene nucleoside **3a** and the free terthiophene **7** are about the same as the reported quantum yield for free terthiophene [27], and the quantum yield of pyrene nucleoside **3d** is similar to that of pyrene-1-butanoic acid. However, the quantum yield we measured is *ca.* 20-fold smaller than the quantum yield reported by *Telser et al.* [28] for pyrenebutanoate. They used a different value for the quantum yield of quinine sulfate (0.70 instead of 0.55) and measured the quantum yield in an aqueous buffer, but those differences do not seem sufficient to explain this difference. The sharpness of the absorption and emission lines for this compound may have caused difficulties in accurately measuring a maximum value. The quantum yield we measured for the stilbene nucleoside **3e** is *ca.* 50% smaller than the quantum yield reported by *Lewis et al.* [29] for a stilbenedicarboxamide at the excitation wavelength of 330 nm, which was determined in a 4 : 1 aqueous EtOH solution, using a phenanthrene standard. The quantum yields for the benzoterthiophene and terphenyl nucleosides **3b** and **3c**,

Table 1. *Absorption and Emission Data and Quantum Yields (Φ_f) for Nucleosides* **3a** – **e** *in Methanol*

	Absorption maxima [nm]	Extinction coeff. [$M^{-1} cm^{-1}$]	Emission maxima [nm]	Φ_f (excitation) [nm]
3a	358	31400	432	0.059 (358)
3b	437	18300	536	0.67 (440)
3c	285	40100	345	0.42 (290)
3d	343	34400	377	0.025 (344)
3e	301	21100	356	0.055 (298)
Pyrene-1-butanoic acid	343	33800	377	0.027 (344)
2,2′: 4′,2″-Terthiophene (**7**)[a]	355	29900	431	0.063 (358)[b]

[a]) CH_2Cl_2 solvent. [b]) The reported quantum yield is 0.055 [27].

respectively are high compared with the other nucleosides, but no literature reference could be found for comparison.

We expect that some of these nucleoside analogues will have a range of uses. There are several colors to choose from, which allows for possible multiplex use in diagnostics studies. In addition, some of the free fluorophores incorporated here are known to undergo excimer formation when two are present at high local concentrations. We expect that excimer, exciplex, and other forms of energy transfer might be observed for some of these nucleosides when placed in proximity with one another. Such studies are currently underway.

We thank the *U.S. Army Research Office* for support, and the *Swiss National Science Foundation* for a 'Stipendium für angehende Forschende' to *C.S.* We also thank *Joseph Langenhan* for early work with the cyclohexene derivative.

Experimental Part

General. Solvents used as reaction media were purified and dried by distillation over CaH_2 (pyridine, MeCN, and CH_2Cl_2), Na (THF), or MeONa (MeOH) before use. Chemicals were purchased from *Acros*, *Aldrich*, *Alfa-Aesar*, *Lancaster*, *Fisher*, or *J. T. Baker*. Flash chromatography (FC): silica gel *Merck 60*, 0.040 – 0.063 mm. ¹H-NMR (400 MHz) and ¹³C-NMR (100 MHz): in CDCl₃ unless otherwise stated; *Bruker-Avance-400* spectrometer; chemical shifts in ppm rel. to SiMe₄, coupling constants *J* in Hz. High resolution mass spectral analyses (HR-MS) were performed by the University of California-Riverside mass spectrometry facility. EI-MS: *HP 5973* mass selective detector. Abbreviations: DMAP = 4-(dimethylamino)pyridine. $(MeO)_2Tr = 4,4'$-dimethoxytrityl.

General Procedure A (G.P. A). A soln. of the aryl bromide **1** in dry THF was slowly added to Mg turnings in dry THF. To start the *Grignard* reaction, a few drops of 1,2-dibromoethane were added, and the mixture was slightly heated. Afer complete addition of the aryl bromide soln., the mixture was stirred for 2 h at 50°. $CdCl_2$ was then added and the mixture stirred for 2 h under reflux. The mixture was cooled to r.t., and a soln. of 2-deoxy-3,5-di-*O*-(*p*-toluoyl)-D-ribofuranosyl chloride [22] in THF was added. After stirring for 16 h at r.t., the solvent was evaporated and the residue suspended in CH_2Cl_2 and washed twice with 10% NH_4Cl soln. The aq. layers were extracted with CH_2Cl_2 and the org. layers dried (MgSO₄) and evaporated. Purification by FC (hexanes/AcOEt 6:1) gave the pure α-D-anomers **2** (the β-D-anomers as minor products were not isolated).

General Procedure B (G.P. B). Freshly prepared 0.5M NaOMe in MeOH was added to a soln. of the protected nucleoside **2** in MeOH/CH_2Cl_2 1:1. After stirring for 4 h at r.t., crystalline NH_4Cl was added, and the solvent was evaporated. Purification by FC (AcOEt) gave the pure deprotected nucleosides.

General Procedure C (G.P. C). The deprotected nucleoside **3** was co-evaporated twice with pyridine and then dissolved in pyridine/CH_2Cl_2. Then, 4,4′-dimethoxytrityl chloride ($(MeO)_2$Tr-Cl), ⁱPr₂EtN, and a catalytic amount of 4-(dimethylamino)pyridine (DMAP) were added, and the mixture was stirred for 4 – 8 h at r.t. Then the solvents were evaporated, and the residue was purified by FC (hexanes/AcOEt 4:1 → 1.5:1, preequilibrated with hexanes containing 5% Et_3N): **4**.

General Procedure D (G.P. D). To a soln. of the (MeO)$_2$Tr-protected nucleoside **4** in CH_2Cl_2, 2-cyanoethyl diisopropylphosphoramidochloridite and $^iPr_2EtN$ were added, and the mixture was stirred for 5 h at r.t. The solvent was evaporated and the residue purified by FC (hexanes/AcOEt 3:1, preequilibrated with hexanes/AcOEt 3:1 containing 5% Et_3N): **5**.

1. *Terthiophene Nucleoside. 5-Bromo-2,2'-bithiophene* (**6**). A soln. of 2-bromothiophene (4.967 g, 24.92 mmol) in dry THF (5 ml) was added dropwise to Mg turnings (752 mg, 30.94 mmol) and a small I_2 crystal in dry THF (25 ml). When *ca.* 1 ml of the bromothiophene soln. was added, the reaction started, and the mixture was kept under reflux. After the addition, the mixture was stirred for 1 h under reflux. The mixture was then transferred with a syringe to an addition funnel and added slowly during 3 h to an ice-cooled mixture of 2,5-dibromothiophene (8.206 g, 33.91 mmol) and [1,1'-bis(diphenylphosphino)ferrocene]dichloropalladium(II) ([PdCl$_2$(dppf)$_2$]; 250 mg, 0.306 mmol; 1:1 complex with CH_2Cl_2 in dry THF (50 ml). This mixture was stirred for 2 h at 0° and for 16 h at r.t. The solvent was evaporated and the residue suspended in AcOEt and washed with sat. NaHCO$_3$ soln. and brine. The aq. layers were extracted with AcOEt and the org. layers dried (MgSO$_4$) and evaporated. Purification by FC (hexanes) gave 4.229 g (57%) of **6** and 1.216 g (16%) of *2,2':5',2''-terthiophene* (**7**) as yellow solids.

Data of **6**: ^{1}H-NMR: 7.22 (*dd, J* = 5.2, 0.8, 1 H); 7.11 (*dd, J* = 3.6, 0.8, 1 H); 7.00 (*dd, J* = 5.2, 3.6, 1 H); 6.96 (*d, J* = 3.9, 1 H); 6.91 (*d, J* = 3.9, 1 H). ^{13}C-NMR: 138.9, 136.4 (2s); 130.6, 127.8, 124.8 (3d); 124.3 (s); 124.0, 123.8 (2d). EI-MS: 246 (100, M^+ (^{81}Br)), 244 (88, M^+ (^{79}Br)), 165 (39), 121 (53).

5-Bromo-2,2':5',2''-terthiophene [23] (**1a**). *From* **6**: A soln. of **6** (3.052 g, 12.45 mmol) in dry THF (20 ml) was added dropwise to Mg turnings (332 mg, 13.66 mmol) and a small I_2 crystal in dry THF (10 ml). To start the *Grignard* reaction, the mixture had to be heated under reflux. After the addition, the mixture was stirred for 2 h under reflux. Then it was cooled to r.t., transferred with a syringe to an addition funnel, and added dropwise during 2 h to a mixture of 2,5-dibromothiophene (3.031 g, 12.53 mmol) and [PdCl$_2$(dppf)$_2$] (113 mg, 0.138 mmol; 1:1 complex with CH_2Cl_2) in dry THF (50 ml) at −20°. This mixture was stirred for 2 h at −20° and then for 14 h at r.t. The solvent was evaporated and the residue dissolved in CH_2Cl_2 and washed with 5% HCl soln. and brine. The aq. layers were extracted with CH_2Cl_2 and the org. layers dried (MgSO$_4$) and evaporated. Purification by FC (hexane) gave 1.597 g (39%) of **1a**.

From **7**: *N*-Bromosuccinimide (3.147 mg, 17.681 mmol) was added in portions during 5 h to a soln. of **7** (4.315 g, 17.373 mmol) in DMF (10 ml) at −20°. After *ca.* 90 min, a precipitate was formed. After stirring for 14 h at r.t., the mixture was dissolved in CH_2Cl_2 (300 ml) and washed with 1N HCl (2 × 100 ml). The aq. layers were extracted with CH_2Cl_2 and the org. layers dried (MgSO$_4$) and evaporated. Purification by FC (hexanes) gave 5.222 g (92%) of **1a**. Yellow solid. ^{1}H-NMR: 7.23 (*dd, J* = 5.2, 1.0, 1 H); 7.17 (*dd, J* = 3.6, 1.0, 1 H); 7.07 (*d, J* = 3.7, 1 H); 7.02 (*dd, J* = 5.3, 3.6, 1 H); 7.01 (*d, J* = 3.6, 1 H); 6.98 (*d, J* = 3.9, 1 H); 6.91 (*d, J* = 3.8, 1 H). ^{13}C-NMR (CDCl$_3$): 138.6; 136.8; 136.7; 135.0; 130.7; 127.9; 124.7; 124.5; 124.3; 123.9; 123.7; 111.0. EI-MS: 328 (100, M^+ (^{81}Br)), 326 (88, M^+, (^{79}Br)), 247 (15), 203 (32).

(1S)-1,4-Anhydro-1,2-dideoxy-([2,2':5',2''-terthiophen]-5-yl)-3,5-di-O-(p-toluoyl)-D-erythro-pentitol (**2a**). According to *G.P. A* with **1a** (1.809 g, 5.528 mmol) in THF (15 ml), Mg turnings (141 mg, 5.802 mmol) in THF (2 ml), CdCl$_2$ (1.021 g, 5.569 mmol) and *Hoffer*'s chlorosugar (2.144 g, 5.514 mmol) in THF (10 ml); 1.816 g (55%) of **2a**. Yellow foam. ^{1}H-NMR: 7.97, 7.83 (2d, J = 8.2, 4 arom. H); 7.24−7.16 (*m*, 6 arom. H); 7.07 (*d, J* = 3.8, 1 arom. H); 7.03−7.01 (*m*, 3 arom. H); 6.91 (*d, J* = 3.5, 1 arom. H); 5.62 (*m*, H−C(3)); 5.57 (*dd, J* = 7.5, 4.9, H−C(1)); 4.70 (*m*, H−C(4)); 4.58 (*m*, 2 H−C(5)); 2.96, 2.47 (2m, 2 H−C(2)); 2.41, 2.39 (2s, 2 MeC$_6$H$_4$). ^{13}C-NMR: 166.3, 166.1 (2 C=O); 145.5, 144.0, 143.9, 137.1, 136.7, 136.1 (6s, 7 arom. C); 129.73, 129.71, 129.1, 129.0, 127.9 (5d, 9 arom. CH); 127.0, 126.8 (2s, 2 arom. C); 125.0, 124.5, 124.3, 124.1, 123.6, 123.2 (6d, 6 arom. CH); 82.1, 76.8, 76.2 (3d, CH(1), CH(3), CH(4)); 64.4, 40.3 (2t, CH(5), CH$_2$(2)); 21.7 (q, CH$_3$C$_6$H$_4$). HR-MS: 600.1109 (C$_{33}$H$_{28}$O$_5$S$_3^+$, M^+; calc. 600.1099).

(1S)-1,4-Anhydro-1,2-dideoxy-1-([2,2':5',2''-terthiophen]-5-yl)-D-erythro-pentitol (**3a**). According to *G.P. B* with **2a** (535 mg, 0.891 mmol) and 0.5M NaOMe in MeOH (1 ml, 0.5 mmol) in MeOH/CH$_2$Cl$_2$ 1:1 (5 ml): 235 mg (72%) of **3a**. Yellow crystals. ^{1}H-NMR: 7.25−6.93 (*m*, 7 arom. H); 5.35 (*m*, H−C(1)); 4.49, 4.10 (2m, H−C(3), H−C(4)); 3.86, 3.76 (2m, 2 H−C(5)); 2.75, 2.26 (2m, 2 H−C(2)). ^{13}C-NMR (D$_5$)pyridine): 148.2, 137.4, 136.9, 136.3 (4s); 128.6, 125.4, 125.1, 124.7, 124.4, 123.9 (6d, 7 arom. CH); 87.7, 76.1, 72.8 (3d, CH(1), CH(3), CH(4)); 63.0, 44.8 (2t, CH(5), CH(2)). HR-MS: 364.0273 (C$_{17}$H$_{16}$O$_3$S$_3^+$, M^+; calc. 364.0262).

(1S)-1,4-Anhydro-1,2-dideoxy-5-O-(4,4'-dimethoxytrityl)-1-([2,2':5',2''-terthiophen]-5-yl)-D-erythro-pentitol (**4a**). According to *G.P.* with C, (MeO)$_2$Tr-Cl (269 mg, 0.794 mmol), $^iPr_2EtN$ (115 μl, 0.672 mmol), a spatula tip of DMAP, **3a** (161 mg, 0.442 mmol), and pyridine/CH$_2$Cl$_2$ 1:1 (8 ml) 6 h. Purification gave 268 mg (91%) of **9a**.

Yellowish foam. ^{1}H-NMR: 7.50 (d, $J = 7.5$, 2 arom. H); 7.41–7.19, 7.10–7.04 (2m, 13 arom. H); 6.95 (d, $J = 3.5$, 1 arom. H); 6.88 (d, $J = 8.8$, 4 arom. H); 5.36 (t, $J = 7.0$, H–C(1)); 4.45 (m, H–C(3)); 4.21 (m, H–C(4)); 3.82 (s, 2 MeO); 3.38, 3.27 (2m, 2 H–C(5)); 2.75, 2.21 (2m, 2 H–C(2)). ^{13}C-NMR: 158.5, 145.9, 144.7, 137.1, 136.7, 136.2, 136.1, 135.9 (8s, 10 arom. C); 130.0, 128.1, 127.9, 126.9, 125.0, 124.5, 124.3, 124.2, 123.7, 123.3, 113.2 (11d, 20 arom. CH); 86.4 (s, 1 C); 84.4, 75.9, 74.8 (3d, CH(1), CH(3), CH(4)); 64.5 (t, CH$_2$(5)); 55.2 (q, 2 MeO); 42.9 (t, CH$_2$(2′)). HR-MS: 689.1430 (C$_{38}$H$_{34}$NaO$_5$S$_3$$^+$, [$M$ + Na]$^+$; calc. 689.1466).

(1S)-1,4-Anhydro-1,2-dideoxy-5-O-(4,4′-dimethoxytrityl)-1-([2,2′:5′,2″-terthiophen]-5-yl)-D-erythro-pentitol 3-(2-(Cyanoethyl Diisopropylphosphoramidite) (**5a**). According to *G.P. D*, with 2-cyanoethyl diisopropylphosphoramidochloridite (172 mg, 0.726 mmol), ^{i}Pr$_2$EtN (350 µl, 2.044 mmol), **4a** (313 mg, 0.469 mmol), and CH$_2$Cl$_2$ (5 ml). Purification gave 321 mg (79%) of **5a**. Yellowish foam. ^{1}H-NMR (mixture of 2 diastereoisomers): 7.53–7.50, 7.42–7.17, 7.09–7.01, 6.93–6.91, 6.86–6.82 (5m, 20 arom. H); 5.43 (m, H–C(1)); 4.57, 4.34 (2m, H–C(3), H–C(4)); 3.81, 3.80 (2s, 2 MeO); 3.72–3.48, 3.38–3.27, 3.17–3.13 (3m, 2 Me$_2$CHN), OCH$_2$CH$_2$CN, 2 H–C(5)); 2.75 (m, 1 H–C(2)); 2.56–2.51 (m, OCH$_2$CH$_2$CN); 2.37 (m, 1 H–C(2)); 1.17–1.03 (m, 2 Me$_2$CHN). ^{13}C-NMR: 158.4; 146.1; 144.9, 137.1, 136.4, 136.1, 136.0 (7s, 10 arom. C); 130.1, 128.2, 127.9, 127.8, 126.7, 124.9, 124.4, 124.3, 124.0, 123.6, 123.2 (11d, 16 arom. CH); 117.5 (s, CN); 113.1 (d, 4 arom. CH); 86.1 (s, 1 C); 84.7, 76.3 (2d, CH(1), CH(4)); 75.0, 74.8 (2d, CH(3)); 63.8 (t, CH$_2$(5)); 58.2, 58.1 (2t, OCH$_2$CH$_2$CN); 55.2 (q, 2 MeO); 43.2, 43.1 (2d, 2 Me$_2$CHN); 42.4 (t, CH$_2$(2)); 24.6, 24.5, 24.4, 24.3 (4q, 2 Me$_2$CHN); 20.1, 20.0 (2t, OCH$_2$CH$_2$CN). HR-MS: 889.2532 (C$_{47}$H$_{51}$N$_2$NaO$_6$PS$_3$$^+$, [$M$ + Na]$^+$; calc. 889.2545).

2. Benzoterthiophene Nucleoside. *(1S)-1,4-Anhydro-1,2-dideoxy-1-[5-[3-(thiophen-2-yl)benzo[c]thiophen-1-yl]thiophen-2-yl]-3,5-di-O-(p-toluoyl)-D-erythro-pentitol* (**2b**). The reaction was performed in the absence of light (flasks wrapped in Al-foil). To a soln. of 1,3-di(thiophen-2-yl)benzo[c]thiophene [22] in THF (40 ml) at −68°, 2.5M BuLi in hexane (1 ml, 2.5 mmol) was slowly added and stirred for 90 min. CdCl$_2$ (281 mg, 1.533 mmol) was added at −68° and the mixture allowed to warm up to r.t. and stirred for 2 h at r.t. A soln. of *Hoffer*'s chlorosugar (831 mg, 2.137 mmol) in THF (20 ml) was added and stirred for 16 h. Then the solvent was evaporated, the residue dissolved in CH$_2$Cl$_2$, the soln. washed twice with 10% NH$_4$Cl soln. The aq. layers were extracted with CH$_2$Cl$_2$ and the org. layers dried (MgSO$_4$) and evaporated. Purification by FC (hexane/AcOEt 7:1) gave 678 mg (49%) of **2b** and 231 mg (37%) of recovered unreacted 1,3-di(thiophen-2-yl)benzo[c]thiophene. **2b**: ^{1}H-NMR: 8.01–7.85 (m, 6 arom. H); 7.41–7.06 (11m, 13 arom. H); 5.68–5.64 (m, H–C(1), H–C(3)); 4.77, 4.62 (2m, H–C(4), 2 H–C(5)); 3.02 (m, H–C(2)); 2.45 (s, MeC$_6$H$_4$); 2.57 (m, H–C(2)); 2.44 (s, MeC$_6$H$_4$); 2.38 (s, MeC$_6$H$_4$). ^{13}C-NMR: 166.4, 166.1 (2 C=O); 143.9, 143.8, 141.6, 140.6, 140.1, 139.7 (6s, arom. C); 129.7, 129.6, 129.1, 129.0, 128.8, 127.5, 127.4, 127.1, 127.0 (10d, 19 arom. CH); 126.8 (s, arom. C); 126.1 (d, 2 arom. CH); 82.2, 80.0, 76.4 (3d, CH(1), CH(3), CH(4)); 64.6, 40.3 (2t, CH$_2$(5), CH$_2$(2)); 21.67, 21.63 (2q, 2 CH$_3$C$_6$H$_4$). HR-MS: 651.1358 (C$_{37}$H$_{31}$O$_5$S$_3$$^+$, [$M$ + H]$^+$; calc. 651.1334).

(1S)-1,4-Anhydro-1,2-dideoxy-1-[5-[3-(thiophen-2-yl)benzo[c]thiophen-1-yl]thiophen-2-yl]-D-erythro-pentitol (**3b**). According to *G.P. B*, in the absence of light (flasks wrapped in Al-foil), with **2b** (548 mg, 0.842 mmol) and 0.5M NaOMe in MeOH (850 µl, 0.425 mmol) in MeOH/CH$_2$Cl$_2$ 1:1 (5 ml): 243 mg (70%) of **3b**. Dark yellow solid. ^{1}H-NMR ((D$_8$)THF): 7.97–7.91 (m, 2 arom. H); 7.48 (d, $J = 4.9$, 1 arom. H); 7.38 (d, $J = 3.2$, 1 arom. H); 7.24–7.21, 7.15–7.10, 7.02–7.00 (3m, 5 arom. H); 5.28 (t, $J = 7.4$, H–C(1)); 4.41, 4.29, 3.90 (3m, H–C(3), H–C(4), 2 OH); 3.67–3.55 (m, 2 H–C(5)); 2.66, 2.08 (m, 2 H–C(2)). ^{13}C-NMR ((D$_8$)THF): 146.8, 133.4, 133.3, 133.1, 132.3 (5s, 5 arom. C); 125.9 (d, 1 arom. CH); 124.9, 124.0 (2s, 2 arom. C); 123.7, 123.5, 123.0, 122.8, 122.7, 122.4, 119.5, 119.3 (8d, 8 arom. CH); 84.5, 73.7, 70.1 (3d, CH(1), CH(3), CH(4)); 60.3, 42.1 (2t, CH$_2$(5), CH$_2$(2)).

(1S)-1,4-Anhydro-1,2-dideoxy-5-O-(4,4′-dimethoxytrityl)-1-[5-[3-(thiophen-2-yl)benzo[c]thiophen-1-yl]thiophen-2-yl]-D-erythro-pentitol (**4b**). According to *G.P. C*, in the absence of light (flasks wrapped in Al-foil), with (MeO)$_2$Tr-Cl (210 mg, 0.620 mmol), ^{i}PrEtN (180 µl, 1.051 mmol), a spatula tip of DMAP, **3b** (104 mg, 0.251 mmol), and pyridine/CH$_2$Cl$_2$ 1:1 (5 ml), 3 days. Purification gave 94 mg (52%) of **4b**. Dark yellow foam. ^{1}H-NMR: 7.99–7.95 (m, 2 arom. H); 7.49–7.15 (m, 15 arom. H); 7.06 (d, $J = 3.6$, 1 arom. H); 6.87 (d, $J = 8.8$, 4 arom. H); 5.42 (t, $J = 7.1$, H–C(1)); 4.47, 4.21 (2m, H–C(3), H–C(4)); 3.82 (s, 2 MeO); 3.46 (dd, $J = 9.5, 4.4$, H–C(5)); 4.21 (dd, $J = 9.5, 6.0$, H–C(5)); 2.81 (m, 1 H–C(2)); 2.31–2.24 (m, 1 H–C(2), 1 OH); 2.03 (d, $J = 5.0$, 1 OH). ^{13}C-NMR: 158.5, 146.8, 144.7, 135.9, 135.6, 135.3, 135.2, 135.1 (8s, 10 arom. C); 130.0, 128.1, 127.9, 126.5 (4d, 9 arom. CH); 126.6, 126.4 (2s, 2 arom. C); 125.6, 125.5, 125.2, 125.1, 124.8, 121.6, 121.5, 113.2 (8d, 13 arom. CH); 86.4 (s, 1 C); 84.3, 75.9, 74.9 (3d, CH(1), CH(3). CH(4)); 64.5 (t, CH$_2$(5)); 55.2 (q, 2 MeO); 42.9 (t, CH$_2$(2)).

(1S)-1,4-Anhydro-1,2-dideoxy-5-O-(4,4′-dimethoxytrityl)-1-[5-[3-(thiophen-2-yl)benzo[c]thiophen-1-yl]thiophen-2-yl]-D-erythro-pentitol 3-(2-cyanoethyl Diisopropylphosphoramidite) (**5b**). According to *G.P. D*, in the absence of light (flasks wrapped in Al-foil), with 2-cyanoethyl diisopropylphosphoramidochloridite (166 mg,

0.701 mmol), ^{i}Pr$_2$EtN (300 µl, 1.75 mmol), **4b** (318 mg, 0.444 mmol), and CH$_2$Cl$_2$ (10 ml). Purification gave 302 mg (74%) of **5b**. Dark yellow foam. ^{1}H-NMR (mixture of 2 diastereoisomers): 7.73 – 7.67, 7.59 – 7.25, 6.91 – 6.87 (3*m*, 26 arom. H); 5.33, 4.65, 4.43 (3*m*, H–C(1), H–C(3), H–C(4)); 3.84, 3.83 (2*s*, 2 MeO); 3.66 – 3.35, 3.26 – 3.23 (2*m*, 2 Me$_2$C*H*N, OC*H*$_2$CH$_2$CN, 2 H–C(5)); 2.81 (*m*, 1 H–C(2)); 2.50 – 2.18 (*m*, OCH$_2$C*H*$_2$CN, 1 H–C(2)); 1.20 – 1.06 (*m*, 2 *Me*$_2$CHN). ^{13}C-NMR: 158.4, 144.9, 142.5, 142.2, 140.7, 140.0, 139.9, 139.7, 139.5, 136.1 (10*s*, 10 arom. C); 130.1, 128.8, 128.3, 127.8, 127.5, 127.40, 127.36, 127.31, 127.0, 126.9, 126.7, 126.5, 126.4, (14*d*, 22 arom. CH); 117.6, 117.5 (2*s*, CN); 113.1 (*d*, 4 arom. CH); 86.1 (*s*, 1 C); 84.8, 79.9 (2*d*, CH(1), CH(4)); 75.7, 75.2 (2*d*, CH(3)); 64.3, 64.0 (2*t*, CH$_2$(5)); 58.3, 58.1 (2*t*, OCH$_2$CH$_2$CN); 55.2 (*q*, 2 MeO); 43.2, 43.1 (2*d*, 2 Me$_2$CHN); 42.5 (*t*, CH$_2$(2)); 24.5, 24.4, 24.3 (3*q*, 2 *Me*$_2$CHN); 20.2, 20.1 (2*t*, OCH$_2$CH$_2$CN). HR-MS: 916.2831 (C$_{53}$H$_{57}$N$_2$NaO$_6$P$^+$, [M + Na]$^+$; calc. 916.2803).

3. Terphenyl Nucleoside. (1S)-1,4-Anhydro-1,2-dideoxy-1-([1,1':4',1''-terphenyl]-4-yl)-3,5-di-O-(p-toluoyl)- D-erythro-*pentitol* (**2c**). A soln. of 1,2-dibromoethane (2.3 ml, 26.7 mmol) in THF (50 ml) was slowly added to a mixture of Mg turnings (1.131 g, 46.54 mmol) in THF (100 ml) at r.t. After addition of *ca.* 5 ml of the dibromoethane soln., the *Grignard* reaction started. Then 4-bromo-1,1':4',1''-terphenyl (**1c**, 5.433 g, 17.57 mmol) was added to the suspension, and the rest of the dibromoethane solution was added slowly during 40 min at 50°. After stirring for 3 h, CdCl$_2$ (2.213 g, 12.07 mmol) was added and the mixture was stirred for 2 h under reflux. After cooling to r.t., a soln. of the chlorosugar (6.832 g, 17.57 mmol) in THF (50 ml) was added, and the mixture was stirred for 16 h at r.t. Then CH$_2$Cl$_2$ (200 ml) was added and the mixture washed twice with 10% NH$_4$Cl soln. The aq. layers were extracted with CH$_2$Cl$_2$/THF 1:1 and the org. layers dried (MgSO$_4$) and evaporated. Purification by FC (hexane/AcOEt 7:1) gave 2.339 g (23%) of **2c**. White powder. ^{1}H-NMR: 8.02 (*d*, *J* = 8.0, 2 arom. H); 7.73 – 7.66, 7.57 – 7.48, 7.42 – 7.39 (3*m*, 13 arom. H); 7.28, 7.17 (2*d*, *J* = 8.9, 4 arom. H); 5.66, 5.48 (2*m*, H–C(1), H–C(3)); 4.77, 4.64 (2*m*, H–C(4), 2 H–C(5)); 3.01 (*m*, H–C(2)); 2.45 (*s*, MeC$_6$H$_4$); 2.43 (*m*, H–C(2)); 2.40 (*s*, MeC$_6$H$_4$). ^{13}C-NMR: 166.4, 166.1 (2 C=O); 143.9, 143.8, 141.6, 140.6, 140.1, 139.7 (6*s*, arom. C); 129.7, 129.6, 129.1, 129.0, 128.8, 127.5, 127.4, 127.1, 127.0 (10*d*, 19 arom. CH); 126.8 (*s*, arom. C); 126.1 (*d*, 2 arom. CH); 82.2, 80.0, 76.4 (3*d*, CH(1), CH(3), CH(4)); 64.6, 40.3 (2*t*, CH$_2$(5), CH$_2$(2)); 21.67, 21.63 (2*q*, 2 CH$_3$C$_6$H$_4$). HR-MS: 582.2427 (C$_{39}$H$_{34}$O$_5$$^+$, M^+; calc. 582.2406).

*(1S)-1,4-Anhydro-1,2-dideoxy-1-([1,1':4',1''-terphenyl]-4-yl)-*D-erythro-*pentitol* (**3c**). According to *G.P. B*, with **2c** (427 mg, 0.732 mmol) and 0.5M NaOMe in MeOH (730 µl, 0.365 mmol) in MeOH/CH$_2$Cl$_2$ 1:1 (5 ml): 201 mg (79%) of **3c**. White powder. ^{1}H-NMR ((D$_8$)THF): 7.70 – 7.61, 7.48 – 7.39, 7.33 – 7.28 (3*m*, 13 arom. H); 5.05 (*m*, H–C(1)); 4.37, 4.16, 3.90, 3.82, 3.65, 3.60 (6*m*, H–C(3), H–C(4), 2 H–C(5), 2 OH)); 2.61, 1.86 (2*m*, 2 H–C(2)). ^{13}C-NMR ((D$_5$)pyridine): 144.1, 140.9, 140.2, 139.6 (4*s*, arom. C); 129.4, 127.9, 127.8, 127.3, 127.2, 127.1 (6*d*, arom. CH); 88.0, 79.8, 73.1 (3*d*, CH(1), CH(3), CH(4)); 63.3, 45.1 (2*t*, CH$_2$(5), CH$_2$(2)). HR-MS: 346.1553 (C$_{23}$H$_{22}$O$_3$$^+$, M^+; calc. 346.1569).

*(1S)-1,4-Anhydro-1,2-dideoxy-5-O-(4,4'-dimethoxytrityl)-1-[[1,1':4',1''-terphenyl]-4-yl)-*D-erythro-*pentitol* (**4c**). According to *G.P. C* (MeO)$_2$Tr-Cl (135 mg, 0.398 mmol), ^{i}Pr$_2$EtN (70 µl, 0.409 mmol), a spatula tip of DMAP, **3c** (94 mg, 0.271 mmol), and pyridine/CH$_2$Cl$_2$ 1:1 (4 ml), for 5 h. Purification gave 144 mg (82%) of **4c**. White foam. ^{1}H-NMR: 7.72 – 7.66, 7.55 – 7.25 (2*m*, 22 arom. H); 6.89 (*d*, *J* = 8.9, 4 arom. H); 5.23 (*t*, *J* = 7.4, H–C(1)); 4.50, 4.26 (2*m*, H–C(3), H–C(4)); 3.83 (*s*, 2 MeO); 3.45 (*dd*, *J* = 9.5, 4.6, H–C(5)); 3.30 (*dd*, *J* = 9.5, 6.0, H–C(5)); 2.77, 2.12 (2*m*, 2 H–C(2)). ^{13}C-NMR: 158.5, 144.8, 142.1, 140.7, 140.1, 139.8, 139.7, 135.9 (8*s*, 10 arom. C); 130.0, 128.8, 128.1, 127.9, 127.5, 127.4, 127.3, 127.1, 127.0, 126.8, 126.3, 113.2 (12*d*, 26 arom. CH); 86.4 (*s*, 1 C); 84.5, 79.5, 75.2 (3*d*, CH(1), CH(3), CH(4)); 64.7 (*t*, CH$_2$(5)); 55.2 (*q*, 2 MeO); 43.1 (*t*, CH$_2$(2)). HR-MS: 671.2786 (C$_{44}$H$_{40}$NaO$_5$$^+$, [$M$ + Na]$^+$; calc. 671.2773).

*(1S)-1,4-Anhydro-1,2-dideoxy-5-O-(4,4'-dimethoxytrityl)-1-([1,1':4',1''-terphenyl]-4-yl)-*D-erythro-*pentitol 3-(2-Cyanoethyl Diisopropylphosphoramidite)* (**5c**). According to *G.P. D*, with 2-cyanoethyl diisopropylphosphoramidochloridite (276 mg, 1.166 mmol) and ^{i}Pr$_2$EtN (500 µl, 2.92 mmol) were reacted with **4c** (494 mg, 0.761 mmol) in CH$_2$Cl$_2$ (10 ml). Purification gave 554 mg (86%) of **5c**. White foam. ^{1}H-NMR (mixture of 2 diastereoisomers): 7.73 – 7.67, 7.59 – 7.25, 6.91 – 6.87 (3*m*, 26 arom. H); 5.33, 4.65, 4.43 (3*m*, H–C(1), H–C(3), H–C(4)); 3.84, 3.83 (2*s*, 2 MeO); 3.66 – 3.35, 3.26 – 3.23 (2*m*, 2 Me$_2$C*H*N, OC*H*$_2$CH$_2$CN, 2 H–C(5)); 2.81 (*m*, 1 H–C(2)); 2.50 – 2.18 (*m*, OCH$_2$C*H*$_2$CN, 1 H–C(2)); 1.20 – 1.06 (*m*, 2 *Me*$_2$CHN). ^{13}C-NMR: 158.4, 144.9, 142.5, 142.2, 140.7, 140.0, 139.9, 139.7, 139.5, 136.1 (10*s*, 10 arom. C); 130.1, 128.8, 128.3, 127.8, 127.5, 127.40, 127.36, 127.31, 127.0, 126.9, 126.7, 126.5, 126.4 (14*d*, 22 arom. CH); 117.6, 117.5 (2*s*, CN); 113.1 (*d*, 4 arom. CH); 86.1 (*s*, 1 C); 84.8, 79.9 (2*d*, CH(1), CH(4)); 75.7, 75.2 (2*d*, CH(3)); 64.3, 64.0 (2*t*, CH$_2$(5)); 58.3, 58.1 (2*t*, OCH$_2$CH$_2$CN); 55.2 (*q*, 2 MeO); 43.2, 43.1 (2*d*, 2 Me$_2$CHN); 42.5 (*t*, CH$_2$(2)); 24.5, 24.4, 24.3 (3*q*, 2 *Me*$_2$CHN); 20.2, 20.1 (2*t*, OCH$_2$CH$_2$CN). HR-MS: 871.3846 (C$_{53}$H$_{57}$N$_2$NaO$_6$P$^+$, [M + Na]$^+$; calc. 871.3852).

4. Stilbene Nucleoside. (1S)-1,4-Anhydro-1,2-dideoxy-1-[4-(2-phenylethenyl)phenyl]-3,5-di-O-(p-toluoyl)- D-erythro-*pentitol* (**2e**). A soln. of *p*-bromostilbene (= 1-bromo-4-(2-phenylethenyl)benzene; 1.444 g,

5.572 mmol) in THF (25 ml) was slowly added to Mg turnings (234 mg, 9.63 mmol). After the addition of 2 ml of the *p*-bromostilbene soln., I_2 crystals, and 2 drops of dibromoethane were added to start the reaction. The rest of the *p*-bromostilbene was added slowly, then the soln. was heated under reflux for 2 h. Subsequently, $CdCl_2$ (1.22 g, 6.67 mmol) was added and the mixture heated for another 2 h under reflux. After cooling to r.t., a soln. of chlorosugar (2.731 g, 6.686 mmol) was added slowly *via* a dropping funnel and stirred for 16 h. The soln. was dried *in vacuo* to remove the THF, the residue dissolved in CH_2Cl_2, and the soln. washed twice with aq. sat. NH_4Cl soln., dried ($MgSO_4$), and evaporated. Purification by FC (hexanes/AcOEt 7 : 1) gave 2.279 g (77%) of **2e**. White powder. ^{1}H-NMR: 8.00 (*d*, *J* = 5.1, 4 H); 7.70 (*d*, *J* = 5.1, 2 H); 7.54 (*m*, 3 H); 7.42 (*m*, 3 H); 7.28 (*m*, 5 H); 7.27 (*m*, 2 H); 5.42 (*m*, 1 H); 5.40 (*t*, *J* = 4.2, 1 H); 4.68 (*m*, 1 H); 4.60 (*d*, *J* = 4.2, 2 H); 2.94 (*q*, *J* = 4.2, 1 H); 2.44 (*s*, 3 H); 2.42 (*s*, 3 H); 1.62 (*s*, 1 H); 1.28 (*t*, *J* = 4.5, 1 H). ^{13}C-NMR: 144.7; 138.6; 137.1; 129.3 (2*s*); 128.9; 128.1; 127.2; 127.1; 126.8; 87.6; 80.1; 73.2; 45.3; 25.7; 25.5; 25.3; 25.1; 24.9. HR-MS: 532.3067 ($C_{35}H_{32}O_5^+$, M^+; calc. 532.2250).

(1S)-1,4-Anhydro-1,2-dideoxy-1-[4-(2-phenylethenyl)phenyl]-D-erythro-pentitol (**3e**). According to *G.P. B*, with **2e** (203 mg, 0.381 mmol) and 0.5M NaOMe in MeOH (0.76 ml, 0.38 mmol) in MeOH/CH_2Cl_2 2.5 : 1 (5 ml): 610 mg (75%) of **3e**. ^{1}H-NMR: 7.42 (*m*, 4 H); 7.24 (*dm*, 4 H); 7.08 (*m*, 2 H); 4.86 (*q*, *J* = 4.1, 1 H); 4.22 (*q*, *J* = 4.0, 1 H); 3.70 (*q*, *J* = 2.7); 3.68 (*q*, *J* = 2.7); 3.45 (*s*, 6 H); 2.46 (*q*, *J* = 4.1, 1 H); 1.72 (*m*, 2 H); 1.17 (*s*, 1 H). ^{13}C-NMR: 166.4; 166.1; 143.9; 143.8; 141.8; 137.3; 136.5; 129.7; 129.6; 129.1; 129.0; 128.7; 128.6; 128.3; 127.6; 127.0; 126.8; 126.5; 126.4; 126.2; 126.0; 82.1; 81.9; 80.0; 76.3; 75.4; 65.1; 64.6; 55.2; 40.3; 39.2; 21.6. HR-MS: 296.1409 ($C_{19}H_{20}O_3^+$, M^+; calc. 296.1412).

(1S)-1,4-Anhydro-1,2-dideoxy-5-O-(4,4'-dimethoxytrityl)-1-[4-(2-phenylethenyl)phenyl]-D-erythro-pentitol (**4e**). According to *G.P. C*, with $(MeO)_2Tr$-Cl (841 mg, 2.46 mmol), $^iPr_2EtN$ (0.54 ml, 3.09 mmol), a spatula tip of DMAP, **3e** (610 mg, 2.058 mmol), and dry pyridine (10 ml), for 4 h. Purification by FC (silica gel preequilibrated with 5% Et_3N in hexanes; hexanes/AcOEt 6 : 1 → 1 : 2): 1.037 mg (84%) of **4e**. Light yellow foam. ^{1}H-NMR: 7.54 (*m*, 6 H); 7.45 (*m*, 13 H); 6.88 (*s*, 2 H); 6.86 (*s*, 2 H); 5.16 (*t*, *J* = 4.6, 1 H); 4.46 (*q*, *J* = 2.9, 1 H); 4.22 (*q*, *J* = 3.6, 1 H); 3.82 (*s*, 6 H); 3.43 (*q*, *J* = 2.9, 1 H); 3.27 (*dd*, *J* = 3.8, 2.1, 1 H); 2.72 (*q*, *J* = 3.9, 1 H); 1.97 (*d*, *J* = 2.8, 1 H); 1.61 (*s*, 1 H). ^{13}C-NMR: 158.5; 144.8; 142.3; 137.3; 135.9; 130.0; 128.6; 128.5; 128.3; 128.1; 127.9; 127.6; 126.8; 126.6; 126.4; 126.1; 113.1; 86.3; 84.4; 79.5; 75.2; 64.7; 55.2; 43.0.

(1S)-1,4-Anhydro-1,2-dideoxy-5-O-(4,4'-dimethoxytrityl)-1-[4-(2-phenylethenyl)phenyl]-D-erythro-pentitol 3-(2-Cyanoethyl Diisopropylphosphoramidite) (**5e**). According to *G.P. D*, with 2-cyanoethyl diisopropylphosphoramidochloridite (0.578 ml, 2.59 mmol), $^iPr_2EtN$ (1.2 ml, 6.9 mmol), **4e** (1.034 mg, 1.727 mmol) and CH_2Cl_2 (20 ml), for 4 h. Purification by FC (hexanes/AcOEt 3 : 1) gave 966 mg (70%) of **5e**. White foam. ^{1}H-NMR: 7.54 (*m*, 3 H); 7.41 (*m*, 7 H); 7.13 (*s*, 1 H); 6.86 (*dd*, *J* = 1.9, 3.7, 2 H); 5.16 (*t*, *J* = 4.6, 1 H); 4.60 (*dm*, 1 H); 4.39 (*q*, *J* = 2.6, 1 H); 3.81 (*s*, 6 H); 3.52 (*m*, 2 H); 3.29 (*m*, 4 H); 2.85 (*m*, 1 H); 2.20 (*m*, 1 H); 1.59 (*s*, 2 H); 1.15 (*m*, 6 H); 1.11 (*d*, *J* = 4.2); 1.05 (*d*, *J* = 4.2). ^{13}C-NMR: 158.3; 144.8; 142.3; 137.3; 135.9; 130.1; 128.6; 128.5; 128.3; 128.1; 127.9; 127.6; 126.8; 126.6; 126.4; 126.1; 113.0; 86.3; 84.4; 80.0; 75.2; 64.7; 55.1; 42.7; 24.6.

5. *Cyclohexene Nucleoside*. *(1S)-1,4-Anhydro-1-(cyclohex-1-en-1-yl)-1,2-dideoxy-3,5-di-O-(p-toluoyl)-D-erythro-pentitol* (**2f**). A soln. of 1-*di*-O-(p-*tol*)-bromocyclohexene (**1f**; 2.090 g, 12.98 mmol) and 1,2-dibromoethane (900 µl, 10.44 mmol) in THF (40 ml) was slowly added to Mg turnings (647 mg, 26.62 mmol) in THF (10 ml). To start the *Grignard* reaction, the mixture was slightly heated. After complete addition of the bromo compound, the mixture was stirred for 2 h at 50°. $ZnCl_2$ (901 mg, 6.61 mmol) was added and the mixture stirred for 2 h under reflux. Then, the mixture was cooled to r.t. and a soln. of *Hoffer*'s chlorosugar (5.076 g, 13.05 mmol) in THF (25 ml) added. After stirring for 16 h at r.t., the solvent was evaporated, the residue suspended in CH_2Cl_2 and washed twice with 10% NH_4Cl soln. The aq. layers were extracted with CH_2Cl_2 and the org. layers dried ($MgSO_4$) and evaporated. Purification by FC (hexanes/AcOEt 6 : 1) gave 2.313 g (41%) of **2f**. Colorless oil which contained *ca*. 13% of a double-bond isomer as an impurity. ^{1}H-NMR: 7.92 (*m*, 4 arom. H); 7.22 (*m*, 4 arom. H); 5.80 (*m*, 1 H); 5.50 (*m*, H−C(1)); 4.61, 4.53−4.44 (2*m*, H−C(3), H−C(4), 2 H−C(5)); 2.62 (*m*, 1 H−C(2)); 2.41, 2.39 (2*s*, 2 MeC_6H_4); 2.15−2.01, 1.67−1.54 (2*m*, 1 H−C(2), 8 H of chx). ^{13}C-NMR: 166.3, 166.1 (2 C=O); 143.9, 143.7 (2*s*, 2 arom. C); 136.7 (*s*, 1 C of chx); 129.7, 129.6, 129.1, 129.0 (4*d*, 8 arom. CH); 127.1, 127.0 (2*s*, 2 arom. C); 123.6 (1*d*, 1 CH of chx); 82.5, 81.2, 76.3 (3*d*, CH(1), CH(3), CH(4)); 64.7, 36.4 (2*t*, $CH_2(5)$, $CH_2(2)$); 24.9, 23.7, 22.5 (3*t*, 4 CH_2 of chx); 21.6 (*q*, $CH_3C_6H_4$). HR-MS: 435.2185 ($C_{27}H_{31}O_5^+$, $[M + H]^+$; calc. 435.2171).

(1S)-1,4-Anhydro-1-(cyclohex-1-en-1-yl)-1,2-dideoxy-D-erythro-pentitol (**3f**). According to *G.P. B*, with **2f** (276 mg, 0.635 mmol) and 0.5M NaOMe in MeOH (630 µl, 0.315 mmol) in MeOH/CH_2Cl_2 1 : 1 (5 ml): 101 mg (80%) of **3f**. Colorless oil. ^{1}H-NMR: 5.75 (br. *s*, 1 H of chx); 4.38, 4.29 (2*m*, H−C(1), H−C(3)); 3.83, 3.73, 3.65 (3*m*, H−C(4), 2 H−C(5)); 2.92, 2.68 (br. *s*, 2 OH); 2.29 (1*m*, 1 H−C(2)); 1.96−1.87, 1.67−1.54 (2*m*, 1 H−C(2),

8 CH of chx). ^{13}C-NMR: 137.6 (1s, 1 C); 123.5 (1d, 1 CH); 84.7, 81.7, 72.6 (3d, CH(1), CH(3′), CH(4′)); 62.3, 38.9 (2t, CH$_2$(5′), CH$_2$(2′)); 24.9, 23.5, 22.4 (3t, 4 CH$_2$ of chx). HR-MS: 198.1252 (C$_{11}$H$_{18}$O$_3^-$; calc. 198.1256).

(1S)-1,4-Anhydro-1-(cyclohex-1-en-1-yl)-1,2-dideoxy-5-O-(4,4′-dimethoxytrityl)-D-erythro-pentitol (**4f**). According to *G.P. C*, with (MeO)$_2$Tr-Cl (179 mg, 0.528 mmol), ^{i}Pr$_2$EtN (150 μl, 0.876 mmol), a spatula tip of DMAP, **3f** (85 mg, 0.428 mmol), and pyridine/CH$_2$Cl$_2$ 1:1 (5 ml), for 3 days. Purification gave 182 mg (85%) of **4f**. Colorless oil. ^{1}H-NMR: 7.35 (*m*, 2 arom. H); 7.25 – 7.09 (*m*, 7 arom. H); 6.74 (*m*, 4 arom. H); 5.67 (*m*, 1 H of chx); 4.32 (*t*, *J* = 7.4, H–C(1)); 4.18, 3.91 (2*m*, H–C(3), H–C(4)); 3.69 (*s*, 2 MeO); 3.23 (*dd*, *J* = 9.4, 4.6, 1 H–C(5)); 3.03 (*dd*, *J* = 9.4, 6.4, 1 H–C(5)); 2.21 (*m*, 1 H–C(2)); 2.11 – 1.93 (*m*, 4 CH of chx); 1.81 (*m*, 1 H–C(2)); 1.64 – 1.44 (*m*, 4 CH of chx). ^{13}C-NMR: 158.4, 144.8, 138.1, 136.0 (4s, 5 arom. C, 1 C of chx); 130.0, 128.1, 127.8, 126.7, 122.9, 113.1 (6d, 13 arom. CH, 1 CH of chx); 86.2 (*s*, 1 C); 83.9, 81.8, 75.1 (3d, CH(1), CH(3′), CH(4′)); 64.9 (*t*, CH$_2$(5)); 55.2 (*q*, 2 MeO); 40.7 (*t*, CH$_2$(2)); 24.9, 23.9, 22.5 (3t, 4 CH$_2$ of chx). HR-MS: 523.2480 (C$_{32}$H$_{36}$NaO$_5^+$, [*M* + Na]$^+$; calc. 523.2460).

(1S)-1,4-Anhydro-1-(cyclohex-1-en-1-yl)-1,2-dideoxy-5-O-(4,4′-dimethoxytrityl)-D-erythro-pentitol 3-(2-Cyanoethyl Diisopropylphosphoramidite) (**5f**). According to *G.P. D*, with 2-cyanoethyl diisopropylphosphoramidochloridite (433 mg, 1.829 mmol), ^{i}Pr$_2$EtN (800 μl, 4 mmol), **4f** (586 mg, 1.170 mmol), and CH$_2$Cl$_2$ (10 ml). Purification gave 662 mg (81%) of **5f**. White foam. ^{1}H-NMR (mixture of 2 diastereoisomers): 7.53 – 7.50, 7.42 – 7.17, 7.09 – 7.01, 6.93 – 6.91, 6.86 – 6.82 (5*m*, 20 arom. H); 5.43 (*m*, H–C(1)); 4.57, 4.34 (2*m*, H–C(3), H–C(4)); 3.81, 3.80 (2*s*, 2 MeO); 3.72 – 3.48, 3.38 – 3.27, 3.17 – 3.13 (3*m*, 2 Me$_2$C*H*N, OC*H*$_2$CH$_2$CN, 2 H–C(5)); 2.75 (*m*, 1 H–C(2)); 2.56 – 2.51 (*m*, OCH$_2$C*H*$_2$CN); 2.37 (*m*, 1 H–C(2)); 1.17 – 1.03 (*m*, 2 Me$_2$CHN). ^{13}C-NMR: 158.4, 146.1, 144.9, 137.1, 136.4, 136.1, 136.0 (7s, 10 arom. C); 130.1, 128.2, 127.9, 127.8, 126.7, 124.9, 124.4, 124.3, 124.0, 123.6, 123.2 (11d, 16 arom. CH); 117.5 (*s*, CN); 113.1 (*d*, 4 arom. CH); 86.1 (*s*, 1 C); 84.7, 76.3 (2d, CH(1), CH(4)); 75.0, 74.8 (2d, CH(3′)); 63.8 (*t*, CH$_2$(5′)); 58.2, 58.1 (2t, OCH$_2$CH$_2$CN); 55.2 (*q*, 2 MeO); 43.2, 43.1 (2d, 2 Me$_2$CHN); 42.4 (*t*, CH$_2$(2)); 24.6, 24.5, 24.4, 24.3 (4q, 2 Me$_2$CHN); 20.1, 20.0 (2t, OCH$_2$CH$_2$CN). HR-MS: 723.3528 (C$_{41}$H$_{53}$N$_2$NaO$_6$P$^+$, [*M* + Na]$^+$; calc. 723.3539).

6. *Absorption Spectra*. Absorption spectra were measured for 10 μM solns. in MeOH at r.t. on a *Cary*-I-UV/VIS spectrometer.

7. *Emission Spectra and Quantum Yields*. Fluorescence spectra of 10 μM solns. in MeOH (except for terthiophene **7**, in CH$_2$Cl$_2$) were measured on a *SPEX-1680* double spectrometer at r.t. The solvents were deoxygenated by bubbling N$_2$ through the solvent for 2 h. The spectra were corrected for instrument response. All slits were set to 2 mm, resulting in a *ca.* 3.4-nm resolution. Fluorescence quantum yields (Φ_f) were calculated by the equation $\Phi_f = (FA_s\eta^2\Phi_s)/(AF_s\eta_0^2)$, where the subscript s refers to the standard and Φ is the quantum yield, F the corrected, integrated fluorescence, A the absorption at the excitation wavelength, η the refractive index of MeOH (or CH$_2$Cl$_2$), and η_0 the refractive index of H$_2$O [30]. Quantum-yield standards were quinine sulfate (*Aldrich*, 99 + %, used without further purification), 10 μM in 1N H$_2$SO$_4$ ($\Phi_f = 0.55$ [31][32]), and fluorescein (*Aldrich*, recrystallized from 1N NaOH by adding 10% AcOH/H$_2$O), 10 μM in 0.1N NaOH ($\Phi_f = 0.90$ [31]).

REFERENCES

[1] a) C. A. Royer, *Methods Mol. Biol.* **1995**, *40*, 65; b) P. Wu, L. Brand, *Anal. Biochem.* **1994**, *218*, 1.

[2] A. Holzwarth, *Methods Enzymol.* **1995**, *246*, 334; b) D. P. Millar, *Curr. Opin. Struct. Biol.* **1996**, *6*, 637.

[3] a) S. L. Beaucage, R. P. Iyer, *Tetrahedron* **1993**, *49*, 1925; b) R. E. Cunningham, *Methods Mol. Biol.* **1999**, *115*, 271.

[4] a) T. Heyduk, Y. Ma, H. Tang, R. H. Ebright, *Methods Enzymol.* **1996**, *274*, 492; b) J. R. Lundblad, M. Laurance, R. H. Goodman, *Mol. Endocrinol.* **1996**, *10*, 607.

[5] a) J. R. Barrio, J. A. Secrist III, N. J. Leonard, *Biochem. Biophys. Res. Commun.* **1972**, *46*, 597; b) S. C. Srivastava, S. K. Raza, R. Misra, *Nucleic Acids Res.* **1994**, *22*, 1296.

[6] D. C. Ward, E. Reich, L. Stryer, *J. Biol. Chem.* **1969**, *244*, 1228.

[7] B. W. Allan, N. O. Reich, J. M. Beechem, *Biochemistry* **1999**, *38*, 5308.

[8] M. R. Otto, L. B. Bloom, M. F. Goodman, J. M. Beechem, *Biochemistry* **1998**, *37*, 10156.

[9] W. Bujalowski, M. M. Klonowska, *Biochemistry* **1994**, *33*, 4682.

[10] R. X. F. Ren, N. C. Chaudhuri, P. L. Paris, S. Rumney, E. T. Kool, *J. Am. Chem. Soc.* **1996**, *118*, 7671.

[11] S. Moran, R. X.-F. Ren, C. J. Sheils, S. Rumney, E. T. Kool, *Nucleic Acids Res.* **1996**, *24*, 2044.

[12] R. S. Coleman, M. L. Madaras, *J. Org. Chem.* **1998**, *63*, 5700.

[13] J. P. Erzberger, D. Barsky, O. D. Scharer, M. E. Colvin, D. M. Wilson, *Nucleic Acids Res.* **1998**, *26*, 2771.

[14] P. L. Paris, J. M. Langenhan, E. T. Kool, *Nucleic Acids Res.* **1998**, *26*, 3789.

[15] K. M. Guckian, B. A. Schweitzer, R. X. F. Ren, C. J. Sheils, P. L. Paris, D. C. Tahmassebi, E. T. Kool, *J. Am. Chem. Soc.* **1996**, *118*, 8182.

[16] T. J. Matray, E. T. Kool, *J. Am. Chem. Soc.* **1998**, *120*, 6191.

[17] T. J. Matray, E. T. Kool, *Nature (London)* **1999**, *399*, 704.

[18] N. C. Chaudhuri, E. T. Kool, *Tetrahedron Lett.* **1995**, *36*, 1795.

[19] N. C. Chaudhuri, R. X. F. Ren, E. T. Kool, *Synlett* **1997**, 341.

[20] F. Morvan, B. Rayner, J.-L. Imbach, *Anticancer Drug Des.* **1991**, *6*, 521.

[21] M. Manoharan, K. L. Tivel, M. Zhao, K. Nafisi, T. L. Netzel, *J. Phys. Chem.* **1995**, *99*, 17461.

[22] A. K. Mohanakrishnan, M. V. Lakshmikantham, C. McDougal, M. P. Cava, J. W. Baldwin, R. M. Metzger, *J. Org. Chem.* **1998**, *63*, 3105.

[23] a) A. McKillop, D. Bromley, E. C. Taylor, *J. Org. Chem.* **1972**, *37*, 88; b) R. Rossi, A. Carpita, M. Ciofalo, V. Lippolis, *Tetrahedron* **1991**, *39*, 8443; c) P. Bäuerle, F. Würthner, G. Götz, F. Effenberger, *Synthesis* **1993**, 1099.

[24] L. Cardona, I. Fernandez, B. Garcia, J. R. Pedro, *Tetrahedron* **1986**, *42*, 2725.

[25] G. Ndebeka, S. Raynal, P. Caubére, *J. Org. Chem.* **1980**, *45*, 5394.

[26] M. Hoffer, *Chem. Ber.* **1960**, *93*, 2777.

[27] P. Garcia, J. M. Pernaut, P. Hapiot, V. Wintgens, P. Valat, F. Garnier, D. Delabouglise, *J. Phys. Chem.* **1993**, *97*, 513.

[28] J. Telser, K. A. Cruickshank, L. E. Morrison, T. L. Netzel, *J. Am. Chem. Soc.* **1989**, *111*, 6966.

[29] F. D. Lewis, T. F. Wu, E. L. Burch, D. M. Bassani, J. S. Yang, S. Schneider, W. Jager, R. L. Letsinger, *J. Am. Chem. Soc.* **1995**, *117*, 8785.

[30] D. J. Simpson, C. J. Unkefer, T. W. Whaley, B. L. Marrone, *J. Org. Chem.* **1991**, *56*, 5391.

[31] J. N. Demas, G. A. Crosby, *J. Phys. Chem.* **1971**, *75*, 991.

[32] J. Olmsted, *J. Phys. Chem.* **1979**, *83*, 2581.

2-(Glucosylthio)ethyl Groups as Potential Biolabile Phosphate-Protecting Groups of Mononucleotides

by Nathalie Schlienger[a]), Christian Périgaud[a])*, Anne-Marie Aubertin[b]), Christine Thumann[b]), Gilles Gosselin[a]), and Jean-Louis Imbach[a])

[a]) Laboratoire de Chimie Organique Biomoléculaire de Synthèse, UMR CNRS 5625, Case courrier 008, Université Montpellier II, Place Eugène Bataillon, F-34095 Montpellier Cedex 5 (fax: (+ 33) 467042029; e-mail: perigaud@univ-montp2.fr)
[b]) Institut de Virologie de la Faculté de Médecine de Strasbourg, Unité INSERM 74, 3 rue Koeberlé, F-67000 Strasbourg

The *in vitro* anti-HIV effects and the stability studies of mononucleoside phosphotriester derivatives **1–3** of 3′-azido-3′-deoxythymidine (AZT) containing 2-(glucosylthio)ethyl moieties as potential biolabile phosphate-protecting groups are reported. The results of the anti-HIV evaluation demonstrate that the described compounds act *via* the release of the free nucleoside analogue and cannot be considered as mononucleotide prodrugs (pronucleotides). These data can be related to the lack of substrate affinity of these derivatives towards target-enzymes as corroborated by decomposition studies in various media and experiments with a purified β-D glucosidase.

Introduction. – In order to mask the negative charges of the phosphate function of 5′-mononucleotides (NMPs), a wide variety of protecting groups has been considered [1–3]. The resulting prodrugs (pronucleotides) were expected to cross cell membranes by passive diffusion and to liberate intracellularly the corresponding NMPs. Recent examples of pronucleotides that display *in vitro* anti-HIV-1 activity include mononucleoside phosphotriesters incorporating enzyme-labile transient phosphate-protecting groups. For instance, 2-(acylthio)ethyl (usually called SATE) groups have already been studied as esterase-mediated biolabile phosphate protections [4]. On the basis of several nucleoside models, it was demonstrated that bis[2-(acylthio)ethyl] phosphotriester derivatives were able to liberate the corresponding NMPs inside the cell [5][6]. Moreover, *in vitro* cytotoxicity studies indicated that neither the (acylthio)ethyl promoieties nor their degradation products induced additional toxicity compared to the parent nucleosides [7]. The proposed decomposition pathways of (acylthio)ethyl pronucleotides (*Scheme*) involve an esterase-mediated activation process leading to unstable phosphotriesters, which decompose spontaneously to afford the corresponding phosphodiesters. These intermediates are converted to NMPs by a similar pathway and/or phosphodiesterase activity.

However, esterases are widely distributed in organs, tissues, and body fluids of mammalian species [8][9]. In this respect, most of the ester prodrugs are rapidly hydrolyzed in human after oral absorption, since the gastrointestinal lumen, mucosal cells, and the liver are rich in these enzymes. Such a presystemic metabolism, preventing the delivery of the prodrugs to other tissues or organs, constitutes the major

Scheme. *Simplified Mechanisms Proposed for the Decomposition of (Acylthio)ethyl (SATE) Pronucleotides and Expected Decomposition Pathways for the Mononucleoside Bis[(glucopyranosylthio)ethyl] Phosphotriesters* **1–3**

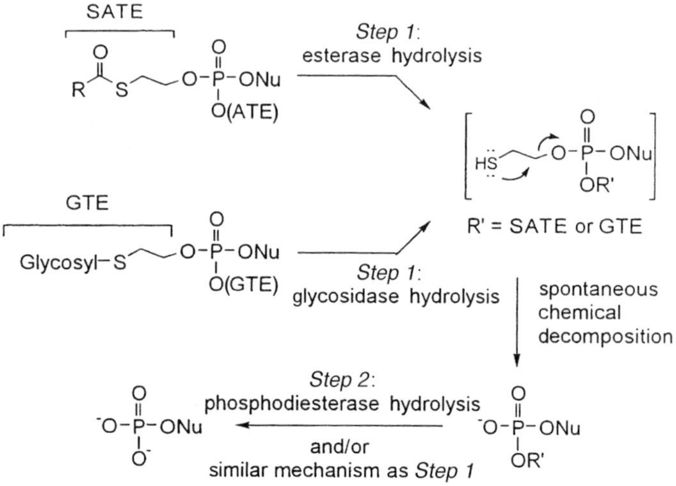

limitation of the *in vivo* development of pronucleotides designed to promote a site-specific delivery by an esterase-mediated activation process [10].

These considerations led us to explore the potential of new biolabile phosphate-protecting groups, namely 2-(glycosylthio)ethyl (GTE) groups, which may involve a more specific enzymatic activation system. The resulting bis[2-(glycosylthio)ethyl] phosphotriester derivatives (*Scheme*) were designed to be hydrolyzed by glycosidases. Glycosidases being intracellular enzymes [11], a more selective delivery of the corresponding NMP should be induced. Furthermore, the carbohydrate moieties could increase the water solubility of the pronucleotides and be used not only as a site-directing moiety toward glycosyl-binding proteins [12] on cell membranes of some cells (like macrophages [13][14]), but also as substrate for monosaccharide-facilitated diffusion-transport systems at the blood/brain barrier [15].

We report herein the biological evaluation in various HIV-1-infected cell lines and the stability studies in several biological media of phosphotriester derivatives **1–3** of AZT (= 3′-azido-3′-deoxythymidine) (*Fig. 1*), which incorporate β-D-glucopyranosyl moieties associated with a thioethyl linker, as a first model of glycosyl-modified phosphate-protecting groups.

Results and Discussion. – *Antiviral Activity.* The (glucopyranosylthio)ethyl phosphotriesters **1–3** (*Fig. 1*) were evaluated, in comparison with AZT and the corresponding (acylthio)ethyl pronucleotide (^tBuSATE = ^tBuC(O)SCH₂CH₂) [16], for their *in vitro* inhibitory effects on the replication of HIV-1 in established human lymphoblastoid cell lines CEM-SS and MT-4 (*Table 1*). This antiretroviral evaluation was extended to human peripheral blood mononuclear cells (PBMC) and monocyte-derived macrophages (M/M), which are most physiologically relevant to predict *in vivo* anti-HIV activity [18]. In these four cell-culture systems, phosphotriester derivatives

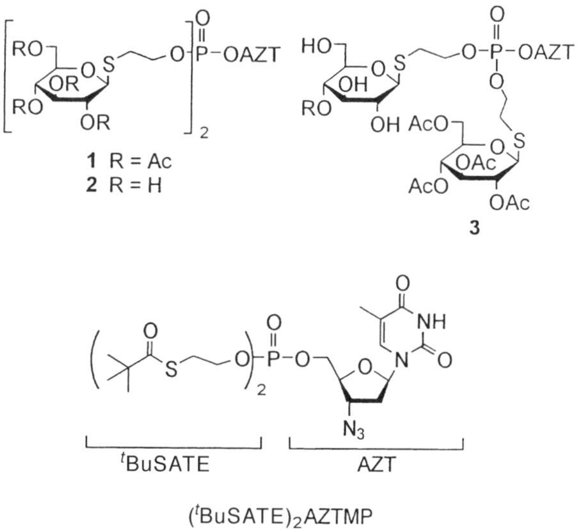

1 R = Ac
2 R = H

3

('BuSATE)₂AZTMP

Fig. 1. *Structures of the (glucopyranosylthio)ethyl phosphotriester derivatives* **1–3** *of AZT and of the corresponding (acylthio)ethyl pronucleotide* ('BuSATE = ^{t}BuC(O)SCH$_2$CH$_2$)

Table 1. *Antiviral Activity* [μM] *of the (Glucopyranosylthio)ethyl Derivatives* **1–5** *Compared to AZT and Its (Acylthio)ethyl Pronucleotide* ('BuSATE = ^{t}BuC(O)SCH$_2$CH$_2$) *in Various Cell Lines Infected with HIV-1*

	CEM-SS		MT-4		PBM		M/M	
	IC_{50}	CC_{50}	IC_{50}	CC_{50}	IC_{50}	CC_{50}	IC_{50}	CC_{50}
1	0.03	> 10	0.3	> 10	0.06	> 10	0.2	> 10
2	0.01	> 10	0.04	> 10	0.01	> 10	0.4	> 10
3	0.001	> 10	0.06	> 10	0.01	> 10	0.04	> 10
4	0.01	> 10	0.05	> 10	0.01	> 10	0.2	> 10
5	0.004	> 10	0.05	> 10	0.02	> 10	0.5	> 10
('BuSATE)₂AZTMP[a])	0.015	> 10	0.81	> 10	0.062	> 10	0.0001	> 100
AZT	0.006	> 100	0.01	75	0.001	75	0.0005	> 100

[a]) Previously published data [16][17] (^{t}BuC(O)SCH$_2$CH$_2$ = 'BuSATE).

1–3 significantly inhibited the multiplication of HIV-1 with IC_{50} values similar to or lower than those observed for AZT. This result was in agreement with data obtained from several bis[(acylthio)ethyl] phosphotriester derivatives of AZT [16], and may be related to the peculiar metabolism of this nucleoside analogue [19][20]. Studies on the metabolism of AZT in both uninfected and HIV-infected cells have shown that this nucleoside analogue, after cell uptake by a non-facilitated diffusion process (passive transport), is efficiently metabolized to AZTMP by cytosolic thymidine kinase (TK). The second phosphorylation step of AZT is catalyzed by thymidine monophosphate kinase (TMPK), but this cellular enzyme phosphorylates AZTMP inefficiently [21][22]. Thus, the intracellular concentration of AZTMP is 2–3 orders of magnitude higher than the one of AZTTP. This is consistent with the idea that TK catalyzes the rate limiting step in the anabolism of AZT to its 5'-triphosphate. Consequently, a mononucleotide prodrug (pronucleotide) of AZT cannot exhibit a higher *in vitro* anti-HIV activity than AZT in cells where cytosolic TK activity is expressed. In fact, contrary to the desired antiviral effect, an intracellular accumulation of AZTMP by means of a pronucleotide approach would lead to an inhibition of TMPK resulting in a blockage of the subsequent phosphorylation steps [23–25]. Concomitant cellular excretion of AZTMP [23][26–28] or dephosphorylation of AZTMP by endogenous pyrimidine nucleotidases and subsequent release of AZT by the cells [29][30] may also be related to the decreased anti-HIV-1 activity observed for an AZT pronucleotide as compared to the parent nucleoside.

More significant differences were found in the antiviral activities of the test compounds in HIV-1-infected CEM/TK⁻ cells (*Table 2*). This cell line is highly deficient in cytosolic TK and should be considered an 'ideal' *in vitro* system to investigate the antiviral activities of nucleotide analogues of AZT that may release the corresponding NMP into the cells. As expected, AZT proved to be completely inactive against HIV-1 replication in CEM/TK⁻ cells, whereas the corresponding 'BuSATE-protected pronucleotide proved to be a potent inhibitor with a IC_{50} value of 0.45 μM. In contrast, the (glucopyranosylthio)ethyl phosphotriesters **1–3** showed no anti-HIV activity in this cell line at concentrations up to 10 μM. This result seems to demonstrate that the phosphotriester derivatives **1–3**, which incorporate (glycosylthio)ethyl (GTE) phosphate-protecting groups, were not able to deliver AZTMP inside the cells.

Table 2. *Antiviral Activity* [μM] *of (Glucopyranosylthio)ethyl Derivatives* **1–5** *Compared to AZT and Its (Acylthio)ethyl Pronucleotide* ('BuSATE = 'BuC(O)SCH₂CH₂), *in CEM/TK⁻ Cell Line Infected with HIV-1*

	1	2	3	4	5	('BuSATE)₂AZTMP[a]	AZT
CEM/TK⁻ IC_{50}	> 10	> 10	> 10	> 10	> 10	0.45	> 100
CC_{50}	> 10	> 10	> 10	> 10	> 10	> 10	> 100

[a]) Previously published data [16] ('BuC(O)SCH₂CH₂ = 'BuSATE).

Table 3. *Half-lives of (Glucopyranosylthio)ethyl Derivatives* **1–5** *in RPMI 1640, Culture Medium, and Total CEM-SS Cell Extract*

	1	2	3	4	5
RPMI	35 h	4.5 h	8.3 h	6.8 d	stable (> 6 d)
Culture medium	30 h	4.5 h	8.3 h	28 h	40 h
Total CEM-SS cell extract	2.3 h	5 h	1.0 h	20 min	stable (> 2 d)

Stability and Decomposition Studies. To explain the inefficiency of the (glucopyranosylthio)ethyl moieties to act as biolabile phosphate-protecting groups of NMP, we studied the stability and the decomposition pathways of the compounds **1–3** in biological media by the improved 'on-line internal-surface reversed-phase (ISRP) cleaning' HPLC method previously developed in our laboratory [16]. This technique allows the direct analysis of biological samples without pretreatment. The stability of the compounds was evaluated in culture medium (RPMI 1640 containing 10% of heat-inactivated foetal calf serum), which is the extracellular medium used for antiviral evaluation in cell-culture systems, and in RPMI 1640, in order to distinguish between chemical and enzyme-controlled hydrolysis. Finally, the total cell extract from CEM-SS

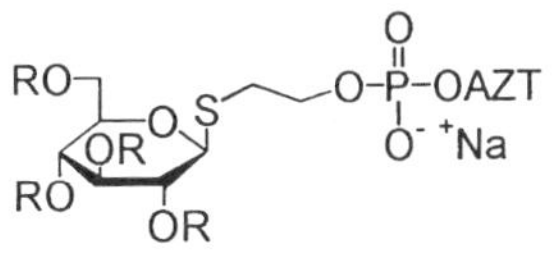

4 R = Ac
5 R = H

Fig. 2. *Structure of the (glucopyranosylthio)ethyl phosphodiester derivatives* **4** *and* **5** *of AZT*

cells was used to mimic the intracellular medium. The phosphodiester derivatives **4** and **5** (*Fig. 2*) were prepared as reference samples for calibration and identification of the decomposition products.

The similar half-lives obtained in culture medium and RPMI for the decomposition of the phosphotriesters **1**–**3** seem to indicate that the hydrolysis proceeds by a chemical mechanism [31]. The faster disappearance of the phosphotriesters **1** and **3** in the total cell extract than in culture medium was essentially due to rapid cleavage of the acetyl-protecting groups by esterases, which are present in higher amounts in the total cell extract.

In the total cell extract, the phosphotriesters **1**–**3** led to a time-dependent accumulation of the phosphodiester **5**. The decomposition of phosphotriester **2** showed that phosphodiester **5** was the only detectable product; in the case of the acetylated phosphotriesters **1** and **3**, the phosphodiester **4** was the main observed metabolite, which was further deacetylated by esterases present in this medium to give **5**. Surprisingly, the phosphodiester **5** was very stable (no decomposition observed after 2 days). This might be related to either low phosphodiesterase and/or β-D-glucosidase activity in the total cell extract. In contrast, the phosphodiesters **4** and **5** were significantly hydrolyzed in culture medium, probably by phosphodiesterase activation, to give AZTMP, which was further slowly dephosphorylated ($t_{1/2}$ *ca.* 106 h) to AZT. This observation could explain the anti-HIV-1 activities observed for the compounds **4** and **5** in TK$^+$ cell lines, which are similar to those of the corresponding phosphotriester derivatives **1**–**3**.

To explain the stability of the phosphodiester **5** in the total cell extract, further studies were performed. An evaluation of the β-D-glucosidase activity with 4-nitrophenyl β-D-glucopyranoside as substrate showed complete absence of β-D-glucosidase activity in this medium. This may be due either to the loss of β-D-glucosidase activity during the preparation of the cell extract, like previously reported for cell homogenates [32], or to intrinsically low β-D-glucosidase activity in CEM-SS cells. Substrate-specificity studies of the phosphodiester **5** towards purified sweet-almond β-D-glucosidase, an enzyme which catalyzes the hydrolysis of a wide range of glucosides [33], showed no detectable formation of AZTMP. The phosphodiesterase activity of the cell extract was evidenced with deoxythymidine (4-nitrophenyl phosphate) as a substrate, which was cleaved with a half-life of 2 h (data not shown). Owing to its stability in the cell extract, the phosphodiester **5** seems to have no substrate affinity towards the phosphodiesterase activity present in this medium.

In contrast to (acylthio)ethyl derivatives, bis[(glucopyranosylthio)ethyl] phospho-triesters cannot be considered as pronucleotides, since they are unable to deliver the corresponding NMP inside the cells. This fact is evidenced by the lack of antiviral activities of the AZT derivatives **1**–**3** in CEM/TK⁻ cells. The anti-HIV-1 activity observed for these compounds in TK-expressing cell lines could be related to the release of the parent nucleoside in the extracellular medium. The hydrolysis of bis-[(glucopyranosylthio)ethyl] phosphotriesters may involve, as a first step, a chemical decomposition process [31] giving rise to the formation of the phosphodiester **5**. This metabolite, which exhibited a great stability in the total cell extract, could subsequently be converted to AZTMP in the culture medium (through a phosphodiesterase activation) and further to AZT, the latter being able to cross the cell membrane.

The results presented above do not preclude the potential interest in a pronucleo-tide approach with glycosyl-modified moieties as transient protecting groups of NMP analogues; but they reveal the importance of the affinity of mononucleotide prodrugs (and corresponding expected metabolites) for intracellular target enzymes. The presence of the O-atom at the anomeric position of the sugar residue appears to be a major requirement to preserve enzymatic recognition. Thus, the design of new kinds of glycosyl-modified pronucleotides will involve necessarily another linker between the P-atom and the glycosyl moiety and, as a consequence, new decomposition processes. In this respect, we have previously reported that mononucleoside *S,S'*-bis[(2-acyl)ethyl phosphorodithioates] allow the efficient intracellular delivery of their parent 5'-mononucleotides [34]. The synthesis and study of new potential gylcosyl-modified pronucleotides incorporating a 2-oxyethyl linker are currently in progress in our laboratory.

These investigations were supported by grants from *CNRS* and the *Agence Nationale de Recherches sur le SIDA* (*ANRS*, France). *N.S.* was supported by an *ANRS* fellowship.

Experimental Part

Chemicals. The synthesis of the phosphotriester and phosphodiester derivatives **1**–**5** has been reported elsewhere [31]. The synthesis of 3'-azido-3'-deoxy-5'-thymidylic acid bis{2-[(2,2-dimethyl-1-oxopropyl)-thio]ethyl} ester ((ᵗBuSATE)₂AZTMP) was published previously [16]. The synthesis of AZTMP was carried out following a general procedure with phosphoryl chloride (= phosphoric trichloride) [35]. 4-Nitrophenyl β-D-glucopyranoside and 5'-thymidylic acid 4-nitrophenyl ester (=thymidine 5'-(4-nitrophenyl hydrogen phos-phate) were purchased from *Aldrich* and *Sigma*, resp.

Virology. The origin of the viruses and the techniques used for measuring inhibition of virus multiplication have been previously described [36]. Briefly, in MT-4 cells, the determination of the antiviral activity of the pronucleotides was based on a reduction of HIV-1-IIIB-induced cytopathogenicity, the metabolic activity of the cells being measured by the property of mitochondrial dehydrogenases to reduce the yellow 3-(4,5-dimethylthiazol-2-yl)-2,5-diphenyl-2*H*-tetrazolium bromide (MTT) to a blue formazan. For CEM-SS or CEM/TK⁻ cells, the production of virus HIV-LAI was measured by quantification of the reverse-transcriptase activity associated with the virus particle released in the culture supernatant [37]. Cells, MT-4, and CEM, were incubated with a *TCID*₅₀ of 50 or 100 or viruses during 30 min; after virus adsorption, unbound particles were eliminated by two washes, and cells were cultured in the presence of different concentrations of test compounds for 5 days, or 6 days in the case of CEM/TK⁻ cells, before virus-production determination. The IC_{50} was derived from the computer-generated median effect plot of the dose/effect data [38]. In parallel experiments, cytotoxicity of the test compounds for uninfected cells was measured after an incubation of 5 or 6 days in their presence by means of the colorimetric MTT test. The CC_{50} is the concentration at which OD_{540} was reduced by one-half and was calculated by means of the program mentioned above. Human peripheral blood mononuclear cells (PBMC) were obtained from HIV-1 seronegative donors. Isolation and culture of human monocytes and infection of monocyte-derived macrophages (M/M) with HIV-1 BAL were performed as previously described [17]. Culture and infection with HIV-1-LAI of human PBMC have been reported elsewhere [36]. IC_{50} Values were calculated on the basis of RT activity in the supernatant of the cell culture. Cytotoxicity of the compounds on uninfected cells was measured with the colorimetric MTT reaction.

Stability Studies. RPMI 1640 was purchased from *GibcoBRL*, and the CEM-SS cell extract was prepared according to published procedures [36]. The stability studies were performed by means of the 'on-line internal-surface reversed-phase (ISRP) cleaning' HPLC method previously described [16]. HPLC Analyses: *Waters* unit equipped with a model *600E* system controller, a model *486* detector (detection at 265 nm), and a *Millennium* data workstation; reversed-phase anal. column (*Macherey-Nagel, Nucleosil, C$_{18}$*, 150 × 4.6 mm, 5 μm), protected by a prefilter and a precolumn (*Guard-Pak* insert, *Delta-Pak C$_{18}$*, 100 Å) hold in a *Guard-Pak* holder; eluents *A* (50 mM triethylammonium acetate buffer (pH 7) and *B* (*A* containing 80% of MeCN); injection of the crude sample (80 μl, initial concentration 50 μM) onto the precolumn, eluent *A* for 5 min, then connection to the column and linear gradient *A* → *B* within 35 min; calibration and identification of the decomposition products by coinjection with authentic samples (AZTMP, AZT, phosphodiesters **4** and **5**). HPLC/MS Coupling experiments: negative electrospray mode; detection of some analytes presumed to result from deacetylation of **1**, **3**, and **4**. The rate constants of disappearance of the pronucleotides were calculated according to pseudo-first-order kinetic models [16], which were in accordance with the experimental data.

Enzyme Assays. Sweet-almond β-D-glucosidase (EC 3.2.1.21) was purchased from *Fluka*. One unit of activity was determined as the amount of enzyme which liberated 1 μmol/min of 4-nitrophenol in 0.1M phosphate buffer (pH 6.8) at 37° with 4-nitrophenyl β-D-glucopyranoside as substrate. Specific activity of the enzyme was 5.6 units/mg, determined spectrophotometrically by following the release of 4-nitrophenol. A suitable enzyme concentration with 4-nitrophenyl β-D-glucopyranoside (50 μM) as substrate was found to be 3.5 μg/ml and led to 10% of hydrolysis after 30 min. Substrate-affinity determination with phosphodiester **5** under the same conditions and also with a 100-fold higher enzyme concentration was carried out. Hydrolysis of the compound was estimated with the HPLC technique described above.

REFERENCES

[1] A. A. Arzumanov, N. B. Dyatkina, *Russ. J. Bioorg. Chem.* **1996**, *22*, 777.
[2] C. Périgaud, J.-L. Girardet, G. Gosselin, J.-L. Imbach, 'Comments on Nucleotide Delivery Forms', in Antiviral Drug Design', Ed. E. De Clercq, JAI Press, London, 1996, pp. 147–172.
[3] J. P. Krise, V. J. Stella, *Adv. Drug Delivery Rev.* **1996**, *19*, 287.
[4] C. Périgaud, G. Gosselin, J.-L. Imbach, 'Minireview: from the Pronucleotide Concept to the SATE Phosphate Protecting Groups', in 'Current Topics in Medicinal Chemistry', Ed. J. C. Alexander, Blackwell Science Ltd., Oxford, 1997, p. 15.
[5] C. Périgaud, G. Gosselin, J.-L. Imbach, *Antiviral Ther.* **1996**, *1* (*Suppl. 4*), 39.
[6] G. Gosselin, J.-L. Girardet, C. Périgaud, S. Benzaria, I. Lefebvre, N. Schlienger, A. Pompon, J.-L. Imbach, *Acta Biochim. Pol.* **1996**, *43*, 195.
[7] C. Périgaud, J.-L. Girardet, I. Lefebvre, M.-Y. Xie, A.-M. Aubertin, A. Kirn, G. Gosselin, J.-L. Imbach, J.-P. Sommadossi, *Antiviral Chem. Chemother.* **1996**, *7*, 338.
[8] E. Heymann, 'Carboxylesterases and Amidases', in 'Enzymatic Basis of Detoxification', Eds. W. B. Jacoby and R. R. Bend, Academic Press, New York, 1980, pp. 291–323.
[9] T. Satoh, *Rev. Biochem. Toxicol.* **1987**, *8*, 155.
[10] D. Egron, C. Périgaud, G. Gosselin, A.-M. Aubertin, J.-L. Imbach, *Bull. Soc. Chim. Belg.* **1997**, *106*, 461.
[11] A. Esen, 'b-Glucosidases', in 'β-Glucosidases, Biochemistry and Molecular Biology', Ed. A. Esen, American Chemical Society, Washington DC, 1993, p. 1.
[12] N. Sharon, H. Lis, *Science (Washington, D.C.)* **1989**, *246*, 227.
[13] P. D. Stahl, J. Somsel Rodman, M. J. Miller, P. H. Schlesinger, *Proc. Natl. Acad. Sci. U.S.A.* **1978**, *75*, 1399.
[14] M. Monsigny, A.-C. Roche, P. Midoux, R. Mayer, *Adv. Drug Delivery Rev.* **1994**, *14*, 1.
[15] J. N. Abbott, I. A. Romero, *Mol. Medicine Today* **1996**, *2*, 106.
[16] I. Lefebvre, C. Périgaud, A. Pompon, A.-M. Aubertin, J.-L. Girardet, A. Kirn, G. Gosselin, J.-L. Imbach, *J. Med. Chem.* **1995**, *38*, 3941.
[17] C. Thumann-Schweitzer, G. Gosselin, C. Périgaud, S. Benzaria, J.-L. Girardet, I. Lefebvre, J.-L. Imbach, A. Kirn, A.-M. Aubertin, *Res. Virol.* **1996**, *147*, 155.
[18] E. R. Kern, 'Preclinical Evaluation of Antiviral Agents: *in vitro* and Animal Model Testing', in 'Antiviral Agents and Viral Diseases of Man', Eds. G. J. Galasso, R. J. Whitley, and T. C. Merigan, Raven Press, New York, 1990, pp. 87–123.

[19] G. J. Veal, D. J. Back, *Gen. Pharmacol.* **1995**, *26*, 1469.

[20] J. Balzarini, *Pharm. World Sci.* **1994**, *16*, 113.

[21] A. Lavie, I. Schlichting, I. R. Vetter, M. Konrad, J. Reinstein, R. S. Goody, *Nat. Med. (N.Y.)* **1997**, *3*, 922.

[22] A. Lavie, I. R. Vetter, M. Konrad, R. S. Goody, J. Reinstein, I. Schlichting, *Nat. Struct. Biol.* **1997**, *4*, 601.

[23] L. W. Frick, D. J. Nelson, M. H. St Clair, P. A. Furman, T. A. Krenitsky, *Biochem. Biophys. Res. Commun.* **1988**, *154*, 124.

[24] P. A. Furman, J. A. Fyfe, M. H. St Clair, K. Weinhold, J. L. Rideout, G. A. Freeman, S. Nusinoff Lehrman, D. P. Bolognesi, S. Broder, H. Mitsuya, D. W. Barry, *Proc. Natl. Acad. Sci. U.S.A.* **1986**, *83*, 8333.

[25] J. A. Harrington, W. H. Miller, T. Spector, *Biochem. Pharmacol.* **1987**, *36*, 3757.

[26] R. P. Agarwal, A. M. Mian, *Biochem. Pharmacol.* **1991**, *42*, 905.

[27] R. P. Agarwal, *Int. J. Purine Pyrimidine Res.* **1991**, *2*, 1.

[28] A. Fridland, M. C. Connely, R. Ashmun, *Mol. Pharmacol.* **1990**, *37*, 665.

[29] A. De Flora, E. Zocchi, L. Guida, C. Polvani, U. Benatti, *Proc. Natl. Acad. Sci. U.S.A.* **1988**, *85*, 3145.

[30] M. Magnani, M. Bianchi, L. Rossi, V. Stocchi, *Biochem. Biophys. Res. Commun.* **1989**, *164*, 446.

[31] N. Schlienger, C. Périgaud, G. Gosselin, J.-L. Imbach, *J. Org. Chem.* **1997**, *62*, 7216.

[32] L. B. Daniels, R. H. Glew, 'β-D-Glucosidases in Tissue', in 'Methods of Enzymatic Analysis. Enzymes 2: Esterases, Glycosidases, Lyases, Ligases', Ed. H. U. Bergmeyer, J. Bergmeyer, and M. Grassl, Verlag Chemie, Weinheim, 1974, p. 217.

[33] M. P. Dale, H. E. Ensley, K. Kern, K. A. R. Sastry, L. D. Byers, *Biochemistry* **1985**, *24*, 3530.

[34] N. Schlienger, C. Périgaud, G. Gosselin, I. Lefebvre, A. Pompon, A.-M. Aubertin, A. Kirn, J.-L. Imbach, *Nucleosides Nucleotides* **1997**, *16*, 1321.

[35] L. A. Slotin, *Synthesis* **1977**, 737.

[36] F. Puech, G. Gosselin, I. Lefebvre, A. Pompon, A.-M. Aubertin, A. Kirn, J.-L. Imbach, *Antiviral Res.* **1993**, *22*, 155.

[37] C. Moog, A. Wick, P. Le Ber, A. Kirn, A.-M. Aubertin, *Antiviral Res.* **1994**, *24*, 275.

[38] J. Chou, T.-C. Chou, in 'Dose-Effect Analysis with Microcomputers: Quantitation of ED_{50}, LD_{50}, Synergism, Antagonism, Low-Dose Risk, Receptor Binding, and Enzyme Kinetics; Computer Software for Apple II Series and IBM-PC and Instruction Manual', Ed. Elsevier-Biosoft, Elsevier Science Publishers, Cambridge, U.K., 1985, p. 19.

Nucleotides

Part LXIII[1])

New 2-(4-Nitrophenyl)ethyl(Npe)- and 2-(4-Nitrophenyl)ethoxycarbonyl(Npeoc)-Protected 2'-Deoxyribonucleosides and Their 3'-Phosphoramidites – Versatile Building Blocks for Oligonucleotide Synthesis

by Holger Lang, Margarete Gottlieb, Michael Schwarz, Silke Farkas, Bernd S. Schulz, Frank Himmelsbach, Ramamurthy Charubala, and Wolfgang Pfleiderer*

Fakultät für Chemie, Universität Konstanz, Postfach 5560, D-78434 Konstanz

A series of new base-protected and 5'-O-(4-monomethoxytrityl)- or 5'-O-(4,4'-dimethoxytrityl)-substituted 3'-(2-cyanoethyl diisopropylphosphoramidites) and 3'-[2-(4-nitrophenyl)ethyl diisopropylphosphoramidites] **52–66** and **67–82**, respectively, are prepared as potential building blocks for oligonucleotide synthesis (see *Scheme*). Thus, 3',5'-di-O-acyl- and N^2,3'-O,5'-O-triacyl-2'-deoxyguanosines can easily be converted into the corresponding O^6-alkyl derivatives **6, 8, 10, 12, 14**, and **16** by a *Mitsunobu* reaction using the appropriate alcohol. Mild hydrolysis removes the acyl groups from the sugar moiety (→ **9, 11, 13, 15**, and **19** (*via* **18**), resp.) which can then be tritylated (→ **38–42**) and phosphitylated (→ **57–61**) in the usual manner. N^2-[2-(4-nitrophenyl)ethoxycarbonyl]-substituted and N^2-[2-(4-nitrophenyl)ethoxycarbonyl]-O^6-[2-(4-nitrophenyl)ethyl]-substituted 2'-deoxyguanosines **5** and **7**, respectively, are synthesized as new starting materials for tritylation (→ **28, 35**, and **37**) and phosphitylation (→ **54, 56, 70**, and **78**). Various O^4-alkylthymidines (see **20–24**) are also converted to their 5'-O-dimethoxytrityl derivatives (see **43–47**) and the corresponding phosphoramidites (see **62–66** and **79–82**).

1. Introduction. – Blocking groups [2] play a crucial role in synthetic approaches using polyfunctional molecules such as nucleosides and nucleotides. A broad variety of protecting strategies [3–7] have been recommended for the buildup of oligonucleotides which can nowadays easily be synthesized by machine-aided methods on solid-support materials [8–11]. Despite the fact that the use of acid- and base-labile protecting groups give very good results, we stressed for many years the idea of applying the 2-(4-nitrophenyl)ethyl (npe) and the 2-(4-nitrophenyl)ethoxycarbonyl (npeoc) group [12–14] as a versatile alternative providing an uniform protection of the amino, amide, hydroxy, mercapto, carboxy, and phosphate functions and the advantage of simultaneous cleavage under aprotic conditions by a β-elimination process [13]. Over the years, a large number of new npe- and npeoc-protected 2'-deoxyribonucleosides and their corresponding 3'-phosphoramidites have been synthesized and proven to be valuable intermediates for oligonucleotide synthesis. Furthermore, the synthesis of similar O^4-alkylthymidine and O^6-alkyl-2'-deoxyguanosine derivatives have been included in these studies. We will report in this paper about the syntheses and the

---

[1]) Part LXII [1].

characterizations of these building blocks starting from thymidine (**1**) and 2'-deoxyguanosine (**2**).

2. Syntheses. – Suitable protection of 2'-deoxyguanosine (**2**) can be seen as the most crucial presupposition for an effective and homogenous synthesis of oligo-2'-deoxyribonucleotides. Thus, 2'-deoxy-N^2-[2-(4-nitrophenyl)ethoxycarbonyl]guanosine (**5**) was synthesized by the transient protection method [14] using trimethylsilyl chloride for intermediary blocking of the sugar OH groups, followed by acylation with 2-(4-nitrophenyl)ethyl carbonochloridate (= 2-(4-nitrophenyl)ethyl chloroformate) (*cf.* the corresponding **3** and **4**) [12]. The preparation of 2'-deoxy-O^6-[2-(4-nitrophenyl)ethyl]-N^2-[2-(4-nitrophenyl)ethoxycarbonyl]guanosine (**7**) could be achieved in a one-pot reaction starting from 3',5'-di-O-acetyl-2'-deoxyguanosine [15] which was first subject to a *Mitsunobu* reaction [16] leading, under O^6-alkylation, to 3',5'-di-O-acetyl-2'-deoxy-O^6-[2-(4-nitrophenyl)ethyl]guanosine. Subsequent acylation with 2-(4-nitrophenyl)ethyl carbonochloridate, followed by treatment with ammonia, led to **7** in an overall yield of 66% (*cf.* also **6**). In a similar manner, 2'-deoxy-O^6-methyl-N^2-[2-(4-nitrophenyl)ethoxycarbonyl]guanosine (**19**) was synthesized *via* its 3',5'-di-O-acetyl derivative **18**, but in this case, besides O^6-methylation (→ **16**) also N^1-methylation to **17** took place in a substantial amount. Nevertheless, the *Mitsunobu* alkylation [17] is superior to nucleophilic displacement reactions of the activated amide function in **2** [18][19] and works also very well with 2'-deoxy-N^2,3'-O,5'-O-triisobutyrylguanosine [12][20] leading with MeOH, EtOH, ^{i}PrOH, and BuOH to the O^6-methyl-, O^6-ethyl-, O^6-isopropyl-, and O^6-butyl derivative **8**, **10**, **12**, and **14**, respectively, of which **10** and **12** were isolated as intermediates before hydrolysis to **11** and **13**, whereas **8** and **14** were converted without isolation into 2'-deoxy-N^2-isobutyryl-O^6-methylguanosine (**9**) and the corresponding O^6-butyl derivative **15**.

In the thymidine series, the *Mitsunobu* reaction leads, unfortunately, to alkylation at N(3) so that the synthesis of the O^4-alkyl derivatives **20–24** was achieved either by reaction with alkyl halide/silver carbonate [21][22] or *via* the triazolide method [22–25] and the use of O^4-[(2,4,6-triisopropylphenyl)sulfonyl] intermediates [26].

Tritylation reactions with 4-monomethoxy- and 4,4'-dimethoxytrityl chloride, respectively, proceeded very well by the conventional method and led, under selective substitution at the primary 5'-OH group, to **28** and **33–47** in yields > 85% (see also **25** [15], **26** and **27** [12], and **29–31** [15]). The synthesis by different routes of some of these compounds, *i.e.* **38**, **39**, **41**, and **43–46**, have already been described [26–29].

Phosphitylations were also performed by conventional methods applying either 2-cyanoethyl diisopropylphosphoramidochloridite (= chloro(2-cyanoethoxy)(diisopropylamino)phosphane; **48**) [30][31] with *Hünig*'s base as acid scavenger or with 2-cyanoethyl tetraisopropylphosphorodiamidite (= (2-cyanoethoxy)bis(diisopropylamino)phosphane; **49**) [32–35] and 1*H*-tetrazole in CH$_2$Cl$_2$ to convert the 5'-O-(4,4'-dimethoxytrityl)-2'-deoxyribonucleosides **33–47** into the corresponding 3'-(2-cyanoethyl diisopropylphosphoramidites) **52–66**. In a second series of reactions, we evaluated the 2-(4-nitrophenyl)ethyl group as a phosphate protecting group [36–38] and synthesized, from **25–36**, **43–45**, and **47**, with the phosphitylating agents 2-(4-nitrophenyl)ethyl diisopropylphosphoramidochloridite (= chloro(diisopropylamino)-[2-(4-nitrophenyl)ethoxy]phosphane; **50**) [39] and 2-(4-nitrophenyl)ethyl tetraisopro-

Scheme

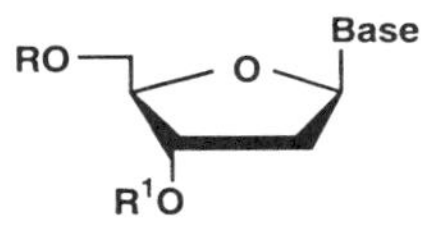

Table (compounds 1–24) — sugar with RO, Base, R¹O, substituents R and R¹

	Base	R	R¹
1	Thy	H	H
2	Gua	H	H
3	Ade^{npeoc}	H	H
4	Cyt^{npeoc}	H	H
5	Gua^{npeoc}	H	H
6	Gua^{npe6}_{ibu}	H	H
7	Gua^{npe6}_{npeoc}	H	H
8	Gua^{me6}_{ibu}	ibu	ibu
9	Gua^{me6}_{ibu}	H	H
10	Gua^{et6}_{ibu}	ibu	ibu
11	Gua^{et6}_{ibu}	H	H
12	Gua^{ipr6}_{ibu}	ibu	ibu
13	Gua^{ipr6}_{ibu}	H	H
14	Gua^{nbu6}_{ibu}	ibu	ibu
15	Gua^{nbu6}_{ibu}	H	H
16	Gua^{me1}	Ac	Ac
17	Gua^{me6}	Ac	Ac
18	Gua^{npeoc}	Ac	Ac
19	Gua^{me6}_{npeoc}	H	H
20	Thy^{me4}	H	H
21	Thy^{et4}	H	H
22	Thy^{ipr4}	H	H
23	Thy^{nbu4}	H	H
24	Thy^{npe4}	H	H

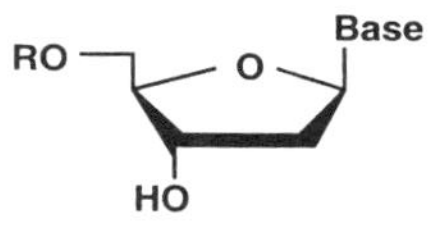

Table (compounds 25–47) — sugar with RO, Base, HO

	Base	R
25	Thy	MeOTr
26	Ade^{npeoc}	MeOTr
27	Cyt^{npeoc}	MeOTr
28	Gua^{npeoc}	MeOTr
29	Thy	$(MeO)_2Tr$
30	Ade^{bz}	$(MeO)_2Tr$
31	Cyt^{bz}	$(MeO)_2Tr$
32	Gua^{ibu}	$(MeO)_2Tr$
33	Ade^{npeoc}	$(MeO)_2Tr$
34	Cyt^{npeoc}	$(MeO)_2Tr$
35	Gua^{npeoc}	$(MeO)_2Tr$
36	Gua^{npe6}_{ibu}	$(MeO)_2Tr$
37	Gua^{npe6}_{npeoc}	$(MeO)_2Tr$
38	Gua^{me6}_{ibu}	$(MeO)_2Tr$
39	Gua^{et6}_{ibu}	$(MeO)_2Tr$
40	Gua^{ipr6}_{ibu}	$(MeO)_2Tr$
41	Gua^{nbu6}_{ibu}	$(MeO)_2Tr$
42	Gua^{me6}_{npeoc}	$(MeO)_2Tr$
43	Thy^{me4}	$(MeO)_2Tr$
44	Thy^{et4}	$(MeO)_2Tr$
45	Thy^{ipr4}	$(MeO)_2Tr$
46	Thy^{nbu4}	$(MeO)_2Tr$
47	Thy^{npe4}	$(MeO)_2Tr$

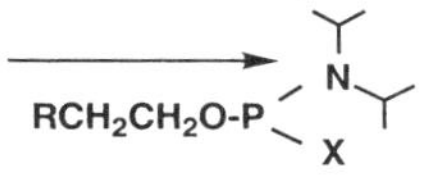

Phosphitylating reagents — $RCH_2CH_2O-P(-N(ipr))-X$

	R	X
48	CN	Cl
49	CN	$N(ipr)_2$
50	$O_2N-\langle\rangle$	Cl
51	$O_2N-\langle\rangle$	$N(ipr)_2$

Table (compounds 52–66) — $(MeO)_2TrO$ sugar with phosphoramidite $-O-P(-N(ipr)_2)-OCH_2CH_2CN$

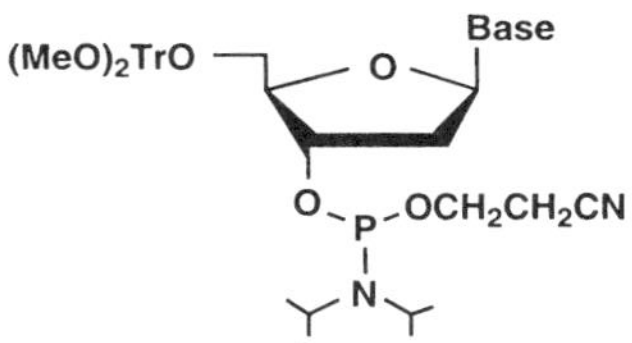

	Base
52	Ade^{npeoc}
53	Cyt^{npeoc}
54	Gua^{npe6}
55	Gua^{npe6}_{ibu}
56	Gua^{npe6}_{npeoc}
57	Gua^{me6}_{ibu}
58	Gua^{et6}_{ibu}
59	Gua^{ipr6}_{ibu}
60	Gua^{nbu6}_{ibu}
61	Gua^{me6}_{npeoc}
62	Thy^{me4}
63	Thy^{et4}
64	Thy^{ipr4}
65	Thy^{nbu4}
66	Thy^{npe4}

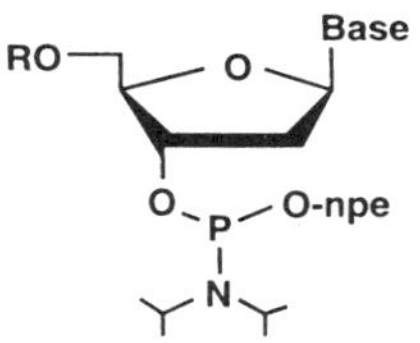

Table (compounds 67–82) — RO sugar with phosphoramidite $-O-P(-N(ipr)_2)-O-npe$

	Base	R
67	Thy	MeOTr
68	Ade^{npeoc}	MeOTr
69	Cyt^{npeoc}	MeOTr
70	Gua^{npeoc}	MeOTr
71	Thy	$(MeO)_2Tr$
72	Ade^{bz}	$(MeO)_2Tr$
73	Cyt^{bz}	$(MeO)_2Tr$
74	Gua^{ibu}	$(MeO)_2Tr$
75	Ade^{npeoc}	$(MeO)_2Tr$
76	Cyt^{npeoc}	$(MeO)_2Tr$
77	Gua^{npeoc}	$(MeO)_2Tr$
78	Gua^{npe6}_{npeoc}	$(MeO)_2Tr$
79	Thy^{me4}	$(MeO)_2Tr$
80	Thy^{et4}	$(MeO)_2Tr$
81	Thy^{ipr4}	$(MeO)_2Tr$
82	Thy^{npe4}	$(MeO)_2Tr$

Legend:

npe = 2-(4-nitrophenyl)ethyl
npeoc = [2-(4-nitrophenyl)ethoxy]carbonyl
ibu = isobutyryl = 2-methyl-1-oxopropyl
ipr = Me_2CH
nbu = $Me(CH_2)_3$
MeOTr = (4-methoxyphenyl)diphenylmethyl
$(MeO)_2Tr$ = bis(4-methoxyphenyl)phenylmethyl

pylphosphorodiamidite (= bis(diisopropylamino)[2-(4-nitrophenyl)ethoxy]phosphane; **51**) [39], respectively, the 3'-[2-(4-nitrophenyl)ethyl diisopropylphosphoramidites] **67–82** by common procedures. Workup and purification were achieved by silica gel column chromatography using efficiently AcOEt/Et$_3$N or toluene/AcOEt in the elution process.

3. Physical Data. – The purity and structures of the newly synthesized compounds were established by chromatographical means and by UV, ^{1}H-NMR, and ^{31}P-NMR spectra, as well as by C,H,N analyses.

Experimental Part

General. TLC: precoated silica-gel thin-layer sheets *60 F 254* from *Merck*. Prep. column chromatography (CC): silica gel (*Merck 60*, 63–200 μm). M.p.: *Gallenkamp* melting-point apparatus; no correction. UV/VIS: *Perkin-Elmer Lambda 15*; λ_{max} in nm (log ε). IR: *Perkin-Elmer FTIR-1600*; $\tilde{\nu}$ in cm^{-1}. ^{1}H-NMR: *Bruker WM-250*; δ in ppm rel. to SiMe$_4$. ^{31}P-NMR: *Jeol JM GX-400*; δ in ppm rel. to 85% H$_3$PO$_4$ soln.

1. *2'-Deoxy-N^2-[2-(4-nitrophenyl)ethoxycarbonyl]guanosine* (**5**). The 2'-deoxyguanosine (**2**) (0.267 g, 1 mmol) was co-evaporated several times with dry pyridine (3 × 2 ml), then suspended in pyridine (20 ml), and treated with Me$_3$SiCl (0.543 g, 0.64 ml, 5 mmol) by stirring at r.t. for 30 min. After cooling in an ice-bath, 2-(4-nitrophenyl)ethyl carbonochloridite [12] (0.459 g, 2 mmol) in CHCl$_3$ (4 ml) was added dropwise within 30 min and stirred at 0° for 1 h and at r.t. for 2 days. MeOH (10 ml) was then added to the clear soln. and stirred at r.t. for 5 h. The solvent was evaporated, the residue treated overnight in a 10% NaHCO$_3$ soln., and the resulting colorless precipitate washed with Et$_2$O and dried at 50° *in vacuo*: 0.23 g (50%) of **5**. TLC (CHCl$_3$/MeOH 4:1): R_f 0.36. M.p. 185–190°. UV (MeOH): 272 (sh, 4.34), 258 (4.41), 251 (sh, 4.37). ^{1}H-NMR ((D$_6$)DMSO): 11.46, 11.32 (2 s, NH); 8.21 (s, H–C(8)); 8.17 (d, 2 H o to NO$_2$); 7.63 (d, 2 H m to NO$_2$); 6.21 (t, H–C(1')); 5.32 (d, OH–C(3')); 4.95 (t, OH–C(5')); 4.48 (t, OCH$_2$CH$_2$ (npeoc)); 4.36 (m, H–C(3')); 3.82 (m, H–C(4')); 3.59–3.45 (m, 2 H–C(5')); 3.14 (t, OCH$_2$CH$_2$ (npeoc)); 2.99 (m, 1 H–C(2')); 2.55 (m, 1 H–C(2')). Anal. calc. for C$_{18}$H$_{20}$N$_6$O$_8$ (460.4): C 49.57, H 4.38, N 18.25; found: C 49.21, H 4.19, N 18.20.

2. *2'-Deoxy-N^2-[2-(4-nitrophenyl)ethoxycarbonyl]-O^6-[2-(4-nitrophenyl)ethyl]guanosine* (**7**). A soln. of 3',5'-di-O-acetyl-2'-deoxyguanosine (0.69 g, 2 mmol) [15], triphenylphosphine (0.84 g, 3.2 mmol), and 2-(4-nitrophenyl)ethanol [40] (0.50 g, 3 mmol) in dry dioxane (40 ml) was stirred at r.t. for a few min and then treated with diethyl diazenedicarboxylate (0.558 g, 3.2 mmol) with stirring for 1 h. The clear soln. was evaporated and co-evaporated with dry pyridine (40 ml). The residue was taken up in pyridine (10 ml), cooled in an ice-bath, and then treated dropwise with 2-(4-nitrophenyl)ethyl carbonochloridate [12] (1.38 g, 6 mmol) in dry CHCl$_3$ (10 ml). After 1 h, stirring was continued at r.t. for 3 h. Then the mixture was diluted with H$_2$O (100 ml) and extracted with CHCl$_3$ (4 × 50 ml), the org. phase dried (Na$_2$SO$_4$), evaporated, and co-evaporated with toluene, and the residue purified by CC (silica gel, CH$_2$Cl$_2$, then CHCl$_3$). The main fraction was evaporated and the resulting residue treated with dioxane (25 ml) and 25% NH$_3$ soln. (25 ml) by keeping the mixture in the icebox for 20 h. After evaporation, the residue was recrystallized from MeOH (80 ml): 0.81 g (66%) of **7**. TLC (CHCl$_3$/MeOH 9:1): R_f 0.53. M.p. 179–182°. UV (MeOH): 269 (4.54), 216 (4.63). ^{1}H-NMR ((D$_6$)DMSO): 10.33 (s, NH); 8.40 (t, H–C(8)); 8.17 (d, 2 H o to NO$_2$); 7.64 (d, 2 H m to NO$_2$); 6.30 (t, H–C(1')); 5.32 (d, OH–C(3')); 4.89 (t, OH–C(5')); 4.73 (t, OCH$_2$CH$_2$, npe); 4.41 (m, H–C(3')); 4.37 (t, OCH$_2$CH$_2$ (npeoc)); 3.83 (m, H–C(4')); 3.67–3.42 (m, 2 H–C(5')); 3.30 (t, OCH$_2$CH$_2$ (npeoc)); 3.11 (t, OCH$_2$CH$_2$ (npe)); 2.71 (m, 1 H–C(2')); 2.25 (m, 1 H–C(2')). Anal. calc. for C$_{27}$H$_{27}$N$_7$O$_{10}$·0.5H$_2$O (618.5): C 52.42, H 4.56, N 15.85; found: C 52.32, H 4.67, N 15.63.

3. *2'-Deoxy-N^2-isobutyryl-O^6-methylguanosine* (= *2'-Deoxy-O^6-methyl-N^2-(2-methyl-1-oxopropyl)guanosine*; **9**) [21]. A soln. of 2'-deoxy-N^2,3'-O,5'-O-triisobutyrylguanosine (5 g, 10.47 mmol) [20], triphenylphosphine (3.44 g, 13.09 mmol), and MeOH (0.63 ml, 15.71 mmol) in dry dioxane (40 ml) was stirred at r.t. for a few min and then treated with diethyl diazenedicarboxylate (2.2 g, 2 ml, 13.09 mmol) and stirred further for 24 h. The clear soln. was evaporated and the residue dissolved in CH$_2$Cl$_2$ and purified by CC (silica gel, (48 × 3 cm), Et$_2$O/petroleum ether (3:1, Et$_2$O). The product fractions were collected and then evaporated to give **8**. To a soln. of this solid in dry MeOH (40 ml), 0.5M NaOMe/MeOH (8 ml) was added and stirred at r.t. for 1 h. The mixture was neutralized with 1M AcOH (5 ml) and evaporated. The residue was dissolved in MeOH, and the soln. mixed with silica gel (20 g) and evaporated. The dried powder was applied to CC (silica gel (10 × 3.5 cm),

CHCl$_3$, then CHCl$_3$/MeOH 19:1): 1.3 g (35%) of **9**. TLC (CHCl$_3$/MeOH 9:1): R_f 0.38. M.p. 183–184°. UV (MeOH): 268 (4.21), 218 (4.31). ^{1}H-NMR ((D$_6$)DMSO): 9.52 (*s*, NH); 7.57 (*s*, H–C(8)); 5.48 (*t*, H–C(1')); 4.46 (*d*, OH–C(3')); 4.07 (*t*, OH–C(5')); 3.56 (*m*, H–C(3')); 3.21 (*s*, MeO–C(6)); 3.09 (*m*, H–C(4')); 2.75 (*m*, 2 H–C(5')); 2.03 (*m*, 1 H–C(2')); 1.83 (*m*, 1 H–C(2')); 1.46–1.37 (*m*, Me$_2$CHCO); 0.24–0.22 (*d*, Me$_2$CHCO). Anal. calc. for C$_{15}$H$_{21}$N$_5$O$_5$ (351.4): C 51.27, H 6.02, N 19.93; found: C 51.30, H 6.22, N 19.66.

4. *2'-Deoxy-O^6-ethyl-N^2,3'-O,5'-O-triisobutyrylguanosine* (= *2'-Deoxy-O^6-ethyl-N^2-(2-methyl-1-oxopropyl)-guanosine 3',5'-Bis(2-methylpropanoate)*; **10**) [21]. A soln. of 2'-deoxy-N^2, 3'-O,5'-O-triisobutyrylguanosine (5 g, 10.47 mmol) [20], triphenylphosphine (3.44 g, 13.09 mmol), and EtOH (0.9 ml, 15.7 mmol) in dry dioxane (40 ml) was stirred at r.t. for a few min and then treated with diethyl diazenedicarboxylate (2.2 g, 2.0 ml, 13 mmol) by stirring for 24 h. The clear soln. was evaporated and purified by CC (silica gel, Et$_2$O): 3.78 g (71%) of **10**. Amorphous solid. TLC (Et$_2$O): R_f 0.25. UV (MeOH): 267 (4.23), 218 (4.35). ^{1}H-NMR (CDCl$_3$): 7.95 (*s*, NH); 7.93 (*s*, H–C(8)); 6.36 (*t*, H–C(1')); 5.38 (*m*, H–C(3')); 4.60 (*t*, MeCH$_2$O–C(6)); 4.49–4.28 (*m*, H–C(4'), 2 H–C(5')); 3.00–2.90 (*m*, 2 Me$_2$CHCO); 2.62–2.49 (*m*, 1 Me$_2$CHCO, 2 H–C(2')); 1.50 (*t*, MeCH$_2$O–C(6)); 1.27–1.11 (*m*, 3 Me$_2$CHCO). Anal. calc. for C$_{24}$H$_{35}$N$_5$O$_7$ (505.6): C 57.01, H 6.98, N 13.85; found: C 56.76, H 7.23, N 13.51.

5. *2'-Deoxy-O^6-ethyl-N^2-isobutyrylguanosine* (= *2'-Deoxy-O^6-ethyl-N^2-(2-methyl-1-oxopropyl)guanosine*; **11**) [21]. To a soln. of **10** (8.09 g, 16 mmol) in dry EtOH (40 ml), 0.5M NaOEt/EtOH (8 ml) was added. The mixture was stirred at r.t. for 1 h, then neutralized with 1M AcOH (5 ml), and evaporated . The residue was dissolved in EtOH, mixed with silica gel (20 g) and evaporated. The dried powder was applied to CC (silica gel 10 × 3.5 cm), CHCl$_3$, then CHCl$_3$/MeOH 19:1): 5.6 g (95%) of **11**. TLC (CHCl$_3$/MeOH 9:1): R_f 0.45. M.p. 160–161°. UV (MeOH): 269 (4.24), 219 (4.35). ^{1}H-NMR ((D$_6$)DMSO): 9.48 (*s*, NH); 7.57 (*s*, H–C(8)); 5.48 (*t*, H–C(1')); 4.48 (*d*, OH–C(3')); 4.05 (*t*, OH–C(5')); 3.76 (*q*, MeCH$_2$O–C(6)); 3.58 (*m*, 2 H–C(3')); 3.00 (*m*, H–C(4')); 2.73 (*m*, 2 H–C(5')); 2.03 (*m*, 1 H–C(2')); 1.83 (*m*, 1 H–C(2')); 1.44 (*m*, Me$_2$CHCO); 0.58 (*t*, MeCH$_2$O–C(6)); 0.25–0.23 (*d*, Me$_2$CHCO). Anal. calc. for C$_{16}$H$_{23}$N$_5$O$_5$ (365.4): C 52.59, H 6.35, N 19.61; found: C 52.30, H 6.45, N 19.00.

6. *2'-Deoxy-N^2,3'-O,5'-O-triisobutyryl-O^6-isopropylguanosine* (= *2'-Deoxy-O^6-(2-methylethyl)-N^2-(2-meth-yl-1-oxopropyl)guanosine 3',5'-Bis(2-methylpropanoate)*; **12**) [21]. As described for **10**, with 2'-deoxy-N^2,3'-O,5'-O-triisobutyrylguanosine (5 g, 10.47 mmol) [20], triphenylphosphine (3.44 g, 13.1 mmol), ^{i}PrOH (0.95 g, 1.2 ml, 16 mmol), and dioxane (40 ml): 4.40 g (81%) of **12**. UV (MeOH): 267 (4.23), 218 (4.37). ^{1}H-NMR (CDCl$_3$): 7.94 (*s*, NH); 7.91 (*s*, H–C(8)); 6.36 (*t*, H–C(1')); 5.57 (*m*, Me$_2$CHO–C(6)); 5.38 (*m*, H–C(3')); 4.48–4.35 (*m*, H–C(4')); 4.29 (*m*, H–C(5')); 2.99–2.89 (*m*, Me$_2$CHCOO, 1 H–C(2')); 2.56 (*m*, Me$_2$CHCON, 1 H–C(2')); 1.43 (*d*, Me$_2$CHO–C(6)); 1.26–1.10 (*m*, Me$_2$CHCO). Anal. calc. for C$_{23}$H$_{37}$N$_5$O$_7$ (519.6): C 57.78, H 7.18, N 13.47; found: C 57.58, H 7.33, N 13.23.

7. *2'-Deoxy-N^2-isobutyryl-O^6-isopropylguanosine* (= *2'-Deoxy-O^6-(2-methylethyl)-N^2-(2-methyl-1-oxopro-pyl)guanosine*; **13**) [21]. As described for **11**, with **12** (8.3 g, 16 mmol): 5.84 g (96%) of **13**. TLC (CHCl$_3$/MeOH 95:5): R_f 0.29 . UV (MeOH): 268 (4.23), 218 (4.35). ^{1}H-NMR ((D$_6$)DMSO): 10.63 (*s*, NH); 7.57 (*s*, H–C(8)); 5.47 ('*t*', H–C(1')); 4.73 (*m*, Me$_2$CHO–C(6)); 4.46 (*d*, OH–C(3')); 4.06 (*t*, OH–C(5')); 3.57 (*m*, H–C(3')); 3.01 (*m*, H–C(4')); 2.70 (*m*, 2 H–C(5')); 2.02, 1.99 (2 *m*, 2 H–C(2')); 1.54 (*m*, Me$_2$CHCO); 0.54 (*d*, Me$_2$CHO–C(6)); 0.26–0.22 (*d*, Me$_2$CHCO). Anal. calc. for C$_{17}$H$_{25}$N$_5$O$_5$ (379.4): C 53.81, H 6.64, N 18.45; found: C 53.01, H 6.90, N 18.03.

8. *O^6-Butyl-2'-deoxy-N^2-isobutyrylguanosine* (= *O^6-Butyl-2'-deoxy-N^2-(2-methyl-1-oxopropyl)guanosine*; **15**). As described for **9**, with 2'-deoxy-N^2,3'-O,5'-O-triisobutyrylguanosine (5 g, 10.47 mmol) [20] and butan-1-ol (1.4 ml, 15.71 mmol) *via* the intermediate **14**: 2.9 g (71%) of **15**. TLC (CHCl$_3$/MeOH 95:5): R_f 0.29. UV (MeOH): 268 (4.30), 218 (4.43). ^{1}H-NMR (CDCl$_3$): 7.86 (*s*, NH, H–C(8)); 6.28 (*t*, H–C(1')); 4.84 (*d*, OH–C(3')); 4.65 (*t*, OH–C(5')); 4.48 (*t*, MeCH$_2$CH$_2$CH$_2$O); 4.10 (*m*, H–C(3')); 3.86 (*m*, 2 H–C(5')); 3.11 (*m*, H–C(4')); 2.89 (*m*, Me$_2$CHCO, 1 H–C(2')); 2.35 (*m*, H–C(2')); 1.77 (*m*, MeCH$_2$CH$_2$CH$_2$O); 1.45 (*m*, MeCH$_2$CH$_2$CH$_2$O); 1.22 (*d*, Me$_2$CHCO); 0.90 (*t*, MeCH$_2$CH$_2$CH$_2$O). Anal. calc. for C$_{18}$H$_{27}$N$_5$O$_5$ (393.3): C 54.95, H 6.92, N 17.80; found: C 54.01, H 6.95, N 17.52.

9. *3',5'-Di-O-acetyl-2'-deoxy-O^6-methylguanosine* (**16**) *and 3',5'-Di-O-acetyl-2'-deoxy-N^1-methylguanosine* (**17**). A suspension of 3',5'-di-O-acetyl-2'-deoxyguanosine [15] (20 g, 57 mmol), triphenylphosphine (28.3 g, 107 mmol), and dry MeOH (4.4 ml, 3.3 g, 110 mmol) in dry dioxane (750 ml) was stirred at r.t. for a few min. Then diethyl diazenedicarboxylate (17.5 ml, 18.6 g, 3.2 mmol) was added and the mixture stirred further for 24 h. The clear soln. was evaporated and the residue purified by FC (Et$_2$O, then AcOEt and AcOEt/MeOH 9:1): 5.88 g (25%) of **16**, followed by 9.0 g (38%) of **17**, after drying in a vacuum desiccator.

Data of **16**: TLC (CHCl$_3$/MeOH 19:1): R_f 0.50. UV (MeOH): 280 (3.99), 248 (4.03), 209 (4.38). ^{1}H-NMR (CDCl$_3$): 7.72 (*s*, H–C(8); 6.26 (*dd*, H–C(1')); 5.42 (*m*, H–C(3')); 4.88 (*s*, NH$_2$); 4.46–4.31 (*m*, H–C(4'),

2 H−C(5′)); 4.05 (s, MeO−C(6)); 2.97 (m, 1 H−C(2′)); 2.55 (m, 1 H−C(2′)); 2.22, 2.07 (2 s, 2 Ac). Anal. calc. for C$_{15}$H$_{19}$N$_5$O$_6$ (365.4): C 49.32, H 5.25, N 19.17; found: C 48.77, H 5.52, N 18.41.

Data of **17**: TLC (CHCl$_3$/MeOH 19:1): R$_f$ 0.16. UV (MeOH): 272 (sh, 4.02), 257 (4.15), 203 (4.25). ^{1}H-NMR (CDCl$_3$): 7.62 (s, H−C(8)); 6.20 (dd, H−C(1′)); 5.41 (m, H−C(3′)); 5.03 (s, NH$_2$); 4.53−4.29 (m, H−C(4′), 2 H−C(5′)); 3.50 (s, Me−N(1)); 2.95 (m, 1 H−C(2′)); 2.51 (m, 1 H−C(2′)); 2.11, 2.08 (2 s, 2 Ac). Anal. calc. for C$_{15}$H$_{19}$N$_5$O$_6$·0.2 CHCl$_3$ (389.2): C 46.91, H 4.97, N 17.99; found: C 47.00, H 5.24, N 17.95.

10. *3′,5′-Di-O-acetyl-2′-deoxy-O^6-methyl-N^2-[2-(4-nitrophenyl)ethoxycarbonyl]guanosine* (**18**). To a soln. of **16** (5.7 g, 15.6 mmol) in dry pyridine (70 ml) cooled to 0° in an ice-bath, 2-(4-nitrophenyl)ethyl carbono-chloridate [12] (5.12 g, 22.3 mmol) was added dropwise slowly. After stirring for 1 h and another 20 h at r.t., the mixture was evaporated and co-evaporated with toluene (50 ml), and the residue was purified by FC (silica gel, CHCl$_3$/MeOH 98:2): 7.59 g (87%) of **18**. Foam. TLC (CHCl$_3$/MeOH 19:1): R$_f$ 0.61. UV (MeOH): 267 (4.23), 256 (sh, 4.17), 216 (4.39). ^{1}H-NMR (CDCl$_3$): 8.16 (d, 2 H o to NO$_2$); 7.91 (s, H−C(8)); 7.50 (s, NH); 7.41 (d, 2 H m to NO$_2$); 6.34 (t, H−C(1′)); 5.48 (m, H−C(3′)); 4.49−4.28 (m, 5 H, OCH$_2$CH$_2$ (npeoc), H−C(4′), 2 H−C(5′)); 4.09 (s, MeO−C(6)); 3.11 (t, OCH$_2$CH$_2$ (npeoc)); 3.07 (m, 1 H−C(2′)); 2.58 (m, 1 H−C(2′)); 2.10, 2.04 (2 s, Ac). Anal. calc. for C$_{24}$H$_{26}$N$_6$O$_{10}$·1.25 H$_2$O (581.0): C 49.61, H 4.94, N 14.46; found: C 49.63, H 4.43, N 14.30.

11. *2′-Deoxy-O^6-methyl-N^2-[2-(4-nitrophenyl)ethoxycarbonyl]guanosine* (**19**). To a soln. of **18** (7 g, 12.5 mmol) in dioxane (25 ml), 25% NH$_3$ soln. (25 ml) was added and stirred at 5° for 20 h. The solvent was evaporated and co-evaporated with MeOH/toluene 1:1 and then the product crystallized from AcOEt/MeOH (50 ml): 5.23 g (88%) of **19**. TLC (CHCl$_3$/MeOH 19:1): R$_f$ 0.13. UV (MeOH): 267 (4.29), 256 (sh, 4.23), 216 (4.43). ^{1}H-NMR ((D$_6$)DMSO): 10.29 (s, NH); 8.40 (s, H−C(8)); 8.17 (d, 1 H o to NO$_2$); 7.62 (d, 1 H m to NO$_2$); 6.31 (dd, H−C(1′)); 5.30 (d, OH−C(3′)); 4.88 (d, OH−C(5′)); 4.40 (m, H−C(3′)); 4.30 (t, OCH$_2$CH$_2$ (npeoc)); 4.03 (s, MeO−C(6)); 3.85 (m, H−C(4′)); 3.63−3.38 (m, 2 H−C(5′)); 3.09 (t, OCH$_2$CH$_2$ (npeoc)); 2.77 (m, 1 H−C(2′)); 2.30 (m, 1 H−C(2′)). Anal. calc. for C$_{20}$H$_{22}$N$_6$O$_8$·0.7 H$_2$O (487.4): C 49.32, H 4.84, N 17.26; found: C 49.33, H 4.49, N 17.24.

12. *O^4-Butylthymidine* (**23**). a) To a soln. of NaOBu (prepared from Na (0.08 g, 3.0 mmol) in BuOH (20 ml), 3′,5′-di-O-acetyl-O^4-(1H-triazol-1-yl)thymidine [23] (1.13 g, 3.0 mmol) was added and stirred at r.t. for 1 h. MeOH (10 ml) was then added, the mixture neutralized with 1M HCl, evaporated, and co-evaporated with MeOH, and the residue purified by CC (silica gel, CHCl$_3$/MeOH 20:1): 0.67 g (75%) of **23**. Colorless foam.

b) A mixture of 3′,5′-di-O-acetyl-O^4-(1H-triazol-1-yl)thymidine (2.83 g, 7.5 mmol), 4-(dimethylamino)-pyridine (2.75 g, 22.5 mmol), BuOH (7.5 ml, 82 mmol), and MeCN (40 ml) was heated under reflux for 27 h. After evaporation, the residue was purified by CC (silica gel, AcOEt). The product fractions were evaporated, and the residue was taken up in MeOH (10 ml) and treated with 25% NH$_3$ soln. (5 ml). After stirring at r.t. for 15 h, the solvent was evaporated and co-evaporated with MeOH. Purification was achieved by CC (CHCl$_3$/MeOH 100:5): 1.75g (78%) of **23**. TLC (CHCl$_3$/MeOH 9:1): R$_f$ 0.68. UV (MeOH): 283 (3.81), 220 (sh, 4.01), 206 (4.25). ^{1}H-NMR ((D$_6$)DMSO): 8.00 (s, H−C(6)); 6.13 (t, H−C(1′)); 5.23 (d, OH−C(3′)); 5.06 (t, OH−C(5′)); 4.24 (m, H−C(3′), MeCH$_2$CH$_2$CH$_2$O); 3.79 (m, H−C(4′)); 3.59 (m, 2 H−C(5′)); 2.18 (m, 1 H−C(2′)); 1.99 (m, 1 H−C(2′)); 1.86 (s, Me−C(5)); 1.65 (m, MeCH$_2$CH$_2$CH$_2$O); 1.37 (m, MeCH$_2$CH$_2$CH$_2$O); 0.90 (t, MeCH$_2$CH$_2$·CH$_2$O). Anal. calc. for C$_{14}$H$_{22}$N$_2$O$_5$ (298.4): C 56.36, H 7.43, N 9.39; found: C 56.26, H 7.31, N 9.47.

13.   *2′-Deoxy-5′-O-(4-monomethoxytrityl)-N^2-[2-(4-nitrophenyl)ethoxycarbonyl]-O^6-[2-(4-nitrophenyl)-ethyl]guanosine* (**28**). To a soln. of dried **7** (4.27 g, 7 mmol) in dry pyridine (35 ml), 4-monomethoxytrityl chloride (2.81 g, 9.1 mmol) was added and the mixture stirred at r.t. for 16 h. After completion of the reaction, MeOH (10 ml) was added and the mixture concentrated to *ca.* 3/4 of the volume, diluted with CHCl$_3$ (100 ml), and washed with H$_2$O (2 × 20 ml). The org. layer was dried (Na$_2$SO$_4$), evaporated, and co-evaporated with toluene. Purification by CC (silica gel, CHCl$_3$/MeOH 100:1) gave 5.55 g (90%) of **28**. Colorless foam. TLC (CHCl$_3$/MeOH 95:5): R$_f$ 0.49. UV (MeOH): 269 (4.55), 236 (4.37). ^{1}H-NMR (CDCl$_3$): 8.12−8.01 (2 d, 4 H o to NO$_2$); 7.96 (s, H−C(8)); 7.51 (s, NH); 7.48−7.10 (m, 12 arom. H, 4 H m to NO$_2$); 6.73 (d, 2 H o to MeO); 6.56 (t, H−C(1′)); 4.73 (m, H−C(3′), OCH$_2$CH$_2$ (npeoc)); 4.34 (t, OCH$_2$CH$_2$ (npe)); 4.20 (m, H−C(4′)); 3.72 (s, MeO); 3.57 (m, OH−C(3′)); 3.35 (m, 1 H−C(5′)); 3.25 (t, OCH$_2$CH$_2$ (npeoc)); 3.02 (t, OCH$_2$CH$_2$ (npe)); 2.70 (m, 1 H−C(2′)); 2.56 (m, 1 H−C(2′)). Anal. calc. for C$_{47}$H$_{43}$N$_7$O$_{11}$ (881.9): C 64.01, H 4.91, N 11.12; found: C 63.89, H 5.05, N 10.92.

14. *General Procedure A* (*G.P. A*): *Synthesis of 5′-O-(4,4′-Dimethoxytrityl)- and Base-Protected 2′-Deoxyribonucleosides*. To a soln. of dried compounds **3**−**7**, **9**, **11**, **13**, **15**, **19**, and **20**−**24** (45 mmol) in pyridine (200 ml) 4,4′-dimethoxytrityl chloride (18.3 g, 53 mmol) was added and stirred at r.t. for 2 to 20 h (TLC control). After completion of the reaction, MeOH (10 ml) was added and the mixture evaporated to *ca.* 1/4 of the volume. The mixture was diluted with CH$_2$Cl$_2$ (200 ml) and washed with H$_2$O (2 × 100 ml). The org. layer was dried

(Na$_2$SO$_4$), evaporated, and co-evaporated with toluene. Purification by CC or FC (silica gel) using the appropriate solvents gave the required compound as a colorless foam.

15. *2'-Deoxy-5'-O-(4,4'-dimethoxytrityl)-N^6-[2-(4-nitrophenyl)ethoxycarbonyl]adenosine* (**33**). From **3** [12] by *G.P. A.* FC (CHCl$_3$/MeOH 100:4). Yield 84%. TLC (CHCl$_3$/MeOH 95:5): R_f 0.34. UV (MeOH): 276 (sh, 4.41), 268 (4.47), 236 (4.43). ^{1}H-NMR (CDCl$_3$): 8.66 (*s*, H–C(8)); 8.36 (*br. s*, NH); 8.16 (*d*, 1 H *o* to NO$_2$); 8.11 (*s*, H–C(2)); 7.44 – 7.15 (*m*, 9 arom. H); 6.80 (*d*, 2 H *o* to MeO); 6.44 (*m*, H–C(1')); 4.71 (*m*, H–C(3')); 4.53 (*t*, OCH$_2$CH$_2$ (npeoc)); 4.24 (*m*, H–C(4')); 3.77 (*s*, 2 MeO); 3.52 – 3.40 (*m*, 2 H–C(5')); 3.14 (*t*, OCH$_2$CH$_2$ (npeoc)); 2.91 – 2.51 (*m*, 2 H–C(2')). Anal. calc. for C$_{40}$H$_{38}$N$_6$O$_9$·0.5 H$_2$O (755.8): C 63.57, H 5.20, N 11.12; found: C 63.62, H 5.21, N 10.95.

16. *2'-Deoxy-5'-O-(4,4'-dimethoxytrityl)-N^2-[2-(4-nitrophenyl)ethoxycarbonyl]guanosine* (**35**). From **5** by *G.P. A.* Purification by CC (CHCl$_3$/Et$_3$N 100:0.2 → CHCl$_3$/MeOH/Et$_3$N 90:10:0.2). Yield 79%. TLC (CHCl$_3$/MeOH 95:5): R_f 0.40. UV (MeOH): 272 (sh, 4.38), 258 (4.42), 249 (4.42), 237 (4.46). ^{1}H-NMR (CDCl$_3$): 11.27 (*s*, NH); 9.76 (*s*, NH); 8.00 (*d*, 2 H *o* to NO$_2$); 7.75 (*s*, H–C(8)); 7.32 – 7.03 (*m*, arom. H); 6.67 (*d*, 2 H *o* to MeO); 6.15 (*t*, H–C(1')); 4.80 (*m*, H–C(3')); 4.44 (*m*, OH–C(3'), OCH$_2$CH$_2$ (npeoc)); 4.15 (*m*, H–C(4')); 3.65 (*s*, MeO); 3.30 (*m*, 2 H–C(5')); 3.01 (*m*, OCH$_2$CH$_2$ (npeoc)); 2.60 – 2.46 (*m*, 2 H–C(2')). Anal. calc. for C$_{40}$H$_{38}$N$_6$O$_{10}$ (762.8): C 62.99, H 5.02, N 11.02; found: C 62.43, H 5.14, N 10.93.

17. *2'-Deoxy-5'-O-(4,4'-dimethoxytrityl)-N^2-[2-(4-nitrophenyl)ethoxycarbonyl]-O^6-[2-(4-nitrophenyl)ethyl]-guanosine* (**37**). From **7** by *G.P. A.* Purification by CC (CHCl$_3$/Et$_3$N 100:0.2 → CHCl$_3$/MeOH/Et$_3$N 99:1:0.2). Yield 85%. TLC (CHCl$_3$/MeOH 95:5): R_f 0.43. UV (MeOH): 268 (4.55), 235 (4.47). ^{1}H-NMR (CDCl$_3$): 8.15 – 8.11 (*m*, NH, 2 H *o* to NO$_2$); 7.96 (*s*, H–C(8)); 7.50 – 7.13 (*m*, 9 arom. H); 6.77 (*d*, 4 H *o* to MeO); 6.48 (*m*, H–C(1')); 4.81 – 4.74 (*t, m*, H–C(3'), OCH$_2$CH$_2$ (npe)); 4.45 (*t*, 2 OCH$_2$CH$_2$ (npeoc)); 4.16 (*m*, H–C(4')); 3.75 (*s*, 2 MeO); 3.56 – 3.25 (*m*, 2 H–C(5'), OCH$_2$CH$_2$ (npe)); 3.09 – 3.03 (*t, br. s*, OH–C(3'), OCH$_2$CH$_2$ (npeoc)); 2.82 – 2.70 (*m*, 1 H–C(2')); 2.55 (*m*, 1 H–C(2')). Anal. calc. for C$_{48}$H$_{45}$N$_7$O$_{12}$ (911.9): C 63.22, H 4.97, N 10.75; found: C 63.08, H 5.14, N 10.67.

18. *2'-Deoxy-5'-O-(4,4'-dimethoxytrityl)-N^2-isobutyryl-O^6-methylguanosine* (**38**). From **9** by *G.P. A.* FC (CHCl$_3$/MeOH 19:1). Yield 93%. TLC (CHCl$_3$/MeOH 19:1): R_f 0.52. UV (MeOH): 280 (sh, 4.32), 269 (4.38), 234 (4.49), 219 (sh, 4.76), 206 (4.94). ^{1}H-NMR (CDCl$_3$): 8.26 (*s*, NH); 8.11 (*s*, H–C(8; 7.50 – 7.24 (*m*, 9 arom. H); 6.85 (*d*, 4 H *o* to MeO); 6.78 (*t*, H–C(1')); 4.81 (*m*, H–C(3')); 4.39 (*s*, OH–C(3')); 4.37 (*d*, H–C(4')); 4.18 (*s*, MeO–C(6)); 3.83 (*s*, 2 MeO); 3.51 – 3.40 (*m*, 2 H–C(5')); 2.77 – 2.72 (*m*, Me$_2$CHCO, 2 H–C(2')); 1.29 – 1.25 (*d*, Me$_2$CHCO). Anal. calc. for C$_{36}$H$_{39}$N$_5$O$_7$ (653.7): C 66.14, H 6.01, N 11.01; found: C 65.78, H 6.08, N 10.81.

19. *2'-Deoxy-5'-O-(4,4'-dimethoxytrityl)-N^2-isobutyryl-O^6-ethylguanosine* (**39**). From **11** by *G.P. A.* FC (CHCl$_3$/MeOH 19:1). Yield 95%. TLC (CHCl$_3$/MeOH 19:1): R_f 0.54. UV (MeOH): 280 (sh, 4.10), 269 (4.26), 234 (sh, 4.37), 216 (sh, 4.57), 203 (4.80). ^{1}H-NMR (CDCl$_3$): 8.02 (*s*, H–C(8)); 8.01 (*s*, NH); 7.44 – 7.19 (*m*, 9 arom. H); 6.81 – 6.78 (*d*, 4 H *o* to MeO); 6.65 (*t*, H–C(1')); 4.75 (*m*, H–C(3')); 4.64 (*q*, MeCH$_2$O–C(6)); 4.24 (*d*, H–C(4')); 3.78 (*s*, 2 MeO); 3.46 – 3.33 (*m*, 2 H–C(5')); 3.00 (*br. s*, OH–C(3')); 2.75 – 2.64 (*m*, Me$_2$CHCO, 2 H–C(2')); 1.51 (*t*, MeCH$_2$O–C(6)); 1.26 (*m*, Me$_2$CHCO). Anal. calc. for C$_{37}$H$_{41}$N$_5$O$_7$ (667.7): C 66.55, H 6.18, N 10.48; found: C 66.14, H 6.30, N 10.30.

20. *2'-Deoxy-5'-O-(4,4'-dimethoxytrityl)-N^2-isobutyryl-O^6-isopropylguanosine* (**40**). From **13** by *G.P. A.* FC (CHCl$_3$/MeOH 19:1). Yield 95%. TLC (CHCl$_3$/MeOH 19:1): R_f 0.56. UV (MeOH): 280 (sh, 4.07), 265 (4.27), 234 (sh, 4.36), 203 (4.83). ^{1}H-NMR (CDCl$_3$): 7.95 (*s*, H–C(8)); 7.76 (*s*, NH); 7.41 – 7.18 (*m*, 9 arom. H); 6.78 (*d*, 4 H *o* to MeO); 6.49 (*t*, H–C(1')); 5.57 (*q*, Me$_2$CHO–C(6)); 4.72 (*m*, H–C(3')); 4.14 (*d*, H–C(4')); 3.77 (*s*, 2 MeO); 3.46 – 3.32 (*m*, 2 H–C(5')); 3.00 (*br. s*, OH–C(3')); 2.81 – 2.73 (*m*, Me$_2$CHCO, 1 H–C(2')); 2.59 (*m*, 1 H–C(2')); 1.46 (*d*, Me$_2$CHO–C(6)); 1.25 – 1.19 (*2 d*, Me$_2$CHCO). Anal. calc. for C$_{38}$H$_{43}$N$_5$O$_7$·0.5 H$_2$O (690.8): C 66.07, H 6.42, N 10.13; found: C 66.16, H 6.50, N 9.88.

21. *O^6-Butyl-2'-deoxy-5'-O-(4,4'-dimethoxytrityl)-N^2-isobutyrylguanosine* (**41**). From **15** by *G.P. A.* FC (CHCl$_3$/MeOH 98:2). Yield 85%. TLC (CHCl$_3$/MeOH 95:5): R_f 0.82. UV (MeOH): 280 (sh, 4.14), 269 (4.30), 234 (sh, 4.41). ^{1}H-NMR (CDCl$_3$): 8.02 (*2 s*, H–C(8), NH); 7.41 – 7.20 (*m*, 9 arom. H); 6.77 (*d*, 4 H *o* to MeO); 6.63 (*t*, H–C(1')); 4.73 (*m*, H–C(3')); 4.53 (*t*, MeCH$_2$CH$_2$CH$_2$O); 4.23 (*d*, H–C(4')); 3.75 (*s*, 6 MeO); 3.40 – 3.32 (*m*, 2 H–C(5')); 2.91 (*br. s*, OH–C(3')); 2.69 – 2.60 (*2 m*, Me$_2$CHCO, 2 H–C(2')); 1.85 (*m*, MeCH$_2$CH$_2$CH$_2$O); 1.52 (*m*, MeCH$_2$CH$_2$CH$_2$O); 1.21 – 1.17 (*2 d*, Me$_2$CHCO); 0.96 (*t*, MeCH$_2$CH$_2$CH$_2$O). Anal. calc. for C$_{39}$H$_{45}$N$_5$O$_7$ (695.8): C 67.32, H 6.52, N 10.07; found: C 67.30, H 6.56, N 10.10.

22. *2'-Deoxy-5'-O-(4,4'-dimethoxytrityl)-O^6-methyl-N^2-[2-(4-nitrophenyl)ethoxycarbonyl]guanosine* (**42**). From **19** by *G.P. A.* CC (CHCl$_3$/MeOH/Et$_3$N 495:4:1). Yield 89%. TLC (CHCl$_3$/MeOH 19:1): R_f 0.52. UV (MeOH): 268 (4.13), 236 (4.15), 204 (4.58). ^{1}H-NMR (CDCl$_3$): 8.17 (*d*, 2 H *o* to NO$_2$); 7.93 (*s*, H–C(8)); 7.41 – 7.17 (*m*, 12 arom. H, NH); 6.77 (*m*, 4 H *o* to MeO); 6.42 (*dd*, H–C(1')); 4.81 (*m*, H–C(3')); 4.43 (*t*, OCH$_2$CH$_2$ (npeoc)); 4.12 – 4.09 (*m*, MeO–C(6), H–C(4')); 3.77 (*s*, 2 MeO); 3.38 (*m*, 2 H–C(5')); 3.09 (*t*, OCH$_2$CH$_2$

(npeoc)); 2.79 (m, 1 H−C(2′)); 2.52 (m, 1 H−C(2′)). Anal. calc. for $C_{41}H_{40}N_6O_{10} \cdot H_2O$ (794.9): C 61.95, H 5.19, N 10.57; found: C 61.85, H 5.21, N 10.52.

23. *5′-O-(4,4′-Dimethoxytrityl)-O⁶-methylthymidine* (**43**). From **20** [22][23] by *G.P. A*. Purification by FC (AcOEt). Yield 84%. TLC (CHCl₃/MeOH 95:5): R_f 0.60. UV (MeOH): 281 (3.91), 278 (sh, 3.90), 230 (sh, 4.33). ¹H-NMR (CDCl₃): 7.86 (s, H−C(6)); 7.30 (m, 9 arom. H); 6.83 (d, 4 H o to MeO); 6.40 (t, H−C(1′)); 4.55 (m, H−C(3′)); 4.10 (m, H−C(4′)); 3.97 (s, MeO−C(4)); 3.79 (s, 2 MeO); 3.43 (m, 2 H−C(5′)); 2.64 (d, OH−C(3′)); 2.50 (m, 1 H−C(2′)); 2.27 (m, 1 H−C(2′)); 1.56 (s, Me−C(5)). Anal. calc. for $C_{32}H_{34}N_2O_7$ (558.6): C 68.80, H 6.13, N 5.01; found: C 68.14, H 6.45, N 4.75.

24. *5′-O-(4,4′-Dimethoxytrityl)-O⁶-ethylthymidine* (**44**) [26][27]. From **21** [22] by *G.P. A*. Purification by FC (AcOEt). Yield 91%. TLC (CHCl₃/MeOH 95:5): R_f 0.50. UV (MeOH): 281 (3.93), 278 (sh, 3.92), 230 (sh, 4.35). ¹H-NMR (CDCl₃): 7.82 (s, H−C(6)); 7.30 (m, 9 arom. H); 6.80 (d, 4 H o to MeO); 6.40 (t, H−C(1′)); 4.54 (m, H−C(3′)); 4.44 (q, MeCH₂O−C(4)); 4.09 (m, H−C(4′)); 3.80 (s, 2 MeO); 3.43 (m, 2 H−C(5′)); 2.61 (m, 1 H−C(2′)); 2.24 (m, 1 H−C(2′)); 2.19 (d, OH−C(3′)); 1.57 (s, Me−C(5)); 1.36 (t, Me*CH₂O−C(4)). Anal. calc. for $C_{33}H_{35}N_2O_7$ (572.7): C 69.21, H 6.34, N 4.89; found: C 68.63, H 6.43, N 4.70.

25. *5′-O-(4,4′-Dimethoxytrityl)-O⁴-isopropylthymidine* (**45**). From **22** [22] by *G.P. A*. Purification by FC (AcOEt). Yield 85%. TLC (CHCl₃/MeOH 95:5): R_f 0.65. UV (MeOH): 281 (3.93), 278 (sh, 3.92), 230 (sh, 4.35). ¹H-NMR (CDCl₃): 7.82 (s, H−C(6)); 7.30 (m, 9 arom. H); 6.83 (d, 4 H o to MeO); 6.42 (t, H−C(1′)); 5.49 (m, Me₂CHO−C(4)); 4.55 (m, H−C(3′)); 4.10 (m, H−C(4′)); 3.79 (s, 2 MeO); 3.44 (m, 2 H−C(5′)); 2.63 (m, 1 H−C(2′), OH−C(3′)); 2.26 (m, 1 H−C(2′)); 1.54 (s, Me−C(5)); 1.32 (m, Me₂CHO−C(4)). Anal. calc. for $C_{34}H_{38}N_2O_7$ (586.7): C 69.61, H 6.53, N 4.77; found: C 69.09, H 6.73, N 4.58.

26. *O⁴-Butyl-5′-O-(4,4′-dimethoxytrityl)thymidine* (**46**). From **24** by *G.P. A*. Purification by FC (AcOEt). Yield 85%. TLC (CHCl₃/MeOH 95:5): R_f 0.65. UV (MeOH): 281 (3.94), 230 (sh, 4.34). ¹H-NMR (CDCl₃): 7.85 (s, H−C(6)); 7.42−7.22 (m, 9 arom. H); 6.83 (d, 4 H o to MeO); 6.42 (t, H−C(1′)); 4.56 (m, H−C(3′)); 4.36 (t, MeCH₂CH₂CH₂O); 4.12 (m, H−C(4′)); 3.79 (s, 2 MeO); 3.47 (m, 2 H−C(5′)); 3.00 (m, OH−C(3′)); 2.63 (m, 1 H−C(2′)); 2.26 (m, 1 H−C(2′)); 1.70 (m, MeCH₂CH₂CH₂O); 1.55 (s, Me−C(5)); 1.42 (m, MeCH₂CH₂CH₂O); 0.94 (t, Me*CH₂CH₂CH₂O). Anal. calc. for $C_{35}H_{40}N_2O_7$ (600.7): C 69.98, H 6.71, N 4.66; found: C 69.63, H 6.43, N 4.70.

27. *5′-O-(4,4′-Dimethoxytrityl)-O⁴-[2-(4-nitrophenyl)ethyl]thymidine* (**47**). From **24** [22] by the *G.P. A*. Purification by FC (CHCl₃/MeOH 100:2). Yield 95%. TLC (CHCl₃/MeOH 95:5): R_f 0.82. UV (MeOH): 280 (sh, 4.25), 275 (4.26), 232 (sh, 4.40). ¹H-NMR (CDCl₃): 8.14 (d, 2 H o to NO₂); 7.92 (s, H−C(6)); 7.30 (m, 9 arom. H); 6.81 (d, 4 H o to MeO); 6.40 (t, H−C(1′)); 4.61 (m, H−C(3′), OCH₂CH₂ (npe)); 4.16 (m, H−C(4′)); 3.77 (s, 2 MeO); 3.73 (d, OH−C(3′)); 3.41 (m, 2 H−C(5′)); 3.15 (t, OCH₂CH₂ (npe)); 2.66 (m, 1 H−C(2′); 2.23 (m, 1 H−C(2′)); 1.46 (s, Me−C(5)). Anal. calc. for $C_{39}H_{38}N_3O_9 \cdot 2H_2O$ (728.7): C 64.28, H 5.81, N 5.76; found: C 64.30, H 5.46, N 5.81.

28. *General Procedure B (G.P. B): Synthesis of 5′-O-(4,4′-Dimethoxytrityl)- and Base-Protected 3′-(2-Cyanoethyl Diisopropylphosphoramidites)* **52−66**. *Method I*: A mixture of protected nucleoside **35−40, 43−45, 47** (1 mmol), 2-cyanoethyl diisopropylphosphoramidochloridite (**48**) [28][29] (1.2 mmol) and ¹Pr₂EtN (0.69 ml, 3.79 mmol) in acid-free dry CH₂Cl₂ (4 ml) was stirred at r.t under Ar for 30 min (TLC control). The soln. was then diluted with CH₂Cl₂ (100 ml) and extracted with sat. NaHCO₃ soln. (2 × 50 ml), the aq. phase reextracted with CH₂Cl₂, the combined org. phase dried (Na₂SO₄) and evaporated, and the residue purified by CC or FC (silica gel) using the appropriate solvent systems to give the product as a diastereomer mixture which was co-evaporated with CH₂Cl₂: solid foam.

Method II: A mixture of protected nucleoside **33, 34, 41, 42, 46** (1 mmol), 2-cyanoethyl tetraisopropyl-phosphordiamidite (**49**) [30−33] (1.5 mmol), and 1*H*-tetrazole (0.05 mmol) in acid-free CH₂Cl₂ (13 ml) was stirred at r.t under Ar for 2 1/2 h. Workup as described above in *Method I*.

29. *2′-Deoxy-5′-O-(4,4′-dimethoxytrityl)-N⁶-[2-(4-nitrophenyl)ethoxycarbonyl]adenosine 3′-(2-Cyanoethyl Diisopropylphosphoramidite)* (**52**). From **33** by *G.P. B, Method II*. FC (toluene/AcOEt 3:7 and 1:1). Yield 92%. TLC (hexane/AcOEt/Et₃N 3:7:1): R_f 0.25, 0.17. UV (MeOH): 275 (sh, 4.41), 266 (4.47), 236 (4.45). ¹H-NMR (CDCl₃): 8.68 (s, H−C(8)); 8.25−8.15 (m, 2 H o to NO₂); 8.01 (m, NH); 7.36−7.34 (d, 2 H m to NO₂); 7.35−7.18 (m, 11 arom. H); 6.74 (d, 4 H o to MeO); 6.45 (t, H−C(1′)); 4.74 (m, H−C(3′)); 4.51 (t, OCH₂CH₂ (npeoc)); 4.29 (m, H−C(4′)); 3.75 (2 s, m, OCH₂CH₂CN, 2 MeO); 3.52−3.43 (m, 2 Me₂CHN); 3.31 (m, 2 H−C(5′)); 3.13 (t, OCH₂CH₂ (npeoc)); 2.82 (m, 1 H−C(2′)); 2.60 (m, 1 H−C(2′)); 2.59, 2.44 (2 t, OCH₂CH₂CN); 1.27−1.08 (m, 2 Me₂CHN). ³¹P-NMR (CDCl₃): 149.6, 149.5. Anal. calc. for $C_{49}H_{55}N_8O_{10}P$ (947.0): C 62.14, H 5.85, N 11.83; found: C 61.93, H 6.06, N 11.32.

30. *2′-Deoxy-5′-O-(4,4′-dimethoxytrityl)-N⁶-[(2-(4-nitrophenyl)ethoxycarbonyl]cytidine 3′-O-(2-Cyano-ethyl Diisopropylphosphoramidite)* (**53**). From **34** [41] by *G.P. B, Method II*. FC (toluene/AcOEt 2:1 and 1:1).

Yield 85%. TLC (hexane/AcOEt/Et$_3$N 3 : 7 : 1): R_f 0.24, 0.13. UV (MeOH): 280 (sh, 4.40), 274 (4.42), 236 (4.56). ^{1}H-NMR (CDCl$_3$): 8.20 (*m*, 2 H *o* to NO$_2$); 8.20 (2 *s*, H−C(6)); 7.51 – 7.12 (*m*, 11 H, H *m* to NO$_2$, H *m* to MeO, Ar); 6.74 (*d*, 4 H *o* to OMe); 6.35 (*m*, H−C(1'), H−C(5)); 4.80 (*t*, OCH$_2$CH$_2$); 4.80 (*m*, H−C(3')); 4.43 (*t*, OCH$_2$CH$_2$ (npeoc)); 4.21 (*m*, H−C(4')); 3.90 (*m*, OCH$_2$CH$_2$ (npeoc)); 3.73 (2 *s*, 2 MeO); 3.70 – 3.47 (*m*, 2 Me$_2$CH*N*, OCH$_2$CH$_2$CN); 3.32 – 3.30 (2 *m*, 2 H−C(5')); 3.08 (*m*, OCH$_2$CH$_2$ (npeoc)); 2.92 (*m*, 1 H−C(2')); 2.75 (*m*, 1 H−C(2')); 2.60, 224 (2 *t*, OCH$_2$CH$_2$CN); 1.10 – 0.91 (*m*, 2 Me$_2$CH*N*). ^{31}P-NMR (CDCl$_3$): 150.0; 149.9. Anal. calc. for C$_{48}$H$_{55}$N$_6$O$_{11}$P (923.0): C 62.46, H 6.00, N 9.10; found: C 62.46, H 6.08, N 8.83.

31. *2'-Deoxy-5'-O-(4,4'-dimethoxytrityl)-N^2-2-[(4-nitrophenyl)ethyoxycarbonyl]guanosine 3'-(2-Cyanoethyl Diisopropylphosphoramidite)* (**54**). From **35** by *G.P. B*, *Method I*. CC (CH$_2$Cl$_2$/MeOH/Et$_3$N 97.5 : 2 : 0.5 → 91.5 : 8 : 0.5). Yield 60%. TLC (CHCl$_3$/MeOH/Et$_3$N 95 : 5 : 2): R_f 0.52. UV(MeOH): 281 (sh, 4.31), 272 (4.37), 258 (4.42), 249 (4.42), 237 (4.46). ^{1}H-NMR (CDCl$_3$): 8.15 – 8.11 (*m*, 2 H *o* to NO$_2$); 7.76, 7.72 (2 *s*, H−C(8)); 7.38 – 7.18 (*m*, 11 H, H *m* to NO$_2$, H *m* to MeO, Ar); 6.77 – 6.71 (*d*, 4 H *o* to MeO); 6.16 (*m*, H−C(1')); 4.74 – 4.61 (*m*, H−C(3')); 4.44 (*t*, OCH$_2$CH$_2$ (npeoc)); 4.24 (*m*, H−C(4')); 3.74, 3.73 (2 *s*, 2 MeO); 4.14 – 3.45 (*m*, OCH$_2$CH$_2$CN, 2 Me$_2$CH*N*); 3.34 – 3.20 (*m*, 2 H−C(5')); 3.09 (*t*, OCH$_2$CH$_2$ (npeoc)); 2.81 – 2.32 (*m*, 2 H−C(2'), OCH$_2$CH$_2$CN); 1.30 – 1.06 (*m*, 2 Me$_2$CH*N*). ^{31}P-NMR (CDCl$_3$): 149.48; 148.89. Anal. calc. for C$_{49}$H$_{55}$N$_8$O$_{11}$P (963.0): C 61.12, H 5.76, N 11.64; found: C 60.80, H 5.90, N 11.40.

32. *2'-Deoxy-5'-O-(4,4'dimethoxytrityl)-N^2-isobutyryl-O^6-2'-[2-(4-nitrophenyl)ethyl]guanosine 3'-(2-Cyanoethyl Diisopropylphosphoramidite)* (**55**). From **36** [18][19] by *G.P. B*, *Method I*. FC (AcOEt/Et$_3$N 99 : 1). Yield 70%. TLC (AcOEt/Et$_3$N 99 : 1): R_f 0.80, 0.72. UV (MeOH): 280 (sh, 4.53), 270 (4.47), 235 (4.43). ^{1}H-NMR (CDCl$_3$): 8.15 – 8.11 (*m*, H$_o$ to NO$_2$); 7.98, 7.97 (2 *s*, H−C(8)); 7.81, 7.67 (2 *s*, H−N(2)); 7.49 (*d*, 2 H *m* to NO$_2$); 7.42 – 7.14 (*m*, 9 H, H *m* to MeO, Ar); 6.80 – 6.73 (*d*, 4 H *o* to MeO); 6.36 (*m*, H−C(1')); 4.83 – 4.66 (*m*, *t*, H−C(3'), OCH$_2$CH$_2$ (npe)); 4.24 (*m*, H−C(4')); 3.74 (*s*, 2 MeO); 3.88 – 3.49 (*m*, 2 Me$_2$CH*N*, OCH$_2$CH$_2$CN); 3.43 – 3.27 (*m*, 2 H−C(5'), OCH$_2$CH$_2$ (npe)); 2.89 – 2.50 (*m*, 2 H−C(2'), Me$_2$CH*CO*); 2.62 – 2.41 (2 *t*, OCH$_2$CH$_2$CN); 1.30 – 0.96 (*m*, 3 Me$_2$CH)). ^{31}P-NMR (CDCl$_3$): 149.6. Anal. calc. for C$_{52}$H$_{61}$N$_8$O$_{10}$P (989.1): C 63.15, H 6.22, N 11.33; found: C 62.58, H 6.35, N 11.04.

33. *2'-Deoxy-5'-O-(4,4'dimethoxytrityl)-N^2-[2-(4-nitrophenyl)ethoxycarbonyl]-O^6-[2-(4-nitrophenyl)ethyl]-guanosine 3'-(2-Cyanoethyl Diisopropylphosphoramidite)* (**56**). From **37** by *G.P. B*, *Method I*. CC (AcOEt/Et$_3$N 99 : 1). Yield 86%. TLC (AcOEt/Et$_3$N 99 : 1): R_f 0.81, 0.76. UV (MeOH): 269 (4.55), 236 (4.48). ^{1}H-NMR (CDCl$_3$): 8.17 – 8.11 (*m*, 2 H *o* to NO$_2$); 7.96, 7.95 (2 *s*, H−C(8)); 7.51 – 7.13 (*m*, 11 arom. H); 6.74 (*d*, 4 H *o* to MeO); 6.36 (*m*, H−C(1')); 4.82 – 4.70 (*m*, *t*, 3 H−C(3'), OCH$_2$CH$_2$ (npe)); 4.42 (*t*, OCH$_2$CH$_2$ (npeoc)); 4.24 (*m*, H−C(4')); 3.74 (2 *s*, 2 MeO); 3.86 – 3.48 (*m*, 2 Me$_2$CH*N*, OCH$_2$CH$_2$CN); 3.37 – 3.26 (*m*, 2 H−C(5'), OCH$_2$CH$_2$ (npe)); 3.09 (*t*, OCH$_2$CH$_2$ (npeoc)); 2.87 – 2.55 (*m*, 2 H−C(2')); 2.62, 2.41 (2 *t*, OCH$_2$CH$_2$CN); 1.26 – 1.07 (*m*, 2 Me$_2$CH*N*). ^{31}P-NMR (CDCl$_3$): 149.59; 149.38. Anal. calc. for C$_{57}$H$_{62}$N$_9$O$_{13}$P (1112.2): C 61.56, H 5.62, N 11.33; found: C 61.76, H 5.65, N 11.05.

34. *2'-Deoxy-5'-O-(4,4'-dimethoxytrityl)-N^2-isobutyryl-O^6-methylguanosine 3'-(2-Cyanoethyl Diisopropylphosphoramidite)* (**57**). From **38** [27] by *G.P. B*, *Method I*. FC (AcOEt/Et$_3$N 96 : 4). Yield 91%. TLC (AcOEt/Et$_3$N 96 : 4): R_f 0.71. UV(MeOH): 280 (sh, 4.09), 269 (4.25), 234 (sh, 4.36), 221 (sh, 4.52). ^{1}H-NMR (CDCl$_3$): 8.01 (*s*, H−C(8)); 7.73 (br. *s*, NH); 7.45 – 7.20 (*m*, 9 arom. H); 6.80 (*d*, 4 H *o* to MeO); 6.42 (*t*, H−C(1')); 4.73 (*m*, H−C(3')); 4.18 (*s*, MeO−C(4)); 3.80 (*s*, 2 MeO); 3.89 – 3.57 (*m*, 2 Me$_2$CH*N*, OCH$_2$CH$_2$CN); 3.38 (*m*, 2 H−C(5')); 3.20 (*m*, Me$_2$CH*CO*); 2.90 (*m*, 1 H−C(2')); 2.64 (*m*, *t*, 2 H, OCH$_2$CH$_2$CN, 1 H−C(2')); 2.50 (*t*, 1 H, OCH$_2$CH$_2$CN); 1.33 – 1.13 (*m*, 3 Me$_2$CH). ^{31}P-NMR (CDCl$_3$): 149.44; 149.37. Anal. calc. for C$_{45}$H$_{56}$N$_7$O$_8$P · H$_2$O (871.9): C 61.98, H 6.70, N 11.24; found: C 61.81, H 6.65, N 11.22.

35. *2'-Deoxy-5'-O-(4,4'-dimethoxytrityl)-O^6-ethyl-N^2-isobutyrylguanosine 3'-(2-Cyanoethyl Diisopropylphosphoramidite)* (**58**). From **39** [27] by *G.P. B*, *Method I*. FC (AcOEt/Et$_3$N 96 : 4). Yield 91%. TLC (AcOEt/Et$_3$N 96 : 4): R_f 0.75. UV(MeOH): 280 (sh, 4.10), 269 (4.26), 234 (sh, 4.37), 221 (sh, 4.54). ^{1}H-NMR (CDCl$_3$): 7.95, 7.93 (2 *s*, H−C(8)); 7.74, 7.67 (2 *s*, NH); 7.41 – 7.5 (*m*, 9 arom. H); 6.77 – 6.73 (*d*, 4 H *o* to MeO); 6.40 (*m*, H−C(1')); 4.75 (*m*, H−C(3')); 4.60 (*q*, MeCH$_2$O−C(6)); 4.27 (*m*, H−C(4')); 3.75 (2 *s*, 2 MeO); 3.83 – 3.44 (*m*, 2 Me$_2$CH*N*, OCH$_2$CH$_2$CN); 3.35 – 3.30 (*m*, 2 H−C(5')); 3.10 (*m*, Me$_2$CH*CO*); 2.82 (*m*, 1 H−C(2')); 2.64 (*m*, *t*, 2 H, OCH$_2$CH$_2$CN, 1 H−C(2')); 2.42 (*t*, 1 H, OCH$_2$CH$_2$CN); 1.48 (*t*, MeCH$_2$O−C(6)); 1.26 – 1.04 (*m*, 3 Me$_2$CH). ^{31}P-NMR (CDCl$_3$): 149.46; 149.39. Anal. calc. for C$_{46}$H$_{58}$N$_7$O$_8$P · 1.5 H$_2$O (895.0): C 61.73, H 6.87, N 10.95; found: C 61.70, H 6.98, N 10.93.

36. *2'-Deoxy-5'-O-(4,4'-dimethoxytrityl)-O^6-isopropyl-N^2-isobutyrylguanosine 3'-(2-Cyanoethyl Diisopropylphosphoramidite)* (**59**). From **40** [27] by *G.P. B*, *Method I*. FC (toluene/AcOEt 2 : 1 → 1 : 1). Yield 93%. TLC (toluene/AcOEt 2 : 1): R_f 0.7. UV (MeOH): 280 (sh, 4.10), 269 (4.26), 233 (sh, 4.36), 221 (sh, 4.53). ^{1}H-NMR (CDCl$_3$): 8.02 (2 *s*, H−C(8)); 7.86, 7.79 (2 *s*, NH); 7.46 – 7.21 (*m*, 9 arom. H); 6.83 – 6.79 (*d*, 4 H *o* to MeO); 6.42 (*t*, H−C(1')); 5.64 – 5.59 (*m*, Me$_2$CH*O*−C(6)); 4.83 (*m*, H−C(3')); 4.32 (*m*, H−C(4')); 3.79 – 3.60 (*m*, *s*,

2 Me$_2$CH*N*, 2 MeO, 1 H of C*H*$_2$CH$_2$CN); 3.30–3.26 (*m*, 2 H–C(5′)); 3.10 (*m*, Me$_2$C*H*CO); 2.90 (*m*, 1 H–C(2′)); 2.67 (*t*, *m*, 1 H of CH$_2$C*H*$_2$CN, 1 H–C(2′)); 2.42 (*t*, 1 H, OCH$_2$C*H*$_2$CN); 1.50 (*m*, *Me*$_2$CHCO); 1.27–1.15 (*m*, *Me*$_2$CHO–C(6)); 1.15–1.02 (*m*, 2 *Me*$_2$CH*N*). ^{31}P-NMR (CDCl$_3$): 149.47; 149.36. Anal. calc. for C$_{47}$H$_{60}$N$_7$O$_8$P · 1.5H$_2$O (909.0): C 62.10, H 6.98, N 10.79; found: C 61.86, H 7.00, N 10.83.

37. *2′-Deoxy-5′-O-(4,4′-dimethoxytrityl)-O⁶-butyl-N²-isobutyrylguanosine 3′-(2-Cyanoethyl Diisopropylphosphoramidite)* (**60**). From **41** [27] by *G.P. B, Method I*. FC (toluene/AcOEt 2:1 → 1:1). Yield 72%. TLC (toluene/AcOEt 2:1): R_f 0.7. UV (MeOH): 280 (sh, 4.12), 269 (4.27), 233 (sh, 4.37). ^{1}H-NMR (CDCl$_3$): 7.95, 7.93 (2 *s*, H–C(8)); 7.71, 7.64 (2 *s*, H–N(2)); 7.35–7.13 (*m*, 9 arom. H); 6.77–6.73 (*d*, 4 H *o* to MeO); 6.31 (*t*, H–C(1′)); 4.68 (*m*, H–C(3′)); 4.47 (*t*, MeCH$_2$CH$_2$C*H*$_2$O); 4.19 (*m*, H–C(4′)); 3.69 (2 *s*, 2 MeO); 3.58 (*m*, 2 Me$_2$C*H*N, OC*H*$_2$CH$_2$CN); 3.30–3.26 (*m*, 2 H–C(5′)); 3.10 (*m*, Me$_2$C*H*CO); 2.72 (*m*, 1 H–C(2′)); 2.57 (*t*, *m*, 2 H, OCH$_2$C*H*$_2$CN, 1 H–C(2′)); 2.37 (*t*, 1 H, OCH$_2$C*H*$_2$CN); 1.78 (*m*, MeCH$_2$C*H*$_2$CH$_2$O); 1.46 (*m*, MeC*H*$_2$CH$_2$CH$_2$O); 1.17–1.02 (*m*, 3 *Me*$_2$CH); 0.90 (*t*, Me*CH$_2$*CH$_2$CH$_2$O). ^{31}P-NMR (CDCl$_3$): 149.59; 149.46. Anal. calc. for C$_{48}$H$_{62}$N$_7$O$_8$P (895.9): C 64.34, H 6.98, N 10.94; found: C 63.63, H 7.00, N 10.65.

38. *2′-Deoxy-5′-O-(4,4′-dimethoxytrityl)-O⁶-methyl-N²-[2-(4-nitrophenyl)ethoxycarbonyl]guanosine 3′-(2-Cyanoethyl Diisopropylphosphoramidite)* (**61**). From **42** by *G.P. B, Method I*. FC (toluene/AcOEt 1:1). Yield 76%. TLC (AcOEt): R_f 0.43. UV (MeOH): 268 (4.13), 236 (4.15). ^{1}H-NMR (CDCl$_3$): 8.18 (*d*, 2 H *o* to NO$_2$); 7.91, 7.89 (2 *s*, H–C(8)); 7.53–7.19 (*m*, 11 arom. H); 6.77–6.74 (*d*, 4 H *o* to MeO); 6.38 (*t*, H–C(1′)); 4.74 (*m*, H–C(3′)); 4.44 (*t*, OCH$_2$C*H*$_2$ (npeoc)); 4.26 (*m*, H–C(4′)); 4.14 (*s*, MeO–C(6)); 3.76 (2 *s*, 2 MeO); 3.68 (*m*, OC*H*$_2$CH$_2$CN, 2 Me$_2$C*H*N); 3.38 (*m*, 2 H–C(5′)); 3.11 (*t*, OCH$_2$C*H*$_2$ (npeoc)); 2.87 (*m*, 1 H–C(2′)); 2.61 (*t*, *m*, 2 H, 1 H–C(2′), OCH$_2$C*H*$_2$CN); 2.44 (*t*, 1 H, OCH$_2$C*H*$_2$CN); 1.22–1.08 (*m*, 2 *Me*$_2$CH). ^{31}P-NMR (CDCl$_3$): 149.42; 149.28. Anal. calc. for C$_{50}$H$_{57}$N$_8$O$_{11}$P (977.0): C 61.46, H 5.88, N 11.47; found: C 61.10, H 5.87, N 10.87.

39. *5′-O-(4,4′-Dimethoxytrityl)-O⁴-methylthymidine 3′-(2-Cyanoethyl Diisopropylphosphoramidite)* (**62**). From **43** [22] [23] by *G.P. B, Method I*. CC (AcOEt/Et$_3$N 199:1). Yield 73%. TLC (AcOEt/Et$_3$N 199:1): R_f 0.77, 0.71. UV (MeOH): 281 (3.92), 276 (3.93), 232 (4.33). ^{1}H-NMR (CDCl$_3$): 7.95, 7.92 (2 *s*, H–C(6)); 7.42–7.19 (2 *m*, 9 arom. H); 6.80 (*dd*, 4 H *o* to MeO); 6.42 (*m*, H–C(1′)); 4.64 (*m*, H–C(3′)); 4.16 (*m*, H–C(4′)); 3.97 (*s*, MeO–C(4)); 3.77, 3.76 (2 *s*, 7 H, 2 MeO, OCH$_2$C*H*$_2$CN); 3.60–3.26 (*m*, 4 H, 1 H–C(5′), 2 Me$_2$C*H*N, OC*H*$_2$CH$_2$CN); 3.30 (*m*, 1 H–C(5′)); 2.65–2.56 (*m*, *t*, 2 H, OCH$_2$C*H*$_2$CN, 1 H–C(2′)); 2.39 (*t*, 1 H, OCH$_2$C*H*$_2$CN); 1.49 (2 *s*, Me–C(5)); 1.17–1.02 (*m*, *d*, 2 *Me*$_2$CH). ^{31}P-NMR (CDCl$_3$): 149.81; 149.21. Anal. calc. for C$_{41}$H$_{51}$N$_4$O$_8$P (758.8): C 64.89, H 6.77, N 7.38; found: C 63.85, H 6.81, N 7.53.

40. *5′-O-(4,4′-Dimethoxytrityl)-O⁴-ethylthymidine 3′-(2-Cyanoethyl Diisopropylphosphoramidite)* (**63**). From **44** by *G.P. B, Method I*. CC (AcOEt/Et$_3$N 199:1). Yield 80%. TLC (AcOEt/Et$_3$N 199:1): R_f 0.76, 0.72. UV (MeOH) 281 (3.95), 276 (3.94), 232 (4.35). ^{1}H-NMR (CDCl$_3$): 7.90, 7.84 (2 *s*, H–C(6)); 7.42–7.19 (2 *m*, 9 arom. H); 6.82–6.79 (*dd*, 4 H *o* to MeO); 6.42 (*m*, H–C(1′)); 4.64 (*m*, H–C(3′)); 4.16 (*m*, H–C(4′)); 4.44 (*q*, MeC*H*$_2$MeO–C(4)); 3.77, 3.76 (2 *s*, 7 H, 2 MeO, OCH$_2$C*H*$_2$CN); 3.60–3.26 (*m*, 4 H, 1 H–C(5′), 2 Me$_2$C*H*, OC*H*$_2$CH$_2$CN); 3.30 (*m*, 1 H–C5(′)); 2.65–2.56 (*m*, *t*, 2 H, OCH$_2$C*H*$_2$CN, 1 H–C(2′)); 2.34 (*t*, 1 H, OCH$_2$C*H*$_2$CN); 2.25 (*m*, 1 H–C(2′)); 1.49, 1.47 (2 *s*, Me–C(5)); 1.35 (*t*, MeC*H*$_2$O–C(6)); 1.17–0.99 (*m*, 2 *Me*$_2$CH). ^{31}P-NMR (CDCl$_3$): 149.80; 149.21. Anal. calc. for C$_{42}$H$_{53}$N$_4$O$_8$P (772.8): C 65.27, H 6.91, N 7.25; found: C 64.42, H 6.83, N 7.31.

41. *5′-O-(4,4′-Dimethoxytrityl)-O⁴-isopropylthymidine 3′-(2-Cyanoethyl Diisopropylphosphoramidite)* (**64**). From **45** by *G.P. B, Method I*. CC (AcOEt/Et$_3$N 199:1). Yield 77%. TLC (AcOEt/Et$_3$N 199:1): R_f 0.87, 0.85. UV (MeOH) 281 (3.96), 276 (3.95), 232 (4.34). ^{1}H-NMR (CDCl$_3$): 7.85, 7.78 (2 *s*, H–C(6)); 7.42–7.19 (2 *m*, 9 arom. H); 6.80 (*dd*, 4 H *o* to MeO); 6.42 (*m*, H–C(1′)); 5.47 (*m*, Me$_2$C*H*O–C(4)); 4.64 (*m*, H–C(3′)); 4.16 (*m*, H–C(4′)); 3.77, 3.76 (2 *s*, 7 H, 2 MeO, OCH$_2$C*H*$_2$CN)); 3.60–3.26 (*m*, 4 H, OCH$_2$C*H*$_2$CN, 1 H–C(5′), 2 Me$_2$C*H*N); 3.28 (*m*, 1 H–C(5′)); 2.65–2.56 (*m*, *t*, 2 H, OCH$_2$C*H*$_2$CN, 1 H–C(2′)); 2.34 (*t*, 1 H, OCH$_2$C*H*$_2$CN); 2.25 (*m*, 1 H–C(2′)); 1.45, 1.43 (2 *s*, Me–C(5)); 1.29 (*m*, *Me*$_2$CHO–C(4)); 1.17–0.99 (*m*, 2 Me$_2$CH*N*). ^{31}P-NMR (CDCl$_3$): 149.84; 149.28. Anal. calc. for C$_{43}$H$_{56}$N$_4$O$_8$P (786.87): C 65.63, H 7.05, N 7.12; found: C 64.33, H 7.11, N 7.24.

42. *O⁴-Butyl-5′-O-(4,4′-dimethoxytrityl)thymidine 3′-(2-Cyanoethyl Diisopropylphosphoramidite)* (**65**). From **46** [27] by *G.P. B, Method II*. CC (toluene/AcOEt 1:1). Yield 82%. TLC (toluene/AcOEt 1:1): R_f 0.58, 0.50. UV (MeOH): 281 (3.95), 276 (3.95), 226 (4.39). ^{1}H-NMR (CDCl$_3$): 7.85, 7.78 (2 *s*, H–C(6)); 7.42–7.19 (2 *m*, 9 arom. H); 6.80 (*dd*, 4 H *o* to MeO); 6.42 (*m*, H–C(1′)); 4.64 (*m*, H–C(3′)); 4.36 (*t*, MeCH$_2$CH$_2$C*H*$_2$O); 4.16 (*m*, H–C(4′)); 3.79 (2 *s*, 7 H, 2 MeO, OC*H*$_2$CH$_2$CN); 3.60–3.40 (*m*, 4 H, 1 H–C(5′), 2 Me$_2$C*H*N, OC*H*$_2$CH$_2$CN); 3.30 (*m*, 1 H–C(5′)); 2.65 (*m*, *t*, 2 H, 1 H–C(2′, OCH$_2$C*H*$_2$CN); 2.39 (*t*, 1 H, OCH$_2$C*H*$_2$CN); 2.27 (*m*, 1 H–C(2′)); 1.71 (*m*, MeCH$_2$C*H*$_2$CH$_2$O); 1.46 (*m*, *d*, Me–C(5), MeC*H*$_2$CH$_2$CH$_2$O); 1.17–0.99 (*d*, *m*, 2 *Me*$_2$CH*N*); 0.94 (*t*, Me*CH$_2$*CH$_2$CH$_2$O). ^{31}P-NMR (CDCl$_3$): 148.50; 147.23. Anal. calc. for C$_{44}$H$_{57}$N$_4$O$_8$P (800.8): C 65.98, H 7.17, N 7.00; found: C 65.70, H 7.22, N 7.08.

43. *5'-O-(4,4'-Dimethoxytrityl)-O⁴-[2-(4-nitrophenyl)ethyl]thymidine 3'-(2-Cyanoethyl Diisopropylphosphoramidite)* (**66**). From **47** by *G.P. B, Method I*. FC (toluene/AcOEt 1:1). Yield 74%. TLC (toluene/AcOEt 1:1): R_f 0.42. UV (MeOH): 280 (sh, 4.24), 275 (4.25), 232 (sh, 4.39). ¹H-NMR (CDCl₃): 8.16 (*d*, 2 H *o* to NO₂); 7.90 (*d*, H−C(6)); 7.30 (2 *m*, 11 arom. H); 6.82 (*dd*, 4 H *o* to MeO); 6.38 (*m*, H−C(1')); 4.63 (*m*, H−C(3'), OCH₂CH₂ (npe)); 4.15 (*m*, H−C(4')); 3.78 (2 *s*, 7 H, 2 MeO, OCH₂CH₂CN); 3.54 (*m*, 4 H, 2 Me₂CHN, OCH₂CH₂CN, 1 H−C(5')); 3.32 (*m*, 1 H−C(5')); 3.16 (*t*, OCH₂CH₂ (npe)); 2.64 (*m*, *t*, 2 H, 1 H−C(2'), OCH₂CH₂CN); 2.32 (*m*, *t*, 2 H, 1 H−C(2'), OCH₂CH₂CN); 1.41 (*d*, Me−C(5)); 1.11 (*d*, *m*, 2 Me₂CH). ³¹P-NMR (CDCl₃): 149.71; 149.16. Anal. calc. for C₄₈H₅₆N₅O₁₂P (894.0): C 64.49, H 6.31, N 7.83; found: C 64.45, H 6.54, N 7.31.

44. *General Procedure C: Synthesis of 5'-O-(4,4'-Dimethoxytrityl)- and Base-Protected 3'-[2-(4-Nitrophenyl)ethyl Diisopropylphosphoramidites]* **67−82**. *Method I:* A mixture of the protected nucleoside (1 mmol), 2-(4-nitrophenyl)ethyl diisopropylphosphoramidochloridite (**50**) [39] (1.2 mmol) and ⁱPr₂EtN (0.69 ml, 3.79 mmol) in acid-free dry CH₂Cl₂ (4 ml) was stirred at r.t under Ar for 30 min (TLC control). The soln. was diluted with CH₂Cl₂ (100 ml) and extracted with sat. NaHCO₃ soln. (2 × 50 ml), the aq. phase reextracted with CH₂Cl₂, the combined org. layer dried (Na₂SO₄) and evaporated, and the residue purified by CC or FC (silica gel) using the appropriate solvent systems to give the product as a diastereomer mixture which was co-evaporated with CH₂Cl₂: solid foam.

Method II: Analogously to *Method I*, with 2-(4-nitrophenyl)ethyl tetraisopropylphosphorodiamidite (**51**) [39] (1.5 mmol) and 1*H*-tetrazole (0.05 mmol) in acid-free CH₂Cl₂ (13 ml) by stirring at r.t. under Ar for 2 1/2 h.

45. *5'-O-(4-Monomethoxytrityl)thymidine 3'-[2-(4-Nitrophenyl)ethyl Diisopropylphosphoramidite]* (**67**). From **25** [15] by *G.P. C, Method I*. CC (AcOEt/Et₃N 95:5). Yield 62%. TLC (AcOEt/Et₃N 95:5): R_f 0.72. UV (MeOH): 269 (4.30), 234 (sh, 4.25). ¹H-NMR (CDCl₃): 8.63 (*br. s*, NH); 8.21 (*m*, 2 H *o* to NO₂); 7.61 (*m*, H−C(6)); 7.42−7.21 (*m*, 14 arom. H); 6.75 (*m*, 2 H *o* to MeO); 6.43 (*m*, H−C(1')); 4.64 (*m*, H−C(3')); 4.16 (*m*, H−C(4')); 3.79 (2 *s*, MeO); 3.93−3.62 (*m*, OCH₂CH₂ (npe)); 3.60−3.26 (*m*, 2 H−C(5'), 2 Me₂CHN); 3.05−2.82 (2 *t*, OCH₂CH₂ (npe)); 2.51−2.17 (*m*, 2 H−C(2')); 1.42 (*s*, Me−C(5)); 1.17−0.99 (*m*, 2 Me₂CHN). ³¹P-NMR (CDCl₃): 147.49. Anal. calc. for C₄₄H₅₁N₄O₉P (810.9): C 65.17, H 6.34, N 6.91; found: C 64.06, H 6.50, N 6.89.

46. *2'-Deoxy-5'-O-(4-monomethoxytrityl)-N⁶-[2-(4-nitrophenyl)ethoxycarbonyl]adenosine 3'-[2-(4-Nitrophenyl)ethyl Diisopropylphosphoramidite]* (**68**). From **26** [12] by *G.P. C, Method I*. FC (AcOEt/Et₃N 96:4). Yield 66%. TLC (AcOEt/Et₃N 95:5): R_f 0.65. UV (MeOH): 275 (sh, 4.52), 268 (4.56), 236 (4.34). ¹H-NMR (CDCl₃): 8.69−8.68 (2 *s*, H−C(8)); 8.31 (*br. s*, NH); 8.21−8.07 (*m*, H−C(2), 4 H *o* to NO₂); 7.45−7.16 (*m*, 16 arom. H); 6.77 (*m*, 2 H *o* to MeO); 6.45 (*m*, H−C(1')); 4.70 (*m*, H−C(3')); 4.53 (*t*, OCH₂CH₂ (npeoc)); 4.35 (*m*, H−C(4')); 3.93−3.77 (*m*, OCH₂CH₂ (PO-npe)); 3.76 (2 *s*, MeO); 3.69−3.47 (*m*, 2 Me₂CHN); 3.45−3.27 (*m*, 2 H−C(5')); 3.15 (*t*, OCH₂CH₂ (npeoc)); 3.04−2.87 (2 *t*, OCH₂CH₂ (PO-npe)); 2.87 (*m*, 1 H−C(2')); 2.57 (*m*, 1 H−C(2')); 1.21−1.06 (*m*, 2 Me₂CHN). ³¹P-NMR (CDCl₃): 147.67. Anal. calc. for C₅₃H₅₇N₈O₁₁P (1013.0): C 62.84, H 5.67, N 11.06; found: C 62.33, H 5.82, N 10.76.

47. *2'-Deoxy-5'-O-(4-monomethoxytrityl)-N⁴-[2-(4-nitrophenyl)ethoxycarbonyl]cytidine 3'-[2-(4-Nitrophenyl)ethyl Diisopropylphosphoramidite]* (**69**). From **27** [12] by *G.P. C, Method I*. FC (AcOEt/Et₃N 96:4). Yield 72%. TLC (AcOEt/Et₃N 95:5): R_f 0.69. UV (MeOH): 282 (4.40), 276 (4.41), 236 (4.47). ¹H-NMR (CDCl₃): 8.26−8.20 (*m*, 4 H *o* to NO₂, H−C(6)); 7.60 (*br. s*, NH); 7.44−7.21 (*m*, 16 arom. H); 6.93−6.81 (*m*, 2 H *o* to MeO, H−C(5)); 6.25 (*m*, H−C(1')); 4.55 (*m*, H−C(3')); 4.43 (*t*, OCH₂CH₂ (npeoc)); 4.17 (*m*, H−C(4')); 3.95−3.64 (*m*, OCH₂CH₂ (PO-npe)); 3.73 (2 *s*, MeO); 3.60−3.32 (*m*, 2 Me₂CHN, 2 H−C(5')); 3.14−3.08 (*t*, OCH₂CH₂ (npeoc)); 3.03−2.85 (2 *t*, OCH₂CH₂ (PO-npe)); 2.72 (*m*, 1 H−C(2')); 2.29 (*m*, 1 H−C(2')); 1.27−1.01 (*m*, 2 Me₂CHN). ³¹P-NMR (CDCl₃): 147.85. Anal. calc. for C₅₂H₅₇N₆O₁₂P (989.0): C 63.15, H 5.81, N 8.50; found: C 61.19, H 5.93, N 8.31.

48. *2'-Deoxy-5'-O-(monomethoxytrityl)-N²-[2-(4-nitrophenyl)ethoxycarbonyl]-O⁶-[2-(4-nitrophenyl)ethyl]-guanosine 3'-[2-(4-Nitrophenyl)ethyl Diisopropylphosphoramidite]* (**70**). From **28** [12] by *G.P. C, Method I*. CC (AcOEt/Et₃N 96:4). Yield 78%. TLC (AcOEt/Et₃N 96:4): R_f 0.75, 0.67. UV (MeOH): 270 (4.62), 236 (4.42). ¹H-NMR (CDCl₃): 8.22−8.07 (*m*, 6 H *o* to NO₂); 7.96, 7.95 (2 *s*, H−C(8)); 7.54−7.18 (*m*, 18 arom. H); 6.88−6.74 (*m*, 2 H *o* to MeO); 6.39 (*m*, H−C(1')); 4.82 (*t*, OCH₂CH₂ (npe)); 4.65 (*m*, H−C(3')); 4.45 (*t*, OCH₂CH₂ (npeoc)); 4.21 (*m*, H−C(4')); 3.90 (*m*, OCH₂CH₂ (PO-npe)); 3.76 (2 *s*, MeO); 3.61−3.43 (*m*, 2 Me₂CHN); 3.38−3.30 (2 *m*, 2 H−C(5'), OCH₂CH₂ (PO-npe)); 3.10 (*t*, OCH₂CH₂ (npeoc)); 3.00−2.82 (2 *t*, OCH₂CH₂ (PO-npe)); 2.84−2.51 (2 *m*, 2 H−C(2')); 1.27−1.03 (*m*, 2 Me₂CHN). ³¹P-NMR (CDCl₃): 147.85. Anal. calc. for C₆₁H₆₄N₉O₁₄P (1178.2): C 62.19, H 5.48, N 10.70; found: C 61.62, H 5.64, N 10.64.

49. *5'-O-(4,4'-Dimethoxytrityl)thymidine 3'-[2-(4-Nitrophenyl)ethyl Diisopropylphosphoramidite]* (**71**). From **29** [15] by *G.P. C, Method I*. FC (toluene/AcOEt/Et₃N 70:30:1 → 50:50:1). Yield 82%. By *G.P. C, Method II*. FC (toluene/AcOEt 2:1 → 1:1). Yield 76%. TLC (AcOEt): R_f 0.87. UV (MeOH): 269 (4.32), 234

(4.38). ^{1}H-NMR (CDCl$_3$): 8.29 (s, NH); 8.13 – 8.05 (m, 2 H o to NO$_2$); 7.59 (m, H–C(6)); 7.38 – 7.21 (m, 11 arom. H); 6.80 (m, 4 H o to MeO); 6.38 (m, H–C(1′)); 4.55 (m, H–C(3′)); 4.12 (m, H–C(4′)); 3.75 (2 s, m, 2 MeO, OCH$_2$CH$_2$ (PO-npe)); 3.61 (m, 1 H–C(5′), 2 Me$_2$CHN); 3.32 (m, 1 H–C(5′)); 2.97, 2.82 (2 t, OCH$_2$CH$_2$ (PO-npe)); 2.31 – 2.21 (2 m, 2 H–C(2′)); 1.38 (s, Me–C(5)); 1.23 – 0.99 (m, 2 Me$_2$CHN). ^{31}P-NMR (CDCl$_3$): 148.49; 148.09. Anal. calc. for C$_{45}$H$_{54}$N$_4$O$_{10}$P (841.9): C 64.18, H 6.46, N 6.65; found: C 63.31, H 6.25, N 6.59.

50. N^6-Benzoyl-2′-deoxy-5′-O-(4,4′-dimethoxytrityl)adenosine 3′-[2-(4-Nitrophenyl)ethyl Diisopropylphosphoramidite] (72). From 30 [15] by G.P. C, Method II. FC (petroleum ether/acetone 3 : 1 and 2 : 1). Yield 65%. TLC (AcOEt): R$_f$ 0.70. UV (MeOH): 272 (4.48), 232 (4.49). ^{1}H-NMR (CDCl$_3$): 9.01 (s, NH); 8.03 (2 s, H–C(8)); 8.16 – 8.00 (m, H–C(2), 2 H o to NO$_2$); 7.78 – 7.50 (m, 2 H o to CO); 7.40 – 7.21 (m, 14 arom. H); 6.95 – 6.75 (d, 4 H o to MeO); 6.48 (d, H–C(1′)); 4.91 – 4.83 (m, H–C(3′)); 4.50 (m, H–C(4′)); 3.77 (s, 2 MeO); 3.91 – 3.62 (m, OCH$_2$CH$_2$ (PO-npe)); 3.54 – 3.33 (2 m, 2 H–C(5′), 2 Me$_2$CHN); 2.90, 2.60 (2 t, m, OCH$_2$CH$_2$ (PO-npe), 1 H–C(2′)); 2.46 (m, 1 H–C(2′)); 1.42 – 0.84 (m, 2 Me$_2$CHN). Anal. calc. for C$_{52}$H$_{56}$N$_7$O$_9$P (954.0): C 65.46, H 5.91, N 10.27; found: C 64.93, H 5.88, N 9.70.

51. N^4-Benzoyl-2′-deoxy-5′-O-(4,4′-dimethoxytrityl)cytidine 3′-[2-(4-Nitrophenyl)ethyl Diisopropylphosphoramidite] (73). From 31 [15] by G.P. C, Method II. FC (toluene/AcOEt 2 : 1 and 1 : 1). TLC (AcOEt): R$_f$ 0.72. UV (MeOH): 305 (4.04), 260 (4.50), 234 (4.53). ^{1}H-NMR (CDCl$_3$): 8.46 (s, NH); 8.15 – 8.05 (dd, H–C(6)); 7.93 (m, 2 H o to CO); 7.80 – 7.70 (m, 2 H o to NO$_2$); 7.68 – 7.53 (m, 14 arom. H); 6.99 – 6.82 (d, 4 H o to MeO); 6.27 (d, H–C(1′)); 4.42 (m, H–C(3′)); 4.19 (m, H–C(4′)); 3.80 (2 s, m, 2 MeO, OCH$_2$CH$_2$); 3.70 (2 m, 2 H–C(5′), 2 Me$_2$CHN); 3.18, 2.87 (2 t, OCH$_2$CH$_2$); 2.76 (m, 1 H–C(2′)); 2.32 (m, 1 H–C(2′)); 1.15 – 1.02 (m, 2 Me$_2$CHN). ^{31}P-NMR (CDCl$_3$): 148.81; 148.47. Anal. calc. for C$_{51}$H$_{56}$N$_5$O$_{10}$P (930.0): C 65.86, H 6.06, N 7.53; found: C 65.73, H 6.01, N 6.91.

52. 2′-Deoxy-5′-O-(4,4′-dimethoxytrityl)-N^2-isobutyrylguanosine 3′-[2-(4-Nitrophenyl)ethyl Diisopropylphosphoramidite] (74). From 32 [42] by G.P. C, Method I. FC (AcOEt/Et$_3$N 100 : 1). Yield 85%. TLC (AcOEt/Et$_3$N 99 : 1): R$_f$ 0.45. UV (MeOH): 281 (sh, 4.36), 273 (4.39), 261 (4.42), 253 (sh, 4.41), 236 (4.44). ^{1}H-NMR (CDCl$_3$): 11.93 (s, NH); 8.11 (m, 2 H o to NO$_2$); 7.81 – 7.70 (m, H–C(8), NH); 7.48 – 7.17 (m, 11 arom. H); 6.78 – 6.73 (m, 4 H o to MeO); 6.13 (m, H–C(1′)); 4.64 (m, H–C(3′)); 4.16 (m, H–C(4′)); 3.88 (m, OCH$_2$CH$_2$ (PO-npe)); 3.74 – 3.72 (3 s, 2 MeO); 3.50 – 3.10 (2 m, 2 H–C(5′), 2 Me$_2$CHN); 3.00 – 2.82 (m, Me$_2$CHCO, OCH$_2$CH$_2$ (PO-npe)); 2.42 (m, 1 H–C(2′)); 1.84 (m, 1 H–C(2′)); 1.11 – 0.84 (m, 3 Me$_2$CH). Anal. calc. for C$_{49}$H$_{58}$N$_7$O$_{10}$P (936.0): C 62.88, H 6.25, N 10.47; found: C 62.46, H 6.37, N 10.10.

53. 2′-Deoxy-5′-O-(4,4′-dimethoxytrityl)-N^6-[2-(4-nitrophenyl)ethoxycarbonyl]adenosine 3′-[2-(4-Nitrophenyl)ethyl Diisopropylphosphoramidite] (75). From 33 by G.P. C, Method I. FC (toluene/AcOEt/NEt$_3$ 70 : 30 : 1 → 50 : 50 : 1). Yield 87%. TLC (AcOEt): R$_f$ 0.66. UV (MeOH): 274 (sh, 4.55), 267 (4.58), 236 (4.45). ^{1}H-NMR (CDCl$_3$): 8.68 (s, H–C(8)); 8.31 (br. s, NH); 8.25 (s, H–C(2)); 8.15 (m, 4 H o to NO$_2$); 7.42 – 7.12 (m, 13 arom. H); 6.75 (d, 4 H o to MeO); 6.35 (t, H–C(1′)); 4.65 (m, H–C(3′)); 4.42 (t, OCH$_2$CH$_2$ (npeoc)); 4.21 (m, H–C(4′)); 3.90 – 3.70 (m, OCH$_2$CH$_2$ (PO-npe)); 3.73 (2 s, 2 MeO); 3.52 – 3.43 (m, 2 Me$_2$CHN); 3.31 (2 m, 2 H–C(5′)); 3.08 (t, OCH$_2$CH$_2$ (npeoc)); 3.00 – 2.82 (2 t, m, OCH$_2$CH$_2$ (PO-npe), 1 H–C(2′)); 2.62 (m, 1 H–C(2′)); 1.25 – 1.02 (m, 2 Me$_2$CHN). ^{31}P-NMR (CDCl$_3$): 148.67; 148.37. Anal. calc. for C$_{54}$H$_{59}$N$_8$O$_{12}$P (1043.1): C 62.18, H 5.70, N 10.74; found: C 61.72, H 6.07, N 9.96.

54. 2′-Deoxy-5′-O-(4,4′-dimethoxytrityl)-N^4-[2-(4-nitrophenyl)ethoxycarbonyl]cytidine 3′-[2-(4-Nitrophenyl)ethyl Diisopropylphosphoramidite] (76). From 34 [41] G.P. C, Method I. FC (AcOEt/Et$_3$N 96 : 4). Yield 91%. By G.P. C, Method II. FC (toluene/AcOEt 2 : 1 and 1 : 1). Yield 66%. TLC (AcOEt/Et$_3$N 96 : 4): R$_f$ 0.68. UV (MeOH): 282 (sh, 4.41), 276 (4.41), 236 (4.56). ^{1}H-NMR (CDCl$_3$): 8.46 (s, NH); 8.19 (d, H–C(6)); 8.14 – 8.04 (dd, 4 H o to NO$_2$); 7.42 – 7.02 (m, 13 arom. H); 6.84 (d, H–C(5), 4 H o to MeO); 6.23 (m, H–C(1′)); 4.58 (m, H–C(3′)); 4.42 (t, OCH$_2$CH$_2$ (npeoc)); 4.15 (m, H–C(4′)); 3.88 – 3.61 (2 s, m, 2 MeO, OCH$_2$CH$_2$ (PO-npe)); 3.47 (2 m, 2 H–C(5′), 2 Me$_2$CHN); 3.10 (t, OCH$_2$CH$_2$ (npeoc)); 2.87 (2 t, OCH$_2$CH$_2$ (PO-npe)); 2.69 (m, 1 H–C(2′)); 2.18 (m, 1 H–C(2′)); 1.20 – 1.01 (m, 2 Me$_2$CHN). ^{31}P-NMR (CDCl$_3$): 148.81; 148.47. Anal. calc. for C$_{53}$H$_{60}$N$_6$O$_{13}$P (1019.0): C 62.46, H 5.83, N 8.24; found: C 61.24, H 5.81, N 8.00.

55. 2′-Deoxy-5′-O-(4,4′-dimethoxytrityl)-N^2-isobutyryl-O^6-[2-(4-nitrophenyl)ethyl]guanosine 3′-[2-(4-Nitrophenyl)ethyl Diisopropylphosphoramidite] (77). From 36 [18] by G.P. C, Method I. FC (toluene/AcOEt/Et$_3$N 70 : 30 : 1 → 50 : 50 : 1). Yield 85%. TLC (AcOEt/Et$_3$N 99 : 1): R$_f$ 0.79. UV (MeOH): 280 (sh, 4.50), 270 (4.58), 235 (4.47). ^{1}H-NMR (CDCl$_3$): 8.18 – 8.05 (m, 4 H o to NO$_2$); 7.97, 7.95 (2 s, H–C(8)); 7.66, 7.64 (2 s, NH); 7.51 – 7.12 (m, 13 arom. H); 6.74 (d, 4 H o to MeO); 6.34 (m, H–C(1′)); 4.83 – 4.78 (t, OCH$_2$CH$_2$ (CO-npe)); 4.68 (m, H–C(3′)); 4.21 (m, H–C(4′)); 3.88 (m, OCH$_2$CH$_2$ (PO-npe)); 3.73 (2 s, 2 MeO); 3.59 – 3.41 (m, 2 Me$_2$CHN); 3.36 – 3.22 (2 m, 2 H–C(5′), OCH$_2$CH$_2$ (CO-npe)); 3.01 (t, 1 H, OCH$_2$CH$_2$ (PO-npe)); 2.92 – 2.68 (t, m, 3 H, 1 H–C(2′), Me$_2$CHCO, OCH$_2$CH$_2$ (PO-npe)); 2.48 (m, 1 H–C(2′)); 1.29 – 1.02 (m, 3 Me$_2$CH). ^{31}P-

NMR (CDCl$_3$): 148.21; 148.56. Anal. calc. for C$_{57}$H$_{65}$N$_8$O$_{12}$P (1085.2): C 63.09, H 6.04, N 10.33; found: C 61.36, H 6.09, N 10.02.

56. *2'-Deoxy-5'-O-(4,4'-dimethoxytrityl)-N^2-[2-(4-nitrophenyl)ethoxycarbonyl]-O^6-[2-(4-nitrophenyl)-ethyl]guanosine 3'-[2-(4-Nitrophenyl)ethyl Diisopropylphosphoramidite]* (**78**). From **37** by *G.P. C, Method I.* FC (AcOEt/Et$_3$N 96:4). Yield 91%. By *G.P. C, Method II.* FC (toluene/AcOEt 2:1 and 1:1). Yield 68%. TLC (AcOEt/Et$_3$N 96:4): R_f 0.83. UV (MeOH): 269 (4.60), 237 (4.46). ^{1}H-NMR (CDCl$_3$): 8.17 – 8.05 (*m*, 6 H *o* to NO$_2$); 7.95 (2 *s*, H – C(8)); 7.51 – 7.14 (*m*, 15 arom. H); 6.75 – 6.72 (*d*, 4 H *o* to MeO); 6.35 (*m*, H – C(1')); 4.80 (*t*, OCH_2CH$_2$ (CO-npe)); 4.65 (*m*, H – C(3')); 4.43 (*t*, OCH_2CH$_2$ (npeoc)); 4.20 (*m*, H – C(4')); 3.90 (*m*, OCH$_2$CH (PO-npe)); 3.73 (2 *s*, 2 MeO); 3.52 – 3.43 (*m*, 2 Me$_2$C*H*N); 3.31 (2 *m*, 2 H – C(5'), OCH$_2$CH_2 (CO-npe)); 3.09 (*t*, OCH$_2$CH_2 (npeoc)); 3.00 – 2.84 (2 *t*, OCH$_2$CH_2 (PO-npe)); 2.82 (*m*, 1 H – C(2')); 2.45 (*m*, 1 H – C(2')); 1.25 – 1.02 (*m*, 2 Me$_2$CH*N*). ^{31}P-NMR (CDCl$_3$): 148.37; 148.28. Anal. calc. for C$_{62}$H$_{66}$N$_9$O$_{15}$P (1208.2): C 61.63, H 5.51, N 10.43; found: C 60.77, H 5.75, N 10.20.

57. *5'-O-(4,4'-Dimethoxytrityl)-O^4-methylthymidine 3'-[2-(4-Nitrophenyl)ethyl Diisopropylphosphoramidite]* (**79**). From **43** [27] [28] by *G.P. C, Method I.* FC (toluene/AcOEt 1:2). Yield 73%. TLC (toluene/AcOEt 1:1): R_f 0.56. UV (MeOH): 281 (sh, 4.23), 275 (4.24), 234 (sh, 4.36). ^{1}H-NMR (CDCl$_3$): 8.10 (*m*, 2 H *o* to NO$_2$); 7.86 (*dd*, H – C(6)); 7.30 (*m*, 11 arom. H); 6.80 (*dd*, 4 H *o* to MeO); 6.38 (*m*, H – C(1')); 4.57 (*m*, H – C(3')); 4.13 (*m*, H – C(4')); 3.98 (*s*, MeO); 3.75 (2 *s*, 2 MeO, OCH$_2$CH$_2$ (PO-npe)); 3.49 (*m*, 1 H – C(5'), 2 Me$_2$C*H*N); 3.28 (*m*, 1 H – C(5')); 2.99, 2.84 (2 *t*, OCH$_2$CH_2 (PO-npe)); 2.60 (*m*, 1 H – C(2')); 2.21 (*m*, 1 H – C(2')); 1.50 (*s*, Me – C(5)); 1.06 (*m*, 2 Me$_2$CH*N*). ^{31}P-NMR (CDCl$_3$): 148.31; 148.01. Anal. calc. for C$_{46}$H$_{55}$N$_4$O$_{10}$P (854.9): C 64.63, H 6.48, N 6.55; found: C 63.94, H 6.35, N 6.31.

58. *5'-O-(4,4'-Dimethoxytrityl)-O^4-ethylthymidine 3'-[2-(4-Nitrophenyl)ethyl Diisopropylphosphoramidite]* (**80**). From **44** [22] [26 – 29] by *G.P. C, Method I.* FC (toluene/AcOEt 1:2). Yield 71%. TLC (toluene/AcOEt 1:1): R_f 0.62. UV (MeOH): 280 (sh, 4.25), 275 (4.26), 232 (sh, 4.39). ^{1}H-NMR (CDCl$_3$): 8.10 (*m*, 2 H *o* to NO$_2$); 7.85 (*dd*, H – C(6)); 7.31 (2 *m*, 11 arom. H); 6.80 (*dd*, 4 H *o* to MeO); 6.39 (*m*, H – C(1')); 4.51 (*m*, MeCH_2O – C(4), H – C(3')); 4.13 (*m*, H – C(4')); 3.75 (*m*, 2 MeO, OCH$_2$CH$_2$ (PO-npe)); 3.48 (*m*, 1 H – C(5'), 2 Me$_2$C*H*N); 3.28 (*m*, 1 H – C(5')); 2.99, 2.84 (2 *t*, OCH$_2$CH_2 (PO-npe)); 2.62 (*m*, 1 H – C(2')); 2.21 (*m*, 1 H – C(2')); 1.49 (*s*, Me – C(5)); 1.36 (*m*, MeCH_2O – C(4)); 1.06 (*m*, 2 Me$_2$CH*N*). ^{31}P-NMR (CDCl$_3$): 148.29; 148.00. Anal. calc. for C$_{47}$H$_{57}$N$_4$O$_{10}$P (869.0): C 64.96, H 6.61, N 6.45; found: C 64.44, H 6.58, N 6.39.

59. *5'-O-(4,4'-Dimethoxytrityl)-O^4-isopropylthymidine 3'-[2-(4-Nitrophenyl)ethyl Diisopropylphosphoramidite]* (**81**). From **45** [27] by *G.P. C, Method I.* FC (toluene/AcOEt 1:1). Yield 71%. TLC (toluene/AcOEt 1:1): R_f 0.71. UV (MeOH): 280 (sh, 4.26), 275 (4.27), 232 (sh, 4.39). ^{1}H-NMR (CDCl$_3$): 8.10 (*m*, 2 H *o* to NO$_2$); 7.82 (*d*, H – C(6)); 7.29 (2 *m*, 11 arom. H); 6.81 (*d*, 4 H *o* to MeO); 6.41 (*m*, H – C(1')); 5.50 (*m*, Me$_2$C*H*O – C(4)); 4.57 (*m*, H – C(3')); 4.13 (*m*, H – C(4')); 3.75 (*m*, 2 MeO, OCH$_2$CH$_2$ (PO-npe)); 3.49 (*m*, 1 H – C(5'), 2 Me$_2$C*H*N); 3.28 (*m*, 1 H – C(5')); 2.99, 2.84 (2 *t*, OCH$_2$CH_2 (PO-npe)); 2.60 (*m*, 1 H – C(2')); 2.22 (*m*, 1 H – C(2')); 1.48 (*s*, Me – C(5)); 1.32 (*m*, Me$_2$CHO – C(4)); 1.06 (*m*, 2 Me$_2$CH*N*). ^{31}P-NMR (CDCl$_3$): 148.28; 147.98. Anal. calc. for C$_{48}$H$_{59}$N$_4$O$_{10}$P (883.0): C 65.29, H 6.74, N 6.35; found: C 64.89, H 6.86, N 6.26.

60. *5'-O-(4,4'-Dimethoxytrityl)-O^4-2-[2-(4-nitrophenyl)ethyl]thymidine 3'-[2-(4-Nitrophenyl)ethyl Diisopropylphosphoramidite]* (**82**). From **47** by *G.P. C, Method II.* FC (AcOEt/Et$_3$N 99:1). Yield 88%. By *G.P. C, Method II.* FC (toluene/AcOEt 2:1 → 1:1). Yield 79%. TLC (AcOEt): R_f 0.77. UV (MeOH): 281 (sh, 4.30), 275 (4.32), 234 (sh, 4.33). ^{1}H-NMR (CDCl$_3$): 8.63 (*m*, 4 H *o* to NO$_2$); 7.61 (*m*, H – C(6)); 7.42 – 7.21 (*m*, 13 arom. H); 6.76 (*m*, 4 H *o* to MeO); 6.43 (*m*, H – C(1')); 4.64 (*m*, H – C(3'), OCH$_2$CH$_2$ (npe)); 4.16 (*m*, H – C(4')); 3.79 (2 *s*, 2 MeO); 3.93 – 3.62 (*m*, OCH$_2$CH$_2$ (PO-npe)); 3.60 – 3.26 (*m*, 2 H – C(5'), 2 Me$_2$C*H*N), OCH$_2$CH_2 (npe)); 3.05 – 2.82 (2 *t*, OCH$_2$CH_2 (PO-npe)); 2.51 (*m*, 1 H – C(2')); 2.17 (*m*, 1 H – C(2')); 1.42 (*s*, Me – C(5)); 1.17 – 0.99 (*m*, 2 Me$_2$CH*N*). ^{31}P-NMR (CDCl$_3$): 148.50; 147.23. Anal. calc. for C$_{53}$H$_{60}$N$_5$O$_{12}$P (990.1): C 64.29, H 6.18, N 7.07; found: C 63.92, H 5.96, N 6.39.

REFERENCES

[1] M. Beier, W. Pfleiderer, *Helv. Chim. Acta* **1999**, *82*, 879.

[2] T. W. Greene, P. G. M. Wuts, 'Protective Groups in Organic Synthesis', 3rd edn., John Wiley & Sons, New York, 1999.

[3] V. Amarnath, A. D. Broom, *Chem. Rev.* **1977**, *77*, 183.

[4] J. Engels, E. Uhlmann, *Adv. Biochem. Eng. Biotechnol.* **1988**, *37*, 73.

[5] S. L. Beaucage, R. P. Iyer, *Tetrahedron* **1992**, *48*, 2223.

[6] S. L. Beaucage, R. P. Iyer, *Tetrahedron* **1993**, *49*, 1925.

[7]   S. L. Beaucage, R. P. Iyer, *Tetrahedron* **1993**, *49*, 6123.

[8]   R. L. Letsinger, V. Mahadevan, *J. Am. Chem. Soc.* **1965**, *87*, 3526.

[9]   T. Brown, D. J. S. Brown, in 'Oligonucleotides and Analogues – A Practical Approach', Ed. F. Eckstein, IRL Press, Oxford, 1991, p. 1.

[10] M. A. Dorman, S. A. Noble, L. J. McBride, M. H. Caruthers, *Tetrahedron* **1984**, *40*, 95.

[11] H. Köster, J. Biernat, J. McManus, A. Wolter, A. Stumpe, C. K. Narang, N. D. Sinha, *Tetrahedron* **1984**, *40*, 103.

[12] F. Himmelsbach, B. S. Schulz, T. Trichtinger, R. Charubala, W. Pfleiderer, *Tetrahedron* **1984**, *40*, 59.

[13] K. P. Stengele, W. Pfleiderer, *Tetrahedron Lett.* **1990**, *31*, 2549.

[14] G. S. Ti, B. L. Gaffney, R. Jones, *J. Am. Chem. Soc.* **1982**, *104*, 1316.

[15] H. Schaller, G. Weimann, B. Lerch, H. G. Khorana, *J. Am. Chem. Soc.* **1963**, *85*, 3821.

[16] O. Mitsunobu, *Synthesis* **1981**, 1.

[17] B. S. Schulz, W. Pfleiderer, *Nucleosides Nucleotides* **1987**, *6*, 529.

[18] B. L. Gaffney, R. A. Jones, *Tetrahedron Lett.* **1982**, *23*, 2253, 2257.

[19] B. L. Gaffney, L. A. Marky, R. A. Jones, *Tetrahedron* **1984**, *40*, 3.

[20] H. Büchi, H. G. Khorana, *J. Mol. Biol.* **1972**, *72*, 251.

[21] B. S. Schulz, W. Pfleiderer, *Tetrahedron Lett.* **1983**, *24*, 3587.

[22] L. Kiriasis, S. Farkas, W. Pfleiderer, *Nucleosides Nucleotides* **1986**, *5*, 517.

[23] C. B. Reese, P. A. Skone, *J. Chem. Soc., Perkin Trans. 1* **1984**, 1263.

[24] W. L. Sung, *J. Org. Chem.* **1982**, *47*, 3623.

[25] B. L. Li, C. B. Reese, P. F. Swann, *Biochemistry* **1987**, *26*, 1086.

[26] P. K. Bridson, W. T. Markiewicz, C. B. Reese, *J. Chem. Soc., Chem. Commun.* **1977**, 791.

[27] H. Borowy-Borowsky, R. W. Chambers, *Biochemistry* **1987**, *26*, 2465.

[28] H. Borowy-Borowsky, R. W. Chambers, *Biochemistry* **1989**, *28*, 1471.

[29] D. F. Forner, Y. Palom, S. Ikuta, E. Pedroso, R. Eritja, *Nucleic Acids Res.* **1990**, *18*, 5729.

[30] N. D. Sinha, J. Biernat, H. Köster, *Nucleic Acids Res.* **1984**, *12*, 4539.

[31] G. Zon, K. A. Gallo, C. J. Samson, K. Shao, R. A. Byrd, M. F. Summers, *Nucleic Acids Res.* **1985**, *13*, 8181.

[32] A. D. Barone, J. Y. Tang, M. H. Caruthers, *Nucleic Acids Res.* **1984**, *12*, 4051.

[33] A. Kraszewski, K. E. Norris, *Nucleic Acids Res.* **1987**, *15*, 177.

[34] J. Nielsen, J. E. Marugg, M. Taagard, J. H. van Boom, O. Dahl, *Nucleic Acids Res.* **1986**, *14*, 4051.

[35] J. Nielsen, O. Dahl, *Nucleic Acids Res.* **1987**, *15*, 3626.

[36] E. Uhlmann, W. Pfleiderer, *Tetrahedron Lett.* **1980**, *21*, 481.

[37] W. Pfleiderer, M. Schwarz, H. Schirmeister, *Chem. Scripta* **1986**, *26*, 147.

[38] P. Herdewijn, R. Charubala, W. Pfleiderer, *Helv. Chim. Acta* **1989**, *72*, 1729.

[39] R. W. Sobol, E. E. Henderson, N. Kon, J. Shao, P. Hitzges, E. Mordechai, N. L. Reichenbach, R. Charubala, W. Pfleiderer, R. J. Suhadolnik, *J. Biol. Chem.* **1995**, *270*, 5963.

[40] E. Uhlmann, W. Pfleiderer, *Helv. Chim. Acta* **1981**, *64*, 1688.

[41] S. Waldvogel, W. Pfleiderer, *Helv. Chim. Acta* **1998**, *81*, 46.

[42] K. L. Agarwal, F. Riftina, *Nucleic Acids Res.* **1978**, *5*, 2805.

Synthesis of Conformationally Restricted Carbocyclic Nucleosides: The Role of the O(4′)-Atom in the Key Hydration Step of Adenosine Deaminase

by Victor E. Marquez[a])*, Pamela Russ[a]), Randolph Alonso[a]), Maqbool A. Siddiqui[a]), Susana Hernandez[a]),
Clifford George[b]), Marc C. Nicklaus[a]), Fang Dai[a]), and Harry Ford, Jr.[a])

[a]) Laboratory of Medicinal Chemistry, Division of Basic Sciences, National Cancer Institute, National Institutes
of Health, Bethesda, MD 20892, USA
[b]) Laboratory for the Structure of Matter, Naval Research Laboratory, Washington, DC 20375, USA

Conformationally restricted carbocyclic nucleosides with either a northern(N)-type conformation, i.e., N-type 2′-deoxy-methanocarba-adenosine **8** ((N)MCdAdo), or a southern(S)-type conformation, i.e. S-type 2′-deoxy-methanocarba-adenosine **9**, ((S)MCdAdo), were used as substrates for adenosine deaminase (ADA) to assess the enzyme's preference for a fixed conformation relative to the flexible conformation represented by the carbocyclic nucleoside aristeromycin (**10**). Further comparison between the rates of deamination of these compounds with those of the two natural substrates adenosine (Ado; **1**) and 2′-deoxyadenosine (dAdo; **2**), as well as with that of the conformationally locked nucleoside LNA-Ado (**11**), which, like the natural substrates, has a furanose O(4′) atom, helped differentiate between the roles of the O(4′) anomeric effect and sugar conformation in controlling the rates of deamination by ADA. Differences in rates of deamination as large as 10000 can be attributed to the combined effect of the O(4′) atom and the enzyme's preference for an N-type conformation. The hypothesis proposed is that ADA's preference for N-type substrates is not arbitrary; it is rather the direct consequence of the conformationally dependent O(4′) anomeric effect, which is more efficient in N-type conformers in promoting the formation of a covalent hydrate at the active site of the enzyme. The formation of a covalent hydrate at the active site of ADA precedes deamination. A new and efficient synthesis of the important carbobicyclic template **14a**, a useful intermediate for the synthesis of (N)MCdAdo (**8**) and other conformationally restricted nucleosides, is also reported.

Introduction. – Adenosine deaminase (ADA, EC 3.5.4.4) is a critical enzyme of the purine metabolic pathway that catalyzes the irreversible hydrolytic deamination of adenosine (Ado; **1**) and 2′-deoxyadenosine (dAdo; **2**) – as well as of other exogenous substrates – to their hypoxanthine derivatives **4** and **5**, respectively, and ammonia (*Scheme 1*) [1][2]. Aberrations in the expression and function of ADA have been associated with impaired B- and T-cell-based immunity, whereas elevated levels of ADA have been observed in certain lymphomas and leukemias [3–5]. Thus, a complete understanding of the ADA-catalyzed mechanism of deamination should provide additional means of intervening therapeutically in ADA-related disorders. The rational design of the first transition-state analogue inhibitor of ADA by *Wolfenden* and *Evans* in 1970 [6] was followed four years later by the discovery of the potent fermentation-product inhibitors coformycin (**6**) [7][8] and 2′-deoxycoformycin (**7**) [9], which were shown to function as stable, high-affinity transition-state-analogue inhibitors of ADA [10][11]. Further confirmation of the transition-state mimicry of these analogues came from several crystal structures of ADA with the ligands bound at the active site of the enzyme [12–14]. In the case of the simple purine riboside **3**

(nebularine), covalent hydration at the active site of ADA is able to generate the potent transition-state inhibitor (6S)-1,6-dihydro-9H-purin-6-ol riboside **3a** [11] (*Scheme 1*). However, since the equilibrium constant for this critical hydration step is very small, the apparent inhibition constant of nebularine is only in the micromolar range [11]. On the other hand, the transition-state inhibitors coformycin (**6**) and 2'-deoxycoformycin (**7**), where the 8-hydroxy substituent of the diazepine ring provides a stable mimic of a hydrated ring, are able to inhibit ADA at picomolar levels [10][11]. From these observations, it was concluded that all ADA substrates must form a covalent hydrate (*i.e.* **1a** and **2a** in *Scheme 1*) as a first critical step before deamination. Calculation of relative hydration free energy differences in heteroaromatic bases have revealed the importance of electronic and steric factors in facilitating hydration of the aglycon as a key step in the deamination reaction [15]. For example, a comparison between nebularine (**3**) and its 8-aza analogue showed a 400-fold greater ADA inhibitor potency for the latter, which is attributable to the enhancing effect of hydration caused by the N-atom at the 8-position [15]. The role of the sugar moiety in this key hydration step, on the other hand, has for the most part been ignored and even considered negligible [15]. In the present work, we wish to address this issue by comparing the deamination rates of substrates that differ from Ado and dAdo in that the furanose ring has been replaced by a cyclopentane ring.

Scheme 1

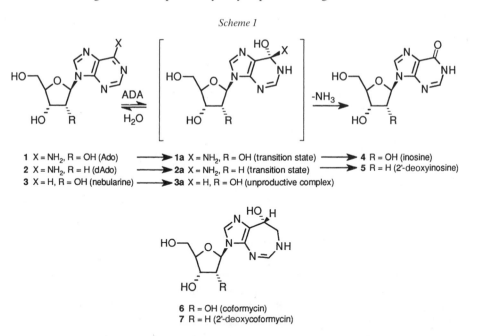

1 X = NH₂, R = OH (Ado) ⟶ **1a** X = NH₂, R = OH (transition state) ⟶ **4** R = OH (inosine)
2 X = NH₂, R = H (dAdo) ⟶ **2a** X = NH₂, R = H (transition state) ⟶ **5** R = H (2'-deoxyinosine)
3 X = H, R = OH (nebularine) ⟶ **3a** X = H, R = OH (unproductive complex)

6 R = OH (coformycin)
7 R = H (2'-deoxycoformycin)

In addition, since ADA-inhibitor complexes [12–14] have revealed a preferred northern(*N*)-type conformation for the bound inhibitor at the active site – and by extension one can assume that substrates will bind with the same preferred conformation – we have chosen to study the deamination rates of carbocyclic adenosine nucleosides locked in antipodal conformations (northern(*N*)- and southern-

(*S*)-type) similar to those of the furanose ring, as defined by the pseudorotation phase angle *P* [16]. Normally, conventional nucleosides in solution undergo rapid equilibration between ranges of *P* defined by two main furanose puckering domains centered around a $^3T_2$ (*N*-type) and a $^2T_3$ (*S*-type) conformation (*Fig. 1*). These conformational modes, in turn, favor particular orientations of the glycosyl torsion angle *χ* (*syn* or *anti*) and the 4-(hydroxymethyl) chain defined by *γ* (+*sc*, − *sc*, and *ap*). Indeed, the range of values of *χ* and *γ* are not uniformly populated, and their preferred distribution is coupled to a particular sugar pucker [17]. The conformationally locked carbocyclic adenosine substrates selected for this study were (*N*)MCdAdo (**8**), as a rigid *N*-type conformer, and (*S*)MCdAdo (**9**) as a rigid *S*-type conformer. The fermentation product aristeromycin (Ari; **10**) was used as a base-line reference for a flexible, carbocyclic nucleoside substrate. Indeed, aristeromycin has been shown to undergo rapid equilibration in solution between a southwestern-type and an *N*-type conformer, with a bias towards the former [18]. Comparison between the rates of deamination of these compounds with those of the two natural substrates Ado (**1**) and dAdo (**2**), as well as with that of the conformationally locked nucleoside LNA-Ado (**11**) [19], helped differentiate the roles of the O(4′) atom's anomeric effect and sugar conformation in controlling the rates of deamination by ADA. As in the natural substrates, the conformationally locked nucleoside LNA-Ado has a O(4′) furanose atom, and its enhanced rate of deamination relative to the conformationally locked carbocyclic nucleosides highlights the importance of the anomeric effect contributed by the O(4′) atom.

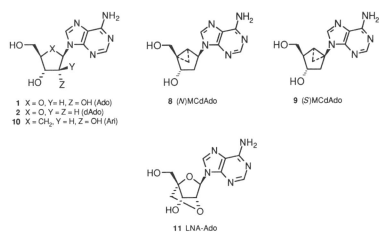

1 X = O, Y = H, Z = OH (Ado)
2 X = O, Y = Z = H (dAdo)
10 X = CH₂, Y = H, Z = OH (Ari)

8 (*N*)MCdAdo

9 (*S*)MCdAdo

11 LNA-Ado

In this work, we also wish to present a new and more efficient six-step synthesis of the critical precursor **14a**, which is an extremely useful intermediate for the convergent synthesis of (*N*)MCdAdo and other *N*-type purine analogues. The overall yield of 35% obtained for **14a** compares favorably with our previous seven-step synthesis reported earlier (25% yield) [20].

Results. – *Synthesis and Structural Analysis of Carbocyclic Nucleosides.* The syntheses of the conformationally locked *N*-type nucleoside (*N*)MCdAdo (**8**), as well

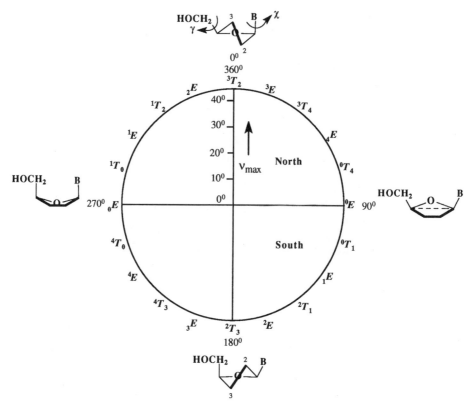

Fig. 1. *Pseudorotational cycle displaying all the puckering modes (P) and maximum puckering amplitude (v_{max})*

as that of the *S*-type analogue (*S*)MCdAdo (**9**), have been reported [20][21]. While the X-ray structure of (*N*)MCdAdo (**8**) was published in conjunction with its synthesis [20][1]), we now report for the first time the crystal structure of (*S*)MCdAdo (**9**)[2]). From comparing the two structures, it is clear that these carbocyclic nucleosides adopt fixed conformations that mimic the ring pucker of a furanose ring (*Figs. 2* and *3*). As discussed in a preliminary communication [22], the crystallographic pseudorotational parameters for the two molecules (A and B) in the unit cell of (*N*)MCdAdo are in complete agreement with the expected values for a *N*-type ($_2E$) conformation ($P = 342.78°$ (A) and $P = 339.25°$ (B)), whereas the equivalent parameters for the antipodal (*S*)MCdAdo correlate with a pure *S*-type ($_3E$) conformation ($P = 199.10°$ (A) and $P = 200.20°$ (B)). Except for the molecule A of (*S*)MCdAdo, all torsion angles χ are in the *anti* range, while the less encumbered angle γ alternates between the *ap* and *+ sc* conformations as it is normally seen in conventional nucleosides [22].

---

[1])   *CCDC* entry NAKZUU.

[2])   Crystallographic data have been deposited with the *Cambridge Crystallographic Data Centre* as deposition No. CCDC-133070. Copies of the data can be obtained, free of charge, by applying to the *CCDC*, 12 Union Road, Cambridge CB2 1EZ, UK (fax: + 44(0)1223-336033; e-mail: deposit@ccdc.com.oc.uk).

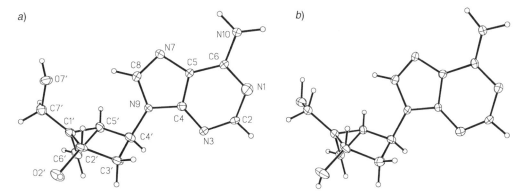

Fig. 2. *Molecular structure of* (N)*MCdAdo* (**8**)*: a) molecule A* ($\chi = -154.8°$ *(anti)*, $\gamma = +54.8°$ *(+sc)) and b) molecule B* ($\chi = -167.6°$ *(anti)*, $\gamma = -177.6°$ *(ap))*

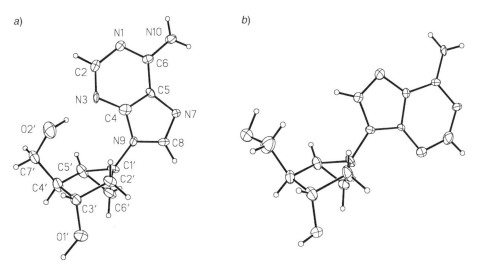

Fig. 3. *Molecular structure of* (S)*MCdAdo* (**9**)*: a) molecule A* ($\chi = +50.9°$ *(syn)*, $\gamma = +54.0°$ *(+sc)) and b) molecule B* ($\chi = -129.8°$ *(anti)*, $\gamma = -171.9°$ *(ap))*

The new approach to the synthesis of (*N*)MCdAdo (**8**) was based on the easily available chiral template (3*R*,4*S*)-4-(phenylmethoxy)-3-[(phenylmethoxy)methyl]cyclopent-1-ene (**12**) [23–25]. This approach compared favorably with respect to the one devised earlier from chiral cyclopentenone **13** [20], particularly in the generation of the key carbobicyclic intermediates **14a,b** (*Scheme 2*). In the same manner as reported previously with the PhSeCl/NaN$_3$ system [25], the double bond in **12** reacted regio- and stereoselectively with PhSeCl/AgCF$_3$CO$_2$ to give the *trans*-2-(phenylseleno)cyclopentan-1-ol intermediate **15** after hydrolysis of the trifluoroacetate ester (*Scheme 3*). Following oxidation of the (phenylseleno) group, unidirectional elimination of PhSeOH gave exclusively the corresponding allylic alcohol **16**. This allylic alcohol was inverted following a *Mitsunobu* esterification which gave the intermediate benzoate **17** and the new allylic alcohol **18** after ester hydrolysis. Finally, a hydroxy-

Scheme 2

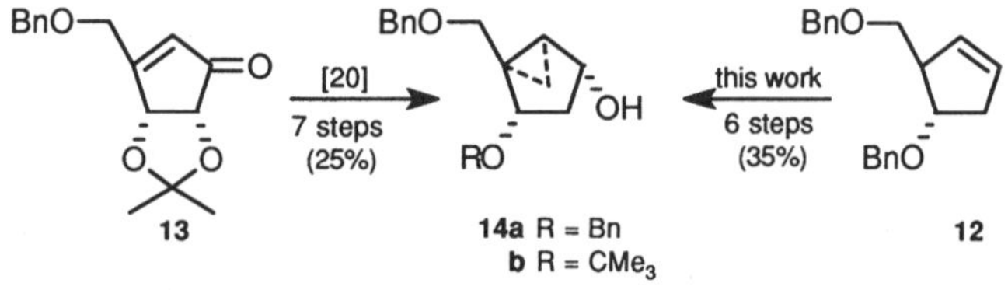

13   14a R = Bn   b R = CMe₃   12

Scheme 3

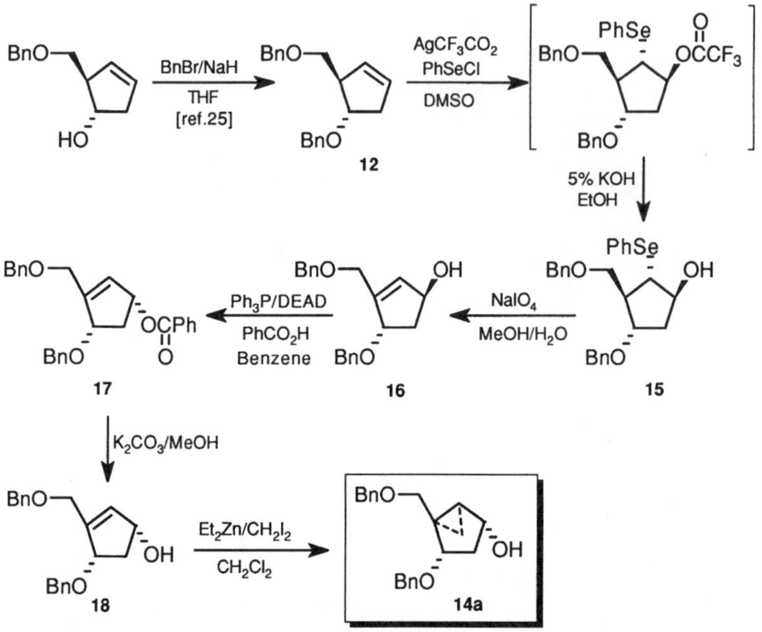

directed cyclopropanation under *Simmons-Smith* conditions gave the key carbobicyclic intermediate **14a** in 35% overall yield. This process required one step less than the previously reported approach [20] and provided a 10% increase in the overall yield of the final product. The synthesis of (*N*)MCdAdo (**8**) from **14a** proceeded essentially as reported before from intermediate **14b** [20], *i.e., via* **19** and **20** (*Scheme 4*).

The conformationally locked, pentofuranose-containing LNA-Ado (**11**), the synthesis of which has been reported earlier [19], was a kind gift from Professor *Wengel*. On conformational grounds, however, the locked 2,5-dioxabicyclo[2.2.1]heptane system does not have the same type of ring pucker as the bicyclo[3.1.0]hexane. The reported X-ray structure of the corresponding LNA-uracil analogue [26] shows that the 2,5-dioxabicyclo[2.2.1]heptane system has a *P* value of 17.4°, which is *ca.* 34–38° away from the pseudorotational angle of the locked *N*-type bicyclo[3.1.0]hexane template of carbocyclic nucleoside **8**. These different envelope conformations, $_2E$ vs. ^3E (see *Fig. 1*), could account for some differences in the rates of deamination discussed below.

Scheme 4

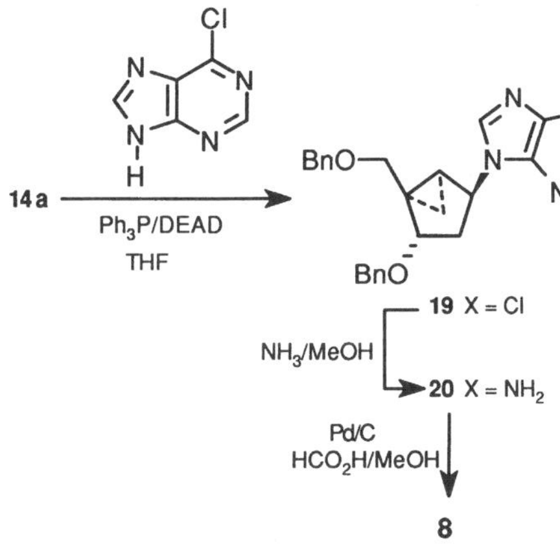

Adenosine Deaminase Activity. The compounds listed in the *Table* can be grouped in two categories depending on their substrate properties. The first three substrates **1, 2**, and **11** fall in the category of good substrates, as indicated by the various kinetic parameters such as K_m, K_{cat}, and the catalytic efficiency (K_{cat}/K_m). The difference between these substrates and the carbocyclic analogues as a group appears substantial in view of the relative hydrolysis rates measured at 50 μM substrate concentration. Despite the poor deamination rates among the locked carbocyclic analogues **8 – 10**, there appears to be a clear preference for the *N*-type conformation. This observation is in agreement with the published X-ray structures of ADA-inhibitor complexes where the sugar moieties of the nucleosides bound at the active site appear puckered as *N*-type sugars [12 – 14]. Interestingly, the rate of deamination of the conformationally flexible aristeromycin (**10**) lies between the rates of deamination of the rigid *N*-type (*N*)MCdAdo (**8**) and the rigid *S*-type (*S*)MCdAdo (**9**) analogues. Obviously, the carbocyclic nucleosides are in a class by themselves as poor substrates of ADA, which appears to be a direct consequence of lacking the furanose O-atom. It has been clearly demonstrated by *Chattopadhyaya* and co-workers that the electronic nature of the aglycon is effectively transmitted to the pentofuranose moiety through the anomeric effect [27][28]. Indeed, the direct correlation of the protonation/deprotonation equilibrium of the aglycon with the two-state *N/S* pseudorotational equilibrium of the sugar moiety permits the facile calculation of the pK_a from the pD-dependent change of $\Delta G°$ in the *N/S* equilibrium [27][28]. On the other hand, a similar pK_a determination is not possible in carbocyclic nucleosides, and specifically for aristero-mycin (**10**), because of the lack of the O(4′) atom and its associated anomeric effect [18][28]. Even though exact kinetic parameters between the two groups could not be compared, the differences in relative rates of deamination revealed some interesting trends worth discussing. For example, a direct comparison between the best ADA substrate, dAdo (**2**), and the best equivalent carbocyclic substrate, (*N*)MCdAdo (**8**), reveals a 122-fold difference in the rate of deamination in favor of the former that can be attributed to a combined effect of the O(4′) atom and the enzyme's preference for binding *N*-type substrates (*Table*). For the other two substrates, Ari (**10**) and

(S)MCdAdo (**9**), this difference can be even greater (200 and 10000, resp.). These larger differences are probably the result of higher energy penalties incurred for having a conformation far removed from the ideal N-type conformation. To gauge the conformational factor alone, a comparison between the rates of deamination of (N)MCdAdo (**8**) and (S)MCdAdo (**9**) would show the enzyme's capacity to effectively discriminate between the two rigid antipodal conformations. The nearly 100-fold difference between the rates of deamination of (N)MDdAdo (**8**) and (S)MCdAdo (**9**) demonstrates the enzyme's clear preference for an N-type conformer. In the real world, however, the effect of conformation is not as dramatic because of the nucleoside's inherent flexibility and capacity to adapt to the conformation demanded by the enzyme, something that would not be possible for (N)MDdAdo (**8**) and (S)MCdAdo (**9**). Indeed, Ado (**1**) and dAdo (**2**), which are known to exist in solution preferentially as S-type conformers [27], are expected to bind to the enzyme as N-type conformers [12–14]. From our study, it can be approximated that the combined anomeric effect of the O(4') atom and the preference for an N-type conformation could account for differences in deamination rates as large as 10000. Since for nucleosides such as Ado (**1**) and dAdo (**2**), the anomeric effect is greater when the conformation of the sugar is of the N-type, the preference shown by ADA for N-type conformers is not a capricious preference; it results from the fact that in an N-type conformation, the anomeric effect stabilizes the protonated base more efficiently, a prerequisite for the formation of the covalent hydrate at the active site of ADA prior to deamination (*Scheme 1*) [11]. In these experiments, the dramatic effect of the O(4') atom was revealed by its capacity to increase the rate of deamination of the conformationally locked nucleoside LNA-Ado (**11**) to *ca.* 20-fold relative to the best carbocyclic substrate (N)MCdAdo (**8**). This makes LNA-Ado only a 4-fold less efficient substrate than Ado (**1**). Such an increase in rate is clearly a direct consequence of the O(4') atom's role in the critical hydration step prior to deamination. However, why did the rate of deamination of LNA-Ado (**11**) not match the level of deamination of Ado (**1**)? We have already discussed in the previous section that the X-ray structures of the locked 2,5-dioxabicyclo[2.2.1]heptane and the bicyclo[3.1.0]hexane systems used as N-type templates differ by 34–38° in pseudorotation (P). This 34–38° difference in P could account for the partial substrate recovery observed for LNA-Ado (**11**), as well as for the fact that there could be possible steric clashes at the receptor site between the structural elements added to constrain the rings and some critical amino-acid side chains. It must be remembered that, while there is a favorable entropic advantage in binding a constrained substrate, the enthalpy penalties could be rather severe when the fit is not perfect. It is possible, therefore, that while LNA-Ado (**11**) exists as an N-type conformer, it does not have the ideal conformation for a perfect fit. Despite this shortcoming, if one assumes that both conformationally rigid compounds, (N)MCdAdo (**8**) and LNA-Ado (**11**), approximate the preferred N-type conformation and are able to bind at the active site of ADA, the role of the O(4') atom appears conclusive. Further confirmation of the importance of the O(4') atom anomeric effect, as a critical determinant for the hydration step catalyzed by ADA and related enzymes, is found in the recent work of *Lindell* and co-workers who report that the 5'-monophosphate of the carbocyclic analogue of nebularine is a very poor inhibitor of the mechanistically equivalent adenosine 5'-monophosphate deaminase (AMPDA) [29].

Table. *Kinetic Constants for Various Substrates of ADA[a]*)

	K_m [μM]	K_{cat} [s⁻¹]	K_{cat}/K_m [μM⁻¹s⁻¹]	Relative rate[b])
Ado (**1**)	33.1 ± 3.7	188	5.7	100
dAdo (**2**)	23.0 ± 2.3	245	10.6	121
LNA-Ado (**11**)	122 ± 12	59.4	0.49	19
(*N*)MCdAdo (**8**)	–	–	–	0.99
Ari (**10**)	–	–	–	0.58
(*S*)MCdAdo (**9**)	–	–	–	0.01

[a]) Kinetic constants were determined by UV at 37° and pH 7.4 with calf-intestine ADA. [b]) Relative rates were measured spectrophotometrically at 37° with 50 μM substrate concentration.

Experimental Part

1. *General.* All chemical reagents were commercially available. M.p.: *MelTemp-II* apparatus, *Laboratory Devices*, USA; uncorrected. Column flash chromatography (FC): silica gel *60* (230–400 mesh; *E. Merck*). ¹H- and ¹³C-NMR Spectra: *Bruker-AC-250* instrument at 250 and 62.9 MHz, resp.; δ in ppm rel. to the solvent in which they were run (7.24 ppm for CDCl₃). Elemental analyses were performed by *Atlantic Microlab, Inc.*, Atlanta, GA.

2. *X-Ray Crystal Structure of (1S,3S,4R,5S)-1-(6-Amino-9H-purin-9-yl)-4-(hydroxymethyl)bicyclo[3.1.0]-hexan-3-ol* (**9**). *Crystal Data*²): 2(C₁₂H₁₅N₅O₂)·3(H₂O), M_r 576.63; crystal size: $0.40 \times 0.08 \times 0.02$ mm³; monoclinic $P2_1$, $a = 11.460(2)$ Å, $b = 7.991(2)$ Å, $c = 15.524(3)$ Å, $\alpha = 90°$, $\beta = 95.55(1)°$, $\gamma = 90°$; $V = 1377.9(5)$ Å³, $Z = 2$, $D_c = 1.390$ mg/m³, $\lambda(CuK_\alpha) = 1.54178$ Å, $\mu = 0.877$ mm⁻¹, $F(000) = 612$, $T = 293(2)$ K. Final R indices $(I > 2\sigma(I_0))$ $R1 = 0.0658$, $wR2 = 0.1395$ for 2136 reflections. Tables of atomic coordinates, bond distances, and angles, and anisotropic thermal parameters have been deposited²).

3. *Kinetic Analysis and Adenosine Deaminase Assays.* Adenosine deaminase (calf intestine) activity was measured spectrophotometrically with a *Shimazdu-UV-PC-2101* spectrophotometer by following the disappearance of adenosine at 265 nm in 1-ml cuvettes with 0.1M phosphate (pH 7.4) at 25°. For the substrates listed in the *Table*, initial hydrolysis rates were determined by following the disappearance of substrate at 265 nm in 1-ml cuvettes with 0.1M phosphate (pH 7.4) at 37°. A similar molar extinction coefficient and a $\Delta\varepsilon = -7900$ was assumed for all adenine analogues examined. Kinetic constants were determined from *Lineweaver-Burk* plots with Graph-Pad Prism V 2.01 (*Graphpad Software, Inc.* San Diego, CA), a personal-computer-based curve-fitting program. A minimum of three determinations were performed. Relative-rate experiments were performed in duplicate using 50 μM substrate concentration at 37°.

4. *Syntheses.* (*3R,4S*)-4-(*Phenylmethoxy*)-3-[(*phenylmethoxy*)*methyl*]*cyclopent-1-ene* (**12**) was prepared in 95% yield as described in [25].

(1S,4S)-4-(Phenylmethoxy)-3-[(phenylmethoxy)methyl]cyclopent-2-en-1-ol (**16**). A mixture of **12** (2.36 g, 8.61 mmol) and phenylselenium chloride (1.98 g, 10.3 mmol) in dry DMSO (8 ml) was stirred under Ar at r.t. until solution occurred. Silver trifluoroacetate (2.28 g, 10.3 mmol) was added, and the mixture was stirred for 1 h. The mixture was then cooled over ice, and hydrolysis of the trifluoroacetate ester was accomplished with 5% KOH/EtOH (15 ml) while stirring for 0.25 h. The mixture was poured into ice-cold H₂O (100 ml) and extracted with Et₂O (3 × 100 ml), the combined Et₂O extract washed with sat. NH₄Cl soln. (35 ml), dried (MgSO₄), and evaporated, and the residue purified by FC (silica gel, step gradient of 10%, 20%, and 50% AcOEt/hexanes): reasonably pure (*1S,2S,3R,4S*)-4-(*phenylmethoxy*)-3-[(*phenylmethoxy*)*methyl*]-2-(*phenylseleno*)*cyclopentan-1-ol* (**15**; 2.98 g, 74%).

A mixture of crude **15** (2.20 g, 4.71 mmol) and NaIO₄ (2.02 g, 9.42 mmol) was stirred in MeOH/H₂O 9:1 (100 ml) for 1 h. After evaporation, the residue was stirred with AcOEt (100 ml), and the insoluble solid was removed by filtration. The filtrate was evaporated and the residue purified by FC (silica gel, step gradient of 25% and 50% AcOEt/hexanes): **16** (1.06 g, 73%). Clear oil. ¹H-NMR (CDCl₃): 7.44–7.30 (*m*, 2 Ph); 6.03 (br. *s*, H–C(2)); 5.10–5.04 (symmetrical *m*, H–C(1)); 4.90–4.84 (symmetrical *m*, H–C(4)); 4.67 (*AB*('*d*', J = 11.9, 1 H, PhCH₂); 4.64 (*AB*('*d*'), J = 11.7, 1 H, PhCH₂); 4.61 (*AB*('*d*'), J = 11.9, 1 H, PhCH₂); 4.52 (*AB*('*d*'), J = 11.9, 1 H, PhCH₂); 4.26 (br. *s*, PhCH₂OCH₂); 2.38 (*ddd*, J = 14.4, 6.8, 3.2, H_a–C(5)); 2.09 (*ddd*, J = 14.1, 6.6, 3.2, H_b–C(5)); 1.91 (br. *s*, OH). ¹³C-NMR (CDCl₃): 145.54; 138.25; 138.03; 132.89; 128.25; 128.24; 127.57; 127.52;

127.48; 82.10; 75.28; 72.70; 71.28; 66.51; 41.16. Anal. calc. for $C_{20}H_{22}O_3 \cdot 0.25\ H_2O$ (314.89): C 76.27, H 7.21; found: C 76.40; H 7.28.

(1R,4S)-4-(Phenylmethoxy)-3-[(phenylmethoxy)methyl]cyclopent-2-enyl Benzoate (**17**). To a stirred soln. of **16** (1.06 g, 3.32 mmol) in dry benzene (50 ml), benzoic acid (0.63 g, 5.13 mmol) and Ph_3P (1.79 g, 6.84 mmol) were added. Diethyl azodicarboxylate (DEAD; 1.08 ml, 6.84 mmol) was added dropwise, and the reaction was allowed to continue at r.t. for 0.5 h. After all the volatiles were removed under reduced pressure, the residue was purified by FC (silica gel, 10% AcOEt/hexanes): **17** (1.27 g, 85%). Thick oil. ^{1}H-NMR (CDCl$_3$): 8.16–8.12, 7.67–7.34 (2*m*, 3 Ph); 6.17 (br. *s*, H–C(2)); 5.92–5.82 (symmetrical *m*, H–C(1)); 4.75–4.60 (*m*, 2 PhCH$_2$, H–C(4)); 4.35 (br. *s*, 2 H, PhCH$_2$OCH$_2$); 3.01 (*dt*, *J* = 14.2, 7.3, H$_a$–C(5)); 2.08 (*dt*, *J* = 14.4, 4.2, H$_b$–C(5)). ^{13}C-NMR: 166.46; 147.62; 138.48; 138.22; 133.07; 130.44; 129.82; 128.57; 128.55; 128.51; 128.47; 127.87; 127.85; 127.80; 80.80; 76.78; 73.11; 71.37; 66.73; 38.22. Anal. calc. for $C_{27}H_{26}O_4$ (414.50): C 78.24, H 6.32; found: C 78.40, H 6.72.

(1R,4S)-4-(Phenylmethoxy)-3-[(phenylmethoxy)methyl]cyclopent-2-en-1-ol (**18**). A stirred soln. of **17** (1.55 g, 3.59 mmol) in MeOH (50 ml) was treated with K_2CO_3 (1.24 g, 8.98 mmol). After 3 h, the solvent was evaporated and the residue purified by FC (silica gel, 5% MeOH/CHCl$_3$): **18** (0.874 g, 80%). Oil. ^{1}H-NMR (CDCl$_3$): 7.54–7.38 (*m*, 2 Ph); 6.03 (br. *s*, H–C(2)); 4.75–4.55 (*m*, 2 × PhCH$_2$O); 4.49–4.54 (*dd*, *J* = 6.8, 3.9, H–C(4)); 4.30 (br. *s*, PhCH$_2$OCH$_2$); 2.74 (*dt*, *J* = 14.2, 7.1, H$_a$–C(5)); 2.68 (br. *s*, OH); 1.82 (*dt*, *J* = 14.2, 3.9, H$_b$–C(5)); ^{13}C-NMR (CDCl$_3$): 144.86; 138.39; 138.20; 133.01; 128.44; 127.82; 127.78; 127.71; 80.99; 73.98; 72.84; 71.48; 66.62; 41.16. Anal. calc. for $C_{20}H_{22}O_3 \cdot 0.5\ H_2O$ (319.39): C 75.21, H 7.26; found: C 75.46, H 7.51.

(1S,2R,4S,5R)-4-(Phenylmethoxy)-5-[(phenylmethoxy)methyl]bicyclo[3.1.0]hexan-2-ol (**14a**). A stirred soln. of alcohol **18** (0.66 g, 2.15 mmol) in dry CH$_2$Cl$_2$ (30 ml) was cooled over an ice-salt bath. Then 1M diethyl zinc in hexane (2.4 ml) was added dropwise, and the reaction was allowed to continue for 0.25 h after the addition. Separately, a soln. of diiodomethane (0.4 ml, 4.74 mmol) in dry CH$_2$Cl$_2$ (10 ml) was prepared, and 5.2 ml of this soln. was added to the mixture. After 5 min stirring, an equal amount of diethylzinc was added dropwise, followed by the remaining soln. of diiodomethane (4.8 ml). The mixture was stirred for a total of 6 h, while the cooling bath gradually warmed to r.t. Then, the mixture was poured into an aq. sat. NH$_4$Cl soln. (50 ml), and extracted with CH$_2$Cl$_2$ (3 × 50 ml). The combined extract was washed with sat. NH$_4$Cl soln. (2 × 50 ml), dried (MgSO$_4$), and evaporated, and the residue purified by FC (silica gel, step gradient of 25% and 50% AcOEt/hexanes): **14a** (0.560 g, 80%). Oil. ^{1}H-NMR (CDCl$_3$): 7.45–7.30 (*m*, 2 Ph); 4.65–4.40 (*m*, 2 PhCH$_2$O, H–C(4)); 4.31 (*t*, *J* = 8.0, H–C(2)); 3.98 (*d*, *J* = 10.5, 1 H, PhCH$_2$OCH$_2$); 3.21 (*d*, *J* = 10.5, 1 H, PhCH$_2$OCH$_2$); 2.37 (*dt*, *J* = 12.9, 7.4, H$_a$–C(3)); 1.70–1.60 (*m*, OH, H–C(3)); 1.40–1.20 (*m*, H–C(1), H$_a$–C(6)); 0.63 (*dd*, *J* = 7.4, 5.6, H$_b$–C(6)). ^{13}C-NMR (CDCl$_3$): 138.84; 138.53; 128.49; 128.47; 127.88; 127.78; 127.72; 127.70; 73.01; 72.13; 71.87; 70.02; 35.59; 32.99; 28.20; 7.00. Anal. calc. for $C_{21}H_{24}O_3$ (324.41): C 77.74, H 7.46; found: C 77.51, H 7.40.

(1R,2S,4S,5S)-4-(6-Amino-9H-purin-9-yl)-2-hydroxybicyclo[3.1.0]hexan-1-methanol (**8**). Diethyl azodicarboxylate (DEAD; 0.63 ml, 3.98 mmol) was added dropwise to a soln. of Ph$_3$P (1.04 g, 3.98 mmol) in dry THF (8.5 ml) previously cooled in an ice-salt bath. After 20 min, this soln. was added dropwise to a soln. of **14a** (0.560 g, 1.73 mmol) and 6-chloro-1*H*-purine (0.431 g, 2.77 mmol) in dry THF (8.5 ml), which was also cooled in an ice-salt bath. The mixture was stirred cold for 0.5 h and then at r.t. for 3 h. The volatiles were removed and the residue purified by FC (silica gel, 25% AcOEt/hexanes): *6-chloro-9-[(1S,2S,4S,5R)-4-(phenylmethoxy)-5-[(phenylmethoxy)methyl]bicyclo[3.1.0]hex-2-yl]-9H-purine* (**19**; 0.741 g, 93%), slightly contaminated with DEAD. ^{1}H-NMR (CDCl$_3$): 9.20 (*s*, H–C(2)); 8.80 (*s*, H–C(8)); 7.50–7.30 (*m*, 2 Ph); 5.35 (*d*, *J* = 5.2, H–C(2′)); 4.75 (*t*, *J* = 8.5, H–C(4′)); 4.65 (br. *s*, PhCH$_2$O); 4.55 (*AB* (*m*), PhCH$_2$O); 4.30–4.50 (DEAD contamination); 4.25 (*d*, *J* = 9.9, 1 H, PhCH$_2$OCH$_2$); 3.25 (*d*, *J* = 9.9, 1 H, PhCH$_2$OCH$_2$); 2.15 (*m*, H$_a$–C(3′)); 1.95 (*m*, H$_b$–C(3′)); 1.80 (*m*, H–C(1′)); 1.45 (DEAD contamination); 1.25 (*m*, H$_a$–C(6′)); 0.95 (*m*, H$_b$–C(6′)).

Since the DEAD contamination was successfuly removed only after the final deprotection step, **19** (0.481 g, 1.04 mmol) was dissolved in a minimum amount of MeOH and added to a sat. NH$_3$/MeOH soln. (90 ml) in a sealed pressure steel bomb which was heated at 70° overnight. After cooling, the soln. was evaporated and co-evaporated twice with MeOH (100 ml) before purification by FC (silica gel, 2% MeOH/CHCl$_3$): *9-[(1S,2S,4S,5R)-4-(phenylmethoxy)-5-[(phenylmethoxy)methyl]bicyclo[3.1.0]hex-2-yl]-9H-purin-6-amine* (**20**; 0.347 g, 75%). Heavy syrup. ^{1}H-NMR (CDCl$_3$): 8.65 (*s*, H–C(2)); 8.40 (*s*, H–C(8)); 7.50–7.30 (*m*, 2 Ph); 6.75 (br. *s*, NH$_2$); 5.25 (*d*, *J* = 6.1, H–C(2′)); 4.75 (*t*, *J* = 8.2, H–C(4′)); 4.65 (br. *s*, PhCH$_2$O); 4.55 (*AB* (*m*), PhCH$_2$O); 4.15–4.35 (*m*, 1 H of PhCH$_2$OCH$_2$, DEAD contamination); 3.25 (*d*, *J* = 9.9, 1 H, PhCH$_2$OCH$_2$); 2.15 (*m*, H$_a$–C(3′)); 1.95 (*m*, H$_b$–C(3′)); 1.70 (*m*, H–C(1′)); 1.40 (DEAD contamination); 1.20 (*m*, H$_a$–C(6′)); 0.90 (*m*, H$_b$–C(6′)).

Compound **20** was used directly for the final deprotection step: A soln. of **20** (0.347 g, 0.787 mmol) in MeOH (96 ml) was treated with 96% formic acid (4 ml) and stirred overnight in the presence of Pd black (0.5 g). The insolubles were removed by filtration with the aid of *Celite*, and the filtrate was evaporated to give a white solid. To this solid, $CHCl_3$ (50 ml) was added and the mixture heated, cooled, and filtered. The remaining solid was recrystallized from $MeOH/Et_2O$: pure **8** (0.160 g, 78%). Solid. M.p. 259–261° (dec.), identical to that of an authentic sample.

REFERENCES

[1] R. Wolfenden, J. Kaufman, J. B. Macon, *Biochemistry* **1969**, *8*, 2412.

[2] B. E. Evans, R. Wolfenden, *Biochemistry* **1973**, *12*, 392.

[3] For a review of ADA deficiency, see M. S. Hershfield, B. S. Mitchell, in 'The Metabolic and Molecular Basis of Inherited Disease', Ed. C. R. Scriver, A. L. Beaudet, W. S. Sly, D. Valle, McGraw Hill, New York, 1995, pp. 1725–1768.

[4] J. K. Lowe, B. Gowans, L. Brox, *Cancer Res.* **1977**, *37*, 3013.

[5] J. M. Wilson, B. S. Mitchell, P. E. Daddona, W. N. Kelley, *J. Clin. Invest.* **1979**, *64*, 1475.

[6] B. Evans, R. Wolfenden, *J. Am. Chem. Soc.* **1970**, *92*, 4751.

[7] M. Ohno, N. Yagisawa, S. Shibahara, S. Kondo, K. Maeda, H. Umezawa, *J. Am. Chem. Soc.* **1974**, *96*, 4326.

[8] H. Nakamura, G. Koyama, Y. Iitaka, M. Ohno, N. Yagisawa, S. Kondo, K. Maeda, H. Umezawa, *J. Am. Chem. Soc.* **1974**, *96*, 4327.

[9] P. W. K. Wood, H. W. Dion, S. M. Lange, L. F. Dahl, L. Durham, *J. Heterocycl. Chem.* **1974**, *11*, 641.

[10] S. Cha, R. P. Agarwal, R. E. Parks Jr., *Biochem. Pharmacol.* **1975**, *24*, 2187.

[11] W. M. Kati, S. A. Acheson, R. Wolfenden, *Biochemistry* **1992**, *31*, 7356.

[12] D. K. Wilson, F. B. Rudolph, F. A. Quiocho, *Science (Washington, D.C.)* **1991**, *252*, 1278.

[13] A. J. Sharf, D. K. Wilson, Z. Chang, F. A. Quiocho, *J. Mol. Biol.* **1992**, *226*, 917.

[14] Z. Wang, F. A. Quiocho, *Biochemistry* **1998**, *37*, 8314.

[15] M. D. Erion, M. R. Reddy, *J. Am. Chem. Soc.* **1998**, *120*, 3295.

[16] C. Altona, M. Sundaralingam, *J. Am. Chem. Soc.* **1972**, *94*, 8205.

[17] For a comprehensive review of these concepts, see W. Saenger, 'Principles of Nucleic Acid Structure', Springer-Verlag, New York, 1984.

[18] C. Thibaudeau, A. Kumar, S. Bekiroglu, A. Matsuda, V. E. Marquez, J. Chattopadhyaya, *J. Org. Chem.* **1998**, *63*, 5447.

[19] A. A. Koshkin, S. K. Singh, P. Nielsen, V. K. Rajwanshi, R. Kumar, M. Meldgaard, C. E. Olsen, J. Wengel, *Tetrahedron* **1998**, *54*, 3607.

[20] M. A. Siddiqui, H. Jr. Ford, C. George, V. E. Marquez, *Nucleosides Nucleotides* **1996**, *15*, 235.

[21] A. Ezzitouni, V. E. Marquez, *J. Chem. Soc., Perkin Trans. 1* **1997**, 1073.

[22] V. E. Marquez, P. Russ, R. Alonso, M. A. Siddiqui, K.-J. Shin, C. George, M. C. Nicklaus, F. Dai, H. Ford Jr., *Nucleosides Nucleotides* **1999**, *18*, 521.

[23] K. Biggadike, A. D. Borthwick, A. M. Exall, B. E. Kirk, S. M. Roberts, P. Youds, A. M. Z. Slawin, D. J. Williams, *J. Chem. Soc., Chem. Commun.* **1987**, 255.

[24] K. Biggadike, A. D. Borthwick, D. Evans, A. M. Exall, B. E. Kirk, S. M. Roberts, L. Stephenson, P. Youds, *J. Chem. Soc., Perkin Trans 1* **1988**, 549.

[25] A. Ezzitouni, P. Russ, V. E. Marquez, *J. Org. Chem.* **1997**, *62*, 4870.

[26] S. Obika, N. Daishu, Y. Hari, K.-I. Morio, Y. In, T. Ishida, T. Imanishi, *Tetrahedron Lett.* **1997**, *38*, 8735.

[27] C. Thibaudeau, J. Plavec, J. Chattopadhyaya, *J. Org. Chem.* **1996**, *61*, 266.

[28] For a comprehensive review of these concepts, see C. Thibaudeau, J. Chattopadhyaya, 'Stereoelectronic Effects in Nucleosides and Nucleotides and their Structural Implications', Uppsala University Press, Uppsala, Sweden, 1999.

[29] S. D. Lindell, B. A. Moloney, B. D. Hewit, C. G. Earnshaw, P. J. Dudfeld, J. E. Dancer, *Bioorg. Med. Chem. Lett.* **1999**, *9*, 1985.

Studies on the Adamantylation of N-Heterocycles and Nucleosides

by Zygmunt Kazimierczuk*, Andrzej Orzeszko, and Agata Sikorska

Institute of Chemistry, Agricultural University, 26/30 Rakowiecka St., 02-528 Warsaw, Poland

Adamantylation of several N-heterocycles and two ribonucleosides (uridine and toyocomycin) was studied. The exact substitution position by the adamantyl carbocation generated from adamantan-1-ol in CF_3COOH depends on the nature of the heterocyclic substrate. Thus, adamantylation of an additional exocyclic amino group (see Scheme 1), N-adamantylation of the heterocycle (Scheme 2), C-adamantylation of the heterocycle (Scheme 3), as well as the formation of heterocyclic N-adamantylcarboxamides via the Ritter reaction (Scheme 4) are possible. The structures of the reaction products were determined by means of elemental analysis and NMR, UV, and IR spectroscopy.

Introduction. – Adamantane derivatives have received considerable attention because of their multifaceted biological activity. For instance, adamantan-1-amine (amantidine) has been found effective in the prophylaxis and treatment of influenza virus A infection [1]. It is also used in the treatment of *Parkinson*'s disease [2]. Many adamantane-moiety-containing compounds show distinct antimicrobial activity [3–5]. The unusual, hydrophobic cage-like structure of adamantane has been employed for modification of many potentially bioactive organic compounds, *e.g.*, nucleosides and polypeptides [6][7]. Substituting heterocyclic compounds with the adamantyl group requires more stringent conditions than those used for derivatizing their aliphatic analogs [8][9]. Recently, a simple, elegant, and promising method has been developed that enables C(5)-adamantylation of barbituric acid (1,2,3,4,5,6-hexahydropyrimidine-2,4,6-trione) as well as N-admantylation of carboxamides and ureas [10]. This method involves heating in CF_3COOH, which somewhat limits its synthetic potential because of the susceptibility of many naturally occurring compounds to the acidic medium. In the present paper, we would like to show some perspectives that have been opened by this reaction in the chemistry of heterocyclic compounds, including certain ribonucleosides.

Results and Discussion. – *Adamantylation of Exocyclic Amino Groups.* Substituted exocyclic amino groups are usually obtained from the corresponding halo, methylthio, and methoxy derivatives. *Via* adamantyl-cation formation from adamantan-1-ol in refluxing CF_3COOH solution, we were able to derivatize the exocyclic NH_2 group of substituted pyrimidin-2- and -4-amines, cytosine, and adenine to form compounds **1a–d**, **2**, and **3**, respectively (Scheme 1). Aniline and phenylalanine did not react under these conditions, which shows that this adamantylation protocol works only with non-protonated NH_2 groups. Heterocyclic bases are protonated at the ring heteroatoms; therefore, the acidic conditions do not hinder the trapping of the adamantyl carbocation by the exocyclic NH_2 group. Because of the relatively high solubility of the heterocyclic base in acidified aqueous medium, the use of an excess of the base allowed product purification without column chromatography. NMR Spectroscopy unequivocally indicated that the substitution took place at the exocyclic NH$_2$ group of pyrimidin-2-amine and cytosine. Determination of the structure of the adamantylation product from adenine was not so simple: the δ(H) of H$-$C(2) and H$-$C(8) of N^6-(adamantan-1yl)adenine (**3**) were separated only at 500 MHz. However, the position of the adamantyl group at the exocyclic amino group of **3** was confirmed by both NOE measurements and comparison with reference compounds obtained according to a previously described procedure [11].

Scheme 1

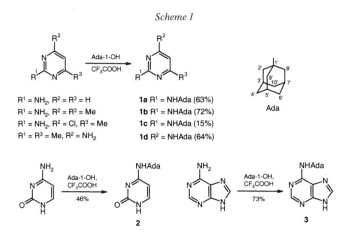

N-*Adamantylation of the Heterocycle.* Another reaction observed under the above-described conditions was adamantylation of the heterocyclic N-atom of certain compounds. This reaction was exemplified by adamantylation of phthalimide, 1,2-dihydropyridazine-3,6-dione, 6-azauracil, and guanine (*Scheme 2*). Phthalimide as well as 1,2-dihydropyridazine-3,6-dione underwent *N*-substitution under rather atypical conditions, giving **4** and **5**, respectively. Usually, *N*-alkylation of these compounds is achieved in an alkaline medium. The results of the present study indicate that the adamantyl cation may also attack the electron-deficient N-atom by replacing the acidic H-atom. The structure of the product was determined spectroscopically in the case of **5** and by comparison with the authentic reference compound obtained by a different method [12] in the case of **4**. Interestingly, adamantylation of 6-azauracil also yielded the *N*(1)-substituted derivative **6** as the sole product, *i.e.*, no *C*-substitution had occurred. The structure of **6** was determined by UV spectroscopy (both neutral and monoanionic species of *N*(1)- and *N*(3)-substituted 6-azauracils display quite different UV spectra [13]) and confirmed by NMR spectroscopy. In compound **7**, the location of the adamantyl substituent at N(9) of the guanine ring was determined based on comparison of NMR and UV spectra of various *N*-substituted guanine derivatives.

C-*Adamantylation of the Heterocycle.* Among the uracil derivatives, 1,3-dimethyl-uracil and uridine gave the respective 5-substituted derivatives **8** and **9** as the main products of the adamantylation (*Scheme 3*). The choice of uridine for nucleoside transformations was justified by the fact that it is the most stable nucleoside in acidic

Scheme 2

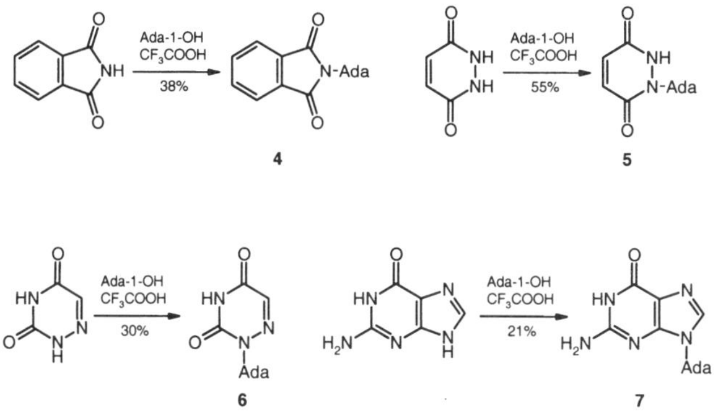

4

5

6

7

Scheme 3

R¹ = R² = Me   **8 (35%)**

R¹ = H,
R² = β-D-ribofuranosyl   **9 (34%)**

10

11

R¹ = H
R² = H

R¹ = H
R² = Me

12a (59%)

12b (63%)

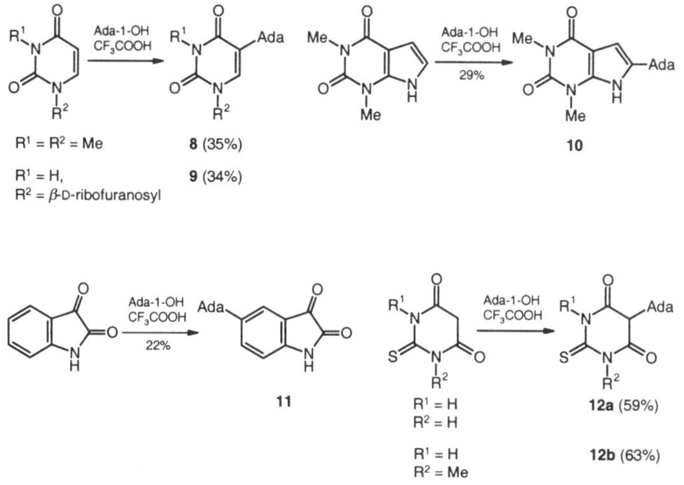

medium. However, prolonged heating of uridine in CF_3COOH gave a more complex mixture (TLC), which was most likely caused by a cleavage of the glycosylic bond and/or anomerization of the sugar moiety. A somewhat unexpected and unexplained result was obtained in the adamantyl-carbocation reaction with uracil. The ^{1}H-NMR spectrum of the product showed the presence of the H−C(6) signal (6.88 ppm) expected for 5-adamantyluracil [14], but also that of two adamantyl-type signal patterns (1.4–2.2 ppm; shifted by 0.5 ppm) instead of one, and of the additional *s* at 6.96 ppm. HPLC Analysis (*Lichrospher-RP-18* column, H_2O/MeCN 9 : 1) revealed that the product consisted of two compounds (t_R 26.79 and 27.25 min). We were unable to isolate individual components of the mixture and to determine the structure of the second product because of its slow solvolysis in MeOH and its chromatographic similarity to 5-adamantyluracil. No product was detected after refluxing thymine, 6-methyluracil, and 6-chlorouracil, respectively, for 2 h with adamantan-1-ol in CF_3COOH. Also, no indication of adamantylation was found in the cases of benzimidazole and 3-deazauridine.

The adamantyl carbocation formed as the intermediate of the reaction under study is a weak electrophile, which explains the 5-substitution of 1,3-dimethyluracil and uridine. In 1,3-dimethyl-7-deazaxanthine, the substitution occurred at C(8) yielding **10** (*Scheme 3*), which is consistent with electrophilic acylation of this ring system [15]. *C*-Substitution also took place with isatin (=1*H*-indole-2,3-dione), which underwent adamantylation at C(5) to give **11**, in analogy to its bromination reaction. In both cases, the substitution position was verified by NOE measurements. The use of the adamantylation reaction for the modification of 2-thiouracil resulted in a highly complex product mixture. By contrast, 2-thiobarbituric acid and 1-methyl-2-thiobarbituric acid gave single products **12a** and **12b**, respectively, under the same conditions. Such behavior has also been observed for barbituric acid itself [10]. However, no products were detected after refluxing 1,3-dimethyl-2-thiobarbituric acid for 24 h under the above-described reaction conditions. It seems that proton dissociation from the heterocyclic nitrogen is necessary for promotion of the reaction in the case of 2-thiobarbituric acids.

It is worth mentioning that adamantylation under the same conditions as described for adamantan-1-ol did not succeed with adamantan-2-ol. This was probably because of the difference in carbocation formation between the tertiary and the corresponding secondary alcohol.

Formation of N-*Adamantylated Carboxamides.* N-(Adamantan-1-yl)amides can be obtained by the *Ritter* reaction of adamantan-1-ol and its derivatives with nitriles [16][17], as well as by *N*-alkylation of amides with adamantan-1-yl halides [18]. In the following, the *Ritter* reaction is exemplified by the transformation of the heterocyclic carbonitriles 5-cyano-1-methyluracil and the antibiotic nucleoside toyocamycin into the substituted amides **13** and **14**, respectively (*Scheme 4*). The cyano group is more nucleophilic than other possible reaction centers, and the formation of diadamantylated derivatives of toyocamycin was negligible. However, TLC revealed the presence of other minor byproducts.

Scheme 4

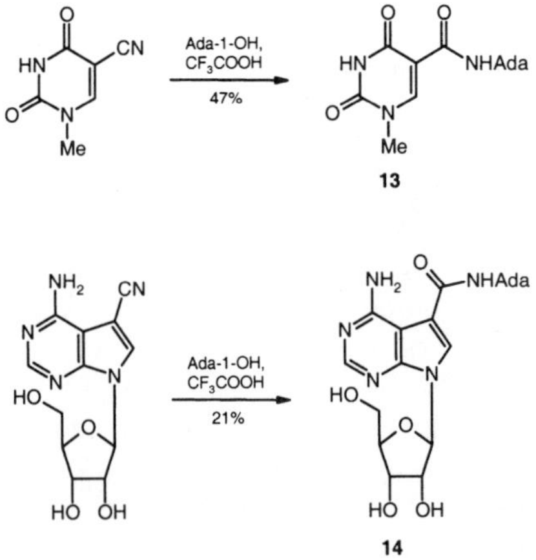

13

14

Conclusion. – The first step of the adamantylation reaction is probably the formation of the corresponding trifluoroacetate of adamantan-1-ol, whereafter a carbenium ion is generated. The rather harsh acidic conditions of this reaction present a major limitation for future use in the derivatization of certain biologically important compounds, *e.g.*, deoxynucleosides and polypeptides. However, the aforementioned examples do not exhaust the list of possible modifications of heterocyclic compounds. Importantly, the above-described reaction provides a facile route for the synthesis of modified bases to be used in nucleoside synthesis. The results of our studies reveal that predicting the actual substitution position at the heterocycle (*i.e.*, *N*- *vs.* *C*-substitution, see *Schemes 2* and *3*) is not an easy task, except in the case of amino- and cyano-substituted *N*-heterocycles (*Schemes 1* and *4*).

This study was supported in part by the *Foundation for the Development of Diagnostics and Therapy*, Warsaw, Poland. The authors are grateful to Drs. *S. Chrapusta* and *H. Rosemeyer* for helpful discussions.

Experimental Part

General. All chemicals, except solvents, were purchased from *Sigma-Aldrich* (Steinheim, Germany). Column flash chromatography (FC): silica gel *60 H* (*Merck*, Germany). Anal. TLC: pre-coated silica gel *60F$_{254}$* (*Merck*, Germany). M.p.: open capillary tubes, *Gallenkamp-5* melting-point apparatus (Leicester, U.K.), uncorrected. UV Spectra: *Kontron Uvikon 940* (Vienna, Austria) spectrometer. FT-IR Spectra (in cm^{-1}): *Perkin-Elmer 2000* apparatus (Beaconsfield, U.K.). NMR Spectra (in ppm): *Varian Gemini 200* MHz and *Varian UNITYplus 500* MHz spectrometers (Palo Alto, CA) in CDCl$_3$ and (D$_6$)DMSO solns. NOE Experiments by the truncated-driven NOE (TOE) method [19].

N-(Adamantan-1-yl)pyrimidin-2-amine (**1a**). A soln. of pyrimidin-2-amine (1.15 g, 11 mmol) and adamantan-1-ol (1.52 g, 10 mmol) in CF$_3$COOH (10 ml) was stirred under reflux for 5 h. The mixture was poured into H$_2$O (50 ml) and brought to pH 7 with conc. aq. NH$_3$ soln. The precipitate formed was filtered and crystallized from EtOH/H$_2$O 1:1: **1** (1.45 g, 63%). Colorless crystals. M.p. 201–203°. TLC (CH$_2$Cl$_2$/MeOH 9:1): R_f 0.78. UV (MeOH/H$_2$O 1:1): 241 (17200), 311 (2500). ^{1}H-NMR ((D$_6$)DMSO): 1.62 (br. *s*, 6 H, Ada); 2.03 (br. *s*, 6 H, Ada); 2.37 (br. *s*, 3 H, Ada); 6.49 (*t*, H−C(5)); 8.80 (*d*, H−C(4), H−C(6)). Anal. calc. for C$_{14}$H$_{19}$N$_3$ (229.33): C 73.33, H 8.35, N 18.32; found: C 73.24, H 8.38, N 18.25.

N-(Adamantan-1-yl)-4,6-dimethylpyrimidin-2-amine (**1b**). As described for **1a**, with 4,6-dimethylpyrimidin-2-amine (1.35 g, 11 mmol), adamantan-1-ol (1.52 g, 10 mmol), and CF$_3$COOH (10 ml). Crystallization from EtOH/H$_2$O 1:1: **1b** (1.85 g, 72%). Colorless crystals. M.p. 164–166°. TLC (CHCl$_3$): R_f 0.40. UV (MeOH/H$_2$O 1:1): 242 (15600), 303 (2900). ^{1}H-NMR (CDCl$_3$): 1.69 (br. *s*, 6 H, Ada); 2.11 (br. *s*, 9 H, Ada); 2.43 (*s*, 6 H, 2 Me); 6.33 (*s*, H−C(5)). Anal. calc. for C$_{16}$H$_{23}$N$_3$ (257.38): C 74.66, H 9.01, N 16.33; found: C 74.70, H 9.08, N 16.22.

N-(Adamantan-1-yl)-4-chloro-6-methylpyrimidin-2-amine (**1c**). As described for **1a**, with 4-chloro-6-methylpyrimidin-2-amine (1.57 g, 11 mmol), adamantan-1-ol (1.52 g, 10 mmol), and CF$_3$COOH (8 ml). The precipitate was adsorbed onto silica gel. FC (silica gel, column 3 × 15 cm, CH$_2$Cl$_2$). Crystallization from EtOH/H$_2$O 1:1: **1c** (415 mg, 15%). Colorless crystals. M.p. 124–126°. TLC (CHCl$_3$): R_f 0.60. UV (MeOH/H$_2$O 1:1): 245 (16200), 304 (3400). ^{1}H-NMR (CDCl$_3$): 1.70 (br. *s*, 6 H, Ada); 2.11 (br. *s*, 9 H, Ada); 2.36 (*s*, 3 H, Me); 6.42 (*s*, H−C(5)). Anal. calc. for C$_{15}$H$_{20}$N$_3$Cl (277.80): C 64.85, H 7.26, N 15.13; found: C 64.85, H 7.31, N 15.23.

N-(Adamantan-1-yl)-2,6-dimethylpyrimidin-4-amine (**1d**). As described for **1a**, with 2,6-dimethylpyrimidin-4-amine (1.35 g, 11 mmol), adamantan-1-ol (1.52 g, 10 mmol), and CF$_3$COOH (10 ml). Crystallization from EtOH/H$_2$O 1:1: **1a** (1.65 g, 64%). Colorless crystals. M.p. 200–201°. TLC (CH$_2$Cl$_2$/MeOH, 9:1): R_f 0.56. UV (MeOH/H$_2$O 1:1): 279 (9800). ^{1}H-NMR (CDCl$_3$): 1.70 (br. *s*, 6 H, Ada); 2.10 (br. *s*, 9 H, Ada); 2.37, 2.57 (2*s*, 6 H, 2 Me); 6.35 (*s*, H−C(5)). Anal. calc. for C$_{16}$H$_{23}$N$_3$ (257.38): C 74.66, H 9.01, N 16.33; found: C 74.77, H 9.06, N 16.28.

N^4-(Adamantan-1-yl)cytosine (**2**). As described for **1a**, with cytosine (330 mg, 2.5 mmol), adamantan-1-ol (610 mg, 4 mmol), and CF$_3$COOH (3.5 ml). The precipitate was washed with a small amount of petroleum ether and crystallized from EtOH: **2** (280 mg, 46%). Colorless crystals. M.p. >320° (dec.). TLC (CH$_2$Cl$_2$/MeOH 8:2): R_f 0.85. UV (MeOH/H$_2$O 2:8): at pH 1: 285 (10900); at pH 7: 268 (7700); at pH 13: 281 (6500). ^{1}H-NMR

((D$_6$)DMSO): 1.61 (br. *s*, 6 H, Ada); 2.03 (br. *s*, 6 H, Ada); 2.48 (overlapped signal, DMSO); 5.57 (*d*, H–C(5)); 7.17 (*d*, H–C(6)). Anal. calc. for C$_{14}$H$_{19}$N$_3$O (245.33): C 68.54, H 7.81, N 17.13; found: C 68.66, H 7.90, N 17.07.

N^6-*(Adamantan-1-yl)adenine* (**3**). A suspension of adenine (1.62 g, 12 mmol) and adamantan-1-ol (1.52 g, 10 mmol) in CF$_3$COOH (15 ml) was stirred under reflux for 8 h. The mixture was poured into water (100 ml) and brought to pH 7 with conc. aq. NH$_3$ soln. The precipitate was filtered off, washed with H$_2$O, and crystallized from EtOH: **3** (1.68 g, 73%), spectrally and chromatographically identical to a sample obtained according to [11]. Colorless powder. M.p. 348–350°. TLC (CH$_2$Cl$_2$/MeOH 8 : 2): R$_f$ 0.90. UV (MeOH/H$_2$O 2 : 8): at pH 1: 261 (13700); at pH 7: 262 (14100); at pH 13: 262 (13500). ^{1}H-NMR ((D$_6$)DMSO): 1.75 (br. *s*, 6 H, Ada); 2.19 (br. *s*, 6 H, Ada); 2.40 (br. *s*, 3 H, Ada); 7.65 (br. *s*, NH–C(6)); 8.19, 8.21 (2*s*, H–C(2), H–C(8)). NOE: irrad. at 8.20 (H–C(2), H–C(8)) → no NOE); irrad. at 1.75 (CH$_2$(2′), CH$_2$(8′), CH$_2$(9′)) → 4% (H–C(6)); 7% (H–C(3′), H–C(5′), H–C(7′)); 7% (CH$_2$(4′), CH$_2$(6′), CH$_2$(10′)). ^{13}C-NMR ((D$_6$)DMSO): 174.0 (C(6)); 150.1 (C(4)); 149.2 (C(2)); 138.8 (C(8)); 119.9 (C(2)); 57.5, 40.7, 35.5, 28.9 (C(Ada)).

N-*(Adamantan-1-yl)phthalimide* (**4**). A soln. of phthalimide (900 mg, 6 mmol) and adamantan-1-ol (910 mg, 6 mmol) in CF$_3$COOH (6 ml) was stirred under reflux for 12 h. The mixture was poured into H$_2$O/ EtOH 8 : 2 (60 ml), the precipitate filtered off, adsorbed onto silica gel, and chromatographed (silica gel, column 3 × 15 cm, hexane/CHCl$_3$ 10 : 2). Crystallization from MeOH yielded **4** (615 mg, 38%). Colorless plates. TLC (hexane/CHCl$_3$ 5 : 2): R$_f$ 0.59. M.p. 138–139° ([12]: 140°). UV (MeOH/H$_2$O 1 : 1): 224 (28600), 297.5 (1900). ^{1}H-NMR (CDCl$_3$): 175 (br. *s*, 6 H, Ada); 2.16 (br. *s*, 6 H, Ada); 2.51 (br. *s*, 3 H, Ada), 7.63–7.78 (*m*, 4 arom. H).

1-(Adamantan-1-yl)-1,2-dihydropyridazine-3,6-dione (**5**). A soln. of pyridazine-3,6-dione (335 mg, 3 mmol) and adamantan-1-ol (910 mg, 6 mmol) in CF$_3$COOH (6 ml) was stirred under reflux for 4 h. The mixture was evaporated and the residue adsorbed onto silica gel and chromatographed (silica gel, column 3 × 12 cm, CHCl$_3$ (200 ml), then CHCl$_3$/MeOH 9 : 1). The product was crystallized from MeOH: **5** (410 mg, 55%). Colorless plates. M.p. 282–284°. TLC (CH$_2$Cl$_2$/MeOH 9 : 1): R$_f$ 0.48. UV (MeOH/H$_2$O 2 : 8): at pH 7: 316 (3200); at pH 12: 225 (7400), 338 (2900). ^{1}H-NMR ((D$_6$)DMSO): 1.65 (br. *s*, 6 H, Ada); 2.09 (br. *s*, 6 H, Ada); 2.26 (br. *s*, 3 H, Ada); 6.73, 6.95 (2*d*, H–C(3), H–C(4)). Anal. calc. for C$_{14}$H$_{18}$N$_2$O$_2$ (246.31): C 68.27, H 7.37, N 11.37; found: C 68.15, H 7.26, N 11.40.

1-(Adamantan-1-yl)-6-azauracil (= *2-(Adamantan-1-yl)-1,2,4-triazine-3,5(2H,4H)-dione*; **6**). As described for **5**, with 6-azauracil (340 mg, 3 mmol), adamantan-1-ol (910 mg, 6 mmol), and CF$_3$COOH (4 ml) (6 h). FC (silica gel, column 3 × 10, CHCl$_3$, then CHCl$_3$/MeOH 8 : 2) and crystallization from MeOH/H$_2$O gave **6** (220 mg, 30%). Colorless plates. M.p. 268–269°. TLC (CH$_2$Cl$_2$/MeOH 9 : 1): R$_f$ 0.80. UV (MeOH/H$_2$O 2 : 8): at pH 7: 275 (7200); at pH 12: 267 (6400). ^{1}H-NMR ((D$_6$)DMSO): 1.64 (br. *s*, 6 H, Ada); 2.11 (br. *s*, 6 H, Ada); 2.18 (br. *s*, 3 H, Ada); 7.39 (*s*, H–C(5)). Anal. calc. for C$_{13}$H$_{17}$N$_3$O$_2$ (247.30): C 63.14, H 6.93, N 16.99; found: C 63.02, H 7.01, N 17.10.

9-(Adamantan-1-yl)guanine (**7**). A soln. of guanine (1.51 g, 10 mmol) and adamantan-1-ol (1.22 g, 8 mmol) in CF$_3$COOH (15 ml) was stirred under reflux for 24 h. The mixture was poured into H$_2$O (150 ml) and brought to pH 8 with conc. aq. NH$_3$ soln. The precipitate was filtered and stirred at 80° in H$_2$O (200 ml). Then, the precipitate was separated and crystallized from DMF to give **7** (0.48 g, 21%). White powder. M.p. >340°. TLC (CH$_2$Cl$_2$/MeOH 9 : 1): R$_f$ 0.37. UV (MeOH/H$_2$O 4 : 1): at pH 7: 251 (13700), 276 (sh, 8400); at pH 1: 254 (12200), 271 (7800); at pH 13: 258–270 (9800). ^{1}H-NMR ((D$_6$)DMSO): 1.76 (br. *s*, 6 H, Ada); 2.15 (br. *s*, 6 H, Ada); 2.31 (br. *s*, 3 H, Ada); 6.28 (br. *s*, 2 H, NH$_2$); 7.67 (*s*, 1 H, H–C(8)); 10.20 (br. *s*, NH). ^{13}C-NMR ((D$_6$)DMSO): 157.0 (C(6)); 152.0, 151.2 (C(2), C(4)); 134.5 (C(8)); 118.2 (C(5)); 56.7, 40.5, 35.5, 28.9 (C(Ada)). Anal. calc. for C$_{15}$H$_{19}$N$_5$O (285.4): C 63.14, H 6.71, N 25.54; found: C 63.00, H 6.81, N 24.41.

5-(Adamantan-1-yl)-1,3-dimethyluracil (**8**). As described for **5**, with 1,3-dimethyluracil (285 mg, 2 mmol) and adamantan-1-ol (610 mg, 4 mmol), and CF$_3$COOH (4 ml) (3 h). FC (silica gel, column 3 × 10, CHCl$_3$, then CHCl$_3$/MeOH 9 : 1) and crystallization from MeOH/H$_2$O gave **8** (190 mg, 35%). Colorless plates. M.p. 208–210°. TLC (CH$_2$Cl$_2$/MeOH 9 : 1): R$_f$ 0.80. UV (MeOH/H$_2$O 1 : 1): 270 (8800). ^{1}H-NMR (CDCl$_3$): 1.74 (br. *s*, 6 H, Ada); 1.95 (br. *s*, 6 H, Ada); 2.05 (br. *s*, 3 H, Ada); 3.32, 3.38 (2*s*, 2 Me); 6.82 (*s*, H–C(6)). Anal. calc. for C$_{16}$H$_{22}$N$_2$O$_2$ (274.31): C 70.04, H 8.08, N 10.21; found: C 70.16, H 8.08, N 10.11.

5-(Adamantan-1-yl)uridine (**9**). A soln. of uridine (970 mg, 4 mmol) and adamantan-1-ol (1.21 g, 8 mmol) in CF$_3$COOH (6 ml) was stirred under reflux for 2 h. The mixture was poured into ice water (50 ml). The precipitate formed was filtered and chromatographed (silica gel, column 4 × 15 cm, CHCl$_3$ (250 ml), then CHCl$_3$/MeOH 9 : 1 (250 ml) and CHCl$_3$/MeOH 8 : 2). The product was crystallized from EtOH: **9** (510 mg, 34%). Colorless crystals. M.p. 178–180°. TLC (CH$_2$Cl$_2$/MeOH 9 : 1): R$_f$ 0.37. UV: at pH 7: 267 (9100); at pH 12: 265 (7100). ^{1}H-NMR ((D$_6$)DMSO): 1.65 (br. *s*, 6 H, Ada); 1.86 (br. *s*, 6 H, Ada); 1.97 (br. *s*, 3 H, Ada); 3.57 (*m*, 2 H–C(5)); 3.7–4.1 (*m*, H–C(2′), H–C(3′), H–C(4′)); 5.81 (*d*, H–C(1′)); 7.53 (*s*, H–C(6)); 11.10 (*s*, NH). ^{13}C-NMR ((D$_6$)DMSO): 162.4 (C(4)); 150.4 (C(2)); 135.5 (C(6)); 121.6 (C(5)); 87.7 (C(1′)); 84.9

(C(4′)); 73.7 (C(2′)); 70.4 (C(3′)); 60.9 (C(5′)); 40.8, 39.7, 34.8, 28.0 (C(Ada)). Anal. calc. for $C_{19}H_{26}N_2O_6$ (378.43): C 60.30, H 6.93, N 7.40; found: C 60.17, H 7.01, N 7.29.

8-(Adamantan-1-yl)-1,3-dimethyl-7-deazaxanthine (=6-*(Adamantan-1-yl)-3,9-dihydro-1,3-dimethyl-1H-pyrrolo[2,3-*d*]pyrimidine-2,4-dione*; **10**). A suspension of 1,3-dimethyl-7-deazaxanthine [20] (560 mg, 3 mmol) and adamantan-1-ol (910 mg, 6 mmol) in CF_3COOH (5 ml) was stirred under reflux for 5 h. The deep-red mixture was evaporated and adsorbed onto silica gel. FC (silica gel, column 3 × 15 cm, $CHCl_3$ (200 ml), then $CHCl_3$/MeOH 95 : 5) and crystallization from a small amount of acetone gave **10** (180 mg, 29%). Colorless crystals. M.p. > 320° (dec. > 260°). TLC (CH_2Cl_2/MeOH 9 : 1): R_f 0.51. UV (MeOH/H_2O 1 : 1): 249 (8400), 283 (6700). ¹H-NMR ((D_6)DMSO): 1.71 (br. *s*, 6 H, Ada); 1.87 (br. *s*, 6 H, Ada); 2.01 (br. *s*, 3 H, Ada); 3.18, 3.46 (2*s*, 2 Me); 5.93 (*s*, H−C(7)); 10.93 (*s*, NH). NOE: irrad. NH → 1% (3.46, Me−N(3)), 2% (1.71, CH_2(2′), CH_2(8′), CH_2(9′)); irrad. Me−N(1) → no NOE; irrad. H−C(7) → 1% (10.93, NH), 3% (1.71, CH_2(2′), CH_2(8′), CH_2(9′); irrad. CH_2(2′), CH_2(8′), CH_2(9′) → 10% (10.93, NH), 6% (5.93, H−C(7)), 7% (1.87, CH_2(4′), CH_2(6′), CH_2(10′), 7% (2.01, H−C(3′), H−C(5′), H−C(7′)). Anal. calc. for $C_{18}H_{23}N_3O_2$ (313.40): C 68.98, H 7.40, N 13.41; found: 69.00, H 7.51, N 13.54.

5-(Adamantan-1-yl)isatin (=5-*(Adamantan-1-yl)-1H-indole-2,3-dione*; **11**). A soln. of isatin (1.47 g, 10 mmol) and adamantan-1-ol (1.21 g, 8 mmol) in CF_3COOH (10 ml) was stirred and refluxed for 6 h. The deep-red mixture was brought to r.t. and diluted with EtOH/H_2O 2 : 8 (30 ml). The red precipitate was filtered off and chromatographed (silica gel, column 4 × 15 cm, $CHCl_3$ (250 ml), then $CHCl_3$/MeOH 95 : 5). The product-containing fractions were evaporated and the residue crystallized from EtOH/H_2O: **11** (620 mg, 22%). Red plates. M.p. 273−275°. TLC (CH_2Cl_2/MeOH 9 : 1): R_f 0.55. UV (MeOH/H_2O 1 : 1): 246 (22900), 305 (3900). ¹H-NMR ($CDCl_3$): 1.74 (br. *s*, 3 H, Ada); 1.79 (br. *s*, 3 H, Ada); 1.86 (br. *s*, 6 H, Ada); 2.11 (br. *s*, 3 H, Ada); 6.92 (*d*, H−C(7)); 7.57 (*m*, H−C(6)); 7.64 (*d*, H−C(4)); 8.83 (*s*, NH). NOE: irrad. NH → 2% (6.92, H−C(7)), 2% (1.79, CH_2(2′), CH_2(8′), CH_2(9′)); irrad. H−C(4) and H−C(6) → 2% (6.92, H−C(7)), 3% (1.86, CH_2(4′), CH_2(6′), CH_2(10′)). Anal. calc. for $C_{18}N_{19}NO_2$ (281.36): C 76.84, H 6.81, N 4.98; found: C 76.75, H 6.69, N 4.97.

5-(Adamantan-1-yl)-2-thiobarbituric Acid (=5-*(Adamantan-1-yl)-1,2,3,4,5,6-hexahydro-2-thioxopyrimidine-2,4,6-(1H,3H,5H)-trione*; **12a**). A suspension of 2-thiobarbituric acid (1.73 g, 12 mmol) and adamantan-1-ol (1.52 g, 10 mmol) in CF_3COOH (15 ml) was stirred under reflux. After 3 h, the suspension was entirely dissolved, and 2 h later, the mixture was poured into H_2O (30 ml). The precipitate formed was filtered and washed with acetone/H_2O 1 : 1 (20 ml). Crystallization from EtOH yielded **12a** (1.63 g, 59%). Pale-yellow plates. M.p. 270°. TLC (CH_2Cl_2/MeOH 9 : 1): R_f 0.70. UV (MeOH/H_2O 1 : 9): at pH 1: 244 (11500), 291 (30400); at pH 7: 239 (15900), 292 (14800). ¹H-NMR ((D_6)DMSO): 1.59 (br. *s*, 12 H, Ada); 1.95 (br. *s*, 3 H, Ada); 2.76 (*s*, H−C(5)); 12.18 (*s*, 2 NH). ¹³C-NMR ((D_6)DMSO): 180.1 (C(2)); 166.9 (C(4), C(6)); 61.8 (C(5)), 28.3, 35.9, 39.0, 39.8 (C(Ada)). Anal. calc. for $C_{14}H_{18}N_2O_2S$ (278.32): C 60.42, H 6.52, N 10.07; found: C 60.45, H 6.63, N 9.97.

5-(Adamantan-1-yl)-1-methyl-2-thiobarbituric Acid (=5-*(Adamantan-1-yl)-1,2,3,4,5,6-hexahydro-1-methyl-2-thioxopyrimidine-2,4,6-trione*; **12b**). As described for **12a**, with 1-methyl-2-thiobarbituric acid (1.58 g, 10 mmol) and adamantan-1-ol (1.22 g, 8 mmol) in CF_3COOH (12 ml). Crystallization from EtOH/H_2O gave **12b** (1.84 g, 63%). Pale-yellow crystals. M.p. 197−198°. TLC (CH_2Cl_2/MeOH 95 : 5): R_f 0.81. UV (MeOH/H_2O 1 : 9): at pH 1: 254 (10800), 290 (285.00), at pH 7: 239 (15900), 292 (14800). ¹H-NMR ((D_6)DMSO): 1.58 (br. *s*, 12 H, Ada); 1.96 (br. *s*, 3 H, Ada); 2.96 (*s*, H−C(5)); 3.45 (*s*, Me); 12.10 (*s*, 1 NH). ¹³C-NMR ((D_6)DMSO): 181.3 (C(2)); 166.7, 165.4 (C(4), C(6)); 62.3 (C(5)); 32.7 C(Me); 28.2, 35.7, 39.1, 39.9 (C(Ada)). Anal. calc. for $C_{15}H_{20}N_2O_2S$ (292.40): C 61.61, H 6.89, N 9.58; found: C 61.49, H 6.83, N 9.67.

5-[[N-(Adamantan-1-yl)amino]carbonyl]-1-methyluracil (=N-*(Adamantan-1-yl)-1,2,3,4-tetrahydro-1-methyl-2,4-dioxopyrimidine-5-carboxamide*; **13**). As described for **5**, with 5-cyano-1-methyluracil [21] (305 mg, 2 mmol), adamantan-1-ol (330 mg, 2.2 mmol), and CF_3COOH (4 ml) (30 min). FC (silica gel, column 2 × 10 cm, $CHCl_3$ (100 ml), then $CHCl_3$/MeOH 95 : 5) and crystallization from EtOH gave **13** (285 mg, 47%). Colorless crystals. M.p. > 320° (dec.). TLC (CH_2Cl_2/MeOH 9 : 1): R_f 0.65. UV (MeOH/H_2O 1 : 9): at pH 7: 225 (9900), 280 (12700): at pH 12: 230 (12000), 279 (10600). FT-IR: 1697 (CONH). ¹H-NMR ((D_6)DMSO): 1.61 (br. *s*, 6 H, Ada); 1.95 (br. *s*, 6 H, Ada); 2.01 (br. *s*, 3 H, Ada); 3.32 (*s*, Me); 8.34 (*s*, H−C(6)); 8.16 (br. *s*, NH). Anal. calc. for $C_{14}H_{18}N_2O_2$ (303.36): C 63.35, H 6.98, N 13.85; found: C 63.40, H 7.01, N 13.74.

N-*(Adamantan-1-yl)-4-amino-7-(β-*D*-ribofuranosyl)-7H-pyrrolo[2,3-*d*]pyrimidine-5-carboxamide* (=N^{amido}-*(Adamantan-1-yl)sangivamycin*; **14**). As described for **5**, with toyocomycin (=4-amino-7-(β-*D*-ribofuranosyl)-7H-pyrrolo[2,3-*d*]pyrimidine-5-carbonitrile; 760 mg, 4 mmol), adamantan-1-ol (1.22 g, 8 mmol), and CF_3COOH (4.5 ml) (1 h). FC (silica gel, column 4 × 15 cm, $CHCl_3$ (300 ml), then $CHCl_3$/MeOH 9 : 1) and crystallization from EtOH/AcOEt gave **14** (380 mg, 21%). Colorless, wool-like crystals. M.p. 248°. TLC

(CH$_2$Cl$_2$/MeOH 9:1): R_f 0.38. UV (MeOH/H$_2$O 1:9): at pH 7: 282 (11800); at pH 1: 237 (13200), 281 (10400). ^{1}H-NMR ((D$_6$)DMSO): 1.66 (br. s, 6 H, Ada); 2.08 (br. s, 6 H, Ada); 2.10 (br. s, 3 H, Ada); 3.57 (m, 2 H−C(5′)); 3.94 (q, H−C(4′)); 4.11 (m, H−C(3′)); 4.35 (t, H−C(2′)); 6.10 (d, H−C(1′)); 7.83 (br. s, NH$_2$); 8.40, 8.45 (2s, H−C(2), H−C(6)). ^{13}C-NMR ((D$_6$)DMSO): 163.1 (C=O); 153.0 (C(4)); 148.7 (C(2)); 145.9 (C(7a)); 126.7 (C(6)); 113.9 (C(5)); 100.9 (C(4a)); 86.7 (C(1′)); 85.5 (C(4′)); 73.8 (C(2′)); 70.1 (C(3′)); 61.4 (C(5′)); 28.7, 39.0, 40.8, 52.3 (C(Ada)). FT-IR: 1692 (CONH). Anal. calc. for C$_{22}$H$_{29}$N$_5$O$_5$ (443.50): C 59.58, H 6.59, N 15.79; found: C 59.47, H 6.63, N 15.68.

REFERENCES

[1] W. L. Davies, R. R. Grunert, R. F. Haff, J. W. McGahen, E. M. Neumayer, M. Paulshock, J. C. Watts, T. R. Wood, E. C. Hermann, C. E. Hoffmann, *Science (Washington, DC)* **1964**, *144*, 862.

[2] R. S. Schwab, A. C. England, D. C. Poskanzer, R. R. Young, *J. Am. Med. Assoc.* **1969**, *208*, 1168.

[3] A. Papadaki-Valiraki, S. Papakonstantinou-Garoufalias, P. Makaros, A. Chytyroglou-Lada, M. Hosoya, J. Balzarini, E. de Clercq, *Farmaco* **1993**, *48*, 1091.

[4] S. Garoufalias, A. Vyzas, G. Fytas, G. B. Foscolos, A. Chytiroglou, *Ann. pharm. franc.* **1988**, *46*, 97.

[5] D. Plachta, B. Starosciak, *Acta Polon. Pharm.* **1994**, *51*, 51.

[6] D. T. Gish, R. C. Kelly, G. W. Camiener, W. J. Wechter, *J. Med. Chem.* **1971**, *14*, 1159.

[7] R. L. Elliott, H. Kopecka, M. J. Bennett, Y.-K. Shue, R. Craig, C.-W. Lin, B. R. Bianchi, T. R. Miller, D. G. Witte, M. A. Stashko, K. E. Asin, A. Nikkel, L. Bednarz, A. M. Nadzan, *J. Med. Chem.* **1994**, *37*, 309.

[8] T. Sasaki, A. Usuki, M. Ohna, *J. Org. Chem.* **1980**, *45*, 3559.

[9] M. E. Gonzalez, B. Alarcon, P. Cabildo, R. M. Claramunt, D. Sanz, J. Elguero, *Eur. J. Med. Chem.* **1985**, *20*, 359.

[10] E. Shokova, T. Mousoulou, Y. Luzikov, V. Kovalev, *Synthesis* **1997**, 1034.

[11] V. M. Cherkasov, G. S. Tret'yakova, L. K. Kirilenko, V. N. Zavatskij, *Ukr. Khim. Zh.* **1967**, *33*, 1316; *Chem. Abstr.* **1968**, *69*, 52102d.

[12] Y. Shibata, M. Shichita, K. Sasaki, K. Nishimura, Y. Hashimoto, S. Iwasaki, *Chem. Pharm. Bull.* **1995**, *43*, 177.

[13] J. Jonas, J. Gut, *Collect. Czech. Chem. Commun.* **1961**, *26*, 1680.

[14] I. Basnak, A. Balkan, P. L. Coe, R. T. Walker, *Nucleosides Nucleotides* **1994**, *13*, 177.

[15] Z. S. Sahir, P. Stegmueller, W. Pfleiderer, *J. Heterocycl. Chem.* **1988**, *25*, 1443.

[16] H. Stetter, J. Mayer, M. Schwarz, C. Wulff, *Chem. Ber.* **1960**, *93*, 1366.

[17] F. Sztaricskai, J. Pelyvas, Z. Dinya, L. Szilagyi, Z. Gyorgydeak, G. Hadhazy, L. Vaczi, R. Bognar, *Pharmazie* **1975**, *30*, 571.

[18] K. Gerson, D. J. Tobias, R. E. Holmes, R. E. Rathbun, R. W. Kattau, *J. Med. Chem.* **1967**, *10*, 603.

[19] G. Wagner, K. Wüthrich, *J. Magn. Reson.* **1979**, *33*, 675.

[20] F. Seela, U. Kretschmer, *J. Heterocycl. Chem.* **1990**, *27*, 479.

[21] G. Shaw, *J. Chem. Soc.* **1955**, 1834.

Six-Membered Carbocyclic Nucleosides with a Purine Base Moiety: Synthesis, Conformational Analysis, and Antiviral Activity

by **Jing Wang** and **Piet Herdewijn***[1])

Rega Institute for Medical Research, K. U. Leuven, Minderbroedersstraat 10, B-3000 Leuven, Belgium

The synthesis, conformational analysis, and antiviral activity of a series of six-membered carbocyclic nucleosides are reviewed. The most interesting compounds are the D- and L-cyclohexene nucleosides **21b** and **22b**, which are both active against a broad range of viruses. It is hypothesized that the reason for this activity is their close structural resemblance with natural furanose nucleosides.

1. Introduction. – Carbocyclic nucleosides are a class of nucleoside derivatives in which the ring O-atom is replaced by a methylene group. This substitution leads to the loss of the anomeric center and hence to a better chemical and enzymatic stability, as compared to the natural nucleosides. These advantages have resulted in continuous interest in the research on carbocyclic nucleosides. A number of carbocyclic nucleosides showing antiviral activity have been described [1][2]. The most striking examples in the series of five-membered carbocyclic nucleosides are carbovir (**1**) [1a][1b], aristeromycin (**2**) [1c–e], neplanocin A (**3**) [1f][1g] and carba-2′-deoxyguanosine (**4**) [1h][1i] (*Fig. 1*). Likewise, three- and four-membered carbocyclic nucleosides have been studied, *e.g.*, **5** and **6** [1j][1k]. Several excellent reviews about their chemical syntheses and biological activities have been published [2].

In contrast to the efforts in the field of five-membered carbocyclic nucleosides, little work has been done on six-membered carbocyclic nucleosides. The main reason for this is that the six-membered ring is much more rigid, and energy barriers between different conformers are higher. Therefore, these compounds were not expected to demonstrate antiviral activity. However, the interest in the chemistry of six-membered carbocyclic nucleosides has increased considerably in recent years after the discovery that the anhydrohexitol nucleosides **7** (*Fig. 2*) demonstrate antiviral activity and that hexitol nucleic acids show antisense activity [3]. Since then, several six-membered carbocyclic nucleosides have been synthesized by us and by other groups (*Fig. 2*) [4]. Unfortunately, none of them showed any biological activity. Our most important efforts in six-membered carbocyclic nucleosides started with the synthesis of the cyclohexane nucleosides **19** (*Fig. 3*), the carbocyclic counterparts of the hexitol nucleosides **7**.

This review summarizes our synthetic work [5–7] on the six-membered carbocyclic nucleosides **19**, **20**, **21**, and **22** (*Fig. 3*). Their conformational analysis and biological activity is also discussed.

---

[1]) Tel. +32-16-337387, Fax. +32-16-337340, e-mail: Piet.Herdewijn@rega.kuleuven.ac.be

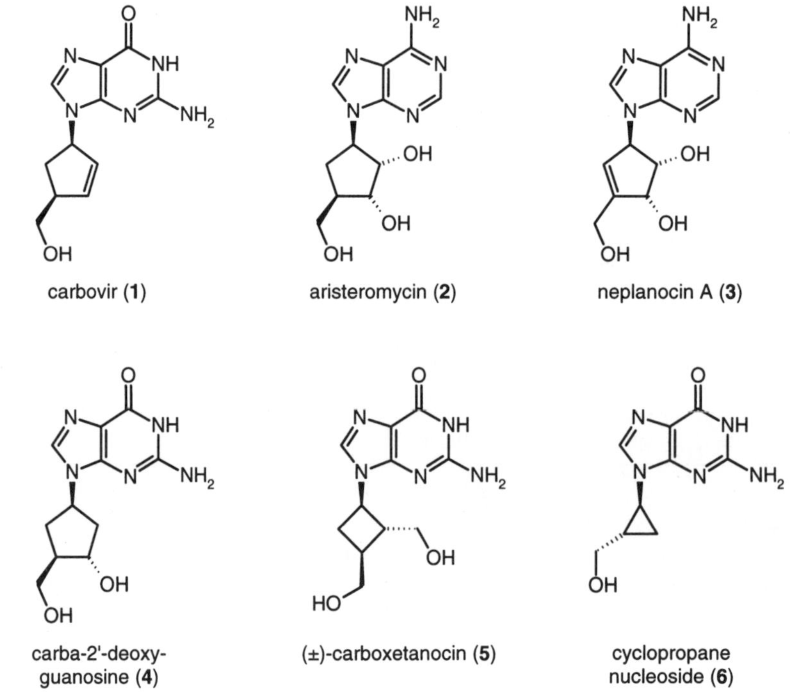

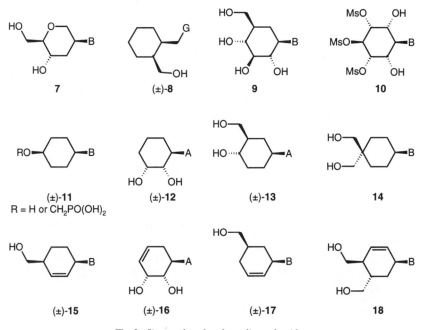

Fig. 1. *Representatives of carbocyclic nucleosides*

Fig. 2. *Six-membered carbocyclic nucleosides*

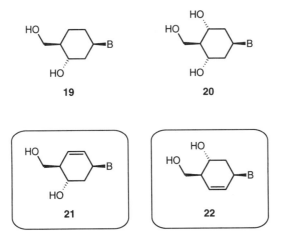

Fig. 3. *Cyclohexane and cyclohexene carbocyclic nucleosides*

2. 3-Hydroxy-(4-hydroxymethyl)cyclohexane Nucleosides 19. – 2.1. *Synthesis.* These compounds [5] are the carbocyclic congeners of the anhydrohexitol nucleosides **7**. The synthetic approach to racemic **19**, based on a conjugate addition reaction [8] of the base moieties to ethyl cyclohexane-1,3-diene-1-carboxylate (**23**) and hydroboration [9] of the cyclohexenyl intermediate, is shown in *Scheme 1*. For the synthesis of adenine

Scheme 1

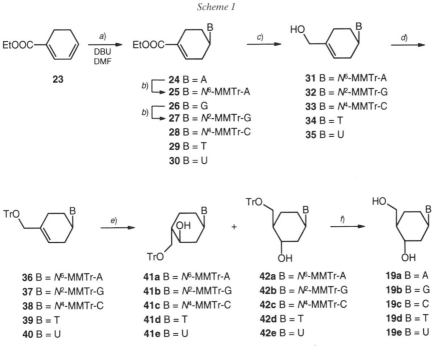

a) For **24**, adenine; for **26**, 1. 2-amino-6-chloropurine, 2. CF$_3$COOH, H$_2$O; for **28**, N^4-MMTr-cytosine, BSA, DMF; for **29**, thymine; for **30**, uracil. *b*) MMTrCl, Py. *c*) DIBAL-H, CH$_2$Cl$_2$. *d*) MMTrCl, Py. *e*) BH$_3$, THF. *f*) 80% aq. AcOH.

compound **19a**, adenine was reacted with **23** in the presence of 1,8-diazabicy-clo[5.4.0]undec-7-ene (DBU) in DMF. The resulting adduct **24** is poorly soluble in organic solvents. Therefore, the adenine moiety was protected as the monomethoxy-trityl ether **25** before carrying out further reactions. Reduction of the ester group with diisobutylaluminium hydride (DIBAL-H) and subsequent protection of the primary hydroxy group as trityl ether afforded **36**. Finally, hydroboration of the double bond with $BH_3 \cdot THF$ gave rise to a 1:1 mixture of **41a** and **42a**, which could easily be separated by chromatography. Complete deprotection of **42a** by treatment with 80% aq. AcOH gave the adenine compound **19a** as a racemate.

The guanine compound **19b** was synthesized in a similar way. 2-Amino-6-chloropurine was used as nucleophile in the *Michael*-type addition, and the chloropurine was converted to the guanine derivative **26**. After protection of the 2-amino group of the guanine base moiety, the ester was reduced to alcohol **32**. Protection of the alcohol and hydroboration of the double bond gave **42b**, which was deprotected to afford the guanine derivative **19b**.

In the case of the pyrimidine derivatives **19c**, **19d**, and **19e**, thymine reacted smoothly with **23** under standard conditions (70°). However, the reaction with uracil had to be carried out at lower temperature (50°) in order to obtain **30** in a moderate yield (28%). For the cytosine derivative, N^4-monomethoxytritylated cytosine was used in the *Michael*-addition reaction to give **28** in 30% yield. Reduction of **28**, **29**, and **30** with DIBAL-H and protection of the generated primary hydroxy group afforded **38**, **39**, and **40**, respectively. The hydroboration reaction had to be carefully controlled due to the sensitivity of the pyrimidine bases under the reaction conditions. With the thymine bases, the reaction time had to be reduced as compared to the purine bases. For the uracil and cytosine derivatives, the reaction was conducted at lower temperature. Chromatographic separation of **42d** from the isomer **41d** and deprotection gave the thymine compound **19d**. Cytosine **19c** and uracil **19e** were obtained by deprotection of the mixtures **41c/42c** and **41e/42e**, respectively, and subsequent chromatographic separation.

The D- and L-enantiomers of **19a** and **19d** were obtained by esterification of the trityl-protected derivatives **42a** and **42d** with (R)-$(-)$-O-methylmandelic acid, followed by chromatographic separation of the resulting diastereoisomeric esters **43a** and **44a**, and **43d** and **44d**, respectively (*Scheme 2*). Deacylation of **43a/43d** using KOH in MeOH, followed by treatment with 80% AcOH to remove the trityl group, gave rise to D-**19a/19d**. Likewise, L-**19a/19d** was obtained from **44a/44d**.

The absolute configurations of D-**19a** and L-**19a** were established by chromatography on a chiral column with optically pure D-**19a** as reference. The latter was synthesized in an enantioselective manner as described below (see *Sect. 4.1*). D-**19a/19d** and L-**19a/19d** were used as building blocks which were incorporated into the oligonucleotides. The conformation and hybridization properties of the resulting cyclohexane nucleic acids were studied [5b].

2.2. *Conformational Analysis and Antiviral Activity*. – NMR Studies and X-ray analysis clearly indicate that the cyclohexane nucleosides **19a–e** exist in a chair conformation similar to the $^4C_1$ of a pyranose nucleoside, with the base moieties oriented equatorially, both in solution and in the solid state. This conformation is opposite to the conformation observed for the anhydrohexitol nucleosides **7**. While the

Scheme 2

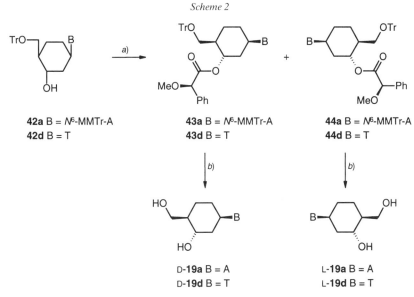

42a B = N⁶-MMTr-A
42d B = T

43a B = N⁶-MMTr-A
43d B = T

44a B = N⁶-MMTr-A
44d B = T

D-**19a** B = A
D-**19d** B = T

L-**19a** B = A
L-**19d** B = T

a) (−)-(*R*)-PhCH(OMe)COOH, DCC, DMAP, CH₂Cl₂. *b*) 1. 0.1M KOH, MeOH; 2. 80% AcOH.

latter show considerable antiviral activity, the cyclohexane nucleosides **19a – e** were devoid of any activity. We expected at that time that this different conformational behavior might be responsible for their different biological activities.

3. 3,5-Dihydroxy-(4-hydroxymethyl)cyclohexane Nucleosides 20. – 3.1. *Synthesis.* In order to further study the structure-activity relationship of cyclohexane nucleosides, an additional 5α-hydroxy group was introduced onto the six-membered ring (→ **20**) in an attempt to invert the preferred conformation to the ¹C₄ form [6]. Although compounds **20** themselves are not chiral, the synthetic strategy had to be designed in such a way that other carbocyclic nucleosides, such as **21** and **22**, could be synthesized enantioselectively from the common intermediate **53** (*Fig. 4*). Inexpensive (*R*)-(−)-carvone (**45**) was chosen as chiral starting material [10].

The synthesis of intermediates **53a – c** is presented in *Scheme 3*. Regio- and stereospecific epoxidation of (*R*)-(−)-carvone (**45**) gave epoxide **46** after distillation, and stereoselective reduction of the carbonyl group using L-*Selectride* and protection of the resulting hydroxy group afforded **47** in high yield. **47** was then converted to acetate **48** *via* oxidative cleavage of the double bond using OsO₄ and NaIO₄, followed by a *Baeyer-Villiger* oxidation [11] with 3-chloroperbenzoic acid (*m*-CPBA). The latter reaction had to be carried out in a buffer solution (pH 8) to avoid ring opening of the epoxide. After hydrolysis of the acetate under basic conditions (→ **49**), the hydroxy group was protected under standard conditions to yield either (*tert*-butyl)dimethylsilyl (TBDMS) ether **50** or Bn ether **51**. Regioselective ring opening of epoxides **50** and **51** by treatment with lithium tetramethylpiperidide (LiTMP)/Et₂AlCl [12] at − 65° afforded **52a** and **52c**, respectively, in quantitative yield, and subsequently, the 5-hydroxy group in **52a** was protected as the Bn ether **52b** using BnBr, NaH, and

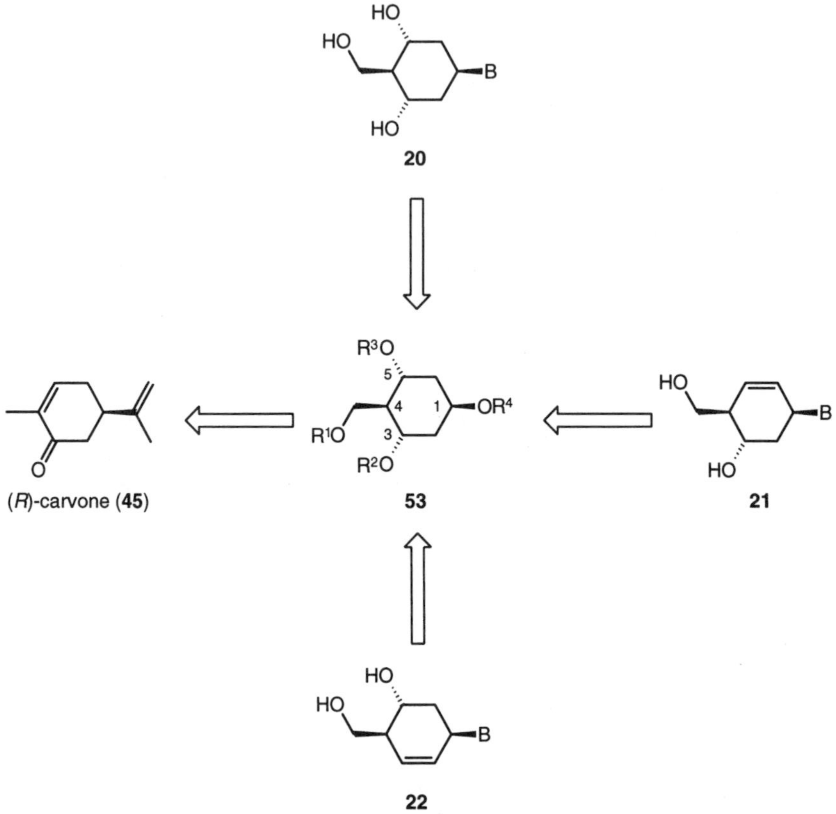

Fig. 4. *Retrosynthetic analysis*

tetrabutylammonium iodide (TBAI). The key reaction of this synthetic scheme is the transformation of the exocyclic double bond into a β-configured hydroxymethyl group. According to the literature, hydroboration of an exocyclic double bond with 9-bora-bicyclo[3.3.1]nonane (9-BBN) proceeds with good regioselectivity, but very poor stereo-selectivity [13]. However, upon treatment of **52b** with 9-BBN at room temperature, followed by oxidation with H_2O_2, the desired β-isomer **53b** was obtained as major product (63% starting from **52a**), together with a small amount of the α-isomer **54b**. The two isomers could easily be separated by chromatography. Compounds **52a** and **52c** reacted similarly. The absolute configurations of **53a – c** were established by NMR and later confirmed by X-ray analysis of compound **57**.

Starting with the key intermediate **53b**, the further synthetic strategy leading to the target compounds **20** involved the formation of a cyclic acetal, inversion of the configuration at C(1) and introduction of the base moieties *via Mitsunobu* reaction [14] (*Scheme 4*). Selective cleavage of the C(3)-OTBDMS group of **53b** was accomplished by treatment with 1 equiv. of tetrabutylammonium fluoride (TBAF), giving diol **55** as the sole product in high yield. This regioselectivity is probably due to neighboring-group participation of the C(4)–CH_2OH moiety (intramolecular hydrogen bonding

Scheme 3

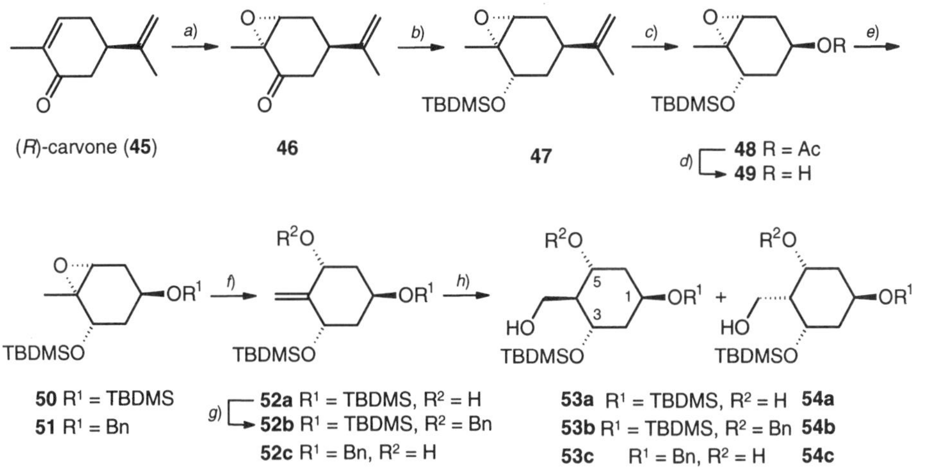

(*R*)-carvone (45)   **46**   **47**   48 R = Ac
                                                            d) 49 R = H

50 R¹ = TBDMS   **52a** R¹ = TBDMS, R² = H   **53a** R¹ = TBDMS, R² = H **54a**
51 R¹ = Bn   *g)* **52b** R¹ = TBDMS, R² = Bn   **53b** R¹ = TBDMS, R² = Bn **54b**
                    52c R¹ = Bn, R² = H   **53c** R¹ = Bn, R² = H **54c**

a) H₂O₂, NaOH, MeOH; 83%. *b*) 1. L-*Selectride*, THF, −65°; 2. TBDMS-Cl, imidazole, DMF; 93%. *c*) 1. OsO₄, KIO₄, THF, H₂O, r.t.; 83%; 2. *m*-CPBA, CHCl₃, pH 8, r.t.; 73%. *d*) K₂CO₃, MeOH; 73%. *e*) for **50**, TBDMS-Cl, imidazole, DMF; 80%; for **51**, NaH, BnBr, TBAI, THF, r.t. *f*) LiTMP, Et₂AlCl, PhH/PhMe, 0°; 100%. *g*) NaH, BnBr, TBAI, THF, r.t. *h*) 9-BBN, THF, r.t.

that accelerates the cleavage of the Si−O bond). Diol **55** was then protected as the cyclic benzylidene acetal **56**, and the C(1)-OTBDMS group was cleaved with excess TBAF to give crystalline **57**.

Inversion of the configuration at C(1) was carried out by a *Mitsunobu*-type reaction, *i.e.*, by treatment of **57** with diethyl azodicarboxylate (DEAD), PhCOOH, and Ph₃P, followed by base hydrolysis. The α-configured alcohol **58** was obtained in good yield. Introduction of the bases was then achieved by a second *Mitsunobu* reaction. Alcohol **58** was treated with adenine in the presence of DEAD and Ph₃P in dioxane to give **59a**, followed by acid hydrolysis of the benzylidene acetal to afford the adenine derivative **60a** (70% yield after the two steps). The guanine derivative **60b** was obtained in a similar way: **58** was reacted with 2-amino-6-chloropurine, followed by treatment with CF₃COOH/H₂O 3:1 at room temperature for 4 days.

Final hydrogenolysis to remove the Bn protecting groups (C(5)−OBn) of **60a** and **60b** was carried out with Pd(OH)₂ and cyclohexene in refluxing MeOH to give the desired cyclohexane nucleosides **20a** and **20b**, respectively (*Scheme 4*).

3.2. *Conformational Analysis and Antiviral Activity.* NMR Studies showed that both adenine **20a** and guanine **20b** exist in a ⁴C₁ chair conformation. They have equatorially oriented base moieties, despite the presence of three other axial substituents. These two compounds did not show any antiviral activity.

4. D- and L-3-Hydroxy-(4-hydroxymethyl)cyclohex-5-ene Nucleosides 21 and 22 [7]. – The lack of antiviral activity of the cyclohexanes **19** and **20**, mimicking the sugar moiety of the active anhydrohexitol nucleosides **7**, was initially explained by the observation that these two families exist in opposite chair conformations. In solution and in the solid state, hexitol nucleosides **7** were found to adopt a chair conformation with the base in an axial position. This conformation can be considered as a mimic of a

Scheme 4

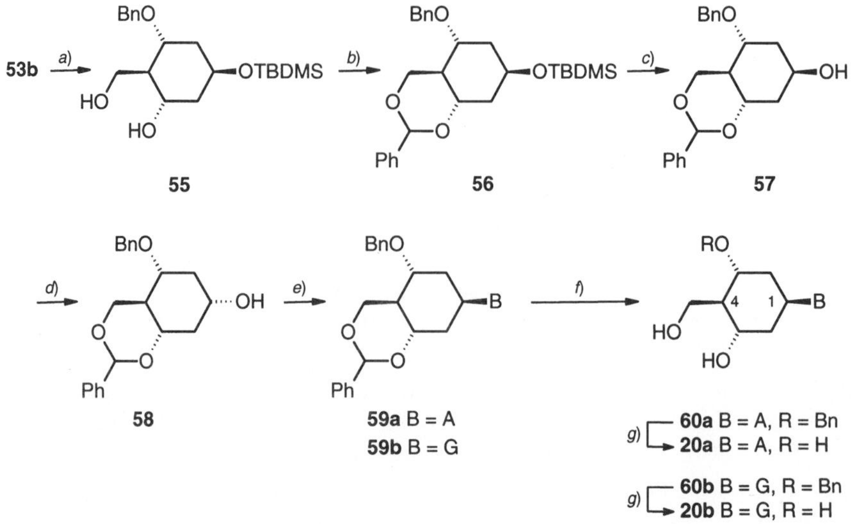

a) TBAF (1 equiv.), THF, r.t.; 93%. *b*) PhCH(OMe)$_2$, TsOH, dioxane; 90%. *c*) TBAF, THF; 93%. *d*) 1. PhCOOH, DEAD, Ph$_3$P, dioxane; 2. K$_2$CO$_3$, MeOH; 72%. *e*) for **59a**, adenine, DEAD, Ph$_3$P, dioxane; for **59b**, 1. 2-amino-6-chloropurine, DEAD, Ph$_3$P, dioxane; 2. CF$_3$COOH/H$_2$O 3 : 1, r.t. *f*) 80% AcOH, 60°. *g*) Pd(OH)$_2$/C, cyclohexene, MeOH, reflux.

furanose nucleoside frozen in the 3′-*endo* conformation. However, co-crystallization experiments with **7** in the active site of herpes virus type-1 thymidine kinase [3c] showed that anhydrohexitol nucleosides are bound in the inverted-chair conformation, *i.e.*, they adopt a conformation with an equatorially oriented base moiety. However, oligonucleotides containing **7** hybridize with RNA and DNA only when the monomeric nucleotides adopt the 3′-*endo* conformation [3d–f]. This finding led to the hypothesis that this conformation might be important for the hexitol nucleosides to be recognized by viral DNA polymerases. It could thus be advantageous for antiviral activity if a nucleoside can adopt several conformations that are interconvertible. In this way, the nucleoside can conformationally adapt to different steric requirements.

In order to substantially reduce the energy difference between the conformations of a six-membered carbocyclic nucleoside, we envisaged **21** and **22**, which contain a double bond in the cyclohexane ring. Molecular-mechanics calculations showed that the energy difference between the lowest-energy conformations of **21** is small (1.6 kJ/mol) and that the half-chair conformation with the base moiety in a pseudoaxial position is preferred. Moreover, the electron-rich double bond may function as a mimic of the furanose O-atom. In other words, cyclohexenes **21** and **22** seemed to be promising six-membered-ring mimics of a natural furanose nucleoside and worth synthesizing.

4.1. D-*3-Hydroxy-(4-hydroxymethyl)cyclohex-5-ene Nucleosides* **21**. Several synthetic methodologies to introduce the base moiety, including the widely applied Pd-coupling reaction, have been investigated for the preparation of D-cyclohexene nucleosides **21**, but they proved tedious and rather unsuccessful. A *Mitsunobu*-type reaction was found to be the best choice, and the enantioselective synthesis of **21** is outlined in *Schemes 5* and *6*.

Diol intermediate **53c** (see *Scheme 3*) was selectively monoprotected as TBDMS ether **61**, and the C(5)–OH group was converted to the corresponding mesylate **62** under standard conditions. Hydrogenolytic cleavage of the C(1)–OBn group with 20% Pd(OH)$_2$ and cyclohexene gave rise to a complex reaction mixture and afforded alcohol **63** in low yield, but fortunately, the reaction proceeded smoothly with 10% Pd/C and HCOONH$_4$. Oxidation of **63** with pyridinium dichromate (PDC) gave a mixture of ketone **64** and enone **65** in a combined yield of 39%. However, with MnO$_2$ as oxidant, an incomplete but clean reaction took place: enone **65** was obtained in 48% yield, and recovered **63** (47%) could be recycled (ketone **64** was not present in the reaction mixture). Reduction of enone **65** was carried out with NaBH$_4$ in the presence of CeCl$_3$ · 7 H$_2$O [15] in MeOH and gave the α-configured alcohol **66** as a single isomer in nearly quantitative yield. The configuration of the C(1)–OH group was assigned by NMR-spectral data.

Scheme 5

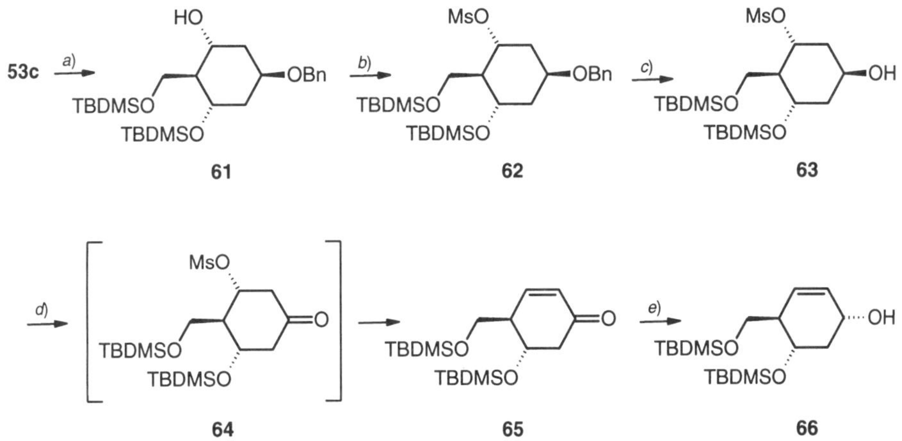

a) TBDMS-Cl (1.2 equiv.), imidazole (1.5 equiv.), DMF, r.t.; 70%. b) MsCl, Et$_3$N, CH$_2$Cl$_2$, 0°; 100%. c) Pd/C (10%), HCOONH$_4$, MeOH, reflux; 76%. d) MnO$_2$, CH$_2$Cl$_2$, r.t.; 48% (47% recovery of **63**). e) NaBH$_4$, CeCl$_3$ · 7 H$_2$O, MeOH, 0° → r.t.

The base moieties were introduced under *Mitsunobu*-type reaction conditions (*Scheme 6*). Upon treatment of **66** with adenine in the presence of DEAD and Ph$_3$P in dioxane, **67** with the desired configuration at C(1) was isolated in 66% yield, together with 17% of its N^7-isomer. Cleavage of the two TBDMS protecting groups proved troublesome. Using excess TBAF, the reaction proceeded quantitatively, but the product was contaminated with a large amount of tetrabutylammonium salts, which could not be removed by standard chromatographic techniques, while deprotection under basic conditions (*t*-BuOK in DMF [16]) resulted in a complex mixture. Therefore, a mild and simple deprotection procedure was applied, and treatment of **67** with TFA/H$_2$O 3:1 at room temperature gave the desired cyclohexene nucleoside **21a** in 54% overall yield starting from **66**. For the guanine compound **21b**, 2-amino-6-chloropurine was used in the *Mitsunobu* reaction of **67**, and the so-obtained chloropurine **68** was treated with TFA/H$_2$O 3:1. These reaction conditions converted the chloropurine group to the corresponding guanine and simultaneously removed the two TBDMS protecting groups, giving cyclohexene nucleoside **21b** in 46% overall yield starting from **66**.

Reduction of the double bond of **21a** with 10% Pd/C in MeOH gave enantiomerically pure D-**19a**. The enantiomeric purity was examined by HPLC on a chiral column and was found to be 99%, which also established the high enantiomeric purity of **21a**.

Scheme 6

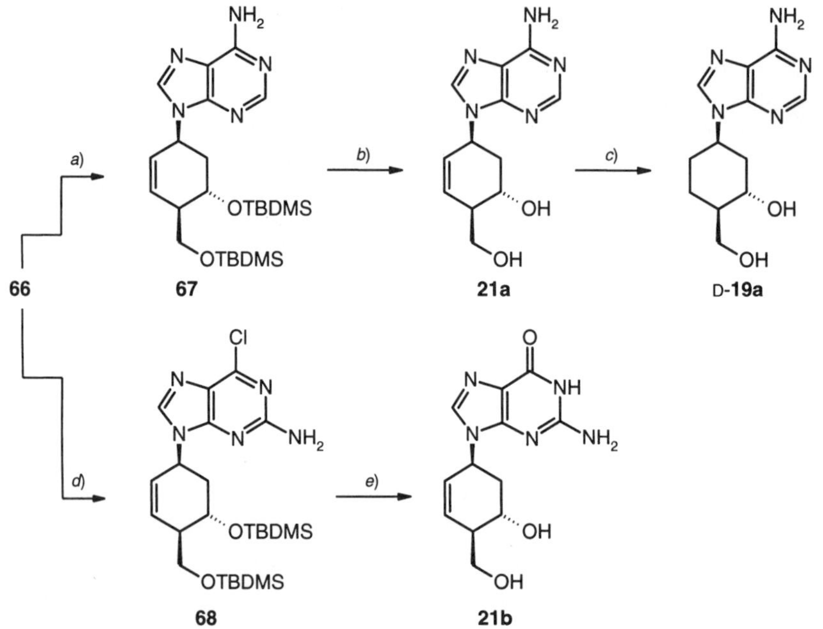

a) Adenine, DEAD, Ph$_3$P, dioxane. *b*) CF$_3$COOH/H$_2$O 3:1, r.t.; 54% from **66**. *c*) Pd/C (10%), H$_2$, MeOH, r.t.; 75%. *d*) 2-Amino-6-chloropurine, DEAD, Ph$_3$P, dioxane. *e*) CF$_3$COOH/H$_2$O 3:1.

4.2. L-3-Hydroxy-(4-hydroxymethyl)cyclohex-5-ene Nucleosides 22.

In the field of antiviral nucleosides, it is recognized that D- and L-nucleosides might have different activity and toxicity spectra, and that the study of the difference in biological activity of enantiomers could provide valuable information. As described in *Sect. 4.1*, our enantioselective approach starting from (*R*)-carvone gave access to the D-cyclohexene nucleosides **21**. As shown in *Fig. 4*, the synthetic pathway included intermediate **53**, which possesses four chiral centers and, disregarding the nature of the protecting groups R^2 and R^3, a plane of symmetry. The present strategy, therefore, also allowed the enantioselective synthesis of the corresponding L-cyclohexene nucleosides **22**, *via* selective transformation of (in this case) OR^2 into an enol function, followed by introduction of the base moiety at C(1).

The diol intermediate **53a** (see *Scheme 3*) was protected as the dibenzoate **69** (*Scheme 7*) in high yield. The equatorial C(3)−OTBDMS group was selectively cleaved using one equivalent of TBAF (74%), and the C(3)−OH group was converted to the corresponding mesylate **71**. The TBDMS group at C(1) was removed with excess TBAF to give alcohol **72**, which was oxidized with PDC to yield **73**. However, the latter reaction was accompanied by simultaneous elimination of the C(3)−OMs group to afford directly the desired enone **74**. Consecutive reduction with NaBH₄ and CeCl₃ · 7 H₂O gave enol **75** as the sole product, with the correct configuration at C(1). The overall yield starting from **53a** was 31%.

Scheme 7

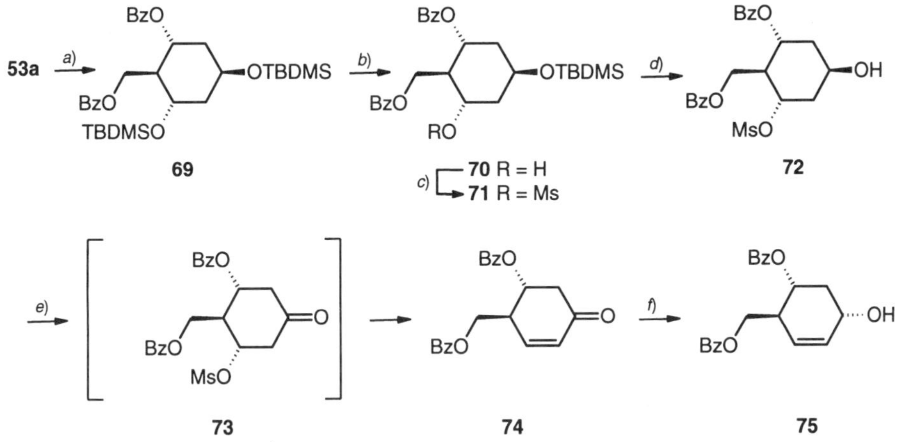

a) Bz₂O, DMAP, CH₂Cl₂, 0°; 98%. *b*) TBAF (1 equiv.), THF, r.t.; 74%. *c*) MsCl, Et₃N, CH₂Cl₂, 0°; 98%. *d*) TBAF, THF, r.t.; 86%. *e*) PDC, CH₂Cl₂, r.t.; 86%. *f*) NaBH₄, CeCl₃·7 H₂O, MeOH, r.t.; 75%.

Similar to the case of **21**, the base moieties were introduced by the *Mitsunobu* methodology (*Scheme 8*). Upon treatment of **75** with adenine under standard *Mitsunobu*-reaction conditions, the desired adenine derivative **76** was isolated in a moderate yield (40%), together with 15% of its *N*⁷-isomer. Final removal of the protecting groups with K₂CO₃ in MeOH gave the L-adenine cyclohexene nucleoside **22a** in 72% yield. Similarly, reaction of **75** with 2-amino-6-chloropurine and conversion of the chloropurine **77** to the guanine derivative with TFA/H₂O 3:1 afforded **78**. Full deprotection was carried out by heating **78** in an NH₃/MeOH solution to give pure L-guanine cyclohexene nucleoside **22b** (42% starting from **75**) after reversed-phase HPLC purification.

4.3. *Conformational Analysis.* Conformational analysis of **21** was carried out by computational methods and ¹H-NMR. The calculated preference for the 3'-*endo* half-chair conformation with pseudoaxial base over the 3'-*exo* conformation by only 1.6 kJ/mol, corresponding to a 7:3 ratio, was confirmed. The experimental coupling constants are in good agreement with the calculated values. Thus, upon introduction of a double bond into the six-membered ring of cyclohexane nucleoside **19**, the conformational preference changes from an exclusively 3'-*exo* conformation with an equatorially oriented base (see *Sect. 2.2*) to a predominantly 3'-*endo* conformation, with the base

Scheme 8

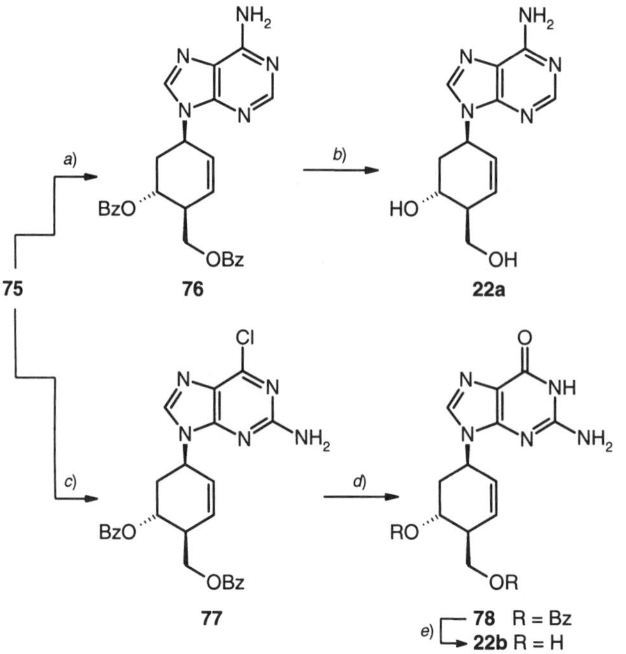

a) Adenine, DEAD, Ph₃P, dioxane, r.t. b) K₂CO₃, MeOH, r.t.; 77%. c) 2-Amino-6-chloropurine, DEAD, Ph₃P, dioxane, r.t. d) TFA/H₂O 3:1, r.t. 48 h; 58% from **75**. e) NH₃/MeOH, 80°, 48 h, 73%.

moiety in a pseudoaxial position, for **21**. According to theoretical considerations, one of the stereoelectronic factors that drives this conformational preference is the $\pi \rightarrow \sigma^*_{C1'-N}$ interaction involving the antibonding orbital of the $C(1')-N$ bond [7a][17]. Such an effect is similar to the anomeric effect observed in natural furanose nucleosides.

In summary, the preference for a 3'-*endo* conformation, the easy interconversion between the 3'-*endo* and 3'-*exo* forms, together with the presence of a double bond mimicking the furanose O-atom, might be the reason for the strong antiviral activity of **21** and **22**.

4.4. *Antiviral Activity*. The adenine nucleosides **21a** and **22a** showed only low antiviral activity and are not further discussed here. However, enantiomeric D-cyclohexene G **21b** and L-cyclohexene G **22b** showed a strikingly similar broad-spectrum antiviral activity and did not show toxicity in four different cell lines (HeLa, Vero, E₆SM, and HEL), indicating that their mode of action is virus-specific. They displayed the same activity against HSV-1 and HSV-2. Against HSV-1, they were as active as acyclovir and brivudin, while against HSV-2, their activity was very similar to that of acyclovir. The activity of D-**21b** was nearly two-fold higher than that of L-**22b**. Both guanine nucleosides retained activity against the TK⁻ strains of HSV-1 and VZV, although the activity was reduced as compared to the activity against the wild type. The activity of D-**21b** against TK⁺ and TK⁻ VZV strains was higher than the respective activities of acyclovir and brivudin. Finally, D-**21b** showed the same potency against CMV as ganciclovir.

In conclusion, the activity spectrum of the two newly synthesized cyclohexene guanine nucleosides **21b** and **22b** is very similar to that of the known antiviral compounds possessing a guanine base moiety (acyclovir, ganciclovir). Both enantiomers show antiviral activity. The high selectivity indices observed for D- and L-cyclohexene G warrant further *in vivo* studies with these compounds against herpesvirus infections.

REFERENCES

[1] a) R. Vince, M. Hua, J. Brownell, S. Daluge, F. Lee, W. M. Shannon, G. C. Lavelle, J. Qualls, O. S. Weislow, R. Kiser, P. G. Canonico, R. H. Schultz, V. L. Narayana, J. G. Mayo, R. H. Shoemaker, M. R. Boyd, *Biochem. Biophys. Res. Commun* **1988**, *156*, 1046; b) R. Vince, M. Hua, *J. Med. Chem.* **1990**, *33*, 17; c) A. Guranowski, J. A. Montgomery, G. L. Cantoni, P. K. Chiang, *Biochemistry*, **1981**, *20*, 100; d) E. De Clercq, *Biochem. Pharmacol.* **1987**, *36*, 2567; e) P. Herdewijn, E. De Clercq, J. Balzarini, H. Vandehaeghe, *J. Med. Chem.* **1985**, *28*, 1385; f) S. Yaginuma, N. Muto, M. Tsujino, Y. Sudate, M. Hayashi, M. Otani, *J. Antibiot.* **1981**, *23*, 359; g) R. Borchardt, B. T. Keller, U. Patel-Thrombre, *J. Biol. Chem.* **1984**, *259*, 5353; h) Y. F. Shealy, C. A. O'Dell, W. M. Shannon, G. Arnett, *J. Med. Chem.* **1984**, *27*, 1416; i) P. M. Price, R. Banerjee, G. Acs, *Proc. Natl. Acad. Sci. U.S.A* **1989**, *86*, 8541; j) D. W. Norbeck, S. Kern, S. Hiyashi, W. Rosenbrook, H. Sham, T. Herrin, J. J. Plattner, J. Erickson, J. Clement, R. Swanson, N. Shipkowitz, D. Hardy, K. Marsh, G. Arnett, W. Shannon, S. Broder, H. J. Mitsuya, *J. Med. Chem.* **1990**, *33*, 1281; k) R. Csuk, Y. Von Scholz, *Tetrahedron* **1994**, *61*, 10431.

[2] V. E. Marquez, M.-I. Lim, *Med. Res. Rev.* **1986**, *6*, 1; L. Agrofoglio, E. Suhas, A. Farese, R. Condom, S. R. Challend, R. A. Earl, R. Guedj, *Tetrahedron* **1994**, *50*, 10611; A. D. Borthwick, K. Biggadike, *Tetrahedron* **1992**, *48*, 571; E. De Clercq, *J. Antimicrob. Chemother.* **1993**, *32, Suppl. A.* 121; P. Herdewijn, *Drug Discovery Today* **1997**, *2*, 235.

[3] I. Verheggen, A. Van Aerschot, S. Toppet, R. Snoeck, G. Jannsen, J. Balzarini, E. De Clercq, P. Herdewijn, *J. Med. Chem.* **1993**, *36*, 2033; I. Verheggen, A. Van Aerschot, L. Van Meervelt, J. Rozenski, L. Wiebe, R. Snoeck, G. Andrei, J. Balzarini, P. Claes, E. De Clercq, P. Herdewijn, *J. Med. Chem.* **1995**, *38*, 826; T. Ostrowski, B. Wroblowski, R. Busson, J. Rozenski, E. De Clercq, M. S. Bennett, J. N. Champness, W. C. Summers, M. R. Sanderson, P. Herdewijn, *J. Med. Chem.* **1998**, *41*, 4343; H. De Winter, E. Lescrinier, A. Van Aerschot, P. Herdewijn, *J. Am. Chem. Soc.* **1998**, *120*, 5381; C. Hendrix, H. Rosemeyer, I. Verheggen, F. Seela, A. Van Aerschot, P. Herdewijn, *Chem. Eur. J.* **1997**, *3*, 110; C. Hendrix, H. Rosemeyer, B. De Bouvere, A. Van Aerschot, F. Seela, P. Herdewijn, *Chem. Eur. J.* **1997**, *3*, 1513.

[4] N. Katagiri, Y. Ito, T. Shiraishi, T. Maruyama, Y. Sato, C. Kaneko, *Nucleosides Nucleotides* **1996**, *15*, 631; M. J. Konkel, R. Vince, *Tetrahedron* **1996**, *52*, 799; M. J. Konkel, R. Vince, *Nucleosides Nucleotides* **1995**, *14*, 2061; H. J. Schaeffer, R. Vince, *J. Med. Chem.* **1968**, *11*, 15; S. Halazy, M. Kenny, J. Dulworth, A. Eggenspiller, *Nucleosides Nucleotides* **1992**, *11*, 1595; I. Kitagawa, B. C. Cha, T. Nakae, Y. Okaichi, Y. Takinama, M. Yoshikawa, *Chem. Pharm. Bull.* **1989**, *37*, 542; F. Calvani, M. Macchia, A. Rossello, M. R. Gismondo, L. Drago, M. C. Fassina, M. Cisternino, P. Domiano, *Bioorg. Med. Chem. Lett.* **1995**, *5*, 2567; S. N. Mikhailov, N. Blaton, J. Rozenski, J. Balzarini, E. De Clercq, P. Herdewijn, *Nucleosides Nucleotides* **1996**, *15*, 867; K. Ramesh, M. S. Wolfe, Y. Lee, D. S. Vander Velde, R. T. Borchardt, *J. Org. Chem.* **1992**, *57*, 5861; Å. Rosenquist, I. Kvarnström, B. Classon, B. Samuelsson, *J. Org. Chem.* **1996**, *61*, 6282; J. H. Arango, A. Geer, J. Rodriguez, P. E. Young, P. Scheiner, *Nucleosides Nucleotides* **1993**, *12*, 773; Y. A. Nsiah, R. L. Tolman, J. D. Karkas, F. Rapp, *Antimicrob. Agents Chemoth.* **1990**, *34*, 1551; M. J. Pérez-Pérez, J. Rozenski, R. Busson, P. Herdewijn, *J. Org. Chem.* **1995**, *60*, 1531; G. J. Aquilar, M. E. Gelpi, R. A. Cadenas, *J. Heterocycl. Chem.* **1992**, *29*, 401.

[5] Y. Maurinsh, J. Schraml, H. De Winter, N. Blaton, O. Peeters, E. Lescrinier, J. Rozenski, A. Van Aerschot, E. De Clercq, R. Busson, P. Herdewijn, *J. Org. Chem.* **1997**, *62*, 2861; Y. Maurinsh, H. Rosemeyer, R. Esnouf, A. Medvedovici, J. Wang, G. Ceulemans, E. Lescrinier, C. Hendrix, R. Busson, P. Sandra, F. Seela, A. Van Aerschot, P. Herdewijn, *Chem. Eur. J.* **1999**, *5*, 2139.

[6] J. Wang, R. Busson, N. Blaton, J. Rozenski, P. Herdewijn, *J. Org. Chem.* **1998**, *63*, 3051.

[7] J. Wang, P. Herdewijn, *J. Org. Chem.* **1999**, *64*, 7820; J. Wang, M. Froeyen, C. Hendrix, G. Andrei, R. Snoeck, E. De Clercq, P. Herdewijn, *J. Med. Chem.* **2000**, *43*, 736.

[8] J. H. Arango, A. Geer, J. Rodriguez, P. E. Young, P. Scheiner, *Nucleosides Nucleotides* **1993**, *12*, 773.

[9] M. Koga, S. J. Schneller, *Org. Chem.* **1993**, *58*, 6471.

[10] K. Weinge, H. Reichert, U. Huber-Patz, H. Irngartinger, *Liebigs Ann. Chem.* **1993**, *4*, 403.

[11] F. Delay, G. Ohloff, *Helv. Chim. Acta* **1979**, *62*, 2168; 'Modern Synthetic Reactions', Ed. H. O. House, Benjamin, Inc., 1972; p. 324.

[12] S. Czernecki, C. Georgoulis, C. Provelenghiou, *Tetrahedron Lett.* **1976**, 3535; K. Kanai, I. Sakamoto, S. Ogawa, T. Suami, *Bull. Chem. Soc. Jpn.* **1987**, *60*, 1529.

[13] D. A. Evans, G. C. Fu, A. H. Hoveyda, *J. Am. Chem. Soc.* **1988**, *110*, 6917; D. A. Evans, G. C. Fu, A. H. Hoveyda, *J. Am. Chem. Soc.* **1992**, *114*, 6671.

[14] O. Mitsunobu, *Synthesis* **1981**, 1.

[15] A. L. Gemal, J. L. Luche, *J. Am. Chem. Soc.* **1981**, *103*, 5454.

[16] G. Megron, F. Vasquezy, G. Galderon, R. Cruz, R. Gavino, G. Islas, *Synth. Commun.* **1998**, *26(16)*, 3021.

[17] B. Doboszewski, N. Blaton, P. Herdewijn, *J. Org. Chem.* **1995**, *60*, 7909; M. Polak, B. Doboszewski, P. Herdewijn, J. Plavec, *J. Am. Chem. Soc.* **1997**, *119*, 9782.

Individual Isomers of Dinucleoside Boranophosphates as Synthons for Incorporation into Oligonucleotides: Synthesis and Configurational Assignment

by **Zinaida A. Sergueeva**[a]), **Dmitri S. Sergueev**[a]), **Anthony A. Ribeiro**[b]), **Jack S. Summers**[a]),
and **Barbara Ramsay Shaw***[a])

[a]) Department of Chemistry, Duke University, Durham, NC 27708, USA
[b]) Duke NMR Spectroscopy Center and Department of Radiology, Duke University, Durham, NC 27708, USA

Individual isomers of the protected boranophosphates **5a** and **5b**, *i.e.*, the N^6-benzyl-2'-deoxy-5'-O-(4,4'-dimethoxytrityl)adenosin-3'-yl 2'-deoxy-4-O-(4-nitrophenyl)uridin-5'-yl boranophosphates, were synthesized *via* stereospecific silylation and boronation of their H-phosphonate precursors. 2D-NMR Spectroscopic studies yielded an initial assignment of the isomer configuration, which was further confirmed unambiguously by a parallel chemical synthesis. Deprotection of the 'dimers' **5a** and **5b** yielded the individual [$P(R)$]- and [$P(S)$]-isomers **7a** and **7b**, respectively, *i.e.*, the 2'-deoxyadenosin-3'-yl 2'-deoxycytidin-5'-yl boranophosphates. Their substrate properties toward phosphodiesterase I were identical to those of the previously characterized isomers of dithymidine boranophosphate. The protected 'dimers' **5a** and **5b** can be used as synthons to incorporate the boranophosphate linkage with a defined configuration to selected positions of an oligonucleotide chain.

Introduction. – In recent years, much attention has been drawn to applications of oligonucleotides as diagnostic tools and therapeutic remedies. Among numerous modified oligonucleotides that were tested [1], several showed considerable promise, *e.g.* phosphorothioates [2], peptide nucleic acids [3], morpholinos [4], and methyl-phosphonates [5]. Oligonucleoside boranophosphates (BH_3^--ODNs), in which borane (BH_3) is substituted for one of the non-bridging phosphate O-atoms ($= P$-borano-phosphate) [6], could be a useful addition to this arsenal of oligonucleotide tools. The boranophosphate modification maintains the negative charge of the normal phosphate and imparts valuable properties to oligonucleotides. The BH_3^--ODNs exhibit high nuclease resistance [6][7], increased hydrophobicity [6][7a,d], altered metal affinities [8], and have potential for use in boron neutron-capture therapy [9]. The usefulness of the boranophosphate modification is demonstrated in a one-step direct PCR sequencing method that exploits the excellent substrate properties of deoxynucleoside P^α-boranotriphosphates and the nuclease resistance of the boranophosphate linkage [10]. Further, the ability of a BH_3^--ODN to support RNase H mediated cleavage of the complementary RNA suggests they might find application in antisense technology [11].

The demand for better performing oligonucleotides in therapeutic and diagnostic applications has led to the design of a next generation of oligonucleotides, comprised of mixed or alternating backbones. Such oligonucleotides have demonstrated enhanced nuclease resistance, improved hybridization properties, and fewer immune-stimulatory side effects [2][12]. In view of this, oligonucleotides containing alternating borano-phosphate linkages deserve attention as potentially useful molecules. In this account, we describe the first synthesis of a protected dinucleoside boranophosphate, which can be further used as a synthon for incorporation into an oligonucleotide chain.

NMR Spectroscopy has been growing as a tool for probing the structure of nucleic acids and their interactions with proteins [13]. The determination of nucleic-acid tertiary structures by NMR can be limited by the small number and redundancy of available protons and their limited exposure to exterior contacts. BH_3^--ODNs may find use as probes in NMR spectroscopy if NOE contacts to the borane H-atoms can be measured. The measurement of NOEs to H-atoms at the phosphonyl positions in boranophosphate-substituted nucleic acids will enhance the reliability and resolution of NMR structures. Since boranophosphates are chiral at the P-atom, comparison of the NOE contacts for the two stereoisomers of a particular boranophosphate linkage could potentially yield important structural information. Introduction of the linkage requires synthesis of diastereoisomerically pure dinucleoside synthons. Here we describe the preparation of the individual isomers of N^6-benzyl-2'-deoxy-5'-O-(4,4'-dimethoxytrityl)-adenosin-3'-yl 2'-deoxy-4-O-(nitrophenyl)uridin-5'-yl boranophosphates in high stereotopic purity and the assignment of their configuration.

Results and Discussion. – *Synthesis and Separation of Isomers.* We have shown that the synthesis of BH_3^--ODNs *via* their phosphonate precursors proceeds smoothly and in high yield [7d][14]. In addition, the two dinucleoside *H*-phosphonate stereoisomers can be conveniently separated by silica gel chromatography [15]. We recently demonstrated that individual isomers of dithymidine *H*-phosphonate ($= P$-deoxythymidinyl-$(3' \rightarrow 5')$-thymidine) could be converted to their boranophosphate analogs stereospecifically with retention of configuration at the P-atom [16]. Based on these results, our strategy for the present synthesis included *H*-phosphonate condensation followed by separation of the isomers, stereospecific silylation and boronation, and finally hydrolysis of trimethylsilyl ester and 3'-OH group deprotection (see below, *Scheme 2*).

The choice of sugar and especially nucleobase protecting groups is very important to the successful synthesis of BH_3^--ODNs with bases other than thymine. We have discovered that acyl base-protecting groups in nucleosides such as benzoyl, isobutyryl, and acetyl are readily reduced by various borane-amine complexes to give N-alkylated nucleosides [17] that could not be easily deprotected. While our synthetic strategy was to avoid the use of amides for base-protection, we were able to employ N^6-benzoyl-2'-deoxyadenosine in our synthesis, as the reduced N^6-benzyl-2'-deoxyadenosine can be deprotected by several methods [18]. Thus, commercially available N^6-benzoyl-2'-deoxy-5'-O-(4,4'-dimethoxytrityl)adenosine 3'-*H*-phosphonate was chosen here as the 5'-nucleotide component for the condensation reaction. The convertible nucleosides, *i.e.* 4-O-substituted 2'-deoxyuridines, are convenient intermediates to various N^4-substituted 2'-deoxycytidine derivatives [19]. Nucleophilic substitution at the 4-position of such 2'-deoxyuridines by ammonia results in 2'-deoxycytidine as the only product [19a,c,d]. We selected for our synthesis 2'-deoxy-4-O-(4-nitrophenyl)-uridine (**1**), described by *Miah et al.* [19c], which is quite stable under mild alkaline conditions, yet is completely converted to 2'-deoxycytidine during standard oligonucleotide deprotection by concentrated aqueous ammonia. Conversion of **1** to nucleoside component **3** followed a standard procedure depicted in *Scheme 1*. Protection of the 5'-OH group with the 4,4'-dimethoxytrityl (MeO)$_2$Tr group resulting in 2'-deoxy-5'-(4,4'-dimethoxytrityl)-4-*O*-(4-nitrophenyl)uridine (**2**) proceeded smoothly with 78% yield. Subsequent 3'-OH acylation with levulinic anhydride ((Lev)$_2$O) and *in situ* 5'-OH deprotection provided 2'-deoxy-3'-*O*-levulinoyl-4-*O*-(4-nitrophenyl)uridine (**3**; 80%) as a slightly yellow solid. The levulinoyl group was preferred over the (*tert*-butyl)dimethylsilyl group ('BuMe$_2$Si) or other acyl groups for 3'-OH protection because of the very mild conditions required for its removal. Treatment with sufficient hydrazine hydrate for 3'-OH deprotection did not cause significant substitution of 4-nitrophenol in **3**, whereas alkaline deprotection of acyl groups or treatment with fluoride ion to remove the 'BuMe$_2$Si group could cause hydrolysis of the aryl group.

Scheme 1. *Synthesis of the 3'-Protected Convertible 2'-Deoxyuridine* **3**

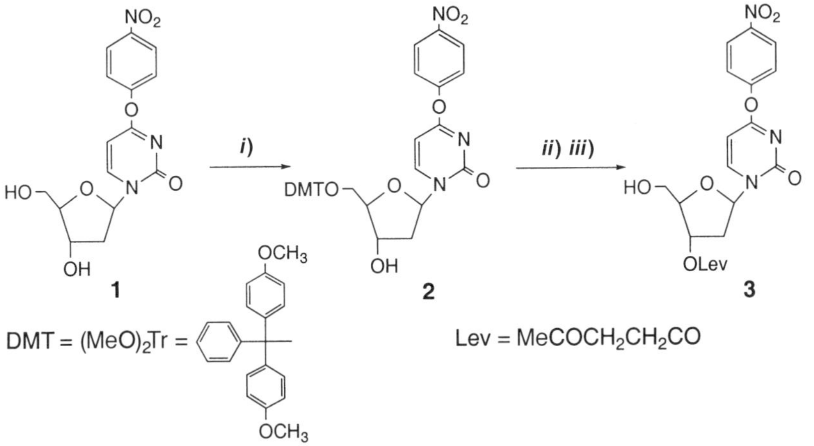

i) (MeO)$_2$TrCl. ii) (Lev)$_2$O; iii) H$^+$.

Condensation of *N*[6]-benzoyl-2'-deoxy-5'-*O*-(4,4'-dimethoxytrityl)adenosine 3'-phosphonate and **3** with pivaloyl chloride (=2,2-dimethylpropanoyl chloride; PivCl) resulted in a *ca.* 1:1 diastereoisomer mixture of *N*[6]-benzoyl-2'-deoxy-5'-*O*-(4,4'-dimethoxytrityl)adenosin-3'-yl 2'-deoxy-3'-*O*-levulinoyl-4-*O*-(4-nitrophenyl)uridin-5'-yl *H*-phosphonate (**4**) according to ^{31}P-NMR (*Scheme 2*). After extraction with saturated NaHCO$_3$ solution, the isomer mixture was separated by HPLC (silica gel, isocratic mode). The fast migrating isomer **4a** was isolated in 98% diastereoisomer purity, while the diastereomer purity of the slower migrating isomer **4b** was 84%, as judged by ^{31}P-NMR. The combined yield of **4a** and **4b** was 73%. The enriched *H*-phosphonate isomer preparations **4a,b** were converted to boranophosphates **5a,b** without isolation of the protected intermediates.

As was shown previously [16], conversion of the *H*-phosphonate diester to the boranophosphate diester proceeds stereospecifically with retention of the configuration. Therefore, we anticipated that diastereoisomeric purity will not change throughout the transformation. Silylation and boronation reactions proceeded near quantitatively, as determined by ^{31}P-NMR analysis. The excess of the silylating and boronating

Scheme 2. *Synthesis of the Individual Isomers* **5a** *and* **5b** *of Protected Dinucleoside Boranophosphate*

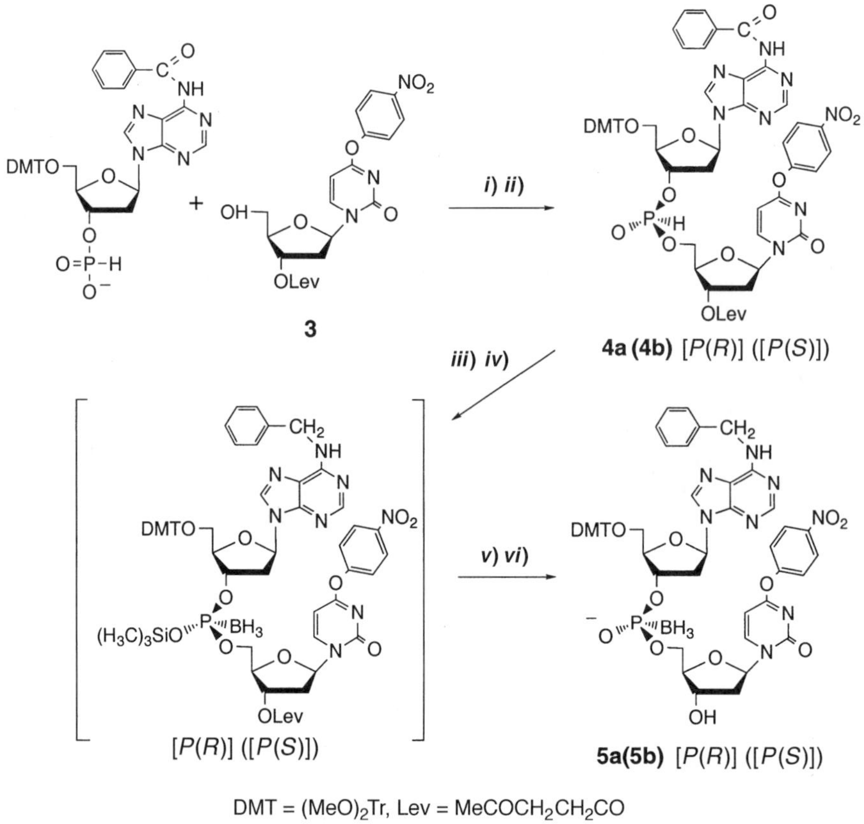

DMT = (MeO)$_2$Tr, Lev = MeCOCH$_2$CH$_2$CO

i) PivCl. *ii*) Direct-phase HPLC separation of the isomers. *iii*) BSA. *iv*) BH$_3$·^{i}Pr$_2$EtN. *v*) H$_2$O. *vi*) NH$_2$NH$_2$·H$_2$O.

agents were removed by precipitation in hexane, and the residue was treated with pyridine/H$_2$O to hydrolyze the boranophosphate trimethylsilyl ester. Removal of the 3'-*O*-levulinoyl protection group proceeded readily with hydrazine hydrate. During the reaction there was an appearance of a faint yellow color, indicating some release of 4-nitrophenol. The estimated extent of the hydrolysis and/or substitution of 4-nitrophenol (based on the color intensity, TLC, and reversed-phase HPLC analysis) was less than 5%. The resulting N^6-benzyl-2'-deoxy-5'-*O*-(4,4'-dimethoxytrityl)adenosin-3'-yl 2'-deoxy-4-*O*-(4-nitrophenyl)uridin-5'-yl boranophosphates (**5a** and **5b**; enriched isomer preparations) were isolated from the crude reaction mixtures by reversed-phase HPLC (*Fig. 1*). Analytical reversed-phase HPLC assessments of the diastereomer purity of the boranophosphates **5** in the crude reaction mixtures (>98% for **5a** and 80–85% for **5b**) agree well with the ^{31}P-NMR assessments of the phosphonates **4a** and **4b**[1]), confirming the stereospecific character of the boronation procedure. The

---

[1])   We could not assess the diastereoisomer purity of the boranophosphates **5** by ^{31}P-NMR due to broad overlapping signals of the boranophosphate isomers.

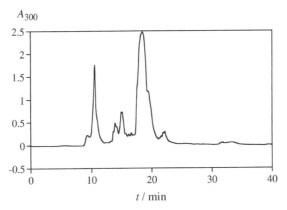

Fig. 1. *Preparative reversed-phase HPLC isolation of* **5b** *from the reaction mixture* (*Scheme 1*). The main-peak shoulder corresponds to **5a** impurity. Conditions: *Microsorb RP-18*. 21.4×250 mm column, elution with gradient 40–80% MeCN in 20 mM $(Et_3NH)HCO_3$ (pH 8.0) for 45 min, flow rate 6 ml/min.

elution order of the isomers observed on reversed-phase HPLC was reversed when compared to elution on silica gel. As shown in *Fig. 1*, **5b** elutes first on reversed-phase HPLC, with the isomer impurity **5a** appearing as a shoulder with greater retention time. The diastereomer purity in the final **5a** and **5b** preparations was increased after reversed-phase HPLC separation to 98 and 94%, respectively.

Configuration Assignment of the Isomers **5a** *and* **5b** *by NOE Contacts to the Borane Protons.* The configuration at the P-atom has been found to be important to a number of properties of substituted nucleic acids, such as nuclease resistance, hybridization ability, and hydrophobicity [20]. However, there is not currently available a general method for assigning the configuration at the P-atom of boranophosphate nucleic acids. Two dimensional (2D) NMR has been used to establish the configuration at the P-atom for a number of phosphotriester and methylphosphonate analogs [21], which are structurally similar to BH_3^--ODNs. We hope to develop a reliable physical method for assigning configuration in boranophosphate analogs on measured NOE contacts to the borane protons. Concurrently, we would obtain useful information about the overall conformation of the BH_3^--ODNs.

COSY and NOESY spectra were recorded for each isomer. All ribose and nucleoside proton resonances of isomers **5a** and **5b**, and most of the protecting-group resonances, were unambiguously assigned based on the cross-peaks in the 2D COSY and NOESY plots. The chemical shifts of the base and sugar protons are presented in *Table 1*.

While, in principle, the connection of the ribose rings *via* five single bonds, *i.e.* $C(3')-O-P-O-C(5')-C(4')$, gives rise to a large number of possible states, evidence suggests that the protected isomers **5a** and **5b** reside in a limited number of accessible conformations, which are not unlike *A*- or *B*-form DNA. The program Insight (*Molecular Simulations Inc.*, San Diego, CA) was used to calculate distances between the two non-bridging phosphonyl O-atom positions and the protons of the bases and ribose units by assuming that the 'dimer' adopted either an *A*- or *B*-like conformation. All NOE contacts predicted by the *A*- and *B*-form models that were not obscured by overlapping resonances were observed (including NOE cross-peaks between protons

Table 1. *Resonance-Frequency Assignments for the ^{1}H-NMR Spectra of Isomers* **5a** *and* **5b**

Proton[2])	5a δ [ppm]	5b δ [ppm]	Proton[2])	5a δ [ppm]	5b δ [ppm]
2'-Deoxyuridine:			2'-Deoxyadenosine:		
H−C(5)(U)	6.175	6.188	H−C(2)(A)	8.090	8.087
H−C(6)(U)	8.558	8.538	H−C(8)(A)	8.212	8.215
OH−C(3U)(U)	5.396	5.333	H−C(1')(A)	6.359	6.362
H−C(1')(U)	5.369	6.080	H−C(2')(A)	2.976	2.977
H−C(2')(U) (up)	2.037	1.992	H'−C(2')(A)	2.464	2.47
H'−C(2')(U) (down)	2.230	2.223	H−C(3')(A)	4.936	4.926
H−C(3')(U)	4.246	4.195	H−C(4')(A)	4.199	4.182
H−C(4')(U)	3.919	3.945	H−C(5')(A)	3.234	3.245
H−C(5')(U)	3.960	3.93	H'−C(5')(A)	3.186	3.158
H'−C(5')(U)	3.754	3.750			

from distant parts of the dinucleotide, such as those of the uracil base and the adenine deoxyribose (H−C(**6**)/H−C(3')(A) and 2 H−C(2')(A); H−C(**5**)(**U**)/H−C(3')(A) and 2 H−C(2')(A)), those between the adenine base and deoxyribose H−C(**2**)(**A**)/ H−C(5')(A) and H−C(2')(A); H−C(**8**)/H−C(1')(A), H−C(3')(A), H−C(4')/A), H−C(5')(A), and 2 H−C(2')(A)), and contacts between the two deoxyribose units H−C(5')(**U**) and H−C(4')(**U**)/2 H−C(2')(A); H'−C(5')(**U**)/2 H−C(2')(A); H−C(**1'**)(**A**)/H−C(4')(U) and H−C(5')(U), and H'−C(5')((U))[2]). Only a few observed NOE contacts could not be explained on the basis of the molecular model. A strong observed NOE that was not predicted by either the *A*- or *B*-form model is the NOE cross-peak H−C(4')(A)/H−C(3')(U). The appearance of this cross-peak could be explained if rotation about the backbone allows two riboses to slide across each other. The similarity of the conformations adopted by the dinucleotide and those of the *A*- and *B*-forms of DNA might be due to favorable stacking interactions between the bases and/or protecting groups [22].

The data do not, however, suggest that the dinucleotide has a strong preference for either of the tested two conformations. Plots of observed NOE *vs.* r^{-6} (where *r* is the calculated distance) showed a great deal of scatter for both the distances calculated for the *A* and *B* conformations (data not shown), indicating that neither model accurately reflects the conformation adopted by the compound. Analysis of ribose-proton splitting patterns (which correlate with the conformation of the ring [23]) indicates that neither the uridine nor adenosine deoxyribose of either isomer has a great preference for adopting an *N*- *vs.* *S*-conformation, and suggests a rapid equilibrium between the available states. The data agree with a one-dimensional NOE and *J*-coupling analysis previously carried out for dithymidine boranophosphate isomers [24].

The similarity of the one-dimensional ^{1}H-NMR spectra of the two isomers **5a** and **5b** suggests that the configuration at the P-atom has little effect on the conformation. The chemical shifts and splitting patterns of the ribose-proton resonances for equivalent protons were very similar for the two isomers, indicating similar ribose and linkage conformations. Additional support for the contention that the config-

---

[2])   For convenience, A, U, and C are used instead of A_d, U_d, and C_d, respectively, for the 2'-deoxynucleoside moieties.

uration at the P-atom does not affect the overall conformation of the dinucleotides is provided by a statistical comparison of the integrated volumes of the cross-peaks in their NOESY plots. If the two isomers adopted substantially different conformations, the differences in the NOEs of protons whose proximities are constrained by their being in the same ribose or nucleoside unit would be expected to display less variation than the NOEs of protons that are not constrained. We found that this was not the case. Of the 46 pairs of data used in the analysis, 8 had a difference in NOE intensity greater than the standard deviation in the differences ($|\Delta i_m| > \sigma$, symbols defined in the *Exper. Part*). The data shows that the rate of occurrence of peaks having $|\Delta i_m| > \sigma$ was greatest when the protons were within a single ribose unit (3 of 14 for contacts between two ribose protons of either A or U^2), next greatest if between a base proton and a ribose of the same nucleoside (4 of 16), rarer for contacts between protons of the two ribose units (1 of 9), and were not observed for contacts between the base of one and the ribose of the other (0 of 9). Since the integrated volumes change with r^6, we feel this provides good evidence that conformational behaviors of the two isomers were very similar.

Other authors have assigned configuration at the P-atom to substituted nucleotide 'dimers' using either coupling constants between the P-atom and ^{13}C of the ribose [25], or from NOEs to $H-C(4')$ at the 5'-terminus of the 'dimer' [21b,c][26]. Neither of these approaches worked for our case. Lines in the ^{13}C-NMR spectra of the two isomers were too broad for us to make an accurate comparison of coupling constants. Comparison of the NOEs of the two isomers from the adenosine $H-C(4')$ (as done in the case of the methyl phosphate NMR study) was not possible because of overlap of the $H-C(4')(A)$ resonance of isomer **5b** with the $H-C(3')(U)$ resonance. We have had more success using the NOE contacts to the borane to distinguish the configuration at the P-atom.

Significantly, the NOESY spectrum showed cross-peaks from a variety of the protons to the borane protons (*Table 2*), with integrated cross-peak volumes dependent on the isomer. The differences in the NOEs observed to the borane ^1H-resonances of the two isomers were much greater than the differences observed for other protons, reflecting the different environments of the borane moiety in the two isomers. Of the 46 pairs of non-borane NOE peaks, 38 differed by less than the standard deviation; in contrast, only 3 of the 10 borane contacts differed by less than 1 σ, and 5 of the 10 differed by more than 2 σ (see *Exper. Part*).

Table 2. *Intensity of the NOESY Contacts to the BH_3 H-Atoms*

Borane to H contact[2])	5a	5b	Borane to H contact	5a	5b
2'-Deoxyuridine:			2'-Deoxyadenosine:		
$H-C(5)(U)$	1.245	1.157	$H-C(3')(A)$	3.363	2.775
$H-C(6)(U)$	1.123	1.292	$H-C(4')(A)$	1.693	1.247
$H-C(2')(U)$	0.425	0.556	$H-C(5')(A)$	0.485	0.000
$H-C(3')(U)$	0.546	$-^a)$	$H'-C(5)(A)$	0.666	0.000
$H-C(4')(U)$	0.329	$-^b)$			
$H-C(5')(U)$	0.451	1.072			
$H'-C(5')(U)$	0.752	1.102			

^{a}) Overlaps the $H-C(4')(A)$ resonance. ^{b}) Overlaps with $H-C(5')(U)$ resonance.

Both isomers displayed NOE transfer to the borane resonance from the 2'-deoxyadenosine H−C(1') and H−C(3') and the 2'-deoxyuridine H−C(6). In the spectrum of isomer **5b**, there was a NOE to a position where the signals from the 2'-deoxyadenosine H−C(4') and the 2'-deoxyuridine H−C(3') protons overlapped. In **5a**, the H−C(4')(A) and H−C(3')(U) resonances were resolved, and both contributed NOEs to the spectrum. Significant differences in the cross-peaks to the borane in the two spectra included the NOEs from H−C(4') and H−C(5')(U) and H'−C(5')(U) and from 2 H−C(5')(A). This is clearly demonstrated in the portion of the 2D spectra reproduced in *Fig. 2*. While better information would surely come with a more rigid structure, our NOE data is most consistent with isomer **5a** having the [P(R)] configuration, and isomer **5b** being [P(S)].

*Assignment of the **5a** and **5b** Configuration by Parallel Synthesis of the Borano-phosphate and Phosphorothioate Analogs.* Conversion of a phosphonate diester to a boranophosphate proceeds under retention of the configuration at the P-atom [16]. Thus, based on our 2D-NMR assignments, we expect that fast eluting H-phosphonate **4a** has a [P(R)] configuration and slow eluting **4b** has [P(S)]. In principle, the elution order of protected dinucleoside *H*-phosphonates in silica-gel chromatography strongly correlates with their configuration, being [P(R)] for the fast-eluting and [P(S)] for the slow-eluting isomer [15]. However, as the *H*-phosphonates **4** are new compounds, we could not rely entirely on the elution order to establish their configuration.

Fortunately, we could take advantage of a study by *Seela* and *Kretschmer* [15a,b], who demonstrated that sulfurization of *H*-phosphonates occurred with retention of the configuration. Both the phosphorothioate and the boranophosphate dinucleosides can be prepared with retention of configuration from a common *H*-phosphonate intermediate. Since assignments of dinucleoside phosphorothioate configuration are well-established [20a][27], we could confirm by parallel synthesis the configurational assignment of the dinucleoside boranophosphates. Thus, the *H*-phosphonates **4a** and **4b** were treated as in *Scheme 1*, but the boronation step was replaced by thioation with elemental sulfur. The phosphorothioate analogs of **5a** and **5b** were then deprotected in analogy to the transformations represented in *Scheme 3*. The resulting isomers of 2'-deoxyadenosin-3'-yl 2'-deoxycytidin-5'-yl phosphorothioate were characterized by their relative mobility on reversed-phase HPLC and their resistance to snake venom phosphodiesterase (not shown). The study revealed that the phosphonate **4a** is converted to phosphorothioate 'dimer' having [P(S)] configuration, thereby confirming that the faster migrating phosphonate 'dimer' has indeed an [P(R)] configuration[3]). Therefore, assignment of the boranophosphate isomers made by 2D-NMR fully agrees with the data obtained by the parallel chemical synthesis of the well-characterized phosphorothioate analogs.

*Preparation of the 2'-Deoxyadenosin-3'-yl 2'-Deoxyuridin-5'-yl Boranophosphates **7a** and **7b** and Study of Their Resistance toward Snake Venom Phosphodiesterase (SVP).* Deprotection of **5a** and **5b** (*Scheme 3*) required optimization of the standard conditions. A number of reports have indicated that the acidic removal of a 5'-(MeO)₂Tr

---

[3])   The change in the assignment [P(S)] to [P(R)] is due to the *CIP* rules, which consider the H-atom in the phosphonates as the least preferred, and the S-atom in the phosphorothioates as the most preferred substituent.

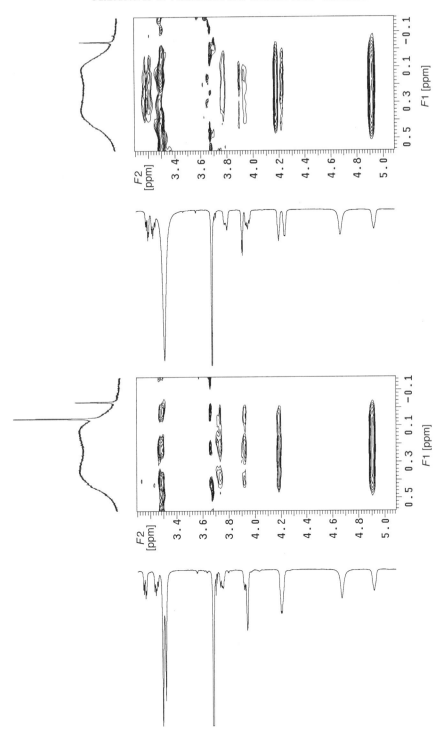

Fig. 2. *Fragments of 2D-NOESY plots of* **5a** *(right) and* **5b** *(left), illustrating interactions of BH₃ and deoxyribose protons*

group caused partial degradation of the boranophosphate bond [28] or loss of a cyanoborane group in base-boronated nucleotides [29]. A detailed study of this phenomena will be published elsewhere. Briefly, we found that the addition of $(MeO)_2Tr$-cation scavengers efficiently suppressed the loss of the boranophosphate group. Borane-pyridine complex was used as a scavenger at 5-fold molar excess over **5a** (**5b**). Without intermediate isolation of the 5′-OH derivative, the reaction mixture was treated with concentrated ammonia for 4 h at room temperature to convert the 2′-deoxy-4-*O*-(4-nitrophenyl)uridine residue to 2′-deoxycytidine. The resulting N^6-benzyl-2′-deoxyadenosin-3′-yl 2′-deoxycytidin-5′-yl boranophosphates **6a** and **6b** were isolated by reversed-phase HPLC with 85% yield.

Scheme 3. *Synthesis of the Deprotected 2′-Deoxyadenosin-3′-yl 2′-Deoxycytidin-5′-yl Boranophosphates* **7a** *and* **7b**

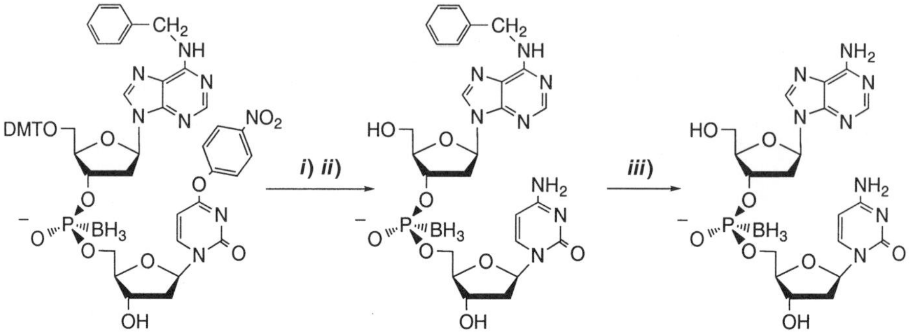

5a (5b) $[P(R)]$ $([P(S)])$ **6a (6b)** $[P(R)]$ $([P(S)])$ **7a (7b)** $[P(R)]$ $([P(S)])$

i) H⁺, BH₃·pyridine. *ii*) NH₄OH. *iii*) $(NH_4)_2S_2O_8$.

We tried a number of methods for N^6-benzyladenine deprotection. Oxidative deprotection of **6a** or **6b** with $NaIO_4$ in the presence of catalytic amounts of RuO_2 [18b,c] resulted in a large amount of by-products and a low yield. The reductive debenzylation with $(NH_4)HCOO$ in the presence of Pd/C [18d] was very inefficient. Reduction through hydrogenolysis as exampled in [18e] was not tried. The most satisfactory results so far were obtained by $(NH_4)_2S_2O_8$ deprotection as described in [18a]. While the original report called for addition of the oxidant in one batch and reaction at 80° for 2 h, we achieved a better yield when the oxidant was added in small portions at 2 h intervals at 60°. Although the yields of the completely deprotected boranophosphate dinucleotide did not exceed 35%, sufficient amounts of the final compounds **7a** and **7b** of 2′-deoxyadenosin-3′-yl 2′-deoxycytidin-5′-yl boranophosphates were isolated by reversed-phase HPLC.

The elution order of the deprotected isomers **7a** and **7b** on the reversed-phase HPLC was the same as for the protected **6a** and **6b**, *i.e.* the $[P(S)]$ isomer was eluted faster than the $[P(R)]$ isomer. It coincided with the elution pattern of the previously synthesized dithymidine boranophosphate isomers [7a][16][24]. The reversed-phase HPLC analysis showed no evidence of boranophosphate-bond racemization during the deprotection procedure.

It was shown previously that oligothymidine boranophosphates are hydrolyzed by snake venom phosphodiesterase (SVP) in a stereospecific manner [7a,b,d]. While the

[$P(S)$] isomer of dithymidine boranophosphate is much more stable (330 times) than its natural counterpart, it can be degraded at high enzyme concentration [7a,b]. In contrast, the [$P(R)$] isomer is completely resistant to the enzyme.

To confirm that the SVP differentiation of boranophosphate isomers is a general phenomenon, we studied the stability of the two isomeric boranophosphates **7a** and **7b** toward SVP hydrolysis. As expected, the [$P(S)$] isomer **7b** was hydrolyzed slowly, giving 2′-deoxyadenosine and 2′-deoxycytidine[4]), and the [$P(R)$] isomer **7a** was completely inert after even prolonged incubation (*Fig. 3*). Even after 65 h of incubation with SVP at 37°, no sign of **7a** hydrolysis was observed. The enzyme in this mixture maintained at least 20% of its original activity at the end of the experiment. The normal d(ApC) was hydrolyzed by SVP at the studied conditions very quickly (*Fig. 3*, insert). Comparison of the initial rates of hydrolysis suggests that the [$P(S)$] boranophosphate isomer is *ca.* $5 \cdot 10^2$ times more resistant to SVP than the natural counterpart. We conclude that the substrate properties of **7a,b** towards SVP are very similar to those of the dithymidine boranophosphates [7a,b] and are in good agreement with those for boranophosphate RNA dimers [30].

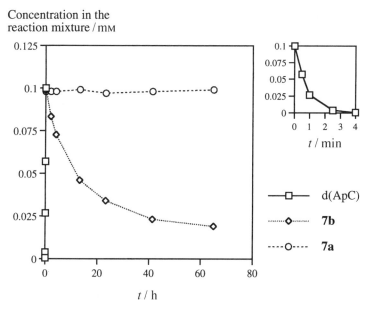

Fig. 3. *Hydrolysis of d(ApC) (insert) and of their boranophosphate analogs* **7a** *and* **7b** *by SVP. Conditions:* 0.1 mM dinucleotide concentration, 0.03 U/ml of SVP activity in 1.25 ml of 0.1M NaCl, 0.1M *Tris* · HCl (9.0), 15 mM MgCl$_2$ buffer, and 0.2 mg/ml of bovine serum albumin at 37°. Hydrolysis of d(ApC) occurred in an appropriate (fast) time scale (insert).

---

[4])   Expected 2′-deoxycytidine 5′-boranophosphate is not stable enough under the buffer conditions and slowly hydrolyzed to 2′-deoxycytidine and boranophosphate [32]. An analogous observation was made for the dithymidine boranophosphate hydrolysis [7a].

Experimental Part

General. 2′-Deoxyuridine, pyrrolidine, phosphorus oxychloride, 4-nitrophenol, 4-(dimethylamino)pyridine (DMAP), Et$_3$N, 4,4′-dimethoxytrityl chloride ((MeO)$_2$TrCl), hydrazine hydrate, *N,O*-bis(trimethylsilyl)acet-amide (BSA), borane-*N,N*-diisopropylethylamine (^{i}Pr$_2$EtN · BH$_3$), borane-pyridine, levulinic acid, and sulfur were purchased from *Aldrich Chemical Co.* Pivaloyl chloride was purchased from *Sigma.* Levulinic anhydride was prepared according to [31]. 3% CHCl$_2$COOH in CH$_2$Cl$_2$ and N^6-benzoyl-2′-deoxy-5′-O-(dimethoxytrityl)-adenosine 3′-*H*-phosphonate were purchased from *Glen Research* (USA). High-performance liquid chroma-tography (HPLC): *Waters-600E* controller system equipped with a *991* photodiode array detector; direct-phase HPLC on a 19 × 300 mm μ-*Porasil* column (*Waters*) using the isocratic mixture AcOEt/MeOH/CH$_2$Cl$_2$ 40:2:58; anal. reversed-phase HPLC on a 3.9 × 300 mm *DeltaPakC18* (15 μm) column, linear gradient 0–30% MeCN in 0.02M (Et$_3$NH)HCO$_3$ (pH 7.5) (system *A*) or isocratic 8% MeCN in 0.02M KH$_2$PO$_4$ (pH 6.5) (system *B*), 1 ml/min flow rate; prep. reversed-phase HPLC on a 21.4 × 250 mm *Microsorb RP-18* column or 25 × 100 mm *RadialDeltaPak* cartridge, linear gradient 0–25% MeCN in 0.02M (Et$_3$NH)HCO$_3$ (pH 7.5) (system *C*) or linear gradient 40–80% MeCN in 0.02M (Et$_3$NH)HCO$_3$ (pH 8.0) (system *D*), 6 ml/min flow rate. 1D-^{1}H- and ^{31}P-NMR Spectra: *Varian Inova 400* spectrometer at 399.9 and 161.9 MHz, resp., chemical shifts δ referenced for ^{1}H to internal deuterated solvent signals (CDCl$_3$ or D$_2$O) and for ^{31}P to external 85% H$_3$PO$_4$ in H$_2$O as a standard. 1D- and 2D-COSY and -NOESY of **5a** and **5b**: at 16 mM concentration in (D$_6$)DMSO; *Varian-Unity-600-MHz* NMR spectrometer in the Duke NMR Spectroscopy Center; 2D-NOESY experiments with 1.0 s delay time, 0.5 s mixing time, and 6598.5 Hz sweep width. Fast-atom-bombardment mass spectrometry (FAB-MS): *Jeol-300* mass spectrometer, pos. and neg. ion modes; high-resolution FAB MS in neg. ion mode with polyethylene glycol *600* as a matrix.

Statistical Comparison of NOESY Cross-peak Volumes. To determine whether the two isomers adopt similar conformations in solution, cross-peak volumes from the NOESY experiment of the first isomer were compared to those from the NOESY of the second isomer. The volumes were measured using VNMR *Varian* software version 6.1A. Only the negative NOE peaks were considered, and peaks from scalar coupled protons, and those involving contacts to the borane protons were omitted from the analysis. To account for small diferences in the concentrations of the two isomers, the integrated volumes of the cross-peaks from the second isomer were scaled so that the summed intensities of all spots of the two isomers were the same. The variation in the measured peak volumes between the first isomer and the scaled values of the second isomer was investigated for the non-borane peaks. The differences between the volumes of a given cross-peak of the two isomers is Δi_m ($\Delta i_m = i_{1,m} - i_{2,m}$, where i represents the integrated volume of the peak indicated by its subscript; the subscript m refers to a specific pair of protons, such as H–C(1′)(U)/H–C(3′)(U), and the subscripts 1 and 2 refer to the spectra of isomer 1 and isomer 2, resp.). Since the intensities of the isomer-2 peaks were scaled to resemble those of isomer 1, the average of the differences in the volumes of the peaks of the two isomers, Δi_{avg} (calculated as $\Delta i_{avg} = (1/n)\Sigma\Delta i_m$ where n is the number of peaks considered) is, by definition, zero. The distribution of Δi values about the mean value (zero) was characterized by its standard deviation σ (calculated as $\sigma^2 = (1/n - 1))\Sigma(\Delta i_m)^2$), and the differences in each pair of peaks was compared to the standard deviation: $|\Delta i_m|/\sigma$. The 46 peaks included in the analysis were[2]: H–C(6)(U)/H–C(3′)(U), 2 H–C(5′)(U), 2 H–C(2′)(U), H–C(3′)(A), 2 H–C(2′)(A); H–C(5)(U)/H–C(3′)(U), H–C(4′)(U), H–C(5′)(A), H′–C(2′)(U), H–C(3′)(A), 2 H–C(2′)(A); H–C(1′)(U)/H–C(3′)(U), H–C(5′)(U), H–C(4′)(U); H–C(5′)(U) and H–C(4′)(U)/2 H–C(2′)(U), 2 H–C(2′)(A); H′–C(5′)(U)/2 H–C(2′)(U), 2 H–C(2′)(A); H–C(2)(A)/ H–C(5′)(A), H–C(2′)(A); H–C(8)(A)/H–C(1′)(A), H–C(3′)(A), H–C(4′)(A), H–C(5′)(A), 2 H–C(2′)(A); H–C(1′)(A)/H–C(3′)(A), H–C(4′)(A), 2 H–C(5′)(A), H–C(4′)(U) and H–C(5′)(U), H–C(5′)(U); 2 H–C(2)(A)/2 H–C(5′)(A); H–C(4′)(A)/2 H–C(2′)(A). The value of σ was 24% of the average value of the volumes of the 92 peaks.

2′-Deoxy-4-O-(4-nitrophenyl)uridine (**1**) was synthesized according to [19c].

2′-Deoxy-5′-O-(dimethoxytrityl)-4-O-(4-nitrophenyl)uridine (**2**). Nucleoside **1** (0.7 g, 2 mmol) was dried by repeated co-evaporation with dry pyridine and finally dissolved in dry pyridine (10 ml). DMAP (12 mg, 0.1 mmol) and Et$_3$N (0.35 ml, 2.4 mmol) were added to the soln., followed by (MeO)$_2$TrCl (0.68 g, 2 mmol). After 2 h stirring, the mixture was diluted with MeOH (3 ml), evaporated, and co-evaporated with toluene (to remove pyridine), and the residue was applied to column chromatography (100 ml of silica gel, 0–10% MeOH/ CH$_2$Cl$_2$). The product fractions (TLC (10% MeOH/CH$_2$Cl$_2$): R_f 0.52) gave **2** (1.0 g, 78%). Yellowish foam. ^{1}H-NMR (CDCl$_3$): 2.26 (*m*, 1 H–C(2′)); 2.59 (*m*, 1 H–C(2′)); 3.42–3.56 (*m*, 2 H–C(5′)); 3.78 (*s*, 2 MeO); 4.03 (*m*, H–C(4′)); 4.49 (*m*, H–C(3′)); 5.81 (*d*, *J* = 7.2, H–C(5)); 6.19 (*t*, *J* = 6.0, H–C(1′)); 6.83 (*m*, 2 arom. H);

7.26–7.38 (m, 13 H, (MeO)$_2$*Tr*); 8.25 (m, 2 arom. H); 8.33 (d, $J = 7.6$, H–C(6)). HR-FAB-MS (pos.): 652.2 ($C_{36}H_{34}N_3O_9^+$, $[M+H^+]$; calc. 651.66).

2'-Deoxy-3'-O-levulinoyl-4-O-(4-nitrophenyl)uridine (**3**). Starting **2** (0.42 g, 0.6 mmol) was dried over P$_2$O$_5$ *in vacuo* overnight and dissolved in dry pyridine (10 ml). A cat. amount of DMAP (12 mg, 0.1 mmol) and levulinic anhydride (0.3 ml, 1.2 mmol) was added and the resulting soln. stirred at r.t. The reaction was completed within 2 h (TLC (10% MeOH/CH$_2$Cl$_2$)) and quenched by adding H$_2$O (2 ml). The mixture was evaporated and the residue treated with 3% CHCl$_2$COOH in CH$_2$Cl$_2$ (20 ml) for 4 min and washed with aq. NaHCO$_3$ soln. Evaporation of the org. layer gave an orange gum, which was purified by chromatography (silica gel, 0–10% MeOH/CH$_2$Cl$_2$): **3** (0.21 g, 80%). Slightly yellow solid. ^{1}H-NMR (CDCl$_3$): 2.17 (s, Me); 2.34 (m, 1 H–C(2')); 2.56 (m, CH$_2$); 2.64 (m, 1 H–C(2')); 2.75 (m, CH$_2$); 3.97–3.85 (m, 2 H–C(5')); 4.14 (m, H–C(4')); 5.30 (m, H–C(3')); 6.16 (d, $J = 7.6$, H–C(5)); 6.19 (t, $J = 6.4$, H–C(1')); 7.32 (m, 2 arom. H); 8.11–8.35 (m, H–C(6), 2 arom. H). HR-FAB-MS (pos.): 448.11 ($C_{20}H_{21}N_3O_9^+$, MH^+; calc. 448.40).

N^6-*Benzoyl-2'-deoxy-5'-O-(4,4'-dimethoxytrityl)adenosin-3'-yl 2'-Deoxy-3'-O-levulinoyl-4-O-(4-nitrophenyl)-uridin-5'-yl H-Phosphonates* **4a** *and* **4b**. A mixture of N^6-benzoyl-2'-deoxy-5'-O-(4,4'-dimethoxytrityl)adenosine 3'-phosphonate (400 mg, 0.48 mmol) and **3** (250 mg, 0.250 mmol) was dried over P$_2$O$_5$ under vacuum overnight and then dissolved in 10 ml of dry pyridine. Pivaloyl chloride (0.2 ml, 1.44 mmol) was added at r.t., and the mixture was stirred for 10 min. CH$_2$Cl$_2$ (10 ml) was added and the soln. was extracted with aq. sat. NaHCO$_3$ soln. (10 ml). The combined org. layer and washings were dried and evaporated. The resulting mixture contained a 1:1 mixture **4a/4b** (^{31}P-NMR: δ 9.18 and 10.32 ppm). Separation of diastereoisomers **4a/4b** by HPLC (μ-*Porasil* (19 × 300 mm column), isocratic (AcOEt/MeOH/CH$_2$Cl$_2$ 40:2:58) gave faster migrating **4a** ($[P(R)]$; 190 mg, 34%; 98% diastereoisomer purity) and slower migrating **4b** ($[P(S)]$; 220 mg, 39%; 84% diastereoisomer purity).

Sulfurization of Diastereoisomers **4a** *and* **4b**. Starting **4a** or **4b** (1 mg, *ca.* 1 μmol of each) was dried over P$_2$O$_5$ under vacuum overnight and then dissolved in dry pyridine (0.1 ml). BSA (5 μl, 0.02 mmol) and Et$_3$N (3 μl, 0.02 mmol) were added, and the soln. was stirred at r.t. for 1 h. Sulfur (2 mg, 0.5 mmol) was added and the mixture stirred for another 16 h. After filtration, the soln. was evaporated and the residue treated with 0.2 ml of 3% CHCl$_2$COOH in CH$_2$Cl$_2$. After 5 min, the acid was neutralized with 20 μl of Et$_3$N, and the mixture was evaporated to oil. The oil was dissolved in NH$_4$OH/MeOH 1:1 (1 ml), the soln. incubated for 48 h at r.t. and then evaporated, and the residue redissolved in H$_2$O and applied to reversed-phase HPLC (*DeltaPakC18* column, 3.9 × 300 mm, linear gradient *A*). The resulting *2'-deoxyadenosin-3'-yl 2'-deoxycytidin-5'-yl phosphoro-thioate* obtained upon sulfurization of **4a** had a t_R of 22.5 min ($[P(S)]$ isomer), and that obtained upon sulfurization of **4b** a t_R of 20.9 min ($[P(R)]$ isomer).

N^6-*Benzyl-2'-deoxy-5'-O-(4,4'-dimethoxytrityl)adenosin-3'-yl 2'-Deoxy-4-O-(4-nitrophenyl)uridin-5'-yl Bora-nosphosphate* **5a** *and* **5b**. Starting **4a** or **4b** (200 mg, 0.17 mmol of each) was dried over P$_2$O$_5$ under vacuum overnight and dissolved in THF (5 ml). BSA (212 μl, 0.85 mmol) was added, and the soln. was stirred at r.t. for 1 h. Then boronating agent BH$_3$·^{i}Pr$_2$EtN (220 μl, 1.7 mmol) was added, and the mixture was stirred for another 30 min. The mixture was cooled in a freezer and added dropwise to cold hexane. The precipitate formed was washed with hexane, dried, and dissolved in pyridine (2 ml), and H$_2$O (18 μl, 1 mmol) was added. The mixture was immediately evaporated to an oil, redissolved in pyridine/AcOH 4:1 (4 ml) and treated with hydrazine hydrate (0.25 ml). After 5 min, pentane-2,4-dione (2 ml) was added to stop the reaction. The mixtures were evaporated and co-evaporated with toluene to remove traces of pyridine, and the residue dissolved in MeCN/aq. 0.2M KH$_2$PO$_4$ (pH 7.5) 1:1 and purified by reversed-phase HPLC (*Microsorb RP-18* column, gradient *D*).

Data of **5a**: 54 mg (27%). ^{1}H-NMR: *Table 1*. ^{31}P-NMR (CD$_3$CN): 93.84. HR-FAB-MS (neg.): 1051.30 ($C_{53}H_{53}BN_8O_{13}P_1^-$, M^-; calc. 1051.82).

Data of **5b**: 32 mg (18%). ^{1}H-NMR: *Table 1*. ^{31}P-NMR (CD$_3$CN): 93.28. HR-FAB-MS (neg.): 1051.40 ($C_{53}H_{53}BN_8O_{13}P_1^-$, M^-, calc. 1051.82).

N^6-*Benzyl-2'-deoxyadenosin-3'-yl 2'-Deoxycytidin-5'-yl Boranophosphates* **6a** *and* **6b**. Protected **5a** or **5b** (10 mg, 10 μmole of each) was dissolved in MeOH/H$_2$O 1:1 (0.2 ml) and treated with AcOH (0.8 ml) for 30 min. Pyridine-borane (5 μl, 0.05 mmol) was added during treatment to scavenge the (MeO)$_2$Tr cation. Then the soln. was diluted with NH$_4$OH soln. (20 ml), stirred for 4 h at r.t., and evaporated and the residue purified by reversed-phase HPLC (*Microsorb RP-18* column, gradient *C*). Product fractions were evaporated and co-evaporated several times with H$_2$O to remove (Et$_3$NH)HCO$_3$.

Data of **6a**: 5.1 mg (85%). ^{1}H-NMR (D$_2$O)^2): 8.08 (s, H–C(2)(A)); 7.99 (s, H–C(8)(A)); 7.66 (d, $J = 7.6$, H–C(6)(C)); 7.15–7.26 (m, 5 H, Bn); 6.24 (t, $J = 6.4$, 1 H, H–C(1')); 6.04 (t, $J = 6.4$, 1 H, H–C(1')); 5.67 (d, $J = 7.6$, H–C(5)(C)); 4.72 (s, 2 H, CH$_2$); 4.32–4.36 (m, 1 H); 4.12 (q, 1 H); 3.95–4.01 (m, 2 H); 3.94–3.89 (m, 1 H); 3.61–3.71 (m, 2 H); 2.74–2.66 (m, 1 H); 2.56–2.48 (m, 1 H); 2.26–2.19 (m, 1 H); 2.10–2.03 (m, 1 H); −0.1–0.6 (br. m, BH$_3$). HR-FAB-MS (neg.): 627.30 ($C_{26}H_{33}N_8BO_8P_1^-$; M^-; calc. 627.38).

Data of **6b**: 5.0 mg (84%). ^{1}H-NMR ($D_2O)^2$): 8.09 (*s*, H−C(2)(A)); 7.97 (*s*, H−C(8)(A)); 7.64 (*d*, *J* = 7.2, H−C(6)(C)); 7.27−7.15 (*m*, 5 H, Bn); 6.22 (*t*, *J* = 6.4, 1 H, H−C(1′)); 6.01 (*t*, *J* = 6.4, 1 H, H−C(1′)); 5.60 (*d*, *J* = 7.2, H−C(5)(C)); 4.82 (*s*, 2 H, CH₂); 4.36−4.31 (*m*, 1 H); 4.12−4.09 (*m*, 1 H); 4.00−3.89 (*m*, 2 H); 3.76−3.63 (*m*, 2 H); 2.71−2.66 (*m*, 1 H); 2.53−2.48 (*m*, 1 H); 2.22−2.17 (*m*, 1 H); 2.10−2.06 (*m*, 1 H); −0.1 to 0.6 (br. *m*, BH₃). HR-FAB-MS (neg.): 627.21 ($C_{26}H_{33}N_8BO_8P_1^-$, *M*⁻; calc. 627.38).

2′-Deoxyadenosin-3′-yl 2′-Deoxycytidin-5′-yl Boranophosphates **7a** *and* **7b**. The benzyl-protected **6a** or **6b** (5.0 mg, 8 µmoles) was dissolved in 10 ml of 1M KH₂PO₄/MeCN 3 : 2 (10 ml), heated to 60°, and treated with (NH₄)₂S₂O₈ (2 mg) every 2 h during an 8 h period. Isolation was performed by reversed-phase HPLC (*DeltaPak C18 Radial* 25 × 100 mm cartridge, 8% MeCN in 0.02M KH₂PO₄ (pH 6.5)).

Data of **7a**: 1.5 mg (35%). HPLC: t_R 14.4 min. ^{1}H-NMR ($D_2O)^2$): 8.08 (*s*, H−C(2)(A)); 7.96 (*s*, H−C(8)(A)); 7.65 (*d*, *J* = 7.6, H−C(6)); 6.20 (*t*, *J* = 6.8, 1 H, H−C(1′)); 6.02 (*t*, *J* = 6.4, 1 H, H−C(1′)); 5.67 (*d*, *J* = 6.9, H−C(5)); 4.82 (*m*, 1 H); 4.34 (*m*, 1 H); 3.85−4.00 (*m*, 2 H); 3.61−3.71 (*m*, 2 H); 2.88 (*m*, 1 H); 2.66 (*m*, 1 H); 2.22 (*m*, 1 H); 2.06 (*m*, 1 H); −0.25−0.65 (br. *m*, BH₃). ^{31}P-NMR (D_2O): 93.80. HR-FAB-MS: 537.1779 ($C_{19}H_{27}N_8BO_8P_1^-$, *M*⁻; calc. 537.2559).

Data of **7a**: 1.4 mg (33%). HPLC: t_R 11.4 min. ^{1}H-NMR ($D_2O)^2$): 8.11 (*s*, H−C(2)(A)); 7.97 (*s*, H−C(8)(A)); 7.64 (*d*, *J* = 7.2, H−C(6)(C)); 6.20 (*t*, *J* = 6.4, 1 H, H−C(1′)); 6.00 (*t*, *J* = 6.0, 1 H, H−C(1′)); 5.59 (*d*, *J* = 7.2, 1 H, H−C(5)(C)); 4.34 (*m*, 1 H); 4.11 (*m*, 1 H); 3.88−4.00 (*m*, 2 H); 3.64−3.77 (*m*, 2 H); 2.68 (*m*, 1 H); 2.51 (*m*, 1 H); 2.19 (*m*, 1 H); 2.07 (*m*, 1 H); −0.3−0.65 (br. *m*, BH₃). ^{31}P-NMR (D_2O): 93.90. HR-FAB-MS: 537.1811 ($C_{19}H_{27}BN_8O_8P_1^-$, *M*⁻; calc. 537.2559).

Enzymatic Hydrolysis of **7a**, **7b**, *and d(ApC)* was carried out at 0.1 mM concentration of dinucleotide and 1 mg/ml (0.03 U/ml) of SVP (*Sigma*) in 1.25 ml of 0.1M NaCl, 0.1M *Tris* · HCl (9.0), 15 mM MgCl₂ buffer, 0.2 mg/ml of bovine serum albumin (*Sigma*) at 37°. At appropriate times, 0.25 ml aliquots were withdrawn, the hydrolysis was stopped by addition of AcOH (50 µl), and the mixture analyzed by anal. reversed-phase HPLC (gradient *A*). Control samples were incubated at identical conditions without SVP. For the reaction mixture containing **7a**, the enzyme activity was checked after 65 h of the hydrolysis at 37° by addition of d(ApC) (0.1 mM). The rate of d(ApC) hydrolysis was determined by reversed-phase HPLC.

This work was supported by *NIH* grant *R01-GM-57693* and *DOE* grant *DE-FG 02-97ER62376* to *B.R.S.* The Duke NMR Spectroscopy Center is partially supported by *NIH* grant *P30-CA-14236*. NMR Instrumentation in the Duke NMR Spectroscopy Center was funded by the *NSF*, the *NIH*, the *NC Biotechnology Center*, and Duke University. We thank Prof. *Terry Oas* for help in preparing the molecular models of *A*- and *B*-form DNA.

REFERENCES

[1] S. Verma, F. Eckstein, *Annu. Rev. Biochem.* **1998**, *67*, 99.

[2] S. Agrawal, Q. Zhao, *Curr. Opin. Chem. Biol.* **1998**, *2*, 519.

[3] B. Hyrup, P. E. Nielsen, *Bioorg. Med. Chem.* **1996**, *4*, 5.

[4] J. Summerton, D. Weller, *Antisense Nucleic Acid Drug Dev.* **1997**, *7*, 187.

[5] R. K. Delong, P. S. Miller, *Antisense Nucleic Acid Drug Dev.* **1996**, *6*, 273.

[6] B. R. Shaw, D. S. Sergueev, K. He, K. W. Porter, J. S. Summers, Z. A. Sergueeva, V. K. Rait, in 'Methods in Enzymology', Ed. M. I. Phillips, Academic Press, Orlando, FL, 1999, Vol. 313 Part A, p. 226; B. R. Shaw, J. Madison, A. Sood, B. F. Spielvogel, *Methods Mol. Biol.* **1993**, *20*, 225.

[7] a) F. Huang, 'Ph. D. Thesis', Duke University, 1994; b) F. Huang, A. Sood, B. F. Spielvogel, B. R. Shaw, *J. Biomol. Struct. Dyn.* **1993**, *10*, a078; c) H. Li, K. Porter, F. Huang, B. R. Shaw, *Nucleic Acids Res.* **1995**, *23*, 4495; d) D. S. Sergueev, B. R. Shaw, *J. Am. Chem. Soc.* **1998**, *120*, 9417.

[8] J. S. Summers, D. Roe, P. D. Boyle, M. Colvin, R. R. Shaw, *Inorg. Chem.* **1998**, *37*, 4148.

[9] B. F. Spielvogel, A. Sood, B. R. Shaw, I. H. Hall, R. G. Fairchild, B. H. Laster, C. Gordon, in 'Progress in Neutron Capture Therapy for Cancer', Eds. B. J. Allen, D. E. Moore, B. V. Harrington, Plenum Press, New York, 1992, p. 211.

[10] K. W. Porter, D. J. Briley, B. R. Shaw, *Nucleic Acids Res.* **1997**, *25*, 1611.

[11] K. V. Rait, B. R. Shaw, *Antisense Nucleic Acid Drug Dev.* **1999**, *9*, 53; K. V. Rait, D. S. Sergueev, J. S. Summers, B. R. Shaw, *Nucleosides Nucleotides* **1999**, *18*, 1379.

[12] S. Agrawal, *TIBTECH* **1996**, *14*, 376; D. Yu, R. P. Iyer, D. R. Shaw, J. Lisziewicz, Y. Li, Z. Jiang, A. Roskey, S. Agrawal, *Bioorg. Med. Chem.* **1996**, *4*, 1685; T. Hamma, P. S. Miller, *Biochemistry* **1999**, *38*, 15333.

[13] F. Aboulela, G. Varani, *Curr. Opin. Biotech.* **1995**, *6*, 89; B. H. Robinson, C. Mailer, G. Drobny, *Annu. Rev. Biophys. Biomol. Struct.* **1997**, *26*, 629.

[14] D. Sergueev, A. Hasan, M. Ramaswamy, B. R. Shaw, *Nucleosides Nucleotides* **1997**, *16*, 1533.

[15] a) F. Seela, U. Kretschmer, *J. Chem. Soc., Chem. Commun.* **1990**, 1154; b) F. Seela, U. Kretschmer, *J. Org. Chem.* **1991**, *56*, 3861; c) H. Almer, J. Stawinski, R. Stromberg, M. Thelin, *J. Org. Chem.* **1992**, *57*, 6163.

[16] Z. A. Sergueeva, D. S. Sergueev, B. R. Shaw, *Tetrahedron Lett.* **1999**, *40*, 2041.

[17] Z. A. Sergueeva, D. S. Sergueev, B. R. Shaw, *Nucleosides, Nucleotides, Nucleic Acids* **2000**, *19*, 275.

[18] a) M. Sako, H. Ishikura, K. Hirota, Y. Maki, *Nucleosides Nucleotides* **1994**, *13*, 1239; b) A. Henderson, J. Riseborough, C. Bleasdale, W. Clegg, M. Eselgood, B. Golding, *J. Chem. Soc., Perkin Trans. 1* **1997**, 3407; c) S. Sarfati, V. Kansal, *Tetrahedron* **1988**, *44*, 6367; d) S. Ram, L. D. Spicer, *Tetrahedron Lett.* **1987**, *28*, 515; e) M. Saady, L. Lebeau, C. Mioskowski, *Tetrahedron Lett.* **1995**, *36*, 2239.

[19] a) C. B. Reese, A. Ubasawa, *Nucleic Acids Res., Symp. Ser.* **1980**, *7*, 5; b) K. J. Divakar, C. B. Reese, *J. Chem. Soc., Perkin Trans. 1* **1982**, 1171; c) A. Miah, C. B. Reese, Q. Song, *Nucleosides Nucleotides* **1997**, *16*, 53; d) C. R. Allerson, S. L. Chen, G. L. Verdine, *J. Am. Chem. Soc.* **1997**, *119*, 7423.

[20] a) F. Eckstein, *Annu. Rev. Biochem.* **1985**, *54*, 367; b) P. S. Miller, *Biotechnology* **1991**, *9*, 358.

[21] a) M. F. Summers, C. Powell, W. Egan, A. R. Byrd, D. W. Wilson, G. Zon, *Nucleic Acids Res.* **1986**, *14*, 7421; b) M. Bower, M. F. Summers, C. Powell, K. Shinozuka, J. B. Regan, G. Zon, D. W. Wilson, *Nucleic Acids Res.* **1987**, *15*, 4915; c) Loschner, T., Engels, J. W., *Nucleic Acids Res.* **1990**, *18*, 5083.

[22] W. Saenger, in 'Principles of Nucleic Acid Structure', Springer-Verlag, New York, 1983.

[23] J. van Wijk, B. D. Huckriede, J. H. Ippel, C. Altona, *Methods Enzymol.* **1992**, *211*, 286; L. J. Rinkel, C. Altona, *J. Biomol. Struct. Dyn.* **1987**, *4*, 621.

[24] H. Li, F. Huang, B. R. Shaw, *Bioorg. Med. Chem.* **1997**, *5*, 787.

[25] D. Machytka, E. Gacs-Baitz, Z. Tegyey, *Nucleosides Nucleotides* **1998**, *17*, 2311.

[26] L. S. Kan, D. M. Cheng, P. S. Miller, J. Yano, P. O. P. Ts'o, *Biochemistry* **1980**, *19*, 2122.

[27] P. M. Burger, F. Eckstein, *Biochemistry* **1979**, *18*, 592; P. M. Burger, B. K. Sathyanarayana, W. Saenger, F. Eckstein, *Eur. J. Biochem.* **1979**, *100*, 585; F. Eckstein, *Angew. Chem., Int. Ed.* **1983**, *22*, 423.

[28] A. Sood, B. R. Shaw, B. F. Spielvogel, E. S. Hall, L. K. Chi, I. H. Hall, *Pharmazie* **1992**, *47*, 833; A. P. Higson, A. Sierzchala, H. Brummel, Z. Zhao, M. H. Carithers, *Tetrahedron Lett.* **1998**, *39*, 3899; Y. Jin, G. Just, *Tetrahedron Lett.* **1998**, *39*, 6429.

[29] A. Hasan, H. Li, J. Tomasz, B. R. Shaw, *Nucleic Acids Res.* **1996**, *24*, 2150; A. Hasan, J. Tomasz, B. R. Shaw, *Bioconjugate Chem.* **1997**, *8*, 813.

[30] Y.-Q. Chen, F.-C. Qu, Y.-B. Zhang, *Tetrahedron Lett.* **1995**, *36*, 745; K. He, D. S. Sergueev, Z. A. Sergueeva, B. R. Shaw, *Tetrahedron Lett.* **1999**, *40*, 4601; K. He, B. R. Shaw, *Nucleic Acids Res., Symp. Ser.* **1999**, *41*, 99.

[31] A. Hassner, G. Strand, *J. Am. Chem. Soc.* **1975**, *97*, 1614.

[32] H. Li, C. Hardin, B. R. Shaw, *J. Am. Chem. Soc.* **1996**, *118*, 6606.

Synthesis of Boron-Containing ADP and GDP Analogues: Nucleoside 5'-(P^α-Boranodiphosphates)

by Jinlai Lin, Kaizhang He, and Barbara Ramsay Shaw*

Department of Chemistry, Duke University, Durham, NC 27708-0346, U.S.A.

New 5'-(P^α-boronated) analogues of the naturally occurring nucleoside diphosphates ADP and GDP were synthesized in good yields, i.e., adenosine 5'-(P^α-boranodiphosphate) (ADPαB; **5a**) and guanosine 5'-(P^α-boranodiphosphate) (GDPαB; **5b**). Their diastereoisomers were successfully separated by reversed-phase HPLC, and chemical structures were established via spectroscopic methods. The isoelectronic substitution of borane (BH$_3$) for one of the non-bridging O-atoms in phosphate diesters should impart an increase in lipophilicity and change in polarity in ADPαB and GDPαB. The boranated nucleoside diphosphates could be employed for investigations of the stereochemical course and metal requirements of enzymatic reactions involving ADP and GDP, and as carriers of ^{10}B in boron neutron-capture therapy (BNCT) for the treatment of cancer.

1. Introduction. – Analogues of naturally occurring nucleotides like cyclic-AMP, ADP, and ATP have a firm place in the array of tools biochemists use to unravel enzyme functions and mechanisms [1]. For example, substrate analogues have been constructed as reversible and irreversible inhibitors, transition-state analogues, and spectroscopic probes. A variety of nucleotides with modifications in the base, sugar, or phosphate residues have been synthesized. Of the latter, the most studied phosphate-modified nucleotides (*Fig.*) are nucleoside phosphorothioates [1] in which a non-bridging O-atom of a phosphate group is replaced by an S-atom. Nucleoside phosphorothioates have found wide applications in biochemistry and molecular biology [1–3]. Another type of modified nucleotide, in which one of the non-bridging O-atoms of a phosphate group is replaced with BH$_3$, are the P-boronated nucleotides (designated nucleoside boranophosphate or borane phosphonates) [2–18].

Figure. *Phosphate-modified nucleotides*

The borane moiety (BH$_3$) in the boranophosphate is isoelectronic with the O-atom in a phosphodiester. At neutral pH, the P-boronated nucleotides carry the same net negative charge as phosphodiesters and phosphorothioates (see *Fig.*), thus making boranophosphates soluble in aqueous solutions. However, distribution of charge density and consequently the polarity of the modified P-moiety would differ as would the metal binding and selectivity. Furthermore, since the BH_3 group in boranophosphates is isoelectronic as well as isosteric with the CH_3^+ group, it might be expected that *P*-boronated nucleotides would also exhibit some desirable properties of the nucleoside methylphosphonates [19], such as increased lipophilicity and resistance to nucleases [10][14]. Additionally, *P*-boronated nucleotides may offer a unique advantage over other modified congeners because the former can be used for boron neutron-capture therapy (BNCT) [20], a radiation therapy that can selectively destroy cells that have preferentially taken up boron.

The synthesis and properties of some P^α-boranonucleotides and oligonucleoside boranophosphates have been reported [4–18]. For example, the nucleoside triphosphate analogues, nucleoside 5′-P^α-boranotriphosphates (dNTPαB and NTPαB), are good substrates for DNA and RNA polymerases [7][9]. Yet, after incorporation, the boranophosphate linkages in DNA or RNA are more resistant to endo- and exonucleases than normal phosphodiesters [6][10]. Here, we investigated methods to synthesize new diphosphate analogues of ADP and GDP. Substitution of BH_3 for the phosphate O-atom introduces a center of chirality into the nucleotide, so that a P^α-boronated nucleoside diphosphate (NDPαB) exists as a pair of diastereoisomers. The two isomers could exhibit quite different behavior toward a particular enzyme. For example, nucleoside 5′-(P^α-thiodiphosphates) (NDPαS) and nucleoside 5′-(P^β-thiodiphosphates) (NDPβS) have been widely used in stereochemical studies of phosphotransferases and NDP-dependent synthetases [1–3]. In the presence of acetate kinase and Mg^{II}, the $[P(R)]$ isomer of ATPβS is obtained from β-prochiral ADPβS, while the $[P(S)]$ isomer of ATPβS is synthesized predominantly in the presence of pyruvate kinase and Cd^{II} [21–23]. Nucleoside *P*-boranodiphosphates should show interesting behavior and special properties that differ from those of their biologically functional analogues, the closely related nucleoside diphosphates (NDP) and nucleoside P^α-thiodiphosphates (NDPαS). In this paper, we report the synthesis of two diphosphate analogues: adenosine 5′-(P^α-boranodiphosphate) (ADPαB) and guanosine 5′-(P^α-boranodiphosphate) (GDPαB), and the isolation of their diastereoisomers.

2. Synthesis and Discussion. – The phosphoramidite approach (*Scheme 1*) for the preparation of nucleoside 5′-(P^α-boranodiphosphates) (NDPαB; N = A, G) is a modification of our previous procedure for the synthesis of thymidine 5′-(P^α-boranotriphosphate) [5]. Phosphitylation of 2′,3′-di-O-acetylribonucleoside **1** by 2-cyanoethyl tetraisopropylphosphorodiamidite in the presence of diisopropylammonium 1*H*-tetrazolide [24] gave phosphoramidite **2**. Intermediate **2** was treated *in situ* with excess borane-*N*,*N*-diisopropylethylamine complex to yield the borane-phosphoramidite complex **3**. Subsequent treatment of **3** with a mixture of ammonium hydroxide and MeOH gave compound **4**, which was identified by a typical broad peak centered at 94 ppm in the ^{31}P-NMR spectrum of the reaction mixture. After removing solvents, compound **4** was precipitated from dry AcOEt (further purification of compound **4** by ion-exchange chromatography (*QA-52* cellulose, HCO_3^-) resulted in lower yields due to some loss of the diisopropylamino group during the chromatography, as recorded afterward in ^{31}P- and ^{1}H-NMR spectra). The precipitated compound **4** was treated with excess tetrabutylammonium dihydrogen phosphate and 1H-tetrazole at 55° to yield the two diastereoisomers of nucleoside 5′-(P^α-boranodiphosphate) **5**. They were identified by the appearance of ^{31}P-NMR signals in the crude reaction mixture at *ca.* -8 ppm (*d*) for P(β) and 84 ppm (broad peak) for P(α). The crude mixture was purified by ion-exchange chromatography to give compound **5** as the ammonium salt with 32 – 36% overall yields (from **1** to **5**). Successful separation of the two diastereoisomers of compound **5** was achieved by reversed-phase HPLC. The chemical structures of the two diastereoisomers were characterized by ^{31}P-NMR, ^{1}H-NMR, MS, and HR-MS.

Scheme 1. *Synthesis of Nucleoside 5′-(P^α-Boranodiphosphates) ADPαB (**5a**) and GDPαB (**5b**)*

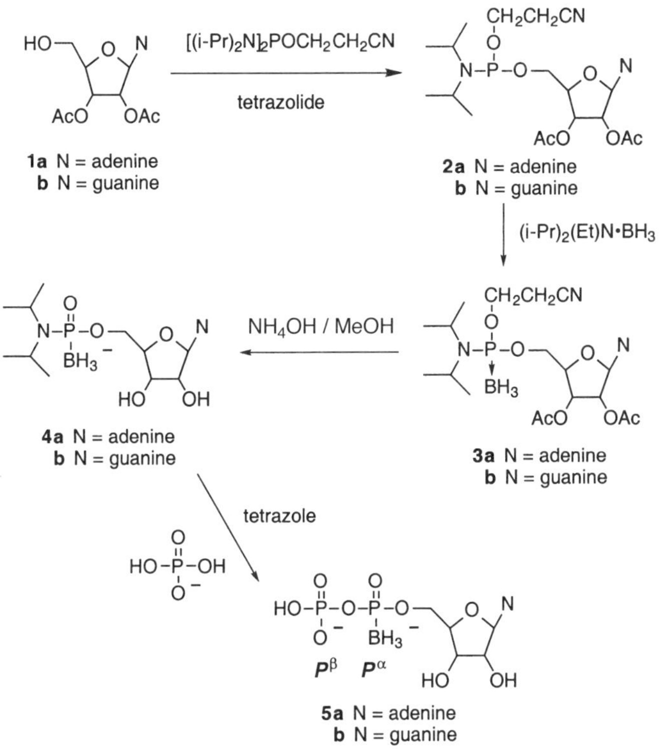

In our synthetic approach, base-unprotected nucleosides were chosen as starting materials. In the first step, we used 2-cyanoethyl tetraisopropylphosphorodiamidite ((^{i}Pr$_2$N)$_2$P(OCH$_2$CH$_2$CN)) as the preferred phosphitylating reagent. We avoided previously used 2-cyanoethyl diisopropylphosphoramidochloridite (^{i}Pr$_2$N-P(OCH$_2$CH$_2$CN)Cl) [5], which reacted with the unprotected NH$_2$ group in the nucleobases used here. With tetrazolide (*i.e.*, diisopropylammonium 1H-tetrazolide) as catalyst, only one of the two diisopropylamino groups of (^{i}Pr$_2$N)$_2$P(OCH$_2$CH$_2$CN) will be substituted by the free 5′-OH group of the nucleoside. By contrast, with 1H-tetrazole as catalyst, both diisopropylamino groups reacted with 5′-OH, resulting in the formation of undesirable by-products such as dinucleoside phosphite triester and a lower overall yield of NDPαB. The key step for the synthesis of NDPαB **5** was the phosphorylation of compound **4**. Catalyzed by 1*H*-tetrazole, the phosphorylation of intermediate **4** by tetrabutylammonium dihydrogen phosphate occurred intermolecularly to produce the desired NDPαB **5** as the major product. We did not observe any by-products arising from the intramolecular cyclization of the diisopropylamino group with either the 2'- or 3'-OH group. The presence of a borane group replacing a nonbridging O-atom at P(α) in NDPαB **5** produced a pair of diastereoisomers that could be resolved by HPLC. Like phosphorothioate NDPαS, the two diastereoisomers of NDPαB are anticipated to have different substrate properties towards nucleosidyl transferases and hydrolases, and should be useful for investigating the roles of phosphate and metal ions in biological processes to elucidate the stereochemical and metal requirements of the enzymatic reactions involving nucleoside diphosphates.

Based on our previous studies showing that boranophosphates are more lipophilic than natural phosphates [14], the introduction of BH_3 at P(α) or P(β) of nucleoside diphosphates is likely to enhance their lipophilicity and assist in their crossing the plasma membrane while maintaining the negative charge of the phosphate. The BH_3 group in an NDPαB is expected to form neither classical H-bonds nor coordinate metal ions as well as the O-atom in the parental ADP and GDP. Thus, the novel combination of high lipophilicity, reasonable water solubility, and nuclease resistance could be extremely useful for drug design [14].

In summary, a simple and efficient synthetic approach to the synthesis of nucleoside 5'-(P^α-boranodiphosphates) (ADPαB and GDPαB) was developed. The individual diastereoisomers were separated by reversed-phase HPLC and characterized by spectroscopic methods. Further investigations of the potential applications of nucleoside 5'-(P^α-boranodiphosphates) as mechanistic probes for enzymatic reactions involving natural nucleoside diphosphates, as substrates for enzymatic synthesis of boranophosphate polyribonucleotides [25], and as carriers of ^{10}B in boron neutron-capture therapy (BNCT) [20] for the treatment of cancer are under consideration.

Experimental Part

General. All solvents and reagents were of anal. grade and used without further purification unless otherwise indicated. ^1H-Tetrazolide [24] was prepared from 1*H*-tetrazole purchased from *ChemGenes*, and 2',3'-di-*O*-acetyladenosine, 2',3'-di-*O*-acetylguanosine, and borane-*N,N*-diisopropylethylamine complex were purchased from *Aldrich*. Ion-exchange chromatography: column packed with *QA-52* cellulose (HCO_3^-) from *Whatman*; linear gradient of 0.05M and 0.2M ammonium hydrogen carbonate (pH 9.6) as eluent. Reversed phase HPLC: *Waters* dual pump system in combination with a *Waters-600E* system controller, a *991* photodiode array UV detector, and a *NEC-Powermate-386* computer; the diastereoisomer mixture of NDPαB was loaded on a *DeltaPackC18* reversed-phase column and eluted with buffers containing 100 mM (Et_3NH)OAc (made from Et_3N and AcOH) (pH 6.8) and MeOH; t_R in min. ^1H- and ^{31}P-NMR Spectra: *Varian Inova-400* spectrometer at 400.0 and 161.9 MHz, resp., δ in ppm downfield from the internal $SiMe_4$ ($=0$ ppm) standard, *J* in Hz (^{31}P with broad-band decoupling); spectra of intermediates were recorded after addition of $CDCl_3$ or (D_6)DMSO to the sample from the reaction soln.

Adenosine 5'-(P^α-Boranodiphosphate) (**5a**) *and Guanosine 5'-(P^α-Boranodiphosphate)* (**5b**): *General Procedure.* To a soln. of 2',3'-di-*O*-acetyladenosine (**1a**; 1 mmol) or 2',3'-di-*O*-acetylguanosine (**1b**; 1 mmol) in anh. dimethylformamide (DMF; 1.6 ml for **1a**; 4.0 ml for **1b**), 2-cyanoethyl tetraisopropylphosphorodiamidite (1.1 mmol, 0.36 ml) and diisopropylammonium 1*H*-tetrazolide (1 mmol, 171 mg) were added under Ar. The reaction was continued for 30 min (for **1a**) or 4 h (for **1b**) at r.t., after which excess borane-*N,N*-diisopropylethylamine complex (0.8 ml) was added *in situ*, and the boronation was allowed to proceed for

2 h. After evaporation, the residue was extracted with AcOEt (40.0 ml) and H_2O (8.0 ml, 2×4.0 ml), the combined org. layer evaporated, and the residue treated with $NH_4OH/MeOH$ 1:1 (σ/σ) (5 ml) at r.t. for 4 h. The solvents were removed under vacuum, and dry AcOEt was added. Without purification, the precipitate **4a** or **4b** (*ca.* 0.5 mmol) was dissolved in anh. DMF (2.0 ml). Excess tetrabutylammonium dihydrogen phosphate (0.8 mmol, 272 mg) and 1*H*-tetrazole (2 mmol, 140 mg) were added, and the reaction was continued for 30 min at 55°. The mixture was extracted with H_2O (10.0 ml) and AcOEt (20.0 ml), the org. layer washed with H_2O (2×3.0 ml), the combined aq. layer concentrated, and the residue purified by ion-exchange column chromatography (15×300 mm LC column, *QA-52* cellulose, linear gradient of 0.005 and 0.2M ammonium hydrogen carbonate buffer (pH 9.56; 700 ml each)). The desired fractions were concentrated, and excess salts were removed by repeated lyophilization with deionized H_2O to yield the ammonium salt of adenosine 5′- (P^α-boranodiphosphate) (**5a**) and guanosine 5′-(P^α-boranodiphosphate) (**5b**). The two diastereoisomers of **5a** or **5b**, resp., were separated by ion-pair chromatography (reversed-phase column *Waters DeltaPakC18*, 3.9×300 mm, 15μ, 100 Å; isocratic conditions: 92% 100 mM ($Et_3NH)OAc$ (pH 6.8) and 8% MeOH as buffers for **5a**: 94% ($Et_3NH)OA$ (pH 6.8) and 6% MeOH as buffers for **5b**). After HPLC purification, the solvents were evaporated. The buffer components, ($Et_3NH)OAc$ and MeOH, were removed by repeated lyophilization.

Adenosine 5′-(P^α-Boranodiphosphate) (ADPαB; **5a**). Overall yield (from **1a** to **5a**), 36%. ^{31}P-NMR (D_2O): 80–84 (br., P(α)); -4.76, -4.93 (d, P(β)).

ADPαB, Isomer I: t_R 11.81. ^{1}H-NMR (D_2O): 8.43 (*s*, H−C(8)); 8.08 (*s*, H−C(2)); 5.97 (unresolved *m*, H−C(1′)); 4.44 (unresolved *m*, H−C(3′)); 4.21 (unresolved *m*, H−C(4′)); 4.09, 3.98 (2*m*, 2 H−C(5′)); 0.40, 0.15 (2 br., BH_3). ^{31}P-NMR (D_2O): 80–84 (br., P(α)); -7.09, -7.31 (d, $J = 35.13$, P(β)). FAB-MS: 424.06 (M^-). HR-MS: 424.0599 ($C_{10}H_{17}BN_5O_9P_2^-$, M^-; calc. 424.0593).

ADPαB, Isomer II: t_R 18.72. ^{1}H-NMR (D_2O): 8.41 (*s*, H−C(8)); 8.08 (*s*, H−C(2)); 5.97 (unresolved *m*, H−C(1′)); 4.36 (unresolved *m*, H−C(3′)); 4.22 (unresolved *m*, H−C(4′)); 4.09, 3.99 (2*m*, 2 H−C(5′)); 0.38, 0.12 (2br., BH_3). ^{31}P-NMR (D_2O): 80–84 (br., P(α)); -7.71, -7.90 (d, $J = 30.44$, P(β)). FAB-MS: 424.06 (M^-). HR-MS: 424.0588 ($C_{10}H_{17}BN_5O_9P_2^-$, M^-; calc. 424.0593).

Guanosine 5′-(P^α-Boranodiphosphate) (GDPαB; **5b**). Overall yield (from **1b** to **5b**), 32%. ^{31}P-NMR (D_2O): 80–84 (br., P(α)); -4.83 (*m*, P(β)).

GDPαB, Isomer I: t_R 7.94. ^{1}H-NMR (D_2O): 8.01 (*s*, H−C(8)); 5.76 (*d, J* = 6.0, H−C(1′)); 4.56 (*m*, H−C(2′)); 4.38 (*dd, J* = 3.2, 4.8, H−C(3′)); 4.17 (*d, J* = 2.8, H−C(4′)); 4.02–3.97 (*m*, 2 H−C(5′)); 0.44, 0.12 (2br., BH_3). ^{31}P-NMR (D_2O): 82–86 (br., P(α)); -8.96, -9.14 (d, $J = 29.14$, P(β)). FAB-MS: 440.09 (M^-). HR-MS: 440.0543 ($C_{10}H_{17}BN_5O_{10}P_2^-$, M^-; calc. 440.0544).

GDPαB, Isomer II: t_R 13.06. ^{1}H-NMR (D_2O): 8.00 (*s*, H−C(8)); 5.76 (*d, J* = 6.0, H−C(1′)); 4.57 (unresolved *m*, H−C(2′)); 4.32 (*dd, J* = 3.2, 5.2, H−C(3′)); 4.17 (*d, J* = 2.8, H−C(4′)); 4.05–3.97 (*2m*, 2 H−C(5′)); 0.41, 0.13 (2br., BH_3). ^{31}P-NMR (D_2O): 82–86 (br., P(α)); -8.97, -9.15 (d, $J = 29.14$, P(β)). FAB-MS: 440.07 (M^-). HR-MS: 440.0542 ($C_{10}H_{17}BN_5O_{10}P_2^-$, M^-; calc. 440.0544).

We thank Drs. *Ahmad Hasan*, *Vladimir Rait*, and *Bozenna K. Krzyzanowska* for their comments. This work was supported by *NIH* grant 1R01-GM57693 to *B. R. S.*

REFERENCES

[1] F. Eckstein, *Annu. Rev. Biochem.* **1985**, *54*, 367.
[2] H.-T. Ho, P. A. Fery, *Biochemistry* **1984**, *23*, 1978.
[3] F. Eckstein, R. S. Goody, *Biochemistry* **1976**, *15*, 1685.
[4] A. Sood, B. R. Shaw, B. F. Spielvogel, *J. Am. Chem. Soc.* **1990**, *112*, 9000.
[5] J. Tomasz, B. R. Shaw, K. Porter, B. F. Spielvogel, A. Sood, *Angew. Chem., Int. Ed.* **1992**, *31*, 1373.
[6] B. R. Shaw, J. Madison, A. Sood, B. F. Spielvogel, *Methods Mol. Biol.* **1993**, *20*, 225.
[7] H. Li, K. Porter, F. Huang, B. R. Shaw, *Nucleic Acids Res.* **1995**, *21*, 4495.
[8] H. Li, C. Hardin, B. R. Shaw, *J. Am. Chem. Soc.* **1996**, *118*, 6606.
[9] K. W. Porter, J. D. Briley, B. R. Shaw, *Nucleic Acids Res.* **1997**, *25*, 1611.
[10] D. S. Sergueev, B. R. Shaw, *J. Am. Chem. Soc.* **1998**, *120*, 9417.
[11] B. R. Shaw, D. Sergueev, K. He, K. Porter, J. Summers, Z. Sergueeva, V. Rait, *Methods Enzymol.: Antisense Technol.* **1999**, *313*, 226.
[12] Z. A. Sergueeva, D. S. Sergueev, B. R. Shaw, *Tetrahedron Lett.* **1999**, *40*, 2041.
[13] K. He, D. S. Sergueev, Z. A. Sergueeva, B. R. Shaw, *Tetrahedron Lett.* **1999**, *40*, 4601.
[14] J.-L. Lin, B. R. Shaw, *Chem. Commun.* **1999**, *16*, 1517.

[15] Y.-Q. Chen, F.-C. Qu, Y.-B. Zhang, *Tetrahedron Lett.* **1995**, *36*, 745.

[16] J. Zhang, T. Terhorst, M. D. Matteucci, *Tetrahedron Lett.* **1997**, *38*, 4957.

[17] Y. Jin, G. Just, *Tetrahedron Lett.* **1998**, *39*, 6429.

[18] S. Sarfati, C. Guerreiro, B. Canard, N. Dereuddre-Bosquet, P. Clayette, D. Dormont, *Nucleosides Nucleotides* **1999**, *18*, 1023.

[19] P. S. Miller, R. P. Reddy, A. Murakami, K. R. Blake, S. Lin, C. H. Agris, *Biochemistry* **1986**, *25*, 5092.

[20] M. F. Hawthorne, *Angew. Chem., Int. Ed.* **1993**, *32*, 950.

[21] E. K. Jaffe, M. Cohn, *J. Biol. Chem.* **1978**, *253*, 4823.

[22] E. K. Jaffe, M. Cohn, *J. Biol. Chem.* **1979**, *254*, 10839.

[23] E. K. Jaffe, J. Nick, M. Cohn, *J. Biol. Chem.* **1982**, *257*, 7650.

[24] A. D. Barone, J.-Y. Tang, M. H. Caruthers, *Nucleic Acids Res.* **1984**, *12*, 4051.

[25] F. Eckstein, H. Gindl, *Eur. J. Biochem.* **1970**, *13*, 558.

Isonucleosides with Exocyclic Methylene Groups

by **Sanjib Bera** and **Vasu Nair***

Department of Chemistry, The University of Iowa, Iowa City, IA 52242, U.S.A. (fax: (319) 353-2621; e-mail: vasu-nair@uiowa.edu)

Synthesis of isonucleosides **13**, **14**, **16**, and **17**, bearing an exocyclic methylidene group at the sugar moiety, starting from a 3-keto sugar is described. The keto compound was converted to the methylene-sugar **10b** (*Scheme 1*), which was coupled with nucleobases by means of the *Mitsunobu* reaction. The coupling reaction with adenine and 8-azaadenine produced both the N^9- and N^3-nucleosides (see **13** and **14**, resp.; *Scheme 2*). The structures of **13a** and **14a** were confirmed by single-crystal X-ray data. Synthesis of the pyrimidine compounds was also approached from the β-amino sugar **20** that was prepared using a *Gabriel*-synthesis methodology (*Scheme 3*).

Introduction. – For a number of years, we have been investigating the synthesis, enzymology, and antiviral studies of various isomeric nucleosides in our laboratory [1]. For example, ($2S,4S$)-4-(6-amino-9H-purin-9-yl)tetrahydrofuran-2-methanol ((S,S)-isoddA; **1**), an isomeric dideoxynucleoside synthesized by us, exhibits activity against HIV-1 and HIV-2 [2]. Our interest in synthesizing analogs of (S,S)-isoddA (**1**) led us to the design of isomeric dideoxynucleosides **3** with an exocyclic methylene group. This synthetic design was supported by the observation of the *anti*-HIV and *anti*-HBV activity of carbocyclic nucleoside **4** [3] and the *anti*-HIV activity of the dioxolanyl compounds **5** and **6** [4]. It is likely that the antiviral activity of **4** is associated with its ability to function as a structural analog of 2′-deoxyguanosine. The exocyclic C=C bond appears to act as a functional replacement for the O-atom at that position. If this is so then our target molecules **3** (and mirror images) are related to both **4**, in which the endocyclic O-atom has replaced the exocyclic OH, and to compounds **5** and **6**. This paper reports the synthesis and antiviral studies of novel isonucleosides with exocyclic methylene groups.

Results and Discussion. – Synthesis of the methylene-isonucleosides was approached from the corresponding carbohydrate precursor bearing the exocyclic methylene group. The synthesis commenced with the protected D-xylose **7** [5]. Oxidation of the 3-OH group with DMSO and P_2O_5 [6] and subsequent *Wittig* reaction [7][8] on the resulting ketone with NaH and (Ph$_3$PMe)Br in DMSO produced the alkene **8**. Acid-catalyzed methanolysis [9] of the acetonide group afforded the α- and β-D-glycosides **9**. Reductive demethoxylation was carried out by refluxing **9** with hexamethyldisilazane (HMDS) followed by treatment with Et$_3$SiH and Me$_3$SiOSO$_2$CF$_3$ [9–11] in MeCN at room temperature to give **10a**. The primary OH group of **10a** was selectively protected [12] with the (*tert*-butyl)diphenylsilyl group to yield **10b**, which was the key starting material for the synthesis of both isodideoxypurine and pyrimidine nucleosides (*Scheme 1*).

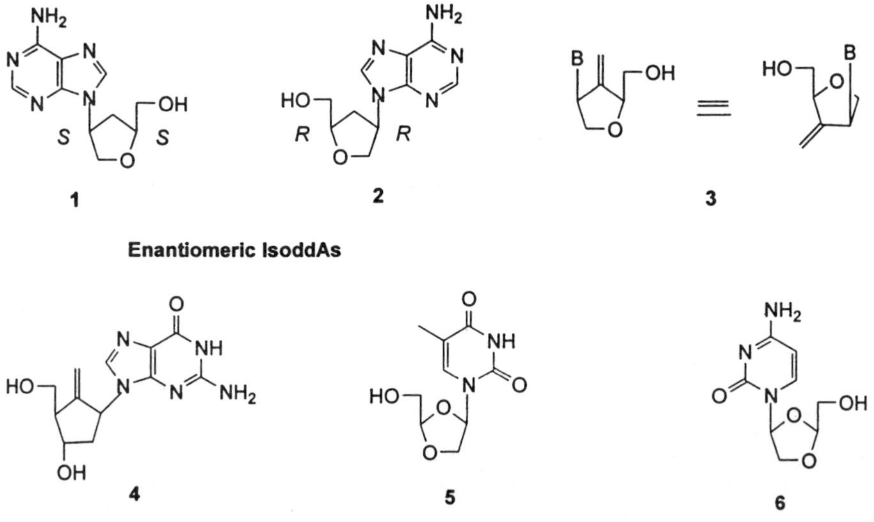

1　　　**2**　　　**3**

Enantiomeric IsoddAs

4　　　**5**　　　**6**

Scheme 1

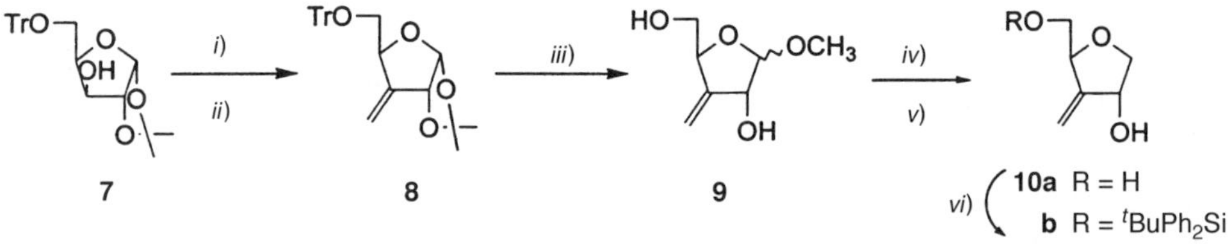

7　　　**8**　　　**9**　　　**10a** R = H
　　　　　　　　　　　　　　　　　　　　　　　　　vi) 　**b** R = 'BuPh₂Si

i) DMSO, P₂O₅, r.t. *ii*) NaH, DMSO, (Ph₃PMe)Br. *iii*) HCl, MeOH. *iv*) Me₃SiCl, HMDS. *v*) Et₃SiH, Me₃SiOSO₂CF₃, MeCN, r.t. *vi*) 'BuPh₂SiCl, pyridine, 4°.

Purine isonucleosides having adenine or 8-azaadenine bases were synthesized *via* the *Mitsunobu* reaction [13–15]. Interestingly, treatment of **10b** with adenine in the presence of DEAD (diethyl diazenedicarboxylate) and Ph₃P in dioxane at room temperature produced the N^9-isomer **11a** and its N^3 isomer **12a** (*Scheme 2*). Similarly, compounds **11b** and **12b** were obtained when **10b** was treated with 8-azaadenine under *Mitsunobu* conditions. Desilylation of compounds **11** and **12** with NH₄F in MeOH afforded compounds **13** and **14**, respectively. The structures of both the N^9- and N^3-isomers of the coupling products were confirmed by their multinuclear NMR and HR-MS data and particularly by their UV spectra and single-crystal X-ray data. For example, the UV spectrum of compound **13a** showed a λ_{max} at 261 nm (ε 14000), whereas compound **14a** exhibited a λ_{max} at 275 nm (ε 13300). The X-ray structure of **13a** (*Fig. 1*) confirmed unequivocally its structure and absolute configuration and showed the glycosidic bond attachment of N(9) of the base, the presence of the exocyclic methylene group, the *anti*-conformation about the glycosidic bond, and the *endo* puckering of the sugar ring at the O-atom. The X-ray structure of compound **14a** also confirmed its structure and absolute configuration but showed the attachment of the base at N(3) (*Fig. 2*). Interestingly, the crystal structure of compound **14a** showed the presence of two independent molecules in an asymmetric unit. Each formed a dimer *via* two H-bonds with a symmetry equivalent molecule of the other (*i.e.*, molecule A was dimerized with molecule B, and *vice versa*). The two molecules differ in conformation mainly at the primary OH group of the sugar moiety.

Scheme 2

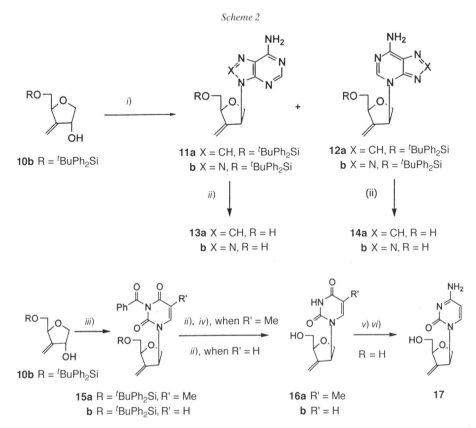

i) Ph$_3$P, DEAD, purine base, dioxane. *ii*) NH$_4$F, MeOH. *iii*) 3-benzoylthymine or uracil, Ph$_3$P, DEAD, THF. *iv*) MeONa, MeOH. *v*) Ac$_2$O, pyridine. *vi*) ^{i}Pr$_3$SiCl, 4-(dimethylamino)pyridine, Et$_3$N, MeCN, then NH$_4$OH

Isodideoxypyrimidine nucleosides having an exocyclic methylene group were also synthesized by the *Mitsunobu* reaction. Thus, compound **10b**, on treatment with 3-benzoyluracil or 3-benzoylthymine under *Mitsunobu* conditions gave the desired products **15a,b** (*Scheme 2*). The protecting groups of **15a,b** were removed by treatment with NH$_4$F or by successive treatment with NH$_4$F and NaOMe to produce **16a,b**. It is interesting to note that in the case of the uracil derivative **15b**, only NH$_4$F treatment was required to completely deprotect to product **16b**. Compound **16b** was converted to the corresponding cytosine derivative **17** according to a previously reported procedure [16].

Because of the relatively low yields obtained in the *Mitsunobu* coupling procedure with pyrimidine nucleosides, we also developed an approach to these compounds from the appropriate amino sugar **20** which was prepared from **10b** (*Scheme 3*). When compound **10b** was treated with phthalimide under *Mitsunobu*-reaction conditions, no

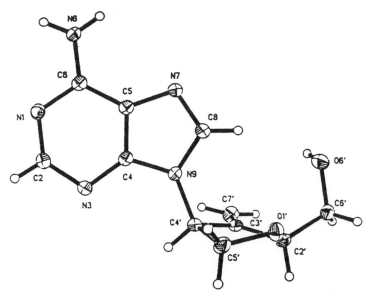

Fig. 1. *ORTEP Plot of single-crystal X-ray structure of* **13a**

phthalimido derivative **19** was obtained. However, the reaction of mesylate **18**, prepared from **10b**, with potassium phthalimide in DMF produced stereospecifically the substituted product **19**. It should be mentioned that treatment of mesylate **18** with NaN₃ did not give the S_N2 product but compound **22** resulting from the S_N2'

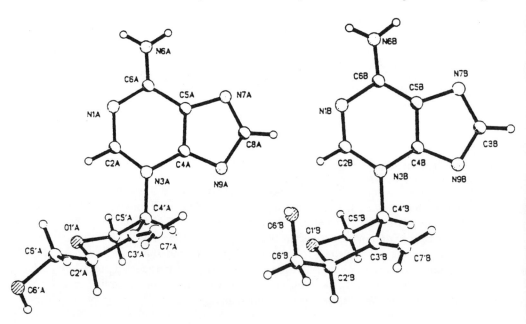

Fig. 2. *ORTEP Plot of single-crystal X-ray structure of* **14a**

displacement reaction by the azido group. The amino derivative **20** was obtained by the deprotection of **19** with hydrazine [17]. Treatment of **20** with 3-methoxy-2-methylprop-2-enoyl isocyanate [10][18][19] produced **21**, which, on acid-catalyzed cyclization, produced the thymine derivative **16a**.

Scheme 3

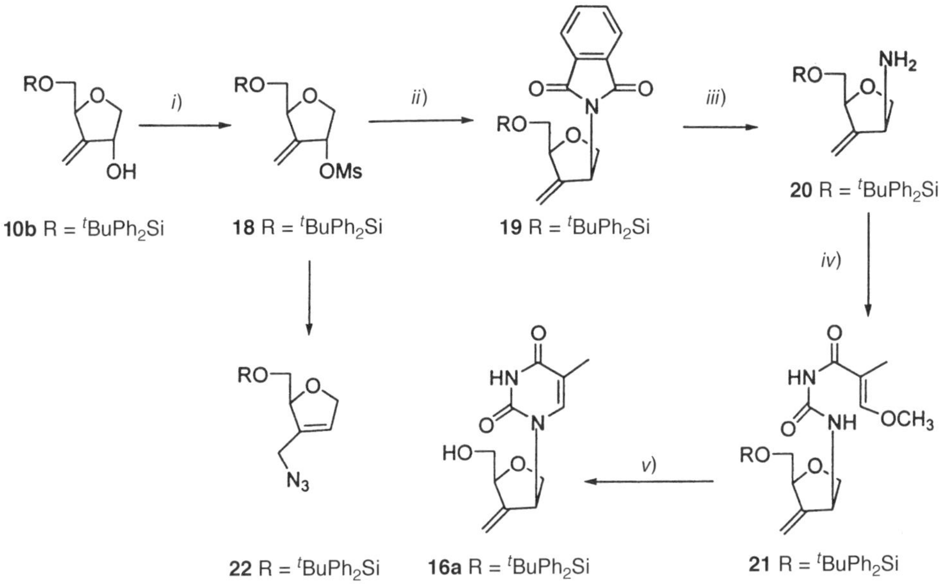

10b R = ^{t}BuPh$_2$Si   18 R = ^{t}BuPh$_2$Si   19 R = ^{t}BuPh$_2$Si

20 R = ^{t}BuPh$_2$Si

22 R = ^{t}BuPh$_2$Si   16a R = ^{t}BuPh$_2$Si   21 R = ^{t}BuPh$_2$Si

i) MsCl, Et$_3$N, CH$_2$Cl$_2$, 0°. *ii*) Potassium phthalimide, DMF, 110°. *iii*) NH$_2$NH$_2$, EtOH, reflux. *iv*) 3-Methoxy-2-methylprop-2-enoyl isocyanate, DMF, Et$_2$O, 0° to r.t. *v*) 2N H$_2$SO$_4$, dioxane, 100°. *vi*) NaN$_3$, DMF, 90°.

In summary, synthesis of the methylene-isonucleosides **13**, **14**, **16**, and **17** was achieved starting from a 3-keto sugar prepared from inexpensive D-xylose. The keto sugar was converted to an appropriate methylene-bearing carbohydrate precursor that was coupled, *via* the *Mitsunobu* reaction, with nucleobases. Synthesis of the pyrimidine compound **16a** was also approached from the *β*-amino sugar **20** which was prepared by means of a *Gabriel*-synthesis methodology. In the case of the direct coupling reaction with adenine and 8-azaadenine, both the N^9- and N^3-nucleosides (see **13** and **14**) were produced. The structure of **13a** and **14a** was unequivocally confirmed by single-crystal X-ray data. Preliminary antiviral evaluations suggested that the target compounds did not show significant *anti*-HIV activity *in vitro* in infected CEM-SS cells. Further biological studies are in progress.

We are grateful to Dr. *Dale Swenson* for the single-crystal X-ray structure determinations and Dr. *Lynn Teesch* and Ms. *Diane Herschberger* for the HR-MS data. This research work was supported by the *N.I.H.* (*National Institute of Allergy and Infectious Diseases*). We thank the *National Cancer Institute* for the *anti*-HIV evaluations.

Experimental Part

General. M.p.s.: uncorrected. *Electrothermal-Engineering-Ltd.* melting-point apparatus. UV Spectra: *Cary-3* UV-Visible spectrophotometer. ^{1}H- and ^{13}C-NMR Spectra: *Bruker-AC-300* instrument at 300 and 75 MHz, resp., δ in ppm referenced to internal SiMe$_4$ for ^{1}H and to solvent CDCl$_3$, (D$_6$)DMSO, (D$_6$)acetone, or CD$_3$OD) for ^{13}C. Column chromatography (CC): 230–400 mesh silica gel. High-resolution FAB-MS: *VG-ZAB-HF* high-resolution mass spectrometer.

3-Deoxy-1,2-O-isopropylidene-3-methylidene-5-O-trityl-α-D-ribofuranose (**8**). To a soln. of **7** [5] (4.3 g, 10 mmol) in DMSO (30 ml), P$_2$O$_5$ (2.6 g, 18.3 mmol) was carefully added while maintaining the temp. at 20–30°. The mixture was stirred at 30° for 20 h. The brown soln. was diluted with CHCl$_3$ (120 ml) and washed with ice-cold H$_2$O until the washings were neutral. The org. phase was dried (Na$_2$SO$_4$) and evaporated and the residue purified by CC: keto compound as white solid foam (2.9 g, 67%).

NaH (0.72 g, 15.1 mmol) was suspended in DMSO (30 ml) and the mixture heated at 75° for 45 min under N$_2$. After the soln. was cooled to r.t., methyltriphenylphosphonium bromide (5.8 g, 16.3 mmol) was added, and the soln. was stirred at r.t. for 30 min. To this mixture a soln. of the keto compound in DMSO (20 ml) was added dropwise. After the addition, the mixture was stirred for an additional 2 h and poured into H$_2$O (100 ml). The aq. soln. was extracted with Et$_2$O (3 × 60 ml), the combined org. phase washed with H$_2$O (2 × 75 ml), dried (Na$_2$SO$_4$), and evaporated, and the gummy residue purified by CC: **8** (2.3 g, 89%). White solid. M.p. 159–160°. ^{1}H-NMR (CDCl$_3$): 7.46–7.17 (*m*, 3 Ph); 5.95 (*d*, *J* = 4.1, H–C(1)); 5.34 (*t*, *J* = 1.1, 1 H, =CH$_2$); 5.04 (*t*, *J* = 1.4, 1 H, =CH$_2$); 4.94–4.87 (*m*, H–C(2), H–C(4)); 3.28 (*dd*, *J* = 9.9, 4.0, H$_a$–C(5)); 3.19 (*dd*, *J* = 9.9, 4.5, H$_b$–C(5)); 1.52 (*s*, 3 H, Me$_2$C); 1.39 (*s*, 3 H, Me$_2$C). ^{13}C-NMR (CDCl$_3$): 147.0 (C(3)); 143.7 (Ph); 128.6, 127.7, 127.0 (Ph); 112.4 (Me$_2$C); 111.8 (=CH$_2$); 104.8 (C(1)); 86.6 (Ph$_3$C); 81.8 (C(2)); 79.1 (C(4)); 65.8 (C(5)); 27.4, 27.1 (Me$_2$C). HR-FAB-MS: 451.1885 (C$_{28}$H$_{28}$NaO$_4$Si$^+$, [*M* + Na]$^+$; calc. 451.1885).

(3R,5S)-5-{[[(tert-Butyl)diphenylsilyl]oxy}methyl}tetrahydro-4-methylidenefuran-3-ol (**10b**). To a soln. of **8** (5.7 g, 13.3 mmol) in MeOH (150 ml) 4N HCl in dioxane (0.27 ml) was added, and the mixture was heated under reflux for 2 h. The acid was neutralized with K$_2$CO$_3$, the suspension adsorbed on silica gel, and the mixture purified by CC: **9** (1.66 g, 78%). Oil.

A mixture of **9** (0.54 g, 3.4 mmol) and Me$_3$SiCl (0.25 ml) in hexamethyldisilazane (HMDS; 25 ml) was heated under reflux for 17 h. Unreacted HMDS was evaporated to afford a pale yellow liquid. Dry MeCN was added to the residue followed by the addition of Et$_3$SiH (1.6 ml, 10.2 mmol) and Me$_3$SiOSO$_2$CF$_3$ (1.8 ml, 10.2 mmol), and the mixture was stirred overnight at r.t. It was then quenched with H$_2$O (5 ml), stirred for 1 h, and then neutralized with *Dowex* resin (OH$^-$ form). The soln. was filtered, the filtrate evaporated, and the residue purified by CC: *(2S,4R)-tetrahydro-4-hydroxy-3-methylidenefuran-2-methanol* (**10a**; 78%). Oil. ^{1}H-NMR (CD$_3$OD): 5.29 (*t*, *J* = 2.2, 1 H, =CH$_2$); 5.11 (*t*, *J* = 2.0, 1 H, =CH$_2$); 4.55–4.48 (*m*, H–C(2), H–C(4)); 4.04 (*dd*, *J* = 9.0, 6.1, H$_a$–C(5)); 3.61 (*dd*, *J* = 11.8, 3.5, 1 H, CH$_2$); 3.50 (*m*, 2 H, H$_b$–C(5), CH$_2$). ^{13}C-NMR (CD$_3$OD): 152.2 (C(3)); 108.6 (=CH$_2$); 82.2 (C(2)); 74.2 (C(4)); 73.5 (C(5)); 65.4 (CH$_2$).

^{t}BuPh$_2$SiCl (0.57 ml, 2.2 mmol) was added to a soln. of **10a** (0.26 g, 2 mmol) in pyridine (20 ml) at 0°. After the addition, the mixture was stirred at 0° for 6 h and at 4° overnight. The reaction mixture was quenched with H$_2$O (2 ml). Pyridine was evaporated and the conc. soln. co-evaporated with toluene. The residue was purified by CC: **10b** (0.62 g, 84%). Gum. ^{1}H-NMR (CDCl$_3$): 7.68–7.38 (*m*, 2 Ph); 5.34 (*s*, 1 H, =CH$_2$); 5.08 (*s*, 1 H, =CH$_2$); 4.60 (*m*, H–C(3), H–C(5)); 4.1 (*dd*, *J* = 9.3, 5.6, H$_a$–C(2)); 3.68 (*m*, H$_b$–C(2), CH$_2$); 1.04 (*s*, Me$_3$C). ^{13}C-NMR (CDCl$_3$): 151.3 (C(4)); 135.5, 135.4, 133.2, 133.1, 129.6, 129.5, 127.5 (Ph); 108.6 (=CH$_2$); 80.6 (C(3)); 73.9, 72.9 (C(2), C(5)); 66.4 (CH$_2$); 26.6 (Me$_3$C); 19.1 (Me$_3$C). HR-FAB-MS: 391.1695 (C$_{22}$H$_{28}$NaO$_3$Si$^+$, [*M* + Na]$^+$; calc. 391.1705).

(2S,4S)-4-(6-Amino-9H-purin-9-yl)tetrahydro-3-methylidenefuran-2-methanol (**13a**) *and (2S,4S)-4-(6-Amino-3H-purin-3-yl)tetrahydro-3-methylidenefuran-2-methanol* (**14a**). To a suspension of **10b** (0.61 g, 1.66 mmol), Ph$_3$P (0.65 g, 2.5 mmol), and adenine in dioxane (80 ml) at r.t., DEAD (0.4 ml, 2.6 mmol) was added dropwise. After the addition, the mixture was stirred at r.t. for 22 h. The solvent was then evaporated and the residue purified by CC: **11a** (0.34 g, 42%) and **12a** (0.15 g, 19%).

Data of **11a**: M.p. 170°. ^{1}H-NMR (CDCl$_3$): 8.33 (*s*, H–C(8)); 7.90 (*s*, H–C(2)); 7.69–7.35 (*m*, 2 Ph); 5.87 (*s*, NH$_2$); 5.64 (br. *s*, H–C(4')); 5.27 (*m*, =CH$_2$); 4.20 (*dd*, *J* = 10.0, 6.0, H$_a$–C(5')); 4.14 (*dd*, *J* = 10.0, 4.0, H$_b$–C(5')); 3.93 (*d*, *J* = 4.7, CH$_2$). ^{13}C-NMR (CDCl$_3$): 155.5 (C(6)); 152.9 (C(2)); 150.0 (C(4)); 146.9 (C(3')); 138.9 (C(8)); 135.6, 132.9, 129.8, 127.8 (2 Ph); 119.3 (C(5)); 112.1 (=CH$_2$); 81.3 (C(2')); 72.1 (C(5')); 65.6 (CH$_2$); 56.6 (C(4')); 26.8 (Me$_3$C); 19.2 (Me$_3$C).

Data of **12a**: M.p. 160–161°. ^{1}H-NMR (CDCl$_3$): 8.36 (*s*, H–C(8)); 8.02 (*s*, H–C(2)); 7.70–7.26 (*m*, 2 Ph); 6.05 (br. *s*, H–C(4')); 5.50 (br. *s*, 1 H, =CH$_2$); 5.40 (br. *s*, 1 H, =CH$_2$); 4.50 (*m*, H–C(2')); 4.25 (*d*, *J* = 4.0,

2 H−C(5′)); 4.02 (*dd*, *J* = 11.5, 3.9, 1 H, CH$_2$); 3.97 (*dd*, *J* = 11.4, 4.4, 1 H, CH$_2$); 1.07 (*s*, Me$_3$C). ^{13}C-NMR (CDCl$_3$): 154.4 (C(6)); 153.7 (C(2)); 150.6 (C(4)); 145.8 (C(3′)); 140.8 (C(8)); 135.5, 132.8, 129.9, 127.8 (Ph); 120.5 (C(5)); 113.9 (=CH$_2$); 81.5 (C(2′)); 72.5 (C(5′)); 64.9 (CH$_2$); 61.3 (C(4′)); 26.8 (Me$_3$C); 19.1 (*Me*$_3$C).

To a soln. of **11a** (0.32 g, 0.66 mmol) in MeOH (30 ml), NH$_4$F (0.4 g) was added and the soln. heated under reflux for 3 h. The soln. was then adsorbed on silica gel and purified by CC: **13a** (0.14 g, 85%). White solid. M.p. 190°. UV (MeOH): 260.6 (14000). ^{1}H-NMR ((D$_6$)DMSO): 8.15 (2*s*, H−C(2), H−C(8)); 7.25 (br. *s*, NH$_2$); 5.56 (*t*, H−C(4′)); 5.30 (br. *s*, 1 H, =CH$_2$); 5.15 (*t*, *J* = 2.0, 1 H, =CH$_2$); 5.06 (*t*, *J* = 5.6, OH); 4.44 (br. *s*, H−C(2′)); 3.73 (*m*, 2 H−C(5′)); 3.31 (*t*, *J* = 6.7, CH$_2$). ^{13}C-NMR (CD$_3$OD): 157.3 (C(6)); 153.7 (C(2)); 150.5 (C(4)); 148.6 (C(3′)); 141.4 (C(8)); 119.9 (C(5)); 112.7 (=CH$_2$); 83.1 (C(2′)); 72.9 (C(5′)); 64.2 (CH$_2$); 58.7 (C(4′)). HR-FAB-MS: 270.0959 (C$_{11}$H$_{13}$N$_5$NaO$_2^+$, [*M* + Na]$^+$; calc. 270.0966).

Compound **12a** was similarly deprotected to give **14a** (86%). White solid. M.p. 256°. UV (MeOH): 274.5 (13300). ^{1}H-NMR (CD$_3$OD): 8.62 (*s*, H−C(8)); 7.90 (*s*, H−C(2)); 5.85 (*m*, H−C(4′)); 5.49 (*m*, CH$_2$); 4.52 (br. *s*, H−C(2′)); 4.29 (*dd*, *J* = 10.6, 2.7, H$_a$−C(5′)); 4.22 (*dd*, *J* = 10.6, 5.9, H$_b$−C(5′)); 4.00 (*dd*, *J* = 12.5, 2.9, 1 H, CH$_2$); 3.90 (*dd*, *J* = 12.4, 3.3, 1 H, CH$_2$). ^{13}C-NMR (CD$_3$OD): 156.6 (C(6)); 152.9 (C(2)); 150.5 (C(4)); 147.4 (C(3′)); 144.3 (C(8)); 120.8 (C(5)); 114.4 (=CH$_2$); 83.3 (C(2′)); 73.1 (C(5′)); 63.7 (CH$_2$); 63.6 (C(4′)). HR-FAB-MS: 270.0954 (C$_{11}$H$_{13}$N$_5$NaO$_2^+$, [*M* + Na]$^+$; calc. 270.0966).

(2S,4S)-4-(7-Amino-3H-1,2,3-triazolo[4,5-d]pyrimidin-3-yl)tetrahydro-3-methylidenefuran-2-methanol (**13b**) *and (2S,4S)-4-(7-Amino-4H-1,2,3-triazolo[4,5-d]pyrimidin-4-yl)tetrahydro-3-methylidenefuran-2-methanol* (**14b**). Compounds **11b** and **12b** were synthesized as described for **13a** and **13b** in 50 and 23% yield, resp.

Data of **11b**: M.p. 168°. ^{1}H-NMR (CDCl$_3$): 8.37 (*s*, H−C(5)); 7.73 – 7.34 (*m*, 2 Ph); 7.06 (br. *s*, NH$_2$); 5.93 (*t*, *J* = 6.3, H−C(4′)); 5.25 (*s*, 1 H, =CH$_2$); 5.04 (*s*, 1 H, =CH$_2$); 4.74 (*t*, *J* = 5.4, H−C(2′)); 4.50 (*dd*, *J* = 9.4, 6.0, H$_a$−C(5′)); 4.33 (*dd*, *J* = 9.4, 7.2, H$_b$−C(5′)); 4.10 (*dd*, *J* = 10.8, 6.6, 1 H, CH$_2$); 3.91 (*dd*, *J* = 10.8, 5.1, 1 H, CH$_2$); 1.07 (*s*, Me$_3$C). ^{13}C-NMR (CDCl$_3$): 156.5 (C(7)); 156.1 (C(5)); 149.1 (C(3a)); 146.2 (C(3′)); 135.6, 133.4, 133.2, 129.6, 127.6 (2 Ph); 124.4 (C(7a)); 111.4 (=CH$_2$); 81.5 (C(2′)); 69.6 (C(5′)); 65.8 (CH$_2$); 60.0 (C(4′)); 26.7 (Me$_3$C); 19.1 (*Me*$_3$C).

Data of **12b**: M.p. 74 – 75°. ^{1}H-NMR (CDCl$_3$): 8.49 (*s*, H−C(5)); 7.70 – 7.32 (*m*, 2 Ph); 5.81 (*m*, H−C(4′)); 5.39 (*s*, 1 H, =CH$_2$); 5.36 (*s*, 1 H, =CH$_2$); 4.68 (*m*, H−C(2′)); 4.68 (*dd*, *J* = 10.1, 4.0, H$_a$−C(5′)); 4.35 (*dd*, *J* = 10.1, 6.5, H$_b$−C(5′)); 3.97 (*dd*, *J* = 10.8, 6.3, 1 H, CH$_2$); 3.85 (*dd*, *J* = 15.1, 5.1, 1 H, CH$_2$). ^{13}C-NMR (CDCl$_3$): 157.8 (C(7)); 156.6 (C(3a)); 156.4 (C(5)); 145.5 (C(3′)); 135.5, 135.4, 133.2, 133.0, 129.6, 127.5 (2 Ph); 126.0 (C(7a)); 113.3 (=CH$_2$); 81.4 (C(2′)); 70.9 (C(5′)); 68.9 (C(4′)); 65.9 (CH$_2$); 26.6 (Me$_3$C); 19.0 (*Me*$_3$C).

Compounds **11b** and **12b** were deprotected as described for **11a** to give **13b** and **14b** in 81 and 88% yield, resp.

Data of **13b**: White solid powder. M.p. 243°. UV (MeOH): 279.5 (12400). ^{1}H-NMR ((D$_6$)DMSO): 8.45 (br. *s*, 1 H, NH$_2$); 8.29 (*s*, H−C(5)); 8.11 (br. *s*, 1 H, NH$_2$); 5.93 (*m*, H−C(4′)); 5.26 (*s*, 1 H, =CH$_2$); 4.95 (*m*, 2 H, =CH$_2$, −OH); 4.51 (br. *s*, H−C(2′)); 4.38 (*dd*, *J* = 9.4, 5.4, H$_a$−C(5′)); 4.25 (*dd*, *J* = 9.4, 8.2, H$_b$−C(5′)); 3.74 (*m*, 1 H, CH$_2$); 3.59 (*m*, 1 H, CH$_2$). ^{13}C-NMR ((D$_6$)DMSO): 156.7 (C(7)); 156.2 (C(5)); 148.6 (C(3a)); 147.1 (C(3′)); 123.9 (C(7a)); 110.4 (=CH$_2$); 81.5 (C(2′)); 68.9 (C(5′)); 63.4 (CH$_2$); 59.6 (C(4′)). HR-FAB-MS: 271.0902 (C$_{10}$H$_{12}$N$_6$NaO$_2^+$, [*M* + Na]$^+$; calc. 271.0919).

Data of **14b**: White powder. M.p. 201°. UV (MeOH): 295.3 (10900). ^{1}H-NMR (CD$_3$OD): 8.28 (*s*, H−C(5)); 5.93 (*m*, H−C(4′)); 5.41 (*d*, *J* = 4.5, =CH$_2$); 4.65 (*dd*, *J* = 10.3, 3.3, H$_a$−C(5′)); 4.58 (*m*, H−C(2′)); 4.33 (*dd*, *J* = 10.3, 6.3, H$_b$−C(5′)); 3.82 (*dd*, *J* = 11.7, 6.7, 1 H, CH$_2$); 3.73 (*dd*, *J* = 11.7, 4.1, CH$_2$). ^{13}C-NMR (CD$_3$OD): 158.7 (C(7)); 158.6 (C(3a)); 157.7 (C(5)); 147.6 (C(3′)); 127.1 (C(7a)); 113.6 (=CH$_2$); 83.2 (C(2′)); 72.1 (C(5′)); 70.6 (C(4′)); 65.1 (CH$_2$). HR-FAB-MS: 271.0919 (C$_{10}$H$_{12}$N$_6$NaO$_2^+$, [*M* + Na]$^+$; calc. 271.0919).

3-Benzoyl-1-[(3S,5S)-5-[[[(tert-butyl)diphenylsilyl]oxy]methyl]tetrahydro-4-methylidenefuran-3-yl]-5-methyl pyrimidine-2,4(1H,3H)-dione (**15a**). To a soln. of Ph$_3$P (0.3 g, 0.92 mmol) in anh. THF (10 ml), DEAD (0.17 ml, 1.1 mmol) was added. The mixture was stirred at 0° for 30 min and then cooled to −45°. To this soln., N^3-benzoylthymine (0.21 g, 0.92 mmol) and **10b** (0.17 g, 0.46 mmol) in THF (10 ml) were added within 1 h. The mixture was stirred overnight at −45°. The soln. was directly adsorbed on silica gel and purified by CC: **15a** (0.08 g, 30%). Low-melting solid. ^{1}H-NMR (CD$_3$OD): 7.92 – 7.25 (*m*, H−C(6), 3 Ph); 5.66 (br. *s*, H−C(3′)); 5.34 (*d*, *J* = 13.6, =CH$_2$); 4.47 (br. *s*, H−C(5′)); 4.14 (*dd*, *J* = 10.2, 6.8, 1 H); 4.04 (*dd*, *J* = 11.0, 3.7, 2 H); 3.94 (*dd*, *J* = 11.3, 4.5, 1 H); 1.64 (*s*, Me); 1.10 (*s*, Me$_3$C). ^{13}C-NMR (CDCl$_3$): 168.9 (C=O); 162.5 (C(4)); 150.3 (C(2)); 146.3 (C(4′)); 136.8 (C(6)); 112.1 (=CH$_2$); 111.6 (C(5)); 81.1 (C(5′)); 71.1 (C(2′)); 65.4 (CH$_2$); 57.3 (C(3′)); 26.9 (Me$_3$C); 19.3 (*Me*$_3$C); 12.2 (Me). HR-FAB-MS: 603.2305 (C$_{34}$H$_{36}$N$_2$NaO$_5$Si$^+$, [*M* + Na]$^+$; calc. 603.2291).

5-Methyl-1-[(3S,5S)-tetrahydro-5-(hydroxymethyl)-4-methylidenefuran-3-yl]pyrimidine-2,4(1H,3H)-dione (**16a**). To a soln. of **15a** (0.07 g, 0.13 mmol) in MeOH (10 ml), NH$_4$F (0.065 g, 2 mmol) was added, and the soln. was heated under reflux for 3 h. The soln. was cooled to r.t., adsorbed on silica gel, and purified by CC to give the $N(3)$-benzoylated derivative, which was dissolved in MeOH (5 ml) and treated with NaOMe (0.03 g). The mixture was stirred at r.t. for 8 h, neutralized with 10% AcOH/H$_2$O, and evaporated. The residue was taken up in MeOH, adsorbed on silica gel, and purified by CC: **16a** (0.018 g, 62% for two steps). Hygroscopic foam. UV (MeOH): 270.9 (9400). ^{1}H-NMR ((D$_6$)acetone): 10.14 (br. s, NH); 7.62 (s, H$-$C(6)); 5.56 (m, H$-$C(3′)); 5.36 ($t, J = 2.1$, 1 H, $=$CH$_2$); 5.30 ($t, J = 2.3$, 1 H, $=$CH$_2$); 4.40 (br. s, H$-$C(5′)); 4.18$-$3.84 (m, H$-$C(2′), H$-$C(3), CH$_2$); 1.76 (s, Me). ^{13}C-NMR (CD$_3$OD): 166.3 (C(4)); 153.2 (C(2)); 148.5 (C(4′)); 140.3 (C(6)); 112.2 ($=$CH$_2$); 111.6 (C(5)); 82.8 (C(5′)); 72.2 (C(2′)); 638 (CH$_2$); 58.9 (C(3′)); 12.3 (Me). HR-FAB-MS: 261.0855 (C$_{11}$H$_{14}$N$_2$NaO$_4^+$, $[M + Na]^+$; calc. 261.0851).

1-[(3S,5S)-Tetrahydro-5-(hydroxymethyl)-4-methylidenefuran-3-yl]pyrimidine-2,4(1H,3H)-dione (**16b**). To a soln. of Ph$_3$P (0.84 g, 3.18 mmol) in anh. THF, DEAD (0.49 ml, 3.12 mmol) was added. The mixture was then stirred at 0° for 30 min and cooled to -70°. A suspension of 3-benzoyluracil (0.54 g, 3.12 mmol) and **10b** (0.46 g, 1.25 mmol) was added to the soln. within 30 min. The mixture was stirred overnight at -40° and then quenched with H$_2$O (1 ml) and evaporated. The gummy residue was purified by CC. The product was dissolved in MeOH (40 ml), NH$_4$F (0.5 g) added, and the resulting soln. heated under reflux for 3 h. The soln. was directly adsorbed on silica gel and purified by CC: **16b** (0.16 g, 57%). White solid. M.p. 178°. UV (MeOH): 266.6 (11800). ^{1}H-NMR (CD$_3$OD): 7.80 ($d, J = 8.1$, H$-$C(6)); 5.64 ($d, J = 8.1$, H$-$C(5)); 5.57 (m, H$-$C(3′)); 5.35 (m, $=$CH$_2$); 4.82 (br. s, H$-$C(5′)); 4.07 ($dd, J = 10.3, 6.5$, H$_a$$-$C(2′)); 4.00 ($dd, J = 10.2, 3.6$, H$_b$$-$C(2′)); 3.90 ($dd, J = 12.3, 2.9$, 1 H, CH$_2$); 3.80 ($dd, J = 12.3, 4.2$, 1 H, CH$_2$). ^{13}C-NMR (CD$_3$OD): 166.2 (C(4)); 153.2 (C(2)); 148.5 (C(4′)); 144.6 (H$-$C(6)); 112.5 ($=$CH$_2$); 102.7 (C(5)); 82.9 (C(5′)); 72.3 (C(2′)); 63.9 (CH$_2$); 59.3 (C(3′)). HR-FAB-MS: 247.0703 (C$_{10}$H$_{12}$N$_2$NaO$_4^+$, $[M + Na]^+$; calc. 247.0694).

4-Amino-1-[(3S,5S)-tetrahydro-5-(hydroxymethyl)-4-methylidenefuran-3-yl]pyrimidin-2(1H)-one (**17**). Ac$_2$O (0.25 ml, 2.65 mmol) was added to a soln. of **16b** (0.12 g, 0.53 mmol) in pyridine (10 ml) at r.t. After the addition, the mixture was stirred at r.t. overnight and then quenched with H$_2$O (1 ml). The solvent was evaporated and the conc. soln. co-evaporated with toluene. The residue was purified by CC to give the acetylated derivative (0.13 g, 92%) as pale yellow foamy solid. Et$_3$N (0.1 ml) was added to a soln. of the acetylated derivative (0.09 g, 0.33 mmol) in MeCN (10 ml) containing 2,4,6-^{i}Pr$_3$C$_6$H$_2$SO$_2$Cl (0.23 g, 0.75 mmol) and 4-(dimethylamino)pyridine (0.092 g, 0.75 mmol) at 0°. The mixture was stirred at r.t. for 4 h. NH$_4$OH soln. (28%, 6 ml) was added, and the soln. was stirred at r.t. for 24 h. After evaporation, the residue was purified by CC: **17** (0.035 g, 47% in two steps). Hygroscopic solid. UV (MeOH): 276.4 (8650). ^{1}H-NMR (CD$_3$OD): 7.80 ($d, J = 7.4$, H$-$C(6)); 5.83 ($d, J = 7.5$, H$-$C(5)); 5.63 (m, H$-$C(3′)); 5.35 ($t, J = 2.1$, 1 H, $=$CH$_2$); 5.30 ($t, J = 2.2$, 1 H, $=$CH$_2$); 4.40 (br. s, H$-$C(5′)); 4.08 ($dd, J = 10.1, 6.5$, H$_a$$-$C(2′)); 3.96 ($dd, J = 10.1, 3.6$, H$_b$$-$C(2′)); 3.90 ($dd, J = 12.3, 3.0$, 1 H, CH$_2$); 3.81 ($dd, J = 12.3, 4.2$, 1 H, CH$_2$). ^{13}C-NMR (CD$_3$OD): 167.4 (C(2)); 159.2 (C(4)); 149.0 (C(4′)); 144.8 (C(6)); 112.2 ($=$CH$_2$); 96.1 (C(5)); 83.1 (C(5′)); 72.9 (C(2′)); 63.9 (CH$_2$); 60.3 (C(3′)). HR-FAB-MS: 224.1040 (C$_{10}$H$_{14}$N$_3$O$_3^+$, $[M + H]^+$; calc. 224.1035).

2-[(3S,5S)-5-{[[(tert-Butyl)diphenylsilyl]oxy}methyl]tetrahydro-4-methylidenefuran-3-yl]-1H-isoindol-1,3(2H)-dione (**19**). To a soln. of **10b** (1.1 g, 3 mmol) in CH$_2$Cl$_2$ (30 ml), Et$_3$N (1.7 ml, 12 mmol) and MsCl (0.7 ml, 9 mmol) were added at 0°. After the addition, the mixture was stirred at 0° for 3 h. H$_2$O (75 ml) was added and extracted with CH$_2$Cl$_2$ (3 × 30 ml). The combined CH$_2$Cl$_2$ part was washed with H$_2$O, dried (Na$_2$SO$_4$), and evaporated. The oily residue was purified by CC: **18** (1.11 g, 83%). Pale yellow low-melting solid. ^{1}H-NMR (CDCl$_3$): 7.70 (m, 4 H, Ph); 7.42 (m, 6 H, Ph); 5.64 (s, 1 H, $=$CH$_2$); 5.51 (m, H$-$C(3′)); 5.33 (s, 1 H, $=$CH$_2$); 4.65 (br. s, H$-$C(5)); 4.18 ($dd, J = 10.6, 4.8$, H$_a$$-$C(2′)); 4.06 ($dd, J = 10.6, 2.8$, H$_b$$-$C(2′)); 3.82 ($dd, J = 11.0, 3.7$, 1 H, CH$_2$); 3.71 ($dd, J = 11.0, 4.6$, 1 H, CH$_2$); 3.04 ($s$, Me); 1.06 ($s$, Me$_3$C). ^{13}C-NMR (CDCl$_3$): 145.4 (C(4′)); 114.3 ($=$CH$_2$); 80.9, 80.2 (C(3), C(5)); 72.1 (C(2′)); 66.3 (CH$_2$); 39.2 (Me); 26.9 (Me$_3$C); 19.3 (Me$_3$C).

To a soln. of **18** (0.3 g, 0.67 mmol) in DMF (30 ml), potassium phthalimide (0.74 g, 4 mmol) was added, and the mixture was heated at 110° overnight. The soln. was cooled and filtered, and the filtrate was evaporated. The residue was taken up in AcOEt (80 ml) and washed with H$_2$O (2 × 50 ml). The AcOEt part was dried (Na$_2$SO$_4$), evaporated, and purified by CC: **19** (0.17 g, 51%). Low-melting solid. ^{1}H-NMR (CDCl$_3$): 7.84$-$7.30 (m, 14 arom. H); 5.34 (m, H$-$C(3′)); 5.03 ($t, J = 2.2$, 1 H, $=$CH$_2$); 4.87 ($t, J = 2.4$, 1 H, $=$CH$_2$); 4.70 (m, H$-$C(5′)); 4.15 ($dd, J = 11, 7.8$, 1 H, CH$_2$); 4.04 ($d, J = 8.7$, 2 H$-$C(2′)); 3.86 ($dd, J = 11.1, 4.6$, 1 H, CH$_2$); 1.11 (s, Me$_3$C). ^{13}C-NMR (CDCl$_3$): 167.1 (C$=$O); 145.8 (C(4′)); 135.7, 135.6, 134.1, 133.8, 133.5, 131.7, 129.5, 127.6, 123.3 (arom. C); 107.0 ($=$CH$_2$); 81.8 (C(5′)); 66.0, 65.9 (C(2′), CH$_2$); 51.9 (C(3′)); 26.8 (Me$_3$C); 19.2 (Me$_3$C). HR-FAB-MS: 520.1930 (C$_{30}$H$_{31}$NNaO$_4$Si$^+$, $[M + Na]^+$; calc. 520.1920).

(3S,5S)-5-{[[(tert-Butyl)diphenylsilyl]oxy}methyl]tetrahydro-4-methylidenefuran-3-amine (**20**). To a soln. of **19** (0.16 g, 0.32 mmol) in EtOH (20 ml), hydrazine hydrate (0.12 g) was added, and the mixture was heated under reflux for 3 h and then allowed to cool to r.t. The white precipitate was filtered and washed with EtOH.

The combined EtOH part was evaporated and the residue purified by CC: **20** (0.09 g, 77%). ^{1}H-NMR (CDCl$_3$): 7.71–7.25 (*m*, 2 Ph); 5.29 (br. *s*, 1 H, =CH$_2$); 4.48 (*m*, H–C(5)); 3.99–3.64 (*m*, 2 H–C(2), H–C(3), CH$_2$); 1.06 (*s*, Me$_3$C). ^{13}C-NMR (CDCl$_3$): 153.3 (C(4)); 135.6, 133.3, 133.2, 129.6, 127.6 (Ph); 106.2 (=CH$_2$); 81.0 (C(5)); 74.4 (C(2)); 66.4 (CH$_2$); 55.5 (C(3)); 26.8 (Me$_3$C); 19.1 (*Me$_3$*C). HR-FAB-MS: 390.1870 (C$_{22}$H$_{29}$NNaO$_2$Si$^+$, [*M* + Na]$^+$; calc. 390.1865).

5-Methyl-1-[(3S,5S)-tetrahydro-5-(hydroxymethyl)-4-methylidenefuran-3-yl]pyrimidine-2,4(1H,3H)-dione (**16a**). A soln. of 3-methoxy-2-methylprop-2-enoyl isocyanate (prepared by heating the corresponding acid chloride (0.11 g, 0.8 mmol) and silver cyanate (0.22 g) in toluene for 0.5 h) was added dropwise to a soln. of **20** (0.09 g, 0.25 mmol) in DMF (10 ml) and Et$_2$O (5 ml) at – 10°. After the addition, the mixture was allowed to warm to r.t. and stirred for 16 h. The volatile materials were evaporated. EtOH was added to the residue with stirring and then evaporated. The gummy residue was purified by CC to give the acryloylurea derivative **21**. This compound was dissolved in dioxane (5 ml) and heated at 100° for 4 h with 2N H$_2$SO$_4$ (1.0 ml). The mixture was allowed to cool to r.t. and then neutralized with 2N aq. NaOH and evaporated. The residue was purified by CC: **16a** (0.04 g, 50% in two steps), identical to the compound prepared by direct coupling and deprotection (see above).

Single-Crystal X-Ray Structure Determination of Compounds **13a** *and* **14a**. A colorless prismatic crystal (0.21 × 0.24 × 0.36 mm and 0.25 × 0.25 × 0.31 mm, resp.) was isolated from the sample and mounted with grease on the tip of a glass capillary epoxied to a brass pin and placed on the diffractometer with the long crystal dimension (*a*-axis and *AB* dimension, resp.) approximately parallel to the diffractometer ϕ-axis. Data were collected on an *Enraf-Nonius-CAD4* diffractometer (MoK$_\alpha$ radiation, graphite monochromator) at 200 K (cold N$_2$ gas cooling) using θ-2θ scans. Intensity standards were measured at 2 h intervals. Net intensities were obtained by profile analysis of the 4469 and 5428 data, resp. *Lorentz* and polarization corrections were applied. No change in the intensity standards was detected. Absorption was minimal and no correction was applied. Equivalent data were averaged yielding 2018 unique data (R-int = 0.023, 1926 > 4 σ (F)) in the case of **13a** and 3817 unique data (R-int = 0.072, 3042 > 4σ(F)) in the case of **14a**. Based on preliminary examination of the data, the space group $P2(1)2(1)2(1)$ was assigned to **13a** and $P2(1)$ to **14a**. The computer programs from the MoLEN package were used for data reduction. The preliminary model of the structure was obtained with XS, a direct-methods program. Least-squares refining of the model *vs.* the data was performed with the XL computer program. Illustrations were made with the XP program, and tables were made with the XCIF program. All are in the SHELXTL v5.0 package. Thermal ellipsoids are drawn at the 35% level unless otherwise noted. All non-H-atoms were refined with anisotropic thermal parameters. All H-atoms were included with a riding model based on default values. In the case of **14a**, there are two independent molecules in the asymmetric unit. Each forms a dimer *via* two H-bonds with a symmetry-equivalent molecule of the other (*i.e.*, molecule A dimerizes with molecule B, and *vice versa*). The two molecules differ in conformation mainly at the sugar CH$_2$OH group.

REFERENCES

[1] V. Nair, T. S. Jahnke, *Antimicrob. Agents Chemother.* **1995**, *39*, 1017; V. Nair, Z. M. Nuesca, *J. Am. Chem. Soc.* **1992**, *114*, 7951.

[2] V. Nair, M. St. Clair, J. E. Reardon, H. C. Krasny, R. J. Hazen, M. T. Paff, L. R. Boone, M. Tisdale, I. Najera, R. E. Dornsife, D. R. Everett, K. Borroto-Esoda, J. L. Yale, T. P. Zimmerman, J. L. Rideout, *Antimicrob. Agents Chemother.* **1995**, *39*, 1993.

[3] G. S. Bisacchi, S. T. Chao, C. Bachard, J. P. Daris, S. Innaimo, G. A. Jacobs, O. Kocy, P. Lapointe, A. Martel, Z. Merchant, W. A. Slusarchyk, J. E. Sundeed, M. G. Young, R. Colonno, R. Zahler, *Bioorg. Med. Chem. Lett.* **1997**, *7*, 127.

[4] H. O. Kim, S. K. Ahn, A. J. Alves, J. W. Beach, L. S. Jeong, B. G. Choi, P. Van Roey, R. F. Schinazi, C. K. Chu, *J. Med. Chem.* **1992**, *35*, 1987; C. K. Chu, S. K. Ahn, H. O. Kim, J. W. Beach, A. J. Alves, L. S. Jeong, Q. Islam, P. Van Roey, R. F. Schinazi, *Tetrahedron Lett.* **1991**, *32*, 3791.

[5] P. A. Levene, A. L. Raymond, *J. Biol. Chem.* **1933**, *102*, 331; V. Nair, D. J. Emanuel, *J. Am. Chem. Soc.* **1977**, *99*, 1571.

[6] K. Onodera, S. Hirano, N. Kashimura, *Carbohydrate Res.* **1968**, *6*, 276.

[7] A. Rosenthal, D. A. Baker, *Tetrahedron Lett.* **1969**, 397.

[8] A. Rosenthal, M. Sprinzl, *Can. J. Chem.* **1969**, *47*, 3941.

[9] P. J. Bolon, T. B. Sells, Z. M. Nuesca, D. F. Purdy, V. Nair, *Tetrahedron* **1994**, *50*, 7747.

[10] D. F. Purdy, L. B. Zintek, V. Nair, *Nucleosides Nucleotides* **1994**, *13*, 109.

[11] J. A. Bennek, G. R. Gary, *J. Org. Chem.* **1987**, *52*, 892.

[12] T. Wenzel, V. Nair, *Bioconjugate Chem.* **1998**, *9*, 683.

[13] C. Bannal, C. Chavis, M. Lucas, *J. Chem. Soc., Perkin Trans 1* **1994**, 1401.

[14] T. F. Jenny, J. Horlacher, N. Previsani, S. A. Benner, *Helv. Chim. Acta* **1992**, *75*, 1944.

[15] L. S. Jeong, S. J. Yoo, *Bioorg. Med. Chem. Lett.* **1998**, *8*, 847; L. S. Jeong, S. J. Yoo, H. R. Moon, M. W. Chun, C.-K. Lee, *Nucleosides Nucleotides* **1999**, *18*, 655.

[16] A. Kakefuda, S. Shuto, T. Nagahata, J. Seki, T. Sasaki, A. Matsuda, *Tetrahedron* **1994**, *50*, 10167.

[17] O. Mitsunobu, *Synthesis* **1981**, 1.

[18] G. Shaw, R. N. Warrener, *J. Chem. Soc.* **1958**, 157.

[19] Y. F. Shealy, C. A. O'Dell, M. C. Thorpe, *J. Heterocycl. Chem.* **1981**, *18*, 383.

Oligonucleotides Functionalized by Fluorescein and Rhodamine Dyes: *Michael* Addition of Methyl Acrylate to 2'-Deoxypseudouridine

by **Natalya Ramzaeva**[a]), **Helmut Rosemeyer**[a]), **Peter Leonard**[a]), **Klaus Mühlegger**[b]), **Frank Bergmann**[b]), **Herbert von der Eltz**[b]), and **Frank Seela**[a])*

[a]) Laboratorium für Organische und Bioorganische Chemie, Institut für Chemie, Universität Osnabrück, Barbarastr. 7, D-49069 Osnabrück

[b]) *Roche Diagnostics GmbH*, Nonnenwald 2, D-82372 Penzberg

The 2'-deoxypseudouridine (**5**) was functionalized at N(1) with methyl acrylate by *Michael* addition. The resulting methyl 2'-deoxypseudouridine-1-propanoate (**6**) was converted to the phosphoramidite **8** and to the amino-functionalized derivative **9**, which was transformed into the fluorescein-labeled phosphoramidites **14** and **16**. Fluorescent oligonucleotides were synthesized either from these building blocks or by post-synthetic modification of oligomers containing 2'-deoxypseudouridine subunits. The stability of oligonucleotide duplexes was determined from the melting profiles, measured by UV- or VIS-light absorbance, as well as from the fluorescence emission spectra. While small spacer residues did not affect the thermal stability of the 2'-deoxypseudouridine-containing duplexes, large dye residues led to destabilization.

Introduction. – Pseudouridine (*Ψ*; **1**), which was isolated in 1957 from tRNA [1], is a pyrimidine *C*-nucleoside. Its structure is related to that of uridine and was established in 1962 [2][3]. Meanwhile, it is known that *Ψ* is present ubiquitously in transfer RNA in the T-*Ψ*-C-Pu loop [4]. Other pyrimidine *C*-nucleosides were isolated from different natural sources and represent antibiotics or exhibit anticancer and/or antiviral activity [5]. Also 2'-deoxypseudouridine (*Ψ*$_d$; **5**) and related pyrimidine 2'-deoxy-*C*-nucleosides have been prepared [6][7]. In 1991, *Piccirilli et al.* incorporated N^1-methyl-2'-deoxypseudouridine (**3**; m^1*Ψ*$_d$) *via* its phosphoramidite into DNA templates [8]. In 1995, *Bhattacharya et al.* incorporated the same compound into G-rich triplex-forming oligonucleotides [9]. *Rosenberg et al.* [10] used the H-phosphonate approach for the synthesis of different oligonucleotides containing 2'-deoxypseudouridine and its *N*-methyl derivatives.

1　　　　　2　　　　　3　　　　　4

As the N(1) atom of 2′-deoxypseudouridine (**5**) is located in the same position as the 5-methyl group of T_d and thereby directed into the major groove of a B-DNA, the introduction of reporter groups such as fluorescent dyes at that position of Ψ_d was anticipated to be favorable. Until now, mainly fluorophore-labeled 2′-deoxyuridine has found application in DNA diagnostics and analysis [11–13]. An alternative nucleoside carrier for reporter groups is 2′-deoxypseudouridine (**5**). The ribonucleoside **1** has been derivatized by regioselective cyanoethylation at N(1) under mild reaction conditions (buffer, pH 8.5, 37°), both on the nucleoside as well as on the tRNA level [14–16].

This report describes the derivatization of 2′-deoxypseudouridine (**5**) by methyl acrylate (= methyl prop-2-enoate) on the nucleoside and on the oligonucleotide level and its functionalization by linkers and fluorescent dyes. The fluorescent oligonucleotides were obtained either by solid-phase DNA synthesis from fluorophore-labeled 2′-deoxypseudouridine phosphoramidites or by post-modification at the oligonucleotide level.

Results. – *Monomers.* The 2′-deoxypseudouridine (= 5-(2-deoxy-β-D-*erythro*-pentofuranosyl)uracil; **5**; Ψ_d) was synthesized from pseudouridine (**1**) according to [7]. Reaction of **5** with methyl acrylate in buffer (pH 8.5) yielded the N^1-alkylated compound **6** in 89% yield (*Scheme 1*). Only traces of by-products were detected. A similar reaction has been performed with acrylonitrile [14]. This is in line with early findings that in slightly alkaline medium, as it is used for the alkylation reaction here,

Scheme 1

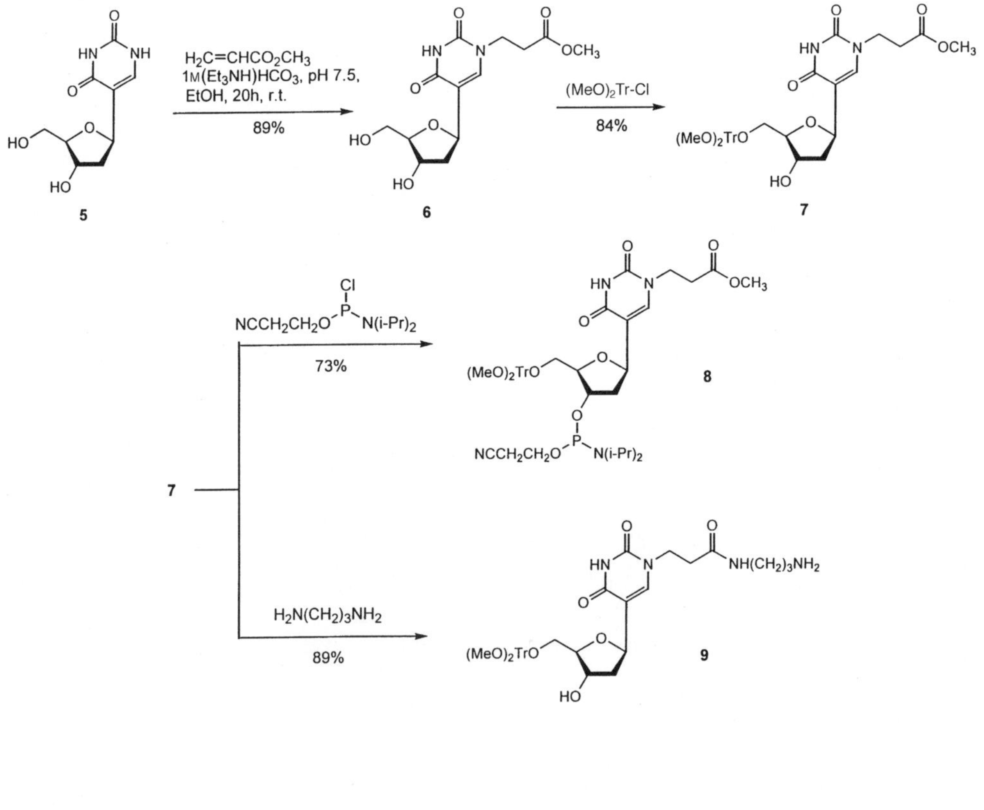

Table 1. *Selected ^{13}C-NMR Data of 2'-Deoxypseudouridine Derivatives in $(D_6)DMSO$ at 303 K*

	C(2)	C(4)	C(5)	C(6)	C(1')	C(2')	C(3')	C(4')	C(5')	CH₂	COOMe	COOMe (or CONHR)	MeO (Tr)
5	151.2	163.5	113.2	137.8	73.2	40.8	72.1	87.0	62.2	–	–	–	–
6	150.4	162.8	113.6	141.7	73.1	40.7	72.0	87.1	62.2	44.1, 32.6	51.5	171.1	–
7	150.4	162.6	113.7	141.2	72.9	40.6	72.2	85.1	64.2	44.3, 32.4	51.4	171.0	55.0
9	150.4	162.6	113.3	141.4	73.0	40.6	72.1	85.1	64.2	45.1, 38.3, 35.9, – 34.2, 31.1		169.3	55.0

pseudouridine exists as a mixture of monoanions ($pK_a^{N(1)}$: 9.3; $pK_a^{N(3)}$: 9.6) in which the N(1) anion predominates significantly over the N(3) anion [14][17].

The assignment of the alkylation position is based on the similarity of the UV spectrum of **6** (λ_{max} 270 nm) with that of N^1-methylpseudouridine (λ_{max} 271 nm) [18], while that of the parent nucleoside **5** exhibits a λ_{max} of 263 nm [19]. Moreover, the similar ^{13}C-NMR chemical shifts of the uracil base of **6** and those of the corresponding resonances of N^1-methylpseudouridine establish N^1-alkylation (*Table 1*) [20].

Protection of compound **6** at the 5'-OH group with a 4,4'-dimethoxytrityl residue ($(MeO)_2Tr$) afforded compound **7** [21]. This was either converted into the phosphoramidite **8** by reaction with 2-cyanoethyl diisopropylphosphoramidochloridite [22] or into the amino-functionalized derivative **9** by reaction with an excess of propane-1,3-diamine (*Scheme 1*). New compounds were characterized by ^1H-, ^{13}C-, and ^{31}P-NMR spectra as well as by elemental analyses (*Table 1* and *Exper. Part*). The assignment of the ^{13}C-NMR resonances was made on the basis of gated-decoupled ^{13}C- and $^1H,^{13}C$-heteronuclear spectra as well as DEPT-135 spectra.

Reaction of compound **9** with esters **10** and **11** derived from N-hydroxysuccinimide and di-O-pivaloyl-protected fluorescein-6-carboxylic acid led to compounds **13** and **15** which both showed strong fluorescence in a spot test upon treatment with ammonia and which were characterized by ^1H-NMR spectroscopy. The compounds differ in the length of the spacer between the fluorophore and the nucleobase (**13**: 10 bonds; **15**: 17 bonds). Reaction of **13** and **15** with 2-cyanoethyl diisopropylphosphoramidochloridite gave the phosphoramidites **14** and **16**, which were characterized by ^{31}P-NMR spectroscopy. The last two reaction steps were performed under the exclusion of light. Ester **12** derived from N-hydroxysuccinimide and a rhodamine dye was used for post-synthetic oligonucleotide labeling (see below).

Oligonucleotides: Syntheses. The syntheses of the oligonucleotides were performed by three different routes: *i*) by solid-phase synthesis *via* phosphoramidites of Ψ_d derivatives, *ii*) by post-modification of oligonucleotides containing a reactive Ψ_d derivative, and *iii*) by a *Michael* reaction on a Ψ_d-containing oligonucleotide.

i) The synthesis of the oligomers **19**, **22** (*Scheme 2*), **24–26**, **28**, **29**, and **36–38** (*Table 2*) was performed with the phosphoramidites **8**, **14**, and **16–18**. The methodology followed the standard protocols of solid-phase technique [23]. The coupling efficiency of the modified building blocks was between 95 and 100% and therewith similar to those determined for the regular compounds (trityl monitoring). The oligonucleotides were detritylated and purified as described in the *Exper. Part*. The composition of the test oligomer **24** (5'-d(Ψ*AGGTCAATACT)-3' with Ψ_d^* = fluorescein-labeled 2'-deoxypseudouridine without spacer) was confirmed by tandem

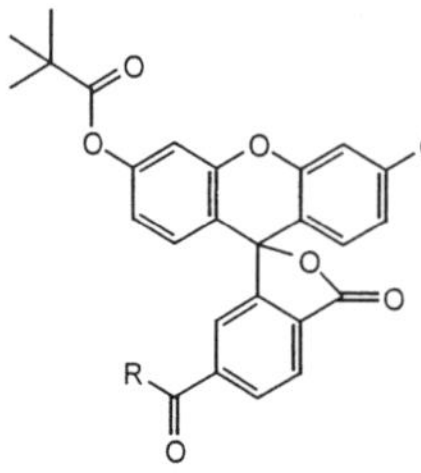

Rhodamin Dye JA-133

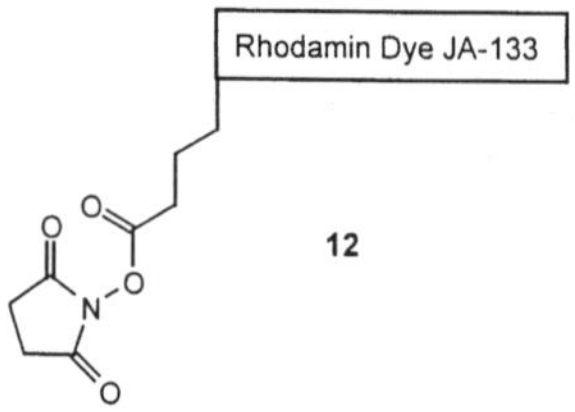

12

10: R=

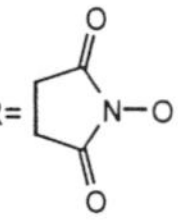

11: R=

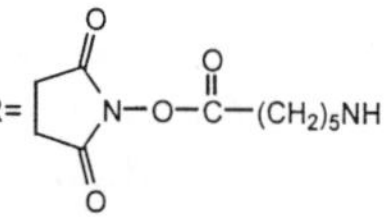

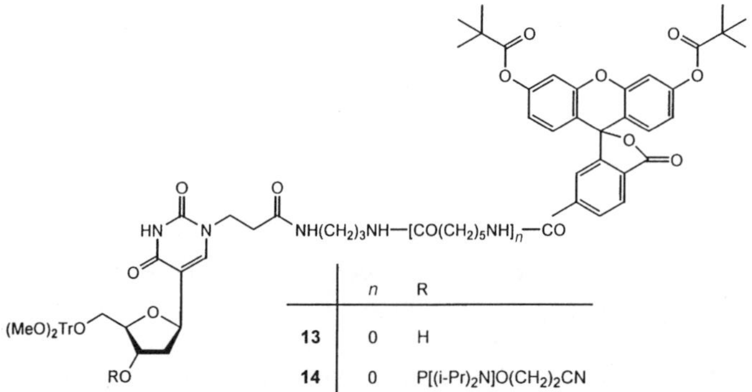

	n	R
13	0	H
14	0	P[(i-Pr)$_2$N]O(CH$_2$)$_2$CN
15	1	H
16	1	P[(i-Pr)$_2$N]O(CH$_2$)$_2$CN

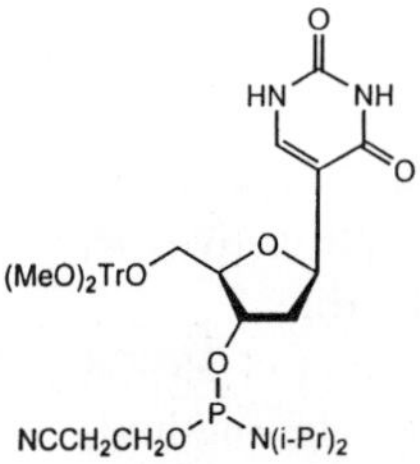

17

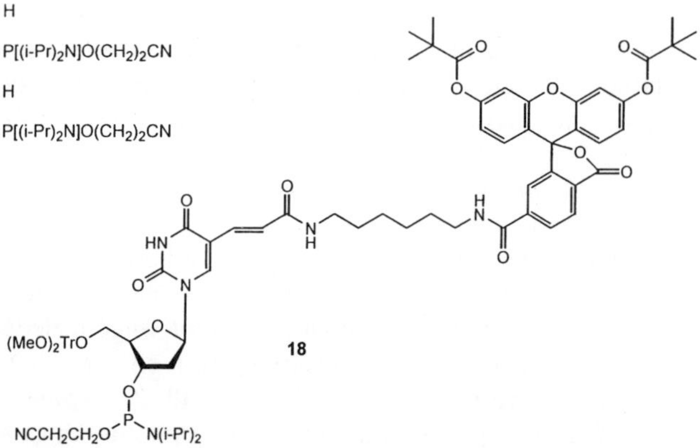

18

hydrolysis with snake-venom phosphodiesterase and alkaline phosphatase as described in the *Exper. Part*. Reversed-phase HPLC (*RP-18*) of the enzymatic digest (*Fig. 1*) was used to estimate the ε_{260} of fluorescein-labeled 2′-deoxypseudouridine (28000). MALDI-TOF Mass spectra were taken to determine the molecular masses of the other modified oligomers **28–35** and **38–40** (*Table 3*).

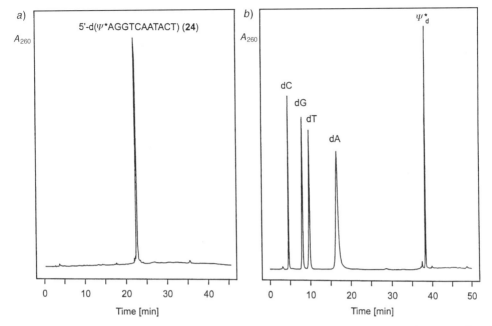

Fig. 1. a) *HPLC Profile of the oligonucleotide* **24** (5'-d(*Ψ**AGGTCAATACT)-3', with *Ψ*$_d^*$ = fluorescein-labeled 2'-deoxypseudouridine); b) *HPLC profile of the reaction products obtained after enzymatic hydrolysis of the oligonucleotide* **24** *by snake-venom phosphodiesterase and alkaline phosphatase in* 1M Tris · HCl *buffer* (pH 8.1) *at 37°. Column: RP-18* (250 × 4 mm); *gradient: 0 – 25 min 100% 0.1M* (Et₃NH)OAc/MeCN 95 : 5, pH 7.5 (= U); 25 – 70 min 0 – 100% MeCN in U.

ii) All other oligonucleotides listed in *Table 2* were prepared by post-synthetic reactions. From the phosphoramidite **8**, the controlled-pore-glass(CPG)-bound, fully-protected oligomer **19** was assembled carrying one 2'-deoxypseudouridine derivative with an ester function in the side chain. Treatment of **19** with 25% aq. ammonia (60°) gave the oligomer **20** (*Scheme 2*) bearing one N^1-(3-amino-3-oxopropyl)-2'-deoxy-pseudouridine residue. Cleavage of the 5'-(4,4'-dimethoxytriphenylmethyl) group of **20** with 80% AcOH/H₂O yielded the oligomer **30** (*Table 2*). In an analogous way, the doubly-modified oligonucleotide **31** (*Table 2*) was prepared. Reaction of **19** with 50% aqueous propane-1,3-diamine for 24 h at room temperature led to compound **21** which – upon detritylation – gave the amino-functionalized oligonucleotide **32** (*Table 2*). This latter oligomer was later used for further post-labeling reactions.

For the synthesis of the oligomer **35** (*Table 2*) bearing one 2'-deoxypseudouridine subunit with an N^1-{3-{[3-(1*H*-imidazol-4-yl)propyl]amino}-3-oxopropyl} side chain, a different strategy was chosen. For this purpose, phosphoramidites of the regular nucleosides with protecting groups suitable for 'ultramild deprotection chemistry' [24] were used together with compound **8**. After assembly of the CPG-linked oligomer **22** (*Scheme 2*), it was treated with a 1M soln. of 1*H*-imidazole-4-propanamine (24 h, room temp.; → **23**) [25] and subsequently detritylated to yield oligomer **35** (*Table 2*). Such oligodeoxyribonucleotides bearing an imidazole tether might be useful as artificial

Scheme 2

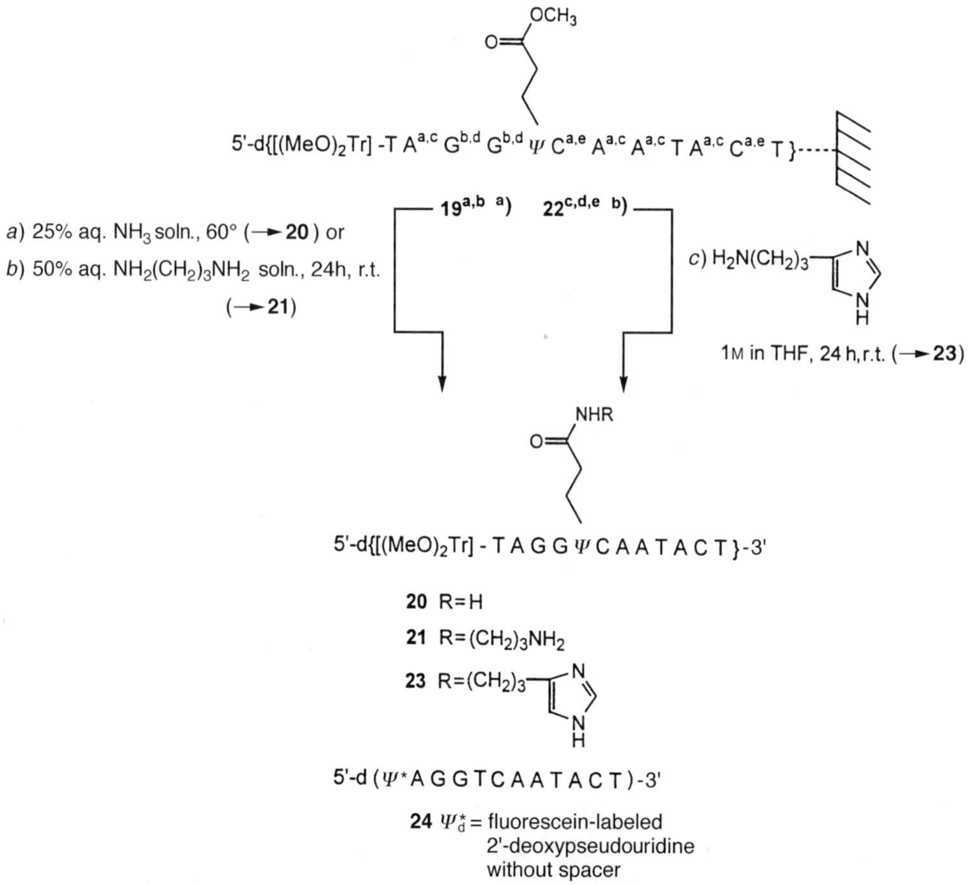

a) 25% aq. NH$_3$ soln., 60° (→ **20**) or

b) 50% aq. NH$_2$(CH$_2$)$_3$NH$_2$ soln., 24h, r.t.

(→ **21**)

c) H$_2$N(CH$_2$)$_3$—[imidazole]

1M in THF, 24 h,r.t. (→ **23**)

5'-d{[(MeO)$_2$Tr] - T A G G Ψ C A A T A C T}-3'

20 R=H

21 R=(CH$_2$)$_3$NH$_2$

23 R=(CH$_2$)$_3$—[imidazole]

5'-d (Ψ* A G G T C A A T A C T)-3'

24 Ψ$_d^*$ = fluorescein-labeled
2'-deoxypseudouridine
without spacer

a, bz (benzoyl); b, ibu (isobutyryl); c, pac (phenoxyacetyl); d, i-Prpac (4-isopropyl)phenoxyacetyl);
e, ac (acetyl); Tr, triphenylmethyl

a) Protected by a and b. *b*) Protected by c, d, and e.

RNases, as it has been recently shown [26] that they promote a sequence-specific strand scission of a complementary RNA strand in the presence of Zn^{2+} ions.

The amino-functionalized oligonucleotide **32** was post-synthetically labeled with the esters **11** and **12** derived from N-hydroxysuccinimide. Both reactions were performed in 0.1M Na-borate buffer (pH 8)/DMF at room temperature [27]. After purification by reversed-phase HPLC (*RP-18*), the labeled oligonucleotides **33** and **34** were obtained (see *Exper. Part* and *Table 2*). The oligonucleotides were characterized by MALDI-TOF mass spectra (*Table 3*).

iii) In a third approach, the oligonucleotide **28** (*Fig. 2,a*) bearing a single 2'-deoxypseudouridine residue was reacted overnight with an excess of methyl acrylate in 1M (Et$_3$NH)HCO$_3$/EtOH 1:1 at 37°. Subsequent analysis of the reaction mixture by reversed-phase HPLC (*RP-18*) showed a relatively clean reaction (*Fig. 2,b*). The main

Table 2. T_m *Values and* ΔG^0_{298} *Data of Duplex Formation of Oligonucleotides*[a])

Oligomer	Modified 2′-deoxynucleoside Ψ_d, Ψ^*_d or T^*_d	T_m [°C]	ΔG^0_{298} [kcal/mol]
5′-d(TAGGTCAATACT)-3′ **25** 3′-d(ATCCAGTTATGA)-5′ **26**	–	46	− 10.4
5′-d(TAGGTCAATACT)-3′ **25** 3′-r(AUCCAGUUAUGA)-5′ **27**	–	45	− 10.2
5′-d(TAGGΨCAATACT)-3′ **28** 3′-d(ATCCAGTTATGA)-5′ **26**	2′-deoxypseudouridine (**5**)	45	− 9.6
5′-d(TAGGΨCAATACT)-3′ **28** 3′-r(AUCCAGUUAUGA)-5′ **27**	2′-deoxypseudouridine (**5**)	46	− 10.7
5′-d(TAGGΨCAAΨACT)-3′ **29** 3′-d(ATCCAGTTATGA)-5′ **26**	2′-deoxypseudouridine (**5**)	42	− 9.0
5′-d(TAGGΨCAAΨACT)-3′ **29** 3′-r(AUCCAGUUAUGA)-5′ **27**	2′-deoxypseudouridine (**5**)	43	− 9.1
5′-d(TAGGΨ*CAATACT)-3′ **30** 3′-d(ATCCAGTTATGA)-5′ **26**	N^1-(3-amino-3-oxopropyl)-2′-deoxypseudouridine	43	− 9.6
5′-d(TAGGΨ*CAATACT)-3′ **30** 3′-r(AUCCAGUUAUGA)-5′ **27**	N^1-(3-amino-3-oxopropyl)-2′-deoxypseudouridine	43	− 9.6
5′-d(TAGGΨ*CAAΨ*ACT)-3′ **31** 3′-d(ATCCAGTTATGA)-5′ **26**	N^1-(3-amino-3-oxopropyl)-2′-deoxypseudouridine	41	− 8.7
5′-d(TAGGΨ*CAAΨ*ACT)-3′ **31** 3′-r(AUCCAGUUAUGA)-5′ **27**	N^1-(3-amino-3-oxopropyl)-2′-deoxypseudouridine	41	− 8.8
5′-d(TAGGΨ*CAATACT)-3′ **32** 3′-d(ATCCAGTTATGA)-5′ **26**	N^1-{3-[(3-aminopropyl)amino]-3-oxopropyl}-2′-deoxypseudouridine	44	− 9.4
5′-d(TAGGΨ*CAATACT)-3′ **32** 3′-r(AUCCAGUUAUGA)-5′ **27**	N^1-{3-[(3-aminopropyl)amino]-3-oxopropyl}-2′-deoxypseudouridine	43	− 9.9
5′-d(TAGGΨ*CAATACT)-3′ **33** 3′-d(ATCCAGTTATGA)-5′ **26**	2′-deoxy-N^1-{3-{{3-{{6-{[(fluorescein-6-yl)carbonyl]amino}-1-oxohexyl}amino}propyl}amino}-3-oxopropyl}pseudouridine	38 (37)[b]	− 8.2
5′-d(TAGGΨ*CAAΨ*ACT)-3′ **38** 3′-d(ATCCAGTTATGA)-5′ **26**	2′-deoxy-N^1-{3-{{3-{{6-{[(fluorescein-6-yl)carbonyl]amino}-1-oxohexyl}amino}propyl}amino}-3-oxopropyl}pseudouridine	32	n.m.
5′-d(TAGGΨ*CAATACT)-3′ **33** 3′-r(AUCCAGUUAUGA)-5′ **27**	2′-deoxy-N^1-{3-{{3-{{6-{[(fluorescein-6-yl)carbonyl]amino}-1-oxohexyl}amino}propyl}amino}-3-oxopropyl}-pseudouridine	36	− 8.2
5′-d(TAGGΨ*CAATACT)-3′ **34** 3′-d(ATCCAGTTATGA)-5′ **26**	2′-deoxy-N^1-{3-{{3-{[4-(JA-133)-1-oxobutyl]amino}propyl}amino}-3-oxopropyl}pseudouridine	37	− 8.1
5′-d(TAGGΨ*CAATACT)-3′ **35** 3′-d(ATCCAGTTATGA)-5′ **26**	2′-deoxy-N^1-{3-{[3-(1H-imidazol-4-yl)propyl]amino}-3-oxopropyl}pseudouridine	43	− 9.6

Table 2. (cont.)

Oligomer	Modified 2'-deoxynucleoside Ψ_d, Ψ_d^* or T_d^*	T_m [°C]	ΔG_{298}^0 [kcal/mol]
5'-d(TAGGΨ*CAATACT)-3' **35** 3'-r(AUCCAGUUAUGA)-5' **27**	2'-deoxy-N^1-{3-{[3-(1H-imidazol-4-yl)propyl] amino}-3-oxopropyl}pseudouridine	42	−9.3
5'-d(TAGGT*CAATACT)-3' **36** 3'-d(ATCCAGTTATGA)-5' **26**	5-{3-{{6-{[(fluorescein-6-yl)carbonyl]amino}hexyl}amino}- 3-oxoprop-1-enyl}thymidine	42	−9.1
5'-d(TAGGT*CAAT*ACT)-3' **37** 3'-d(ATCCAGTTATGA)-5' **26**	5-{3-{{6-{[(fluorescein-6-yl)carbonyl]amino}hexyl}amino}- 3-oxoprop-1-enyl}thymidine	39	−8.7
5'-d(TAGGT*CAATACT)-3' **36** 3'-r(AUCCAGUUAUGA)-5' **27**	5-{3-{{6-{[(fluorescein-6-yl)carbonyl]amino}hexyl}amino}- 3-oxoprop-1-enyl}thymidine	39	−8.6
5'-d(TAGGT*CAAT*ACT)-3' **37** 3'-r(AUCCAGUUAUGA)-5' **27**	5-{3-{{6-{[(fluorescein-6-yl)carbonyl]amino}hexyl}amino}- 3-oxoprop-1-enyl}thymidine	38	−9.0
5'-d(TAGGΨ*CAATACT)-3' **39** 3'-d(ATCCAGTTATGA)-5' **26**	Compound **6**	43	−9.6
5'-d(TAGGΨ*CAAΨ*ACT)-3' **40** 3'-d(ATCCAGTTATGA)-5' **26**	Compound **6**	39	−8.1

[a]) Measured in 10 mM Na-cacodylate, 10 mM MgCl$_2$, and 100 mM NaCl, pH 7.0, 5 + 5 μM of single strands.
[b]) Measured from an oligomer sample synthesized from compound **16**.

Table 3. *Molecular Masses of Oligonucleotides Determined by MALDI-TOF Mass Spectra*

Oligonucleotide	Modified 2'-deoxynucleoside Ψ_d or Ψ_d^*	M^+ (calc.)	M^+ (found)
5'-d(TAG GΨC AAT ACT)-3' **28**	2'-deoxypseudouridine	3628.6	3629.3
5'-d(TAG GΨCAAΨACT)-3' **29**	2'-deoxypseudouridine	3614.6	3615.2
5'-d(TAG GΨ*C AAT ACT)-3' **30**	N^1-(3-amino-3-oxopropyl)-2'-deoxypseudouridine	3703.0	3703.0
5'-d(TAG GΨ*C AAΨ* ACT)-3' **31**	N^1-(3-amino-3-oxopropyl)-2'-deoxypseudouridine	3760.0	3756.6
5'-d(TAG GΨ*C AAT ACT)-3' **32**	N^1-{3-[(3-aminopropyl)amino]-3-oxopropyl}-2'- deoxypseudouridine	3755.7	3755.7
5'-d(TAG GΨ*C AAT ACT)-3' **35**	N^1-{3-[(3-aminopropyl)amino]-3-oxopropyl}-2'- deoxypseudouridine	3808.6	3809.0
5'-d(TAG GΨ*C AAT ACT)-3' **33**	2'-deoxy-N^1-{3-{3-{6-{[(fluorescein-6-yl)carbonyl]amino}- 1-oxohexyl}amino}propyl}amino}-3-oxopropyl}pseudouridine	4229.5	4228.2 [a]) 4230.3 [b])
5'-d(TAG GΨ*C AAΨ*ACT)-3' **38**	2'-deoxy-N^1-{3-{{3-{6-{[(fluorescein-6-yl)carbonyl]amino}- 1-oxohexyl}amino}propyl}amino}-3-oxopropyl}pseudouridine	4814.5	4817.5
5'-d(TAG GΨ*C AAT ACT)-3' **34**	2'-deoxy-N^1-{3-{3-{[4-(JA-133)-1-oxobutyl]amino}propyl}- amino}-3-oxopropyl}pseudouridine	4437.9	4437.5
5'-d(TAG GΨ*C AAT ACT)-3' **39**	Compound **6**	3715.6	3716.6
5'-d(TAG GΨ*C AAΨ* ACT)-3' **40**	Compound **6**	3786.7	3790.7

[a]) Oligomer prepared by post-synthetic modification. [b]) Oligomer prepared by solid-phase synthesis from compound **16**.

Scheme 3

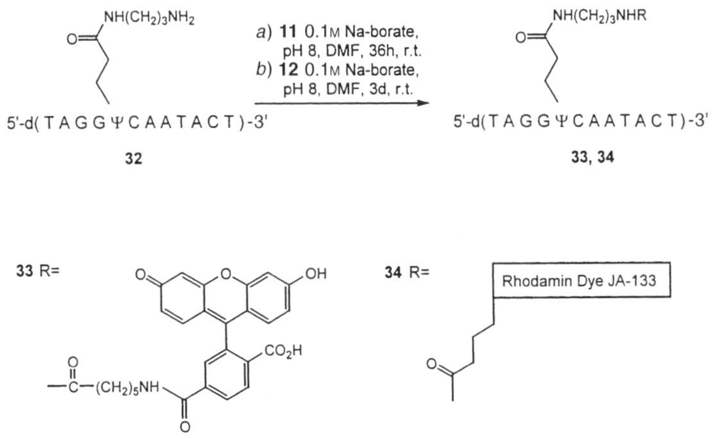

NH(CH₂)₃NH₂

O=

5'-d(TAGG Ψ CAATACT)-3'

32

a) **11** 0.1M Na-borate, pH 8, DMF, 36h, r.t.
b) **12** 0.1M Na-borate, pH 8, DMF, 3d, r.t.

NH(CH₂)₃NHR

O=

5'-d(TAGG Ψ CAATACT)-3'

33, 34

33 R= ...OH ...CO₂H

O=C—(CH₂)₅NH—

34 R= Rhodamin Dye JA-133

product **39** (*Table 2*), which migrates slightly slower than **28** (*Fig. 2,c*), was purified (*Fig. 2,d*) and characterized by MALDI-TOF mass spectrometry (*Table 3*) as well as by enzymatic tandem hydrolysis with snake-venom phosphodiesterase and alkaline phosphatase. It was shown that a single modification had occurred at the Ψ_d subunit of **28** under formation of the ester **6**, which co-migrates with dA in reversed-phase HPLC under various elution conditions. The other canonical nucleosides gave almost no reaction with methyl acrylate on the oligomer level under the above-mentioned conditions. This was demonstrated by reaction of methyl acrylate with *i*) the parent oligonucleotide **25**, which remained unmodified (reversed-phase HPLC) and *ii*) by reaction with an artificial mixture of G_d, C_d, A_d, and T_d. In the latter case, reversed-phase HPLC analysis of the reaction mixture showed that only T_d was partly alkylated (data not shown) [14b,c]. Analogously, the oligonucleotide **29**, which carries two Ψ_d residues, was reacted with methyl acrylate to give the doubly labeled oligomer **40**, which was purified and characterized as described for **39** (*Table 3*).

Duplex Formation of Oligonucleotides With 2'-Deoxypseudouridine or Functionalized Derivatives Thereof. The 2'-deoxypseudouridine represents a '*Janus*'-type molecule [28][29] exhibiting two identical but discernable *Watson-Crick* base-pairing sites **I** and **II**. The base-pair motif **I**, in which both bases adopt the *anti* conformation, involves HN(3)/O=C(2) of Ψ_d as donor and acceptor atoms. In motif **II** the atoms HN(1) and O=C(2) of Ψ_d are used for the pairing with A_d. In the latter case, Ψ_d adopts the *syn* conformation.

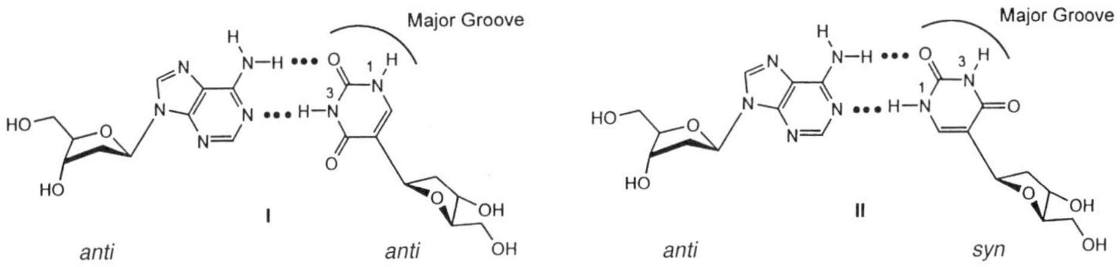

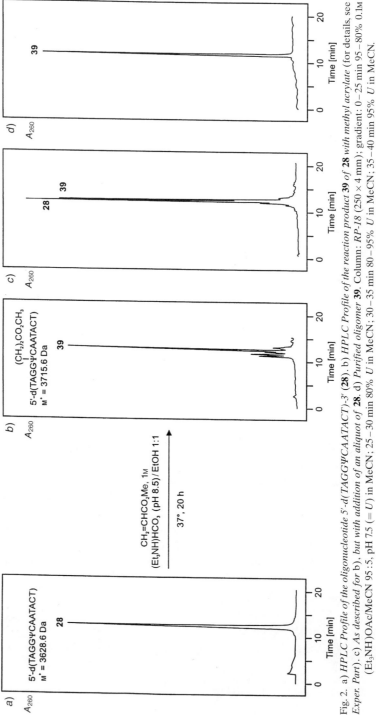

Fig. 2. a) *HPLC Profile of the oligonucleotide 5′-d(TAGGΨCAATACT)-3′* (**28**). b) *HPLC Profile of the reaction product* **39** *of* **28** *with methyl acrylate (for details, see Exper. Part).* c) *As described for* b), *but with addition of an aliquot of* **28**. d) *Purified oligomer* **39**. *Column: RP-18* (250 × 4 mm); *gradient: 0–25 min 95–80% 0.1M (Et₃NH)OAc/MeCN 95:5, 25–30 min 80% U in MeCN; 30–35 min 80–95% U in MeCN; 35–40 min 95% U in MeCN.*

Next, the conformation of 2′-deoxypseudouridine was studied in solution. Irradiation of H−C(6) of compound **5** results in an NOE of 3.8 % at H−C(1′), establishing a preferred *anti* conformation at the *C*-glycosidic bond. This is in line with results obtained by *Bhattacharya et al.* [9] with the N^1-methyl derivative **3**, and it is typical also for regular pyrimidine *N*-nucleosides such as T_d or U_d [30]. Contrary, 2′-deoxypseudouridine and 2′-deoxyuridine show significant differences with respect to their sugar puckering. From a pseudorotational analysis of the ^3J(H,H) coupling constants of vicinal sugar protons [31] (*Table 4*), a constrained *S*-type conformer population ($_{2'}T^3$) of 85% can be calculated, compared to only 60% for U_d (*Table 4*). Due to the lack of an anomeric effect in the cases of *C*-nucleosides, the $N \rightleftharpoons S$ conformer equilibrium of the *C*-nucleoside is strongly biased towards the *S*-type puckered pseudorotamer [32][33].

Table 4. $^3J(H,H)$ Coupling Constants of 2′-Deoxypseudouridine (**5**) and 2′-Deoxyuridine (U_d)

^3J(H,H) [Hz][a]							Sugar conformation					Pseudorotational parameters			
$J(1',2')$	$J(1',2'')$	$J(2',3')$	$J(2'',3')$	$J(3',4')$	$J(4',5')$	$J(4',5'')$	%N	%S	%γ^{g+}	%γ^t	%γ^{g-}	P_N	P_S	$\Psi_{m(N)}$	$\Psi_{m(S)}$
5 10.10	6.05	5.90	2.25	2.70	4.20	5.40	15	85	38	38	24	26	143.7	36	37.8
U_d 6.70	6.85	6.40	4.25	3.95	3.50	5.10	40	60	49	34	17	26	161.8	36	32.0

[a]) Solvent, D_2O; r.m.s. ≤ 0.4 Hz; $|\Delta J_{max}| \leq 0.4$ Hz. Primed and doubly primed locants are used to distinguish between the two protons at C(2′) and C(5′).

Only very few reports appeared until now that deal with the incorporation of Ψ_d or its derivatives into oligonucleotides and their influence on the duplex or triplex stability. *Bhattacharya et al.* [9] reported that the incorporation of 2′-deoxy-N^1-methylpseudouridine ($m^1\Psi_d$; **3**) into G-rich, triplex-forming oligonucleotides does not improve the stability of triple helices, neither in the antiparallel nor in the parallel hybridization mode. *Rosenberg et al.* [10] showed that the incorporation of the same compound $m^1\Psi_d$ (**3**) into oligonucleotide duplexes also leads to a significant decrease in the T_m value. This is surprising, as it has been shown that oligonucleotide duplexes containing T_d in place of U_d exhibit an enhanced thermal stability of 0.5−0.8°/modification [34]. To the best of our knowledge, no report appeared until now describing the effect that unmodified 2′-deoxypseudouridine exerts on the structure and thermal stability of oligonucleotide duplexes. Therefore, 2′-deoxypseudouridine (**5**) as well as its N^1-substituted derivatives described above were incorporated into the non-self-complementary DNA · DNA duplex **25**·**26** (T_m 46°, *Table 2*) and the corresponding DNA · RNA hybrid (**25**·**27**; T_m 45°, *Table 2*). For this purpose, either one or two of the modified nucleotides were incorporated into the single strand **25** at position 5 as well as in positions 5 and 9. The modified oligonucleotides were hybridized with the corresponding complementary DNA (**26**) or RNA (**27**) strand, and the T_m values were determined by temperature-dependent UV measurements in a 10 mM Na-cacodylate, 10 mM $MgCl_2$, and 100 mM NaCl solution (pH 7.0) (*Table 2*). As can be seen from *Table 2*, the incorporation of one or two Ψ_d (**5**) residues leads to an only slight decrease ($\Delta T_m = -1-2°$/modification) of the thermal stability of **25**·**26** and **25**·**27**.

Next, the T_m values of DNA · DNA and DNA · RNA duplexes containing in one DNA strand either one or two N^1-(3-amino-3-oxopropyl)-2′-deoxypseudouridine subunits (*Table 2*, **30** · **26**, **30** · **27**, **31** · **26**, **31** · **27**) were measured. From *Table 2* it can be seen that the introduction of a short side chain into 2′-deoxypseudouridine that does not protrude from the major groove of the double helix leads to a further slight decrease of the thermal stability. An analogous thermal stability was found for the duplex **39** · **26**, containing one 2′-deoxypseudouridine subunit with a methyl propanoate side chain, *i.e.*, **6**. The duplex **40** · **26**, carrying two corresponding Ψ_d^* residues, shows a T_m value of 39°.

An elongation of the side chain to an N^1-{3-[(3-aminopropyl)amino]-3-oxopropyl} residue does not further influence the T_m values, neither of the DNA · DNA nor of the DNA · RNA duplex (**32** · **26**, **32** · **27**, *Table 2*). Also the introduction of one 2′-deoxypseudouridine derivative with an N^1-{3-{[3-(1*H*-imidazol-4-yl)propyl]amino}-3-oxopropyl} side chain (see **35**) only slightly influences the thermal stability of duplexes with either the complementary DNA or RNA strand (**35** · **26**, **35** · **27**, *Table 2*).

Duplex Formation and Spectral Properties of Oligonucleotides With Fluorophor-Tethered 2′-Deoxypseudouridine Subunits. The results described above were good auspices for the introduction of bulky fluorophores into DNA · DNA and DNA · RNA duplexes *via* a 2′-deoxypseudouridine carrier. This was realized by two methods: *i*) First, the phosphoramidite **14** was used to synthesize a series of 31-mers (not shown), each carrying one fluorescein-labeled 2′-deoxypseudouridine residue without 6-amino-hexanoic acid spacer. Between the nucleobase and the fluorophore, a 10-bond spacer is inserted. The T_m values were determined by temperature-dependent UV-measurements and compared with those of the unmodified duplexes. In all cases, the introduction of only one fluorescein-labeled 2′-deoxypseudouridine residue leads to a significant decrease of the thermal stability ($\Delta T_m = -5$ and $-8°$) [35] – a finding that is in contrast to our expectations.

Therefore, the oligonucleotides **33** and **38** carrying one or two fluorescein-labeled residues, with a 17-bond distance between the uracil base and the fluorophore, were prepared using the phosphoramidite **16** and hybridized to the counter strand **26** (*Table 2*). T_m Measurements, however, revealed that – regardless of the longer spacer length – the decrease of the thermal stability of the duplexes compared to that of the unmodified standard-oligomer duplex (**25** · **26**) is similar to that where the phosphor-amidite **14** without 6-aminohexanoic acid spacer had been used (**33** · **26**: T_m 37°; **38** · **26**: T_m 32°). A doubly fluorescein-labeled duplex (**38** · **26**) exhibits a melting temperature that lies 15° below that of the standard duplex (*Table 2*).

ii) In a second approach, we used the amino-functionalized oligonucleotide **32** for a post-synthetic labeling reaction [36] with the ester **11** derived from *N*-hydroxysuc-cinimide to yield the oligomer **33**, which has already been described above. Moreover, compound **32** was reacted with the ester **12** yielding oligomer **34** which carries one 2′-deoxypseudouridine derivative with the cationic rhodamine dye JA-133 [37] in a 13-bond distance to the base. Both oligomers were hybridized to their complementary DNA strand (→ **33** · **26**, **34** · **26**), and their T_m values were determined by temperature-dependent UV, VIS, and fluorescence spectra. *Fig. 3,a* displays the UV-melting profile of **33** · **26** from which a T_m value of 38° can be taken, which means a T_m decrease of $-8°$ compared to the parent duplex **25** · **26**. A plot of A_{492} – the λ_{max} of fluorescein – *vs.* the temperature (*Fig. 3,b*) as well as a plot of the fluorescence emission intensity at 521 nm *vs.* the temperature (*Fig. 3,c*) do not show any cooperativity. The change in fluorescence emission between 15 and 60° amounts to 15% [27].

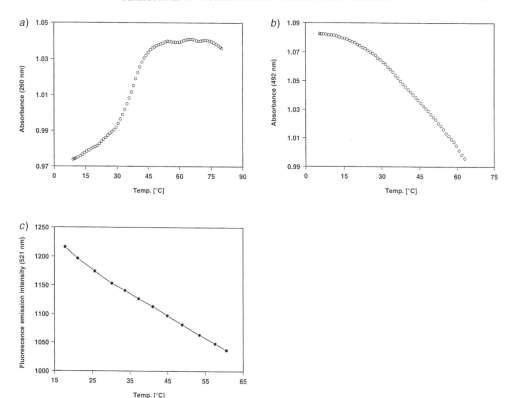

Fig. 3. a) A$_{260}$ vs. *temperature plot of the duplex* **33·26**, b) A$_{492}$ vs. *temperature plot of the duplex* **33·26**, *and* c) *fluorescence emission intensity at 521 nm* (excitation at 492 nm) *as a function of temperature of* **33·26**. In 10 mM Na-cacodylate, 100 mM NaCl, and 10 mM MgCl$_2$ (pH 7); oligomer concentration, 5 μM of each strand in *a*) and *b*), and 0.5 μM of each strand in *c*).

The oligonucleotide duplex **34 · 26** containing one cationic rhodamine dye (JA-133) labeled 2′-deoxypseudouridine subunit, exhibits a similar UV-melting temperature (T_m 37°, *Fig. 4,a*). In contrast to the duplex **33 · 26**, however, the T_m value of **34 · 26** can be verified by measuring the temperature-dependent absorbance at 615 nm, the λ_{max} of the rhodamine dye JA-133 (*Fig. 4,b*), as well as the temperature-dependent fluorescence emission at 639 nm (*Fig. 4,c*). The change in fluorescence emission between 15 and 70° amounts to 18% which is similar to that of **33 · 26**. In both cases, a melting temperature of 37° can be determined from the inflection points. Also, plots of temperature-dependent λ_{max} values of the excitation and emission spectra exhibit cooperativity (data not shown).

For comparison, two oligonucleotides **36** and **37** were synthesized containing one or two 5-{3-{{6-{[(fluorescein-6-yl)carbonyl]amino}hexyl}amino}-3-oxoprop-1-enyl}thymidine subunits from the phosphoramidite **18**. In this case, fluorescein and carrier base are separated by a spacer with 13 bonds. When these oligomers were hybridized with their corresponding DNA (**36 · 26**, **37 · 26**, *Table 2*) and RNA strands (**36 · 27**, **37 · 27**, *Table 2*), the resulting duplexes exhibit T_m values that are 3.5 – 4.0°/modification

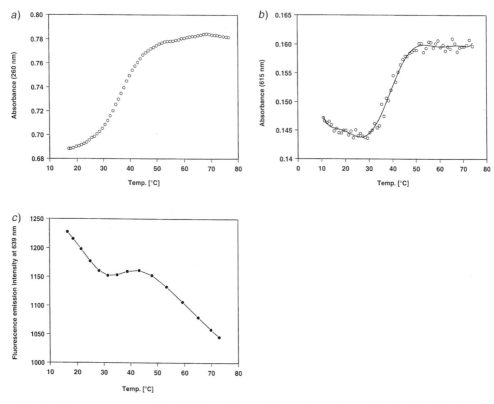

Fig. 4. a) A$_{260}$ vs. temperature plot of the duplex **34·26**, b) A$_{615}$ vs. temperature plot of the duplex **34·26**, and c) fluorescence emission intensity at 639 nm (excitation at 615 nm) as a function of temperature of **34·26**. In 10 mM Na-cacodylate, 100 mM NaCl, and 10 mM MgCl$_2$ (pH 7); oligomer concentration, 5 μM of each strand in a) and b), and 0.5 μM of each strand in c).

lower than those of the parent duplexes **25 · 26** and **25 · 27** (*Table 2*). This means that the fluorescein-labeled thymidine residue reduces the T_m value of the duplex only half as much as the fluorescein-labeled 2′-deoxypseudouridine, regardless of the shorter spacer length. *Fig. 5* shows exemplarily the melting profiles of **36·26** determined by three different methods, all displaying a T_m value of 42°.

Discussion and Conclusion. – A comparison of the T_m data presented above clearly indicates that 2′-deoxypseudouridine is a suitable nucleotide carrier for reporter groups such as fluorescein or the cationic rhodamine dye JA-133 [37]. However, the T_m values of oligonucleotides containing fluorophor-labeled Ψ_d are lower than corresponding duplexes carrying fluorescein-labeled 2′-deoxyuridine. The reduction of the T_m, however, is not due to the incorporation of the *C*-nucleoside or amino-functionalized 2′-deoxypseudouridine analogues *per se*; there is almost no decrease in the T_m prior to labeling with the fluorescent dyes. No significant difference in the melting temperature can be observed between a 13-bond and a 17-bond linker between the dye and the base carrier. Moreover, the charge of the dye ligand has no significant influence on the

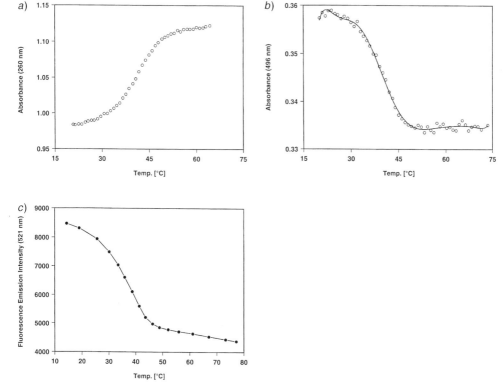

Fig. 5. a) A_{260} vs. *temperature plot of the duplex* **36·26**, b) A_{496} vs. *temperature plot of the duplex* **36·26**, *and* c) *fluorescence emission intensity at 521 nm* (excitation at 496 nm) *as a function of temperature of* **36·26**. In 10 mM Na-cacodylate, 100 mM NaCl, and 10 mM MgCl$_2$ (pH 7); oligomer concentration, 5 μM of each strand in a) and b), and 0.5 μM of each strand in c).

thermal stability of the duplex. A striking difference between the dye-labeled oligonucleotides **33·26** and **34·26** (*Table 2*), however, is the finding that only in the second case, cooperative melting can be observed by temperature-dependent monitoring of the optical properties of the dye, namely fluorescence emission and absorbance. It is assumed that this is due to the fact that in the first case (**33·26**), the fluorescein tether protrudes stretched-out from the major groove of the double helix and does not fold back because of charge repulsion between the negatively charged dye and the phosphodiester backbone, so that a dissociation of the DNA strands does not change the electronic environment of the fluorescent dye.

This implies, on the other hand, that in the case of **34·26** (*Table 2*), in which the cationic rhodamine dye (JA-133 [37]) is located at a distance of only 13 bonds from the uracil carrier, the fluorophore must be in close contact with the double helix, augmented by charge attraction between the positively charged dye and the phosphodiester backbone, so that a strand dissociation influences the electronic environment of the fluorophore.

The main advantage of 2'-deoxypseudouridine over the hitherto used 2'-deoxyuridine as a carrier nucleoside for reporter groups within an oligonucleotide chain is the fact that it can be functionalized under very mild reaction conditions, which allows a post-synthetic cascade of modifications on the oligonucleotide level.

We gratefully acknowledge excellent technical assistance by Mrs. *Eva-Maria Albertmann*, Mrs. *Elisabeth Feiling* (Osnabrück), Mr. *Andreas Huber*, and Mr. *Gerhard Lassonczyk* (Penzberg). We also thank Dr. *Thomas Wenzel, Bruker Saxonia Analytic GmbH*, Leipzig, for the measurement of the MALDI-TOF mass spectra. Moreover, financial support by the *Deutsche Forschungsgemeinschaft* is gratefully acknowledged.

Experimental Part

General. The phosphoramidites **17** and **18** were purchased from *Glen Research* (USA); the esters **10**–**12** of N-hydroxysuccinimide were generously provided by Dr. *H.-P. Josel* (*Roche Diagnostics GmbH*, Penzberg, Germany). The oligoribonucleotide **27** was a generous gift of Dr. *L. Beigelman* (*Ribozyme*, USA). Flash chromatography (FC): at 0.5 bar with silica gel *60* (*Merck*, Darmstadt, Germany). Solvent systems for FC and TLC: CH_2Cl_2/MeOH 9:1 (*A*), CH_2Cl_2/acetone 1:1 (*B*), CH_2Cl_2/acetone 7:3 (*C*), and i-PrOH/NH_3/H_2O 4:1:1 (*D*). Samples were collected with an *UltroRac II* fractions collector (*LKB Instruments*, Sweden). UV Spectra: *U3200* spectrophotometer (*Hitachi*, Japan). Fluorescence spectra: in 10 mM sodium cacodylate, 100 mM NaCl, and 10 mM $MgCl_2$ (pH 7.0) solns.; *F-4500* fluorescence spectrometer (*Hitachi*, Japan). NMR Spectra: *Avance-DPX-250* and *AMX-500* spectrometers (*Bruker*, Germany); δ values rel. to Me_4Si or external H_3PO_4. Microanalyses were performed by *Mikroanalytisches Laboratorium Beller* (Göttingen, Germany).

Oligonucleotide Synthesis and Purification. Oligonucleotides were synthesized with an *ABI-392* DNA synthesizer (*Applied Biosystems*, Weiterstadt, Germany) according to a standard protocol recommended by the manufacturer either in the 'trityl-off' or the 'trityl-on' mode with the modified phosphoramidites **8**, **14**, and **16**–**18** on a 1-μmol scale together with the phosphoramidites of $[(MeO)_2Tr]T_d$, $[(MeO)_2Tr]ib^2G_d$, $[(MeO)_2Tr]b-z^6A_d$, and $[(MeO)_2Tr]bz^4C_d$. The coupling yields of the modified phosphoramidites **8**, **14**, and **16**–**18** were generally > 95%.

In the cases of unmodified oligonucleotides as well as in the cases in which compounds **14** and **16** were used, the 5'-*O*-(dimethoxytrityl)-oligonucleotides ('trityl-on' synthesis) were cleaved from the solid support and subsequently deprotected in 25% NH_3/H_2O (12–18 h at 60°). The 5'-*O*-(dimethoxytrityl)-oligonucleotides were purified by reversed-phase HPLC (*RP-18*) with the following apparatus and procedure: 250 × 4 mm *RP-18* column (*Merck*, Germany); *Merck-Hitachi* HPLC apparatus, with a *655 A-12 D-2000* chromato-integrator (*Merck-Hitachi*, Darmstadt, Germany); eluents 0.1M (Et_3NH)OAc (pH 7.0)/MeCN 95:5 (*U*) and MeCN (*V*); gradient *Ia:* 0–30 min 0–40% *V* in *U*, flow rate 1 ml min⁻¹; gradient *Ib:* 0–60 min 0–100% *V* in *U*, flow rate 1 ml min⁻¹. Removal of the 4,4'-dimethoxytrityl residues was performed by treating the purified oligomers with 80% aq. AcOH soln. for 20 min at r.t. After neutralization with Et_3N and evaporation, followed by co-evaporation with MeOH, the oligomers were again purified by HPLC (*RP-18*, above-mentioned device; gradient *II:* 0–25 min 5–20% *V* in *U*, 25–30 min 20% *V* in *U*, 30–35 min 20–5% *V* in *U*, 35–40 min 5% *V* in *U*, flow rate 1 ml min⁻¹). Subsequent desalting of all oligonucleotides was performed by HPLC (*RP-18*, 100 × 4 mm, apparatus as described above; solvent for adsorption H_2O; solvent for desorption, MeOH/H_2O 3:2).

Oligonucleotide syntheses with the commercially available phosphoramidites **17** and **18** were made in the 'trityl-off' mode. After cleavage from the solid support and evaporation, the crude detritylated oligonucleotides were purified as described above. Upon purification of the Ψ_d-containing oligonucleotides **28** and **29**, a slightly slower-migrating by-product appearing as a shoulder was cut off from the corresponding main peak.

The phosphoramidite **8** was used to prepare the fully protected polymer-bound oligonucleotide **19** in the 'trityl-on' mode. Treatment of **19** with NH_3/H_2O (16 h, 60°) gave the base-deprotected 5'-*O*-(dimethoxytrityl)-oligomer **20**, which was purified, detritylated, and again purified as described above to give **30**. In an analogous manner, the oligomer **31** was prepared containing two N^1-(3-amino-3-oxopropyl)-modified 2'-deoxypseudouridine subunits.

Treatment of **19** with propane-1,3-diamine/H_2O 1:1 (24 h, r.t.) gave the 5'-*O*-(dimethoxytrityl)-oligomer **21**, which was purified, detritylated, and again purified as described above to give **32**. This oligonucleotide was subsequently used for further post-modification reactions with fluorescent-dye-labeled esters of N-hydroxy-succinimide as described below.

For the preparation of the oligomer **22**, the phosphoramidite **8** was used together with the phosphoramidites of $[(MeO)_2Tr]T_d$, $[(MeO)_2Tr]pac^6A_d$, $[(MeO)_2Tr]ac^4C_d$, and $[(MeO)_2Tr]i$-Pr-pac^2G_d in the 'trityl-off' mode. After the assembly of the oligomer **22** was complete, it was reacted in heterogeneous phase with 1M 1*H*-

imidazole-4-propanamine in THF at r.t. for 24 h. After filtration, the soln. was evaporated in a *Speedvac* concentrator at elevated temperature. The resulting oligonucleotide **23** was then detritylated and purified by HPLC (*RP-18*, 250 × 4 mm, gradient *II* as described above) and then desalted to give **35** (*Table 2*).

The enzymatic hydrolysis of the oligonucleotide 5′-d(Ψ*AGGTCAATACT) (**24**) was performed as described [38], except that the incubation time was 5 h. HPLC (*RP-18*, 250 × 4 mm); gradient: 0 – 25 min 100% *U*, 25 – 70 min 100 – 0% *U* in *V*; flow rate 0.7 ml min^{-1}) separated the nucleoside constituents. On the basis of the following extinction coefficients ε_{260}: G$_d$ 11700, C$_d$ 7300, A$_d$ 15400, T$_d$ 8800 as well as of the given oligonucleotide sequence, the ε_{260} of fluorescein-labeled 2′-deoxypseudouridine was estimated to be 28000 (*Fig. 1,a* and *b*). All other oligonucleotides were characterized by MALDI-TOF spectra (*Table 3*).

Post-Synthetic Reaction of the Amino-Functionalized Oligonucleotide **32** *with the Esters* **11** *and* **12** *of* N-*Hydroxysuccinimide.* Two portions of the oligonucleotide **32** (5 A_{260} units, 37 µmol, each), both dissolved in 0.1M Na-borate buffer (pH 8, 200 µl), were reacted with either the fluorescein derivative **11** (5 mg, 6.8 µmol) [36] dissolved in DMF (100 µl) for 36 h (r.t.) or with compound **12** (2 mg, 2.5 µmol) dissolved in DMF (100 µl) for 3 d (r.t.) under exclusion of light. Both reaction mixtures were evaporated in a *Speedvac* concentrator at elevated temp. and then redissolved in the HPLC buffer *U* (500 µl). Both labeled oligonucleotides **33** and **34** were purified by HPLC (*RP-18*, 250 × 4 mm); gradient: 0 – 50 min 0 – 50% *V* in *U*; flow rate of 1 ml min^{-1}. The t_R of fluorescein-labeled **33**, 19.5 min; t_R of (JA-133)-labeled **34**, 33 min. Integration of the HPLC peaks allowed for both labeling reactions a rough estimation of a 50% yield.

Post-Synthetic Alkylation of the Oligomer **28** *with Methyl Acrylate.* Oligonucleotide **28** (5 A_{260} units, 46 nmol) was dissolved in 1M aq. (Et$_3$NH)HCO$_3$/EtOH 1:1 (100 µl), and methyl acrylate (3µl, 33µmol) was added. After 20 h at 37°, the mixture was lyophilized in a *Speedvac* concentrator (*Savant*, Farmingdale, USA), the residue redissolved in H$_2$O, and the mixture purified by reversed-phase HPLC (*RP-18*, 250 × 4 mm), gradient *II* as described above, flow rate 0.5 ml/min). The main zone (t_R 12.3 min) was collected and lyophilized: 2.5 A_{260} units of 5′-d(TAG G6C AAT ACT)-3′ (**39**; 47% by HPLC peak integration). The product was characterized by MALDI-TOF mass spectrometry as well as by enzymatic tandem hydrolysis with snake venom phosphodiesterase and alkaline phosphatase as described [38].

Determination of Melting Curves and Thermodynamics: a) Absorbance *vs.* temp. profiles were measured on *Cary-1* or *Cary-1E* spectrophotometers (*Varian*, Australia) with a *Cary* thermoelectrical controller. The T_m values were measured in the reference cell with a Pt-100 resistor. Thermodynamic data (ΔG^0_{298}) were calculated with the program MeltWin 3.0 [39].

b) Fluorescence emission or excitation as a function of temperature were recorded on an *F-4500* fluorescence spectrophotometer (*Hitachi*, Japan) in a thermostatted 1-cm-quartz cuvette using an *RC6* thermostate (*MWG Lauda*, Germany).

*Methyl 5-(2′-Deoxy-β-*D*-erythro-pentofuranosyl)-1,2,3,4-tetrahydro-2,4-dioxopyrimidine-1-propanoate* (**6**). To a soln. of 2′-deoxypseudouridine (**5**; 1.2 g, 5.26 mmol) in 1M (Et$_3$NH)HCO$_3$ (TBK; pH 8.5, 50 ml) and EtOH (60 ml), methyl acrylate (15 ml, 14.34 g, 166.6 mmol) was added slowly. After stirring for 20 h at r.t., the soln. was evaporated and co-evaporated twice with 50% aq. EtOH and then with toluene. FC (silica gel, column 13 × 5 cm, *A*) afforded 1.5 g (89%) of **6**. Colorless foam. TLC (silica gel, *A*): R_f 0.42. UV (H$_2$O): 270 (8600). ^{1}H-NMR ((D$_6$)DMSO): 11.30 (*s*, NH); 7.62 (*s*, H–C(6)); 4.97 (*d*, *J* = 3.8, OH–C(3′)); 4.81 (*t*, *J* = 5.8, H–C(1′)); 4.66 (*t*, *J* = 3.8, OH–C(5′)); 4.14 (*m*, H–C(3′)); 3.91 (*t*, *J* = 6.80, CH$_2$); 3.71 (*m*, H–C(4′)); 3.57 (*s*, MeO); 3.44 (*t*, *J* = 5.2, CH$_2$(5′)); 2.70 (*s*, *J* = 6.8, CH$_2$); 2.03 (*m*, H$_\beta$–C(2′)); 1.77 (*m*, H$_\alpha$–C(2′)). Anal. calc. for C$_{13}$H$_{18}$N$_2$O$_7$ (314.3): C 49.68, H 5.77, N 8.91; found: C 49.39, H 5.70, N 8.82.

*Methyl 5-[2′-Deoxy-5′-O-(4,4′-dimethoxytriphenylmethyl)-β-*D*-erythro-pentofuranosyl]-1,2,3,4-tetrahydro-2,4-dioxopyrimidine-1-propanoate* (**7**). Compound **6** (960 mg, 3.05 mmol) was dried by repeated co-evaporation from anh. pyridine and then dissolved in pyridine (40 ml). 4,4′-Dimethoxytriphenylmethyl chloride (1.14 g, 3.29 mmol) was added, and the soln. was stirred for 4 h at r.t. MeOH (4 ml), and 10 min later, a 5% aq. NaHCO$_3$ soln. (60 ml) were added. The aq. layer was extracted with CH$_2$Cl$_2$ (3 × 100 ml) and the org. layer dried (Na$_2$SO$_4$) and evaporated. FC (silica gel, column 15 × 5 cm, *B*) afforded 1.6 g (84%) of **7**. Colorless foam. TLC (silica gel, *B*): R_f 0.5. ^{1}H-NMR ((D$_6$)DMSO): 11.30 (*s*, NH); 7.47 (*s*, H–C(6)); 5.06 (*d*, *J* = 4.2, OH–C(3′)); 4.84 (*dd*, *J* = 5.0, 8.8, H–C(1′)); 4.11 (*m*, H–C(3′)); 3.86 (*m*, CH$_2$(5′)); 3.73 (*m*, H–C(4′)); 3.75 (*s*, 2 MeO); 3.53 (*s*, COOMe); 3.10 (*m*, CH$_2$); 2.62 (*m*, CH$_2$); 2.11 (*m*, H$_\beta$–C(2′)); 1.82 (*m*, H$_\alpha$–C(2′)). Anal. calc. for C$_{34}$H$_{36}$N$_2$O$_9$ (616.7): C 66.22, H 5.88, N 4.54; found: C 66.13, H 5.91, N 4.46.

*Methyl 5-[2′-Deoxy-5′-O-(4,4′-dimethoxytriphenylmethyl)-β-*D*-erythro-pentofuranosyl]-1,2,3,4-tetrahydro-2,4-dioxopyrimidine-1-propanoate 3′-(2-Cyanoethyl Diisopropylphosphoramidite]* (**8**). A soln. of **7** (400 mg, 0.65 mmol) in anh. CH$_2$Cl$_2$ (20 ml) was preflushed with Ar. Then, (i-Pr)$_2$EtN (215 µl, 1.3 mmol) and 2-cyanoethyl diisopropylphosphoramidochloridite (258 µl, 1.12 mmol) were added. After stirring for 20 min at r.t., an ice-cold 5% aq. NaHCO$_3$ soln. (20 ml) was added and the mixture extracted with CH$_2$Cl$_2$ (2 × 30 ml). The org. layer was dried (Na$_2$SO$_4$) and evaporated. FC (silica gel, column 10 × 5 cm, *C*) afforded a diastereoisomer mixture **8** (400 mg, 75%). TLC (silica gel, *C*): R_f 0.75, 0.83. ^{31}P-NMR (CDCl$_3$): 149.1; 149.6.

N^1-*(3-Aminopropyl)-5-[2′-deoxy-5′-O-(4,4′-dimethoxytriphenylmethyl)-β-D-erythro-pentofuranosyl]-1,2,3,4-tetrahydro-2,4-dioxopyrimidine-1-propanamide* (**9**). To a soln. of **7** (500 mg, 0.81 mmol) in i-PrOH (25 ml), propane-1,3-diamine (8 ml, 96 mmol) was added during 1 h at r.t. After stirring for 48 h at r.t., the soln. was evaporated and co-evaporated from anh. pyridine and toluene. Precipitation from toluene gave **9**. Colorless solid (474 mg, 89%). TLC (silica gel, *D*): R_f 0.8. The product showed a positive ninhydrin reaction. ^{1}H-NMR ((D$_6$)DMSO): 11.30 (*s*, NH); 8.00 (*t*, *J* = 10.8, N*H*CH$_2$); 7.42 (*s*, H−C(6)); 4.82 (br. *s*, 9 H, H−C(1′), OH−C(3′), NH$_3^+$, 2H$_2$O); 4.07 (*m*, H−C(3′)); 3.80 (*m*, CH$_2$(5′)); 3.73 (*s*, 2 MeO); 3.67 (*m*, H−C(4′)); 3.04 (*m*, 2 CH$_2$); 2.54 (*m*, CH$_2$); 2.30 (*m*, CH$_2$); 2.07 (*m*, H$_\beta$−C(2′)); 1.81 (*m*, H$_\alpha$−C(2′)); 1.50 (*m*, CH$_2$). Anal. calc. for C$_{36}$H$_{42}$N$_4$O$_8$ (658.75): C 65.64, H 6.43, N 8.51; found: C 65.55, H 6.25, N 8.42.

N^1-*[3-[({3′,6′-Bis(2,2-dimethyl-1-oxopropoxy)-3-oxospiro[isobenzofuran-1(3H),9′-[9H]xanthen]-6-yl}carbonyl)amino]propyl]-5-[2′-deoxy-5′-O-(4,4′-dimethoxytriphenylmethyl)-β-D-erythro-pentofuranosyl]-1,2,3,4-tetrahydro-2,4-dioxopyrimidine-1-propanamide* (**13**). To a soln. of **9** (400 mg, 0.64 mmol) in anh. pyridine (20 ml) 1-{[(3,6-di-*O*-pivaloyl)fluorescein-6-yl]oxy}pyrrolidine-2,5-dione (**10**; 411 mg, 0.64 mmol) [40] was added. After stirring at r.t. for 2 h, an ice-cold 5% aq. NaHCO$_3$ soln. (30 ml) was added, the aq. layer extracted (3 × 60 ml) with CH$_2$Cl$_2$, and the org. layer dried (Na$_2$SO$_4$) and evaporated. FC (silica gel, *B*) afforded 415 mg (56%) of **13**. Colorless solid. TLC (silica gel, *B*): R_f 0.23. A spot test with aq. NH$_3$ gave a positive fluorescein reaction. ^{1}H-NMR (CDCl$_3$): 9.7 (br. *s*, NH); 7.55 (*t*, *J* = 7.6, N*H*CH$_2$); 7.41 (*s*, H−C(6)); 4.83 (*dd*, *J* = 5.5, 9.4, H−C(1′)); 4.24 (*m*, H−C(3′)); ca. 3.7 (*m*, CH$_2$(5′), 2 MeO); 3.02 (*m*, 2 CH$_2$); 2.60−2.15 (*m*, 2 CH$_2$, H$_\beta$−C(2′)); ca. 1.8 (*m*, H$_\alpha$−C(2′)); ca. 1.5 (*m*, CH$_2$).

N^1-*[3-[({3′,6′-Bis(2,2-dimethyl-1-oxopropoxy)-3-oxospiro[isobenzofuran-1(3H),9′-[9H]xanthen]-6-yl}carbonyl)amino]propyl]-5-[2′-deoxy-5′-O-(4,4′-dimethoxytriphenylmethyl)-β-D-erythro-pentofuranosyl]-1,2,3,4-tetrahydro-2,4-dioxopyrimidine-1-propanamide 3′-(2-Cyanoethyl Diisopropylphosphoramidite)* (**14**). A soln. of **13** (296.5 mg, 0.25 mmol) in anh. CH$_2$Cl$_2$ (15 ml) was preflushed with Ar. Then, (i-Pr)$_2$EtN (84 μl, 0.50 mmol) and 2-cyanoethyl diisopropylphosphoramidochloridite (84 μl, 0.37 mmol) were added. After stirring for 20 min at r.t., the soln. was diluted with CH$_2$Cl$_2$ (30 ml) and extracted with 5% aq. NaHCO$_3$ soln. (20 ml). The org. layer was dried (Na$_2$SO$_4$), filtered, and evaporated. FC (silica gel, *B*) gave the diastereoisomer mixture **14** (270 mg, 78%). Colorless foam. TLC (silica gel, *B*): R_f 0.8. ^{31}P-NMR (CDCl$_3$): 149.11, 149.21.

N^1-*[3-({6-[({3′,6′-Bis(2,2-dimethyl-1-oxopropoxy)-3-oxospiro[isobenzofuran-1(3H),9′-[9H]xanthen]-6-yl}carbonyl)amino]-1-oxohexyl}amino)propyl]-5-[2′-deoxy-5′-O-(4,4′-dimethoxytriphenylmethyl)-β-D-erythro-pentofuranosyl]-1,2,3,4-tetrahydro-2,4-dioxopyrimidine-1-propanamide* (**15**). Compound **9** (80 mg, 0.13 mmol) was reacted with the ester **11** (80 mg, 0.13 mmol) and worked up as described for compound **13**: 90 mg (53%) of **15**. Colorless foam. TLC (silica gel, *B*): R_f 0.2. A spot test with aq. NH$_3$ gave a positive fluorescein reaction. ^{1}H-NMR (CDCl$_3$): 9.5 (br. *s*, NH); 4.90 (*dd*, *J* = 5.4, 9.2, H−C(1′)); 4.30 (*m*, 1 H, H−C(3′)).

N^1-*[3-({6-[({3′,6′-Bis(2,2-dimethyl-1-oxopropoxy)-3-oxospiro[isobenzofuran-1(3H),9′-[9H]xanthen]-6-yl}carbonyl)amino]-1-oxohexyl}amino)propyl]-5-[2′-deoxy-5′-O-(4,4′-dimethoxytriphenylmethyl)-β-D-erythro-pentofuranosyl]-1,2,3,4-tetrahydro-2,4-dioxopyrimidine-1-propanamide 3′-(2-Cyanoethyl Diisopropylphosphoramidite]* (**16**). Compound **15** (200 mg, 0.15 mmol) was reacted with 2-cyanoethyl diisopropylphosphoramidochloridite (64 μl, 0.28 mmol) and worked up as described for **14**: 150 mg (67%) of **16**. Colorless foam. TLC (silica gel, *B*): R_f 0.7. ^{31}P-NMR (CDCl$_3$): 149.5; 149.1.

REFERENCES

[1] F. F. Davis, F. W. Allen, *J. Biol. Chem.* **1957**, *227*, 907; W. E. Cohn, *Fed. Proc.* **1957**, *16*, 166.

[2] R. Shapiro, R. W. Chambers, *J. Am. Chem. Soc.* **1961**, *83*, 3920.

[3] A. M. Michelson, W. E. Cohn, *Biochemistry* **1962**, *1*, 490; R. W. Chambers, *Prog. Nucleic Acid Res. Mol. Biol.* **1966**, *5*, 349.

[4] R. Cortese, H. O. Kammen, S. J. Spengler, B. N. Ames, *J. Biol. Chem.* **1974**, *249*, 1103.

[5] K. A. Watanabe, in 'Chemistry of Nucleosides and Nucleotides', Ed. L. B. Townsend, Plenum Press, New York, London, 1994, p. 421.

[6] C. K. Chu, U. Reichman, K. A. Watanabe, J. J. Fox, *J. Heterocycl. Chem.* **1977**, *14*, 1119.

[7] K. Pankiewicz, A. Matsuda, K. A. Watanabe, *J. Org. Chem.* **1982**, *47*, 485.

[8] J. A. Piccirilli, S. E. Moroney, S. A. Benner, *Biochemistry* **1991**, *30*, 10350.

[9] B. K. Bhattacharya, R. V. Devivar, G. R. Revankar, *Nucleosides Nucleotides* **1995**, *14*, 1269.

[10] I. Rosenberg, J. F. Soler, Z. Tocik, W.-Y. Ren, L. A. Ciszewski, P. Kois, K. H. Pankiewicz, M. Spassova, K. A. Watanabe, *Nucleosides Nucleotides* **1993**, *12*, 381.

[11] J. Haralambidis, M. Chai, G. W. Tregear, *Nucleic Acids Res.* **1987**, *15*, 4857.

[12] P. Hagmar, M. Bailey, G. Tong, J. Haralambidis, W. H. Sawyer, B. E. Davidson, *Biochim. Biophys. Acta* **1995**, *1244*, 259.

[13] J. A. Brumbaugh, L. R. Middendorf, D. L. Grone, J. L. Ruth, *Proc. Natl. Acad. Sci. U.S.A.* **1988**, *85*, 5610.

[14] R. W. Chambers, *Biochemistry* **1965**, *4*, 219; A. Wilk, A. Grajkowski, L. R. Phillips, S. A. Beaucage, *J. Org. Chem.* **1999**, *64*, 7515; M. Mag, J. W. Engels, *Nucleic Acids Res.* **1988**, *16*, 3525.

[15] M. Yoshida, T. Ukita, *Biochim. Biophys. Acta* **1968**, *157*, 455.

[16] A. V. Rake, G. M. Tener, *Biochemistry* **1966**, *5*, 3992.

[17] I. Luyten, K. W. Pankiewicz, K. A. Watanabe, J. Chattopadhyaya, *J. Org. Chem.* **1998**, *63*, 1033.

[18] R. A. Earl, L. B. Townsend, *J. Heterocycl. Chem.* **1977**, *14*, 699.

[19] A. Matsuda, C. K. Chu, U. Reichman, K. Pankiewicz, K. A. Watanabe, J. J. Fox, *J. Org. Chem.* **1981**, *46*, 3603.

[20] A. D. Argoudelis, S. A. Mizsak, *J. Antibiotics* **1976**, *29*, 818.

[21] H. Schaller, G. Weimann, B. Lerch, H. G. Khorana, *J. Am. Chem. Soc.* **1963**, *85*, 3821.

[22] S. L. Beaucage, M. Caruthers, *Tetrahedron Lett.* **1981**, *22*, 1859.

[23] B. C. Froehler, 'Protocols for Oligonucleotides and Analogs', in 'Methods in Molecular Biology', Ed. E. S. Agrawal, Humana Press, Totowa, N.J., 1993, Vol. 20, p. 63.

[24] J. C. Schulhof, D. Molko, R. Teoule, *Nucl. Acids Res.* **1987**, *15*, 397 (manufactures information: http:// www.glenres.com).

[25] S. Kohgo, K. Shinozuka, H. Ozaki, H. Sawai, *Tetrahedron Lett.* **1998**, *39*, 4067.

[26] J. Hovinen, A. Guzaev, E. Azhayeva, A. Azhayev, H. Lönnberg, *J. Org. Chem.* **1995**, *60*, 2205.

[27] J. B. Randolph, A. S. Waggoner, *Nucleic Acids Res.* **1997**, *25*, 2923.

[28] T. L. Trapane, M. S. Christopherson, C. D. Roby, P. O. P. Ts'o, D. Wang, *J. Am. Chem. Soc.* **1994**, *116*, 8412.

[29] N. Branda, G. Kurz, J.-M. Lehn, *J. Chem. Soc., Chem. Commun.* **1996**, 2443.

[30] H. Rosemeyer, G. Toth, B. Golankiewicz, Z. Kazimierczuk, W. Bourgeois, U. Kretschmer, H.-P. Muth, F. Seela, *J. Org. Chem.* **1990**, *55*, 5784.

[31] J. van Wijk, C. Altona, 'PSEUROT 6.2 – A Program for the Conformational Analysis of the Five-Membered Rings', University of Leiden, July, 1993.

[32] C. Thibaudeau, J. Plavec, K. A. Watanabe, J. Chattopadhyaya, *J. Chem. Soc., Chem. Commun.* **1994**, 537.

[33] H. Rosemeyer, M. Zulauf, N. Ramzaeva, G. Becher, E. Feiling, K. Mühlegger, I. Münster, A. Lohmann, F. Seela, *Nucleosides Nucleotides* **1997**, *16*, 821.

[34] F. Seela, Y. He, unpublished results.

[35] Dr. Frank Bergmann (*Roche Diagnostics GmbH*, Penzberg), personal communication.

[36] G. T. Hermanson, 'Bioconjugate Techniques', Academic Press, New York, 1996, p. 670.

[37] R. Hermann, H.-P. Josel, K.-H. Drexhage, J. Arden-Jacob, European Patent EP 0 567 622 B1, 24.04.1996.

[38] F. Seela, S. Lampe, *Helv. Chim. Acta* **1991**, *74*, 1790.

[39] J. A. McDowell, D. H. Turner, *Biochemistry* **1996**, *35*, 14077.

[40] F. M. Rossi, J. P. Y. Kao, *Bioconjugate Chem.* **1997**, *8*, 495.

Base-Pairing Properties of 8-Aza-7-deazaadenine Linked *via* the 8-Position to the DNA Backbone

by Frank Seela*, Matthias Zulauf, and Harald Debelak

Laboratorium für Organische und Bioorganische Chemie, Institut für Chemie, Universität Osnabrück,
Barbarastr. 7, D-49069 Osnabrück, Germany (fax: +49(541)969-2370; email: Fraseela@rz.uni-osnabrueck.de)

The base-pairing properties of oligonucleotides containing the unusual N^8-linked 8-aza-7-deazaadenine 2'-deoxyribonucleoside (**2a**) as well as its 7-bromo derivative **2b** are described. The oligonucleotides were prepared by solid-phase synthesis employing phosphoramidite chemistry. Compound **2a** forms a strong base pair with T_d for which a reverse *Watson-Crick* pair is suggested (*Fig. 9*). Compound **2a** displays a lower *N*-glycosylic-bond stability than its N^9-nucleoside and shows strong stacking interactions when incorporated into oligonucleotides. The replacement of 2'-deoxyadenosine by **2a** does not significantly influence the duplex stability. However, this behavior depends on the position of the incorporation.

Introduction. – The current knowledge of usual DNA structures such as hairpins, cruciforms, triplexes, tetraplexes, or pentaplexes or left-handed Z-DNA is related to special sequence motifs [1]. Also, changes in the environmental conditions, such as counter ions, or the interaction with high-molecular-weight proteins that bind to DNA can alter the nucleic acid structure [2]. The variety of DNA structures is increased by those of backbone-modified nucleic acids (PNA, hexose-DNA, bicyclo-DNA) or DNA containing modified nucleobases (7-deazapurines, isoguanine) [3–7].

Earlier, it was shown that purine N^7-(2'-deoxyribofuranosides) related to 2'-deoxyadenosine or 2'-deoxyguanosine form stable base pairs in duplex DNA [8][9]. It was also found that the N^8-glycosylated 8-azaadenine (pyrazolo[3,4-*d*]pyrimidin-4-amine; purine numbering is used throughout the *General Part*) forms a rather stable base pair with thymidine when it replaces 2'-deoxyadenosine in self-complementary duplexes such as 5'-d(A-T)$_6$-3' or 5'-d(C-T-G-G-A-T-C-C-A-G)-3' [10]. As this

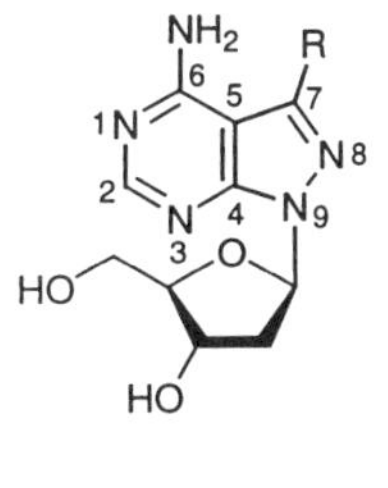

1a R = H
 b R = Br
 c R = I

Purine numbering

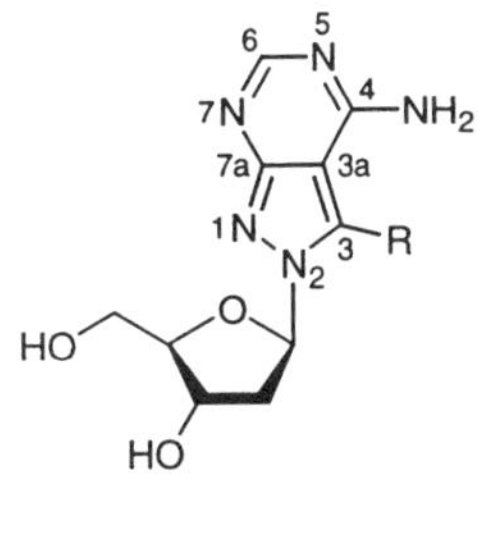

2a R = H
 b R = Br
 c R = I

Systematic numbering

observation was unexpected, we decided to investigate this subject in more detail. For this purpose, we incorporated the N^8-linked nucleoside **2a** [10][11] into a series of self-complementary and non-self-complementary oligonucleotides and studied their base-pairing properties. Furthermore, the influence of bulky substituents at the 7-position (see **2b**) were examined. For comparison, hybridization experiments were performed with oligonucleotides containing the regular (N^9-linked) 8-aza-7-deazaadenine 2′-deoxyribonucleoside (**1a**) [10][11].

Results and Discussion. – 1. *Monomers.* The glycosylation of 8-aza-7-deaza-6-methoxypurine (**3a**) as well as of the 7-bromo or 7-iodo derivatives **3b,c** with 2′-deoxyribofuranosyl chloride **4** gave the N^9-nucleosides **6a**–**c** as main products, while the N^8-compounds **5a**–**c** were formed as the minor components [11–13]. The regularly linked nucleosides **1a**–**c** were prepared as described earlier [11][13]. Detoluoylation of the intermediates **5b,c** (methanolic ammonia) furnished the deprotected nucleosides **2b,c** under simultaneous displacement of the MeO group by the NH_2 function (*Scheme 1*). The halogen substituents could be retained under these conditions, and a cleavage of the N-glycosylic bond was not observed. The chromatographic mobility of

Scheme 1

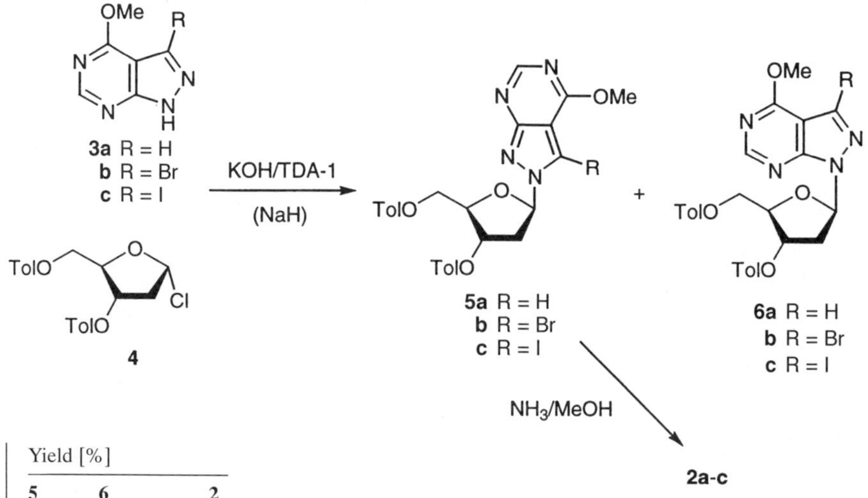

	Yield [%]		
	5	6	2
a	25	56	89
b	8 (15)	44 (34) [13]	44
c	11 (18)	38 (38) [13]	48

the N^8-glycosylated compounds **2a**–**c** (HPLC (*RP-18*); *Fig. 1*) points to a higher polarity of the N^8-nucleosides compared to their N^9-counterparts **1a**–**c**. As the corresponding adenine N^7-(2′-deoxyribonucleosides) *vs.* their N^9-isomers do not show such behavior [9], the presence of the *o*-quinoid moiety in **2a**–**c** might be responsible for the increased polarity. The 7-iodo derivative **1c** served as starting material for the

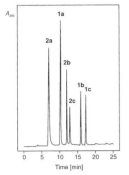

Fig. 1. *HPLC Profiles of the 8-aza-7-deazaadenine N^9-deoxyribonucleosides* **1a–c** *and of their* N^8*-regioisomers* **2a–c**

Table 1. *Half-life Values (τ) of Proton-Catalyzed Glycosylic-Bond Hydrolysis of the N^8-Linked 8-Aza-7-deazaadenine 2'-Deoxyribonucleosides* **2a–c** *and of Their N^9-Derivates* **1a–c**

	τ [min][a]	λ [nm]
$N^8c^7z^8A_d$ (**2a**) [15]	16	268, 280
$N^8Br^7c^7z^8A_d$ (**2b**)	<1 (13)[b]	271, 293
$N^8I^7c^7z^8A_d$ (**2c**)	3 (51)[b]	275, 300
$c^7z^8A_d$ (**1a**) [15]	61	258
$I^7c^7z^8A_d$ (**1c**)	300	233
A_d [15]	3.5	

[a]) Measured UV-spectrophotometrically in 1.0N HCl at 25°. [b]) Measured UV-spectrophotometrically in 0.1N HCl.

palladium(0)-catalyzed cross-coupling reaction with hex-1-yne [12]. Although the 7-alkynylated N^9-nucleosides were obtained under those conditions [12], the coupling reaction failed in the case of the N^8-linked iodo isomer **2c**. Instead, the deiodinated N^8-nucleoside **2a** was obtained as established by NMR spectroscopy.

Next, the glycosylic-bond stability of **2a–c** was investigated in acidic medium. The reaction was followed UV-spectrophotometrically in 1.0M and 0.1M HCl at 25°. The acid stability of the regularly linked nucleoside **1a** is significantly higher than that of the N^8-nucleoside **2a** or 2'-deoxyadenosine [11][12] (*Table 1*). Similarly, the N^8-nucleosides **2b,c** carrying a halogen atom at C(7) are more labile than the regularly linked nucleosides **1b,c** while the N-glycosylic bond of the latter is stabilized by the halogen substituent at C(7). The behavior of the halogenated N^8-nucleosides is similar to that of the 8-halogenated purine N^9-nucleosides, which are always less stable than their non-halogenated counterparts [14].

It is expected that the nucleosides **2a–c** show a dynamic conformational equilibrium in solution as it is found for the canonical DNA constituents. The conformational states are described by *i*) the puckering of the pentofuranosyl moiety ($N \leftrightarrow S$, $^3T_{2'} \leftrightarrow {}_{3'}T^{2'}$), *ii*) the rotational equilibrium about the C(4')–C(5') bond ($\gamma^{g+} \leftrightarrow \gamma^t \leftrightarrow \gamma^{g-}$), as well as by *iii*) the *syn-anti* equilibrium of the base around the N-glycosylic bond [16]. The nucleobases linked to the anomeric sugar C-atom drive the two-state ($N \leftrightarrow S$) pseudorotational equilibrium in nucleosides by two counteracting contributions [17]: *i*) the anomeric effect (stereoelectronic interactions between O(4') and the nucleobase N-atom at C(1')), which places the aglycone in the pseudoaxial orientation and *ii*) the inherent steric effect of the nucleobase, which opposes the anomeric effect by its tendency to take up the pseudoequatorial position (*Fig. 2*). The latter is sterically favored in the *S*-type conformation.

The sugar conformation of the nucleosides **1a–c** and **2a–c** was determined from the vicinal ^3J(H,H) coupling constants of the ^{1}H-NMR spectra measured in D$_2$O (*Table 2*) by the PSEUROT program [18][19]. In the case of 2'-deoxyadenosine, the preferred conformation is *S* (72%, *Table 3*) [20]. This *S*-conformer population is decreased in the case of the pyrazolo[3,4-*d*]pyrimidine nucleosides **1a–c**. A further

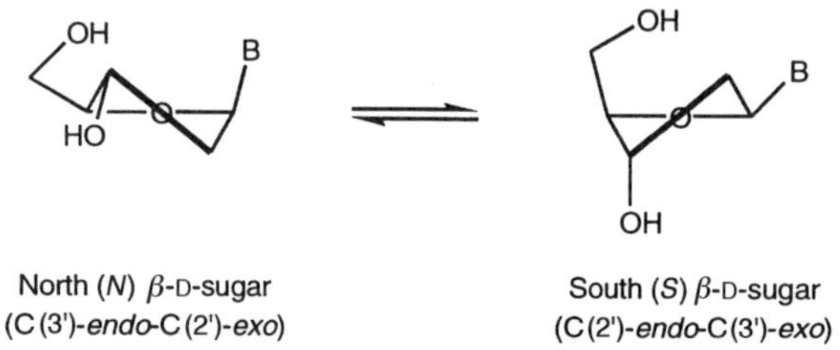

North (*N*) β-D-sugar   South (*S*) β-D-sugar

(C(3')-*endo*-C(2')-*exo*)   (C(2')-*endo*-C(3')-*exo*)

Fig. 2. *Two-state* ($N \leftrightarrow S$) *pseudorotational equilibrium of a 2'-deoxynucleoside*

Table 2. ^3J *Coupling Constants* [Hz] *of the Sugar Protons of the* N^8-*Linked 8-Aza-7-deazaadenine 2'-Deoxyribonucleosides* **2a**–**c** *and for Comparison of Their* N^9-*Isomers* **1a**–**c**

	$^3J(1',2')$	$^3J(1',2'')$	$^3J(2',3')$	$^3J(2'',3')$	$^3J(3',4')$	$^3J(4',5')$	$^3J(4',5'')$
A_d[a])	7.2	6.5	6.5	3.3	3.3	3.5	4.3
1a[a])	6.6	6.7	6.5	4.0	3.7	4.0	5.9
b[a])	6.4	6.4	6.6	4.5	3.3	4.4	6.0
c[a])	6.3	6.5	6.6	4.1	3.4	4.8	6.0
2a[a])	6.0	6.4	6.0	4.6	4.1	3.8	5.9
b[b])	5.0	6.9	6.4	4.9	4.4	4.1	6.3
c[b])	5.9	6.6	5.7	5.0	4.4	4.1	6.1
z^8A_d [21][a])	6.5	6.6	5.7	5.2	5.2	3.7	5.8
$N^8z^8A_d$ [21][a])	4.1	6.8	6.5	6.6	4.3	3.9	6.5

[a]) Measured in D_2O at 20°. [b]) Measured in $D_2O/(D_6)DMSO$ 95:5 at 35°.

decrease occurs in the case of the N^8-nucleosides **2a**–**c**, which show an almost equal population of *S*- and *N*-conformers. This trend proceeds when the conformer population of 8-aza-2'-deoxyadenosine (*Table 3*) is examined. This behavior indicates that nucleobases with electron-attracting properties drive the equilibrium from *S*- towards the *N*-conformation. Concerning the conformation about the C(4')–C(5') bond, the following relationship is observed: ($\gamma^{g+} \leftrightarrow \gamma^t \leftrightarrow \gamma^{g-}$) (*Fig. 3, Table 3*). While

Table 3. *Calculated Pseudorotational Parameters and the Rotational Equilibrium about the* C(4')–C(5') *Bond of the 8-Aza-7-deazaadenine 2'-deoxyribonucleosides* **1a**–**c** *and* **2a**–**c**[a])

	%*N*	%*S*	% $\gamma^{g+}(+sc)$	% $\gamma^t(-sc)$	% $\gamma^{g-}(ap)$
A_d	28	72	59	25	16
1a	37	63	35	43	22
b	39	61	29	44	25
c	37	63	25	44	31
2a	44	56	38	42	20
b	48 ·	52	29	48	23
c	47	53	31	46	23
z^8A_d [22]	50	50	39	42	19
$N^8z^8A_d$ [22]	59	41	29	50	21

[a]) R.m.s. ≤ 0.4 for all calculations; $|\Delta J_{max}| \leq 0.5$ Hz.

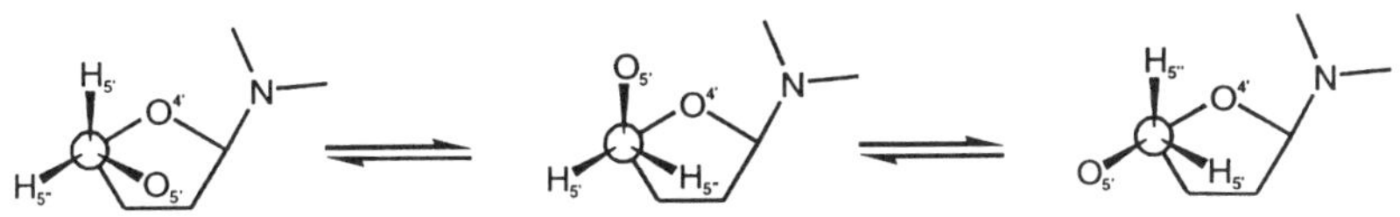

$+sc\,(g^+)$ $ap\,(g^-)$ $-sc\,(t)$

Fig. 3. *Conformations about the C(4')–C(5') bond of* **1a–c** *and* **2a–c**

A_d shows a $+sc$ rotamer population of 59%, it decreases to 38% for the N^8-glycosylated nucleoside **2a**. This phenomenon is similar to that found for the related pyrazolo[3,4-d]pyrimidine and triazolo[4,5-d]pyrimidine nucleosides but opposite to that of 7-substituted 7-deazapurine nucleosides [21][22].

Single-crystal X-ray analyses show that the regularly linked pyrazolo[3,4-d]-pyrimidine nucleosides of type **1** show a rather different N-glycosylic-bond conformation (*high-anti*) compared to the purine nucleosides [23]; these changes are also observed in the CD spectra [23]. *Fig. 4,a*, shows that a change of the glycosylation position from N^9 to N^8 as in **1a–c** *vs.* **2a–c** reverses the sign of the *Cotton* effect in the CD spectra and shifts the signal to longer wavelength. According to *Fig. 4,b*, the UV maximum of **1a** (*ca.* 280 nm) is shifted to a longer wavelength compared to that of 2'-deoxyadenosine (260 nm); in **2a,b** the quinoid structure gives rise to a further shift of the UV-maximum.

The phosphoramidites **9** and **10** were synthesized as the starting materials for the oligonucleotide synthesis. Compound **9** was prepared as described earlier [10]. The (dimethylamino)ethylidene residue was chosen as amino protecting group of the 7-brominated nucleoside ($\rightarrow$ **7**). The half-life value of the deprotection of compound **7** which was determined UV-spectrophotometrically in 25% aqueous NH_3 solution was 36 min at 20°. Subsequently, the 4,4'-dimethoxytriphenylmethyl group was introduced furnishing the derivative **8** (*Scheme 2*). Phosphitylation of compound **8** with 2-cyanoethyl diisopropylphosphoramidochloridite furnished the phosphoramidite **10** (63% yield).

All compounds were characterized by ^1H-, ^{13}C-, and ^{31}P-NMR spectroscopy (for ^{13}C-NMR, see *Table 4*; for ^1H-NMR, see *Exper. Part*) as well as by elemental analyses or FAB-MS. The ^{13}C-NMR signals of the 8-aza-7-deaza-2'-deoxyadenosine derivatives **2b,c**, **7**, and **8** were assigned by gated-decoupled ^{13}C-NMR or heteronuclear $^1H/$$^{13}C$-NMR correlation spetra. The ^{13}C-NMR data indicate a significant upfield shift of the C(7) signal when the glycosylic position is changed from N(9) to N(8). The introduction of the acetamidine protecting group ($\rightarrow$ **7**) has a strong impact on the ^{13}C-NMR chemical shifts of the heterocyclic system.

2. Oligonucleotides. 2.1. Synthesis. To investigate the base-pairing properties of the N^8-nucleosides **2a,b**, a series of oligonucleotides was synthesized. Automated solid-phase synthesis was performed with the methyl phosphoramidite **9** [10] or the corresponding cyanoethyl phosphoramidite, described elsewhere [24], as well as the cyanoethyl phosphoramidite **10**. The coupling yields were always higher than 95%. Deprotection was performed with 25% aqueous NH_3 solution and purification by OPC cartridges [26] or by reversed-phase HPLC (see *Exper. Part*). The oligonucleotides that were synthesized with the methyl phosphoramidite were demethylated with thiophe-

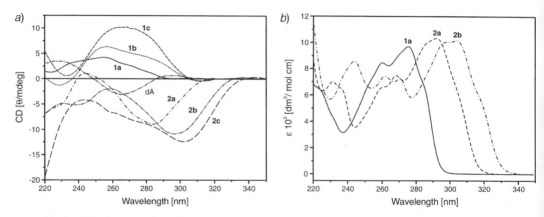

Fig. 4. a) *CD Spectra of the nucleosides* **1b,c** *and* **2b,c**; b) *UV spectra of the* N⁹-*nucleoside* **1a**, *the* N⁸-*nucleoside* **2a**, *and its 7-bromo derivative* **2b**. Measured at 10° in bidistilled water with 10 mM nucleoside concentration.

Scheme 2

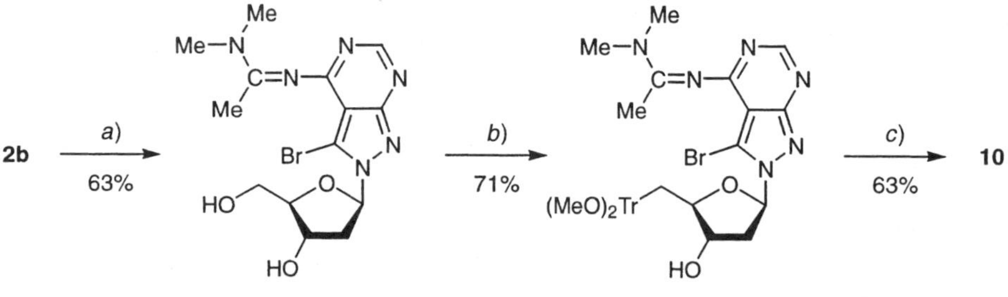

a) Me₂NC(Me)(OMe)₂, MeOH, 50°, 3 h. b) (MeO)₂Tr-Cl, pyridine, 50°, 2 h. c) (i-Pr)₂NP(Cl)O(CH₂)₂CN, CH₂Cl₂, 30 min, r.t.

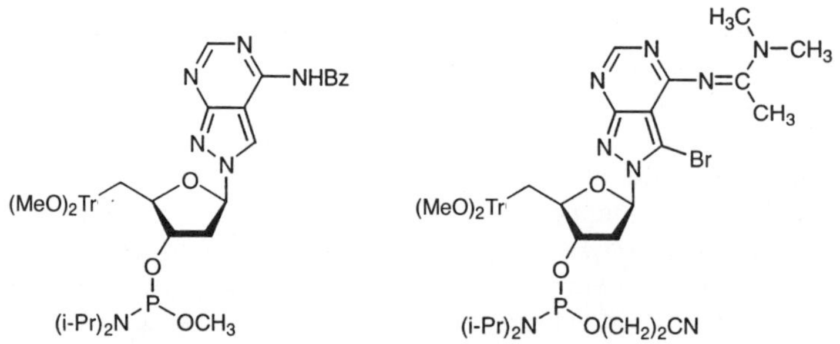

nol/Et₃N/dioxane 1:1:2 [10][25]. The homogeneity of the obtained oligonucleotides **11–39** (see below, *Tables 5–9*), was established by reversed-phase HPLC. MALDI-TOF-MS were measured (see *Exper. Part Table 10*), and the nucleoside composition

Table 4. ^{13}C-NMR Chemical Shifts of the 8-Aza-7-deazaadenine 2'-Deoxyribonucleoside Derivatives[a])

	C(7)[b]) C(3)[c])	C(5)[b]) C(3a)[c])	C(6)[b]) C(4)[c])	C(2)[b]) C(6)[c])	C(4)[b]) C(7a)[c])	C(1')[b])	C(2')[b])
1a	133.2	100.6	158.2	156.2	153.8	84.2	38.1
b	118.9	99.8	157.3	156.9	154.5	84.0	37.8
c	91.0	103.5	157.6	156.2	154.0	84.0	37.9
2a	124.0	101.4	159.5	156.7	159.6	90.5	[d])
b	107.5	101.8	158.7	158.8	157.2	88.5	38.4
c	81.2	106.0	159.5	159.2	156.3	89.6	38.4
7	109.2	107.8	162.8	163.2	159.2	88.5	[d])
8	109.0	108.1	163.0	163.1	163.1	87.4	[d])

	C(3')	C(4')	C(5')	MeO	Me₂N	C=N	Me
1a	71.2	87.7	62.6				
b	70.8	87.7	62.3				
c	70.9	87.7	62.3				
2a	70.7	88.4	62.1				
b	70.8	87.5	62.2				
c	70.9	88.4	62.2				
7	70.9	87.5	62.2			34.3, 42.2	157.0	16.9
8	70.0	85.0	63.8	54.9	[d])	157.7	16.8

[a]) Measured in (D₆)DMSO. [b]) Purine numbering. [c]) Systematic numbering. [d]) Superimposed by DMSO.

was determined after digestion of the oligonucleotides with snake-venom phospho-diesterase followed by alkaline phosphatase. Representative examples confirming the nucleoside composition are shown in *Fig. 5*.

2.2. *Stability of Duplexes with Antiparallel Chain Orientation. Tables 5 – 9* summarize the T_m values and the thermodynamic data of a series of oligonucleotide duplexes. The data were determined by curve-shape analysis of the melting profiles (MeltWin; version 3.0) [27]. The duplexes 5'-d(A-T)₆ that contain 8-aza-7-deaza-2'-deoxyadeno-sine (**1a**) or its N^8-glycosylated isomer **2a** [10] in place of A$_d$ are more stable than the parent 5'-d(A-T)₆ (see **12·12** and **13·13** vs. **11·11**) [10] (*Table 5*). While the T_m increase of the duplex **12·12** over that of the parent **11·11** is rather small ($\Delta T_m = 3°$), the duplex **13·13** is very stable ($\Delta T_m = 16°$). The homomeric duplexes **16·15** or **17·15** (*Table 5*) show slightly lower T_m values than the parent duplex **14·15**. Only duplexes that are composed of tracts of N^8-linked 'purines' and pyrimidines within the same strand are not as stable as those with a regular glycosylation position (compare **20·20** with **18·18** or **19·19**).

The CD spectrum of the duplex **13·13** formed by the alternating residues **2a** and T_d is different to that of the duplex **11·11** formed by **1a** and T_d (*Fig. 6,a*). The modified duplex shows a negative *Cotton* effect at 305 nm: the negative *Cotton* effect of 5'-d(A-T)₆-3' at 245 nm is absent. Similar CD changes are observed in the case of oligonucleotide duplexes of homomeric and block oligonucleotides (*Fig. 6,b* and *c*). While the CD spectra of the oligonucleotides **16·15** and **19·19** containing N^9-glycosylated 8-aza-7-deaza-2'-deoxyadenosine (**1a**) show similarities to the parent 'adenine' oligonucleotides **14·15** or **18·18**, respectively, the duplexes **17·15** and **20·20** containing the N^8-glycosylated compound **2a** are bathochromically shifted. To show the

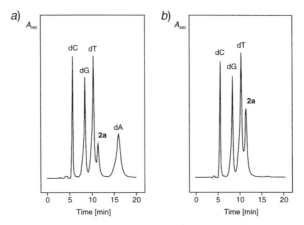

Fig. 5. *Reversed-phase HPLC profiles of the hydrolysis products of* a) *5′-d(T-A-G-G-T-C-**2a-2a**-T-A-C-T)-3′* (**24**) *and* b) *5′-d(T-**2a**-G-G-T-C-**2a-2a**-T-**2a**-C-T)-3′* (**25**) *formed by snake-venom phosphodiesterase followed by alkaline phosphatase. Conditions, see* Exper. Part.

Table 5. T_m *Values and Thermodynamic Data of the Oligonucleotides Containing 8-Aza-7-deaza-2′-deoxyade-nosine* (**1a**) *and the* N^8*-Glycosylated Nucleoside* **2a**

	T_m [°][a])	ΔH^0 [kcal/mol]	ΔS^0 [cal/mol · K]	ΔG^0_{298} [kcal/mol]
5′-d[(A-T-A-T-A-T-A-T-A-T)]$_2$-3′ (**11·11**)	33 (26)	−45 (−44)	−125 (−127)	−6.3 (−5.5)
5′-d[(**1a**-T-**1a**-T-**1a**-T-**1a**-T-**1a**-T)]$_2$-3′ (**12·12**) [10]	36	−63	−180	−7.2
5′-d[(**2a**-T-**2a**-T-**2a**-T-**2a**-T-**2a**-T)]$_2$-3′ (**13·13**) [10]	49 (39)	−74 (−62)	−207 (−175)	−9.6 (−7.3)
	41[c]) (22)[d])	−62 (−23)	−175 (−54)	−7.8 (−5.9)
5′-d(A$_{12}$)-3′ · 3′-d(T$_{12}$)-5′ (**14·15**) [28]	44 (37)	−84 (−91)	−238 (−267)	−9.8 (−7.9)
5′-d(**1a**$_{11}$-A)-3′ · 3′-d(T$_{12}$)-5′ (**16·15**) [28]	38 (32)	−91 (−65)	−266 (−186)	−8.4 (−6.8)
5′-d(**2a**$_{12}$)-3′ · 3′-d(T$_{12}$)-5′ (**17·15**)	38 (31)	−68 (−78)	−195 (−232)	−7.8 (−6.3)
5′-d[(A$_6$-T$_6$)]$_2$-3′ (**18·18**) [28]	46 (40)	−81 (−75)	−232 (−219)	−9.1 (−7.4)
5′-d[(**1a**$_6$-T$_6$)]$_2$-3′ (**19·19**) [28]	44 (39)	−49 (−58)	−133 (−163)	−8.3 (−7.1)
5′-d[(**2a**$_6$-T$_6$)]$_2$-3′ (**20·20**)	25 (17)	[b])	[b])	[b])

[a]) Measured at 260 nm. Data without parentheses are measured in 1M NaCl containing 100 mM MgCl$_2$ and 60 mM Na-cacodylate (pH 7.0) with 10 μM oligonucleotide concentration. Data in parentheses are measured in 100 mM NaCl, 10 mM MgCl$_2$, and 10 mM Na-cacodylate (pH 7.0) with 10 μM oligonucleotide concentration. [b]) Not determined. [c]) 1M NaCl, 10 mM Na-phosphate, and 0.1 mM EDTA (pH 7.0). [d]) 10 mM NaCl, 10 mM Na-phosphate, and 0.1 mM EDTA (pH 7.0).

structural characteristics of oligonucleotide exclusively due to residue **2a**, the CD spectrum of homomer **17** was measured and compared with that of d(A$_{12}$) [28]; the significant differences in these spectra are due to the conformational properties of the monomers. When the single-stranded oligomer **17** was hydrolyzed with snake-venom phosphodiesterase and alkaline phosphatase, the UV spectrum of the monomer **2a** was obtained (*Fig. 7,a*). According to the pronounced differences observed in the CD spectra of the homomer **17** and the monomer **2a** (*Fig. 7,b*), the single strands have a

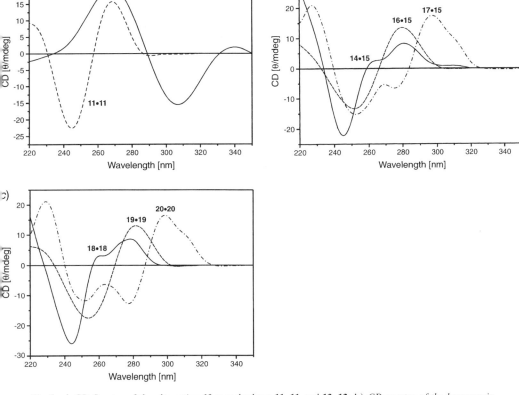

Fig. 6. a) *CD Spectra of the alternating 12-mer duplexes* **11·11** *and* **13·13**; b) *CD spectra of the homomeric duplexes* **14·15**, **16·15**, *and* **17·15**; c) *CD spectra of the block oligomers* **18·18**, **19·19**, *and* **20·20**. Spectra were measured at 10° in 1M NaCl, 100 mM MgCl₂, and 60 mM Na-cacodylate (pH 7.0) with 10 μM oligomer concentration. For sequences, see *Table 5*.

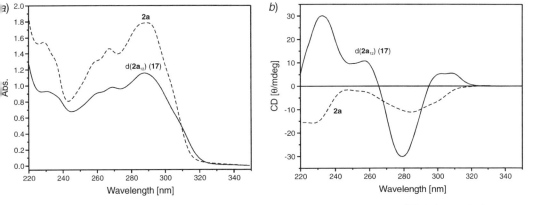

Fig. 7. a) *UV Spectra and* b) *CD spectra of the homomeric oligonucleotide* d(**2a**₁₂) (**17**) *before and after enzymatic digestion*. Measured at 10° in bidistilled water.

well-organized structure. This is due to stacking interactions between the base residues reflected by a hypochromicity of 35% (289 nm) determined by the enzymatic hydrolyses of the oligomer d(**2a**)$_{12}$.

Next, a series of oligonucleotides with random base composition was synthesized. They are derived from the oligomers 5′-d(T-A-G-G-T-C-A-A-T-A-C-T)-3′ (**21**) and 3′-d(A-T-C-C-A-G-T-T-A-T-G-A)-5′ (**22**), which are commonly used in our laboratory to study the influence of modified bases on the duplex stability. When two 2′-deoxyadenosine residues of the duplex **21·22** are replaced by compound **2a**, a decrease of the T_m value of only 1° per incorporated N^8-nucleoside residue results (see **23·22**), while a decrease of 3° per residue is observed in the case of duplex **24·22** (see *Table 6*). This indicates the dependence of the duplex stability on the incorporation position of residue **2a**. Furthermore, the incorporation at some positions of duplex **21·22** is more sensitive than at others. When two residues **2a** replace A$_d$, resulting in the duplexes **23·22**, **24·22**, and **21·28**, the decrease of the T_m value changes from 2° to 12° when compared with the parent duplex **21·22**. From the T_m data of *Table 6*, it can also be seen that the duplex stability is strongly affected when both strands are modified (see **23·28**, **24·28**, and **25·28**). Nevertheless, the destabilization caused by the incorporation of the N^8-linked nucleoside **2a** in one strand is rather small compared to the effect of residues causing real mismatches (ΔT_m/mismatch > 10°). The CD spectra of some duplexes of *Table 6* are given in *Fig. 8*.

The T_m values were also measured in Mg^{2+}-free buffer solution at various concentrations of NaCl. According to *Table 7*, the salt dependence of the T_m values of a duplex containing A$_d$ (**21·22**) with those containing the modified nucleoside **2a** shows an almost identical behavior. The stability of the unmodified DNA·RNA hybrid **21·38** (T_m 46°, *Table 8*) is slightly lower than that of the DNA·DNA duplex **21·22** (T_m 47°, *Table 7*). However, when compound **2a** is replacing 2′-deoxyadenosine (see **23·38**, **24·38**, and **25·38**), the T_m decrease is rather important (*Table 8*).

2.3. *Stability of Parallel-Stranded Duplexes.* The influence of the N^8-linked nucleoside **2a** on duplexes with parallel-strand (ps) orientation was also studied. For this purpose, the oligodeoxynucleotide **39**, wherein guanine is replaced by isoguanine and cytosine by 5-methylisocytosine, was chosen (*Table 9*) [29]. According to *Table 9*, the incorporation of the nucleoside **2a** reduces the T_m values (see **23·39** and **24·39** vs. **21·39**), as it is observed for the duplexes with antiparallel-strand (aps) orientation. Interestingly, the ΔT_m between modified and unmodified duplexes is larger in the case of ps-hybrids (*Table 9*) than in the case of aps duplexes (see **23·22** and **24·22** vs. **21·22** in *Table 6*). A comparison of the ΔG^0_{298} values of the modified aps or ps duplexes with their unmodified counterparts, determined in 1M NaCl, 100 mM MgCl$_2$, 60 mM Na-cacodylate, indicates a decrease of $|\Delta G^0_{298}|$ upon modification; *e.g.*, the aps duplex **23·22** or the ps duplex **23·39**, compared with the unmodified duplexes **21·22** or **21·39**, respectively, show a $\Delta\Delta G^0_{298}$ of 1.3 and 1.9 kcal/mol, respectively. If another sequence is chosen but the number of substitutions is the same (*e.g.* 2), different $\Delta\Delta G^0_{298}$ values are calculated; 2.3 kcal/mol for **24·22** in comparison to **21·22** and 2.8 kcal/mol for **24·39** in comparison to **21·39**. This indicates that the decrease of ΔG^0 upon incorporation of **2a** depends on the position of the modification. However, this dependence seems to be rather similar in aps- and ps-DNA, a finding that was not expected.

Table 6. T_m *Values and Thermodynamic Data of the Oligonucleotides Containing* N^8-*Linked 8-Aza-7-deaza-adenine 2′-Deoxyribonucleoside* **2a** *and its 7-Bromo Derivative* **2b**

	T_m [°][a])	ΔH^0 [kcal/mol]	ΔS^0 [cal/mol · K]	ΔG^0_{298} [kcal/mol]
5′-d(T-A-G-G-T-C-A-A-T-A-C-T)-3′ (**21**) 3′-d(A-T-C-C-A-G-T-T-A-T-G-A)-5′ (**22**)	50 (47)	−90 (−89)	−252 (−253)	−11.8 (−10.9)
5′-d(T-**2a**-G-G-T-C-A-A-T-**2a**-C-T)-3′ (**23**) 3′-d(A-T-C-C-A-G-T-T-A-T-G-A)-5′ (**22**)	48 (44)	−80 (−78)	−224 (−222)	−10.5 (−9.6)
5′-d(T-A-G-G-T-C-**2a**-**2a**-T-A-C-T)-3′ (**24**) 3′-d(A-T-C-C-A-G-T-T-A-T-G-A)-5′ (**22**)	44 (41)	−79 (−82)	−224 (−236)	−9.5 (−8.9)
5′-d(T-**2a**-G-G-T-C-**2a**-**2a**-T-**2a**-C-T)-3′ (**25**) 3′-d(A-T-C-C-A-G-T-T-A-T-G-A)-5′ (**22**)	41 (39)	−78 (−78)	−222 (−224)	−8.8 (−8.5)
5′-d(T-A-G-G-T-C-A-A-T-A-C-T)-3′ (**21**) 3′-d(A-T-C-C-**2a**-G-T-T-A-T-G-A)-5′ (**26**)	46 (42)	−77 (−81)	−217 (−231)	−9.8 (−9.1)
5′-d(T-A-G-G-T-C-A-A-T-A-C-T)-3′ (**21**) 3′-d(A-T-C-C-A-G-T-T-**2a**-T-G-A)-5′ (**27**)	45 (42)	−70 (−72)	−194 (−204)	−9.5 (−8.8)
5′-d(T-A-G-G-T-C-A-A-T-A-C-T)-3′ (**21**) 3′-d(A-T-C-C-**2a**-G-T-T-**2a**-T-G-A)-5′ (**28**)	38 (34)	−57 (−50)	−158 (−138)	−7.8 (−7.0)
5′-d(T-**2a**-G-G-T-C-A-A-T-**2a**-T-T)-3′ (**23**) 3′-d(A-T-C-C-**2a**-G-T-T-**2a**-T-G-A)-5′ (**28**)	33 (27)	−48 (−37)	−122 (−99)	−6.8 (−6.3)
5′-d(T-A-G-G-T-C-**2a**-**2a**-T-A-C-T)-3′ (**24**) 3′-d(A-T-C-C-**2a**-G-T-T-**2a**-T-G-A)-5′ (**28**)	27 (24)	−45 (−40)	−126 (−110)	−6.2 (−5.7)
5′-d(T-**2a**-G-G-T-C-**2a**-**2a**-T-**2a**-C-T)-3′ (**25**) 3′-d(A-T-C-C-**2a**-G-T-T-**2a**-T-G-A)-5′ (**28**)	25 (21)	−48 (−56)	−136 (−168)	−5.7 (−4.2)
5′-d(C-G-A-A-C-T-G-G-C-G-T-C)-3′ (**29**) 3′-d(G-C-T-T-G-A-C-C-G-C-A-G)-5′ (**30**)	62 (61)	−99 (−96)	−270 (−263)	−15.1 (−14.6)
5′-d(C-G-A-A-C-T-G-G-C-G-T-C)-3′ (**29**) 3′-d(G-C-T-T-G-**2a**-C-C-G-C-**2a**-G)-5′ (**31**)	57 (57)	−97 (−100)	−267 (−279)	−13.7 (−13.8)
5′-d(C-G-**2a**-**2a**-C-T-G-G-C-G-T-C)-3′ (**32**) 3′-d(G-C-T-T-G-A-C-C-G-C-A-G)-5′ (**30**)	54 (53)	−72 (−71)	−195 (−192)	−11.4 (−11.1)
5′-d(C-G-**2a**-**2a**-C-T-G-G-C-G-T-C)-3′ (**32**) 3′-d(G-C-T-T-G-**2a**-C-C-G-C-**2a**-G)-5′ (**31**)	50 (50)	−81 (−86)	−225 (−240)	−11.0 (−11.3)
5′-d(T-**2b**-G-G-T-C-A-A-T-**2b**-C-T)-3′ (**33**) 3′-d(A-T-C-C-A-G-T-T-A-T-G-A)-5′ (**22**)	46	−75	−210	−9.8
5′-d(T-A-G-G-T-C-A-A-T-A-C-T)-3′ (**21**) 3′-d(A-T-C-C-**2b**-G-T-T-**2b**-T-G-A)-5′ (**34**)	32	−40	−107	−6.9
5′-d(T-**2b**-G-G-T-C-A-A-T-**2b**-C-T)-3′ (**33**) 3′-d(A-T-C-C-**2b**-G-T-T-**2b**-T-G-A)-5′ (**34**)	24	−30	−78	−5.9
5′-d(T-**1b**-G-G-T-C-**1b**-**1b**-T-**1b**-C-T)-3′ (**35**) [27] 3′-d(A-T-C-C-**2b**-G-T-T-**2b**-T-G-A)-5′ (**34**)	28	−42	−114	−6.5
5′-d(T-**1c**-G-G-T-C-**1c**-**1c**-T-**1c**-C-T)-3′ (**36**) [27] 3′-d(A-T-C-C-**2b**-G-T-T-**2b**-T-G-A)-5′ (**34**) [27]	37	−45	−119	−7.7
5′-d(T-**1b**-G-G-T-C-**1b**-**1b**-T-**1b**-C-T)-3′ (**35**) 3′-d(A-T-C-C-**1b**-G-T-T-**1b**-T-G-A)-5′ (**37**)	61	[b])	[b])	[b])

[a]) Measured at 260 nm in 1M NaCl, 100 mM MgCl$_2$, and 60 mM Na-cacodylate (pH 7.0) with 5 µM single-strand concentration. Data in parentheses are measured in 100 mM NaCl, 10 mM MgCl$_2$, and 10 mM Na-cacodylate (pH 7.0) with 10 µM oligonucleotide concentration. [b]) Not determined.

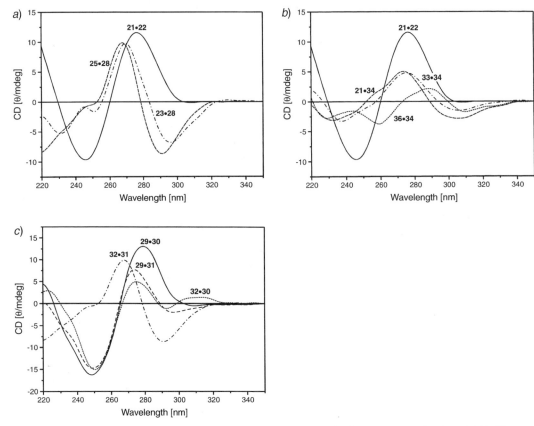

Fig. 8. a) *CD Spectra of the oligonucleotides* **21** · **22**, **23** · **28**, *and* **25** · **28**; b) *CD spectra of the oligomers* **21** · **34**, **33** · **34**, *and* **36** · **34**; c) *CD spectra of the oligonucleotides* **29** · **30**, **29** · **31**, **32** · **30**, *and* **32** · **31**. Spectra were measured in 1M NaCl, 100 mM MgCl₂, and 60 mM Na-cacodylate (pH 7.0) with 5 + 5 μM oligonucleotide concentration.

Table 7. T_m *Values of Oligonucleotides Containing* N^8-*Linked-8-Aza-7-deazaadenine 2'-Deoxyribonucleoside* **2a**
Measured in Mg^{2+}-*Free Buffers*[a])

	T_m [°][b])	T_m [°][c])	T_m [°][d])
5'-d(T-A-G-G-T-C-A-A-T-A-C-T)-3' (**21**) 3'-d(A-T-C-C-A-G-T-T-A-T-G-A)-5' (**22**)	53	43	40
5'-d(T-**2a**-G-G-T-C-A-A-T-**2a**-C-T)-3' (**23**) 3'-d(A-T-C-C-A-G-T-T-A-T-G-A)-5' (**22**)	49	38	34
5'-d(T-A-G-G-T-C-**2a**-**2a**-T-A-C-T)-3' (**24**) 3'-d(A-T-C-C-A-G-T-T-A-T-G-A)-5' (**22**)	43	34	30
5'-d(T-**2a**-G-G-T-C-**2a**-**2a**-T-**2a**-C-T)-3' (**25**) 3'-d(A-T-C-C-A-G-T-T-A-T-G-A)-5' (**22**)	40	31	27
5'-d(T-A-G-G-T-C-A-A-T-A-C-T)-3' (**21**) 3'-d(A-T-C-C-**2a**-G-T-T-**2a**-T-G-A)-5' (**28**)	37	24	20

[a]) Measured at 260 nm. [b]) 1M NaCl, 10 mM Na-phosphate, and 0.1 mM EDTA. [c]) 100 mM NaCl, 10 mM Na-phosphate, and 0.1 mM EDTA. [d]) 50 M NaCl, 10 mM Na-phosphate, and 0.1 mM EDTA.

Table 8. T_m Values and Thermodynamic Data of the DNA/RNA Hybrids Containing N^8-Linked 8-Aza-7-deaza-adenine 2'-Deoxyribonucleoside **2a**[a])

	T_m [°][a])	ΔH^0 [kcal/mol]	ΔS^0 [cal/mol·K]	ΔG^0_{298} [kcal/mol]
5'-d(T-A-G-G-T-C-A-A-T-A-C-T)-3' (**21**) [28] 3'-(A-U-C-C-A-G-U-U-A-U-G-A)-5' (**38**)	46	−82	−230	−10.1
5'-d(T-**2a**-G-G-T-C-A-A-T-**2a**-C-T)-3' (**23**) 3'-(A-U-C-C-A-G-U-U-A-U-G-A)-5' (**38**)	37	−66	−189	−7.1
5'-d(T-A-G-G-T-C-**2a**-**2a**-T-A-C-T)-3' (**24**) 3'-(A-U-C-C-A-G-U-U-A-U-G-A)-5' (**38**)	33	−61	−178	−7.1
5'-d(T-**2a**-G-G-T-C-**2a**-**2a**-T-**2a**-C-T)-3' (**25**) 3'-(A-U-C-C-A-G-U-U-A-U-G-A)-5' (**38**)	22	−40	−112	−5.9

[a]) Measured at 260 nm in 0.1M NaCl, 10 mM MgCl$_2$, and 10 mM Na-cacodylate (pH 7.0) with 5 μM single-strand concentration.

Table 9. T_m Values and Thermodynamic Data of the Oligonucleotides Containing N^8-Linked 8-Aza-7-deazaadenine 2'-Deoxyribonucleoside **2a** in Parallel-Stranded DNA[a]). d(iC) = m^{5}iC$_d$.

	T_m [°]	ΔH^0 [kcal/mol]	ΔS^0 [cal/mol·K]	ΔG^0_{298} [kcal/mol]
5'-d(T-A-G-G-T-C-A-A-T-A-C-T)-3' (**21**) 5'-d(A-T-iC-iC-A-iG-T-T-A-T-iG-A)-3' (**39**) [29]	39	−74	−211	−8.8
5'-d(T-**2a**-G-G-T-C-A-A-T-**2a**-C-T)-3' (**23**) 5'-d(A-T-iC-iC-A-iG-T-T-A-T-iG-A)-3' (**39**)	33	−50	−138	−6.9
5'-d(T-A-G-G-T-C-**2a**-**2a**-T-A-C-T)-3' (**24**) 5'-d(A-T-iC-iC-A-iG-T-T-A-T-iG-A)-3' (**39**)	28	−49	−140	−6.0

[a]) Measured at 260 nm in 1M NaCl containing 100 mM MgCl$_2$ and 60 mM Na-cacodylate (pH 7.0) with 10 μM oligonucleotide concentration.

3. *Base-Pair Motifs and Conclusion.* The nucleoside **2a** forms a rather strong base pair with thymidine in aps-DNA, while that in ps-DNA is less stable. The most likely base-pair motifs are a reverse *Watson-Crick* base pair for **2a** with T_d in aps-DNA (motif **I**) and a *Watson-Crick* pair in ps-DNA (motif **III**; *Fig. 9*). The distances of the two anomeric centers within these base pairs are rather similar (motif **I** *vs.* **II** or motiv **III** *vs.* **IV**) [30]. However, the glycosylic-bond angles are different in motif **I** (O(4')−C(1')−N(8)−N(9)) compared to motif **II** (O(4')−C(1')−N(9)−C(4)). The glycosylic-torsion angle of the N^8-nucleoside **2a** in base pair **I** is in the *syn* range. The amino group of motif **I** points towards the minor groove and not towards the major groove as observed for *Watson-Crick* base pairs (see motif **II**). Consequently, substituents at C(7) also point towards the minor groove (motif **I**) and not towards the major groove (motif **II**). According to the limitation of space of the minor groove, bulky 7-substituents should destabilize the duplex. This is actually the case, as the incorporation of the bromo nucleoside **2b** destabilizes the duplex in comparison to that containing **2a** (*Table 6*).

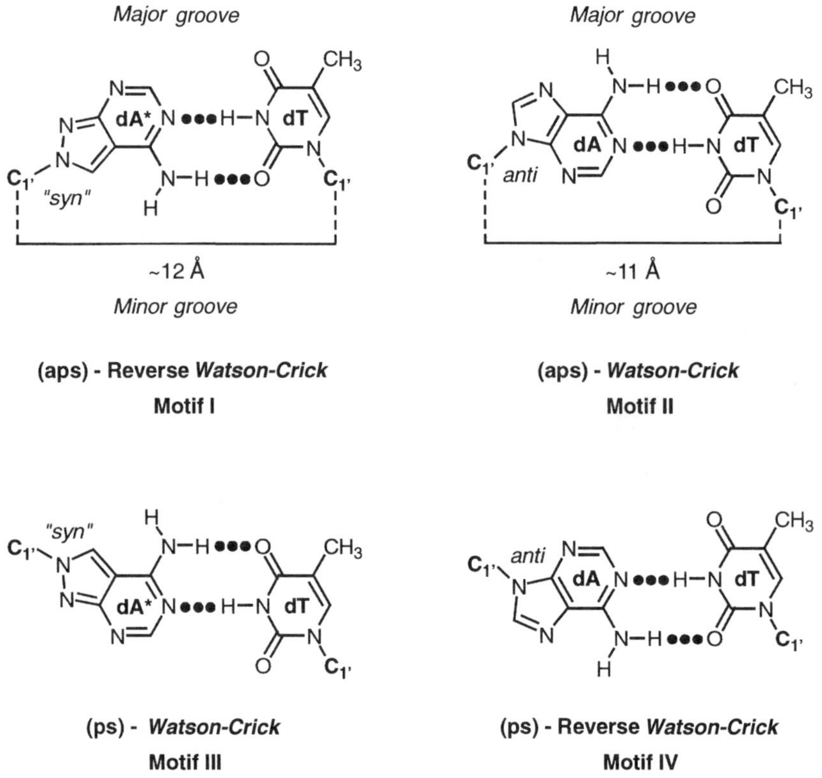

Fig. 9. *Comparison of the base-pair motifs of* **2a** (dA*) *with those of regular* A_d (dA) *in antiparallel-strand* (aps) *and parallel-strand orientation* (ps)

Experimental Part

General. All chemicals were purchased from *Aldrich, Sigma,* or *Fluka* (*Sigma-Aldrich Chemie GmbH,* Deisenhofen, Germany). Solvents were of laboratory grade. TLC: aluminum sheets, silica gel *60 F$_{254}$*, 0.2 mm layer (*Merck*, Germany). Column flash chromatography (FC): silica gel 60 (*Merck*, Germany) at 0.4 bar (4 · 10^4 Pa); solvent systems: CH$_2$Cl$_2$/MeOH 9:1 (*A*), CH$_2$Cl$_2$/MeOH 8:2 (*B*), petroleum ether/acetone 1:2 (*C*), CH$_2$Cl$_2$/acetone 8:2 (*D*), CH$_2$Cl$_2$/acetone 9:1 (*E*); sample collection with an *UltroRac II* fractions collector (*LKB Instruments*, Sweden). M.p.: *Büchi-SMP-20* apparatus (*Büchi*, Switzerland); uncorrected. UV Spectra: *U-3200* spectrometer (*Hitachi*, Japan). NMR Spectra: *Avance-250* or *AMX-500* spectrometers (*Bruker*, Karlsruhe, Germany), at 250.13 and 500.14 MHz for ^{1}H and at 125.13 MHz for ^{13}C; δ in ppm rel. to SiMe$_4$ as internal standard, *J* values in Hz. Positive-ion fast-atom-bombardment (FAB) mass spectra were provided by Dr. *M. Sauer* (Universität Heidelberg, Germany) using 3-nitrobenzyl alcohol (3-NOBA) as matrix. Elemental analyses were performed by *Mikroanalytisches Laboratorium Beller* (Göttingen, Germany).

3-Bromo-2-[2-deoxy-β-D-erythro-pentofuranosyl]-2H-pyrazolo[3,4-d]pyrimidin-4-amine (**2b**). Compound **5b** (1.0 g, 1.72 mmol) [13] was stirred at 80° for 6 h with sat. (0°) NH$_3$/MeOH soln. (200 ml) in an autoclave. The soln. was evaporated and the residue applied to FC (column 12 × 3 cm, solvent *B*): **2b** (259 mg, 44%). Colorless amorphous powder. TLC (*A*): 0.17. UV (MeOH): 244 (9200), 262 (8200), 303 (9200). ^{1}H-NMR ((D$_6$)DMSO): 2.34 (*m*, 1 H–C(2')); 2.91 (*m*, 1 H–C(2')); 3.45 (*m*, 2 H–C(5')); 3.85 (*m*, H–C(4')); 4.50 (*m*, H–C(3')); 4.76 (*t, J* = 5.7, OH–C(5')); 5.36 (*d, J* = 4.5, OH–C(3')); 6.44 (*'t', J* = 5.7, H–C(1')); 7.29 (br. *s*, NH$_2$); 8.19 (*s*, H–C(6)). Anal. calc. for C$_{10}$H$_{12}$BrN$_5$O$_3$ (330.14): C 36.38, H 3.66, N 21.21; found: C 36.48, H 3.52, N 21.17.

2-[2-Deoxy-β-D-erythro-pentofuranosyl]-3-iodo-2H-pyrazolo[3,4-d]pyrimidin-4-amine (**2c**). As described for **2b**, with **5c** (200 mg, 0.32 mmol) [13] and NH_3/MeOH soln. (200 ml). After FC, crystallization from MeOH/ H_2O furnished **2c** (60 mg, 48%). Colorless solid. M.p. 208–211°. TLC (*A*): 0.15. UV (MeOH): 248 (6900), 266 (7900), 274 (8000), 307 (9200). ^{1}H-NMR ((D$_6$)DMSO): 2.33 (*m*, 1 H–C(2′)); 2.90 (*m*, 1 H–C(2′)); 3.42 (*m*, 2 H–C(5′)); 3.87 (*m*, H–C(4′)); 4.50 (*m*, H–C(3′)); 4.79 ('*t*', *J* = 5.6, OH–C(5′)); 5.36 (*d*, *J* = 4.6, OH–C(3′)); 6.43 ('*t*', *J* = 5.8, H–C(1′)); 7.30 (br. *s*, NH$_2$); 8.19 (*s*, H–C(6)). Anal. calc. for C$_{10}$H$_{12}$IN$_5$O$_3$ (377.14): C 31.85, H 3.21, N 18.57; found: C 32.08, H 3.13, N 18.44.

3-Bromo-2-(2-deoxy-β-D-erythro-pentofuranosyl)-4-[(dimethylamino)ethylidene]-2H-pyrazolo[3,4-d]pyri-midin-4-amine (**7**). A soln. of **2b** (200 mg, 0.61 mmol) in MeOH (10 ml) was stirred with *N,N*-dimethylacet-amide dimethyl acetal (2.0 g) for 3 h at 50°. After evaporation, the residue was applied to FC (column 12 × 3 cm, solvent *B*): **7** (152 mg, 62%). Colorless foam. TLC (*A*): 0.28. UV (MeOH): 275 (8500), 329 (14900). ^{1}H-NMR ((D$_6$)DMSO): 2.29 (*s*, 1 Me); 2.36 (*m*, 1 H–C(2′)); 2.87 (*m*, 1 H–C(2′)); 3.19 (*s*, MeN); 3.21 (*s*, MeN); 3.49 (*m*, 2 H–C(5′)); 3.86 (*m*, H–C(4′)); 4.51 (*m*, H–C(3′)); 4.78 ('*t*', *J* = 5.4, OH–C(5′)); 5.37 (*d*, *J* = 4.6, OH–C(3′)); 6.49 ('*t*', *J* = 5.7, H–C(1′)); 8.41 (*s*, H–C(6)).

3-Bromo-2-[2-deoxy-5-O-(4,4′-dimethoxytriphenylmethyl)-β-D-erythro-pentofuranosyl]-4-[(dimethylamino)-ethylidene]-2H-pyrazolo[3,4-d]pyrimidin-4-amine (**8**). To a soln. of **7** (140 mg, 0.35 mmol) in dry pyridine (0.5 ml), 4,4′-dimethoxytriphenylmethyl chloride (130 mg, 0.38 mmol) was added. After stirring at 50° for 2 h, the mixture was poured into an ice-cold 3% aq. NaHCO$_3$ soln. (5 ml) and extracted with CH$_2$Cl$_2$ (2 × 75 ml). The combined org. layers were dried (Na$_2$SO$_4$) and evaporated. The residue was applied to FC (column 15 × 2 cm, solvent *A*): **8** (175 mg, 71%). Colorless foam. TLC (*A*): 0.51. UV (MeOH): 275 (10900), 335 (12700). ^{1}H-NMR ((D$_6$)DMSO): 2.26 (*s*, 1 CH$_3$); 2.38 (*m*, 1 H–C(2′)); 2.90 (*m*, 1 H–C(2′)); 3.03 (*m*, 2 H–C(5′)); 3.17 (*s*, MeN); 3.23 (*s*, MeN); 3.68 (*s*, MeO); 3.69 (*s*, MeO); 3.93 (*m*, H–C(4′)); 4.64 (*m*, H–C(3′)); 5.40 (*d*, *J* = 5.1, OH–C(3′)); 6.58 (*dd*, *J* = 5.4, H–C(1′)); 6.75 (*m*, (MeO)$_2$Tr); 7.10–7.38 (*m*, (MeO)$_2$Tr); 8.40 (*s*, H–C(6)). Anal. calc. for C$_{35}$H$_{37}$BrN$_6$O$_5$ (701.62): C 59.92, H 5.32, N 11.98; found: C 60.21, H 4.97, N 11.60.

3-Bromo-2-[2-deoxy-5-O-(4,4′-dimethoxytriphenylmethyl)-β-D-erythro-pentofuranosyl]-4-[(dimethylamino)-ethylidene]-2H-pyrazolo[3,4-d]pyrimidin-4-amine 3′-(2-Cyanoethyl Diisopropylphosphoramidite) (**10**). To a stirred soln. of dry **8** (150 mg, 0.21 mmol) and anh. ^{i}Pr$_2$EtN (80 mg, 0.63 mmol) in dry THF (1 ml), 2-cyanoethyl diisopropylphosphoramidochloridite (65 mg, 0.28 mmol) was added under Ar. The mixture was stirred for 30 min and then filtered. The filtrate was diluted with AcOEt (80 ml) and extracted twice with an ice-cold 3% aq. NaHCO$_3$ soln. (2 × 10 ml) and H$_2$O (2 × 10 ml). The org. phase was dried (Na$_2$SO$_4$) and evaporated. The residue was applied to FC (column 12 × 2 cm, solvent *C*): **10** (120 mg, 63%). Colorless oil. TLC (*C*) 0.3, 0.4. ^{31}P-NMR (CDCl$_3$): 150.4.

Table 10. *Molecular Masses Determined by MALDI-TOF Mass Spectroscopy of the Modified Oligonucleotides Containing $c^7z^8A_d^*$ (**2a**) and its 7-Bromo-Derivative $Br^7c^7z^8A_d^*$ (**2b**)*

	M^+ (calc.)	M^+ (found)
5′-d(**2a**-T-**2a**-T-**2a**-T-**2a**-T-**2a**-T)-3′ (**13**)	3640.7	3638.8
5′-d(T-**2a**-G-G-T-C-A-A-T-**2a**-C-T)-3′ (**23**)	3642.7	3641.7
5′-d(T-A-G-G-T-C-**2a**-**2a**-T-A-C-T)-3′ (**24**)	3642.7	3642.2
5′-d(T-**2a**-G-G-T-C-**2a**-**2a**-T-**2a**-C-T)-3′ (**25**)	3642.7	3642.1
5′-d(A-G-T-**2a**-T-T-G-**2a**-C-C-T-A)-3′ (**28**)	3642.7	3641.5
5′-d(G-**2a**-C-G-C-C-**2a**-G-T-T-C-G)-3′ (**31**)	3644.7	3643.3
5′-d(C-G-**2a**-**2a**-C-T-G-G-G-C-G-C-T)-3′ (**32**)	3644.7	3643.5
5′-d(T-**2b**-G-G-T-C-A-A-T-**2b**-C-T)-3′ (**33**)	3798.5	3800.4
5′-d(A-G-T-**2b**-T-T-G-**2b**-C-C-T-A)-3′ (**34**)	3798.5	3802.5

Synthesis and Purification of the Oligonucleotides **11–39**. The oligonucleotide synthesis was performed on an *ABI-392-08* DNA synthesizer (*Applied Biosystems*, Weiterstadt, Germany) on a 1-μmol scale with the phosphoramidites **9** and **10** and those of the regular 2′-deoxynucleosides (*Applied Biosystems*, Weiterstadt, Germany) according to the synthesis protocol for 3′-phosphoramidites [31]. The crude oligonucleotides were purified and detritylated on oligonucleotide purification cartridges by the standard protocol for purification [26]. The oligonucleotides were lyophilized on a *Speed-Vac* evaporator to yield colorless solids which were stored frozen at −18°. The enzymatic hydrolysis of the oligomers was performed as described in [32]. Quantification of the constituents was made on the basis of the peak areas, which were divided by the extinction coefficients of the nucleoside (ε_{260} values: A_d 15400, C_d 7300, G_d 11400, T_d 8800, **2a** 6600, **2b** 7300). Snake-venom phophodiesterase (EC 3.1.15.1., *Crotallus durissus*) and alkaline phosphatase (EC 3.1.3.1., *E. coli*) were generous gifts of the *Roche Diagnostics GmbH* (Penzberg, Germany). MALDI-TOF-MS were provided by Dr. *J. Gross* (Universität Heidelberg, Germany). Some selected MALDI-TOF data of modified oligonucleotides are shown in *Table 10*.

Oligonucleotide analysis was carried out by reversed-phase HPLC with a *Merck-Hitachi* HPLC: 250 × 4 mm *RP-18* column; gradients of 0.1M (Et₃NH)OAc (pH 7.0)/MeCN 95:5 (*A*) and MeCN (*B*); gradient I: 50 min 0–50% *B* in *A*, flow rate 1 ml/min; gradient II: 20 min 0–25% *B* in *A*, flow rate 0.7 ml/min, then 30 min 25–40% *B* in *A*, flow rate 1 ml/min; gradient III: 20 min 0–25% *B* in *A*, flow rate 1 ml/min.

Determination of T_m values and Thermodynamic Data. Absorbance *vs.* temp. profiles were measured on a *Cary-1/1E* UV/VIS spectrophotometer (*Varian*, Australia) equipped with a *Cary* thermoelectric controller. The T_m values were measured in the reference cell with a *Pt-100* resistor and the thermodynamic data (ΔH^0, ΔS^0, ΔG^0_{298}) were calculated by means of the 'MeltWin 3.0' program package [27]. CD Spectra: *Jasco-600* (*Jasco*, Japan) spectropolarimeter with a thermostatically (*Lauda RCS-6* bath, Germany) controlled 1-cm cuvette.

We thank Mr. *Yang He* for the NMR spectra, Dr. *Helmut Rosemeyer* for the PSEUROT calculations, and Mrs. *Elisabeth Feiling* for the oligonucleotide syntheses. Financial Support by the *Deutsche Forschungsgemein-schaft* and the *Roche Diagnostics GmbH*, Penzberg, Germany, is gratefully acknowledged.

REFERENCES

[1] W. Saenger, 'Principles of Nucleic Acid Structure', in 'Springer Advanced Texts in Chemistry', Ed. C. R. Cantor, Springer Verlag, New York-Berlin-Heidelberg-Tokyo, 1984.

[2] C. R. Cantor, P. R. Schimmel, 'Biophysical Chemistry', W. H. Freeman & Co., San Francisco, CA, 1980.

[3] E. Uhlmann, A. Peyman, G. Breipohl, D. W. Will, *Angew. Chem., Int. Ed.* **1998**, *37*, 2796, and refs. cit. therein.

[4] K. Groebke, J. Hunzicker, W. Fraser, L. Peng, U. Diederichsen, K. Zimmermann, A. Holzner, C. Leumann, A. Eschenmoser, *Helv. Chim. Acta* **1998**, *81*, 375.

[5] M. Bolli, H. U. Trafelet, C. Leumann, *Nucleic Acids Res.* **1996**, *24*, 4660.

[6] F. Seela, H. Thomas, *Helv. Chim. Acta* **1995**, *78*, 94.

[7] F. Seela, B. Gabler, *Helv. Chim. Acta* **1994**, *77*, 622.

[8] F. Seela, H. Winter, *Helv. Chim. Acta* **1994**, *77*, 597.

[9] F. Seela, P. Leonard, *Helv. Chim. Acta* **1998**, *81*, 2244.

[10] F. Seela, K. Kaiser, *Helv. Chim. Acta* **1988**, *71*, 1813.

[11] F. Seela, H. Steker, *Helv. Chim. Acta* **1985**, *68*, 563.

[12] F. Seela, M. Zulauf, *J. Chem. Soc., Perkin Trans. 1* **1998**, 3233.

[13] F. Seela, M. Zulauf, G. Becher, *Nucleosides Nucleotides* **1997**, *16*, 305.

[14] F. Jordan, H. Niv, *Nucleic Acids Res.* **1977**, *4*, 697.

[15] H. Steker, Dissertation, Universität Paderborn, 1998.

[16] O. Röder, H.-D. Lüdemann, E. von Goldhammer, *Eur. J. Biochem.* **1975**, *53*, 517.

[17] J. Plavec, W. Tong, J. Chattopadhyaya, *J. Am. Chem. Soc.* **1993**, *115*, 9734.

[18] C. A. G. Haasnoot, F. A. A. M. de Leeuw, C. Altona, *Tetrahedron* **1980**, *36*, 2783.

[19] J. van Wijk, C. Altona, 'PSEUROT 6.2, a Program for the Conformational Analysis of Five-Membered Rings', University of Leiden, July, 1993 (licensee: C. Altona, Gorlaeus Laboratories, Leiden, The Netherlands).

[20] H. Rosemeyer, M. Zulauf, N. Ramzaeva, G. Becher, E. Feiling, K. Mühlegger, I. Münster, A. Lohmann, F. Seela, *Nucleosides Nucleotides* **1997**, *16*, 821.

[21] H. Rosemeyer, F. Seela, *J. Chem. Soc., Perkin Trans. 2* **1997**, 2341.

[22] F. Seela, I. Münster, U. Löchner, H. Rosemeyer, *Helv. Chim. Acta* **1998**, *81*, 1139.

[23] F. Seela, G. Becher, H. Rosemeyer, H. Reuter, G. Kastner, I. A. Mikhailopulo, *Helv. Chim. Acta* **1999**, *82*, 105.

[24] F. Seela, H. Debelak, in preparation.

[25] F. Seela, A. Kehne, *Biochemistry* **1985**, *24*, 7556.

[26] Applied Biosystems, 'Manual for Oligonucleotide Purification Cartridges'.

[27] J. A. McDowell, D. H. Turner, *Biochemistry* **1996**, *35*, 14077.

[28] F. Seela, M. Zulauf, *J. Chem. Soc., Perkin Trans. 1* **1999**, 479.

[29] F. Seela, C. Wei, *Helv. Chim. Acta* **1999**, *82*, 726.

[30] G. M. Blackburn, M. J. Gait, in 'Nucleic Acids in Chemistry and Biology', IRL press, Oxford, Vol. 1, p. 33.

[31] Applied Biosystems, 'Users Manual of the DNA synthesizer', p. 392.

[32] F. Seela, S. Lampe, *Helv. Chim. Acta* **1991**, *74*, 1790.

Nucleic-Acid Analogs with Restricted Conformational Flexibility in the Sugar-Phosphate Backbone ('Bicyclo-DNA')

Part 7[1])

Synthesis and Properties of Oligodeoxynucleotides Containing [(3′S,5′S,6′R)-6′-Amino-2′-deoxy-3′,5′-ethano-β-D-ribofuranosyl]thymine (= (6′R)-6′-Amino-bicyclo-thymidine)

by **Roland Meier**, **Sabine Grüschow**, and **Christian Leumann***

Department of Chemistry and Biochemistry, University of Bern, Freiestrasse 3, CH-3012 Bern

We describe the synthesis of the acetamido- and trifluoroacetamido-functionalized bicyclo-thymidines **11** and **12**, starting from the silyl enol ether **1**, in 6 steps. These nucleosides were converted to the corresponding cyanoethyl phosphoramidite building blocks **16** and **17** and subsequently incorporated into the homo-thymidylate decamers **18**–**22**. Upon deprotection of the oligomers, the trifluoroacetamido functions were cleaved, leaving behind a free amino function in the sugar-phosphate backbone that is protonated at neutral pH, giving rise to partially zwitterionic oligonucleotides. Pairing properties with the complementary DNA oligomer d(A$_{10}$), as determined by UV/melting curves, revealed a slightly increased stability of the duplex d(A$_{10}$) · **20**, in which the decathymidylate sequence shows an alternating arrangement of natural thymidine and amino-bicyclo-thymidine residues, relative to the natural reference duplex. The dependence of T_m on the salt concentration of the medium is reduced in this case. Duplex destabilization occurs if the amino-bicyclo-thymidine residues are replaced by the charge-neutral acetamido-bicyclo-nucleosides (e.g., d(A$_{10}$) · **22**), most probably due to steric interference of the acetamido substituent with the backbone P−O(5′) bond.

1. Introduction. – The design and synthesis of new oligonucleotide analogs with improved pairing properties, enhanced biostability and bioavailability relative to natural DNA and RNA, is of importance with respect to potential applications as antisense agents in human therapy and biotechnology [2], in DNA or RNA diagnostics [3], and in materials and computer science [4]. Furthermore, such oligonucleotides serve as tools to chemically rationalize the supramolecular assembly of oligonucleotide single strands, thus contributing to the understanding of the interrelation between monomeric nucleoside structure and the association properties of oligomers thereof.

In this context, we recently prepared the conformationally constrained DNA analog bicyclo-DNA and explored its pairing properties in detail (*Fig. 1*) [5–8]. As the backbone torsion angle γ in bicyclo-DNA, in contrast to natural A- and B-DNA, is preferentially located in the antiperiplanar (*ap*) and not in the synclinal (*+sc*) arrangement, this provided a unique opportunity to explore the effect of a change of

---

[1]) Part 6: [1].

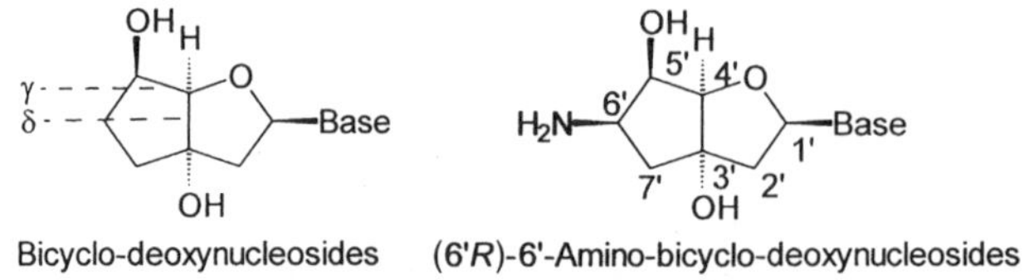

Bicyclo-deoxynucleosides   (6'R)-6'-Amino-bicyclo-deoxynucleosides

Fig. 1. *Structure of the bicyclo-deoxynucleosides and the (6'R)-6'-amino-bicyclo-deoxynucleosides*

this particular torsion angle on the association mode of duplex formation [9]. This conformational study was further extended to oligodeoxynucleotides containing 5'-epi-bicyclo-deoxynucleosides, in which γ is restricted to the $-ac/-sc$ conformational space [1].

In addition to probing the effect of structural preorganization on duplex formation, the carbocyclic ring in the bicyclo-deoxynucleosides is ideally suited for introducing functional groups into well-defined positions at the sugar-phosphate backbone. Such functional groups may serve various purposes, *e.g.*, the adjustment of torsion angle γ by changing the preferred conformation of the carbocyclic ring or the change of the electrostatic environment of the sugar-phosphate backbone. Furthermore, they may be useful as anchor points for the introduction of reporter molecules or chemically reactive groups, or to enhance the catalytic power of DNA enzymes [10]. Here we report on the synthesis of $(6'R)$-6'-amino-bicyclo-thymidine, its incorporation into oligonucleotides, and a first assessment of the base-pairing properties of accordingly modified oligonucleotides.

2. Synthesis of the (6'R)-6'-Amino-bicyclo-deoxysugar Unit. – Of primary interest to us were the $(6'R)$-6'-amino-bicylo-deoxynucleosides, in which the amino function is attached to the concave (β) side of the bicyclic ring system. The (R)-configuration was chosen for two reasons. From the *cis* relationship of the substituents at C(5') and C(6'), we expected a conformational change of the carbocyclic ring, relative to the unsubstituted bicyclo-nucleosides, towards a conformation in which the 5'-OH group is located in the pseudoaxial position. This orients torsion angle γ into the ($+sc$) range and thus mimicks closely the orientation observed in DNA duplexes of the A and B type. Furthermore, from model building, it appeared that attachment to the β face orients the functional group into the direction of the major groove of a DNA duplex, while attachement to the convex α face would orient it straight away from the duplex into the solvent.

The synthesis started from the silyl enol ether **1** (*Scheme 1*), prepared and used previously in the synthesis of tricyclo-DNA [11]. Although the relative configuration at the anomeric center of **1** seemed to be of minor importance with respect to the stereochemical outcome of the envisaged transformations in the carbocyclic part of the molecule, both, the α-D- and β-D-anomers were used separately and in parallel. While we anticipated that direct methods for the introduction of the amino substituent, *e.g. via* aziridination [12] of **1** would yield predominantly products with the amino substituent in α-position, we preferred a classical procedure *via* hydroboration followed by exchange of the resulting OH substituent at C(7) by an NH$_2$ substituent under inversion of configuration.

Scheme 1. *Synthesis of the Amino-Functionalized Bicyclo-sugar* **7**

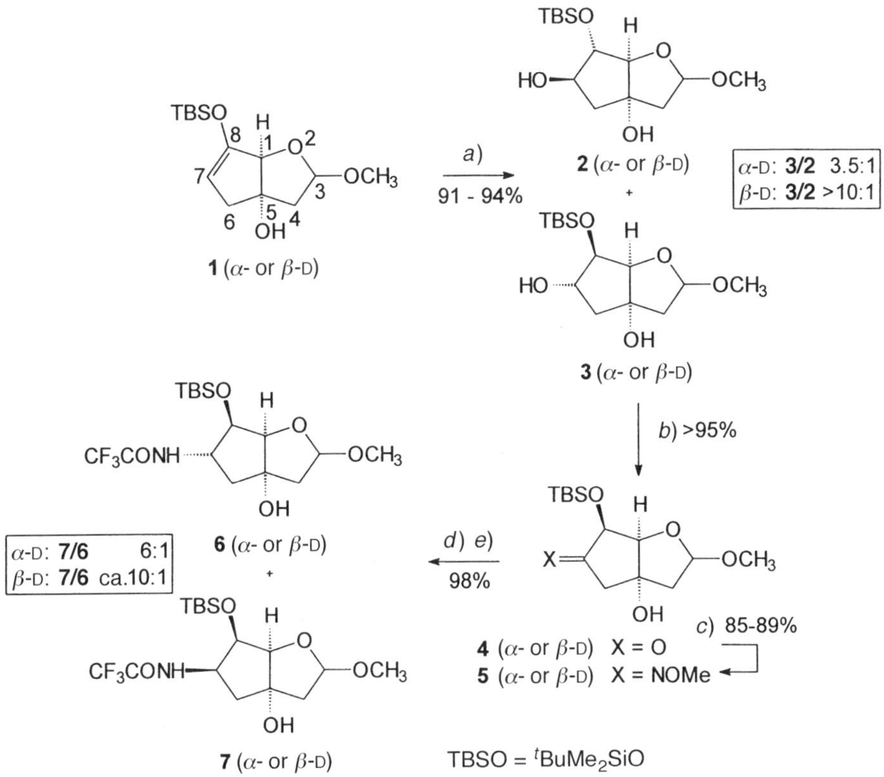

a) BH$_3 \cdot$THF (3 equiv.), THF, $-78° \rightarrow$ r.t., 31 h. b) *Dess-Martin* periodinane (2 equiv.), CH$_2$Cl$_2$, r.t., 2 h. c) (MeONH$_3$)Cl (2 equiv.), NaOAc (1 equiv.), 90% EtOH, r.t., 20 min. d) *Raney*-Ni, EtOH/H$_2$O/conc. NH$_3$ soln. 10 : 4 : 0.8, H$_2$ (10 bar), r.t., 12 h. e) CF$_3$COOEt, Et$_3$N (0.2 equiv.), r.t., 1 h.

Hydroboration of **1** proceeded smoothly and produced the two diastereoisomeric diols **2** and **3** in yields >90% and in ratios **3/2** of 3.5 : 1 in the α-D series and >10 : 1 in the β-D series[2]. The higher selectivity in the β-D series is most likely due to the MeO substituent that sterically obstructs the attack of the borane reagent from the β-face, thus pronouncing the already inherent preference for attack at the convex α face of the bicyclic system. While direct conversion of the secondary-alcohol function to an amino function in **3** *via Mitsunobu* reaction [13] failed in our hands, we had to adopt a redox procedure to introduce the desired functionality. Oxidation of **3** ($\rightarrow$**4**) followed by treatment with *O*-methylhydroxylamine gave the corresponding (*E*)/(*Z*)-oximes **5** in high yields in both the α-D and β-D series. Subsequent hydrogenolysis with *Raney*-Ni

---

[2]   The relative configurations at the centers C(7) and C(8) of all intermediates **2**–**7** were assigned on the basis of ^{1}H-NMR NOE experiments and are discussed in detail in [14].

then produced in almost quantitative yield[3]) the corresponding amines, which were isolated as the corresponding trifluoracetates **6/7**. As expected, the reaction proceeded with considerable stereoselectivity with ratios of **6/7** > 6 : 1 in both the α-D and β-D series.

3. X-Ray Analysis and Solution Structure of α-7[4]). – Suitable crystals of α-**7** were subjected to X-ray analysis (*Fig. 2*)[5]). The structure unequivocally supports the assignment of the relative configuration at the centers C(7) and C(8) and, furthermore, provides insight into the preferred conformational properties of the bicyclic sugar component. As in the unsubstituted bicyclo-deoxynucleosides, the furanose unit adopts an almost perfect 1'-*exo* conformation with a pseudorotation phase angle P of 144.3°. Importantly, the carbocyclic ring exists in a conformation with the silyloxy substituent at C(8) in a pseudoaxial and the trifluoroacetamido substituent at C(7) in a pseudoequatorial position. Its conformation thus deviates substantially from that of the unsubstituted bicyclo-deoxynucleosides in which the corresponding OH substituents were shown to exist invariably in the pseudoequatorial orientation.

Translated into the structural description of the DNA backbone, the introduction of the amido substituent corrects torsion angle γ from the (unnatural) *ap*-orientation (*ca.* 150°) to the (in A- and B-DNA naturally occurring) *sc*-orientation (77.2°), while torsion angle δ (135.5°) remains largely unaffected with values as observed in B-DNA helices [15]. Analysis of the conformation of α-**7** in solution by NMR coupling constant analysis essentially confirms that occurring in the solid state [14].

4. Synthesis of Nucleosides and Building Blocks for DNA Synthesis. – With both anomers α- and β-**7** in hand, we then approached the nucleosidation reaction (*Scheme 2*). Interestingly, the stereochemical outcome of the nucleosidation according to the *Vorbrüggen* procedure [16] was strongly dependent upon the configuration of the anomeric center of **7** and upon the temperature. In the α-D series, ratios of nucleosides α/β-**8**[4]) varied from > 6 : 1 at high temperature to 1 : 1.7 at low temperature. In the β-D series, no reaction took place at low temperatures, while mixtures α/β-**8** 4 : 1 were obtained upon heating. The higher reactivity of α-**7** over β-**7** is in accord with a

---

[3])   During preliminary experiments for the reduction of **5** with *Raney*-Ni, we isolated amine **12** in yields of up to 25%. This product most likely arises from reduction of the imine formed between acetaldehyde and the already formed amine. The source of the acetaldehyde is unknown but most probably arises from the solvent EtOH, from which it is produced in minor amounts by the catalyst. The formation of this by-product could be completely suppressed by addition of ammonia as a competing amine.

12

[4])   The descriptor α or β preceding a key number refers to the configuration at the anomeric center.

[5])   Crystallographic data (excluding structure factors) for α-**7** have been deposited with the *Cambridge Crystallographic Data Centre* as deposition No. CCDC-132022. Copies of the data can be obtained, free of charge, on application to the *CCDC*, 12 Union Road, Cambridge CB2 1EZ, UK (fax: + 44 (1223) 336033; e-mail: deposit@ccdc.can.ac.uk).

a)

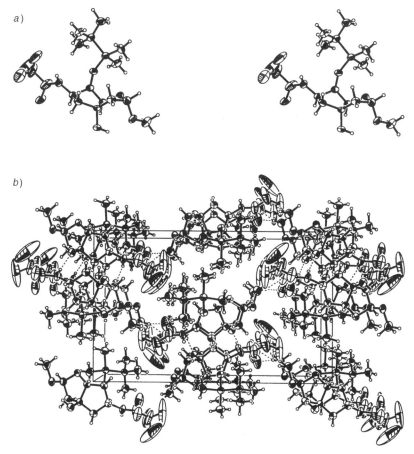

b)

Fig. 2. *ORTEP Plot of crystals of α-**7**:* a) *stereoscopic view* (25% probability thermal ellipsoids) *and* b) *view along the crystallographic* a-*axis*

pronounced kinetic anomeric effect. The preferred formation of the β-D-nucleoside at low temperature is most likely the result of a kinetically controlled reaction, while formation of the α-D-nucleoside at high temperature seems to occur under thermodynamic control[6]).

The further synthetic transformations into the building blocks for oligonucleotide synthesis were straightforward and are summarized in *Scheme 3*. Desilylation of the trifluoroacetyl-protected amine β-**8** afforded **12** in high yield, which could be tritylated to **14** and converted into the phosphoramidite **16** according to typical protocols in nucleotide chemistry. Moreover, the trifluoroacetyl group of **12** could be easily

---

[6]) Support for this interpretation comes from additional equilibration experiments. When β-**8** was subjected to the reaction conditions for nucleosidation at room temperature, equilibration to a mixture α/β-**8** 1.5 : 1 was observed after 23 h. No such equilibration was observed when pure α-**8** was used.

Scheme 2. *One-Pot Nucleosidation Reaction of Thymine with* **7**

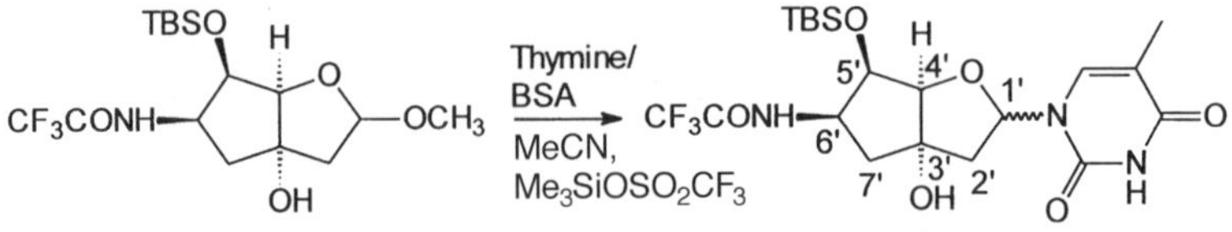
TBSO
H
CF3CONH
OCH3
OH
Thymine/
BSA
MeCN,
Me3SiOSO2CF3
CF3CONH
5'
4'
1'
6'
7'
3'
2'
N
O
N
H

7 **8**

TBSO = ^{t}BuMe$_2$SiO

7[4])	Temp. [°C]	α/β-**8**[4])	Yield [%]
α	<10	1:1.7	84
α	60	6:1	77
β	10	n.r.[a])	n.r.[a])
β	60	4:1	63

[a]) n. r. = no reaction

Scheme 3. *Preparation of Building Blocks* **16** *and* **17** *for Oligonucleotide Synthesis*

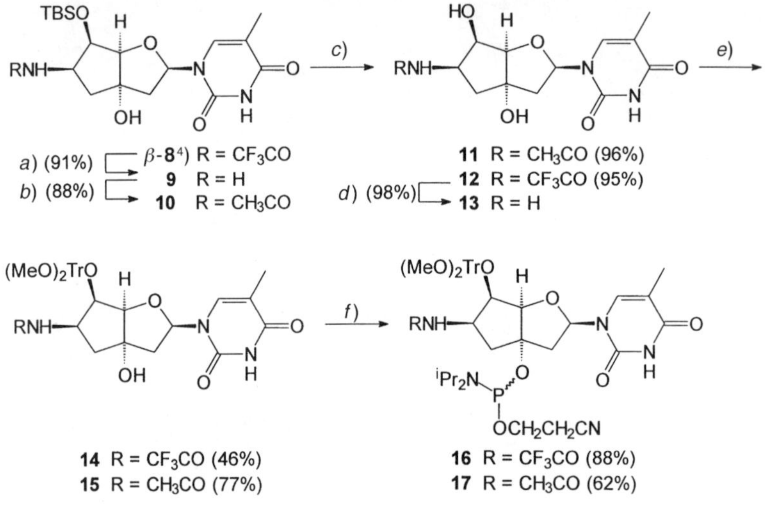
TBSO
RNH
OH
c)
HO
e)
a) (91%)
b) (88%)
β-8[4]) R = CF3CO
9 R = H
10 R = CH3CO
11 R = CH3CO (96%)
12 R = CF3CO (95%)
d) (98%)
13 R = H
(MeO)2TrO
f)
iPr2N
P
OCH2CH2CN
14 R = CF3CO (46%)
15 R = CH3CO (77%)
16 R = CF3CO (88%)
17 R = CH3CO (62%)

TBSO = ^{t}BuMe$_2$Si, (MeO)$_2$Tr = 4,4'-dimethoxytrityl

a) MeOH, conc. NH_3, r.t., 4 h. *b*) Ac_2O (1 equiv.), C_5H_5N, 0°, 2 h. *c*) $Et_3N \cdot 3HF$ (1.3 equiv.), C_5H_5N, r.t., 24 h (**11**); $Bu_4NF \cdot 3H_2O$ (8 equiv.), MeCN, AcOH (10 equiv.), r.t., 2 h (**12**). *d*) Conc. NH_3 soln., MeOH, r.t. 45 min. *e*) $(MeO)_2TrOSO_2CF_3$ (1.5–2.0 equiv.), C_5H_5N, 50°, 16 h. *f*) $^iPr_2NEt$ (4 equiv.), Cl[P(OCH_2CH_2CN)(N^iPr_2)] (2 equiv.), MeCN, r.t., 2 h.

removed (→**13**). A similar deprotection of the amino function in β-**8** (→**9**) and subsequent transformation into the acetamide **10** provided access to the acetamido series. The corresponding phosphoramidite building block **17** was obtained, in analogy to the trifluoroacetyl series, from **10** *via* desilylation (→**11**), tritylation (→**15**), and phosphitylation. While the trifluoroacetyl group will be lost during standard oligonucleotide deprotection to produce a free, protonatable amino function, the acetyl group is stable under these conditions and thus leads to a largely isosteric but charge-neutral residue.

5. Synthesis of Oligonucleotides. – The synthesis of the decamers **18–22** containing (6′R)-6′-amino- and (6′R)-6′-acetamido-bicyclo-deoxythymidine (see *Fig. 3*) were carried out on a *Pharmacia-Gene-Assembler-Plus*® DNA synthesizer on the 1.3-μmol scale in the 'trityl-off' mode (for details, see *Exper. Part*). Commercially available, thymidine-loaded controlled-pore glass (CPG) was used as the starting unit. Coupling yields, as monitored by the on-line trityl assay, amounted on average to 90–98%, when the modified building blocks **16** and **17** were coupled to a natural thymidine unit. The same yields were observed for coupling of a natural thymidine unit to a 6′-amino-bicyclo-thymidine unit. The coupling of two consecutive amino-bicyclo-thymidine building blocks, however, was less efficient and proceeded with coupling yields per cycle of only *ca.* 60%. This low yield, most likely arising from steric interference of the 6′-substituents, precluded the synthesis of completely modified decamers. After completion of chain assembly, the detachment from the solid support and deprotection was carried out by standard treatment with conc. NH₃ solution.

18 d(T-T-T-T-T-t⁺-T-T-T-T)

19 d(T-T-T-T- t⁺- t⁺-T-T-T-T)

20 d(t⁺-T- t⁺-T- t⁺-T- t⁺-T- t⁺-T)

21 d(T-T-T-T-T- tᴬᶜ-T-T-T-T)

22 d(tᴬᶜ-T- tᴬᶜ-T- tᴬᶜ-T- tᴬᶜ-T- tᴬᶜ-T)

23 d(t-T- t-T- t-T- t-T- t-T)

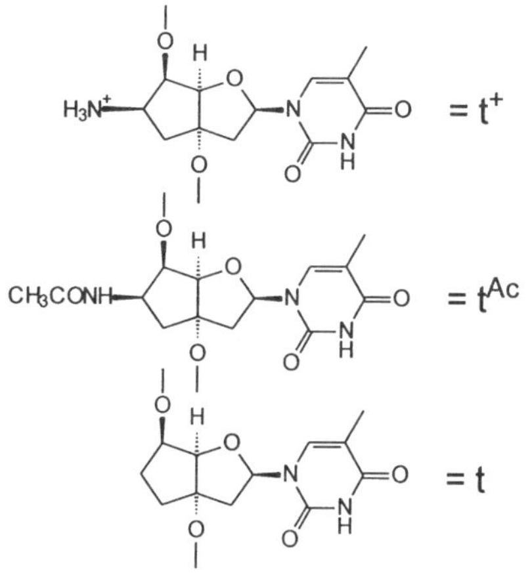

Fig. 3. *Sequences of the oligonucleotides* **18–23**

The crude oligomers were purified by anion-exchange HPLC and the purity of the collected fractions controlled by reversed-phase HPLC. As expected, the retention times during DEAE-HPLC of the decamers decreased with increasing content of amino-bicyclo-deoxynucleotide units, due to the reduced overall negative charge of the oligomers. The integrity of the oligomers was confirmed by matrix-assisted laser-desorption-ionization time-of-flight mass spectrometry (MALDI-TOF-MS) (*Table 2*, *Exper. Part*). The synthesis of sequence **23**, containing unsubstituted bicyclo-deoxythymidine units was described previously [6].

All sequences **18–22** were chemically stable (HPLC and MALDI-ToF-MS control) under the conditions used for recording UV/melting curves (NaCl-containing aqueous buffer, pH 6–10, 0–100°). No enhanced instability of the oligomers towards hydrolysis, due to the presence of the 6′-amino function was observed.

6. Pairing Properties. – Initial pairing properties with the DNA complement d(A₁₀) were obtained from UV/melting-curve analysis in buffer solutions containing 150 mM

Table 1. T_m *Values for Duplex Formation of Sequences* **18**–**23** *with* $d(A_{10})$. Buffer: 10 mм Na-cacodylate (pH 6.0/7.0), 10 mм Na$_2$HPO$_4$ (pH 8.0/9.0), 150 mм NaCl; c(duplex) 3.7–4.0 µм.

	T_m [°][a])			
	pH 6.0	pH 7.0	pH 8.0	pH 9.0
18, d(T-T-T-T-T-t$^+$-T-T-T-T)	22.4	22.2	21.7	21.3
19 d(T-T-T-T-t$^+$-t$^+$-T-T-T-T)	18.9	19.0	18.4	18.2
20 d(t$^+$-T-t$^+$-T-t$^+$-T-t$^+$-T-t$^+$-T)	27.5	27.3	26.1	24.6
21 d(T-T-T-T-T-t^{Ac}-T-T-T-T)	20.0	19.5	20.1	19.6
22 d(t^{Ac}-T-t^{Ac}-T-t^{Ac}-T-t^{Ac}-T-t^{Ac}-T)	13.5	13.5	13.2	13.1
23 d(t-T-t-T-t-T-t-T-t-T)	–	21.0[b])	–	–

[a]) T_m 23° for d(A$_{10}$)·dT($_{10}$) under identical conditions at pH 7.0. [b]) Taken from [6].

NaCl, and in the pH range 6.0–9.0. T_m Data were extracted from the melting curves and are summarized in *Table 1*.

While none of the oligonucleotides **18**–**23** showed any transition without the DNA complement, which excludes self-aggregation of the pyrimidine strands, all formed stable duplexes with d(A$_{10}$)[7]. Compared to the natural reference duplex d(A$_{10}$)· d(T$_{10}$), which under the given conditions melts with a T_m of 23°, the duplex **20**· d(A$_{10}$), with five charged residues (t$^+$) spaced by five natural thymidine residues, displayed slightly enhanced T_m values. As expected, the T_m is dependent on pH and slightly decreases with increasing pH. Replacement of the positively charged t$^+$ by its acetylated, charge-neutral derivative t^{Ac}, as in the duplex **22**·d(A$_{10}$), leads to a noticeable decrease of T_m compared to both, the all-DNA duplex d(A$_{10}$)·d(T$_{10}$) and the duplex d(A$_{10}$)·**23**, containing the underivatized bicyclo-thymidine unit t. As expected, duplex melting in the case of **22**·d(A$_{10}$) is pH insensitive in the pH-range investigated. Duplexes containing the mono- and disubstituted pyrimidine sequences **18**, **19**, and **21** essentially follow the described overall properties.

Duplex formation in the case of **20**·d(A$_{10}$) is less dependent on the salt concentration of the medium, compared to the duplex d(T$_{10}$)·d(A$_{10}$). In the range of 25–600 mм NaCl, a linear dependence of T_m from ln[NaCl] was measured with slopes $\delta T_m/\delta$ln[NaCl] of 4.32 K·м$^{-1}$ for **20**·d(A$_{10}$) and 5.24 K·м$^{-1}$ for d(T$_{10}$)·d(A$_{10}$). The lower dependence of T_m from the electrolyte concentration is again in agreement with the partially zwitterionic nature of the backbone of **20**.

7. Discussion and Conclusions. – Zwitterionic oligonucleotides were prepared in the past either by derivatization of the pyrimidine C(5) position with a flexible aminohexyl chain [17][18] or by introducing a basic 2-(dimethylamino)ethylphosphor-amidate group as the linking unit between nucleoside residues [19]. In both cases, a negligible salt-concentration dependence of duplex formation was reported. However, no benefit in terms of strength of duplex formation arose as a result of the modifications.

---

[7]   All complexes formed between the oligonucleotides **18**–**23** and d(A$_{10}$) at 1:1 stoichiometry of single strands were duplexes and not triple helices. This is deduced from the fact that only monophasic and not biphasic melting curves were observed at 260 nm, and that no transitions at 284 nm occurred. At the latter wavelength, the *Hoogsteen*-strand melting in dT·dA·dT triple helices is visible.

From the experiments presented here, we conclude that, indeed, the positive charge associated with the amino-bicyclo-nucleosides exhibits a slightly stabilizing effect at neutral pH, most probably by neutralizing intrastrand repulsion of two neighboring negatively charged phosphate units. However, this stabilizing electrostatic effect seems in part to be compromised by a repulsive steric interaction between the substituent at position C(6') of the bicyclo-nucleosides and the 5'-phosphate residues. This interpretation is corroborated by the relatively low T_m of duplexes containing the residues t^{Ac} (e.g., **22**·d(A$_{10}$)), in which charge contributions to stability are disentangled from structural contributions. Thus, geometrically constrained and suitably positioned cationic units in the DNA backbone can stabilize DNA duplexes and may do so even more effectively, if the charge-carrying group does not sterically interfere with the preferred backbone conformation.

Although the 6'-substituted (6'R)-bicyclo-deoxynucleosides seem to prefer a conformation favoring the $+sc$ orientation of torsion angle γ as deduced from the X-ray structure of the precursor α-**7**, not enhanced but reduced affinity to the DNA target, compared to unmodified bicyclo-DNA is the result. This is evident from comparison of the T_m data of the duplexes **22**·d(A$_{10}$) and **23**·d(A$_{10}$). Model building suggests that this is due to repulsive nonbonding interactions between the substituent at C(6') and the adjacent O(5')−P ester bond, perturbing the preferred conformation of the latter with respect to the B-DNA backbone structure (too many atoms in the sugar-phosphate backbone(!), see [20]).

Given the relatively weak destabilizing effect upon replacement of one (6'R)-6'-acetamido-bicyclo-deoxynucleoside residue for a thymidine residue within a decamer duplex, we envision the use of this scaffold, as well as of its (6'S)-isomer, in the future for the site-specific introduction of functional units into DNA double and triple helices.

We gratefully acknowledge financial support from the *Swiss National Science Foundation*, from the *Wander-Stiftung*, Bern, and from *Novartis AG*, Basel. We thank the BENEFRI Small Molecule Crystallography Service directed by Prof. *Helen Stoeckli-Evans* for measuring the X-ray data set, Dr. *Eugen Stulz* for structure generation and refinement, and Dr. *Verena Meyer* for help with preparative HPLC.

Experimental Part

General. Solvents for extraction: technical grade, distilled. Solvents for reactions: reagent grade, distilled over CaH$_2$ (MeCN, CH$_2$Cl$_2$, pyridine) or Na (THF). Reagents: if not otherwise stated, from *Fluka*, highest quality available. TLC: *Merck SiL G-25 UV$_{254}$*; non-UV-visible compounds were stained by dipping the plate in a mixture of EtOH (180 ml), 4-methoxybenzaldehyde (10 ml), conc. H$_2$SO$_4$ soln. (10 ml), and AcOH (2 ml), followed by heating with a heat gun. Flash column chromatography (FC): silica gel (30–60 μm) from *Baker*. HPLC: *Pharmacia-LKB-2249* gradient pump attached to an *ABI-Kratos-Spectroflow-757*-UV/VIS detector and a *Tarkan-W + W* recorder *600*; t_R in min. UV/Melting curves: *Varian-Cary-3E*-UV/VIS spectrometer equipped with a temp. controller unit and connected to a *Compaq-ProLinea-3/25-zs* personal computer, temp. gradient 0.5°/min; data-point collection in intervals of ca. 0.3°; at <20°, the cell compartment was flushed with N$_2$ to avoid condensation of H$_2$O on the UV cells; the transition temperature T_m was determined as the maximum of the first derivative of the melting curve using the software package *Origin*™ *V5.0*. M.p.: *Büchi 510*; uncorrected. Optical rotations: *Perkin-Elmer-241* polarimeter; 10-mm cell. IR: *Perkin Elmer FTIR 1600*; $\tilde{\nu}$ in cm^{-1}. NMR: *Bruker AC-300, DRX500*; δ in ppm, ^{13}C multiplicities from DEPT spectra, *J* in Hz. EI-MS: *Varian MAT CH-7A*; ionizing voltage 70 eV; *m/z* (intensity in %). Liquid secondary ion mass spectrometry (LSI-MS): *Micromass Autospec Q*, primary ions Cs$^+$ (25 keV); matrix: dithioerythrol/dithio-DL-threitol. Matrix-assisted laser desorption ionization time-of-flight mass spectrometry (MALDI-TOF-MS) of oligonucleotides was performed as described in [21].

(1R,3S,5S,7R,8S)-8-{[(tert-Butyl)dimethylsilyl]oxy}-3-methoxy-2-oxabicyclo[3.3.0]octan-5,7-diol   (α-**2**)
and (1R,3S,5S,7S,8R)-8-{[(tert-Butyl)dimethylsilyl]oxy}-3-methoxy-2-oxabicyclo[3.3.0]octan-5,7-diol (α-**3**). A
soln. of α-**1** (1.462 g, 5.11 mmol) in THF (abs.) was treated with $BH_3 \cdot THF$ (1M, 3 equiv.) at $-78°$, allowed to
warm up to r.t., and quenched after 31 h with sat. $NaHCO_3$ soln. (26 ml). A soln. of $KHSO_5$ triple salt (12.55 g,
20.4 mmol) in sat. $NaHCO_3$ soln. (125 ml) was added, and the mixture stirred for an additional 2 h. Extraction
with Et_2O (4 × 200 ml) followed by drying the org. phase (Na_2SO_4) and evaporation afforded the crude
products, which were separated by FC (hexane/AcOEt 5 : 1 → 0 : 1): 0.320 g (21%) of α-**2** and 1.14 g (73%) of α-
3, both as white crystals.

*Data of α-***2**: M.p. 87.0–87.5°. TLC (CH_2Cl_2, 7.5% MeOH): R_f 0.39. IR (CCl_4): 3620w, 3561w, 2954s, 2930s,
2858m, 1472w, 1463w, 1447w, 1389w, 1362w, 1316w, 1293w, 1253m, 1197m, 1145m, 1125m, 1100s, 1070s, 1028vs,
1006w, 983w, 940m, 863m, 838s. ^1H-NMR (300 MHz, $CDCl_3$): 0.09, 0.11 (2s, Me_2Si); 0.88 (s, Me_3CSi); 1.87
($dd, J = 14.0, 6.6$, 1 H–C(6)); 1.92 ($d, J = 4.4$, OH); 2.12 ($dd, J = 13.6, 0.7$, 1 H–C(4)); 2.19 ($dd, J = 13.8, 3.9$,
1 H–C(4)); 2.29 ($dd, J = 14.0, 6.6$, 1 H–C(6)); 2.62 (s, OH); 3.33 (s, MeO); 3.78 ($dd, J = 5.2, 4.0$, H–C(8));
4.02 ($d, J = 3.7$, H–C(1)); 4.13–4.21 (m, H–C(7)); 5.07 ($dd, J = 3.9, 0.9$, H–C(3)). ^{13}C-NMR (75 MHz,
$CDCl_3$): $-4.88, -4.70$ ($2q$, Me_2Si); 18.03 (s, Me_3CSi); 25.77 (q, Me_3CSi); 43.64, 48.63 ($2t$, C(4), C(6)); 54.49
(q, MeO); 78.29, 82.69 ($2d$, C(7), C(8)); 84.34 (s, C(5)); 94.66 (d, C(1)); 106.79 (d, C(3)). EI-MS: 273 (11,
$[M - OMe]^+$), 131 (*100*).

*Data of α-***3**: M.p. 95–97°. TLC (CH_2Cl_2 + 7.5% MeOH): R_f 0.32. IR ($CHCl_3$): 3601w, 3566w (br.), 3003m,
2954s, 2931s, 2858m, 1472m, 1464m, 1442w, 1434w, 1390m, 1362m, 1344w, 1314w, 1300m, 1257s, 1224vs, 1206vs,
1142s, 1098s, 1080s, 1050m, 1020m, 986w, 949m, 919m, 882m, 850s, 839s. ^1H-NMR (300 MHz, $CDCl_3$): 0.09
(s, Me_2Si); 0.88 (s, Me_3CSi); 1.70 ($dd, J = 13.4, 7.2$, 1 H–C(6)); 1.94 ($dd, J = 13.8, 4.2$, 1 H–C(4)); 2.11 ($d, J =
13.6$, 1 H–C(4)); 2.21 ($dd, J = 13.6$, 1 H–C(6)); 2.26 (s, OH); 2.91 (s, OH); 3.32 (s, MeO); 3.87–3.98
(m, H–C(7), H–C(8)); 4.16 ($d, J = 5.5$, H–C(1)); 5.06 ($d, J = 4.4$, H–C(3)). ^{13}C-NMR (75 MHz, $CDCl_3$):
$-4.90, -4.73$ ($2q$, Me_2Si); 18.27 (s, Me_3CSi); 25.81 (q, Me_3CSi); 41.85, 48.74 ($2t$, C(4), C(6)); 54.34 (q, MeO);
76.85, 77.80 ($2d$, C(7), C(8)); 83.75 (s, C(5)); 88.48 (d, C(1)); 106.51 (d, C(3)). EI-MS: 273 (5, $[M - OMe]^+$),
215 (*100*).

(1R,3R,5S,7S,8R)-8-{[(tert-Butyl)dimethylsilyl]oxy}-3-methoxy-2-oxabicyclo[3.3.0]octan-5,7-diol   (β-**3**).
From β-**1** (588 mg, 2.05 mmol), as described above. FC (CH_2Cl_2, 2.5% MeOH) yielded β-**3** (566 mg, 91%).
White crystals. M.p. 124–125°. TLC (CH_2Cl_2, 5% MeOH): R_f 0.18. IR ($CHCl_3$): 3602m, 3423w (br.), 2955s,
2931s, 2900m, 2858m, 1558w, 1472m, 1463m, 1450w, 1389w, 1362m, 1310m, 1252s, 1159s, 1109s, 1068s, 1034m,
1007w, 988m, 956m, 877m, 840s, 800vs. ^1H-NMR (300 MHz, $CDCl_3$): 0.09 (s, Me_2Si); 0.90 (s, Me_3CSi); 1.66
($ddd, J = 13.2, 9.1, 1.1$, 1 H–C(6)); 1.93 (br. s, OH); 2.09 ($dd, J = 13.4, 1.7$, 1 H–C(4)); 2.21 (br. s, OH); 2.31
($ddd, J = 13.6, 5.8, 1.6$, 1 H–C(4)); 2.41 ($dd, J = 13.2, 7.0$, 1 H–C(6)); 3.35 (s, MeO); 3.83 ($dd, J = 7.7, 5.9$,
H–C(8)); 4.04 ($d, J = 5.5$, H–C(1)); 4.23 ($dd, J = 16.4, 7.5$, H–C(7)); 5.05 ($dd, J = 5.7, 1.6$, H–C(3)).
^{13}C-NMR (75 MHz, $CDCl_3$): $-4.76, -4.62$ ($2q$, Me_2Si); 18.22 (s, Me_3CSi); 25.84 (q, Me_3CSi); 44.18, 49.15
($2t$, C(4), C(6)); 54.56 (q, MeO); 76.04, 79.05 ($2d$, C(7), C(8)); 83.75 (s, C(5)); 88.40 (d, C(1)); 105.69
(d, C(3)). EI-MS: 273 ($[M - OMe]^+$, 3), 215 (*100*).

(1R,3S,5R,8S)-8-{[(tert-Butyl)dimethylsilyl]oxy}-5-hydroxy-3-methoxy-2-oxabicyclo[3.3.0]octan-7-one (α-
4). To a soln. of α-**3** (2.376 g, 7.81 mmol) in CH_2Cl_2 (36 ml), a soln. of *Dess-Martin* periodinane [22] (6.622 g,
15.58 mmol) in CH_2Cl_2 (32 ml) was added at r.t. After 2 h, the mixture was diluted with ^tBuOMe (200 ml) and
extracted with sat. $NaHCO_3$/20% $Na_2S_2O_3$ soln. 1 : 1 (100 ml) and the org. phase dried ($MgSO_4$) and evaporated:
crude α-**4** (2.515 g, 90% pure by ^1H-NMR). Colorless oil that was directly introduced into the next step. Anal.
data of a FC-purified sample (AcOEt/hexane 1 : 4): TLC (5% MeOH/CH_2Cl_2): IR ($CHCl_3$): 3673w, 3584w,
3509w, 3425w(sh), 3002w, 2933s, 2858m, 1764s, 1605w, 1467m, 1394m, 1364m, 1297m, 1253s, 1218s, 1190m,
1145m, 1076s, 1019m, 949s, 842s. ^1H-NMR (300 MHz, $CDCl_3$): 0.10, 0.13 (2s, Me_2Si); 0.89 (s, Me_3CSi); 2.15
($dd, J = 13.8, 1.7$, 1 H–C(4)); 2.25 ($dd, J = 14.0, 4.8$, 1 H–C(4)); 2.50 ($d, J = 18.8$, 1 H–C(6)); 2.59 ($dd, J = 18.8$,
1.5, 1 H–C(6)); 2.94 (br. s, OH); 3.38 (s, MeO); 4.27 ($d, J = 5.9$, H–C(1) or H–C(8)); 4.37 ($dd, J = 6.1, 1.7$,
H–C(1) or H–C(8)); 5.04 ($dd, J = 5.0, 1.7$, H–C(3)). ^{13}C-NMR (75 MHz, $CDCl_3$): $-5.15, -4.76$ ($2q$, Me_2Si);
18.45 (s, Me_3CSi); 25.71 (q, Me_3CSi); 46.69 (t, C(6)); 48.06 (t, C(4)); 55.01 (q, MeO); 76.60, 84.28 ($2d$, C(1),
C(8)); 80.38 (s, C(5)); 105.28 (d, C(3)); 210.57 (s, C(7)). EI-MS: 271 (2, $[M - OMe]^+$), 213 (*100*).

(1R,3R,5R,8S)-8-{[(tert-Butyl)dimethylsilyl]oxy}-5-hydroxy-3-methoxy-2-oxabicyclo[3.3.0]octan-7-one (β-**4**).
From β-**3** (566 mg, 1.86 mmol), as described above: crude β-**4** (740 mg). White solid that was used in the
next step without further purification. Anal. data of a FC-purified sample (7.5% MeOH/CH_2Cl_2): TLC (7.5%
MeOH/CH_2Cl_2): R_f 0.54. ^1H-NMR (300 MHz, $CDCl_3$): 0.08, 0.13 (2s, Me_2Si); 0.91 (s, Me_3CSi); 2.21 ($d, J = 12.5$,
1 H–C(4)); 2.29 ($ddd, J = 12.5, 4.6, 1.3$, 1 H–C(4)); 2.59 ($d, J = 18.4$, 1 H–C(6)); 2.72 ($d, J = 18.4$, 1 H–C(6));
3.17 (s, MeO); 4.45 ($d, J = 1.8$, H–C(1) or H–C(8)); 4.48 ($d, J = 1.8$, H–C(1) or H–C(8)); 5.00 ($d, J = 4.4$,

H–C(3)). ^{13}C-NMR (300 MHz, CDCl$_3$): – 5.15, – 4.69 (2*q*, Me$_2$Si); 18.36 (*s*, Me$_3$CSi); 25.75 (*q*, *Me$_3$*CSi); 48.26, 49.10 (2*t*, C(4), C(6)); 54.19 (*q*, MeO); 78.13, 85.21 (2*d*, C(1), C(8)); 80.23 (*s*, C(5)); 104.40 (*d*, C(3)); 210.65 (*s*, C(7)).

(1R,3S,5S,7E/Z,8R)-8-*[[(tert-Butyl)dimethylsilyl]oxy]-5-hydroxy-3-methoxy-2-oxabicyclo[3.3.0]octan-6-one O-Methyloxime* (α-**5**). A soln. of crude α-**4** (5.03 g, *ca*. 15 mmol) in abs. EtOH (92 ml) was treated with a (MeONH$_3$)Cl soln. (2.609 g, 31.24 mmol) and anh. NaOAc (1.281 g, 15.62 mmol) in H$_2$O (11 ml) (pH *ca*. 4), and the mixture was stirred for 20 min at r.t. Dilution with *t*BuOMe (400 ml) followed by extraction with sat. NaHCO$_3$ soln. (400 ml), drying of the org. phase (MgSO$_4$), evaporation, and FC of the residual yellowish oil (AcOEt/hexane 4.5 : 1 → AcOEt) afforded α-**5** (4.626 g, 89%; (*E*)/(*Z*) 5 : 1 (tentative assignment) by ^{1}H-NMR) as a yellow solid. Anal. data are from separated isomers.

Data of α-**5** (apolar (*E*)-isomer (tentative)): TLC (2.5% MeOH/CH$_2$Cl$_2$): *R*$_f$ 0.51. IR (CHCl$_3$): 3516*w* (br.), 3001*w*, 2935*m*, 2858*w*, 1657*w*, 1467*w*, 1421*w*, 1394*w*, 1358*w*, 1307*w*, 1253*m*, 1217*vs*, 1126*m*, 1082*s*, 1042*s*, 945*m*, 901*m*, 868*m*, 837*m*. ^{1}H-NMR (300 MHz, CDCl$_3$): 0.06, 0.07 (2*s*, Me$_2$Si); 0.85 (*s*, Me$_3$CSi); 2.06 (*d*, *J* = 13.2, 1 H–C(4)); 2.27 (*dd*, *J* = 13.2, 4.8, 1 H–C(4)); 2.61 (*d*, *J* = 19.1, 1 H–C(6)); 2.78 (*d*, *J* = 19.5, 1 H–C(6)); 3.27 (*s*, OH); 3.37 (*s*, MeO); 3.82 (*s*, MeON=C); 4.25 (*d*, *J* = 5.1, H–C(1) or H–C(8)); 4.36 (*d*, *J* = 5.5, H–C(1) or H–C(8)); 5.11 (*d*, *J* = 4.8, H–C(3)). ^{13}C-NMR (75 MHz, CDCl$_3$): 4.97, – 4.75 (*q*, Me$_2$Si); 18.19 (*s*, Me$_3$CSi); 25.70 (*q*, *Me$_3$*CSi); 35.19, 46.16 (2*t*, C(4), C(6)); 54.91 (*q*, MeO); 61.81, 72.66, 89.93 (2*d*, *q*, C(1), C(8), *Me*ON=C); 84.48 (*s*, C(5)); 108.18 (*d*, C(3)); 160.37 (*s*, C(7)). EI-MS: 331 (1, *M*$^+$), 242 (*100*).

Data of α-**5** (polar (*Z*)-isomer (tentative)): TLC (2.5% MeOH/CH$_2$Cl$_2$): *R*$_f$ 0.35. IR (CHCl$_3$): 3516*w*, 3001*w*, 2934*m*, 2858*w*, 1466*w*, 1428*w*, 1393*w*, 1358*w*, 1311*w*, 1250*m*, 1217*vs*, 1133*m*, 1082*s*, 1046*s*, 1008*w*, 939*m*, 879*m*, 839*m*. ^{1}H-NMR (300 MHz, CDCl$_3$): 0.03, 0.06 (2*s*, Me$_2$Si); 0.83 (*s*, Me$_3$CSi); 2.09 (*d*, *J* = 13.2, 1 H–C(4)); 2.23 (*dd*, *J* = 13.2, 4.4, 1 H–C(4)); 2.52 (*d*, *J* = 18.4, 1 H–C(6)); 2.88 (*d*, *J* = 18.0, 1 H–C(6)); 3.35 (*s*, OH); 3.37 (*s*, MeO); 3.80 (*s*, MeON=C); 4.20 (*d*, *J* = 5.5, H–C(1) or H–C(8)); 4.75 (*dt*, *J* = 0.7, 5.5, H–C(1) or H–C(8)); 5.12 (*d*, *J* = 4.4, H–C(3)). ^{13}C-NMR (75 MHz, CDCl$_3$): *ca*. – 5 (*q*, Me$_2$Si); 18.17 (*s*, Me$_3$CSi); 25.65 (*Me$_3$*CSi); 37.47, 45.73 (2*t*, C(4), C(6)); 54.90 (*q*, MeO); 61.38, 66.60 (2*d*, C(1), C(8)); 83.98 (*s*, C(5)); 90.66 (*q*, *Me*ON=C); 108.29 (*d*, C(3)); 160.60 (*s*, C(7)). EI-MS: 331 (38, *M*$^+$), 210 (*100*).

(1R,3R,5S,7E/Z,8R)-8-*[[(tert-Butyl)dimethylsilyl]oxy]-5-hydroxy-3-methoxy-2-oxabicyclo[3.3.0]octan-6-one O-Methyloxime* (β-**5**). From crude β-**4** (566 mg, 1.86 mmol), as described above. FC (AcOEt/hexane 5 : 1) gave β-**5** (523 mg, 85%; (*E*)/(*Z*) 6 : 1 (tentative assignment) by ^{1}H-NMR) as a colorless oil. Anal. data are from pure isomers:

Data of β-**5** (apolar (*E*)-isomer (tentative)): TLC (5% MeOH/CH$_2$Cl$_2$): *R*$_f$ 0.54. IR (CHCl$_3$): 3600*w*, 3418*w* (br.), 3000*w*, 2956*m*, 2933*s*, 2893*m*, 2853*m*, 1479*w*, 1468*w*, 1448*w*, 1414*w*, 1395*w*, 1368*w*, 1312*w*, 1257*m*, 1168*s*, 1110*m*, 1052*s*, 1024*w*, 985*w*, 971*w*, 944*m*, 888*m*, 864*s*, 844*s*. ^{1}H-NMR (300 MHz, CDCl$_3$): 0.11, 0.12 (2*s*, Me$_2$Si); 0.91 (*s*, Me$_3$CSi); 1.75 (*s*, OH); 2.16–2.27 (*m*, 2 H–C(4)); 2.41 (*d*, *J* = 18.8, 1 H–C(6)); 3.15 (*dd*, *J* = 18.8, 1.1, 1 H–C(6)); 3.29 (*s*, MeO); 3.83 (*s*, MeON=C); 4.09 (*d*, *J* = 5.5, H–C(1) or H–C(8)); 4.62 (*dd*, *J* = 5.5, 1.1, H–C(1) or H–C(8)); 5.13 (*dd*, *J* = 4.6, 2.8, H–C(3)). ^{13}C-NMR (75 MHz, CDCl$_3$): – 4.93, – 4.62 (2*q*, Me$_2$Si); 18.44 (*s*, Me$_3$CSi); 25.82 (*q*, *Me$_3$*CSi); 38.43, 47.73 (2*t*, C(4), C(6)); 54.83 (*q*, MeO); 61.75, 73.68 (2*d*, C(1), C(8)); 88.18 (*q*, *Me*ON=C); 83.25 (*s*, C(5)); 105.82 (*d*, C(3)); 160.30 (*s*, C(7)). EI-MS: 332 (2, [*M* + H]$^+$), 331 (3, *M*$^+$), 242 (*100*).

Data of β-**5** (polar (*Z*)-isomer (tentative)): TLC (5% MeOH/CH$_2$Cl$_2$): *R*$_f$ 0.43. IR (CHCl$_3$): 3600*w*, 3422*w* (br.), 3000*w*, 2967*s*, 2933*s*, 2904*m*, 2862*m*, 1476*m*, 1463*m*, 1445*w*, 1390*w*, 1362*w*, 1312*w*, 1296*w*, 1257*m*, 1129*s*, 1090*m*, 1057*s*, 1028*w*, 985*w*, 943*m*, 887*m*, 863*m*, 844*m*. ^{1}H-NMR (300 MHz, CDCl$_3$): 0.09, 0.11 (2*s*, Me$_2$Si); 0.89 (*s*, Me$_3$CSi); 1.69 (*s*, OH); 2.18 (*dd*, *J* = 12.9, 4.4, H–C(4)); 2.28 (*dd*, *J* = 12.9, 5.1, H–C(4)); 2.41 (*d*, *J* = 17.3, 1 H–C(6)); 3.18 (*d*, *J* = 17.7, 1 H–C(6)); 3.38 (*s*, MeO)); 3.81 (*s*, MeON=C); 4.06 (*d*, *J* = 5.5, H–C(1) or H–C(8)); 4.85 (*d*, *J* = 5.5, H–C(1) or H–C(8)); 5.24 (*dd*, *J* = 5.5, 4.4, H–C(3)). EI-MS: 331 (3, *M*$^+$), 89 (*100*).

N-[(1R,3S,5S,7S,8R)-8-*[[(tert-Butyl)dimethylsilyl]oxy]-5-hydroxy-3-methoxy-2-oxabicyclo[3.3.0]oct-7-yl]-2,2,2-trifluoroacetamide* (α-**6**) and *N-[(1R,3S,5S,7R,8R)*-8-*[[(tert-Butyl)dimethylsilyl]oxy]-5-hydroxy-3-methoxy-2-oxabicyclo[3.3.0]oct-7-yl]-2,2,2-trifluoroacetamide* (α-**7**). In an autoclave, *Raney*-Ni (1.03 g) was suspended in H$_2$O (4 ml) and pretreated with H$_2$ (10 bar) under heavy stirring. After 10 min, a degassed soln. of α-**5** ((*E*)/(*Z*); 873 mg, 2.64 mmol) in EtOH (10 ml), containing 0.8 ml of conc. NH$_3$ soln. was carefully transferred to the autoclave, and the mixture was hydrogenated for 12 h at 10 bar. The mixture was filtered over *Celite* and the filtrate evaporated to give 796 mg (99%) of the crude amines as a colorless oil. This oil was dissolved in CF$_3$COOEt (1.55 ml) containing 70 μl (0.2 equiv.) of Et$_3$N, the soln. stirred for 1 h at r.t. and evaporated, and the residue separated by FC (AcOEt/hexane 2.5 : 1 → 1 : 1): α-**6** (144 mg, 14%) as a colorless oil and α-**7** (888 mg, 84%) as a white solid.

Data of α-6: TLC (8% MeOH/CH$_2$Cl$_2$): R_f 0.50. IR (CHCl$_3$): 3553w, 3426m, 2995w, 2955s, 2932s, 2858m, 1723vs, 1541m, 1471m, 1442m, 1386w, 1347w, 1296w, 1280w, 1252s, 1189s, 1170vs, 1131s, 1105s, 1055m, 1021m, 1005w, 947m, 927s, 886m, 872m, 842s. ¹H-NMR (300 MHz, CDCl$_3$): 0.07, 0.09 (2s, Me$_2$Si); 0.86 (s, Me$_3$CSi); 1.73 (dd, J = 13.6, 6.6, 1 H−C(6)); 2.01 (dd, J = 13.8, 4.2, 1 H−C(4)); 2.14 (d, J = 13.6, 1 H−C(4)); 2.34 (dd, J = 14.0, 5.9, 1 H−C(6)); 3.06 (s, OH); 3.34 (s, MeO); 4.04−4.16 (m, H−C(1), H−C(7), H−C(8)); 5.13 (d, J = 4.4, H−C(3)); 6.62 (d, J = 6.7, NH). ¹³C-NMR (75 MHz, CDCl$_3$): − 5.18, − 4.82 (2q, Me$_2$Si); 18.12 (s, Me$_3$CSi); 25.62 (q, Me$_3$CSi); 39.22, 48.12 (t, C(4), C(6)); 54.49 (q, MeO); 57.47, 75.28, 88.84 (3d, C(1), C(7), C(8)); 85.16 (s, C(5)); 107.34 (d, C(3)); 115.68 (q, J(C,F) = 288.1, CF$_3$); 156.89 (q, J(C,F) = 37.2, COCF$_3$). EI-MS: 368 (14, M − OMe]⁺), 310 (100).

Data of α-7: M.p. 84−85°. TLC (8% MeOH/CH$_2$Cl$_2$): R_f 0.61. IR (CHCl$_3$): 3531w, 3427m, 2998w, 2956w, 2931m, 2859m, 1724s, 1534m, 1472m, 1464m, 1443m, 1427w, 1390w, 1373w, 1362w, 1344w, 1322w, 1303w, 1261s, 1173s, 1092s, 1043s, 966m, 948m, 875m, 838s. ¹H-NMR (300 MHz, CDCl$_3$): 0.05, 0.07 (2s, Me$_2$Si); 0.87 (s, Me$_3$CSi); 1.89−2.01 (m, 1 H−C(4), 1 H−C(6)); 2.15 (d, J = 13.6, 1 H−C(4)); 2.23 (dd, J = 13.8, 7.6, 1 H−C(6)); 2.96 (s, OH); 3.34 (s, MeO); 4.14 (t, J = 4.0, H−C(8)); 4.19 (d, J = 4.4, H−C(1)); 4.53 (ddd, J = 16.8, 8.4, 3.8, H−C(7)); 5.10 (d, J = 4.4, H−C(3)); 6.66 (d, J = 7.7, NH). ¹³C-NMR (75 MHz, CDCl$_3$): − 5.45, − 4.43 (2q, Me$_2$Si); 18.10 (s, Me$_3$CSi); 25.66 (q, Me$_3$CSi); 41.14, 48.24 (2t, C(4), C(6)); 53.34 (d, C(7)); 54.65 (q, MeO); 73.34 (d, C(8)); 84.95 (s, C(5)); 91.05 (d, C(1)); 107.96 (d, C(3)); 111.93 (q, J(C,F) = 286, CF$_3$); 156.24 (q, J(C,F) = 37, CF$_3$CO). EI-MS: 368 (11, [M − OMe]⁺), 312 (100).

X-Ray Analysis of α-7: Suitable crystals were obtained from 0.02M α-7 in hexane at − 24°. Colorless transparent needles (0.57 × 0.27 × 0.19 mm); C$_{16}$H$_{28}$F$_3$NO$_5$Si; orthorhombic space group P2$_1$2$_1$2$_1$. Intensities were measured with a *Stoe-AED*-4-circle diffractometer (MoK$_α$, λ 0.71073 Å). Of the 2265 independent reflections (θ = 2.03−25.47°), 1645 with F > 4σ(F) were used in the refinement. The structure was solved using direct methods with SHELX-86 [23] and refined by full-matrix least-square procedures SHELXL-97. Non-H-atoms were refined anisotropically. The positions of all H-atoms were calculated and adjusted after every least-squares cycle. The refinement converged at R = 0.0627, Rw = 0.1755.

N-*[(1R,3R,5S,7S,8R)-8-[[(tert-Butyl)dimethylsilyl]oxy]-5-hydroxy-3-methoxy-2-oxabicyclo[3.3.0]oct-7-yl]-2,2,2-trifluoroacetamide* (β-6) *and* N-*[(1R,3R,5S,7R,8R)-8-[[(tert-Butyl)dimethylsilyl]oxy]-5-hydroxy-3-me-thoxy-2-oxabicyclo[3.3.0]oct-7-yl]-2,2,2-trifluoroacetamide* (β-7). Procedure not optimized. As described above, from β-5 ((E)/(Z); 495 mg, 1.50 mmol) by hydrogenation (H$_2$/20 bar) over 10% Pd/C (495 mg) in MeOH, followed by trifluoroacetylation. FC (2% MeOH/CH$_2$Cl$_2$) yielded β-6/β-7 (104 mg, 17%; ratio 1:5 by ¹H-NMR) besides 33% of nonconverted β-5.

Data of β-7: TLC (10% MeOH/CH$_2$Cl$_2$): R_f 0.64. IR (CHCl$_3$): 3597w, 3416w, 3265w(sh), 3101w, 2954m, 2931s, 2901m, 2858m, 1720vs, 1548m, 1472m, 1464m, 1448m, 1385m, 1362m, 1314m, 1282m, 1256s, 1166vs, 1118m, 1103m, 1088m, 1071m, 1031m, 1006w, 978m, 959m, 940m, 922m, 865m, 839s. ¹H-NMR (300 MHz, CDCl$_3$): 0.06, 0.10 (2s, Me$_2$Si); 0.87 (s, Me$_3$CSi); 2.13 (s, OH); 2.21−2.26 (m, 3 H, CH$_2$(4), CH$_2$(6)); 2.45 (dd, J = 14.0, 6.3, 1 H, CH$_2$(4), CH$_2$(6)); 3.41 (s, MeO); 4.05 (d, J = 4.8, H−C(1)); 4.30 (dd, J = 6.4, 5.0, H−C(8)); 4.56 (m, H−C(7)); 5.17 (dd, J = 6.6, 2.2, H−C(3)); 8.08 (d, J = 8.1, NH). ¹³C-NMR (75 MHz, CDCl$_3$): − 5.27, − 5.05 (2q, Me$_2$Si); 18.13 (s, Me$_3$CSi); 25.61 (q, Me$_3$CSi); 43.87, 49.03 (2t, C(4), C(6)); 53.03, 56.22 (d + q, C(7), MeO)); 73.09, 90.11 (2d, C(1), C(8)); 84.31 (s, C(5)); 107.39 (d, C(3)); 117.95 (q, CF$_3$); 156.75 (q, CF$_3$CO).

(3'S,5'R,6'R)-1-*[5'-O-[(tert-Butyl)dimethylsilyl]-2'-deoxy-6'-(trifluoroacetamido)-3',5'-ethano-α/β-D-ribofur-anosyl]thymine* (α/β-8). A mixture of α-7 (1.000 g, 2.51 mmol) and dry thymine (631 mg, 5.00 mmol) in abs. MeCN (25 ml) was cooled to 3° and treated with N,O-bis(trimethylsilyl)acetamide (BSA; 3.06 ml, 1.25 mmol) followed by Me$_3$SiCl (158 μl, 1.25 mmol). After 90 min (→ clear soln.), Me$_3$SiOSO$_2$CF$_3$ (1.82 ml, 10.03 mmol) was added and the mixture stirred for 20 h at 3°. Then the mixture was treated with 1M HCl (25 ml) for 5 min, diluted with CH$_2$Cl$_2$ (150 ml), and extracted with sat. NaHCO$_3$ soln. The org. phases were dried (MgSO$_4$), and evaporated and the crude nucleosides purified by FC (0−3% MeOH/CH$_2$Cl$_2$) to give α/β-8 (1.110 g, 90%; ratio 1:1.5 by ¹H-NMR) as a white foam. Separation of the anomers was effected by prep. HPLC (*LiChrosorb SI 60*, 7 μ, 23 × 250 mm; hexane/iPrOH 8:2; t_R 22.5 (β-8), 27.5 (α-8)) to give anal. pure β-8 (577 mg, 47%) and α-8 (332 mg, 27%), both as white foams.

Data of α-8: TLC (hexane/iPrOH 8:2): R_f 0.35. IR (CHCl$_3$): 3426m, 2956m, 2932m, 2900w, 2860m, 1691vs, 1535m, 1472m, 1443m, 1427w, 1411w, 1364m, 1264s, 1237s, 1229s, 1212s, 1200s, 1173s, 1104s, 1051m, 1006m, 967m, 872m, 837s. ¹H-NMR (300 MHz, CDCl$_3$): 0.07, 0.12 (2s, Me$_2$Si); 0.90 (s, Me$_3$CSi); 1.84 (s, Me−C(5)); 2.06 (dd J = 13.6, 9.9, 1 H−C(7')); 2.38 (dd, J = 13.6, 7.7, 1 H−C(7')); 2.49−2.63 (m, 2 H−C(2')); 4.18 (t, J = 3.9, H−C(5')); 4.33 (br. s, OH); 4.53−4.63 (m, H−C(6')); 4.58 (d, J = 4.0, H−C(4')); 6.06 (dd, J = 6.8, 2.8, H−C(1')); 6.66 (d, J = 8.1, CF$_3$CONH); 7.32 (s, H−C(6)); 9.76 (br. s, H−N(3)). ¹³C-NMR (75 MHz, CDCl$_3$):

−5.44, −4.47 (2*q*, Me₂Si); 12.40 (*q*, *Me*−C(5)); 18.00 (*s*, Me₃CSi); 25.68 (*q*, *Me₃*CSi); 43.40, 47.99 (2*t*, C(2′), C(7′)); 52.05 (*d*, C(6′)); 73.47 (*d*, C(5′)); 84.34 (*s*, C(3′)); 91.40 (C(1′)); 93.32 (*d*, C(4′)); 109.95 (*s*, C(5)); 115.67 (*q*, *J*(C,F) = 288, CF₃); 137.58 (*s*, C(6)); 150.56 (*s*, C(2)); 156.52 (*q*, *J*(C,F) = 37, COCF₃); 164.53 (*s*, C(4)). LSI-MS: 646.2 (2, [*M* + matrix]⁺), 532.1 (3, [*M* + K]⁺), 516.2 (2, [*M* + Na]⁺), 494.2 (39, [*M* + 1]⁺), 368.2 (*100*).

*Data of β-**8**:* TLC (hexane/ⁱPrOH 8:2): R_f 0.27. IR (CHCl₃): 3596*w*, 3397*m*, 2957*m*, 2932*m*, 2899*w*, 2860*w*, 1693*vs*, 1532*m*, 1472*m*, 1442*w*, 1427*w*, 1414*w*, 1371*m*, 1318*w*, 1281*m*, 1261*s*, 1238*s*, 1213*s*, 1199*s*, 1175*s*, 1101*s*, 1056*m*, 1012*m*, 977*m*, 931*m*, 910*m*, 839*s*. ¹H-NMR (300 MHz, CDCl₃): 0.07 (*s*, Me₂Si); 0.89 (*s*, Me₃CSi); 1.87 (*s*, Me−C(5)); 1.96 (*dd*, *J* = 13.2, 9.9, 1 H−C(2′)); 2.16−2.30 (*m*, H−C(7′)); 2.55 (*dd*, *J* = 13.6, 5.2, 1 H−C(2′)); 4.10 (*d*, *J* = 5.2, H−C(4′)); 4.38 (*t*, *J* = 5.0, H−C(5′)); 4.44−4.50 (*m*, H−C(6′) OH); 6.32 (*dd*, *J* = 9.6, 5.5, H−C(1′)); 6.98 (*d*, *J* = 6.6, CF₃CONH); 7.21 (*s*, H−C(6)); 10.19 (*s*, H−N(3)). ¹³C-NMR (75 MHz, CDCl₃): −5.40, −4.52 (2*q*, Me₂Si); 12.14 (*q*, Me−C(5)); 18.13 (*s*, Me₃CSi); 25.72 (*q*, *Me₃*CSi); 42.36, 46.58 (2*t*, C(2′), C(7′)); 54.20 (*d*, C(6′)); 71.80 (*d*, C(5′)); 83.75 (*s*, C(3′)); 85.49 (*d*, C(1′)); 88.99 (*d*, C(4′)); 111.82 (*s*, C(5)); 115.72 (*q*, *J*(C,F) = 288, CF₃); 134.91 (*d*, C(6)); 150.87 (*s*, C(2)); 156.74 (*q*, *J*(C,F) = 37, COCF₃); 163.87 (*s*, C(4)). LSI-MS: 646.1 (1, [*M* + matrix]⁺), 532.1 (4, [*M* + K]⁺), 516.1 (2, [*M* + Na]⁺), 494.1 (39, [*M* + 1]⁺), 368.1 (*100*).

(3′S,5′R,6′R)-1-[6′-Amino-5′-O-[(tert-butyl)dimethylsilyl]-2′-deoxy-3′,5′-ethano-α/β-D-ribofuranosyl]thymine (**9**). To a soln. of β-**8** (640 mg, 1.3 mmol) in MeOH (1.3 ml), conc. NH₃ soln. (26 ml) was added. The resulting mixture was stirred for 4 h at r.t. and subsequently evaporated. CC (CH₂Cl₂/MeOH/NH₃ soln. 50:5:3) gave **9** (472 mg, 91%). Slightly yellow oil. TLC (CH₂Cl₂/MeOH/NH₃ soln. 18:2:1): R_f 0.43. UV (MeOH): 264 (9930). IR (KBr): 3419*s* (br.), 3198*s* (br), 3062*s* (br.), 2954*s*, 2930*s*, 2898*s*, 2857*s*, 1698*vs*, 1686*vs*, 1472*m*, 1287*w*, 1262*w*, 1152*w*, 836*w*, 780*w*. ¹H-NMR (300 MHz, CDCl₃): 0.07, 0.10 (2*s*, Me₂Si); 0.91 (*s*, Me₃CSi); 1.84−1.99 (*m*, 2 H−C(7′)); 1.88 (*s*, Me−C(5)); 2.33 (*dd*, *J* = 13.1, 9.6, 1 H−C(2′)); 2.41 (*dd*, *J* = 13.2, 5.5, 1 H−C(2′)); 3.62 (*dd*, *J* = 10.3, 4.3, H−C(6′)); 4.03 (*d*, *J* = 5.7, H−C(4′)); 4.11−4.14 (*m*, H−C(5′)); 6.36 (*dd*, *J* = 9.6, 5.3, H−C(1′)); 7.79 (*d*, *J* = 1.1, H−C(6)). ¹³C-NMR (75 MHz, CDCl₃): −5.00, −4.45 (2*q*, Me₂Si); 12.21 (*q*, Me−C(5)); 18.26 (*s*, Me₃CSi); 25.91 (*q*, *Me₃*CSi); 43.74, 46.72 (2*t*, C(2′), C(7′)); 55.70 (*d*, C(6′)); 73.60 (*d*, C(5′)); 84.55 (*s*, C(3′)); 87.03 (*d*, C(1′)); 89.21 (*d*, C(4′)); 110.92 (*s*, C(5)); 137.79 (*d*, C(6)); 151.13 (*s*, C(2)); 164.27 (*s*, C(4)). HR-LSI-MS (C₁₈H₃₂N₃O₅Si): 398.2111 (calc. 398.2109).

(3′S,5′R,6′R)-1-[6′-Acetamido-5′-O-[(tert-butyl)dimethylsilyl]-2′-deoxy-3′,5′-ethano-β-D-ribofuranosyl}thymine (**10**). To a soln. of **9** (396 mg, 0.99 mmol) in pyridine (3.5 ml), Ac₂O (102 µl, 1 mmol) was added at 0°, and the mixture was stirred for 2 h. Dilution with CH₂Cl₂ (50 ml), extraction with sat. NaHCO₃ soln. (50 ml), and evaporation of the dried (MgSO₄) org. phase, followed by FC (CH₂Cl₂/MeOH 100:8) yielded **10** (387 mg, 88%). White foam. TLC (CH₂Cl₂/MeOH 100:8): R_f 0.48. UV (MeOH): 262 (9480). IR (KBr): 3368*m* (br.), 3068*w* (br.), 2954*m*, 2930*m*, 2892*m*, 2856*m*, 1700*s*, 1540*w*, 1472*m*, 1374*m*, 1284*m*, 1260*m*, 1154*m*, 1046*m*, 980*w*, 883*w*, 836*m*, 779*m*. ¹H-NMR (300 MHz, CD₃OD): 0.11, 0.13 (*s*, Me₂Si); 0.98 (*s*, Me₃CSi); 1.93 (*s*, Me−C(5)); 2.01 (*s*, MeCO); 2.01−2.07 (*m*, 1 H−C(7′)); 2.16−2.26 (*m*, 1 H−C(2′), 1 H−C(7′)); 2.36 (*dd*, *J* = 13.1, 5.2, 1 H−C(2′)); 4.06 (*d*, *J* = 4.8, H−C(4′)); 4.40 (*t*, *J* = 4.4, H−C(5′)); 4.44−4.49 (*m*, H−C(6′)); 6.23 (*dd*, *J* = 10.0, 5.2, H−C(1′)); 7.53 (*d*, *J* = 1.1, H−C(6)). ¹³C-NMR (75 MHz, CDCl₃): −4.48, −3.75 (2*q*, Me₂Si); 12.77 (*q*, *Me*−C(5)); 19.78 (*s*, Me₃CSi); 23.09 (*q*, *Me*CO); 26.91 (*q*, *Me₃*CSi); 43.42, 46.47 (2*t*, C(2′), C(7′)); 55.13 (*d*, C(6′)); 73.72 (*d*, C(5′)); 84.27 (*s*, C(3′)); 88.63 (*d*, C(1′)); 91.42 (*d*, C(4′)); 111.78 (*s*, C(5)); 138.75 (*d*, C(6)); 152.52 (*s*, C(2)); 166.45 (*s*, C(4)); 173.06 (*s*, *Me*CO). HR-LSI-MS (C₂₀H₃₄N₃O₆Si): 440.2217 (calc. 440.2201).

(3′S,5′R,6′R)-1-(6′-Acetamido-2′-deoxy-3′,5′-ethano-β-D-ribofuranosyl)thymine (**11**). To a soln. of **10** (305 mg, 0.69 mmol) in pyridine (1.5 ml), Et₃N · 3 HF (160 µl, 0.98 mmol) was added and the mixture stirred. After 24 h at r.t., excess solid (NH₄)₂CO₃ was added, the mixture was filtered, and the filtrate evaporated. FC (CH₂Cl₂/MeOH 100:8) gave **11** (215 mg, 96%). White foam. TLC (CH₂Cl₂/MeOH 100:8): R_f 0.11. UV (H₂O): 264 (9370). IR (KBr): 3384*s* (br.), 3066*m*, 2930*m*, 2820*w*, 1700*s*, 1684*s*, 1654*m*, 1540*m*, 1472*m*, 1374*m*, 1288*m*, 1264*m*, 1144*m*, 1056*m*, 940*w*, 814*w*, 782*w*, 668*w*, 610*w*, 560*w*. ¹H-NMR (300 MHz, CD₃OD): 1.92 (*d*, *J* = 1.3, Me−C(5)); 2.02 (*s*, MeCO); 2.07−2.23 (*m*, 1 H−C(2′), 2 H−C(7′)); 2.40 (*dd*, *J* = 13.0, 5.2, 1 H−C(2′)); 4.15 (*t*, *J* = 3.8, H−C(5′)); 4.21 (*d*, *J* = 4.2, H−C(4′)); 4.52 (*ddd*, *J* = 11.0, 8.6, 3.3, H−C(6′)); 6.44 (*dd*, *J* = 10.1, 5.2, H−C(1′)); 8.20 (*d*, *J* = 1.1, H−C(6)). ¹³C-NMR (75 MHz, CDCl₃): 12.81 (*q*, *Me*−C(5)); 22.84 (*q*, *Me*CO); 43.21, 47.76 (2*t*, C(2′), C(7′)); 54.05 (*d*, C(6′)); 84.27 (*s*, C(3′)); 89.30, 92.72 (2*d*, C(1′), C(4′)); 111.76 (*s*, C(5)); 139.07 (*d*, C(6)); 152.60 (*s*, C(2)); 166.67 (*s*, C(4)); 173.22 (*s*, *Me*CO). HR-LSI-MS (C₁₄H₂₀N₃O₆): 326.1352 (calc. 326.1358).

(3′S,5′R,6′R)-1-[2′-Deoxy-6′-(trifluoroacetamido)-3′,5′-ethano-β-D-ribofuranosyl]thymine (**12**). To a soln. of β-**8** (993 mg, 2.00 mmol) in MeCN (10 ml), a soln. of Bu₄NF · 3H₂O (5.05 g, 16.02 mmol) in MeCN (10 ml) and AcOH (1.15 ml, 20.01 mmol) was added, and the resulting mixture was stirred for 2 h at r.t. Dilution with AcOEt (100 ml), extraction with sat. NaHCO₃ soln., followed by drying of the org. phase (MgSO₄) gave crude

product that was purified by FC (AcOEt/hexane 3 : 1): **12** (719 mg, 95%). White foam. TLC (8% MeOH/ CH$_2$Cl$_2$): R_f 0.19. UV (H$_2$O): 265 (8910). IR (KBr): 3408s (br.), 3000s (sh), 3072m, 2954w, 2830w, 1698vs, 1562m, 1552m, 1536w, 1475m, 1408w, 1376m, 1289s, 1266s, 1215s, 1188s, 1158s, 1107m, 1054m, 1010w, 968w, 944m, 882w, 852w, 814w, 784w, 756w, 668w, 644w. ^{1}H-NMR (500 MHz, CD$_3$OD): 1.92 (d, $J = 1.3$, Me−C(5)); 2.16 (d, $J = 13.5$, 8.0, 1 H−C(7′)); 2.17 (dd, $J = 13.0$, 10.0, 1 H−C(2′)); 2.34 (dd, $J = 13.4$, 11.8, 1 H−C(7′)); 2.42 (dd, $J = 13.0$, 5.2, 1 H−C(2′)); 4.23−4.25 (m, H−C(4′), H−C(5′)); 4.55 (ddd, $J = 11.4$, 8.4, 2.6, H−C(6′)); 6.47 (dd, $J = 10.1$, 5.2, H−C(1′)); 8.19 (d, $J = 1.3$, H−C(6)). ^{13}C-NMR (125 MHz, CD$_3$OD): 12.79 (q, Me−C(5)); 42.43, 47.65 (2t, C(2′), C(7′)); 54.65 (d, C(6′)); 72.86 (d, C(4′) or C(5′)); 85.71 (s, C(3′)); 89.37 (d, C(1′)); 92.52 (d, C(4′) or C(5′)); 111.83 (s, C(5)); 117.69 (q, J(C,F) $= 287$, CF$_3$); 138.93 (d, C(6)); 152.60 (s, C(2)); 158.77 (q, J(C,F) $= 37$, COCF$_3$); 166.66 (s, C(4)). LSI-MS: 380.0 (18, $[M + H]^+$), 127 (*100*).

(3′S,5′R,6′R)-1-(6′-Amino-2′-deoxy-3′,5′-ethano-β-D-ribofuranosyl)thymine (**13**). A soln. of **12** (20 mg, 53 μmol) in MeOH (0.1 ml) was treated with conc. NH$_3$ soln. (1.5 ml). After 45 min at r.t., the mixture was lyophilized and the residue adsorbed on silica gel (MeOH) and purified by FC (CH$_2$Cl$_2$/MeOH 5 : 1 + 5% conc. NH$_3$ soln.): **13** (15 mg, 98%). Colorless film. TLC (MeCl$_2$/MeOH 4 : 1 + 5% conc. NH$_3$): R_f 0.25. UV (MeOH): 267 (9770). IR (KBr): 3072m (br.), 2963m, 1684s, 1472m, 1436w, 1292m, 1264m, 1203s, 1132s, 1046m, 934w. ^{1}H-NMR (300 MHz, CD$_3$OD): 1.92 (d, $J = 1.1$, Me−C(5)); 2.18−2.37 (m, 1 H−C(2′), 2 H−C(7′)); 2.44 (dd, $J = 13.1$, 5.4, 1 H−C(2′)); 3.86 (ddd, $J = 9.8$, 8.7, 3.7, H−C(6′)); 4.21 (d, $J = 4.2$, H−C(4′)); 4.31 (t, $J = 4.0$, H−C(5′)); 6.40 (dd, $J = 9.9$, 5.1, H−C(1′)); 7.96 (d, $J = 1.1$, H−C(6)). ^{13}C-NMR (75 MHz, CDCl$_3$): 12.8 (q, Me−C(5)); 42.5, 47.4, (2t, C(2′), C(7′)); 54.6 (d, C(6′)); 72.0 (d, C(5′)); 85.9 (s, C(3′)); 90.0, 92.2 (2d, C(1′), C(4′)); 112.1 (s, C(5)); 139.0 (d, C(6)); 152.7 (s, C(2)); 166.6 (s, C(4)). HR-LSI-MS (C$_{11}$H$_{18}$N$_3$O$_5$): 284.1245 (calc. 284.1246).

(3′S,5′R,6′R)-1-[2′-Deoxy-5′-O-[(4,4′-dimethoxytriphenyl)methyl]-6′-(trifluoroacetamido)-3′,5′-ethano-β-D-ribofuranosyl}thymine (**14**). A soln. of **12** (173 mg, 0.46 mmol) in dry pyridine (0.5 ml), containing activated molecular sieves (3 Å), was treated with (MeO)$_2$TrOSO$_2$CF$_3$ (312 mg, 0.69 mmol) and heated to 50°. After 9 h, another portion of (MeO)$_2$TrOSO$_2$CF$_3$ (103 mg, 0.23 mmol) was added. After a total of 16 h, the mixture was diluted with AcOEt (30 ml) and extracted with sat. NaHCO$_3$ soln. (30 ml), the org. phase dried (MgSO$_4$) and evaporated, and the residual oil purified by FC (2.5−10% MeOH/CH$_2$Cl$_2$ + 1% Et$_3$N): **14** (144 mg, 46%), besides 40% of recovered starting material. TLC (8% MeOH/CH$_2$Cl$_2$): R_f 0.46. IR (CHCl$_3$): 3595w, 3393w, 2961w, 2935w, 2912w, 2839w, 1694s, 1628w, 1608m, 1580w, 1528w, 1512m, 1467m, 1443w, 1426w, 1413w, 1370w, 1320w, 1303m, 1281m, 1257s, 1237s, 1216s, 1201vs, 1180s, 1097m, 1057m, 1035m, 1014m, 910w, 872w, 825m. ^{1}H-NMR (300 MHz, CDCl$_3$): 1.76 (dd, $J = 15.3$, 6.4, 1 H−C(7′)); 1.93 (s, Me−C(5)); 2.04 (dd, $J = 13.6$, 9.9, 1 H−C(2′)); 2.35 (d, $J = 15.1$, 1 H−C(7′)); 2.49 (dd, $J = 13.8$, 5.3, 1 H−C(2′)); 3.00 (dd, $J = 10.5$, 5.7, H−C(6′)); 3.4 (br. s, OH); 3.75 (s, MeO); 3.89 (d, $J = 5.5$, H−C(4′)); 4.22 (t, $J = 6.1$, H−C(5′)); 6.19 (dd, $J = 9.6$, 5.5, H−C(1′)); 6.79−6.83 (m, 4 arom. H); 7.16−7.50 (m, 14 H). ^{13}C-NMR (75 MHz, CDCl$_3$): 12.16 (q, Me−C(5)); 42.13, 47.92 (2t, C(2′), C(7′)); 53.89 (d, C(6′)); 55.29 (q, MeO); 72.29 (d, C(5′)); 83.47, 88.10 (2s, C(3′), Ar$_2$CPh); 84.85 (d, C(1′)); 88.69 (d, C(4′)); 115.72 (q, J(C,F) $= 289$, CF$_3$); 112.36 (s, C(5)); 113.57, 113.61, 127.41, 127.53, 128.23, 129.56, 129.67 (7d, arom. C); 134.56 (d, C(6)); 135.30, 135.59, 144.47 (3s, arom. C); 150.52 (s, C(2)); 156.76 (q, J(C,F) $= 37$, COCF$_3$); 159.04, 159.07 (2s, arom. C); 163.39 (s, C(4)). LSI-MS: 720.1 (2, $[M + K]^+$), 681.1 (2, M^+), 303.1 (*100*).

(3′S,5′R,6′R)-1-[6′-Acetamido-2′-deoxy-5′-O-[(4,4′-dimethoxytriphenyl)methyl]-3′,5′-ethano-β-D-ribofurano-syl}thymine (**15**). As described for **14**, from **11** (186 mg, 0.57 mmol) and (MeO)$_2$TrOSO$_2$CF$_3$ (395 mg, 0.87 mmol) in pyridine (0.6 ml). FC (CH$_2$Cl$_2$/MeOH 50 : 1) gave **15** (277 mg, 77%). Slightly brownish foam. TLC (CH$_2$Cl$_2$/MeOH 50 : 4): R_f 0.53. ^{1}H-NMR (300 MHz, CDCl$_3$): 1.70 (dd, $J = 14.8$, 6.3, 1 H−C(7′)); 1.90 (d, $J = 0.9$, Me−C(5)); 1.94 (s, MeCO); 2.21−2.32 (m, 1 H−C(2′), 1 H−C(7′)); 2.41 (dd, $J = 13.7$, 5.8, 1 H−C(2′)); 3.26 (br. s, OH−C(3′)); 3.32 (dd, $J = 11.0$, 6.0, H−C(6′)); 3.65 (d, $J = 5.5$, H−C(4′)); 3.74 (s, MeO); 4.08−4.14 (m, H−C(5′)); 6.01 (dd, $J = 9.4$, 5.9, H−C(1′)); 6.55 (d, $J = 4.6$, CONH); 6.79 (d, $J = 8.6$, 4 arom. H); 7.21 (d, $J = 0.9$, H−C(6)); 7.16−7.50 (3m, 9 arom. H); 9.10 (s, H−N(3)). ^{13}C-NMR (75 MHz, CDCl$_3$): 12.61 (q, Me−C(5)); 23.40 (q, MeCO); 43.33, 47.41 (2t, C(2′), C(7′)); 53.66 (d, C(6′)); 55.25 (q, MeO); 72.94 (d, C(5′)); 83.35, 87.94 (2s, C(3′), Ar$_2$CPh); 85.64 (d, C(1′)); 88.78 (d, C(4′)); 111.40 (s, C(5)); 113.40, 127.21, 127.85, 128.06, 129.84 (5d, arom. C); 135.82 (d, C(6)); 135.94, 144.84 (2s, arom. C); 150.41 (s, C(2)); 158.83 (s, arom. C); 163.57 (s, C(4)); 170.41 (s, CO). LSI-MS: 628 (3, $[M + 1]^+$), 303 (*100*). Anal. calc. for C$_{35}$H$_{37}$N$_3$O$_8$·0.5H$_2$O: C 66.03, H 6.02, N 6.60; found: C 66.24, H 6.25, N 6.56.

(3′S,5′R,6′R)-1-[2′-Deoxy-5′-O-[(4,4′-dimethoxytriphenyl)methyl]-6′-(trifluoroacetamido)-3′,5′-ethano-β-D-ribofuranosyl}thymine 3′-(2-Cyanoethyl Diisopropylphosphoramidite) (**16**). A soln. of **14** (73 mg, 0.11 mmol) in MeCN (1 ml) was treated with ^{i}Pr$_2$NEt (73.5 μl, 0.43 mmol) and [P(OCH$_2$CH$_2$CN)(N^{i}Pr$_2$)]Cl (48 μl, 0.22 mmol) and the resulting mixture stirred for 2 h at r.t. Then additional ^{i}Pr$_2$NEt (37 μl, 0.22 mmol) and

Cl[P(OCH$_2$CH$_2$CN)(N^{i}Pr$_2$)] (24 µl, 0.11 mmol) were added, and the mixture was worked up after 7 h by dilution with AcOEt and extraction with sat. NaHCO$_3$ soln. After drying and evaporation of the org. phase, the residual gum was purified by FC (hexane/AcOEt 2:1→0:1): **16** (83 mg, 88%). White foam. TLC (hexane/AcOEt 1:1): R_f 0.31, 0.26. ^{1}H-NMR (300 MHz, CDCl$_3$): 1.05–1.13 (*m*, 2 *Me*$_2$CH); 1.93, 1.94 (2*d*, *J* = 0.7, Me–C(5)); 2.02–2.20 (*m*, 2 H, CH$_2$(2′) or CH$_2$(7′)); 2.34–2.39 (*m*, 1 H, CH$_2$(2′) or CH$_2$(7′)); 2.50–2.59 (*m*, OCH$_2$CH$_2$CN); 2.84–2.92 (*m*, 1 H, CH$_2$(2′) or CH$_2$(7′)); 3.17–3.27 (*m*, Me$_2$CH); 3.44–3.65 (*m*, OCH$_2$CH$_2$CN); 3.65–3.75 (*m*, H–C(5′)); 3.76 (*s*, MeO); 3.82, 3.99 (2*d*, *J* = 5.3, H–C(4′)); 4.11–4.19 (*m*, H–C(6′)); 5.96–6.04 (*m*, 4 arom. H); 7.12–7.50 (*m*, H–C(6), 9 arom. H); 9.21 (br. *s*, H–N(3)). ^{13}C-NMR (75 MHz, CDCl$_3$): 12.18 (*q*, *Me*–C(5)); 20.10, 20.13, 20.20, 20.23 (4*t*, *J*(C,P) = 2.4, NCCH$_2$CH$_2$O); 24.16, 24.19, 24.27, 24.29, 24.37, 24.40, 24.46, 24.48 (8*q*, *Me*$_2$CH); 41.56, 41.917 (2*t*, *J*(C,P) = 8.9, C(2′), C(7′)); 43.27, 43.43 (2*d*, Me$_2$*C*H); 44.97, 45.81 (2*t*, *J*(C,P) = 9.0, C(2′), C(7′)); 53.20, 53.43 (2*d*, C(6′)); 55.22 (*q*, MeO); 57.66, 57.89 (2*t*, *J*(C,P) = 9.5, NCCH$_2$CH$_2$O); 72.12, 72.30 (2*d*, C(5′)); 85.55, 86.15 (2*d*, *J*(C,P) = 1.2, C(1′)); 86.99, 87.23 (2*s*, *J*(C,P) = 7.9, 8.5, C(3′)); 87.46, 87.69 (2*d*, *J*(C,P) = 5.5, 7.3, C(4′)); 88.06 (*s*, C(3′), Ar$_2$*C*Ph); 111.90, 111.91 (2*s*, C(5)); 113.47, 113.50, 113.53 (3*d*, arom. C); 113.81, 117.64 (2*q*, *J*(C,P) = 288, CF$_3$); 117.52, 117.55 (2*s*, CN); 127.26, 127.29, 127.54, 127.57, 128.14, 128.17, 129.59, 129.63, 129.69, 129.71 (10*d*, arom. C); 135.04, 135.06, 135.55, 135.58 (4*d*, C(6), arom. C); 135.34, 135.42, 144.49, 144.52 (4*s*, arom. C); 150.29, 150.38 (2*s*, C(2)); 156.72 (*q*, *J*(C,P) = 37, COCF$_3$); 158.93, 158.95, 158.97 (3*s*, arom. C); 163.50 (*s*, C(4)). ^{31}P-NMR (81 MHz, CDCl$_3$): 141.6, 142.6.

(*3′S,5′R,6′R*)-*1-{(6′-Acetamido-2′-deoxy-5′-O-[(4,4′-dimethoxytriphenyl)methyl]-3′,5′-ethano-β-D-ribofuranosyl}thymine 3′-(2-Cyanoethyl Diisopropylphosphoramidite)* (**17**). As described for **16**, from **15** (213 mg, 0.34 mmol), ^{i}Pr$_2$NEt (235 µl, 1.37 mmol), and [P(OCH$_2$CH$_2$CN)(N^{i}Pr$_2$)]Cl (155 µl, 0.69 mmol) in MeCN (3 ml). FC (CH$_2$Cl$_2$/MeOH 50:1) gave **17** (176 mg, 62%; *ca.* 1:1 diastereoisomer mixture). White foam. TLC (CH$_2$Cl$_2$/MeOH 25:2): R_f 0.48, 0.65. ^{1}H-NMR (300 MHz, CDCl$_3$): 1.04–1.12 (*m*, 2 *Me*$_2$CH); 1.92, 1.93 (2*s*, Me–C(5)); 1.95, 1.96 (2*s*, 3 MeCO); 1.96–2.02 (*m*, 1 H, CH$_2$(2′) or CH$_2$(7′)); 2.34–2.39 (*m*, 2 H, CH$_2$(2′) or CH$_2$(7′)); 2.49–2.57 (*m*, OCH$_2$CH$_2$CN); 2.77–2.83 (*m*, 1 H, CH$_2$(2′) or CH$_2$(7′)); 3.41–3.75 (*m*, H–C(4′), H–C(5′), OCH$_2$CH$_2$CN, 2 Me$_2$*C*H); 3.77 (*s*, 2 MeO); 4.01–4.11 (*m*, H–C(6′)); 5.83–5.93 (*m*, H–C(1′)); 6.57, 6.71 (2*d*, *J* = 4.8, 5.3, CONH); 6.79–6.83 (*m*, 4 H arom.); 7.18–7.50 (*m*, H–C(6), 9 arom. H); 8.41 (br. *s*, H–N(3)). ^{31}P-NMR (81 MHz, CDCl$_3$): 141.8, 142.4. LSI-MS: 828 (2, [*M* + 1]$^+$), 303 (*100*). Anal. calc. for C$_{44}$H$_{54}$N$_5$O$_9$P: C 63.83, H 6.57, N 8.46; found: C 63.53, H 6.76, N 8.37.

Oligonucleotide Synthesis. Oligonucleotide synthesis was performed on a *Pharmacia Gene Assembler Special* connected to a *Compaq-Pro-Linea-3/25-zs* personal computer. All syntheses were performed using the 1.3-µmol cycle with coupling times of 6–9 min and detritylation times of 60–90 s per unnatural building block. Solvents and solns. were prepared according to the manufacturer's protocol. Phosphoramidite (0.1M in MeCN) and 1*H*-tetrazole (0.45M in MeCN) solns. were equal in concentration to those used for the synthesis of natural oligodeoxynucleotides. For the synthesis of **21** and **22**, the activator 1*H*-tetrazole was replaced by 5-

Table 2. *Synthesis and Analytical Data of Oligonucleotides* **18–22**

Sequence (1.3 µmol)	HPLC	Isolated yield OD(260 nm) ([%])	MALDI-TOF-MS [*M* – H]$^-$	
			m/z (calc.)	*m/z* (found)
18 d(TTTTT t$^+$TTTT)	DEAE^a): 20–35% *B* in 20 min; t_R 16.5 RP^b): 15–22% *B* in 30 min; t_R 11	37 (40)	3020.1	3018.6
19 d(TTTTt$^+$t$^+$TTTT)	DEAE^a): 20–35% *B* in 20 min; t_R 13 RP^b): 15–22% *B* in 30 min; t_R 9.5	15 (17)	3061.1	3059.7
20 d(t$^+$T t$^+$T t$^+$T t$^+$T t$^+$T)	DEAE^a): 10–18% *B* in 13 min; t_R 9 RP^b): 10–12% *B* in 30 min; t_R 11.8	31 (33)	3184.3	3185.9
21 d(TTTTTt^{Ac}TTTT)	DEAE^a): 25–45% *B* in 13 min; t_R 18.1 RP^b): 10–30% *B* in 30 min; t_R 12.5	45 (48)	3061.9	3061.1
22 d(t^{Ac}Tt^{Ac}Tt^{Ac}Tt^{Ac}Tt^{Ac}T)	DEAE^a): 25–55% *B* in 30 min; t_R 17.8 RP^b): 5–20% *B* in 30 min; t_R 21.5	48 (51)	3394.2	3394.1

^{a}) *Nucleogen* DEAE 60-7, 125 × 4.0 mm (*Macherey & Nagel*); *A*: 20 mM KH$_2$PO$_4$ in H$_2$O/MeCN 4:1, pH 6.0; *B*: *A* + 1M KCl; flow 1 ml/min; detection at 260 nm. ^{b}) *Aquapore Rp-300* 220 × 4.6 mm, 7 µm (*Brownlee Labs*); *A*: 0.1M (Et$_3$NH)OAc in H$_2$O, pH 7.0; *B*: 0.1M (Et$_3$NH)OAc in H$_2$O/MeCN 1:4, pH 7.0; flow 1 ml/min; detection at 260 nm.

(benzylthio)-1H-tetrazole (0.25M in MeCN) [24]. Average coupling yields, monitored by the on-line trityl assay, were in the range of 90–98% for **18** and **20–22**. For sequence **19**, the coupling step between the two adjacent trifluoroacetamido-bicyclo-thymidine residues proceeded with only 59% yield. All syntheses were run in the trityl-off mode.

  Deprotection and Purification of Oligonucleotides. Removal of the protecting groups and detachment from the solid support was effected in conc. NH_3 soln. (1–2 ml) at r.t. for 13–18 h. The crude oligomers were purified by DEAE ion-exchange HPLC, desalted over *Sep-Pak* (*Waters*), and their purity controlled by reversed-phase chromatography. *Table 2* contains synthetic and anal. data of the oligonucleotides described here. All natural DNA sequences used in this study were prepared according to standard CED- or PAC-phosphoramidite chemistry and purified by HPLC.

REFERENCES

[1] J. C. Litten, C. Leumann, *Helv. Chim. Acta* **1996**, *79*, 1129.
[2] E. Uhlmann, *Chemie in unserer Zeit* **1998**, *32*, 150.
[3] B. Hyrup, P. E. Nielsen, *Bioorg. Med. Chem.* **1996**, *4*, 5.
[4] C. M. Niemeyer, *Angew. Chem.* **1997**, *109*, 603; ibid., *Int. Ed. Engl.* **1997**, *36*, 585.
[5] M. Bolli, H. U. Trafelet, C. Leumann, *Nucleic Acids Res.* **1996**, *24*, 4660.
[6] M. Tarköy, M. Bolli, C. Leumann, *Helv. Chim. Acta* **1994**, *77*, 716.
[7] M. Bolli, P. Lubini, C. Leumann, *Helv. Chim. Acta* **1995**, *78*, 2077.
[8] M. Bolli, C. Leumann, *Angew. Chem.* **1995**, *34*, 694; ibid., *Int. Ed. Engl.* **1995**, *34*, 694.
[9] M. Bolli, C. Litten, R. Schütz, C. Leumann, *Chem. Biol.* **1996**, *3*, 197.
[10] S. W. Santoro, G. F. Joyce, *Proc. Natl. Acad. Sci. U.S.A.* **1997**, *94*, 4262.
[11] R. Steffens, C. J. Leumann, *J. Am. Chem. Soc.* **1999**, *121*, 3249.
[12] C. Heathcock, J. E. Kropp, M. Lorber, *J. Am. Chem. Soc.* **1970**, *92*, 1326.
[13] O. Mitsunobu, *Synthesis* **1981**, 1.
[14] R. Meier, Dissertation, Universität Bern, 1997.
[15] M. Tarköy, M. Bolli, B. Schweizer, C. Leumann, *Helv. Chim. Acta* **1993**, *76*, 481.
[16] H. Vorbrüggen, K. Krolikiewicz, B. Bennua, *Chem. Ber.* **1981**, *114*, 1234.
[17] H. Hashimoto, M. G. Nelson, C. Switzer, *J. Org. Chem.* **1993**, *58*, 4194.
[18] H. Hashimoto, M. G. Nelson, C. Switzer, *J. Am. Chem. Soc.* **1993**, *115*, 7128.
[19] R. L. Letsinger, C. N. Singman, G. Histand, M. Salunkhe, *J. Am. Chem. Soc.* **1988**, *110*, 4470.
[20] A. Eschenmoser, in 'Proceedings of the Robert A. Welch Foundation, 37th Conference on Chemical Research', Houston, Texas, 1993, p. 219.
[21] U. Pieles, W. Zürcher, M. Schär, H. E. Moser, *Nucleic Acids Res.* **1993**, *21*, 3191.
[22] D. B. Dess, J. C. Martin, *J. Am. Chem. Soc.* **1991**, *113*, 7277.
[23] G. M. Sheldrick, in 'Crystallographic Computing 3: Data Collection, Structure Determination and Databases', Eds. G. M. Sheldrick, C. Krueger, and R. Goddard, Oxford University Press, Oxford, 1985.
[24] X. Wu, S. Pitsch, *Nucleic Acids Res.* **1998**, *26*, 4315.

Oligonucleotides Containing Novel 4'-*C*- or 3'-*C*-(Aminoalkyl)-Branched Thymidines[1])

by **Henrik M. Pfundheller**[a]), **Torsten Bryld**[b]), **Carl E. Olsen**[c]), and **Jesper Wengel**[b])*

[a]) Department of Chemistry, University of Southern Denmark, Odense University, DK-5230 Odense M
[b]) Center for Synthetic Bioorganic Chemistry, Department of Chemistry, University of Copenhagen, Universitetsparken 5, DK-2100 Copenhagen
[c]) Department of Chemistry, Royal Veterinary and Agricultural University, DK-1871 Frederiksberg C

The synthesis of four novel 3'-*C*-branched and 4'-*C*-branched nucleosides and their transformation into the corresponding 3'-*O*-phosphoramidite building blocks for automated oligonucleotide synthesis is reported. The 4'-*C*-branched key intermediate **11** was synthesized by a convergent strategy and converted to its 2'-*O*-methyl and 2'-deoxy-2'-fluoro derivatives, leading to the preparation of novel oligonucleotide analogues containing 4'-*C*-(aminomethyl)-2'-*O*-methyl monomer **X** and 4'-*C*-(aminomethyl)-2'-deoxy-2'-fluoro monomer **Y** (*Schemes 2* and *3*). In general, increased binding affinity towards complementary single-stranded DNA and RNA was obtained with these analogues compared to the unmodified references (*Table 1*). The presence of monomer **X** or monomer **Y** in a 2'-*O*-methyl-RNA oligonucleotide had a negative effect on the binding affinity of the 2'-*O*-methyl-RNA oligonucleotide towards DNA and RNA. Starting from the 3'-*C*-allyl derivative **28**, 3'-*C*-(3-aminopropyl)-protected nucleosides and 3'-*O*-phosphoramidite derivatives were synthesized, leading to novel oligonucleotide analogues containing 3'-*C*-(3-aminopropyl)thymidine monomer **Z** or the corresponding 3'-*C*-(3-aminopropyl)-2'-*O*,5-dimethyluridine monomer **W** (*Schemes 4* and *5*). Incorporation of the 2'-deoxy monomer **Z** induced no significant changes in the binding affinity towards DNA but decreased binding affinity towards RNA, while the 2'-*O*-methyl monomer **Z** induced decreased binding affinity towards DNA as well as RNA complements (*Table 2*).

1. Introduction. – A large number of chemically modified oligonucleotides (ONs) have been synthesized and investigated to improve the binding affinity towards complementary RNA and the stability towards nucleases [1b][2]. As important examples, 2'-deoxy-2'-fluoro-modified ONs [3][4], and 2'-*O*-alkylated ONs, including 2'-*O*-methyl derivatives, have shown increased binding affinity towards RNA [5–8]. The increased binding affinity towards RNA has been attributed to the tendency of these nucleotides to adopt a C(3')-*endo* conformation, giving thermally stable A-type duplexes [9].

ONs containing 4'-*C*-substituted nucleotide monomers have likewise been reported to show, in general, minor decreases in binding affinity towards RNA and small increases in binding affinity towards DNA [10–15], presumably originating from the C(2')-*endo*-like conformation preferentially adopted by these nucleotides [16], in analogy with what is found in B-type DNA · DNA duplexes [9]. Furthermore, ONs containing 2'-*O*-alkylated or 4'-*C*-substituted nucleotides have shown increased

---

[1])   Part of this work has been published in a preliminary form [1a].

resistance towards enzymatic degradation, most pronounced for ONs containing modified nucleotides with an ionizable amino group in the sugar moiety, *e.g.*, 2'-*O*-(aminopropyl)- [17] and 4'-*C*-(aminoalkyl)-substituted thymidines [12][13][15]. Based on the results described above, we decided to evaluate the thermal stability of duplexes formed between ONs containing *2',4'-dimodified* nucleotide monomers and complementary RNA and DNA. Thus, in an attempt to maximize the binding affinity of ONs containing 4'-*C*-substituted nucleotides towards RNA and to increase the nuclease stability, we decided to synthesize and evaluate ONs containing 4'-*C*-(aminomethyl)-2'-*C*-methoxy- and 4'-*C*-(aminomethyl)-2'-fluoro-modified thymidines.

In addition, we and others have been interested in obtaining modified ONs with a DNA-selective binding, *e.g.* for diagnostics applications. Thus, 2'-*β*-ethynyl-modified ONs were recently shown to preferentially hybridize with complementary DNA [18] which was explained by the tendency of these modified nucleotides to adopt a C(2')-*endo*-type conformation. Likewise, we have reported the synthesis and evaluation of ONs containing 3'-*C*-(3-hydroxypropyl)- [19] and 3'-*C*-allyl-branched [20] thymidines showing better binding affinities towards DNA than towards RNA. However, this was not observed for ONs containing similar nucleotides with a C_1 3'-*C*-branch, *e.g.* 3'-*C*-(hydroxymethyl) [21], 3'-*C*-methyl [22][23], and 3'-*C*-(aminomethyl) [23]. A 3'-*C*-branch is expected to be oriented in a pseudoequatorial position, driving the sugar pucker towards a C(2')-*endo* conformation. An explanation for the results described above could be that a larger C_3 3'-*C*-branch is not as well accommodated in the DNA · RNA duplex as a smaller C_1 branch. Herein we report the synthesis and evaluation of ONs containing 3'-*C*-(3-aminopropyl)thymidine and 3'-*C*-(3-aminopropyl)-2'-*O*,5-dimethyluridine in an attempt to improve the thermal stability of duplexes formed between ONs containing 3'-*C*-branched monomers and complementary DNA.

2. Results and Discussion. – 2.1. *2'-Substituted 4'-C-(Aminomethyl) Nucleosides and ONs.* For the synthesis of 4'-*C*-(aminomethyl)-2'-*C*-methoxy- and 4'-*C*-(aminomethyl)-2'-fluoro-derivatized thymidine nucleosides, a strategy starting from the known 3-*O*-benzyl-4-*C*-[(benzyloxy)methyl]-1,2-di-*O*-isopropylidene-*β*-L-lyxofuranose (**2**), which can be synthesized from 3-*O*-benzyl-4-*C*-(hydroxymethyl)-1,2-di-*O*-isopropylidene-*β*-L-lyxofuranose (**1**) *via* regioselective benzylation [24][25], was chosen (*Scheme 1*). In an attempt to improve the regioselectivity of the monobenzylation, furanose **1** was benzylated by the dibutyltin oxide method [26][27]. However, equal amounts of two dibenzylated regioisomers were obtained in moderate yields, and the reported [24][25] regioselective monobenzylation method was therefore used. It was decided to use an azido substituent as a group which could be converted to an amino functionality later at an appropriate step in the synthesis. For the introduction of this azido group, furanose **2** was mesylated under standard conditions to give furanose **3** in 92% yield, as reported [28]. However, treatment of **3** with NaN_3 in hot DMF was unsuccessful resulting in complete recovery of **3** that might be ascribed to the 'neopentylic' character of the mesyloxy group of **3** and its position on the concave face of the oxabicyclo[3.3.0]octane system [24]. Instead, both by isopropylidene group hydrolysis in 80% $AcOH/H_2O$ followed by basic acetylation, and by direct one-pot acidic acetolysis [29], the 1,2-di-*O*-acetyl derivative **4** was obtained in 54% yield. Stereoselective reaction with silylated thymine [30][31] afforded in 75% yield the nucleoside **5**, which was heated under

Scheme 1

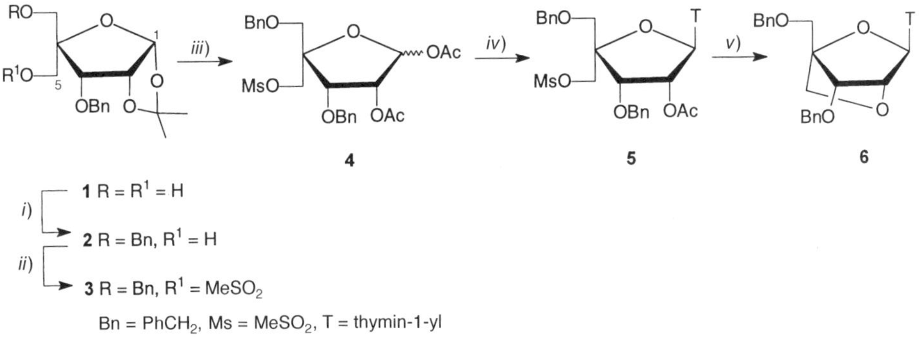

4   **5**   **6**

i) ⌐ **1** R = R^1 = H

ⅱ) └→ **2** R = Bn, R^1 = H

ⅱ) ⌐ **3** R = Bn, R^1 = MeSO$_2$

Bn = PhCH$_2$, Ms = MeSO$_2$, T = thymin-1-yl

i) BnBr, NaH, DMF. *ii*) MsCl, Py. *iii*) 1. 50% AcOH; 2. Ac$_2$O, Py *or* AcOH/Ac$_2$O/conc. H$_2$SO$_4$ soln. 100 : 10 : 0.1.
    iv) Thymine, *N,O*-bis(trimethylsilyl)acetamide (BSA), Me$_3$SiOSO$_2$CF$_3$, MeCN. *v*) LiEt$_3$BH, THF.

reflux in DMF with NaN$_3$ to give several products of which none was identified. This disappointing result is consistent with another report describing the reaction of 4′-*C*-[(mesyloxy)methyl]thymidine derivatives with LiN$_3$ to give the desired 4′-*C*-(azido-methyl)nucleoside in only low yields [32]. At this point, it was realized that it was impossible to use the mesyloxy group as a leaving group with azide as the nucleophile. Instead, we used hydride as the nucleophile in an attempt to obtain a 4′-*C*-methylnucleoside. However, treatment of nucleoside **5** with lithium triethylborohy-dride yielded the 2,5-dioxabicyclo[2.2.1]heptane locked nucleic acid (LNA) monomer **6** [25][33] in 86% yield, presumably *via* reduction of the acetate followed by intramolecular attack of the resulting alkoxide on the 'neopentylic' C-atom.

In an attempt to introduce the azido group while taking advantage of the excellent leaving-group ability of the triflate ion, the 4-*C*-(triflyloxy)methyl derivative **7** was synthesized using standard chemistry and was reacted with NaN$_3$ in DMF at 60° to give furanose **8** in 59% yield (two steps) (*Scheme 2*). Acidic hydrolysis and basic acetylation (→ **9**) followed by coupling with silylated thymine as described above (→ **10**) gave, after deacetylation, nucleoside **11** in 83% yield from **8**. To obtain the 2′-*O*-methyl derivative **12**, nucleoside **11** was chemoselectively *O*-methylated in 90% yield by the reaction conditions described by *Wang* and *Seifert* [12]. We have successfully performed methylation of 3′-*C*-allyl-3′,5′-di-*O*-benzyl-5-methyluridine under similar conditions [34]. The structure of **12** was verified by ^{1}H- and ^{13}C-NMR data (NH of thymine at δ 9.35 and CH$_3$O−C(2′) at δ 59). Reduction of the azido group with *Lindlar* catalyst (→ **13**) followed by trifluoroacetylation of the resulting amino group afforded nucleoside **14** in 73% yield from **12**. Debenzylation to give diol **15** in 92% yield followed by dimethoxytritylation with (MeO)$_2$TrCl to give the 5′-*O*-(4,4′-dimethoxy-trityl)-protected analogue **16** in 91% yield, and subsequent phosphitylation afforded the phosphoramidite derivative **17** in 43% yield.

To obtain the analogous 4′-*C*-(aminomethyl)-2′-deoxy-2′-fluoro phosphoramidite derivative **27**, the configuration at C(2′) of nucleoside **11** was inverted by the anhydro approach [35]. Thus, mesylation of L-*lyxo*-configured **11** followed by reaction with aq. base afforded the L-*xylo*-configured nucleoside **19** in 94% yield from **11** (*Scheme 3*). Several attempts to prepare the 2′-fluoronucleoside **21** directly from **19** with

Scheme 2

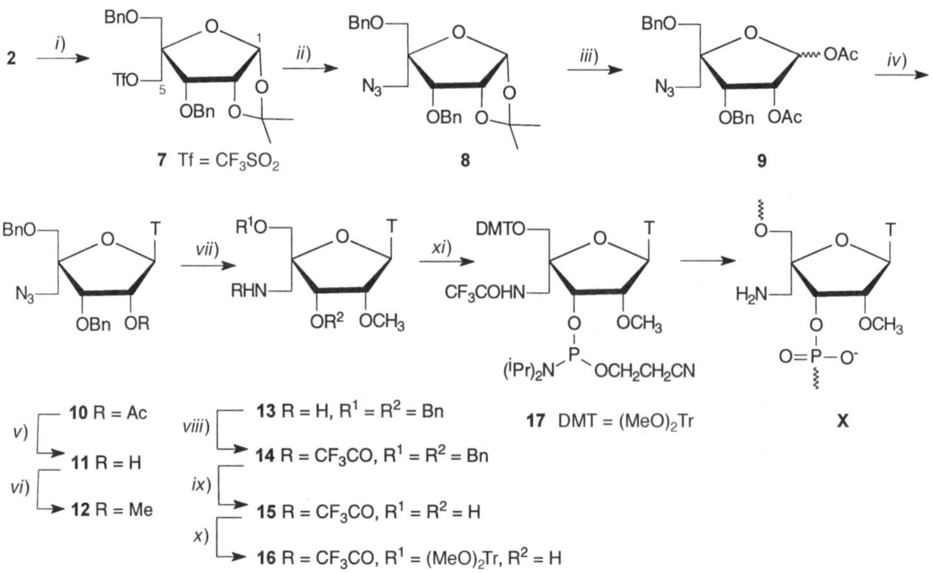

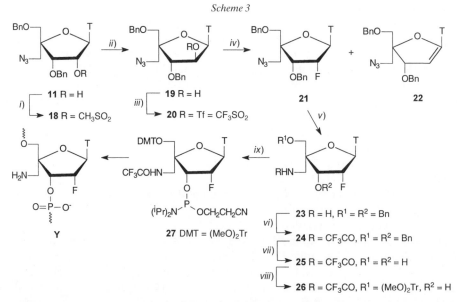

i) $(CF_3SO_2)_2O$, Py, CH_2Cl_2. ii) NaN_3, DMF. iii) 1. 50% AcOH; 2. Ac_2O, Py. iv) Thymine, BSA, $Me_3SiOSO_2CF_3$, MeCN. v) NH_3, MeOH. vi) MeI, NaH, THF. vii), H_2, *Lindlar* catalyst, EtOH. viii) CF_3COOEt, Et_3N, CH_2Cl_2. ix) H_2, 5% Pd/C, EtOH. x) $(MeO)_2TrCl$, Py. xi) $NC(CH_2)_2OP(Cl)N^iPr_2$, $^iPr_2EtN$, CH_2Cl_2.

Scheme 3

i) $MeSO_2Cl$, Py. ii) NaOH, EtOH, H_2O. iii) Tf_2O, 4-(dimethylamino)pyridine (DMAP), Py, CH_2Cl_2. iv) Bu_4NF, THF. v) H_2, *Lindlar* catalyst, dioxane, EtOH. vi) CF_3COOEt, NEt_3, CH_2Cl_2. vii) H_2, 5% Pd/C, EtOH. viii) $(MeO)_2TrCl$, Py. ix) $NC(CH_2)_2OP(Cl)N^iPr_2$, $^iPr_2EtN$, CH_2Cl_2.

(diethylamino)sulfur trifluoride (DAST) [36][37] and different temperatures and solvents (CH$_2$Cl$_2$ or DMF) were unsuccessful. An attempt to react tris(dimethylamino)sulfur (trimethylsilyl)difluoride (TASF) [38][39] with the (trifluoromethyl)sulfonyl derivative **20**, synthesized in 77% yield from **19** under standard conditions, was likewise unsuccessful. However, under strictly anhydrous conditions, 2'-fluoronucleoside **21** was obtained with dried 'anhydrous' tetrabutylammonium fluoride (Bu$_4$NF) [40][41]. Thus, nucleoside **20** was reacted with 10 equiv. of Bu$_4$NF in anh. THF to give **21** in 37% yield and ketene *N,O*-acetal **22** in 30% yield, the latter formed *via trans*-elimination of the 2'-*C*-(triflyloxy) group. The structure of **22** was verified by MS and NMR and by comparison with other structurally similar 1',2'-unsaturated compounds [42][43]. This type of elimination reaction has been reported earlier [44], but the analogous reaction of the 4'-*C*-unbranched 2'-*O*-(triflyloxy) derivative of **20** afforded exclusively the desired 2'-fluoro product in a very high yield [45]. The *gauche* effect of the 2'-*O*-(triflyloxy) group positioned at the α-face of the furanose ring should drive the sugar pucker towards a C(2')-*endo*-like conformation in both **20** and its 4'-*C*-unbranched analogue. However, also the 4'-*C*-(azidomethyl) group shifts the pentofuranose sugar pucker towards a C(2')-*endo* conformation. Accordingly, the $^3J(\mathrm{H-C(1')}, \mathrm{H}_\beta-\mathrm{C(2')})$ coupling constant of 4.3 Hz observed for **20**, corresponding to a H$-$C(1')$-$C(2')$-$H$_\beta$ torsion angle of *ca.* 45°, as obtained from a simplified *Karplus* equation [46], is in accordance with a C(2')-*endo*-like conformation. Thus, H$-$C(1') and the 2'-*C*-(triflyloxy) group exist in a *trans*-diaxial geometry advantageous for *E2* elimination, offering an explanation for the nearly equal amounts of substitution and elimination observed in the reaction. Following similar synthetic procedures as for the 2'-*O*-methylnucleosides (**12** → **16**, *Scheme 1*), 4'-*C*-(azidomethyl)-2'-deoxy-2'-fluoro derivative **21** was reduced to give the 4'-*C*-(aminomethyl)nucleoside **23**, trifluoroacetylated (→ **24**), debenzylated (→ **25**) and 5'-*O*-[(MeO)$_2$Tr]-protected to give **26** in an overall yield of 46% from **21**. Nucleoside **26** was further transformed into the 3'-*O*-phosphoramidite derivative **27** in 76% yield.

The phosphoramidites **17** and **27** were incorporated into ONs as monomers **X** and **Y**, respectively, on an automated DNA synthesizer. The stepwise coupling efficiency obtained for **17** was only 40% when 1*H*-tetrazole was used as activator (2 × 12 min coupling). With pyridinium hydrochloride as activator [47][48], the coupling efficiency was > 90% (10 min coupling). The ONs were deprotected by heating (55° for 12 h) in aqueous NH$_3$ solution followed by reversed-phase cartridge purification. Analysis by capillary gel electrophoresis verified the purity to be > 90%, and the composition of the ONs was confirmed by MALDI-MS. The phosphoramidite **17** was incorporated into three types of ONs: 14-mer oligothymidylates **II** – **IV**, mixed 9-mer sequences **VI** – **VIII**, and a sequence **XIII** containing 2'-*O*-methyl-modified ribonucleotides instead of unmodified 2'-deoxynucleotides. The hybridization properties of these **X**-modified oligonucleotides and of their reference strands **I**, **V**, and **XII** towards complementary DNA and RNA were measured both in a medium salt buffer and in a high salt buffer (*Table 1*).

Likewise, oligonucleotides **IX** – **XI** and **XIV** (*Table 1*), containing the monomer **Y**, were synthesized with a stepwise coupling efficiency for the amidite **27** of 80% with pyridinium hydrochloride as activator (10 min coupling). *Krug et al.* [49] have reported that when 2'-deoxy-2'-fluorouridine is incorporated as the 3'-terminal residue of an ON,

Table 1. *Melting Experiments of Oligonucleotides Containing Monomers* **X** *and* **Y**[a])

		Complementary ssDNA $T_m/°$ or $\Delta T_m/°$ per mod. [b])		Complementary ssRNA $T_m/°$ or $\Delta T_m/°$ per mod. [b])	
		[Na$^+$]: 150 mM[c])	[Na$^+$]: 750 mM[d])	[Na$^+$]: 150 mM[c])	[Na$^+$]: 750 mM[d])
I	5'-T$_{14}$	33.0	n.d.	31.0	n.d.
II	5'-T$_7$**X**T$_6$	-4.0	n.d.	-4.0	n.d.
III	5'-T$_6$**X**T**X**T$_5$	-3.0	n.d.	-3.0	n.d.
IV	5'-T$_{12}$**X**T	±0.0	n.d.	-1.0	n.d.
V	5'-G$_d$TG$_d$A$_d$A$_d$TA$_d$TG$_d$C$_d$	31.0	36.5	29.7	34.0
VI	5'-G$_d$TG$_d$A$_d$A$_d$**X**A$_d$TG$_d$C$_d$	$+1.5$	-0.5	$+2.5$	$+2.5$
VII	5'-G$_d$TG$_d$A$_d$A$_d$**X**A$_d$**X**G$_d$C$_d$	$+1.5$	-0.3	$+1.3$	$+1.3$
VIII	5'-G$_d$**X**G$_d$A$_d$A$_d$**X**A$_d$**X**G$_d$C$_d$	-0.3	-1.7	$+0.8$	$+1.0$
IX	5'-G$_d$TG$_d$A$_d$A$_d$**Y**A$_d$TG$_d$C$_d$	$+1.5$	-0.5	$+0.5$	±0.0
X	5'-G$_d$TG$_d$A$_d$A$_d$**Y**A$_d$**Y**G$_d$C$_d$	$+1.8$	±0.0	$+1.5$	$+0.8$
XI	5'-G$_d$**Y**G$_d$A$_d$A$_d$**Y**A$_d$**Y**G$_d$C$_d$	$+0.3$	-0.8	$+1.0$	$+0.7$
XII	5'-G^{m}U^{m}G^{m}A^{m}U^{m}A^{m}U^{m}G^{m}C	31.5	37.0	42.5	50.5
XIII	5'-G^m**X**G^{m}A^m**X**A^m**X**G^{m}C	24.0	28.0	34.5	40.5
XIV	5'-G^m**Y**G^{m}A^m**Y**A^m**Y**G^{m}C	19.0	23.0	29.0	37.0

[a]) **X** = 1-[4-C-(aminomethyl)-2-O-methyl-α-L-lyxofuranosyl]thymine unit; **Y** = 1-[4-C-(aminomethyl)-2-deoxy-2-fluoro-α-L-lyxofuranosyl)thymine unit, G^m = 2'-O-methylguanosine unit, U^m = 2'-O-methyluridine unit, A^m = 2'-O-methyladenosine unit. [b]) T_m is shown for the oligonucleotides **I**, **V**, and **XII–XIV**. ΔT_m is shown for oligonucleotides **II–IV** with **I** as reference and for oligonucleotides **VI–XI** with **V** as reference. T_m = melting temperature determined as the maximum of the first derivative of the absorbance *vs.* temperature curve; $\Delta T_m/°$ per mod. = change in T_m per modification compared to the unmodified reference strand; n.d. = not determined. [c]) Measured at 260 nm in medium salt buffer: 1 mM EDTA, 10 mM sodium phosphate, 140 mM sodium chloride, pH 7.1. [d]) Measured at 260 nm in high salt buffer: 1 mM EDTA, 10 mM sodium phosphate, 740 mM sodium chloride, pH 7.1.

it is converted to arabinosyluracil *via* internal nucleophilic attack from the pyrimidine nucleobase and ring opening of the 2,2'-anhydro intermediate during the deprotection step with hot aqueous NH$_3$ solution. However, when 2'-deoxy-2'-fluorouridine was incorporated at other positions in an ON, a similar conversion to arabinosyluracil was not observed which is consistent with results reported by others [50]. A similar kind of degradation was reported for oligophosphoramidates containing 2'-deoxy-2'-fluoro-uridine monomers [51]. Furthermore, *Kawasaki et al.* [3] reported a significant degradation of 2'-deoxy-2'-fluoropyrimidine nucleosides under the usual ON deprotection conditions. They suggested the alternative use of methanolic NH$_3$ solution for 24 h at room temperature for ON deprotection. However, when oligonucleotide **XI** was deprotected under the latter conditions, MALDI-MS showed that *ca.* 50% of the ON products still contained the trifluoroacetyl protecting group. Furthermore, the N^2-isobutyryl protecting group of the guanine bases was not removed. Substitution of methanolic NH$_3$ with aqueous NH$_3$ solution did induce removal of the trifluoroacetyl protecting group after 24 h, but only *ca.* 50% of the N^2-isobutyryl protecting groups. Therefore, the usual deprotecting conditions (aqueous NH$_3$ solution at 55° for 12 h) were used in the present study. In fact, MALDI-MS did not indicate any kind of degradation of the ONs, as further verified by capillary gel electrophoresis showing the ONs to be > 90% pure. The MALDI-MS data were determined with an experimental error of less than 0.4 Da with internal standards. The hybridization properties of the modified oligonucleotides **IX – XI** and **XIV** and of their reference strands **V** and **XII** towards complementary DNA and RNA were measured both in a medium salt buffer and in a high salt buffer (*Table 1*).

Upon single and two-fold incorporation of 2'-*O*-methyl monomer **X** in the middle of a 14-mer oligothymidylate (see ONs **II** and **III**, resp.), the stability of the duplexes with DNA and RNA was significantly decreased (ΔT_m/mod. $= -3$ and $-4°$) compared to the unmodified reference strand **I** when measured in a medium salt buffer, whereas the presence of a single **X** towards the 3'-end of sequence **IV** had no significant effect on the binding affinity towards complementary DNA and RNA. These results are consistent with results obtained with ONs containing 4'-*C*-(hydroxymethyl)uridine [52], although this analogue (see **U**) displayed even lower binding affinity towards complementary DNA (ΔT_m/mod. $= -8°$ for the sequence 5'-T$_7$UT$_6$). Due to the disappointing melting results, the hybridization properties of ONs **II – IV** were not evaluated in a high salt buffer.

U

More encouraging results were obtained when monomers **X** and **Y** were incorporated into 9-mer mixed oligonucleotide sequences (*Table 1*, **VI – XI**). Thus, the presence of a single **X** or **Y** (ONs **VI** and **IX**) or of two **X** or **Y** (ONs **VII** and **X**) in the 9-mer caused an increase (ΔT_m/mod. $\approx +1.5°$) in the stability of the resulting duplex with DNA when measured in a medium salt buffer. Incorporation of monomer **X** three times (ON **VIII**) caused a minor decrease (ΔT_m/mod. $= -0.3°$) in the thermal stability of the resulting duplex with DNA compared with the reference strand **V**, while the presence of **Y** in the same sequence (ON **XI**) caused a small increase (ΔT_m/mod. $= +0.3°$). When the corresponding ONs containing 2'-*O*-methyl derivative **X** (ONs **VI – VIII**) were hybridized with complementary RNA in a medium salt buffer, increases in the thermal stability of the resulting duplexes (ΔT_m/mod. $= +2.5, +1.3,$ and $+0.8°$) were observed. Similar results were obtained with ONs containing the 2'-deoxy-2'-fluoro derivative **Y** though the presence of **Y** in the singly modified ON **IX** caused a significant smaller increase in T_m compared to **VI** containing the 2'-*O*-methyl monomer **X**. In relation to the main purpose of this work, namely evaluation of the possibility of maximizing the thermal affinity of *C*-branched ONs, comparison with results obtained with the 4'-*C*-(hydroxymethyl)uridine monomer **U** appeared relevant. In the mixed-sequence 9-mer context, monomer **U** in general induced decreased T_m values compared to the reference ON **V** with both DNA and RNA complements (ΔT_m/mod. against DNA $= 0, -2.0$ and $-2.0°$ and ΔT_m/mod. against RNA $= +2.0, 0$ and $-0.3°$ for the 9-mer sequences containing one, two or three **U** monomers, respectively). It is thus evident that the combined exchange of the HO$-$C(2') and the $HOCH_2-C(4')$ groups with a $F-C(2')/MeO-C(2')$ group and a $NH_2CH_2-C(4')$ group leads to increased thermal stability of duplexes with both DNA and RNA complements. The apparent sequence dependence on the influence of the different modifications on the thermal stability of the resulting duplexes stresses the importance of comparing only in identical sequence contexts and excludes direct comparison of the effect induced by, *e.g.*, monomers **X** and **U** with effects reported in the literature for, *e.g.*, 2'-*O*-alkyl-substituted ONs [5–8] or 4'-*C*-(aminomethyl)-modified ONs [23].

For ONs **VI** – **VIII** containing the 2'-*O*-methyl monomer **X**, ΔT_m/mod. for duplexes with DNA as well as RNA was reduced with increasing number of **X** monomers incorporated (medium salt buffer). The same tendency, though less pronounced, could be seen for ONs **IX** – **XI** containing the 2'-deoxy-2'-fluoro monomer **Y** when hybridized with complementary DNA in a medium salt buffer. The 2'-*O*-methyl group and the 2'-fluoro substituent both direct the sugar pucker towards a $C(3')$-*endo* conformation (indicated by $^3J(H-C(1'), H-C(2'))$ values of *ca.* 1.5–4.5 Hz), which in general is expected to induce an increased thermal stability of duplexes with an RNA complement. However, compared to other 2'-*O*-methyl- and 2'-deoxy-2'-fluorouridines, 2'-modified 4'-*C*-(aminomethyl) nucleotides **X** and **Y** probably display a less pronounced tendency towards adopting a $C(3')$-*endo*-like conformation, as indicated by the relatively high $^3J(H-C(1'), H-C(2'))$ values. Thus, the increase in stability of the duplexes **VI**, **VII**, and **IX** – **XI** with DNA is, most likely, caused by a reduction in the electrostatic repulsion between phosphate moieties due to the presence of the positively charged aminomethyl group under the experimental conditions used. To confirm an effect of the protonated aminomethyl group, the duplex stabilities were measured under conditions of high ionic strength ($[Na^+] = 750$ mм, *Table 1*). As anticipated, the stabilities of the duplexes involving ONs **VI** – **VIII** containing monomer **X** or ONs **IX** – **XI** containing monomer **Y** and complementary DNA were decreased relative to the stability of the duplex involving the unmodified reference strands **V** under high ionic strength conditions. This confirms the positive effect of the protonated aminomethyl group regarding duplex stability with complementary DNA. However, for ONs **VI** and **VII** in experiments towards complementary RNA, the presence of the positively charged aminomethyl group does not seem to affect the duplex stability.

Due to the larger *gauche* effect of the 2'-F-atom compared to the 2'-*O*-methyl group, we anticipated that substitution of the 2'-*O*-methyl monomer **X** with the 2'-deoxy-2'-fluoro monomer **Y** would cause an increase in binding affinity towards RNA. However, comparable binding affinities towards RNA were obtained for ONs **VII** and **VIII** containing 2'-*O*-methyl monomer **X** and ONs **X** and **XI** containing 2'-deoxy-2'-fluoro monomer **Y** ($[Na^+] = 150$ mм). When the hybridization affinities were measured under conditions of high ionic strength, the increases in binding affinity towards RNA for ONs **IX** – **XI** containing **Y** were somewhat reduced compared to the corresponding oligonucleotides **VI** – **VIII** containing **X**.

We also evaluated the effect of the presence of the 4'-*C*-(aminomethyl) group in an otherwise uniformly modified 2'-*O*-methyl-RNA ON. For the 2'-*O*-methyl-RNA reference strand **XII** (*Table 1*), containing uracil as nucleobase instead of thymine, no change in the thermal stability towards a DNA complement was observed compared to the 2′-deoxy counterpart **V** confirming results reported earlier [5] [53]. However, when monomer **X** was incorporated three times in the 9-mer 2′-*O*-methyl-RNA strand (ON **XIII**), the thermal stability of the duplex with DNA was reduced significantly (ΔT_m value −7.5°). Thus, the positive effect of the 4′-*C*-(aminomethyl) substituent in the thymine monomer **X** observed earlier for ONs **VI–VIII** regarding binding affinity towards DNA is not seen with ON **XIII**. In fact, substitution of a thymine base with a uracil base generally results in a decrease in T_m of *ca.* 0.5° per substitution [54], which further emphasizes the detrimental effect of the 4′-*C*-(aminomethyl) group on the T_m value obtained for **XIII**. As expected, 2′-*O*-methyl-modified oligonucleotide **XII** showed increased binding affinity towards RNA ($\Delta T_m/2′$-*O*-methylnucleotide = +1.6°) compared to the 2′-deoxy analog **V**. ON **XIII** showed significantly reduced binding affinity towards RNA in both salt buffers compared to the reference strand **XII**, but still increased binding affinity towards RNA compared to the completely unmodified 2′-deoxy analog **V**. Likewise, the 2′-deoxy-2′-fluoro monomer **Y** was incorporated in an otherwise 2′-*O*-methyl modified ON (**XIV**, *Table 1*). The negative effect of the presence of the 2′-deoxy-2′-fluoro monomer **Y** on the thermal stability of the 2′-*O*-methyl-RNA strand was even more pronounced than observed for the similar oligonucleotide containing **X** with resulting ΔT_m values of *ca.* −13° for duplexes with complementary DNA and RNA. Based on these results, it appears that the incorporation of branched 4′-*C*-alkylnucleotide monomers in a 2′-*O*-methyl-RNA ON causes a detrimental disruption of the regular duplex structure.

2.2. *3′-C-(Aminoalkyl)-Substituted Nucleosides and ONs.* We have recently published the synthesis and evaluation of ONs containing 3′-*C*-(3-hydroxypropyl)thymidine [19]. We have also been interested in evaluating ONs containing the corresponding amino derivative, 3′-*C*-(3-aminopropyl)thymidine, which could potentially increase the stability of the DNA · DNA and DNA · RNA duplexes by lowering the electrostatic repulsion between the two anionic strands. Thus, 3′-*C*-(hydroxypropyl)nucleoside **29** was synthesized *via* hydroboration/oxidation of the 3′-*C*-allyl derivative **28** as reported earlier [19] (*Scheme 4*). Standard *Mitsunobu* reaction [55] afforded the phthaloyl(Phth)-protected primary-amino-substituted nucleoside derivative **30** contaminated with triphenylphosphine oxide. Several attempts to remove the phosphine oxide by-product, *e.g.*, trituration with toluene, precipitation in Et$_2$O and extensive column chromatography, failed. Impure **30** was, therefore, thioacylated to give 2′-*O*-[(pentafluorophenoxy)thiocarbonyl]nucleoside **31** in 27% yield (two steps). Deoxygenation afforded the 2′-deoxynucleoside **32** in disappointing 25% yield (after extensive column-chromatographic purification and prep. TLC to remove tin impurities) and nucleoside **30** in 19% yield. Removal of the benzyl protecting groups of **32** by hydrogenation afforded the debenzylated diol **33** in only 18% yield. TLC showed the formation of several products upon addition of additional Pd(OH)$_2$/C during the reaction. Nevertheless, the small amount of diol **33** was 5′-*O*-(MeO)$_2$Tr-protected to give nucleoside **34**, which was subsequently phosphitylated to give the 3′-*O*-phosphoramidite building block **35** [20] in 23% yield (two steps).

The corresponding 3′-*C*-(3-aminopropyl)-2′-*O*-methylnucleoside derivative **42** was likewise synthesized and evaluated as a monomeric substitution in ONs. As the first step, by the same procedure as for the methylation of **11**, nucleoside **28** was methylated

Scheme 4

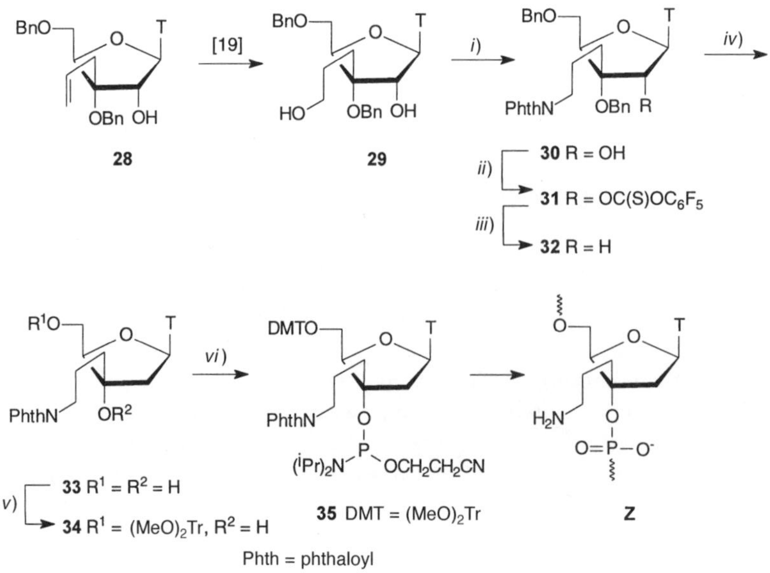

Phth = phthaloyl

i) Phthalimide, PPh$_3$, diethyl diazenedicarboxylate (DEAD), THF. *ii*) C$_6$F$_5$OC(S)Cl, DMAP, CH$_2$Cl$_2$. *iii*) 2,2'-azobis[isobutyronitrile] (AIBN), Bu$_3$SnH, benzene. *iv*) H$_2$, 20% Pd(OH)$_2$/C, EtOH, dioxane. *v*) (MeO)$_2$TrCl, Py. *vi*) NC(CH$_2$)$_2$OP(Cl)N^{i}Pr$_2$, ^{i}Pr$_2$EtN, CH$_2$Cl$_2$.

to give the 2'-*O*-methyl derivative **36** in 84% yield as reported earlier [34] (*Scheme 5*). Nucleoside **36** was subjected to a hydroboration/oxidation reaction sequence to afford the desired primary alcohol **37** in 54% yield. Furthermore, we isolated a mixture of the *Markovnikov* diastereoisomers **38** in 13% yield. *Mitsunobu* reaction of **37** with phthalimide [55] afforded the phthaloyl-protected derivative **39** in 20% yield in addition to a fraction contaminated with triphenylphosphine oxide. Subsequent debenzylation with BCl$_3$ afforded diol **40** whereupon derivative **41** was obtained by standard 5'-*O*-(MeO)$_2$Tr protection in 65% yield (two steps). Eventual phosphitylation yielded the 3'-*O*-phosphoramidite derivative **42** in 33% yield after extensive column-chromatographic purification.

ONs containing 3'-*C*-(3-aminopropyl)thymidine monomer **Z** were synthesized from the phosphoramidite **35**. With pyridinium hydrochloride as activator and *t*-BuOOH [56] [57] as oxidizing agent, **35** was coupled in >95% stepwise yield (10 min coupling time). Standard deprotection with aq. NH$_3$ solution at 55° and cartridge purification afforded the ONs **XVI** and **XIX** (*Table 2*). The purity (>85%) and composition of the ONs was verified as described earlier for the ONs containing monomers **X** and **Y**. To avoid formation of ONs containing partially ring-opened phthaloyl functionalities under standard deprotection conditions, the use of methyl-amine for deprotection has been reported [17]. However, according to MALDI-MS this was not a problem for ONs **XVI** and **XIX**. Instead, a minor peak at $[M + 54]^+$ in the MALDI-MS of oligomer **XVI** was seen, presumably originating from *Michael* addition of the amino group to acrylonitrile during deprotection. Similarly, ONs **XVII** and **XX**

Scheme 5

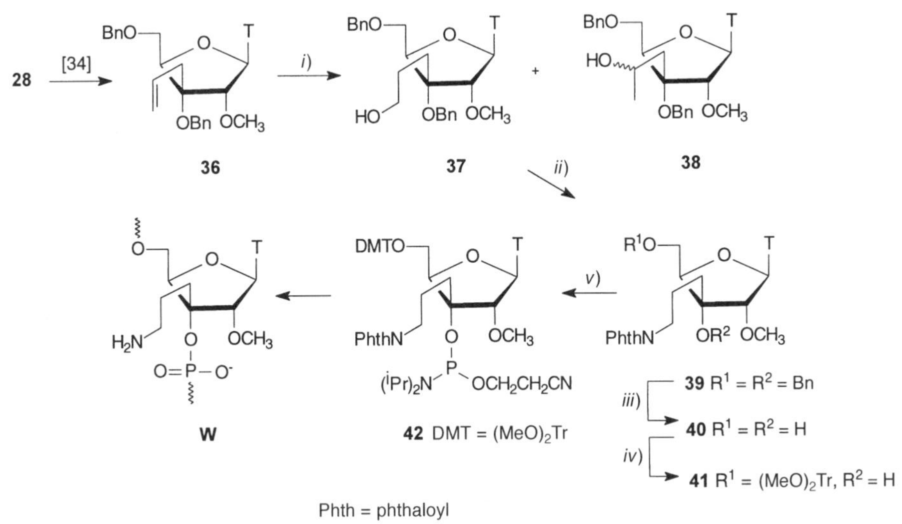

Phth = phthaloyl

i) 1. BH$_3$ in 1,4-oxathiane, THF; 2. aq. NaOH soln., H$_2$O$_2$. *ii*) Phthalimide, PPh$_3$, DEAD, THF. *iii*) BCl$_3$ in hexane, CH$_2$Cl$_2$. *iv*) (MeO)$_2$TrCl, Py. *v*) NC(CH$_2$)$_2$OP(Cl)N(^{i}Pr)$_2$, ^{i}Pr$_2$EtN, CH$_2$Cl$_2$.

containing monomer **W** were synthesized with a stepwise coupling efficiency for the phosphoramidite **42** of > 80% (10 min coupling) with pyridinium hydrochloride as activator and *t*-BuOOH as oxidizing agent. Deprotection, purification, capillary gel electrophoresis, and MALDI-MS were performed as described earlier to yield ONs **XVII** and **XX** with a purity of > 85%.

The hybridization of the modified ONs **XVI** and **XIX** containing the 3′-*C*-(3-aminopropyl) nucleoside monomer **Z** were evaluated in comparison to the reference strands **XV** and **XVIII**, respectively, in a medium salt buffer, and for duplexes involving a DNA complement furthermore in a high salt buffer (*Table 2*). Incorporation of 3′-*C*-

Table 2. *Melting Experiments of Oligonucleotides Containing Monomers* **Z** *and* **W**[a])

		Complementary ssDNA [Na$^+$] = 150 mM[b])		[Na$^+$] = 750 mM[c])		Complementary ssRNA [Na$^+$] = 150 mM[b])	
		$T_m/°$[d])	$\Delta T_m/°$ per mod.[d])	$T_m/°$[d])	$\Delta T_m/°$ per mod.[d])	$T_m/°$[d])	$\Delta T_m/°$ per mod.[d])
XV	5′-T$_{14}$	33.0	reference	42.5	reference	31.0	reference
XVI	5′-T$_7$**Z**T$_6$	33.0	±0.0	41.5	−1.0	28.5	−2.5
XVII	5′-T$_7$**W**T$_6$	23.5	−9.5	n.d.	n.d.	n.d.	n.d.
XVIII	5′-G$_d$TG$_d$A$_d$TA$_d$TG$_d$C$_d$	31.0	reference	36.5	reference	29.5	reference
XIX	5′-G$_d$TG$_d$A$_d$**Z**A$_d$TG$_d$C$_d$	31.5	+0.5	36.5	±0.0	25.5	−4.0
XX	5′-G$_d$TG$_d$A$_d$**W**A$_d$TG$_d$C$_d$	27.5	−3.5	n.d.	n.d.	26.0	−3.5

[a]) **Z** = 3′-*C*-(3-aminopropyl)thymidine unit; **W** = 3′-*C*-(3-aminopropyl)-2′-*O*-methyl-5-methyluridine unit.
[b]) Measured at 260 nm in medium salt buffer: 1 mM EDTA, 10 mM sodium phosphate, 140 mM sodium chloride, pH 7.1.   [c]) Measured at 260 nm in high salt buffer: 1 mM EDTA, 10 mM sodium phosphate, 740 mM sodium chloride, pH 7.1.   [d]) T_m = melting temperature determined as the maximum of the first derivative of the absorbance *vs.* temperature curve; $\Delta T_m/°$ per mod. = change in T_m per modification compared to the unmodified reference strand; n.d. = not determined.

(3-aminopropyl)thymidine monomer **Z** in the middle of an oligothymidylate (ON **XVI**) or in the middle of a mixed sequence 9-mer (ON **XIX**) induced no significant change in the thermal stability involving a DNA, while the presence of **Z** had a destabilizing effect on the duplexes involving complementary RNA (ΔT_m/mod. $= -2.5$ and $-4.0°$) when measured in a medium salt buffer. The hybridization results with DNA are in contrast to the results obtained with ONs containing 3'-*C*-(3-hydroxypropyl)thymidine [19] which showed small decreases in binding affinity towards DNA. An interaction between the positively charged amino group of **Z** and the anionic phosphate backbone could be an immediate explanation of the better hybridization results towards DNA obtained for ONs containing the 3'-*C*-(aminopropyl) branch. We, therefore, measured the T_m values of ONs **XVI** and **XIX** in a buffer with high salt concentration (*Table 2*). As the resulting changes in the T_m values for ONs **XVI** and **XIX** ($+8.5$ and $+5.0°$, resp.) are slightly smaller than for the reference strands **XV** and **XIII** ($+9.5$ and $+5.5°$, resp.), an electrostatic interaction between the amino group in **Z** and the phosphate backbone appears likely. Compared with the few reports describing ONs containing 3'-*C*-branched monomers, ONs containing the 3'-*C*-(aminopropyl)-branched monomer **Z** show quite promising hybridization properties towards DNA. Thus, in general, ONs containing 3'-*C*-(hydroxymethyl)- [21], 3'-*C*-methyl- [22][23], 3'-*C*-(aminomethyl)- [23], and 3'-*C*-allylnucleotide [20] monomers hybridize with complementary DNA with ΔT_m/mod. values of approximately 0 to $-1.0°$ under medium salt conditions.

The significant decrease in binding affinity of mid-modified ONs containing monomer **Z** towards RNA compared to DNA is noticeable (*Table 2*). Similar results were observed with ONs containing 3'-*C*-allylthymidine [20] or 3'-*C*-(3-hydroxypropyl)thymidine [19] and can be explained by the conformation of the sugar moiety. Thus, the presence of a 3'-β-substituent on the pentofuranose ring shifts the sugar pucker towards a C(2')-*endo* type conformation through its preference to adopt a pseudo-equatorial orientation [58]. For monomer **Z**, coupling constants between $H-C(1')$ and $H_\alpha-C(2')/H_\beta-C(2')$ are *ca.* 5.5 and *ca.* 9.5 Hz, respectively, as seen, *e.g.*, for **34**, confirming the C(2')-*endo*-type conformation, as these coupling constants correspond to a $H-C(1')-C(2')-H_\alpha$ torsion angle of *ca.* 41° and a $H-C(1')-C(2')-H_\beta$ torsion angle close to 160° according to a simplified *Karplus* equation [46]. The presence of the monomer **Z** with C(2')-*endo* conformation in the otherwise unmodified DNA strand might cause a rather significant distortion of the structure of the DNA·RNA duplex. For ONs containing 3'-*C*-(hydroxymethyl)- [21][59], 3'-*C*-methyl- [22][23], and 3'-*C*-(aminomethyl)thymidine [23] monomers, smaller decreases in ΔT_m/mod. values were in general observed for duplexes towards complementary RNA. The C_3-alkyl branch in **Z**, 3'-*C*-(3-hydroxypropyl)thymidine [19], and 3'-*C*-allylthymidine [20] monomers might be involved in some unfavorable steric interactions which are less pronounced for the modified monomers with only one C-atom in the 3'-*C*-branch. Thus, apparently a more selective DNA binding is observed when the 3'-*C*-branch contains three C-atoms. However, the limited amount of experimental data should be kept in mind before any general conclusions are drawn.

The hybridization properties of the modified oligonucleotides **XVII** and **XX**, containing monomer **W**, towards complementary DNA and RNA measured in a medium salt buffer are depicted in *Table 2*. Incorporation of a single 3'-*C*-(3-

aminopropyl)-2′-*O*,5-dimethyluridine monomer **W** in a 14-mer oligothymidylate induces a major decrease in the thermal stability of the duplex formed with complementary DNA with a ΔT_m/mod. value of $-9.5°$. More encouraging results were obtained when a single **W** was incorporated in the mixed 9-mer ON **XX**, inducing a ΔT_m/mod. value of $-3.5°$ towards both complementary DNA and RNA. The binding affinities towards complementary DNA obtained for **XVII** and **XX** are in a striking contrast to the results obtained with the 2′-deoxy analogue **Z**. However, ONs containing 3′-*C*-allyl-2′-*O*,5-dimethyluridine [34], 3′-*C*,2′-*O*,5-trimethyluridine [22], the conformationally restricted 2′-*O*-methyl-6,3′-ethanouridine [60], and 2′-*O*-methyl-3′-*C*,2′-*O*-linked bicyclic nucleosides [34] displayed similar pronounced decreases in the thermal stability of duplexes towards DNA and RNA.

Despite the finding that the 2′-*O*-methyl modification in general has shown RNA-affinity-enhancing properties, probably by shifting the pentofuranose pucker towards the C(3′)-*endo* conformation due to the *gauche* effect, the presence of the 2′-*O*-methyl modification in monomer **W** does not lead to an increased affinity of the oligonucleotides towards RNA compared to DNA. An explanation might be found in the conformation of the sugar ring since the coupling constant $^3J(\text{H}-\text{C}(1′),\text{H}-\text{C}(2′))$ for both monomers (as seen, *e.g.*, for nucleoside **41**) is *ca.* 8 Hz, thus indicating a C(2′)-*endo* conformation of the pentofuranose ring and additionally highlighting the dominating effect of the 3′-*β*-branch on the sugar conformation. Consequently, the 2′-*O*-methyl group is oriented in a pseudoequatorial position, probably causing a significant distortion of the structure of the duplexes with DNA as well as RNA, as seen in molecular-modeling studies of oligonucleotides containing 3′-*C*-allyl-2′-*O*,5-dimethyl-uridine [34].

3. Conclusion. – 4′-*C*-(Aminomethyl)-2′-deoxy-2′-fluoro- and 4′-*C*-(aminomethyl)-2′-*O*-methylnucleosides have been efficiently synthesized by a convergent approach involving substitution of a 2′-[(methylsulfonyl)oxy] group or chemoselective 2′-*O*-methylation, respectively. The corresponding 3′-*O*-phosphoramidite derivatives were used in syntheses of partly modified 14-mer oligothymidylates and 9-mer mixed sequences by means of an automated DNA synthesizer. Compared with the corresponding 4′-*C*-branched 2′-deoxy derivatives, increased binding affinities towards complementary DNA and RNA were in general obtained, and a positive effect of the positively charged aminomethyl group was indicated at a physiologically relevant ionic strength. The introduction of the 2′-fluoro or 2′-methoxy substituents induces a conformational shift towards a C(3′)-*endo*-type furanose conformation. However, no additional effect was observed after incorporation of the 2′-substituted 4′-*C*-(amino-methyl) monomers in a 9-mer 2′-*O*-methyl-RNA strand. On the contrary, significantly reduced binding affinities resulted, indicating an unfavorable steric interaction involving the 4′-*C*-(aminomethyl) branch. The 3′-*C*-(3-aminopropyl)-2′-deoxy- and 4′-*C*-(aminomethyl)-2′-*O*-methylnucleosides were likewise synthesized by a convergent approach involving either free-radical deoxygenation at C(2′) or chemoselective 2′-*O*-methylation, respectively. The corresponding 3′-*O*-phosphoramidite derivatives were used for syntheses of partly modified 14-mer oligothymidylates and 9-mer mixed sequences. Compared with the corresponding unmodified reference ONs, the 2′-unsubstituted and 2′-substituted 3′-*C*-(3-aminopropyl)thymidine monomers induce unchanged or decreased thermal stabilities of duplexes towards complementary DNA and decreased thermal stabilities towards RNA. Thus, one of the aims of the present research, namely to develop oligonucleotide analogues displaying DNA-selective hybridization, has been accomplished, but neither here nor earlier has it been possible to obtain significantly increased binding affinities for 3'-C-branched ONs. Noteworthy are, however, the affinity-enhancing properties of the 2'-substituted 4'-C-(amino-methyl)thymidine monomers reported herein. The somewhat troublesome synthetic routes devised for the relevant nucleosides and the nonideal oligomerizations of the corresponding phosphoramidite derivatives on automated synthesizers have so far precluded the evaluation of fully modified 3'-C- and 4'-C-branched ON derivatives. However, 4'-C- and 3'-C-branched ON analogues are likely to find utility as monomeric constituents, e.g., conjugation sites, in antisense oligonucleotides.

Ms. *Britta M. Dahl* is thanked for oligonucleotide synthesis. *The Danish Natural Science Research Council* and *The Danish Technical Research Council* are thanked for financial support.

Experimental Part

General. Reactions were conducted under Ar when anh. solvents were used. Petroleum ether was of distillation range 60–80°. On workup, org. phases were dried (Na_2SO_4) and filtered. Column chromatography (CC): glass columns; silica gel *60* (0.040–0.063 mm). NMR spectra: chemical shift values δ in ppm rel. to $SiMe_4$ as internal reference (^{1}H and ^{13}C) and rel. to 85% H_3PO_4 soln. as external reference (^{31}P). MS: in m/z. Microanalyses were performed at The Microanalytical Laboratory, Department of Chemistry, University of Copenhagen.

3-O-Benzyl-4-C-[(benzyloxy)methyl]-1,2-di-O-isopropylidene-β-L-lyxofuranose (**2**). To a soln. of **1** (570 mg, 1.83 mmol) in anh. MeOH (50 ml), dibutyltin oxide (478 mg, 1.92 mmol) was added. The mixture was heated under reflux for 1 h, and the resulting clear soln. was cooled to r.t. and evaporated to give a yellow/white solid material which was redissolved in anh. toluene (25 ml). BnBr (0.65 ml, 5.5 mmol) was added to the mixture followed by Bu_4NBr (295 mg, 0.92 mmol), and the mixture was heated under reflux for 15 h. Additional Bu_4NBr (325 mg, 1.01 mmol) was added followed by BnBr (0.65 ml, 5.5 mmol), but no further conversion was observed according to anal. TLC. After additional heating under reflux for 24 h, the mixture was cooled to r.t. CH_2Cl_2 (30 ml) was added followed by a sat. aq. $NaHCO_3$ soln. (30 ml). The resulting precipitate was filtered off and the separated org. phase washed with sat. aq. NaCl soln. (3×20 ml), dried (Na_2SO_4), and evaporated. Purification by CC (10–50% AcOEt/petroleum ether, then 5% $MeOH/CH_2Cl_2$) afforded **2** (139 mg, 19%), another dibenzyl derivative (151 mg, 21%), and a tribenzyl derivative (206 mg, 23%) as oils, in addition to starting material **1** (*ca.* 128 mg, *ca.* 23% (contaminated with Sn-alkyls)) as a yellowish solid material. **2**: ^{1}H- and ^{13}C-NMR: in accordance with [24].

1,2-Di-O-acetyl-3-O-benzyl-4-C-[(benzyloxy)methyl]-5-O-(methylsulfonyl)-L-lyxofuranose (**4**). *Method A:* Furanose **3** [28] (630 mg, 1.32 mmol) was heated in 50% aq. AcOH soln. (5.5 ml) at 90° for 3 h and subsequently cooled to r.t. Evaporation followed by co-evaporation with abs. EtOH, toluene, and pyridine afforded an oil which was redissolved in anh. pyridine (3 ml). Ac_2O (2 ml) was added and the mixture stirred under Ar for 12 h at r.t. The mixture was poured into ice-cold H_2O (10 ml), and sat. aq. $NaHCO_3$ soln. (10 ml) was added followed by AcOEt (20 ml). The org. phase was washed several times with sat. aq. $NaHCO_3$ soln. The aq. phase was extracted with AcOEt (2×10 ml) and the combined org. phase dried (Na_2SO_4) and evaporated. Purification by CC (0–2% $MeOH/CH_2Cl_2$) afforded **4** (α/β-L 3:1; 370 mg, 54%).

Method B: To a soln. of **3** [28] (118 mg, 0.25 mmol) in AcOH (2 ml) and Ac_2O (0.2 ml), conc. H_2SO_4 (2 µl) was added. The soln. was stirred at r.t. for 12 h, poured into ice-cold H_2O (5 ml), and extracted with CH_2Cl_2 ($3 \times$ 10 ml). The org. phase was washed successively with sat. aq. $NaHCO_3$ soln. (3×10 ml) and sat. aq. NaCl soln. (10 ml), dried ($MgSO_4$), and evaporated. Purification by CC (0–2% $MeOH/CH_2Cl_2$) afforded furanose **4** (α/β-L 4:1) (70 mg, 54%). Oil. ^{1}H-NMR ($CDCl_3$): 7.37–7.21 (*m*, Bn); 6.38 (*d, J* = 4.7, H−C(1)(β-L)); 6.14 (*s*, H−C(1) (α-L)); 5.35 (*d, J* = 5.0, H−C(2)(α-L)); 5.17 (*dd, J* = 4.6, 6.3, H−C(2)(β-L)); 4.69−4.35 (*m*, Bn, H−C(3)); 4.25 (*d, J* = 11.3, Bn); 3.64, 3.52, 3.45, 3.39 (*4d, J* = 9.7, 10.0, 9.7, 10.0, 2 H−C(5), CH_2−C(4)); 2.98, 2.95 (*2s*, $2MeSO_2$), 2.13, 2.06, 2.04, 1.90 (*4s*, 4 MeCO). ^{13}C-NMR ($CDCl_3$): 170.1, 170.0, 169.6, 169.0 (4 C=O); 137.7,

137.0, 128.6, 128.5, 128.2, 128.1, 128.0, 127.9, 127.8 (Bn); 97.5 (C(1)(α-L)); 94.2 (C(1)(β-L)); 86.3 (C(4)(β-L)); 84.3 (C(4)(α-L)); 79.1, 74.5, 74.1, 73.7, 73.5, 71.6, 70.6, 70.2, 69.8 (C(2), C(3), C(5), CH_2-C(4), Bn); 37.6, 37.4 (2 $MeSO_2$); 21.0, 20.7, 20.6, 20.4 (4 MeCO). FAB-MS: 545 ($[M + Na]^+$). Anal. calc. for $C_{25}H_{30}O_{10}S$: C 57.46, H 5.79; found: C 57.72, H 5.74.

1-(2-O-Acetyl-3-O-benzyl-4-C-[(benzyloxy)methyl]-5-O-(methylsulfonyl)-α-L-lyxofuranosyl)thymine (**5**). To a soln. of **4** (460 mg, 0.88 mmol) in anh. MeCN (2 ml), thymine (0.20 g, 1.59 mmol) was added followed by *N,O*-bis(trimethylsilyl)acetamide (BSA; 1.31 ml, 5.30 mmol). The mixture was heated under reflux until the soln. became clear (5 min). After cooling to 0°, $Me_3SiOSO_2CF_3$ (Me_3SiOTf; 0.27 ml, 1.50 mmol) was added dropwise, and the soln. was stirred for 12 h at r.t. As anal. TLC showed that the reaction was not complete, identical amounts as specified of more thymine and BSA were further added, and the soln. was heated under reflux for 15 min, whereupon Me_3SiOTf (0.54 ml, 3.0 mmol) was added dropwise at 0°. After stirring at r.t. for 1.5 h, sat. aq. $NaHCO_3$ soln. (10 ml) was added, the mixture extracted with CH_2Cl_2 (3 × 20 ml), and the org. phase washed with sat. aq. $NaHCO_3$ soln. (2 × 10 ml), dried (Na_2SO_4), and evaporated. Purification by CC (0–2% $MeOH/CH_2Cl_2$) afforded **5** (386 mg, 75%). White solid material. ^{1}H-NMR ($CDCl_3$): 8.92 (br. *s*, NH); 7.40–7.26 (*m*, 11 H, H–C(6), Bn); 6.14 (*d*, *J* = 5.0, H–C(1')); 5.48 (*t*, *J* = 5.4, H–C(2')); 4.62 (*d*, *J* = 11.1, 1 H, Bn); 4.58 (*m*, 5 H, H–C(3'), 2 H–C(5'), Bn); 4.27 (*d*, *J* = 11.3, 1 H, Bn); 3.89 (*d*, *J* = 10.1, 1 H, CH_2–C(4')); 3.59 (*d*, *J* = 10.2, 1 H, CH_2–C(4')); 2.92 (*s*, $MeSO_2$); 2.10 (*s*, COMe); 1.59 (*d*, *J* = 0.5, Me). ^{13}C-NMR ($CDCl_3$): 170.2 (C=O); 163.6 (C(4)); 150.4 (C(2)); 137.0, 136.9, 136.1, 128.9, 128.8, 128.7, 128.5, 128.3, 128.2, 127.8 (C(6), Bn); 111.6 (C(5)); 87.8, 85.5 (C(1'), C(4')); 77.8, 74.8, 74.4, 73.8, 70.8, 69.0 (C(2'), C(3'), C(5'), CH_2–C(4'), Bn); 37.2 ($MeSO_2$); 20.6 (*Me*CO); 11.9 (Me). FAB-MS: 589 ($[M + H]^+$). Anal. calc. for $C_{28}H_{32}N_2O_{10}S$: C 57.13, H 5.48, N 4.76; found: C 57.17, H 5.38, N 4.36.

(1S,3R,4R,7S)-7-(Benzyloxy)-1-[(benzyloxy)methyl]-3-(thymin-1-yl)-2,5-dioxabicyclo[2.2.1]heptane (**6**) [25]. A soln. of **5** (73 mg, 0.124 mmol) in anh. THF (3 ml) was cooled to 0°. Then, 1.0M $LiEt_3BH$ in THF (0.25 ml, 0.25 mmol) was added dropwise, and the mixture was stirred at r.t. for 3 h and then cooled to 0°. An identical amount as specified above of $LiEt_3BH$ was additionally added dropwise, and the mixture was stirred for 12 h at r.t., and cooled to 0°. The reaction was quenched by slow addition of 35% aq. H_2O_2 soln./2N NaOH 1:1 (0.5 ml). The mixture was diluted with CH_2Cl_2 (10 ml) and washed with sat. aq. NaCl soln. (3 × 10 ml). The org. phase was dried (Na_2SO_4) and evaporated. Purification by CC (1% $MeOH/CH_2Cl_2$) afforded **6** (48 mg, 86%). White solid material. ^{1}H-NMR, ^{13}C-NMR, and FAB-MS: in accordance with [25].

3-O-Benzyl-4-C-[(benzyloxy)methyl]-5-O-[(trifluoromethyl)sulfonyl]-1,2-O-isopropylidene-β-L-lyxofuranose (**7**). To a soln. of **2** (6.34 g, 15.8 mmol) in anh. CH_2Cl_2 (60 ml), anh. pyridine (3.83 ml, 47.5 mmol) was added. The mixture was cooled to −50° and Tf_2O (4.15 ml, 25.3 mmol) added dropwise. The mixture was stirred for 1 h, then allowed to warm up to 0°, and stirred for 1 h, whereupon the reaction was quenched by slow addition of sat. aq. $NaHCO_3$ soln. (15 ml). Additional CH_2Cl_2 (40 ml) and sat. aq. $NaHCO_3$ soln. (50 ml) were added. The org. phase was successively washed with ice-cold 1M HCl (2 × 75 ml) and sat. aq. $NaHCO_3$ soln. (75 ml). The aq. phases were separately extracted with CH_2Cl_2 (25 ml each), and the combined org. phase was dried (Na_2SO_4) and evaporated: **7** as a white solid material which was used directly in the next step without further purification. ^{1}H-NMR ($CDCl_3$): 7.38–7.26 (*m*, 10 H, Bn); 5.74 (*d*, *J* = 3.6, H–C(1)); 5.00 (*d*, *J* = 10.7, 1 H, Bn); 4.81 (*d*, *J* = 10.7, 1 H, Bn); 4.75 (*d*, *J* = 12.1, 1 H, Bn); 4.61–4.54 (*m*, H–C(2), 2 H–C(5), Bn); 4.22 (*d*, *J* = 5.2, H–C(3)); 3.54 (*d*, *J* = 10.0, 1 H, CH_2–C(4)); 3.49 (*d*, *J* = 10.3, 1 H, CH_2–C(4)); 1.64, 1.35 (2*s*, Me_2C). ^{13}C-NMR ($CDCl_3$): 137.8, 137.3, 128.8, 128.7, 128.5, 128.4, 128.3, 128.1, 128.0 (Bn); 118.8 (*q*, *J* = 320, CF_3); 114.1 (Me_2C); 104.5 (C(1)); 84.2 (C(4)); 79.5, 78.8 (C(2), C(3)); 76.4, 74.0, 72.8, 70.5 (C(5), CH_2–C(4), Bn); 26.5, 25.9 (Me_2C). Anal. calc. for $C_{24}H_{27}F_3O_8S \cdot 0.25 H_2O$: C 53.67, H 5.12; found: C 53.73, H 4.91.

5-Azido-3-O-benzyl-4-C-[(benzyloxy)methyl]-5-deoxy-1,2-O-isopropylidene-β-L-lyxofuranose (**8**). To a soln. of **7** in anh. DMF (230 ml), NaN_3 (8.22 g, 0.126 mol) was added in one portion and the mixture stirred at 60° for 6 h. The soln. was cooled to r.t., filtered, and evaporated to give an oil which was redissolved in AcOEt (120 ml). The org. phase was washed with sat. aq. NaCl soln. (3 × 75 ml), dried (Na_2SO_4), and evaporated. Purification by CC (10–20% AcOEt/petroleum ether) afforded **8** (3.99 g, 59% from **2**). Clear oil. IR (NaCl): 2103 (−N_3). ^{1}H-NMR ($CDCl_3$): 7.38–7.26 (*m*, 10 H, Bn); 5.79 (*d*, *J* = 3.8, H–C(1)); 4.76 (*d*, *J* = 12.0, 1 H, Bn); 4.63 (*dd*, *J* = 3.8, 5.1, H–C(2)); 4.58 (*s*, 1 H, Bn); 4.54 (*s*, 1 H, Bn); 4.48 (*d*, *J* = 12.0, 1 H, Bn); 4.20 (*d*, *J* = 5.5, 1 H, H–C(3)); 4.05 (*d*, *J* = 13.4, 1 H, CH_2–C(4)); 3.58 (*d*, *J* = 10.2, 1 H, H_b–C(5)); 3.46 (*d*, *J* = 10.2, 1 H, H_a–C(5)); 3.33 (*d*, *J* = 13.4, 1 H, CH_2–C(4)); 1.67, 1.36 (2*s*, Me_2C). ^{13}C-NMR ($CDCl_3$): 138.0, 137.6, 128.5, 128.4, 128.0, 127.9, 127.7 (Bn); 113.5 (Me_2C); 104.2 (C(1)); 86.4 (C(4)); 78.9, 78.3, 73.6, 72.4, 71.3 (C(2), C(3), C(5), Bn); 52.1 (CH_2–C(4)); 26.6, 25.8 (Me_2C). FAB-MS: 448 ($[M + Na]^+$). Anal. calc. for $C_{23}H_{27}N_3O_5$: C 64.93, H 6.40, N 9.88; found: C 64.76, H 6.32, N 9.72.

1,2-Di-O-acetyl-5-azido-3-O-benzyl-4-C-[(benzyloxy)methyl]-5-deoxy-L-lyxofuranose (**9**). Furanose **8** (5.93 g, 13.9 mmol) was heated in 50% aq. AcOH soln. (120 ml) at 100° for 1 h and cooled to r.t. Evaporation followed by co-evaporation successively with abs. EtOH, toluene, and pyridine afforded an oil which was redissolved in anh. pyridine (30 ml). Ac$_2$O (20 ml) was added, and the mixture was stirred under Ar for 12 h at r.t. The mixture was cooled to 0° and poured into ice-cold H$_2$O (50 ml). Sat. aq. NaHCO$_3$ soln. (50 ml) was added followed by AcOEt (100 ml). The org. phase was washed with sat. aq. NaHCO$_3$ soln., the aq. phase extracted with AcOEt (2 × 50 ml), and the combined org. phase dried (Na$_2$SO$_4$) and evaporated. Purification by CC (CH$_2$Cl$_2$) afforded **9** (α/β-L 1:4; 5.86 g, 90%). Clear oil. ^{1}H-NMR (CDCl$_3$): 7.39 – 7.24 (*m*, Bn); 6.42 (*d*, *J* = 4.7, H–C(1)(α-L)); 6.18 (*s*, H–C(1)(β-L)); 5.35 (*d*, *J* = 4.8, H–C(2)(β-L)); 5.25 (*dd*, *J* = 4.8, 6.6, H–C(2)(α-L)); 4.63 – 4.48 (*m*, Bn); 4.36 (*d*, *J* = 5.0, H–C(3)(β-L)); 4.29 (*d*, *J* = 6.5, H–C(3)(α-L)); 3.74 (*d*, *J* = 13.1, 1 H, CH$_2$–C(4)(α-L)); 3.64 – 3.43 (*m*, H–C(5), 2 H–C(4)); 3.37 (*d*, *J* = 13.3, 1 H, CH$_2$–C(4)(α-L)); 2.13, 2.12, 2.05, 1.90 (4*s*, 4 MeCO). ^{13}C-NMR (CDCl$_3$): 169.7, 169.6, 168.9 (4 C=O); 137.8, 137.3, 128.5, 128.4, 128.0, 127.9, 127.8, 127.7, 127.6 (Bn); 97.5 (C(1)(β-L)); 94.3 (C(1)(α-L)); 87.8 (C(4)(α-L)); 86.1 (C(4')(β-L)); 79.0, 77.4, 74.3, 74.2, 73.7, 73.5, 73.4, 71.6, 71.5, 71.0 (2 C(2), 2 C(3), 2 C(5), Bn); 53.0 (CH$_2$–C(4)(β-L)); 52.7 (CH$_2$–C(4)(α-L)); 21.0, 20.7, 20.6, 20.3 (4 MeCO). FAB-MS: 492 ([*M* + Na]$^+$). Anal. calc. for C$_{24}$H$_{27}$N$_3$O$_7$: C 61.40, H 5.80; N 8.95; found: C 61.64, H 5.81, N 8.96.

1-[2-O-Acetyl-5-azido-3-O-benzyl-4-C-[(benzyloxy)methyl]-5-deoxy-α-L-lyxofuranosyl}thymine (**10**). To a soln. of **9** (5.86 g, 12.5 mmol) in anh. MeCN (50 ml), thymine (3.15 g, 25 mmol) and BSA (24.7 ml, 0.1 mol) were added. The soln. was heated under reflux for 30 min, cooled to 0°, and Me$_3$SiOTf (4.5 ml, 25 mmol) was added dropwise. The mixture was heated under reflux for another 30 min and then cooled to 0°. The reaction was quenched by addition of a sat. aq. NaHCO$_3$ soln. (60 ml). The mixture was filtered and evaporated, the oily residue dissolved in CH$_2$Cl$_2$ (75 ml), and the soln. washed with sat. aq. NaHCO$_3$ soln. (3 × 50 ml), dried (Na$_2$SO$_4$), and evaporated. Purification by CC (0 – 1% MeOH/CH$_2$Cl$_2$) afforded **10** (6.22 g, 93%). White solid material. ^{1}H-NMR (CDCl$_3$): 9.22 (br. *s*, NH); 7.40 (*d*, *J* = 1.3, H–C(6)); 7.39 – 7.26 (*m*, 10 H, Bn); 6.21 (*d*, *J* = 5.1, H–C(1')); 5.46 (*t*, *J* = 5.6, H–C(2')); 4.63 (*d*, *J* = 11.3, 1 H, Bn); 4.56 (*m*, 4 H, H–C(3'), Bn); 3.77 (*d*, *J* = 9.9, H$_b$–C(5')); 3.65 (*d*, *J* = 13.0, 1 H, CH$_2$–C(4')); 3.51 (*d*, *J* = 10.3, H$_a$–C(5')); 3.38 (*d*, *J* = 13.1, 1 H, CH$_2$–C(4')); 2.10 (*s*, COMe); 1.57 (*d*, *J* = 1.1, Me). ^{13}C-NMR (CDCl$_3$): 170.2 (C=O); 163.8 (C(4)); 150.5 (C(2)); 137.2, 137.1, 135.8 (C(6), Bn); 128.7, 128.6, 128.3, 128.2, 128.1, 127.8 (Bn); 111.5 (C(5)); 87.1 (C(1'), C(4')); 77.5, 74.8, 74.6, 73.7, 71.4 (C(2'), C(3'), C(5'), Bn); 52.6 (CH$_2$–C(4')); 20.5 (*Me*CO); 11.9 (Me). FAB-MS: 536 ([*M* + H]$^+$). Anal. calc. for C$_{27}$H$_{29}$N$_5$O$_7$: C 60.55, H 5.46, N 13.08; found: C 60.34, H 5.39, N 12.87.

1-[5-Azido-3-O-benzyl-4-C-[(benzyloxy)methyl]-5-deoxy-α-L-lyxofuranosyl}thymine (**11**). Nucleoside **10** (6.18 g, 11.54 mmol) was stirred in a sat. NH$_3$ soln. in MeOH (100 ml) at r.t., and after 5 h, the mixture was evaporated. Purification by CC (1% MeOH/CH$_2$Cl$_2$) afforded **11** (6.18 g, 97%). White solid material. ^{1}H-NMR (CDCl$_3$): 10.3 (br. *s*, NH); 7.54 (*d*, *J* = 1.1, H–C(6)); 7.38 – 7.26 (*m*, 10 H, Bn); 5.99 (*d*, *J* = 4.4, H–C(1')); 4.91 (*d*, *J* = 11.8, 1 H, Bn); 4.61 (*d*, *J* = 11.5, 1 H, Bn); 4.54 – 4.49 (*m*, 3 H, H–C(2'), Bn); 4.31 (*d*, *J* = 5.5, H–C(3')); 3.81 (*d*, *J* = 13.5, 1 H, CH$_2$–C(4')); 3.80 (*d*, *J* = 10.3, H$_b$–C(5')); 3.52 (*d*, *J* = 10.1, H$_a$–C(5')); 3.31 (*d*, *J* = 13.3, 1 H, CH$_2$–C(4')); 1.50 (*d*, *J* = 0.8, Me). ^{13}C-NMR (CDCl$_3$): 164.0 (C(4)); 151.4 (C(2)); 137.5, 137.3, 136.1 (C(6), Bn); 128.7, 128.5, 128.2, 128.1, 127.8 (Bn); 110.7 (C(5)); 90.3, 87.2 (C(1'), C(4')); 77.9, 74.7, 73.9, 73.6, 71.4 (C(2'), C(3'), C(5'), Bn); 52.9 (CH$_2$–C(4')); 11.9 (Me). FAB-MS: 494 ([*M* + H]$^+$). Anal. calc. for C$_{25}$H$_{27}$N$_5$O$_6$: C 60.84, H 5.51, N 14.19; found: C 60.80, H 5.38, N 14.10.

1-[5-Azido-3-O-benzyl-4-C-[(benzyloxy)methyl]-5-deoxy-2-O-methyl-α-L-lyxofuranosyl}thymine (**12**). A soln. of **11** (0.99 g, 2.01 mmol) in anh. THF (20 ml) was cooled to 0°. NaH (0.241 g of a 60% dispersion in mineral oil, 6.03 mmol) was added and the mixture stirred for 30 min. MeI (0.63 ml, 10.1 mmol) was added dropwise, and the mixture was stirred at 0° for 10 h. Ice-cold H$_2$O (10 ml) was added followed by AcOEt (25 ml). The org. phase was washed with sat. aq. NaHCO$_3$ soln. (3 × 30 ml), dried (Na$_2$SO$_4$), and evaporated. Purification by CC (0.5% MeOH/CH$_2$Cl$_2$) afforded **12** (0.92 g, 90%). White solid material. ^{1}H-NMR (CDCl$_3$): 9.35 (br. *s*, NH); 7.67 (*d*, *J* = 1.3, H–C(6)); 7.41 – 7.22 (*m*, 10 H, Bn); 6.07 (*d*, *J* = 2.4, H–C(1')); 4.74 (*d*, *J* = 11.7, 1 H, Bn); 4.68 (*d*, *J* = 11.4, 1 H, Bn); 4.52 (*s*, 1 H, Bn); 4.49 (*d*, *J* = 11.7, 1 H, Bn); 4.35 (*d*, *J* = 5.5, 1 H, H–C(3')); 3.94 (*d*, *J* = 10.4, H$_b$–C(5')); 3.93 (*d*, *J* = 13.6, 1 H, CH$_2$–C(4')); 3.87 (*dd*, *J* = 2.4, 5.7, H–C(2')); 3.58 (*s*, MeO); 3.55 (*d*, *J* = 10.0, H$_a$–C(5')); 3.34 (*d*, *J* = 13.7, 1 H, CH$_2$–C(6')); 1.48 (*d*, *J* = 1.2, Me). ^{13}C-NMR (CDCl$_3$): 164.0 (C(4)); 150.3 (C(2)); 137.3, 137.2, 135.8 (C(6), Bn); 128.7, 128.6, 128.2, 127.9, 127.8 (Bn); 110.8 (C(5)); 88.5, 87.2, 83.4 (C(1'), C(2'), C(4')); 75.9, 73.6, 73.1, 70.2 (C(3'), C(5'), Bn); 59.2 (MeO); 53.0 (CH$_2$–C(4')); 11.7 (Me). FAB-MS: 508 ([*M* + H]$^+$). Anal. calc. for C$_{26}$H$_{29}$N$_5$O$_6$: C 61.53, H 5.76, N 13.80; found: C 61.26, H 5.78, N 13.62.

1-[5-Amino-3-O-benzyl-4-C-[(benzyloxy)methyl]-5-deoxy-2-O-methyl-α-L-lyxofuranosyl}thymine (**13**). To a soln. of **12** (440 mg, 0.87 mmol) in abs. EtOH (13 ml), *Lindlar* catalyst (130 mg) was added. The mixture was stirred at r.t. under H$_2$ for 8 h, filtered through a thin layer of silica gel, and evaporated. Purification by

CC (2–5% MeOH/CH$_2$Cl$_2$) afforded **13** (316 mg, 75%). White solid material. ^{1}H-NMR (CDCl$_3$): 7.60 (*d*, *J* = 0.9, H–C(6)); 7.38–7.21 (*m*, 10 H, Bn); 6.12 (*d*, *J* = 3.3, H–C(1′)); 4.79 (*d*, *J* = 11.9, 1 H, Bn); 4.56–4.46 (*m*, 2 H, Bn); 4.48 (*d*, *J* = 11.7, 1 H, Bn); 4.33 (*d*, *J* = 5.9, H–C(3′)); 3.88 (*dd*, *J* = 3.6, 5.9, H–C(2′)); 3.82 (*d*, *J* = 10.4, H$_b$–C(5′)); 3.59 (*d*, *J* = 10.5, H$_a$–C(5′)); 3.50 (*s*, MeO); 3.12 (*d*, *J* = 13.8, 1 H, CH$_2$–C(4′)); 2.83 (*d*, *J* = 13.8, 1 H, CH$_2$–C(4′)); 1.49 (*d*, *J* = 0.5, Me). ^{13}C-NMR (CDCl$_3$): 164.0 (C(4)); 150.5 (C(2)); 137.7, 137.4, 135.9 (C(6), Bn); 128.5, 128.1, 128.0, 127.8, 127.7 (Bn); 111.0 (C(5)); 88.0, 87.8, 83.9 (C(1′), C(2′), C(4′)); 76.8, 73.6, 73.4, 71.8 (C(3′), C(5′), Bn); 59.1 (MeO); 44.2 (CH$_2$–C(4′)); 11.8 (Me). FAB-MS: 482 ([*M* + H]$^+$). Anal. calc. for C$_{26}$H$_{31}$N$_3$O$_6$ · 0.25 H$_2$O: C 64.25, H 6.53, N 8.65; found: C 64.27, H 6.38, N 8.55.

1-[3-O-Benzyl-4-C-[(benzyloxy)methyl]-5-deoxy-2-O-methyl-5-[(trifluoroacetyl)amino]-α-L-lyxofuranosyl]-thymine (**14**). To a soln. of nucleoside **13** (300 mg, 0.62 mmol) in anh. CH$_2$Cl$_2$ (8 ml), ethyl trifluoroacetate (0.15 ml, 1.25 mmol) and Et$_3$N (0.086 ml, 0.62 mmol) were added. The mixture was stirred for 12 h at r.t. and then evaporated. Purification by CC (1% MeOH/CH$_2$Cl$_2$) afforded **14** (346 mg, 97%). White solid material. ^{1}H-NMR (CDCl$_3$): 8.86 (br. *s*, NH); 7.52 (*d*, *J* = 1.3, H–C(6)); 7.39–7.18 (*m*, 11 H, NHCOCF$_3$, Bn); 5.90 (*d*, *J* = 1.6, H–C(1′)); 4.72 (*d*, *J* = 11.5, 1 H, Bn); 4.54–4.44 (*m*, 3 H, Bn); 4.43 (*d*, *J* = 6.1, H–C(3′)); 3.95 (*dd*, *J* = 1.7, 6.3, H–C(2′)); 3.77–3.68 (*m*, CH$_2$–C(4′)); 3.55–3.48 (*m*, 2 H–C(5′)); 3.58 (*s*, MeO); 1.51 (*d*, *J* = 1.0, Me). ^{13}C-NMR (CDCl$_3$): 164.1 (C(4)); 157.4 (*q*, *J* = 37, COCF$_3$); 150.2 (C(2)); 136.8, 136.1 (C(6), Bn); 128.6, 128.4, 128.2, 128.1, 127.9, 127.8, 127.7 (Bn); 115.9 (*q*, *J* = 288, COCF$_3$); 110.9 (C(5)); 90.2, 86.8, 83.0 (C(1′), C(2′), C(4′)); 76.0, 73.6, 73.2, 70.8 (C(3′), CH$_2$–C(4′), Bn); 59.3 (MeO); 41.4 (C(5′)); 11.8 (Me). FAB-MS: 578 ([*M* + H]$^+$). Anal. calc. for C$_{28}$H$_{30}$F$_3$n$_3$O$_7$: C 58.23, H 5.23, N 7.28; found: C 58.27, H 5.21, N 7.34.

1-[5-Deoxy-2-O-methyl-5-[(trifluoroacetyl)amino]-α-L-lyxofuranosyl]thymine (**15**). To a soln. of **14** (346 mg, 0.60 mmol) in abs. EtOH (10 ml), 5% Pd/C catalyst (150 mg) was added. The mixture was stirred at r.t. under H$_2$ for 1.5 h, filtered through a thin layer of silica gel, and evaporated. Purification by CC (3% MeOH/CH$_2$Cl$_2$) afforded **15** (223 mg, 92%). White solid material. ^{1}H-NMR (CD$_3$OD): 7.86 (*d*, *J* = 1.1, H–C(6)); 6.05 (*d*, *J* = 4.6, H–C(1′)); 4.51 (*d*, *J* = 5.8, H–C(3′)); 4.03 (*dd*, *J* = 4.8, 5.8, H–C(2′)); 3.70–3.62 (*m*, CH$_2$–C(4′), 2 H–C(5′)); 3.50 (*s*, MeO); 1.49 (*d*, *J* = 1.0, Me). ^{13}C-NMR (CD$_3$OD): 166.3 (C(4)); 157.4 (*q*, *J* = 35, COCF$_3$); 152.4 (C(2)); 138.4 (C(6)); 117.5 (*q*, *J* = 286, COCF$_3$); 111.7 (C(5)); 89.0, 88.9, 84.9 (C(1′), C(2′), C(4′)); 71.1, 64.3 (C(3′), CH$_2$–C(4′)); 59.1 (MeO); 42.1 (C(5′)); 12.4 (Me). EI-MS: 397 (*M*$^+$). Anal. calc. for C$_{14}$H$_{18}$F$_3$N$_3$O$_7$ · 0.25 H$_2$O: C 41.85, H 4.64, N 10.58, found: C 41.93, H 4.42, N 10.40.

1-[5-Deoxy-4-C-[[(4,4′-dimethoxytrityl)oxy]methyl]-2-O-methyl-5-[(trifluoroacetyl)amino]-α-L-lyxofurano-syl]thymine (**16**). To a soln. of **15** (223 mg, 0.56 mmol) in anh. pyridine (8 ml), (MeO)$_2$TrCl (361 mg, 1.07 mmol) was added and the mixture stirred at r.t. for 6 h. MeOH (2 ml) was added and the soln. evaporated. The oily residue was redissolved in CH$_2$Cl$_2$ (20 ml) and the soln. washed with sat. aq. NaCl soln. (3 × 15 ml), dried (Na$_2$SO$_4$), and evaporated. Purification by CC (0.5–1.5% MeOH/CH$_2$Cl$_2$ containing 0.5% pyridine) afforded **16** (382 mg, 91%). Yellowish solid material. ^{1}H-NMR (CDCl$_3$): 9.54 (br. *s*, NH); 7.45 (*d*, *J* = 0.9, H–C(6)); 7.39–7.08 (*m*, 10 H, NHCOCF$_3$, (MeO)$_2$*Tr*); 6.84 (*dd*, *J* = 1.7, 9.0, 4 H, (MeO)$_2$*Tr*); 6.10 (*d*, *J* = 5.6, H–C(1′)); 4.49 (*d*, *J* = 5.1, H–C(3′)); 4.20 (*t*, *J* = 5.8, H–C(2′)); 3.82–3.75 (*m*, CH$_2$–C(4′)); 3.63–3.61 (*m*, 2 H–C(5′)); 3.79, 3.78 (2*s*, (MeO)$_2$*Tr*); 3.42 (br. *s*, OH–C(3′)); 3.36 (*s*, MeO); 1.48 (*d*, *J* = 1.0, Me). ^{13}C-NMR (CDCl$_3$): 164.0 (C(4)); 159.0 ((MeO)$_2$*Tr*); 157.5 (*q*, *J* = 37, COCF$_3$); 150.7 (C(2)); 143.9 ((MeO)$_2$*Tr*); 137.9, 135.7, 134.8, 134.7 (C(6), (MeO)$_2$*Tr*); 130.2, 130.1, 129.1, 128.3, 128.1, 127.4 ((MeO)$_2$*Tr*); 115.7 (*q*, *J* = 288, COCF$_3$); 113.4 ((MeO)$_2$Tr); 112.0 (C(5)); 87.7, 87.2, 87.0, 83.0 (C(1′), C(2′), C(4′), C(Ar$_3$)); 71.1, 65.9 (C(3′), CH$_2$–C(4′)); 59.0 (MeO); 55.1 ((MeO)$_2$*Tr*); 41.3 (C(5′)); 11.6 (Me). FAB-MS: 699 (*M*$^+$). Anal. calc. for C$_{35}$H$_{36}$F$_3$N$_3$O$_9$ · 0.5 C$_7$H$_8$: C 62.01, H 5.41, N 5.63; found: C 62.04, H 5.50, N 5.50.

1-[5-Deoxy-4-C-[[(4,4′-dimethoxytrityl)oxy]methyl]-2-O-methyl-5-[(trifluoroacetyl)amino]-α-L-lyxofurano-syl]thymine 3′-(2-Cyanoethyl Diisopropylphosphoramidite) (**17**). To a soln. of **16** (160 mg, 0.23 mmol) in anh. CH$_2$Cl$_2$ (4 ml), ^{i}Pr$_2$EtN (0.16 ml, 0.92 mmol) was added, and the mixture was cooled to 0°. 2-Cyanoethyl diisopropylphosphoramidochloridite (0.11 ml, 0.46 mmol) was added dropwise, and the mixture was stirred for 4 h at r.t., diluted with AcOEt (25 ml), and washed with sat. aq. NaCl soln. (3 × 10 ml). The org. phase was dried (Na$_2$SO$_4$) and evaporated. After purification by CC (AcOEt/CH$_2$Cl$_2$/petroleum ether/Et$_3$N 15 : 50 : 35 : 1), the product was redissolved in anh. toluene and precipitated in petroleum ether (50 ml) at −30° under vigorous stirring. A second purification by CC (1% Et$_3$N/CH$_2$Cl$_2$) followed by precipitation in petroleum ether as described above afforded phosphoramidite **17** (89 mg, 43%). White solid material. ^{31}P-NMR (CDCl$_3$): 153.0, 152.5. FAB-MS: 900 (*M*$^+$).

1-[5-Azido-3-O-benzyl-4-C-[(benzyloxy)methyl]-5-deoxy-2-O-(methylsulfonyl)-α-L-lyxofuranosyl]thymine (**18**). A soln. of **11** (1.16 g, 2.35 mmol) in anh. pyridine (6 ml) was cooled to 0°, and MsCl (0.36 ml, 4.7 mmol) was added dropwise. The mixture was stirred at r.t. for 2 h, poured into ice-cold H$_2$O (10 ml) and CH$_2$Cl$_2$ (50 ml), and a sat. aq. NaHCO$_3$ soln. (20 ml) was added. The org. phase was separated and washed successively with sat. aq. NaHCO₃ soln. (3 × 25 ml) and ice-cold 1M HCl (3 × 25 ml), dried (Na₂SO₄), and evaporated to give **18** as a red foam which was used directly in the next step without further purification. ¹H-NMR (CDCl₃): 9.28 (br. *s*, NH); 7.52 (*d*, *J* = 1.2, H−C(6)); 7.35−7.22 (*m*, 10 H, Bn); 6.13 (*d*, *J* = 3.9, H−C(1′)); 5.35 (*dd*, *J* = 4.1, 5.5, H−C(2′)); 4.87 (*d*, *J* = 11.4, 1 H, Bn); 4.54−4.47 (*m*, 3 H, Bn); 4.41 (*d*, *J* = 5.7, H−C(3′)); 3.85 (*d*, *J* = 10.4, Hᵦ−C(5′)); 3.73 (*d*, *J* = 13.4, 1 H, CH₂−C(4′)); 3.49 (*d*, *J* = 10.4, Hₐ−C(5′)); 3.39 (*d*, *J* = 13.5, 1 H, CH₂−C(4′)); 3.15 (*s*, MeSO₂); 1.49 (*s*, Me). ¹³C-NMR (CDCl₃): 163.7 (C(4)); 150.7 (C(2)); 136.9, 135.4 (Bn); 128.8, 128.7, 128.5, 128.4, 127.8 (C(6), Bn); 111.6 (C(5)); 87.7, 87.5 (C(1′), C(4′)); 79.5, 76.3, 74.1, 73.7, 70.8 (C(2′), C(3′), C(5′), Bn); 52.7 (CH₂−C(4′)); 38.8 (MeSO₂); 11.8 (Me). FAB-MS: 572 ([*M* + H]⁺).

1-[5-Azido-3-O-benzyl-4-C-[(benzyloxy)methyl]-5-deoxy-α-L-xylofuranosyl]thymine (**19**). Nucleoside **18** was stirred in 90% aq. EtOH soln. (80 ml), and 1N NaOH (5.17 ml, 5.17 mmol) was added. The mixture was heated under reflux for 30 min, cooled to r.t., neutralized with AcOH and evaporated. The resulting oil was redissolved in CH₂Cl₂ (50 ml), and a sat. aq. NaHCO₃ soln. (50 ml) was added. The org. phase was washed with sat. aq. NaHCO₃ soln. (3 × 30 ml), dried (Na₂SO₄), and evaporated. Purification by CC (1.5% MeOH/CH₂Cl₂) afforded **19** (1.09 g, 94%). White solid material. ¹H-NMR (CDCl₃): 10.8 (br. *s*, NH); 7.45 (*d*, *J* = 0.9, H−C(6)); 7.39−7.26 (*m*, 10 H, Bn); 6.24 (*d*, *J* = 3.8, H−C(1′)); 4.97 (*d*, *J* = 5.1, OH−C(2′)); 4.84 (*m*, H−C(2′)); 4.72 (*d*, *J* = 11.8, 1 H, Bn); 4.56 (*s*, 1 H, Bn); 4.55 (*s*, 1 H, Bn); 4.49 (*d*, *J* = 11.5, 1 H, Bn); 4.14 (*d*, *J* = 2.1, H−C(3′)); 3.73 (*d*, *J* = 9.8, Hᵦ−C(5′)); 3.71 (*d*, *J* = 9.8, Hₐ−C(5′)); 3.66 (*d*, *J* = 12.8, 1 H, CH₂−C(4′)); 3.49 (*d*, *J* = 12.7, 1 H, CH₂−C(4′)); 1.61 (*s*, Me). ¹³C-NMR (CDCl₃): 165.7 (C(4)); 150.8 (C(2)); 138.7, 137.5, 137.3 (C(6), Bn); 128.5, 128.4, 128.1, 128.0, 127.7 (Bn); 108.2 (C(5)); 86.6, 86.0, 83.7 (C(1′), C(2′), C(4′)); 73.9, 73.5, 72.0, 69.9 (C(3′), C(5′), Bn); 51.7 (CH₂−C(4′)); 12.0 (Me). FAB-MS: 494 ([*M* + H]⁺). Anal. calc. for C₂₅H₂₇N₅O₆: C 60.84, H 5.51, N 14.19; found: C 60.81, H 5.52, N 14.05.

1-[5-Azido-3-O-benzyl-4-C-[(benzyloxy)methyl]-5-deoxy-2-O-[(trifluoromethyl)sulfonyl]-α-L-xylofuranosyl]thymine (**20**). A soln. of **19** (450 mg, 0.911 mmol) in anh. CH₂Cl₂ (20 ml) was cooled to 0°, 4-(dimethylamino)pyridine (DMAP; 445 mg, 3.64 mmol) and anh. pyridine (0.74 ml, 9.1 mmol) were added, followed by dropwise addition of trifluoromethanesulfonic anhydride (0.33 ml, 2.0 mmol). The mixture was stirred at r.t. for 2 h and poured into an ice-cold sat. aq. NaHCO₃ soln. (10 ml). The org. phase was washed successively with 1M ice-cold HCl (2 × 20 ml) and sat. aq. NaHCO₃ soln. (2 × 20 ml), dried (Na₂SO₄), and evaporated. Purification by CC (silica gel, 1.5% MeOH/CH₂Cl₂) afforded **20** (437 mg, 77%). Yellowish solid material which was used directly in the next step without further purification. ¹H-NMR (CDCl₃): 8.71 (br. *s*, NH); 7.40−7.21 (*m*, 11 H, H−C(6), Bn); 6.39 (*d*, *J* = 4.4, H−C(1′)); 5.54 (*t*, *J* = 4.2, H−C(2′)); 4.81 (*d*, *J* = 12.0, 1 H, Bn); 4.56−4.47 (*m*, 3 H, Bn); 4.52 (*d*, *J* = 11.6, 1 H, Bn); 3.67 (*d*, *J* = 12.7, 1 H, CH₂−C(4′)); 3.54−3.44 (*m*, 2 H−C(5′)); 3.43 (*d*, *J* = 12.9, 1 H, CH₂−C(4′)); 1.76 (*d*, *J* = 1.0, Me). FAB-MS: 626 ([*M* + H]⁺).

1-[5-Azido-3-O-benzyl-4-C-[(benzyloxy)methyl]-2,5-dideoxy-2-fluoro-α-L-lyxofuranosyl]thymine (**21**) *and 1-[4-C-(Azidomethyl)-3,5-di-O-benzyl-2,5-dideoxy-D-erythro-pent-1-enofuranosyl]thymine* (**22**). To a soln. of **20** (437 mg, 0.70 mmol) in anh. THF (10 ml) at 0°, Bu₄NF (2.38 g, dried for 48 h at 0.3 Torr at 40°, *ca.* 9.11 mmol) in anh. THF (4 ml) was added slowly, and the mixture was stirred for 24 h at 5°. The mixture was evaporated, the resulting oil redissolved in AcOEt (20 ml), and the soln. washed with sat. aq. NaCl soln. (3 × 20 ml), dried (Na₂SO₄), and evaporated. Purification by CC (1. 10−25% AcOEt/petroleum ether; 2. 0−2% MeOH/CH₂Cl₂) afforded **21** (130 mg, 37%). White solid material. ¹H-NMR (CDCl₃): 8.28 (br. *s*, NH); 7.43 (*d*, *J* = 1.1, H−C(6)); 7.40−7.22 (*m*, 10 H, Bn); 6.20 (*dd*, *J* = 3.5, 15.6, H−C(1′)); 5.18 (*ddd*, *J* = 3.6, 5.0, 52.9, H−C(2′)); 4.82 (*d*, *J* = 11.7, 1 H, Bn); 4.58−4.49 (*m*, 3 H, Bn); 4.37 (*dd*, *J* = 5.0, 14.9, H−C(3′)); 3.84 (*d*, *J* = 10.5, Hᵦ−C(5′)); 3.69 (*d*, *J* = 13.3, 1 H, CH₂−C(4′)); 3.57 (*d*, *J* = 10.5, Hₐ−C(5′)); 3.43 (*d*, *J* = 13.3, 1 H, CH₂−C(4′)); 1.57 (*s*, Me). ¹³C-NMR (CDCl₃): 163.9 (C(4)); 150.4 (C(2)); 137.0, 136.9, 136.0 (C(6), Bn); 128.7, 128.6, 128.3, 128.2, 128.0, 127.8 (Bn); 111.3 (C(5)); 92.3 (*d*, *J* = 194, C(2′)); 88.3 (*d*, *J* = 34, C(1′)); 87.1 (C(4′)); 76.4 (*d*, *J* = 14.7, C(3′)); 73.7, 73.5, 70.9 (2*s* and *d* (73.5), *J* = 2.1, C(5′), Bn); 52.6 (*d*, *J* = 3.2, CH₂−C(4′)); 11.8 (Me). FAB-MS: ([*M* + H]⁺). Anal. calc. for C₂₅H₂₆FN₅O₅: C 60.60, H 5.29, N 14.13; found: C 60.71, H 5.06, N 13.96.

In addition, the less polar derivative **22** was isolated as a yellowish gum (100 mg, 30%). ¹H-NMR (CDCl₃): 9.33 (br. *s*, NH); 7.50 (*d*, *J* = 0.9, H−C(6)); 7.39−7.26 (*m*, 10 H, Bn); 5.73 (*d*, *J* = 2.4, H−C(2′)); 4.78 (*d*, *J* = 2.7, H−C(3′)); 4.60 (*d*, *J* = 11.6, 1 H, Bn); 4.58 (*s*, 1 H, Bn); 4.56 (*s*, 1 H, Bn); 4.50 (*d*, *J* = 11.9, 1 H, Bn); 3.91 (*d*, *J* = 12.9, 1 H, CH₂−C(4′)); 3.74 (*d*, *J* = 9.5, Hᵦ−C(5′)); 3.68 (*d*, *J* = 13.2, 1 H, CH₂−C(4′)); 3.40 (*d*, *J* = 9.3, Hₐ−C(5′)); 1.94 (*d*, *J* = 0.9, Me). ¹³C-NMR (CDCl₃): 163.0 (C(4)); 148.1, 146.1 (C(2), C(1′)); 137.7, 137.3 (Bn); 134.4 (C(6)); 128.6, 128.5, 128.4, 128.3, 127.9, 127.8, 127.7, 127.6 (Bn); 111.5 (C(5)); 88.5, 88.4, 83.1 (C(2′), C(3′), C(4′)); 73.6, 71.3, 68.5 (C(5′), Bn); 50.3 (CH₂−C(4′)); 12.4 (Me). FAB-MS: 476 ([*M* + H]⁺). Due to the instability of **22**, elemental analysis could not be obtained.

1-[5-Amino-3-O-benzyl-4-C-[(benzyloxy)methyl]-2,5-dideoxy-2-fluoro-α-L-lyxofuranosyl]thymine (**23**). As described for **13**, with **21** (390 mg, 0.79 mmol), abs. EtOH/anh. dioxane 1:4 (11 ml), and *Lindlar* catalyst

(120 mg). Purification by CC (2–5% MeOH/CH$_2$Cl$_2$) afforded **23** (240 mg, 64%). White solid material. ^{1}H-NMR (CDCl$_3$): 7.45 (d, $J = 1.4$, H–C(6)); 7.38–7.23 (m, 10 H, Bn); 6.26 (dd, $J = 4.0$, 15.1, H–C(1′)); 5.28–5.08 (ddd, $J = 4.4$, 5.1, 52.1, H–C(2′)); 4.85 (d, $J = 11.6$, 1 H, Bn); 4.53–4.50 (m, 3 H, Bn); 4.36 (dd, $J = 5.3$, 12.8, H–C(3′)); 3.90 (br. s, NH$_2$); 3.77 (d, $J = 10.5$, H$_b$–C(5′)); 3.61 (d, $J = 10.4$, H$_a$–C(5′)); 3.00 (d, $J = 13.7$, 1 H, CH$_2$–C(4′)); 2.90 (d, $J = 13.9$, 1 H, CH$_2$–C(4′)); 1.58 (d, $J = 0.8$, Me). ^{13}C-NMR (CDCl$_3$): 164.0 (C(4)); 150.6 (C(2)); 137.3, 135.9 (C(6), Bn); 128.7, 128.6, 128.2, 128.1, 127.9, 127.7 (Bn); 111.3 (C(5)); 92.7 (d, $J = 195$, C(2′)); 87.9 (C(4′)); 87.5 (d, $J = 33$, C(1′)); 77.1 (d, $J = 14$, C(3′)); 73.6, 72.2 (C(5′), Bn); 44.1 (CH$_2$–C(4′)); 11.9 (Me). FAB-MS: 470 ([$M + $H]$^+$). Anal. calc. for C$_{25}H_{28}FN_3O_5 \cdot 0.25$ H$_2$O: C 63.35, H 6.06, N 8.86; found: C 63.46, H 5.95, N 8.85.

1-[3-O-Benzyl-4-C-[(benzyloxy)methyl]-2,5-dideoxy-2-fluoro-5-[(trifluoroacetyl)amino]-α-L-lyxofuranosyl]-thymine (**24**). As described for **14**, with **23** (230 mg, 0.49 mmol), anh. CH$_2$Cl$_2$ (6 ml), ethyl trifluoroacetate (0.18 ml, 1.48 mmol), and Et$_3$N (0.14 ml, 0.98 mmol). Purification by CC (1% MeOH/CH$_2$Cl$_2$) afforded **24** (248 mg, 89%). White solid material. ^{1}H-NMR (CDCl$_3$): 9.55 (br. s, NH); 7.40–7.22 (m, 11 H, H–C(6), Bn); 7.10 (t, $J = 6.3$, NHCOCF$_3$); 5.90 (dd, $J = 2.3$, 19.3, H–C(1′)); 5.42–5.21 (ddd, $J = 2.3$, 5.3, 54.0, H–C(2′)); 4.79 (d, $J = 11.5$, 1 H, Bn); 4.54–4.45 (m, 4 H, H–C(3′), Bn); 3.77–3.67 (m, 3 H, 2 H$_b$–C(5′), CH$_2$–C(4′)); 3.57 (d, $J = 10.5$, 1 H, CH$_2$–C(4′)); 1.67 (d, $J = 1.2$, Me). ^{13}C-NMR (CDCl$_3$): 164.0 (C(4)); 157.8 (q, $J = 37$, COCF$_3$); 150.3 (C(2)); 137.4, 136.9, 136.6 (C(6), Bn); 128.7, 128.6, 128.5, 128.4, 128.2, 128.0, 127.8 (Bn); 115.8 (q, $J = 288$, COCF$_3$); 111.5 (C(5)); 92.0 (d, $J = 192$, C(2′)); 91.1 (d, $J = 36$, C(1′)); 86.3 (C(4′)); 77.6 (C(3′)); 73.7, 73.5, 72.1 ($2s$ and d (73.5), $J = 2.0$, CH$_2$–C(4′), Bn); 41.4 (C(5′)); 11.8 (Me). FAB-MS: 566 ([$M + $H]$^+$). Anal. calc. for C$_{27}H_{27}F_4N_3O_6$: C 57.34, H 4.81, N 7.43; found: C 57.37, H 4.74, N 7.37.

1-[2,5-Dideoxy-2-fluoro-5-[(trifluoroacetyl)amino]-α-L-lyxofuranosyl]thymine (**25**). As described for **15**, with **24** (242 mg, 0.43 mmol), abs. EtOH (7 ml), and 5% Pd/C (110 mg). Purification by CC (2–5% MeOH/CH$_2$Cl$_2$) afforded **25** (154 mg, 92%). White solid material. ^{1}H-NMR (CD$_3$OD): 7.82 (d, $J = 1.1$, H–C(6)); 6.17 (dd, $J = 3.1$, 17.2, H–C(1′)); 5.31–5.10 (ddd, $J = 3.2$, 5.4, 53.8, H–C(2′)); 4.64 (dd, $J = 5.6$, 17.2, H–C(3′)); 3.75–3.38 (m, 2 H–C(5′), CH$_2$–C(4′)); 1.67 (d, $J = 1.0$, Me). ^{13}C-NMR (CD$_3$OD): 166.6 (C(4)); 159.5 (q, $J = 37$, COCF$_3$); 152.4 (C(2)); 138.3 (C(6)); 117.6 (q, $J = 287$, COCF$_3$); 111.8 (C(5)); 95.0 (d, $J = 190$, C(2′)); 89.7 (d, $J = 34$, C(1′)); 88.8 (C(4′)); 71.2 (d, $J = 15$, C(3′)); 63.7 (CH$_2$–C(4′)); 41.7 (d, $J = 4.1$, C(5′)); 12.3 (Me). FAB-MS: 386 ([$M + $H]$^+$). Anal. calc. for C$_{13}H_{15}F_4N_3O_6 \cdot 0.25$ H$_2$O: C 40.06, H 4.01, N 10.78; found: C 40.09, H 3.89, N 10.48.

1-[2,5-Dideoxy-5-O-(4,4′-dimethoxytrityl)-2-fluoro-5-[(trifluoroacetyl)amino]-α-L-lyxofuranosyl]thymine (**26**). As described for **16**, with **25** (145 mg, 0.38 mmol), anh. pyridine (4 ml), and (MeO)$_2$TrCl (206 mg, 0.61 mmol). Purification by CC (1% MeOH/CH$_2$Cl$_2$ containing 0.5% of pyridine) afforded **26** (246 mg, 87%). Yellowish solid material. ^{1}H-NMR (CDCl$_3$): 9.69 (br. s, N H); 7.38–7.14 (10 H, (MeO)$_2$Tr, H–C(6)); 7.09 (t, $J = 6.3$, NHCOCF$_3$); 6.82 (dd, $J = 3.6$, 9.2, 4 H, (MeO)$_2$Tr); 6.05 (dd, $J = 4.0$, 16.8, H–C(1′)); 5.44–5.23 (ddd, $J = 4.4$, 4.6, 54.1, H–C(2′)); 4.65 (dd, $J = 5.5$, 11.9, H–C(3′)); 4.30 (br. s, OH–C(3′)); 3.76, 3.77 ($2s$, 2 MeO); 3.72–3.62 (m, 2 H–C(5′)); 3.41 (d, $J = 10.8$, 1 H, CH$_2$–C(4′)); 3.34 (d, $J = 10.5$, 1 H, CH$_2$–C(4′)); 1.60 (s, Me). ^{13}C-NMR (CDCl$_3$): 164.0 (C(4)); 158.9 ((MeO)$_2$Tr); 159.5 (q, $J = 37$, COCF$_3$); 150.6 (C(2)); 144.0 ((MeO)$_2$Tr); 136.9, 136.5, 134.8 (C(6), (MeO)$_2$Tr); 130.1, 129.1, 128.3, 128.1, 125.3 ((MeO)$_2$Tr); 115.7 (q, $J = 288$, COCF$_3$); 113.3 ((MeO)$_2$Tr); 112.0 (C(5)); 92.7 (d, $J = 191$, C(2′)); 89.1 (d, $J = 33$, C(1′)); 87.6, 86.8 (C(4′), C(Ar$_3$)); 71.8 (d, $J = 16$, C(3′)); 66.0 (CH$_2$–C(4′)); 55.1 (2 MeO); 41.1 (C(5′)); 11.6 (Me). FAB-MS: 687 (M^+). Anal. calc. for C$_{34}$H$_{33}$F$_4$N$_3$O$_8 \cdot 0.6$C$_7$H$_8$: C 61.75, H 5.13, N 5.66; found: C 61.96, H 5.16, N 5.89.

1-(5-Deoxy-4-C-[[(4,4′-dimethoxytrityl)oxy]methyl]-2-fluoro-5-[(trifluoroacetyl)amino]-α-L-lyxofuranosyl)-thymine 3′-(2-Cyanoethyl Diisopropylphosphoramidite) (**27**). As described for **17**, with **26** (246 mg, 0.36 mmol), anh. CH$_2$Cl$_2$ (4 ml), ^{i}Pr$_2$EtN (0.16 ml, 0.94 mmol), and 2-cyanoethyl diisopropylphosphoramidochloridite (0.11 ml, 0.47 mmol). After purification by CC (1% Et$_3$N/CH$_2$Cl$_2$), the product was redissolved in anh. toluene and precipitated in vigorously stirred petroleum ether (40 ml) at $-30°$. A second purification by CC (0.5% Et$_3$N/CH$_2$Cl$_2$) afforded **27** (245 mg, 76%). White solid material. ^{31}P-NMR (CDCl$_3$): 154.2 (dd, $J = 9.4$, 18.2); 152.9 (d, $J = 18.7$). FAB-MS: 888 ([$M + $H]$^+$).

1-[3,5-Di-O-benzyl-3-C-[3-(phthalimido)propyl]-β-D-ribofuranosyl]thymine (**30**). To a soln. of **29** [19] (3.47 g, 6.99 mmol) in anh. THF (10 ml), phthalimide (1.23 g, 8.39 mmol) and PPh$_3$ (2.29 g, 8.74 mmol) were added, and the mixture was cooled to 0°. A soln. of diethyl diazenedicarboxylate (1.37 ml, 8.67 mmol) in anh. THF (3 ml) cooled to 0° was added dropwise within 15 min, and the mixture was stirred for 12 h at r.t. AcOEt (60 ml) was added, and the mixture was washed successively with sat. aq. NaCl soln. (2 × 50 ml) and H$_2$O (2 × 50 ml). The aq. phases were extracted with AcOEt (2 × 50 ml), and the combined org. phase was dried (Na$_2$SO$_4$) and evaporated. Purification by CC (30–90% AcOEt/petroleum ether) afforded **30** (4.34 g), contaminated with PPh$_3$(O). ^{1}H-NMR (CDCl$_3$): 8.35 (s, NH); 7.81–7.19 (m, Bn, H–C(6), phth, PPh$_3$(O)); 6.10

($d, J = 7.8$, H–C(1')); 4.51 (m, 4 H, Bn); 4.37 (br. s, H–C(4')); 4.16 ($dd, J = 7.8$, 11.5, H–C(2')); 3.83 ($dd, J = 2.9$, 10.7, H$_b$–C(5')); 3.75 ($t, J = 6.4$, 2 H–C(3'')); 3.52 ($d, J = 9.9$, H$_a$–C(5')); 2.92 ($d, J = 11.4$, OH–C(2')); 2.21–1.88 (m, 2 H–C(1''), 2 H–C(2'')); 1.44 (3 H, $d, J = 0.8$, Me). ^{13}C-NMR (CDCl$_3$): 168.6 (phth); 163.5 (C(4)); 151.1 (C(2)); 137.1, 136.7, 136.1, 134.1, 133.3 (Bn, C(6), phth); 132.2–131.9 (several signals, Bn, phth, PPh$_3$(O)); 128.8–127.4 (several signals, Bn, phth, PPh$_3$(O)); 123.3 (phth); 111.3 (C(5)); 87.3, 82.6, 81.0, 79.1 (C(1'), C(2'), C(3'), C(4')); 73.6, 69.8, 65.6 (Bn, C(5')); 38.1 (C(3'')); 27.6, 22.4 (C(1''), C(2'')); 11.7 (Me). FAB-MS: 626 ([M + H]$^+$). HR-FAB-MS: 626.2584 (C$_{35}$H$_{36}$N$_3$O$_8^+$; calc. 626.2502).

1-[3,5-Di-O-benzyl-2-O-[(pentafluorophenoxy)thiocarbonyl]-3-C-[3-(phthalimido)propyl]-β-D-ribofurano-syl]thymine (**31**). To a soln. of **30** (3.96 g, contaminated with PPh$_3$(O)) in anh. CH$_2$Cl$_2$ (20 ml) under Ar, DMAP (1.39 g, 11.4 mmol) was added and the soln. cooled to $-15°$. *O*-(Pentafluorophenyl) carbonochloridothioate (1.32 ml, 8.2 mmol) was added dropwise, and the soln. was stirred for 12 h at r.t. The mixture was poured into ice-cold H$_2$O (70 ml), stirred for 30 min, and filtered through a thin layer of *Celite*. The mixture was successively washed with H$_2$O (30 ml), sat. aq. CuSO$_4$ soln. (2 × 50 ml), and sat. aq. NaCl soln. (2 × 25 ml). All the aq. phases were extracted with CH$_2$Cl$_2$ (25 ml each), and the combined org. phase was dried (Na$_2$SO$_4$) and evaporated. Purification by CC (0–3% MeOH/CH$_2$Cl$_2$ containing 0.25% of pyridine) afforded **31** (1.23 g, 27%, two steps) as a yellowish solid material. In addition, a fraction (1.78 g) consisting of unreacted starting material and PPh$_3$(O) was isolated. **31**: ^{1}H-NMR (CDCl$_3$): 8.47 (s, NH); 7.78–7.62 (m, 5 H, H–C(6), phth); 7.38–7.24 (m, 10 H, Bn); 6.59 ($d, J = 8.1$, H–C(2')); 6.16 ($d, J = 8.0$, H–C(1')); 4.71 ($d, J = 11.4$, 1 H, Bn); 4.58 ($d, J = 11.5$, 1 H, Bn); 4.54 (s, 2 H, Bn); 4.42 (br. s, H–C(4')); 3.92 ($dd, J = 2.4$, 11.2, H$_b$–C(5')); 3.73–3.61 (m, H$_a$–C(5'), 2 H–C(3'')); 2.10–1.80 (m, 2 H–C(1''), 2 H–C(2'')); 1.35 ($d, J = 0.7$, Me). ^{13}C-NMR (CDCl$_3$): 191.8 (C=S); 168.4 (phth); 163.6 (C(4)); 150.6 (C(2)); 137.4, 136.4, 135.9, 134.1, 131.9, 128.7, 128.6, 128.3, 127.8, 127.4, 127.1 (Bn, C(6), phth); 123.2 (phth); 111.4 (C(5)); 87.6, 83.9, 83.4, 82.6 (C(1'), C(2'), C(3'), C(4')); 73.6, 69.5, 65.1 (Bn, C(5')); 37.7 (C(3'')); 26.9, 22.3 (C(1''), C(2'')); 11.5 (Me). FAB-MS: 852 ([M + H]$^+$).

3',5'-Di-O-benzyl-3'-C-[3-(phthalimido)propyl]thymidine (**32**). Through a soln. of **31** (1.18 g, 1.39 mmol) in anh. benzene (4 ml), Ar was bubbled for 15 min with stirring. Then, 2,2'-azobis[isobutyronitrile] (AIBN; 0.114 g, 0.70 mmol) was added, and bubbling with Ar was continued for further 15 min. The temp. was raised to 90°, Bu$_3$SnH (0.56 ml, 2.1 mmol) added, and the mixture heated under reflux for 1 h, then allowed to cool to r.t., and subsequently evaporated. Purification by CC (1. petroleum ether, then 0–2% MeOH/CH$_2$Cl$_2$; 2. petroleum ether, then 0–1% MeOH/CH$_2$Cl$_2$) and prep. TLC (1. 2% MeOH/CH$_2$Cl$_2$; 2. 4% MeOH/CH$_2$Cl$_2$) afforded **32** (208 mg, 25%). White solid material. ^{1}H-NMR (CDCl$_3$): 8.52 (s, NH); 7.81–7.61 (m, 5 H, H–C(6), phth); 7.36–7.15 (m, 10 H, Bn); 6.39 ($dd, J = 5.3$, 9.5, H–C(2')); 4.50–4.38 (m, 4 H, Bn); 4.27 (br. s, H–C(4')); 3.82 ($dd, J = 3.0$, 11.0, H$_b$–C(5')); 3.74–3.67 (m, 2 H–C(3'')); 3.56 ($d, J = 9.9$, H$_a$–C(5')); 2.53 ($dd, J = 5.0$, 13.0, H$_b$–C(2')); 2.00–1.77 (m, H$_a$–C(2'), 2 H–C(1'')); 1.48 (s, Me). ^{13}C-NMR (CDCl$_3$): 168.5 (phth); 163.8 (C(4)); 150.4 (C(2)); 137.8, 136.9, 136.3, 134.1, 131.9, 128.7, 128.5, 128.1, 127.7, 127.4, 123.3 (Bn, C(6), phth)); 110.7 (C(5)); 86.4, 84.3, 83.1 (C(1'), C(3'), C(4')); 73.5, 70.2, 64.1 (Bn, C(5')); 41.6 (C(2')); 37.8 (C(3'')); 27.6, 23.5 (C(1''), C(2'')); 11.8 (Me). HR-FAB-MS: 610.2505 (C$_{35}$H$_{36}$N$_3$O$_7^+$: calc. 610.2553).

3'-C-[3-(Phthalimido)propyl]thymidine (**33**). To a soln. of **32** (0.326 mg, 0.53 mmol) in abs. EtOH (1 ml) and anh. dioxane (0.5 ml), 20% Pd(OH)$_2$/C (100 mg) was added. The mixture was degassed and H$_2$ introduced *via* a balloon. The black soln. was stirred for 3 days under H$_2$ at r.t. (additional catalyst (100 mg) was added after 24 h and 48 h). Then 10% MeOH/CH$_2$Cl$_2$ was added and the mixture stirred for 12 h. After filtration through a *Celite* pad, which was washed thoroughly with MeOH, the filtrate was evaporated. Purification by CC (0–3% MeOH/CH$_2$Cl$_2$) afforded **33** (42 mg, 19%). White solid material. ^{1}H-NMR (CD$_3$OD): 8.04 ($d, J = 1.0$, H–C(6)); 7.83–7.75 (m, 4 H, phth); 6.30 ($dd, J = 5.3$, 9.3, H–C(1')); 3.86 ($d, J = 3.0$, H–C(4')); 3.74–3.63 (m, 2 H–C(5'), 2 H–C(3'')); 2.19 ($dd, J = 5.4$, 12.7, H$_b$–C(2')); 2.03 ($dd, J = 9.4$, 12.6, H$_a$–C(2')); 1.83 ($d, J = 1.2$, Me); 1.78–1.71 (m, 2 H–C(1''), 2 H–C(2'')). ^{13}C-NMR (CD$_3$OD): 170.1 (phth); 166.6 (C(4)); 152.7 (C(2)); 138.8, 135.5, 133.5 (C(6), phth); 124.2 (phth); 111.4 (C(5)); 90.4, 86.1, 81.8 (C(1'), C(3'), C(4')); 62.5 (C(5')); 44.0 (C(2')); 39.1 (C(3'')); 33.6, 24.6 (C(1''), C(2'')); 12.4 (Me). FAB-MS: 430 ([M + H]$^+$).

5'-O-(4,4'-Dimethoxytrityl)-3'-C-[3-(phthalimido)propyl]thymidine (**34**) [20]. To a soln. of **33** (55 mg, 0.13 mmol) in anh. pyridine (1 ml) at r.t., (MeO)$_2$TrCl (95 mg, 0.28 mmol) was added in one portion, and the mixture was stirred at r.t. for 12 h. MeOH (0.20 ml) was added to quench the reaction, and the mixture was evaporated. The residue was redissolved in CH$_2$Cl$_2$ (5 ml) and the org. phase washed with sat. aq. NaCl soln. (3 × 5 ml), dried (Na$_2$SO$_4$), and evaporated. Purification by CC (0–2% MeOH/CH$_2$Cl$_2$ containing 1% of pyridine) afforded **34** (54 mg, 57%). White solid material. ^{1}H-NMR (CDCl$_3$): 9.06 (br. s, NH); 7.84–7.63 (2m, 5 H, H–C(6), phth); 7.38–7.22 (m, 9 H, (MeO)$_2$Tr); 6.83 ($dd, J = 2.2$, 9.0, 4 H, (MeO)$_2$Tr); 6.49 ($dd, J = 5.0$, 9.6, H–C(1')); 4.05 (br. s, H–C(4')); 3.78 (s, 2 MeO); 3.61 ($dd, J = 3.5$, 10.5, H$_b$–C(5')); 3.58–3.47 (m, 2 H–C(3'')); 3.14 ($dd, J = 2.4$, 10.8, H$_a$–C(5')); 2.38 ($dd, J = 5.3$, 12.8, H$_b$–C(2')); 2.09 ($dd, J = 9.8$, 12.4,

H_a–C(2′)); 1.75–1.45 (m, 2 H–C(1″), 2 H–C(2″)); 1.21 (s, Me). ^{13}C-NMR (CDCl$_3$): 168.6 (phth); 163.9 (C(4)); 158.9 ((MeO)$_2$Tr); 150.7 (C(2)); 143.8 ((MeO)$_2$Tr); 136.2, 136.1, 135.0, 134.8, 134.1, 132.0, 130.3, 130.3, 128.5, 128.0, 127.4, 123.3 (C(6), (MeO)$_2$Tr, phth); 113.3 ((MeO)$_2$Tr); 111.3 (C(5)); 87.9, 87.3, 84.2, 81.0 (C(1′), C(3′), C(4′), C(Ar)$_3$); 62.7 (C(5′)); 55.2 (2 MeO); 44.0 (C(2′)); 37.9 (C(3″)); 31.8, 23.4 (C(1″), C(2″)); 11.2 (Me). FAB-MS: 731 (M$^+$). NMR Data: in accordance with [20].

5′-O-(4,4′-Dimethoxytrityl)-3′-C-[3-(phthalimido)propyl]thymidine 3′-(2-Cyanoethyl Diisopropylphosphoramidite) (**35**). To a mixture of **34** (54 mg, 0.074 mmol), anh. CH$_2$Cl$_2$ (2 ml), and ^{i}Pr$_2$EtN (0.05 ml, 0.30 mmol) at 0°, 2-cyanoethyl diisopropylphosphoramidochloridite (0.036 ml, 0.15 mmol) was added dropwise, and the mixture was stirred for 12 h at r.t. The reaction was quenched with sat. aq. NaHCO$_3$ soln. (0.1 ml) and diluted with AcOEt (5 ml). The mixture was washed with sat. aq. NaHCO$_3$ soln. (4 × 5 ml) and the org. phase dried (Na$_2$SO$_4$) and evaporated. Purification by CC (AcOEt/CH$_2$Cl$_2$/petroleum ether 10:35:1:54) afforded **35** (28 mg, 41%). White solid material. ^{31}P-NMR (CDCl$_3$): 141.0, 140.3. FAB-MS: 932 (M$^+$).

1-[3,5-Di-O-benzyl-3-C-(3-hydroxypropyl)-2-O-methyl-β-D-ribofuranosyl]thymine (**37**) *and* *1-[3,5-Di-O-benzyl-3-C-[(RS)-2-hydroxypropyl]-2-O-methyl-β-D-ribofuranosyl]thymine* (**38**). To a soln. of **36** [34] (703 mg, 1.43 mmol) in anh. THF (6 ml), 7.8M BH$_3$ in 1,4-oxathiane (0.22 ml, 1.72 mmol) was added dropwise at r.t., and stirring was continued for 90 min. After cooling to 0° 2M aq. NaOH (0.86 ml, 1.72 mmol) was dropwise added, followed by the addition of 35% aq. H$_2$O$_2$ soln. (w/w, 0.2 ml). The mixture was allowed to warm to r.t. and, after stirring for 1 h, poured into ice-cold H$_2$O (20 ml). The mixture was filtered and extracted with CH$_2$Cl$_2$ (3 × 10 ml). The combined org. phase was washed successively with H$_2$O (3 × 5 ml) and sat. aq. NaHCO$_3$ soln. (5 ml), dried (Na$_2$SO$_4$), and evaporated. Purification by CC (0–2.5% MeOH/CH$_2$Cl$_2$) afforded as the less polar product **38** (98 mg, 13%) as a diastereoisomer mixture and **37** (400 mg, 54%) as the most polar compound.

Data of mixture **38**: ^{1}H-NMR (CDCl$_3$): 8.91 (br. s); 7.82 (d, J = 1.2); 7.80 (d, J = 1.3); 7.42–7.26 (m); 6.54 (d, J = 7.8); 6.40 (d, J = 7.7); 4.85 (d, J = 11.5); 4.67–4.51 (m); 4.31 (br. s); 4.21–4.17 (m); 4.00 (d, J = 7.7); 3.81–3.77 (m); 3.66 (dd, J = 1.6, 11.0); 3.51–3.34 (m); 3.44 (s, MeO); 3.39 (s, MeO); 3.19 (d, J = 1.5); 2.05–1.91 (m); 1.83 (br. s); 1.62 (d, J = 1.0, Me); 1.57 (d, J = 0.9, Me); 1.23, 1.21 (2s, 2 H–C(3″)). ^{13}C-NMR (CDCl$_3$): 163.8, 163.7 (2 C(4)); 150.6 (2 C(2)); 138.0, 137.8, 136.8, 136.5, 136.2, 136.0 (Bn, 2 (C(6)); 128.9, 128.8, 128.6, 128.5, 128.0, 127.8, 127.7, 127.6, 127.3, 126.9 (Bn); 111.7, 111.5 (2 C(5)); 86.4, 86.3, 85.5, 85.2, 84.7, 84.2, 82.6, 81.5 (2 C(1′), 2 C(2′), 2 C(3′), 2 C(4′)); 73.7, 73.6, 69.8, 69.5, 65.7, 64.9, 64.4, 62.7, 59.0, 58.2 (2 MeO, Bn, 2 C(5′), 2 C(2″)); 40.5, 37.9 (2 C(1″)); 23.6, 23.2, (2 C(3″)); 12.0, 11.9 (2 Me). FAB-MS: 511 ([M + H]$^+$). Anal. calc. for C$_{28}$H$_{34}$N$_2$O$_7$·0.25 H$_2$O: C 65.29, H 6.75, N 5.44; found: C 65.41, H 6.80, N 5.10.

Data of **37**: ^{1}H-NMR (CDCl$_3$): 8.75 (br. s, NH); 7.77 (d, J = 1.0, H–C(6)); 7.42–7.25 (m, 10 H, Bn); 6.36 (d, J = 7.9, H–C(1′)); 4.92 (d, J = 11.9, 1 H, Bn); 4.61 (s, 2 H, Bn); 4.59 (d, J = 9.8, 1 H, Bn); 4.20 (d, J = 2.6, H–C(4′)); 4.00 (d, J = 7.9, H–C(2′)); 3.78 (dd, J = 2.8, 11.0, H$_b$–C(5′)); 3.65 (m, H$_a$–C(5′), 2 H–C(3″)); 3.39 (s, MeO); 2.22–2.16 (m, H$_b$–C(2″)); 2.00–1.78 (m, 2 H–C(1″)); 1.62 (m, H$_a$–C(2″)); 1.57 (d, J = 1.0, Me). ^{13}C-NMR (CDCl$_3$): 163.8 (C(4)); 150.7 (C(2)); 138.9, 137.0, 136.5 (Bn, C(6)); 128.8, 128.4, 128.3, 127.8, 127.4, 127.1 (Bn); 111.3 (C(5)); 87.7, 85.7, 83.6, 83.5 (C(1′), C(2′), C(3′), C(4′)); 73.6, 69.6, 65.6, 62.6 (Bn, C(5′), C(3″)); 59.4 (MeO); 26.2, 26.1 (C(1″), C(2″)); 11.9 (Me). FAB-MS: 511 ([M + H]$^+$). Anal. calc. for C$_{28}$H$_{34}$N$_2$O$_7$·0.25 H$_2$O: C 65.29, H 6.75, N 5.44; found: C 65.36, H 6.66, N 5.32.

1-[3,5-Di-O-benzyl-2-O-methyl-3-C-[3-(phthalimido)propyl]-β-D-ribofuranosyl]thymine (**39**). To a soln. of **37** (560 mg, 1.10 mmol) in anh. THF (2 ml) under Ar, phthalimide (0.194 g, 1.32 mmol) and PPh$_3$ (0.362 g, 1.38 mmol) were added, and the mixture was cooled to 0°. A soln. of diethyl diazenedicarboxylate (0.24 ml, 1.36 mmol) in anh. THF (0.3 ml) cooled to 0° was added dropwise during 15 min, and the mixture was stirred for 12 h at r.t. AcOEt (10 ml) was added and the mixture washed successively with a sat. aq. NaCl soln. (2 × 10 ml) and H$_2$O (2 × 10 ml). The aq. phases were extracted with AcOEt (2 × 5 ml). The combined org. phase was dried (Na$_2$SO$_4$) and evaporated. Purification by CC (40–80% AcOEt/petroleum ether) afforded **39** (140 mg, 20%), a fraction (417 mg) consisting of **39** and PPh$_3$(O), and a fraction (137 mg) consisting of unreacted **37** and PPh$_3$(O). ^{1}H-NMR (CDCl$_3$): 8.69 (br. s, NH); 7.80–7.63 (m, 5 H, H–C(6), phth); 7.36–7.20 (m, 10 H, Bn); 6.36 (d, J = 8.0, H–C(1′)); 4.89 (d, J = 11.6, 1 H, Bn); 4.57 (d, J = 11.6, 1 H, Bn); 4.48 (d, J = 11.6, 1 H, Bn); 4.42 (d, J = 11.5, 1 H, Bn); 4.18 (br. s, H–C(4′)); 3.94 (d, J = 8.0, H–C(2′)); 3.73 (m, H$_b$–C(5′), 2 H–C(3″)); 3.57 (d, J = 10.8, H$_a$–C(5′)); 3.36 (s, MeO); 2.20–1.80 (m, 2 H–C(1″), 2 H–C(2″)); 1.48 (s, Me). ^{13}C-NMR (CDCl$_3$): 168.5 (phth); 163.7 (C(4)); 150.6 (C(2)); 138.6, 136.8, 136.2, 134.1, 131.9, 128.8, 128.4, 128.3, 127.5, 127.4, 127.2 (Bn, C(6), phth); 123.3 (phth); 111.3 (C(5)); 87.5, 85.5, 83.3 (C(1′), C(2′), C(3′), C(4′)); 73.5, 69.8, 65.8 (Bn, C(5′)); 59.3 (MeO); 38.0 (C(3″)); 27.2, 22.5 (C(1″), C(2″)); 11.8 (Me). FAB-MS: 640 ([M + H]$^+$). Anal. calc. for C$_{36}$H$_{37}$N$_3$O$_8$·0.5 H$_2$O: C 66.65, H 5.90, N 6.48; found: C 66.69, H 5.84, N 6.53.

1-[2-O-Methyl-3-C-[3-(phthalimido)propyl]-β-D-ribofuranosyl]thymine (**40**). To a soln. of **39** (120 mg, 0.19 mmol) in anh. CH$_2$Cl$_2$ (5 ml) at −78°, 1M BCl$_3$ in hexane (0.75 ml, 0.75 mmol) was added dropwise during

15 min with stirring. After 2 h, the mixture was allowed to slowly warm up to $-20°$, and after further stirring for 6 h, MeOH (3 ml) was added. The mixture was stirred for 12 h at r.t. followed by evaporation and co-evaporation several times with MeOH. Purification by CC ($0-4\%$ MeOH/CH$_2$Cl$_2$) afforded **40** (73 mg, 85%). White solid material. ^{1}H-NMR ((D$_6$)DMSO): 11.3 (br. s, NH); 7.97 (d, $J = 0.8$, H−C(6)); 7.90−7.81 (m, 4 H, phth); 5.94 (d, $J = 7.9$, H−C(1')); 5.24 (t, $J = 3.8$, OH−C(5')); 4.81 (s, OH−C(3')); 3.76 (d, $J = 7.4$, H−C(2')); 3.73 (br. s, H−C(4')); 3.61−3.53 (m, 2 H−C(5'), 2 H−C(3'')); 3.23 (s, MeO); 1.76 (d, $J = 0.8$, Me); 1.81−1.68 (m, 2 H−C(1''), 2 H−C(2'')). ^{13}C-NMR ((D$_6$)DMSO): 168.4 (phth); 164.0 (C(4)); 151.2 (C(2)); 136.9, 134.7, 132.0, 123.4 (C(6), phth); 110.2 (C(5)); 86.8, 85.6, 85.3, 78.4 (C(1'), C(2'), C(3'), C(4')); 61.0, 58.6 (MeO, C(5')); 38.4 (C(3'')); 31.0, 22.8 (C(1''), C(2'')); 12.7 (Me). HR-FAB-MS: 460.1780 (C$_{22}$H$_{26}$N$_3$O$_8^+$; calc. 460.1720).

1-[5-O-(4,4'-Dimethoxytrityl)-2-O-methyl-3-C-[3-(phthalimido)propyl]-β-D-ribofuranosyl]thymine (**41**). To a soln. of **40** (85 mg, 0.19 mmol) in anh. pyridine (1.5 ml) at r.t., (MeO)$_2$TrCl (125 mg, 0.37 mmol) was added in one portion, and the mixture was stirred for 12 h, whereupon MeOH (0.25 ml) was added to quench the reaction. After stirring for 10 min, the mixture was evaporated, the residue dissolved in CH$_2$Cl$_2$ (5 ml), and the soln. washed with sat. aq. NaCl soln. (3×5 ml), dried (Na$_2$SO$_4$), and evaporated. Purification by CC ($0-1.5\%$ MeOH in CH$_2$Cl$_2$/pyridine 99:1) afforded **41** (117 mg, 77%). Yellowish solid material. ^{1}H-NMR (CDCl$_3$): 8.73 (br. s, NH); 7.83−7.69 (m, 5 H, H−C(6), phth); 7.40−7.16 (m, 9 H, (MeO)$_2$Tr); 6.86 (dd, $J = 1.6$, 9.1, 4 H, (MeO)$_2$Tr); 6.19 (d, $J = 6.6$, H−C(1')); 4.10 (br. s, H−C(4')); 3.91 (d, $J = 6.8$, H−C(2')); 3.80, 3.79 (2s, 2 MeO); 3.70 (dd, $J = 3.0$, 10.9, H$_b$−C(5')); 3.51 (s, MeO); 3.41−3.24 (m, H$_a$−C(5'), 2 H−C(3'')); 2.87 (s, OH−C(3')); 1.69−1.50 (m, 2 H−C(1''), 2 H−C(2'')); 1.14 (d, $J = 0.8$, Me). ^{13}C-NMR (CDCl$_3$): 168.4 (phth); 163.7 (C(4)); 159.1 ((MeO)$_2$Tr); 150.6 (C(2)); 143.5 ((MeO)$_2$Tr); 136.1, 134.6, 134.0, 132.1, 130.5, 129.1, 128.3, 128.1, 127.6, 136.1, 125.4, 123.3 (C(6), (MeO)$_2$Tr, phth); 113.3 ((MeO)$_2$Tr); 111.8 (C(5)); 87.7, 86.2, 85.7, 85.1, 78.7 (C(1'), C(2'), C(3'), C(4'), C(Ar$_3$)); 62.2 (C(5')); 59.1 (MeO); 55.2 (2 MeO); 37.9 (C(3'')); 30.5, 22.6 (C(1''), C(2'')); 11.0 (Me). FAB-MS: 761 (M^+). Anal. calc. for C$_{43}$H$_{43}$N$_3$O$_{10}$·0.6 C$_7$H$_8$: C 69.41, H 5.90, N 5.14; found: C 69.30, H 6.07, N 5.03.

1-[5-O-(4,4'-Dimethoxytrityl)-2-O-methyl-3-C-[3-(phthalimido)propyl]-β-D-ribofuranosyl]thymine 3'-(2-Cyanoethyl Diisopropylphosphoramidite) (**42**). To a soln. of **41** (109 mg, 0.14 mmol) in anh. CH$_2$Cl$_2$ (2 ml) and ^{i}Pr$_2$EtN (0.20 ml, 1.20 mmol) under Ar at 0°, 2-cyanoethyl diisopropylphosphoramidochloridite (0.14 ml, 0.58 mmol) was added dropwise, and the mixture was stirred for 12 h at r.t. The reaction was quenched by addition of sat. aq. NaHCO$_3$ soln. (0.1 ml), and the resulting mixture was diluted with AcOEt (5 ml) and washed with a sat. aq. NaHCO$_3$ soln. (4×5 ml). The org. phase was dried (Na$_2$SO$_4$) and evaporated. After purification by CC (1. $0-1\%$ Et$_3$N/CH$_2$Cl$_2$; 2. AcOEt/CH$_2$Cl$_2$/Et$_3$N/petroleum ether 15:45:1:39; 3. AcOEt/CH$_2$Cl$_2$/Et$_3$N/ petroleum ether 15:35:1:49), the residue was dissolved in anh. toluene (1 ml) and precipitated in vigorously stirred petroleum ether (250 ml) at $-50°$: **42** (45 mg, 34%). White solid material. ^{31}P-NMR (CDCl$_3$): 141.3, 141.2. FAB-MS: 984 ([$M+$Na]$^+$).

Oligonucleotide Synthesis. All ONs were prepared on a *Biosearch 8700* DNA synthesizer by the phosphoramidite approach as described earlier [61] with *tert*-butyl hydroperoxide as oxidant for ONs containing units **Z** or **W**, and with a mixture of I$_2$, pyridine, and H$_2$O as oxidant for ONs containing units **X** or **Y**. After completion of the sequences, deprotection with conc. NH$_3$ in MeOH (32% (w/w), 55°, 12 h) of 5'-O-(MeO)$_2$Tr-on ONs and reversed-phase cartridge purification yielded the final ON products, which by capillary gel electrophoresis were shown to be $>85\%$ pure. The composition of the modified ONs was verified by MALDI-MS analysis.

Thermal Stability Studies. Melting temperatures (T_m values) were determined as described earlier [25]. A medium salt buffer and a high salt buffer with compositions as indicated in *Table 1* were used.

REFERENCES

[1] a) H. M. Pfundheller, J. Wengel, *Bioorg. Med. Chem. Lett.* **1999**, *9*, 2667; b) J. Wengel, *Acc. Chem. Res.* **1999**, *32*, 301.

[2] S. M. Freier, K. H. Altmann, *Nucleic Acids Res.* **1997**, *25*, 4429.

[3] A. M. Kawasaki, M. D. Casper, S. M. Freier, E. A. Lesnik, M. C. Zounes, L. L. Cummins, C. Gonzales, P. D. Cook, *J. Med. Chem.* **1993**, *36*, 831.

[4] B. P. Monia, E. A. Lesnik, C. Gonzales, W. F. Lima, D. M. McGee, C. J. Guinosso, A. M. Kawasaki, P. D. Cook, S. M. Freier, *J. Biol. Chem.* **1993**, *268*, 14514.

[5] H. Inoue, Y. Hayase, A. Imura, S. Iwai, K. Miura, E. Ohtsuka, *Nucleic Acids Res.* **1987**, *15*, 6131.

[6] E. A. Lesnik, C. J. Guinosso, A. M. Kawasaki, H. Sasmor, M. Zounes, L. L. Cummins, D. J. Ecker, P. D. Cook, S. M. Freier, *Biochemistry* **1993**, *32*, 7832.

[7] L. L. Cummins, S. R. Owens, L. M. Risen, E. A. Lesnik, S. M. Freier, D. McGee, C. J. Guinosso, P. D. Cook, *Nucleic Acids Res.* **1995**, *23*, 2019.

[8] M. Grøtli, M. Douglas, R. Eritja, B. S. Sproat, *Tetrahedron* **1998**, *54*, 5899.

[9] W. Saenger, 'Principles of Nucleic Acid Structure', Springer-Verlag, New York, 1984.

[10] J. Fensholdt, H. Thrane, J. Wengel, *Tetrahedron Lett.* **1995**, *36*, 2535.

[11] H. Thrane, J. Fensholdt, M. Regner, J. Wengel, *Tetrahedron* **1995**, *51*, 10389.

[12] G. Wang, W. E. Seifert, *Tetrahedron Lett.* **1996**, *37*, 6515.

[13] Y. Ueno, M. Kanazaki, S. Shuto, A. Matsuda, Poster 281 at the 'XIII International Round Table Nucleosides, Nucleotides and Their Biological Applications', September 6–10, 1998, Montpellier, France.

[14] Y. Ueno, Y. Nagasawa, I. Sugimoto, N. Kojima, M. Kanazaki, S. Shuto, A. Matsuda, *J. Org. Chem.* **1998**, *63*, 1660.

[15] G. Wang, P. J. Middleton, C. Lin, Z. Pietrzowski, *Bioorg. Med. Chem. Lett.* **1999**, *9*, 885.

[16] A. Marx, P. Erdmann, M. Senn, S. Körner, T. Jungo, M. Petretta, P. Imwinkelried, A. Dussy, K. J. Kulicke, L. Macko, M. Zehnder, B. Giese, *Helv. Chim. Acta* **1996**, *79*, 1980.

[17] R. H. Griffey, B. P. Monia, L. L. Cummins, S. M. Freier, M. J. Greig, C. J. Guinosso, E. A. Lesnik, S. M. Manalili, V. Mohan, S. Owens, B. R. Ross, H. Sasmor, E. Wancewicz, K. Weiler, P. D. Wheeler, P. D. Cook, *J. Med. Chem.* **1996**, *39*, 5100.

[18] R. Buff, J. Hunziker, *Bioorg. Med. Chem. Lett.* **1998**, *8*, 521.

[19] H. M. Pfundheller, P. N. Jørgensen, U. S. Sørensen, S. K. Sharma, M. Grimstrup, C. Ströch, P. Nielsen, G. Viswanadham, C. E. Olsen, J. Wengel, *J. Chem. Soc., Perkin Trans. 1* **1998**, 1409.

[20] P. N. Jørgensen, U. S. Sørensen, H. M. Pfundheller, C. E. Olsen, J. Wengel, *J. Chem. Soc., Perkin Trans. 1* **1997**, 3275.

[21] P. N. Jørgensen, P. C. Stein, J. Wengel, *J. Am. Chem. Soc.* **1994**, *116*, 2231.

[22] C. Schmit, M.-O. Bèvierre, A. De Mesmaeker, K. H. Altmann, *Bioorg. Med. Chem. Lett.* **1994**, *4*, 1969.

[23] G. Wang, P. J. Middleton, L. He, V. Stoisavljevic, W. E. Seifert, *Nucleosides Nucleotides* **1997**, *16*, 445.

[24] T. Waga, T. Nishizaki, I. Miyakawa, H. Ohrui, H. Meguro, *Biosci. Biotech. Biochem.* **1993**, *57*, 1433.

[25] A. A. Koshkin, S. Singh, P. Nielsen, V. K. Rajwanshi, R. Kumar, M. Meldgaard, C. E. Olsen, J. Wengel, *Tetrahedron* **1998**, *54*, 3607.

[26] S. David, S. Hanessian, *Tetrahedron* **1985**, *41*, 643.

[27] A. Veyrières, *J. Chem. Soc., Perkin Trans. 1* **1981**, 1626.

[28] P. Nielsen, J. Wengel, *Chem. Commun.* **1998**, 2645.

[29] I. A. Mikhailopulo, N. E. Poopeiko, T. M. Tsvetkova, A. P. Marochkin, J. Balzarini, E. De Clerq, *Carbohydr. Res.* **1996**, *285*, 17.

[30] H. Vorbrüggen, K. Krolikiewicz, B. Bennua, *Chem. Ber.* **1981**, *114*, 1234.

[31] H. Vorbrüggen, G. Höfle, *Chem. Ber.* **1981**, *114*, 1256.

[32] C. O-Yang, W. Kurz, E. M. Eugui, M. J. McRoberts, J. P. H. Verheyden, L. J. Kurz, K. A. M. Walker, *Tetrahedron Lett.* **1992**, *33*, 41.

[33] S. K. Singh, P. Nielsen, A. A. Koshkin, J. Wengel, *Chem. Commun.* **1998**, 455.

[34] H. M. Pfundheller, A. A. Koshkin, C. E. Olsen, J. Wengel, *Nucleosides Nucleotides* **1999**, *18*, 2017.

[35] J. J. Fox, N. C. Miller, *J. Org. Chem.* **1963**, *28*, 936.

[36] W. J. Middleton, *J. Org. Chem.* **1975**, *40*, 574.

[37] Y. Sato, K. Utsumi, T. Maruyama, T. Kimura, I. Yamamoto, D. D. Richman, *Chem. Pharm. Bull.* **1994**, *42*, 595.

[38] K. W. Pankiewics, B. Nawrot, H. Gadler, R. W. Price, K. A. Watanabe, *J. Med. Chem.* **1987**, *30*, 2314.

[39] B. Doboszewski, G. W. Hay, W. A. Szarek, *Can. J. Chem.* **1987**, *65*, 412.

[40] R. K. Sharma, J. L. Fry, *J. Org. Chem.* **1983**, *48*, 2112.

[41] D. P. Cox, J. Terpinski, W. Lawrynowics, *J. Org. Chem.* **1984**, *49*, 3216.

[42] M. J. Robins, E. M. Trip, *Tetrahedron Lett.* **1974**, *38*, 3369.

[43] A. Fraser, P. Wheeler, P. D. Cook, Y. S. Sanghvi, *J. Heterocycl. Chem.* **1993**, *30*, 1277.

[44] R. Ranganathan, *Tetrahedron Lett.* **1977**, *15*, 1291.

[45] H. Ikeda, R. Fernandez, A. Wilk, J. J. Barchi, Jr., X. Huang, V. E. Marquez, *Nucleic Acids Res.* **1998**, *26*, 2237.

[46] D. B. Davies, *Prog. NMR Spectrosc.* **1978**, *12*, 135.

[47] S. M. Gryaznov, R. L. Letsinger, *Nucleic Acids Res.* **1992**, *20*, 1879.

[48] M. Beier, W. Pfleiderer, *Helv. Chim. Acta* **1999**, *55*, 879.

[49] A. Krug, T. S. Oretskaya, E. M. Volkov, D. Cech, Z. A. Shabarova, A. Rosenthal, *Nucleosides Nucleotides* **1989**, *8*, 1473.

[50] D. M. Williams, F. Benseler, F. Eckstein, *Biochemistry* **1991**, *30*, 4001.

[51] R. G. Schultz, S. M. Gryaznov, *Nucleic Acids Res.* **1996**, *24*, 2966.

[52] K. D. Nielsen, F. Kirpekar, P. Roepstorff, J. Wengel, *Bioorg. Med. Chem.* **1995**, *3*, 1493.

[53] M. Majlessi, N. C. Nelson, M. M. Becker, *Nucleic Acids Res.* **1998**, *26*, 2224.

[54] Y. S. Sanghvi, G. D. Hoke, S. M. Freier, M. C. Zounes, C. Gonxales, L. Cummins, H. Sasmor, P. D. Cook, *Nucleic Acids Res.* **1993**, *21*, 3197.

[55] O. Mitsunobu, M. Wada, T. Sano, *J. Am. Chem. Soc.* **1972**, *94*, 679.

[56] C. Scheuer-Larsen, B. M. Dahl, J. Wengel, O. Dahl, *Tetrahedron Lett.* **1998**, *39*, 8361.

[57] W. Bannwarth, A. Trzeciak, *Helv. Chim. Acta* **1987**, *70*, 175.

[58] L. H. Koole, H. M. Buck, J.-M. Vial, J. Chattopadhyaya, *Acta Chem. Scand.* **1989**, *43*, 665.

[59] S. K. Sharma, J. Wengel, University of Copenhagen, unpublished data.

[60] M.-O. Bèvierre, A. De Mesmaeker, R. M. Wolf, S. M. Freier, *Bioorg. Med. Chem. Lett.* **1994**, *4*, 237.

[61] M. Raunkjær, C. E. Olsen, J. Wengel, *J. Chem. Soc., Perkin Trans. 1* **1999**, 2543.

1*H*-Benzotriazole as Synthetic Auxiliary in a Facile Route to N⁶-(Arylmethyl)-2′-deoxyadenosines: DNA Intercalators Inserted into Three-Way Junctions

by **Sherif A. El-Kafrawy**[a]), **Magdy A. Zahran**[b]), **Erik B. Pedersen**[a])*, and **Claus Nielsen**[c])

[a]) Department of Chemistry, University of Southern Denmark, Odense Universitet, DK-5230 Odense M
[b]) Chemistry Department, Faculty of Sciences, Menoufia University, Egypt
[c]) Department of Virology, State Serum Institute, DK-2300 Copenhagen S
(fax: +4566158780, e-mail: EBP@Chem.sdu.dk)

The 2′-deoxy-N⁶-(naphthalen-1-ylmethyl)- (**5a**) and N⁶-(pyren-1-ylmethyl)adenosine (**5b**) were synthesized in two steps from 2′-deoxyadenosine and the adequate arenecarbaldehyde with 1*H*-benzotriazole as a synthetic auxiliary (*Scheme*). When the N⁶-(arylmethyl)-2′-deoxyadenosines were inserted into the junction region of a DNA three-way junction, its thermal stability increased.

1. Introduction. – Carcinogenic polycyclic aromatic hydrocarbons are known to undergo metabolic activation to reactive diol epoxide intermediates that bind covalently to DNA *in vivo* [1–3]. Moreover, the presence of bulky polycyclic aromatic residues derived from the binding of stereoisomeric mutagenic and tumorigenic metabolites of benzo[*a*]pyrene to 2′-deoxyguanosine residues in DNA is known to block replication [4] and to induce mutations [5], and is believed to constitute the critical initial step in carcinogenesis [6]. To gain greater insight into polyaromatic hydrocarbon (PAH) adducts to DNA and the biological consequences of their replication, efficient methods for the site-specific synthesis of PAH-oligodeoxynucleotides, with the PAH attached covalently to the exocyclic amino groups of specific dA and dG bases, are required.

Oligonucleotides complementary to strategic regions of viral or messenger RNAs were first shown by *Zamecnik* and *Stephenson* [7] to inhibit viral replication. The factors that are usually assumed to limit the activity of antisense oligonucleotides are cellular uptake, resistance to nuclease, and the stability of the hybrid formed. Modifications are usually chosen to improve one or more of these properties. However, a steady improvement in activity has been achieved by this rational approach that is encouraging for the design of conjugates to meet specific requirements. The manipulation of sequence-specific protein binding and gene expression with antigene or antisense technology requires oligonucleotides that bind DNA or RNA with high affinity and specificity [8–11]. Oligonucleotides capable of selective recognition of RNA must accommodate, in addition, both the kinetic [12][13] and thermodynamic [13][14] consequences of nonuniform RNA conformation. The structural complexity inherent in RNA complicates recognition because it limits the accessibility of single-stranded regions that must be differentiated for specific binding. It is estimated that

11–15 nucleotides are necessary to define a unique mRNA sequence in eukaryotic cell [15], yet few well-characterized RNA secondary structures contain contiguous single-stranded regions of this size.

The object of this investigation will be focused on the effect of intercalators as a chemical modification on strengthening the binding force in an oligonucleotide three-way junction (TWJ). It is believed that three-way junctions with two short single-stranded arms are interesting target sites for antisense oligonucleotides when contiguous single-stranded regions are not available for binding. However, the stability of the hybridization of oligodeoxynucleotides to TWJ single-stranded arms is lower compared to the corresponding duplex with the same number of base pairs [16][17].

2. Synthesis. – In continuation of our previous work [18], we now report a facile and short direct route to synthesize 2′-deoxy-N^6-(naphthalen-1-ylmethyl)adenosine (**5a**) and 2′-deoxy-N^6-(pyren-1-ylmethyl)adenosine (**5b**) by means of 1*H*-benzotriazole [19] as a facile and versatile reagent, and this in good yield and without any by-products [20][21] arising from direct alkylation at other sites, *e.g.*, at N(1), N(3) or N(7). *Harvey* and co-workers [22] reported recently the synthesis of 2′-deoxy-N^6-(pyren-1-ylmethyl)-adenosine (**5b**), which involves in the key step coupling of an appropriately protected halopurine derivative with the amino derivative of the polyaromatic hydrocarbon, in this case pyrene-1-methanamine. Our synthetic strategy entailed in the key step the condensation of 1*H*-benzotriazole (**2**) with the appropriate polyaromatic aldehydes **3a,b** and with 2′-deoxyadenosine (**1**) by refluxing this mixture in toluene with azeotropic removal of H_2O (*Dean-Stark*) in an acid-catalyzed reaction, *i.e.*, the reaction took place in the presence of a few drops of AcOH (*Scheme*). After *ca.* 6 h, the solvent was evaporated, and the 1*H*-benzotriazole adducts **4a** or **4b** were obtained from the residue after flash-chromatographic (FC) purification in 85 and 80% yield, respectively. The 1*H*-benzotriazole adducts **4a** and **4b** were then reduced by LiAlH$_4$ in dry THF to give **5a** or **5b** in 80 and 70% yield, respectively, after FC purification. The structures of the adducts **4a,b** were elucidated from their ^{1}H- and ^{13}C-NMR and FAB mass spectra. The former spectra were complex due to the presence of the stereogenic C-atom Bt-CH-Ar. The diastereomer mixtures could not be separated by column chromatography.

In our previous study [18], we had the problem of very bad yields, or no reaction at all, when the reaction between 1*H*-benzotriazole (**2**), 2′-deoxyadenosine (**1**), and polycyclic aromatic aldehyde **3** was carried out by reflux in EtOH in the presence of a catalytic amount of AcOH. We believed that this may be due to the steric hindrance around the stereogenic C-atom (N^6-CH) generated by the presence of the 1*H*-benzotriazol-1-yl and the bulky polycyclic aromatic group at the same C-atom. We also thought that the H_2O liberated during the condensation reaction could retard the reaction progress. Therefore, we now modified the reaction conditions by carrying out the reaction in refluxing toluene, using the *Dean-Stark* trap to remove azeotropically the H_2O formed, to force the reaction to proceed to completion.

The reduction of the adducts **4a,b** was performed with LiAlH$_4$ in THF at room temperature to give the N^6-(arylmethyl)deoxyadenosines **5a,b** in 70–80% yield. Attempts to carry out the reduction with NaBH$_4$ resulted in poor yields and required long reaction times. The spectra of 2′-deoxy-N^6-(pyren-1-ylmethyl)adenosine (**5b**) were in good agreement with those reported [24].

Scheme

Naph = naphthalen-1-yl, Pyr = pyren-1-yl, Bt = 1*H*-benzotriazol-1-yl, DMT = 4,4′-dimethoxytrityl = (MeO)$_2$Tr

The 4,4′-dimethoxytrityl ((MeO)$_2$Tr), and the phosphoramidite derivatives of **5a,b**, *i.e.*, **6a,b** and **7a,b**, respectively, were prepared according to *Caruthers*' methodology [23]. For the preparation of **7a,b** the (MeO)$_2$Tr derivatives **6a,b** were treated with 2-cyanoethyl diisopropylphosphoramidochlorite in the presence of ^{i}Pr$_2$EtN. The purity of the obtained phosphoramidites **7a,b** (yield 83 and 85%, resp.) was 94 and 99%, respectively, according to ^{31}P-NMR, the remaining 1–6% being non-phosphoramidite by-products.

The modified and unmodified oligodeoxynucleotides (ODNs) were synthesized on a *Pharmacia Gene-Assembler-Special* synthesizer on a 0.2-μmol scale according to the standard phosphoramidite method [23] with **7** and commercial phosphorami-dites. Phosphorothioates were prepared with the *Beaucage* reagent [25][26] (3*H*-1,2-benzodithiol-3-one 1,1-dioxide) on a 0.2-μmol scale. The ability of ODNs to hybridize to their complementary DNA strand was examined by UV melting measurements. The melting points T_m were determined as the maxima of the first derivative of the melting curves. DNA Three-way junctions were formed from equimolar amounts (3 μM) in each strand at pH 7.0 in a 1 mM EDTA, 10 mM Na$_2$HPO$_4$, and 140 mM NaCl buffer.

3. Stability of Three-Way Junction. – To study the intercalating ability of the prepared modified nucleosides **5**, we chose a three-way junction composed of two dangling single strands and a double-stranded DNA stem rich in C · G base pairs to shift the melting temperature of the stem to a higher temperature to avoid its interference with those obtained when hybridizing the dangling strands (*Fig. 1*).

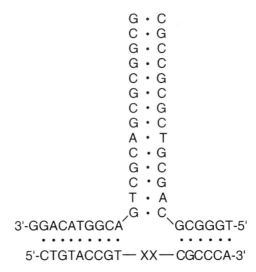

Fig. 1. *Three-way junction I.* XX = Site of insertion. The nucleoside symbols represent 2'-deoxynucleosides.

On inserting compound **5a** and an extra dG as a bulge in the junction region of the three-way junction I (*Table 1, Entry 3*), an increase in T_m of 6.4° was observed compared to the insertion of dA and dG as a reference (*Entry 2*). Insertion of **5a** in both the 5'-end and the junction region had a small further effect (*Entry 7, $\Delta T_m = 6.8°$*) when compared to the reference with dA instead of **5a** (*Entry 4*). When the reference ODN had **5a** added at the 5'-end and dG and dA in the junction region (*Entry 6*), a $\Delta T_m = 3.2°$ was observed when **5a** was inserted instead of dA in the junction region (*Entry 7*). Mismatches in the targeting ODN resulted in remarkable decreases in the

Table 1. *Hybridization Data for the Three-Way Junction I with Inserted dG.* **X = 5a**.

Entry	Targeting ODN			T_m [°C]
1	5'-d(CTGTACCGT		CGCCCA)-3'	41.2
2	5'-d(CTGTACCGT	AG	CGCCCA)-3'	44.8
3	5'-d(CTGTACCGT	**X**G	CGCCCA)-3'	51.2
4	5'-d(ACTGTACCGT	AG	CGCCCA)-3'	44.0
5	5'-d(ACTGTACCGT	**X**G	CGCCCA)-3'	48.4
6	5'-d(**X**CTGTACCGT	AG	CGCCCA)-3'	47.6
7	5'-d(**X**CTGTACCGT	**X**G	CGCCCA)-3'	50.8
8	5'-d(CT<u>T</u>GACCGT	**X**G	CGCCCA)-3'[a])	31.2
9	5'-d(CTGTACCGT	**X**G	C<u>C</u>GCCA)-3'[a])	40.0

[a]) Mismatch site in the targeting ODN is underlined.

stabilization of the three-way junction (*Entries 8 and 9*). This confirms hybridization in both arms of TWJ.

Insertion of a dG in the junction region could lead to base pairing with the first dC in the stem, leading to the disruption of the corresponding first base pair in the stem. In fact, DNA folding calculations [27] show that inserted d(AG) in the junction region leads to disruption of the first two base pairs in the stem (d(GC) and d(AT)). To overcome this problem, another antisense ODN containing dT and dA and dT and **5a** insertions in the junction region was used. The hybridization data shown in *Table 2* showed a $\Delta T_m = 6.0°$ when inserting dT and **5a** in the junction region (*Entry 11*) compared to the reference with d(TA) in the same region (*Entry 13*). A more pronounced effect ($\Delta T_m = 13.6°$) was achieved when two modified nucleosides **5a** (*Entry 12*) were inserted in the junction region compared to the insertion of two dA (*Entry 14*) in the same region. As a preliminary test for the increased binding properties of **5b** over **5a**, a single trial was made with the insertion of dT and **5b** in the junction region, and a $\Delta T_m = 10.4°$ was observed (*Entry 11 vs. 13*). We have recently suggested that intercalators inserted at TWJ junctions stabilize the TWJ by stacking on top of the deflecting arm [28]. Further stabilization by insertion of two intercalators (*Entry 12*) could then be due to stacking of both intercalators at the top of the deflecting arm.

Table 2. *Hybridization Data for the Three-Way Junction I Avoiding Inserted dG.* **X** = **5a**.

Entry	Targeting ODN			T_m [°C]
10	5'-d(CTGTACCGT		CGCCCA)-3'	41.2
11	5'-d(CTGTACCGT	**TX**	CGCCCA)-3'	44.8 (49.2)[a]
12	5'-d(CTGTACCGT	**XX**	CGCCCA)-3'	51.2
13	5'-d(CTGTACCGT	TA	CGCCCA)-3'	38.8
14	5'-d(CTGTACCGT	AA	CGCCCA)-3'	37.6

[a]) Value in parentheses for **X** = **5b**.

The human immunodeficiency virus (HIV) contains at least two structured RNAs that interact with viral protein and mediate novel genetic regularity pathways. One of these is a 234-nucleotide sequence located within the *env* coding region called the Rev response element (RRE). Interaction of the RRE with viral protein Rev [29][30] regulates the appearance of unspliced or singly spliced viral mRNAs in the cytoplasm of infected cells [31–40]. *Cload* and *Schepartz* [41] studied the *in vitro* inhibitory effect of some tethered oligonucleotide probes (TOPs) on the function of RRE. We chose the same antisense sequence (*Fig. 2*) as that reported by *Cload* and *Schepartz* to have potent concentration-dependent inhibition of RRE function *in vitro*. Only the tethering propane-1,3-diol moiety was replaced with an intercalating nucleoside.

Thus, a ΔT_m of 12.2° was observed when inserting **5b** and dG in the junction region (*Table 3, Entry 16*) compared to *Entry 15*. Also mismatching by interchanging two nucleosides in the antisense ODN decreased the stability of the system (*Entries 17 and 18*), indicating that full complementarity took place between the three strands of the system.

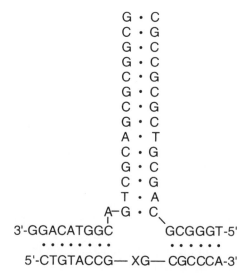

```
          G · C
          C · G
          G · C
          G · C
          C · G
          G · C
          C · G
          G · C
          A · T
          C · G
          G · C
          C · G
          T · A
       A–G · C
       |            \
3'-GGACATGGC          GCGGGT-5'
 · · · · · · · ·      · · · · · ·
5'-CTGTACCG— XG— CGCCCA-3'
```

Fig. 2. *Three-way junction II*. XG = Site of insertion. The nucleoside symbols represent 2'-deoxynucleosides.

Considering the stabilizing effect of the pyrenyl intercalator on DNA TWJ, it was found interesting to test the activity of the corresponding phosphorothioates against HIV-1 in MT-4 cells. The ODNs given in *Table 3* were also synthesized as their phosphorothioate analogues, as well as additional examples with dA, d(AA), or d(XX) insertions in the middle of the oligonucleotide. The HIV-1 strain HTLV-IIIB was propagated in MT-4 cells at 37° with RPMI-1640, 10% heat-inactivated fetal-calf serum (FCS), and antibiotics (growth medium). The cells were preincubated with lipofectin to improve their uptake of the oligonucleotide. Unfortunately, none of the prepared phosphorothioates showed any *anti*-HIV activity. Moreover, they were cytotoxic at 100 μM.

Table 3. *Hybridization Data for the Three-Way Junction II*. **X = 5b**.

Entry	Targeting ODN		T_m [°C]
15	5'-d(CTGTACCG	CGCCCA)-3'	40.8
16	5'-d(CTGTACCG **XG**	CGCCCA)-3'	53.0
17	5'-d(CTGTACCG **XG**	C<u>C</u>GCCA)-3' (m)[a]	33.2
18	5'-d(CT<u>TG</u>ACCG **XG**	CGCCCA)-3' (m)[a]	40.4

[a]) Mismatch site in the targeting ODN is underlined.

Experimental Part

General. Anal. TLC: *Merck* precoated silica gel *60 F₂₅₄* plates. Flash column chromatography (FC): silica gel (0.040–0.063 mm) from *Merck*. NMR Spectra: at 300 (^{1}H), 75.5 (^{13}C), and 121.5 MHz (^{31}P); *Varian-Gemini 2000* 300-MHz spectrometer; δ values in ppm rel. to SiMe₄ as an internal standard (^{1}H, ^{13}C) and rel. to 85% H₃PO₄ as external standard (^{31}P). FAB-MS: positive mode; *Kratos MS-50-RF* spectrometer.

Oligonucleotides. Melting experiments were carried out on a *Perkin-Elmer* UV/VIS spectrometer *Lambda 2* fitted with a *PTP-6-Peltier* temp.-programming element. The UV absorbance at 260 nm as a function of time was recorded while the temp. was raised gradually (1°/min) in a 1-cm cuvette, at a 3 μM concentration of each component. DNA Syntheses were performed on a *Pharmacia Gene-Assembler-Special* DNA synthesizer, on a

0.2-μmol scale. The coupling efficiencies for the modified amidites (20-min couplings) **7a,b** were *ca.* 85% compared to 99% for the commercial ones (2-min coupling). The efficiency of each coupling step was monitored by the release of dimethoxytrityl cation after each step. Removal from solid support and deprotection was carried out at r.t. for 4 days in 25% ammonia and 0.01m EDTA in 25% ammonia for the phosphorothioates. Desalting of the oligonuclotides was accomplished with disposable *NAP-10* columns (*Pharmacia*).

N⁶-[(1H-Benzotriazol-1-yl)(naphthalen-1-yl)methyl]-2′-deoxyadenosine (**4a**). A soln. of 2′-deoxyadenosine (**1**; 2.51 g, 10 mmol), *1H*-benzotriazole (**2**; 1.43 g, 12 mmol), and naphthalene-1-carboxaldehyde (**3a**; 1.6 ml, 12 mmol) was refluxed in toluene in the presence of 5 drops of AcOH as a catalyst with azeotropic removal of H_2O in a *Dean-Stark* apparatus. After 6 h, the solvent was evaporated and the residue submitted to FC (CHCl₃/MeOH 99:1): **4a** (4.3 g, 85%). FAB-MS (CHCl₃ + 3-nitrobenzyl alcohol): 509 ($[M+H]^+$).

N⁶-[(1-H-Benzotriazol-1-yl)pyren-1-yl)methyl]-2′-deoxyadenosine (**4b**): As described for **4a**, with **1** (2.51 g, 10 mmol), **2** (1.43 g, 12 mmol), pyrene-1-carboxaldehyde (**3b**; 2.76 g, 12 mmol), toluene and 5 drops of AcOH: **4b** (4.6 g, 80%). FAB-MS (CHCl₃ + 3-nitrobenzyl alcohol): 583 ($[M+H]^+$).

2′-Deoxy-N⁶-(naphthalen-1-ylmethyl)adenosine (**5a**). A soln. of **4a** (2.55 g, 5 mmol) was stirred with a suspension of LiAlH₄ (1.14 g, 30 mmol) in dry THF at r.t. for 2 h. The mixture was then poured into ice-water, neutralized with AcOH and extracted with CHCl₃. The crude mixture was purified by FC (CHCl₃/MeOH 97:3): **5a** (1.56 g, 80%). ¹H-NMR (CDCl₃): 8.39 (*s*, H−C(8)); 8.06 (*s*, H−C(2)); 7.86−7.36 (*m*, Naph.); 6.12 (*m*, H−C(1′)); 5.25 (br., NHCH₂Ar); 4.69 (*d*, 1 OH); 4.07 (*m*, H−C(3′)); 3.92 (*m*, H−C(4′)); 3.77 (*m*, 2 H−C(5′)); 2.94 (*m*, 1 H−C(2′)); 2.17 (*m*, 1 H−C(2′)). ¹³C-NMR (CDCl₃): 154.86 (C(6)); 152.71 (C(2)); 147.86 (C(4)); 139.47 (C(8)); 133.88, 133.36, 131.53, 128.83, 128.70, 126.63, 126.49, 126.03, 125.42, 123.48 (Naph); 121.17 (C(5)); 89.54 (C(4′); 87.56 (C(1′)); 73.06 (C(3′)); 63.25 (C(5′)); 42.57 (NHCH₂Ar); 40.67 (C(2′)). FAB-MS (CHCl₃ + 3-nitrobenzyl alcohol): 392 ($[M+H]^+$).

2′-Deoxy-N⁶-(pyren-1-ylmethyl)adenosine (**5b**). As described for **5a**, with **4b** (1.16 g, 2 mmol) and LiAlH₄ (0.456 g, 12 mmol) in dry THF: **5b** (0.65 g, 70%). ¹H-NMR ((D₆)DMSO): 8.60 (*s*, H−(C8)); 8.56 (*s*, H−C(2)); 8.42−8.05 (*m*, Pyr); 6.41 (*m*, H−C(1′)); 5.49 (br., NHCH₂Ar); 5.35 (*d*, 1 OH); 4.61 (*t*, 1 OH); 4.46 (*m*, H−C(3′)); 3.94 (*m*, H−C(4′)); 3.61 (*m*, 2 H−C(5′)); 2.78 (*m*, 1 H−C(2′)); 2.32 (*m*, 1 H−C(2′)). ¹³C-NMR (CDCl₃): 154.30 (C(6)); 152.36 (C(2)); 147.79 (C(4)); 139.03 (C(8)); 131.19, 130.51, 130.15, 130.00, 128.10, 127.23, 126.76, 126.62, 125.93, 25.41, 124.62, 124.55, 124.10, 123.90, 122.32 (Pyr); 120.48 (C(5)); 88.74 (C(4′)); 86.33 (C(1′)); 71.61 (C(3′)); 62.50 (C(5′)); 41.58 (NHCH₂Ar), 40.03 (C(2′)). FAB-MS (CHCl₃ + 3-nitrobenzyl alcohol): 466 ($[M+H]^+$).

4,4′-Dimethoxytrityl Derivatives **6**: *General Procedure.* Compound **5a** or **5b** (2 mmol) was stirred in pyridine at r.t. with 2.2 mmol of 4,4′-dimethoxytrityl chloride for 3 h. The pyridine was evaporated and the residue submitted to FC (CHCl₃): yield 90%.

2′-Deoxy-5′-O-(4,4′-dimethoxytrityl)-N⁶-(naphthalen-1-ylmethyl)adenosine (**6a**): ¹H-NMR (CDCl₃): 8.42 (*s*, H−C(8)); 8.06 (*s*, H−C(2)); 7.84−7.13 (*m*, arom. H); 6.42 (*m*, H−C(1′), NH); 5.30 (br., NHCH₂Ar); 4.62 (*m*, OH); 4.10 (*m*, H−C(3′)); 3.80−3.69 (*m*, MeO, H−C(4′)); 3.37 (*m*, 2 H−C(5′)); 2.72 (*m*, 1 H−C(2′)); 2.47 (*m*, 1 H−C(2′)). ¹³C-NMR (CDCl₃): 154.95 (C(6)); 152.25 (C(2)); 149.63 (C(4)); 138.25 (C(8)); 136.37−123.52 (arom. C); 120.04 (C(5)); 113.20 ((MeO)₂*Tr*); 86.01 (C(4′)); 84.20 (C(1′)); 72.22 (C(3′)); 63.64 (C(5′)); 55.11 (MeO); 42.80 (NHCH₂); 40.22 (C(2′)). FAB-MS (CHCl₃ + 3-nitrobenzyl alcohol): 694 ($[M+H]^+$).

2′-Deoxy-5′-O-(4,4′-dimethoxytrityl)-N⁶-(pyren-1-ylmethyl)adenosine (**6b**): ¹H-NMR (CDCl₃): 8.33 (*s*, H−C(8)); 8.30 (*s*, H−C(2)); 8.17−7.95 (*m*, Pyr); 7.37−7.16 (*m*, (MeO)₂*Tr*); 6.25 (*t*, H−C(1′)); 5.53 (br., NHCH₂Ar), 4.55 (*m*, OH); 4.03 (*m*, H−C(3′)); 3.70 (*s*, MeO); 3.30 (*m*, H−C(5′)); 2.60 (*m*, 1 H−C(2′)); 2.33 (*m*, 1 H−C(2′)). ¹³C-NMR (CDCl₃): 154.45 (C(6)); 153.23 (C(2)); 144.59 (C(4)); 138.13 (C(8)); 135.72−122.93 (arom. C); 120.08 (C(5)); 113.16 ((MeO)₂*Tr*); 85.47 (C(4′)); 84.02 (C(1′)); 72.19 (C(3′)); 65.80 (C(5′)); 55.07 (MeO); 42.82 (NHCH₂Ar); 40.00 (C(2′)). FAB-MS (CHCl₃ + 3-nitrobenzyl alcohol): 768 ($[M+H]^+$).

Phosphoramidite Derivatives **7**: *General Procedure.* Compound **6a** or **6b** (0.5 mmol) was co-evaporated with dry MeCN and dissolved in a mixture of ¹Pr₂EtN (0.4 ml) and dry CH₂Cl₂ (2.5 ml). Then, 2-cyanoethyl diisopropylphosphoramidochlorite (0.8 mmol) was added dropwise under Ar. The mixture was stirred at r.t. for 1 h. Then the reaction was quenched by addition of MeOH (0.2 ml). After addition of AcOEt and CH₂Cl₂, the org. phase was washed with sat. NaHCO₃ soln. (3 × 30 ml), dried (Na₂SO₄), and evaporated. The product was purified by CC (AcOEt/CH₂Cl₂/Et₃N 45:45:10). The resulting gum was dissolved in 2 ml of dry toluene, and the soln. was added dropwise to stirred cold petroleum ether (100 ml). The solid product was filtered off: **7a,b**.

2′-Deoxy-5′-O-(4,4′-dimethoxytrityl)-N⁶-(naphthalen-1-ylmethyl)adenosine 3′-(2-Cyanoethyl Diisopropylphosphoramidite) (**7a**): ³¹P-NMR (CDCl₃): 149.951 (94% pure). FAB-MS (CHCl₃ + 3-nitrobenzyl alcohol): 894 ($[M+H]^+$).

2'-Deoxyy-5'-O-(4,4'-dimethoxytrityl)-N⁶-(pyren-1-ylmethyl)adenosine 3'-(2-Cyanoethyl Diisopropylphos-phoramidite) (**7b**): ³¹P-NMR (CDCl₃): 149.98 (99% pure). FAB-MS (CHCl₃ + 3-nitrobenzyl alcohol): 968 ([*M* + H]⁺).

REFERENCES

[1] R. G. Harvey, 'Polycyclic Hydrocarbons and Carcinogenesis', American Chemical Society, Washington, D.C., 1985.

[2] R. G. Harvey, N. E. Geacintov, *Acc. Chem. Res.* **1988**, *21*, 66.

[3] I. B. Weinstein, A. M. Jeffery, K. W. Jennette, S. H. Blobstein, R. G. Harvey, C. Harris, H. Autrup, H. Kasai, K. Nakanishi, *Science (Washington, D.C.)* **1976**, *193*, 592; A. M. Jeffery, K. W. Jennette, S. H. Blobstein, I. B. Weinstein, F. A. Beland, R. G. Harvey, H. Kasai, I. Miura, K. Nakanishi, *J. Am. Chem. Soc.* **1976**, *98*, 5714.

[4] A. M. Hruszkewycz, K. A. Canella, K. Peltonen, L. Kotrappa, A. Dipple, *Carcinogenesis* **1992**, *13*, 2347; D. B. Reardon, A. C. H. Bigger, J. Strandberg, H. Yagi, D. M. Jerina, A. Dipple, *Chem. Res. Toxicol.* **1989**, *2*, 12.

[5] S. J. C. Wei, R. L. Chang, C.-Q. Wong, N. Bhachech, X. X. Cui, E. Hennig, H. Yagi, J. M. Sayer, D. M. Jerina, B. D. Preston, A. H. Conney, *Proc. Natl. Acad. Sci. U.S.A.* **1991**, *88*, 11227; H. Rodriguez, E. L. Loechler, *Biochemistry* **1993**, *32*, 1759; S. Shibutani, L. A. Margulis, N. E. Geacintov, A. P. Grollman, *Biochemistry* **1993**, *32*, 7531.

[6] J. A. Ross, G. B. Nelson, K. H. Wilson, J. R. Rabinowitz, A. Galati, G. D. Stoner, S. Nesnow, M. Moss, *Cancer Res.* **1995**, *55*, 1039; M. S. Greenblatt, W. P. Bennett, M. Hollstein, C. C. Harris, *Cancer Res.* **1994**, *54*, 4855.

[7] P. C. Zamecnik, M. L. Stephenson, *Proc. Natl. Acad. Sci. U.S.A.* **1978**, *75*, 280.

[8] S. T. Crooke, *Biotechnology* **1992**, *10*, 882.

[9] C. Hélène, J.-J. Toulmé, *Biochim. Biophys. Acta* **1990**, *1049*, 99.

[10] E. Uhlmann, A. Peyman, *Chem. Rev.* **1990**, *90*, 543.

[11] H. M. Weintraub, *Sci. Am.* **1990**, *262*, 34.

[12] W. F. Lima, B. P. Monia, D. J. Ecker, S. M. Freier, *Biochemistry* **1992**, *31*, 12055.

[13] S. T. Cload, P. L. Richardson, Y.-H. Huang, A. Schepartz, *J. Am. Chem. Soc.* **1993**, *115*, 5005.

[14] D. J. Ecker, T. A. Vickers, T. W. Bruice, S. M. Freier, R. D. Jenison, M. Manoharan, M. Zounes, *Science (Washington, D.C.)* **1992**, *257*, 958.

[15] T. M. Woolf, D. A. Melton, C. G. B. Jennings, *Proc. Natl. Acad. Sci. USA* **1992**, *89*, 7305.

[16] J.-C. François, N. T. Thuong, C. Hélène, *Nucleic Acids Res.* **1994**, *22*, 3943.

[17] M. A. Rosen, D. J. Patel, *Biochemistry* **1993**, *32*, 6563.

[18] S. A. El-Kafrawy, M. A. Zahran, E. B. Pedersen, *Acta Chem. Scand.* **1999**, *53*, 280.

[19] A. R. Katritzky, X. Lan, J. Z. Yang, O. V. Denisko, *Chem. Rev.* **1998**, *98*, 409.

[20] R. G. Harvey, 'Polycyclic Aromatic Hydrocarbons: Chemistry and Cacinogenicity', Cambridge University Press, Cambridge, England, 1991.

[21] S. C. Cheng, A. S. Prakash, M. A. Pigott, B. D. Hilton, J. Roman, H. Lee, R. G. Harvey, A. Dipple, *Chem. Res. Toxicol.* **1988**, *1*, 216.

[22] H. Lee, E. Luna, M. Hinz, J. J. Stezowski, A. S. Kiselyov, R. G. Harvey, *J. Org. Chem.* **1995**, *60*, 5604.

[23] L. J. McBride, M. H. Caruthers, *Tetrahedron Lett.* **1983**, *24*, 245.

[24] H. Lee, M. Hinz, J. J. Stezowski, R. G. Harvey, *Tetrahedron Lett.* **1990**, *31*, 6773.

[25] R. P. Iyer, W. Egan, J. B. Regan, S. L. Beaucage, *J. Am. Chem. Soc.* **1990**, *112*, 1253.

[26] R. P. Iyer, L. R. Phillips, W. Egan, J. B. Regan, S. L. Beaucage, *J. Org. Chem.* **1990**, *55*, 4693.

[27] J. Jr. SantaLucia, *Proc. Natl. Acad. Sci. U.S.A.* **1998**, *95*, 1460.

[28] O. M. Ali, E. B. Pedersen, *Acta Chem. Scand.* **1999**, *53*, 497.

[29] M. L. Zapp, M. R. Green, *Nature (London)* **1989**, *342*, 714.

[30] T. J. Daly, K. S. Cook, G. S. Gray, T. E. Maione, J. R. Rusche, *Nature (London)* **1989**, *342*, 816.

[31] C. A. Rosen, E. Terwilliger, A. Dayton, J. G. Sodroski, W. A. Haseltine, *Proc. Natl. Acad. Sci. U.S.A.* **1988**, *85*, 2071.

[32] M. H. Malim, J. Hauber, S.-Y. Le, J. V. Maizel, B. R. Cullen, *Nature (London)* **1989**, *338*, 254.

[33] B. K. Felber, M. Hadzopoulou-Cladaras, C. Cladaras, T. Copeland, G. N. Pavlakis, *Proc. Natl. Acad. Sci. U.S.A.* **1989**, *86*, 1495.

[34] M. Emerman, R. Vazeux, K. Peden, *Cell* **1989**, *57*, 1155.

[35] E. T. Dayton, D. M. Powell, A. I. Dayton, *Science (Washington, D.C.)* **1989**, *246*, 1625.

[36] M.-L. Hammarskjöld, J. Heimer, B. Hammarskjöld, I. Sangwan, L. Albert, D. Rekosh, *J. Virol.* **1989**, *63*, 1959.

[37] D. D. Chang, P. A. Sharp, *Cell* **1989**, *59*, 789.

[38] X. Lu, J. Heimer, D. Rekosh, M.-L. Hammarskjöld, *Proc. Natl. Acad. Sci. U.S.A.* **1990**, *87*, 7598.

[39] M. Hadzopoulou-Cladaras, B. K. Felber, C. Cladaras, A. Athanassopoulos, A. Tse, G. N. Pavlakis, *J. Virol.* **1989**, *63*, 1265.

[40] J. Kjems, M. Brown, D. D. Chang, P. A. Sharp, *Proc. Natl. Acad. Sci. U.S.A.* **1991**, *88*, 683.

[41] S. T. Cload, A. Schepartz, *J. Am. Chem. Soc.* **1994**, *116*, 437.

Dipyrido[3,2-a:2',3'-c]phenazine-Tethered Oligo-DNA: Synthesis and Thermal Stability of Their DNA·DNA and DNA·RNA Duplexes and DNA·DNA·DNA Triplexes

by Dimitri Ossipov, Edouard Zamaratski, and Jyoti Chattopadhyaya*

Department of Bioorganic Chemistry, Box 581, Biomedical Center, University of Uppsala, Sweden
(tel: + 4618-4714577; fax: + 4618-554495; e-mail: jyoti@bioorgchem.uu.se)

Dipyrido[3,2-a:2',3'-c]phenazine (dppz) derivatives were conjugated to 9-mer and 18-mer DNA (ODN) at a site without nucleobase, either at the 5'- or 3'-end or at a internucleotide position, via linkers of 7, 12, or 18 atoms lengths. These dppz-linked ODNs were synthesized using novel backbone glycerol phosphoramidites: Glycerol, serving as artificial nucleoside without nucleobase, was modified to amines 10, 23, and 24, which were suitable for the subsequent key reaction with dppz-carboxylic acid 3 (Schemes 2 and 3). The products of these reactions (see 5–7) were then transformed to the standard phosphoramidite derivatives (see 27, 29, and 30) or used for loading on a CPG support (see 28, 31, and 32). The dppz-modified ODNs were subsequently assembled in the usual manner using automated solid-phase DNA synthesis. The 9-mer ODN-dppz conjugates 35–43 were tested for their ability to form stable duplexes with target DNA or RNA strands (D11 (60) or R11 (61)), while the 18-mer ODN-dppz conjugates 48–56 were tested for their ability to form stable triplexes with a DNA target duplex D24·D24 (62) (see Tables 1 and 2). The presence of the conjugated dppz derivative increases the stability of DNA·DNA and DNA·RNA duplexes, typically by a ΔT_m of 7.3–10.9° and 4.5–7.4°, respectively, when the dppz is tethered at the 5'- or 3'-terminal (Table 2). The dppz derivatives also stabilize triplexes when attached to the 5'- or 3'-end, with a ΔT_m varying from 3.8–11.1° (Table 3). The insertion of a dppz building block at the center of a 9-mer results in a considerably poorer stability of the corresponding DNA·DNA duplexes (ΔT_m = 0.5 to 4.2°) and DNA·RNA duplexes (ΔT_m = − 1.5 to 0.9°), while the replacement of one interior nucleotide by a dppz building unit in the corresponding 8-mer ODN does not reveal the formation of any duplex at all. Different types of modifications in the middle of the 18-mer ODN, in general, do not lead to any triplex formation, except when the dppz derivative is tethered to the ODN through a 12-atom-long linker (Entry 9 in Table 3).

1. Introduction. – Oligonucleotides (ODNs) can recognize both single-stranded (ss) DNA or RNA or double-stranded (ds) DNA in a sequence-selective manner. The high specificity of ODN-ssRNA or ODN-dsDNA recognition has led to the development of the antisense strategy [1][2] to arrest the translation of messenger RNA or antigene strategy [3–5] aimed at modulating gene expression. ODN Analogues carrying various functional groups (aromatic systems, polyamines, or metal complexes etc.) have been shown [6–8] to be able to assist in stabilizing the DNA triplex or DNA·DNA (DNA·RNA) duplex, which need to be efficient and specific if such materials are to be developed into an important class of diagnostics or therapeutics. On the other hand, the incorporation of DNA ligands activating the degradation of complementary DNA or RNA strands under appropriate conditions could potentially serve as powerful therapeutics to both arrest undesirable translation or transcription processes [9][10] as well as to destroy the target.

In this paper, we describe the synthesis of dipyrido[3,2-*a*:2′,3′-*c*]phenazine (dppz) [11] derivatives covalently linked to ODNs through linkers of different lengths. The reasons for our synthesis of dppz-tethered ODNs are as follows: *1*) Ru^{II} Complexes with the dppz ligand **1** (like [Ru(phen)$_2$(dppz)]$^{2+}$, phen = 1,10-phenanthroline) exhibit long-lived emissions when intercalated into a DNA duplex but not when free in solution, which could be an important property for designing diagnostics [12]. *2*) It is known that [Ru(phen)$_2$(dppz)]$^{2+}$ intercalatively binds to double-stranded DNA at least 3 orders of magnitude more effectively than the tris-phenanthroline (phen) counterpart [13]. One of the reasons for stronger binding of [Ru(phen)$_2$(dppz)]$^{2+}$ than [Ru(phen)$_3$]$^{2+}$ is most probably due to the fact that the dppz moiety has a higher ability for intercalation into the DNA duplexes in comparison with phen. Although both the phen and the dppz ligands have the same chelating N-atoms for interaction with different metal ions, the advantage of relatively larger specificity and tighter binding by the dppz ligand lies in the fact that it has a relatively larger planar extended aromatic system compared to phen. *3*) As a result, various tetrahedral [14] and octahedral [15–17] complexes of the phen ligand, which oxidatively or in a photoinduced manner cleave the oligonucleotide chain under suitable conditions, would act more specifically and actively if the dppz ligand is used instead.

An important aspect of this work is to develop new specific Ru^{II}-based agents, which are both specific as well as tight binders. Hence, the present work deals with the preparation of a new type of dppz-tethered molecules which would enable us to synthesize octahedral complexes such as [Ru(phen)$_2$(dppz)]$^{2+}$ in which the linker is connected through the dppz moiety instead of the phen ligand. Additionally, we also wish to use the chemistry reported herein to synthesize new types of dppz-tethered derivatives such as [Ru(dppz)$_3$]$^{2+}$, which is also connected to the ODN chain through the dppz ligand.

We report herein our new synthetic procedure to tether a dppz chromophore to the ODN chain using the phosphoramidite chemistry, which allows its multiple introduction at any position of choice. In this study, the dppz moiety was covalently connected to the defined sequences at the 3′- or at the 5′-end or in the middle position of the phosphodiester backbone of an ODN through a linker chain of various lengths (7, 12, and 18 atoms). This permitted us to comprehensively examine the properties of these dppz-tethered ODNs with respect to their ability to stabilize DNA · DNA and DNA · RNA duplexes or triple helices. A non-nucleosidic building block based on a glycerol residue [18][19] was used for the covalent connection to the linker arms to provide maximum flexibility and to prevent stereochemical alterations of the nucleoside conformation (or hybridization properties). Two types of interior attachment of dppz to the probe strand were employed to explore the stability of duplexes and triplexes depending upon the possible configurational orientation of the dppz chromophore at the site without nucleobase: *i*) the addition of the dppz-building unit to the probe strand such that it is fully complementary to the target strand (see below, *Table 1*), while the dppz-glycerol unit is bulged out, or *ii*) the replacement of one of the interior nucleotide units by the dppz-glycerol unit in the same probe strand such that one of the nucleotide units of the target strand has no complementary nucleotide to base-pair with (*Table 1*).

2. Results and Discussion. – 2.1. *Synthesis of Dppz-Tethered Phosphoramidite Blocks* **27**, **29**, *and* **30** *and of the Appropriate 3'-Succinamido-Anchored CPG Supports.* In our initial attempts to prepare ODN-dppz conjugates, we needed to find an appropriate linker. Our attempts to functionalize the parent dppz by allylic bromination of 11-methyl-dppz with *N*-bromosuccinimide (NBS) or nucleophilic substitution in *N*-alkylated-dppz failed under a variety of conditions. The alternative strategy involved the use of dipyridophenazine-11-carboxylic acid **3** (obtained from **2**) [20], which, in the activated form, formed an amide linkage upon reaction with an appropriate amine (*Scheme 1*). The task was arduous due to poor solubility of **3** in organic solvents. We, however, found that heating a suspension of **3** and 1,1'-carbonylbis[1*H*-imidazole] in pyridine at 80° for 15–20 min led to the quantitative formation of 1-acyl-1*H*-imidazole, which, upon immediate treatment with an appropriate amine for 30 min at 80°, followed by cooling to room temperature for 30 min, gave the corresponding amides **4**, **5**, **6**, or **7** in 75, 38, 72, or 44% yield, respectively. It is noteworthy that while the secondary amine gave two isomeric dynamically interconverting tertiary amides **4** at room temperature because of restricted rotation about the amide bond, the primary amines **10** (see below, *Scheme 2*) and **23** and **24** (*Scheme 3*) gave a single secondary amide **5**, **6**, or **7**, as expected.

Scheme 1

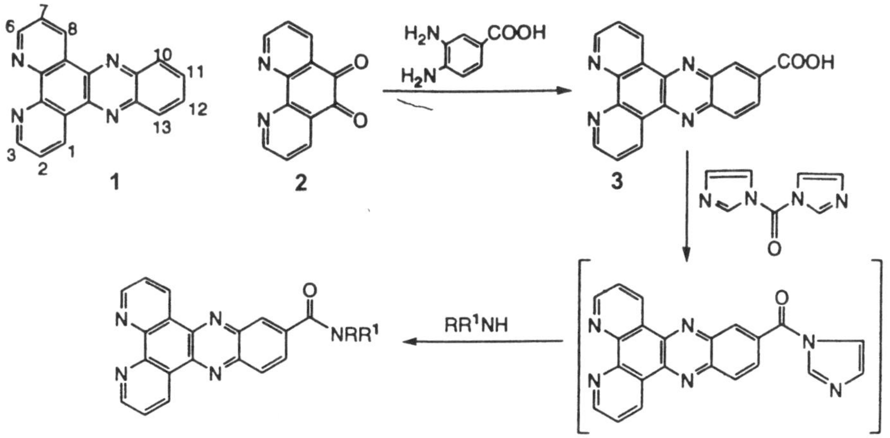

4 R = CH₂CH₂OH, R¹ = Me

5 R = CH₂CH₂CH₂O⟍⟍O, R¹ = H

6 R = CH₂CH₂(OCH₂CH₂)₂OCH₂CH(OH)CH₂OTr(MeO)₂, R¹ = H

7 R = CH₂CH₂(OCH₂CH₂)₄OCH₂CH(OH)CH₂OTr(MeO)₂, R¹ = H

For the design of an achiral linker unit, we chose a simple C₃ *sn*-glycerol backbone because it is cheap, provides maximum flexibility of the backbone to make a bulge, and possesses good H₂O solubility; this C₃ linker also allows multiple introductions of the chromophore into synthetic ODNs at any position of choice through linkers of various lengths. Two synthetic schemes were used for the construction of linker arms of

Scheme 2

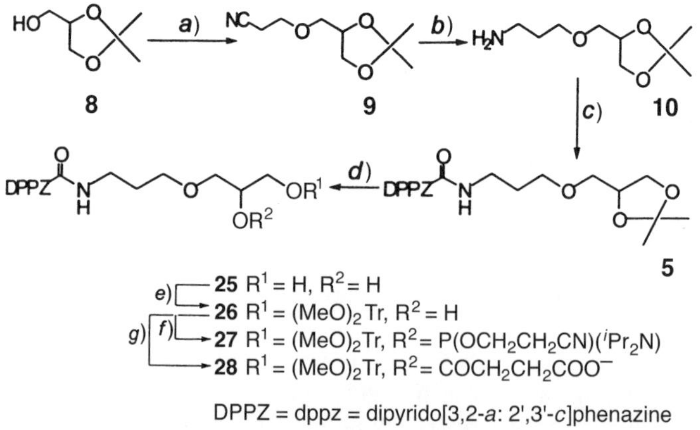

25 R^1 = H, R^2 = H
26 R^1 = (MeO)$_2$Tr, R^2 = H
27 R^1 = (MeO)$_2$Tr, R^2 = P(OCH$_2$CH$_2$CN)(^{i}Pr$_2$N)
28 R^1 = (MeO)$_2$Tr, R^2 = COCH$_2$CH$_2$COO$^-$

DPPZ = dppz = dipyrido[3,2-*a*: 2',3'-*c*]phenazine

a) Acrylonitrile, NaH, 1 h, 20°. *b*) NaBH$_4$, CoCl$_2$·6H$_2$O, MeOH, 1 h, 20°. *c*) **3**, 1,1'-carbonylbis[1*H*-imidazole], pyridine, 0.5 h, 80°, 0.5 h, 20°. *d*) 1M HCl/THF, 1.5 h, 20°. *e*) (MeO)$_2$TrCl; pyridine, 1.5 h, 20°. *f*) PCl(OCH$_2$CH$_2$CN)(^{i}Pr$_2$N)/^{i}Pr$_2$NEt, THF, 1.5 h, 20°. *g*) Succinic anhydride 4-(dimethylamino)pyridine (DMAP), CH$_2$Cl$_2$, 4 h, 20°.

different lengths to the 3-hydroxy group of the glycerol backbone: The shortest propane-1,3-diyl linker arm [18] was prepared from readily available solketal (= 2,2-dimethyl-1,3-dioxolane-4-methanol; **8**), which was quantitatively transformed to the 3-(alkyloxy)propanenitrile, **9** (*Scheme 2*). The latter was then reduced with NaBH$_4$ in the presence of cobalt(II) chloride to afford 3-(alkyloxy)propanamine **10** (55%).

To construct longer arms (see **6** and **7**), flexible, hydrophilic and bifunctional [21] tri- and penta(ethylene glycol) were used to tether to dppz. Because of the low coupling yield of solketal (**8**) to *n*(ethylene glycol) (in form of mesylated *n*(ethylene glycol) monoazide [19], data not shown), we built the glycerol part of the linker unit through the addition of the allyl radical [22]. This procedure involved the preparation of monotosylates **13** and **14** from tri- and penta(ethylene glycol) **11** and **12**, respectively (yield 52 and 51%, resp.; *Scheme 3*). The tosyloxy groups of **13** and **14** were then converted to the corresponding phthalimido groups of **15** (57%) and **16** (77%), which served both as the precursor and protector of the amino function. The reaction of **15** and **16** with allyl bromide in the presence of NaH in THF/DMF 15:1 afforded the corresponding olefins **17** (47%) and **18** (46%), which were then oxidized with permanganate to the respective diols **19** (52%) and **20** (38%). The permanganate treatment was carefully monitored because a prolonged reaction period led also to the removal of the phthalimido group. The primary-alcohol moiety of **19** and **20** was protected with a 4,4'-dimethoxytrityl group yielding **21** (85%) and **22** (87%), respectively. After aminolysis of the latter with methylamine, the target amines **23** and **24** were obtained in 86 and 84% yield, respectively.

The amines **10** (*Scheme 2*) and **23** and **24** (*Scheme 3*) were linked to the dppz moiety through the amide bond under the condition described in *Scheme 1* **10** → **5** (38%), **23** → **6** (72%), **24** → **7** (44%)). The dppz derivative **5** was then treated with 1M HCl/THF 1:1 to remove the isopropylidene group, and the crude product **25** (92%) was

Scheme 3

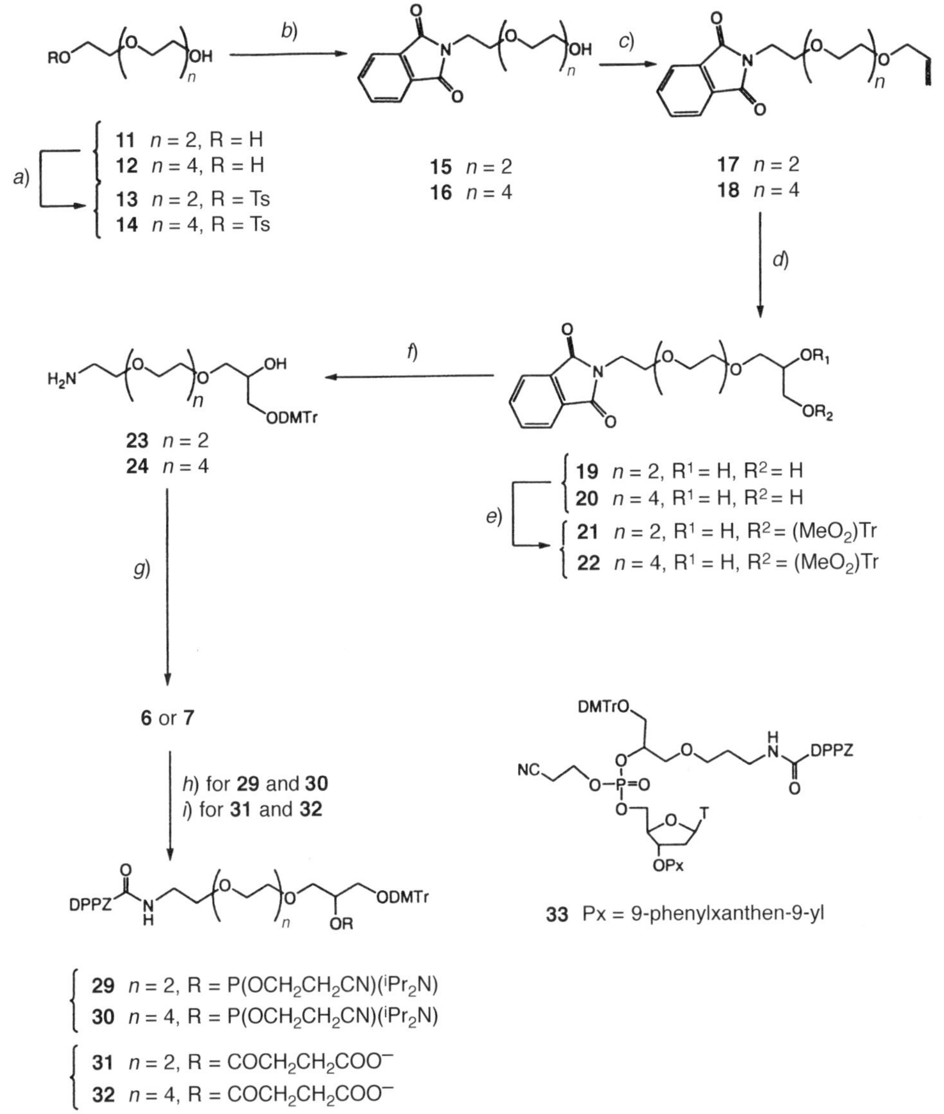

33 Px = 9-phenylxanthen-9-yl

DMTr = (MeO₂)Tr; DPPZ = dppz = dipyrido[3,2-a: 2',3'-c]phenazine

a) TsCl pyridine/CH₂Cl₂, 1 h, 0°, 1.5 h, 20°. b) Phthalimide, DBU, DMF, 18 h, 80°. c) Allyl bromide, NaH, THF/ DMF, 1 h, 0°, 18 h, 20°. d) KMnO₄, acetone/H₂O, 10 min, 20°. e) (MeO)₂TrCl/pyridine, 1.5 h, 20°. f) 40% CH₂NH₂/H₂O/MeOH, 5 h, 55°. g) 3, 1,1'-carbonylbis[1H-imidazole], pyridine, 0.5 h, 80°, 0.5 h, 20°. h) PCl(OCH₂CH₂CN)(ᶦPr₂N), ᶦPr₂EtN, THF, 1.5 h, 20°. i) Succinic anhydride, DMAP, CH₂Cl₂, 4 h, 20°.

tritylated to give **26** (91%) (*Scheme 2*). Finally, compounds **26** and **6** and **7** (*Scheme 3*) were converted in the usual manner [23] to the corresponding phosphoramidite blocks **27**, **29**, and **30** (94, 88, and 43% yield, resp.) for incorporation of these dppz-tethered units into oligodeoxynucleotides at the 5'-end or in the middle of the chain. Compounds **26**, **6**, and **7** were also treated with succinic anhydride and 4-(dimethylamino)pyridine (DMAP) in CH_2Cl_2 to give the corresponding succinate building blocks **28** (81%), **31**, (81%) and **32** (80%), respectively, which were then immobilized onto a 3-amino-propyl-CPG support [24], and were used for incorporation of the dppz-tethered building blocks at the 3'-termini of the oligonucleotides.

2.2. *Oligonucleotide Synthesis.* Prior to the solid-phase synthesis, we studied the stability of the amide linkage in the dppz-tethered dimeric block **33** (*Scheme 3*) under the basic conditions normally employed for the deprotection of the cyanoethyl group from the phosphotriester (ammonia, 17 h 55°). No significant degradation (< 5%) of dppz-tethered thymidine could be observed (as monitored by TLC and NMR) under these conditions.

The ODN analogues **34 – 62** (*Table 1*) used in this study were synthesized by the phosphoramidite methodology [26] with a commercially available DNA/RNA synthesizer (see *Exper. Part*). After deprotection, each dppz-ODN conjugate showed a very broad peak by HPLC analysis, which did not allow an estimation of the ODN purity. The purity of the HPLC-purified oligonucleotides were however, found to be satisfactory on 20% polyacrylamide gel containing 7M urea.

2.3. *Thermal Denaturation Study of the DNA · DNA and DNA · RNA Duplexes.* The duplexes (dppz-ODN) · DNA (*Entries 2 – 13* in *Table 2*) and (dppz-ODN) · RNA (*Entries 15 – 26* in *Table 2*) were generated by hybridization of the dppz-tethered 9-mers **35 – 43** and dppz-tethered 8-mers **44 – 46** with the target 11-mer ODN D11 (**60**) and oligo-RNA R11 (**61**) in a 1 : 1 ratio (1 μM of each strand in 20 mM PO_4^{3-} and 0.1M NaCl buffer at pH 7.3) (see *Table 1* for abbreviations). All the melting curves of these duplexes exhibited a monophasic dissociation. The melting temperatures obtained (*Table 2*) allow to draw the following conclusions: Tethering the dppz derivative at the 5'- or 3'-terminus increases the T_m for both DNA · DNA and DNA · RNA duplexes. A 12-atom linker gives a maximal stability ($\Delta T_m = 10.9°$ for the DNA · DNA duplex (*Entry 3, Table 2*), and $\Delta T_m = 5.4$ for the DNA · RNA duplex (*Entry 16*) for 5'-tethered dppz duplexes compared to 7- and 18-atom linkers (*Entries 2, 4, 15,* and *17*). Conversely, 7-atom linkers (*Entries 5* and *18*) and 18-atoms linkers (*Entries 7* and *20*) show the maximal stability for the 3'-dppz tether compared to the 12-atoms linker (*Entries 6* and *19*) for both DNA · DNA and DNA · RNA duplexes. It is noteworthy that the stability of the DNA · RNA duplex, **40 · 61**; 3'Z9 · R11 $\Delta T_m = 7.4°$ (*Entry 20*), is comparable to the corresponding DNA · DNA duplexes in general.

The dppz incorporation at the center of the unmodified 9-mer **34** results in considerable reduction of stabilization of the DNA · DNA and DNA · RNA duplexes ($\Delta T_m = 0.5 – 4.2°$, *Entries 8 – 10*; $\Delta T_m = -1.5$ to 0.9°, *Entries 21 – 23*; see *Table 2*), probably because of the formation of the bulge and steric interference by the linker arm within the duplex grooves. According to the T_m data, such interference is minor in the case of a 12-atoms linker (as in mY9 · D11 (**42 · 60**), *Entry 9*, and mY9 · R11 (**42 · 61**), *Entry 22*). A comparison of the two types of middle-modified DNA · DNA duplexes (*i.e., Entries 8 – 10* with a bulge, and *Entries 11 – 13* without bulge) shows that the

Table 1. *Synthetic DNA-dppz Conjugates and Their Target Oligo-DNA and Oligo-RNA*

dppz =   X =   Y (n = 2) or Z (n = 4) =

	Oligonucleotide	Linker		Abreviation	
	natural	5'-d(TCCAAACAT)-3'	–	N9	(34)
9-Mer	5'-modified	5'-d(MTCCAAACAT)-3'	M = X	5'X9	(35)
			M = Y	5'Y9	(36)
			M = Z	5'Z9	(37)
	3'-modified	5'-d(TCCAAACATM)-3'	M = X	3'X9	(38)
			M = Y	3'Y9	(39)
			M = Z	3'Z9	(40)
9-Mer	middle-modified	5'-d(TCCAMAACAT)-3'	M = X	mX9	(41)
			M = Y	mY9	(42)
			M = Z	mZ9	(43)
8-Mer	middle-modified	5'-d(TCCAMACAT)-3'	M = X	mX8	(44)
			M = Y	mY8	(45)
			M = Z	mZ8	(46)
18-Mer	natural	5'-d(TTCT$_6$CT$_6$CT)-3'	–	N18	(47)
18-Mer	5'-modified	5'-d(MTTCT$_6$CT$_6$CT)-3'	M = X	5'X18	(48)
			M = Y	5'Y18	(49)
			M = Z	5'Z18	(50)
	3'-modified	5'-d(TTCT$_6$CT$_6$CTM)-3'	M = X	3'X18	(51)
			M = Y	3'Y18	(52)
			M = Z	3'Z18	(53)
18-Mer	middle-modified	5'-d(TTCT$_6$MCT$_6$CT)-3'	M = X	mX18	(54)
			M = Y	mY18	(55)
			M = Z	mZ18	(56)
17-Mer	middle-modified	5'-d(MTTCT$_6$MT$_6$CT)-3'	M = X	mX17	(57)
			M = Y	mY17	(58)
			M = Z	mZ17	(59)
11-Mer	single-strand targets	5'-d(CATGTTTGGAC)-3'	–	D11	(60)
		5'-r(CAUGUUUGGAC)-3'	–	R11	(61)
24-Mer	duplex target	5'-d(GCCAAGA$_6$GA$_6$GACGC)-3'	–	D24·D24	(62)
		3'-d(CGGTTCT$_6$CT$_6$CTGCG)-5'			

absence of the bulge in the middle of the duplex indeed destabilizes the duplexes enormously. This is also true for the DNA·RNA duplexes (compare *Entries 21–23* with *Entries 24–26*). This is presumably due to the fact [27] that when a site without nucleobase forms no bulge (as in *Entries 11–13* and *24–26*) and the chromophore yields a mismatch pair, the duplex becomes energetically very unfavorable because the opposite nucleobase is most probably flipped out of the helix. On the other hand, when the duplex is fully matched (as in *Entries 8–10* or *21–23*) and the site without nucleobase bulges out, the chromophore can be possibly stacked with neighboring base pairs, which is energetically more favorable. In general, DNA·DNA duplexes were more stable and revealed a much more stabilizing effect with dppz-tethering than the corresponding DNA·RNA duplexes.

Table 2. *Thermal Stability* (T_m) *of DNA · DNA and DNA · RNA Duplexes Obtained from a 9-Mer* (or 8-Mer) *and a 11-Mer* (1 : 1 mixture)

Entry	DNA · DNA Duplexes			Entry	DNA · RNA Duplexes		
	Duplex type[a])	T_m [°]	ΔT_m [°]		Duplex type[a])	T_m [°]	ΔT_m [°]
1	N9·D11 (**34·60**)	25.9	–	*14*	N9·R11 (**34·61**)	20.6	–
2	5′X9·D11 (**35·60**)	33.5	7.6	*15*	5′X9·R11 (**35·61**)	25.2	4.6
3	5′Y9·D11 (**36·60**)	36.8	10.9	*16*	5′Y9·R11 (**36·61**)	26.1	5.4
4	5′Z9·D11 (**37·60**)	33.2	7.3	*17*	5′Z9·R11 (**37·61**)	25.5	4.9
5	3′X9·D11 (**38·60**)	34.3	8.4	*18*	3′X9·R11 (**38·61**)	27.1	6.5
6	3′Y9·D11 (**39·60**)	33.2	7.3	*19*	3′Y9·R11 (**39·61**)	25.2	4.5
7	3′Z9·D11 (**40·60**)	35.6	9.7	*20*	3′Z9·R11 (**40·61**)	28.0	7.4
8	mX9·D11 (**41·60**)	29.0	3.1	*21*	mX9·R11 (**41·61**)	19.2	– 1.5
9	mY9·D11 (**42·60**)	30.1	4.2	*22*	mY9·R11 (**42·61**)	21.5	0.9
10	mZ9·D11 (**43·60**)	26.4	0.5	*23*	mZ9·R11 (**43·61**)	21.1	0.4
11	mX8·D11 (**44·60**)	14.7	– 11.2	*24*	mX8·R11 (**44·61**)	n.f.[b])	n.f.[b])
12	mY8·D11 (**45·60**)	14.6	– 11.3	*25*	mY8·R11 (**45·61**)	n.f.[b])	n.f.[b])
13	mZ8·D11 (**46·60**)	n.f.[b])	n.f.[b])	*26*	mZ8·R11 (**46·61**)	n.f.[b])	n.f.[b])

[a]) See *Table 1* for abbreviations. [b]) n.f. = duplex not formed.

2.4. *Thermal Denaturation Studies of Triplexes*. Triplexes (*Entries 2 – 13 in Table 3*) were generated by hybridization of the dppz-tethered 18-mers **48–56** and dppz-tethered 17-mers **57–59** with the target 24-mer duplex D24 · D24 (**62**) [28] (18-mer (or 17-mer)/D24/D24 1 : 1 : 1; 1 μM of each strand in 20 mM PO_4^{3-} and 0.1M NaCl buffer at pH 7.3). Three similar transitions were observed for all melting profiles. The transition occurring in the temperature range 12–35° was attributed to the thermal dissociation of the third strand, the transition observed in the temperature range 42–50° corresponded to the dissociation of the mismatched duplex between the 18-mer and the GA-rich 24-mer oligonucleotide, while the transition observed in the temperature range 60–63° corresponded to the dissociation of the 24-mer duplex, D24 · D24 (**62**). All T_ms are shown in *Table 3*. These results show that all of the triple helixes formed with the oligonucleotides dppz-tethered at the 5′- or 3′-terminal are more stable than those formed with the unmodified ODN N18 (**47**), yet there are significant variations depending on the point of attachment to the ODN (3′- or 5′-) and the length of the

Table 3. *Thermal Stability* (T_m) *of DNA · DNA · DNA Triplexes Obtained from a 18-Mer* (or 17-Mer) *and Two 24-Mers* (1 : 1 : 1 mixture)

Entry	DNA · DNA · DNA Triplexes			Entry	DNA · DNA · DNA Triplexes		
	Triplex type[a])	T_m [°]	ΔT_m [°]		Triplex type[a])	T_m [°]	ΔT_m [°]
1	N18·D24·24 (**47·62**)	13.5	–	*8*	mX18·D24·D24 (**54·62**)	n.d.[b])	n.d.[b])
2	5′X18·D24·D24 (**48·62**)	24.6	11.1	*9*	mY18·D24·D24 (**55·62**)	20.6	7.1
3	5′Y18·D24·D24 (**49·62**)	20.9	7.4	*10*	mZ18·D24·D24 (**56·62**)	n.d.[b])	n.d.[b])
4	5′Z18·D24·D24 (**50·62**)	17.3	3.8	*11*	mX17·D24·D24 (**57·62**)	15.8	2.3
5	3′X18·D24·D24 (**51·62**)	22.6	9.1	*12*	mY17·D24·D24 (**58·62**)	n.d.[b])	n.d.[b])
6	3′Y18·D24·D24 (**52·62**)	20.3	6.8	*13*	mZ17·D24·D24 (**59·62**)	n.d.[b])	n.d.[b])
7	3′Z18·D24·D24 (**53·62**)	20.5	7.0				

[a]) See *Table 1* for abbreviations. [b]) n.d. = not determined.

linker arm used. The most essential stabilization was seen with ODN 5'X18 (**48**; dppz attached to the 5'-end through the 7-atom linker arm). The triplex formed by it shows a T_m value of 24.6°, which is an increase of 11.1° compared with the unsubstituted third strand **47**. In the case of a 5'-conjugation, the increase of the linker length up to 18 atoms leads to a sharp decrease of stabilization (ΔT_ms are changed from 11.1° to 3.8°). The same trend is also observed for a 3'-attachment, albeit more gradually (ΔT_ms are changed from 9.1° to 6.8°), suggesting that, in practical terms, there is no difference between 12-atom and 18-atom linker lengths as far as the triplex stability is concerned. The above data show that the increase of the linker length interferes with the triplex structure more strongly, especially in the case of 5'-modification. The increase of entropy with long linker arms could also play the destabilizing role for the triplex formation. Comparison of the 3'- and 5'-modifications for 7-atoms and 12-atoms linkers show that the 5'-dppz conjugation seems to be more desirable for stabilizing the triple helix. On the contrary, the longest linker arm (18-atoms) is preferable for a 3'-attachment of the dppz group in terms of triplex stability.

Triplexes derived from the third strand containing the dppz incorporation at the central position (mM18·D24·D24 and mM17·D24·D24 (M = X, Y, Z) gave relatively low thermal stabilities compared to a modification at the 5'- or 3'-terminus, except for two cases, namely for mY18·D24·D24 (**55**·**62**; *Entry 9* in *Table 3*) and mX17·D24·D24 (**57**·**62**; *Entry 11*). This observation shows that the optimal linker length of 12-atoms on the dppz chromophore can successfully penetrate into the interior of a triplex containing the unmodified 18-mer N18 (**47**) and thereby provide an excellent stability for the resulting triplex mY18·D24·D24 (**55**·**62**; (ΔT_m = 7.1°). It is noteworthy that any other shorter or longer linkers with a dppz modification in the middle could not stabilize the triplex to the extent of an 18-atom linker. For triplex studies, we also chose a second type of central modification with no site without nucleobase being bulged out (as in *Entries 11–13* in *Table 3*), in which the chromophore, therefore, yields a mismatch pair. The aim was to explore if such a modification has any influence on the triplex stability in comparison with the duplex stabilization. It turned out that this type of modification does not stabilize any triplex significantly.

3. Conclusion. The synthetic protocol of a dppz attachment to the various ODNs was devised at the 3'- or 5'-ends as well as in the middle of an ODN chain to study the charge-transport phenomenon by a Ru^{II}-dppz-tethered ODN through the DNA stacks. The dppz ligand was attached to a non-nucleosidic glycerol-based building block through various linkers of different lengths. The preparation of these ODN conjugates enabled us also, in the first step, to examine how a dppz group tethered through linkers of various sizes actually stabilizes the DNA·DNA and DNA·RNA duplexes as well as DNA·DNA·DNA triplexes. The results reported here show that an ODN with a covalently 5'- or 3'-attached dppz forms more stable DNA·DNA and DNA·RNA duplexes as well as DNA·DNA·DNA triplexes as compared with non-modified counterparts. Among the set of 5'- and 3'-derivatized ODNs, the length of the linker arm and the point of attachment play a considerable role in determining the stability of the duplexes and triplexes formed. The best DNA·DNA duplex stabilization is found for 5'Y9·D11 (**36**·**60**; *Entry 3* in *Table 2*, ΔT_m = 10.9°), while the most stabilized DNA·RNA duplex is 3'Z9·R11 (**40**·**61**; *Entry 20* in *Table 2*, ΔT_m = 7.4°). The triplex

5′X18 · D24 · D24 (**48 · 62**) is the most stabilized one ($\Delta T_m = 11.1°$) among the triple helix hybrids (see *Table 3*). Insertion of a glycerol residue carrying the dppz ligand in the middle of the ODN leads to remarkably reduced DNA · DNA and DNA · RNA duplex stabilities (compare this with 3′- or 5′-modifications in *Table 2*), most probably because of a distortion by the bulged nucleotide without nucleobase. In this particular case, the stabilizing factor of the dppz chromophore competes with the destabilization arising from the bulged moiety without a nucleobase; it is likely that a more appropriate linker arm could help to enhance the stabilizing interactions of the dppz group within the duplex, which is evident from our study of the 12-atom linker arm for triplexes such as mY18 · D24 · D24 (**55 · 62**; $\Delta T_m = 7.1°$, *Entry 9* in *Table 3*). Remarkably, other dppz-tethered linkers incorporated in the middle of the oligonucleotide do not stabilize the triplex formation at all (see *Table 3*).

Work is now in progress to explore the $[Ru(phen)_2(dppz)]^{2+}$-tethered ODN and other [(metal ion)(dppz)]-tethered-directed cleavage reaction of target DNA and RNA.

The authors thank the *Swedish Natural Science Research Council* (*NFR*) and the *Swedish Engineering Research Council* (*TFR*) for generous funding.

Experimental Part

General. Dry pyridine was obtained by distillation over 4-toluenesulfonyl chloride. MeCN and CH_2Cl_2 were distilled from P_2O_5 under Ar. DMF and THF were distilled over CaH_2. Column chromotography (CC): silica gel *Merck G60*. Semi-prep. HPLC: *Spherisorb 5ODS2* (reversed phase); *Gilson* equipment with pump model *303*, manometric module model *802C*, and dynamic mixer *811B* connected to a *Dynamax* computer program for gradient control. TLC: pre-coated silica gel F_{254} plates with fluorescent indicator; eluents $CH_2Cl_2/EtOH$ 95 :5 (*A*), 90 :10 (*B*), and 80 :20 (*C*). Thermal denaturation experiments: PC-interfaced *Perkin-Elmer*-UV/VIS spectrophotometer *Lambda 40* with a *PTP-6 Peltier* temperature controller. NMR Spectra: *JNM-GX-270* spectrometer, at 270 (^1H) or 36 MHz (^{31}P); δ in ppm rel. to $SiMe_4$ as an internal standard (^1H) or 85% phosphoric acid as external standard (^{31}P); *J* in Hz.

Dipyrido[3,2-a:2′,3′-c]phenazine-11-carboxylic acid (**3**). A mixture of 1,10-phenanthroline-5,6-dione (**2**; 10.5 g, 50 mmol) in EtOH (370 ml) and 3,4-diaminobenzoic acid (8.37 g, 55 mmol) in EtOH (180 ml) was boiled to reflux for 10 min. After 1 – 2 min of heating, a grey precipitate was formed. After cooling, the precipitate was filtered and washed with EtOH: 15.35 g (94%) of **3**. ^1H-NMR (CF₃COOD): 9.86 (*dt, J*(3,2) = *J*(6,7) = 8.2, *J*(3,1) = *J*(6,8) = 1.4, H−C(3), H−C(6)); 9.05 (*dd, J*(1,2) = *J*(8,7) = 5.2, *J*(1,3) = *J*(8,6) = 1.4, H−C(1), H−C(8)); 8.95 (*d, J*(10,12) = 1.6, H−C(10)); 8.35 (*dd, J*(12,13) = 8.9, *J*(12,10) = 1.6, H−C(12)); 8.27 (*d, J*(13,12) = 8.9, H−C(13)); 8.09 (*dt, J*(2,3) = *J*(7,6) = 8.2, *J*(2,1) = *J*(7,8) = 5.2, H−C(2), H−C(7)).

3-[(2,2-Dimethyl-1,3-dioxolan-4-yl)methoxy]propan-1-amine (**10**) was synthesized as previously described [18].

2,2′-[Ethane-1,2-diylbis(oxy)]bis[ethanol] Mono(4-methylbenzenesulfonate) (= *Triethylene Glycol Monotosylate*; **13**). Triethylene glycol (**11**; 2.823 g, 18.8 mmol) was co-evaporated with anh. pyridine and dissolved in pyridine (75 ml). The soln. was cooled to − 20° and treated dropwise within 1 h with TsCl (3.942 g, 20.68 mmol) in dry CH_2Cl_2 (25 ml) under vigorous stirring at − 20°. After an additional hour stirring at 0° then 1.5 h at r.t., the reaction was quenched by addition of 1 ml of MeOH and the mixture stirred for 15 min. The crude material was obtained after extraction with aq. sat. $NaHCO_3$ soln./CH_2Cl_2 and drying of the org. phase by filtration through $MgSO_4$. CC (silica gel, 0–4% EtOH/CH_2Cl_2) gave 2.951 g (52%) of **13**. TLC (*B*): R_f 0.50. ^1H-NMR (CDCl₃): 7.73 (*d, J* = 8.0, 2 arom. H); 7.28 (*d, J* = 8.0, 2 arom. H); 4.10 (*t, J* = 4.8, TsOCH₂); 3.64 (*t, J* = 4.8, TsOCH₂CH₂, OCH₂CH₂OH); 3.55 (*s*, OCH₂CH₂O); 3.51 (*t, J* = 5.9, CH₂OH); 2.38 (*s*, Me).

3,6,9,12-Tetraoxatetradecane-1,14-diol Mono(4-methylbenzenesulfonate) (= *Pentaethylene Glycol Monotosylate*; **14**) was obtained from pentaethylene glycol (**12**) as described for **13**: 3.605 g (51%) of **14**. TLC (*B*): R_f 0.62. ^1H-NMR (CDCl₃): 7.80 (*d, J* = 8.3, 2 arom. H); 7.34 (*d, J* = 8.3, 2 arom. H); 4.16 (*t, J* = 4.8, TsOCH₂); 3.71−3.59 (*m*, 18 H); 2.6 (*t, J* = 5.0, OH); 2.45 (*s*, Me).

2-[2-[2-(2-Hydroxyethoxy)ethoxy]ethyl]-1H-isoindole-1,3(2H)-dione (**15**). To a stirred soln. of **13** (2.78 g, 9.14 mmol) in anh. DMF (40 ml), phthalimide (1.41 g, 9.59 mmol) and 1,8-diazabicyclo[5.4.0]undec-7-en (DBu; 1.43 ml, 9.59 mmol) were added. The soln. was heated at 80° for 18 h and then evaporated. The crude yellow oil was purified by FC (silica gel, 0–5% EtOH/CH$_2$Cl$_2$): **13** (1.45 g, 57%). Oil. TLC: R_f 0.77 (*B*), 0.48 (*A*). ^{1}H-NMR (CDCl$_3$): 7.89–7.82 (*m*, 2 arom. H); 7.75–7.68 (*m*, 2 arom. H); 3.92 (*t*, *J* = 5.7, NCH$_2$); 3.76 (*t*, *J* = 5.7, NCH$_2$*CH*$_2$); 3.70–3.59 (*m*, 6 H); 3.54 (*t*, *J* = 5.9, CH$_2$OH).

2-(14-Hydroxy-3,6,9,12-tetraoxatetradec-1-yl)-1H-isoindole-1,3(2H)-dione (**16**). As described for **15**, from **14**: 2.5 g (77%) of **16**. TLC: R_f 0.73 (*B*), 0.38 (*A*). ^{1}H-NMR (CDCl$_3$): 7.88–7.81 (*m*, 2 arom. H); 7.75–7.68 (*m*, 2 arom. H); 3.90 (*t*, *J* = 5.9, NCH$_2$); 3.74 (*t*, *J* = 5.9, NCH$_2$*CH*$_2$); 3.74–3.58 (*m*, 16 H); 2.63 (*t*, *J* = 4.6, OH).

2-[2-[2-[2-(Prop-2-enyloxy)ethoxy]ethoxy]ethyl]-1H-isoindole-1,3(2H)-dione (**17**). Compound **15** (1.45 g, 5.24 mmol), co-evaporated from THF, was redissolved in dry THF (17 ml)/dry DMF (1.15 ml). The soln. was cooled to 0° under N$_2$, allyl bromide (0.67 ml, 7.87 mmol) and then an equimolar amount of solid NaH added, and the opaque yellow suspension stirred at 0° for 1 h and then at r.t. for 18 h. MeOH (6 ml) was added and the mixture evaporated. The crude material was submitted to CC (silica gel, 50–100% AcOEt/cyclohexane): 779 mg (47%) of **17**. TLC: R_f 0.28 (*A*), 0.75 (*B*). ^{1}H-NMR (CDCl$_3$): 7.88–7.81 (*m*, 2 arom. H); 7.74–7.68 (*m*, 2 arom. H); 5.96–5.82 (*m*, C*H*=CH$_2$); 5.29–5.14 (*m*, CH=C*H*$_2$); 3.99 (*d*, *J* = 5.7, OCH$_2$CH=CH$_2$); 3.90 (*t*, *J* = 5.8, NCH$_2$); 3.74 (*t*, *J* = 5.8, NCH$_2$*CH*$_2$); 3.68–3.59 (*m*, 6 H); 3.55–3.51 (*m*, CH$_2$OCH$_2$CH=CH$_2$).

2-(3,6,9,12,15-Pentaoxaoctadec-17-en-1-yl)-1H-isoindole-1,3(2H)-dione (**18**). As described for **17**, from **16**: 1.40 g (46%) of **18**. TLC: R_f 0.53 (*A*), 0.77 (*B*). ^{1}H-NMR (CDCl$_3$): 7.88–7.81 (*m*, 2 arom. H); 7.75–7.68 (*m*, 2 arom. H); 5.99–5.84 (*m*, C*H*=CH$_2$); 5.30–5.15 (*ddq*, *J* = 26.2, 13.85, 1.5, CH=C*H*$_2$); 4.02 (*dt*, *J* = 5.7, 1.5, OCH$_2$CH=CH$_2$); 3.90 (*t*, *J* = 5.6, NCH$_2$); 3.74 (*t*, *J* = 5.6, NCH$_2$*CH*$_2$); 3.67–3.57 (*m*, 16 H).

2-[2-[2-[2-(2,3-Dihydroxypropoxy)ethoxy]ethoxy]ethyl]-1H-isoindole-1,3(2H)-dione (**19**). A soln. of **17** (688 mg, 2.16 mmol) in acetone (3 ml), was cooled to 0°. An aq. soln. (6 ml of H$_2$O) of potassium permanganate (398 mg, 2.52 mmol) was added dropwise under stirring within 10 min. The mixture was stirred for a further 10 min at r.t. The resulting precipitate was filtered off through *Celite*, and the filtrate was evaporated and co-evaporated with abs. EtOH twice. The product was isolated by CC (silica gel, 0–10% EtOH/CH$_2$Cl$_2$): 396 mg (52%) of **19**. TLC: R_f 0.49 (*B*), 0.31 (*A*). ^{1}H-NMR (CDCl$_3$): 7.88–7.82 (*m*, 2 arom. H); 7.75–7.69 (*m*, 2 arom. H); 3.91 (*t*, *J* = 5.7, NCH$_2$); 3.88–3.80 (*m*, HOCH$_2$*CH*(OH)); 3.75 (*t*, *J* = 5.7, NCH$_2$*CH*$_2$); 3.67–3.49 (*m*, 12 H).

2-(17,18-Dihydroxy-3,6,9,12,15-pentaoxaoctadec-1-yl)-1H-isoindole-1,3(2H)-dione (**20**). As described for **19**, from **18**: 523 mg (38%) of **20**. TLC: R_f 0.51 (*B*), 0.14 (*A*). ^{1}H-NMR (CDCl$_3$): 7.88–7.82 (*m*, 2 arom. H); 7.75–7.69 (*m*, 2 arom. H); 3.90 (*t*, *J* = 5.8, NCH$_2$); 3.90–3.81 (*m*, HOCH$_2$*CH*(OH)); 3.74 (*t*, *J* = 5.8, NCH$_2$*CH*$_2$); 3.71–3.53 (*m*, 20 H).

2-[2-[2-[2-[3-[(4,4'-Dimethoxytrityl)oxy]-2-hydroxypropoxy]ethoxy]ethoxy]ethyl]-1H-indole-1,3(2H)-dione (**21**). Compound **19** (590 mg, 1.67 mmol) was dried by co-evaporation from anh. pyridine and then dissolved in dry pyridine (15 ml), and after cooling to 0° (MeO)$_2$Tr−Cl (679 mg, 2.01 mmol) was added. The mixture was stirred for 1.5 h at r.t., then poured into aq. sat. NaHCO$_3$ soln., and extracted with CH$_2$Cl$_2$. The org. phase was evaporated and the residue co-evaporated with toluene/CH$_2$Cl$_2$ and subjected to CC (silica gel, 50–100% AcOEt/cyclohexane): 930 mg (85%) of **21**. TLC (*A*): R_f 0.46. ^{1}H-NMR (CDCl$_3$): 7.85–7.80 (*m*, 2 arom. H (pht)); 7.72–7.66 (*m*, 2 arom. H(pht)); 7.44–7.40 (*m*, 2 arom. H (MeO)$_2$); 7.34–7.09 (*m*, 7 arom. H ((MeO)$_2$/Tr)); 6.84–6.79 (*m*, 4 arom. H ((MeO)$_2$/Tr)); 3.99–3.91 (*m*, (MeO)$_2$TrOCH$_2$*CH*(OH)); 3.88 (*t*, *J* = 5.6, NCH$_2$); 3.78 (*s*, 2 MeO); 3.72 (*t*, *J* = 5.6, NCH$_2$*CH*$_2$); 3.64–3.45 (*m*, 10 H); 3.20–3.10 (*m*, (MeO)$_2$TrOCH$_2$).

2-[18-(4,4'-Dimethoxytrityl)-17-hydroxy-3,6,9,12,15-pentaoxaoctadec-1-yl]-1H-isoindole-1,3(2H)-dione (**22**). As described for **21** from **20**: 771 mg (87%) of **22**. TLC (*A*): R_f 0.35. ^{1}H-NMR (CDCl$_3$): 7.87–7.80 (*m*, 2 arom. H (pht)); 7.73–7.67 (*m*, 2 arom. H (pht)); 7.44–7.40 (*m*, 2 arom. H ((MeO)$_2$Tr)); 7.34–7.17 (*m*, 7 arom. H ((MeO)$_2$Tr)); 6.84–6.79 (*m*, 4 arom. H ((MeO)$_2$Tr)); 4.00–3.92 (*m*, (MeO)$_2$TrOCH$_2$*CH*(OH)); 3.89 (*t*, *J* = 5.5, NCH$_2$); 3.78 (*s*, 2 MeO); 3.73 (*t*, *J* = 5.5, NCH$_2$*CH*$_2$); 3.65–3.48 (*m*, 18 H); 3.22–3.11 (*m*, (MeO)$_2$TrOCH$_2$); 2.82 (*d*, *J* = 4.2, OH).

2-[2-[2-(2-Aminoethoxy)ethoxy]ethoxy]-1-[[(4,4'-dimethoxytrityl)oxy]methyl]ethanol (**23**). Compound **21** (961 mg, 1.47 mmol) was taken up in 40% aq. MeNH$_2$ soln. (10 ml), and MeOH (3 ml) was added to solubilize the starting material. The mixture was heated at 55° for 5 h under stirring and then evaporated and co-evaporated with abs. EtOH. The crude material was purified by CC (silica gel, 0–50% EtOH/CH$_2$Cl$_2$): **23** (662 mg, 86%). Pale yellow oil. TLC (*C*): R_f 0.19. ^{1}H-NMR (CDCl$_3$): 7.43–7.40 (*m*, 2 arom. H ((MeO)$_2$Tr)); 7.32–7.16 (*m*, 7 arom. H ((MeO)$_2$)); 6.84–6.80 (*m*, 4 arom. H ((MeO)$_2$Tr)); 4.08–4.01 (*m*, (MeO)$_2$TrOCH$_2$·*CH*(OH)); 3.77 (*s*, 2 MeO); 3.72–3.50 (overlapping *m*, 12 H); 3.19–3.06 (*m*, (MeO)$_2$TrOCH$_2$, *CH*$_2$CH$_2$NH$_2$).

17-Amino-1-[[(4,4'-dimethoxytrityl)oxy]methyl]-3,6,9,12,15-pentaoxaheptadecan-1-ol (**24**). As described for **23** from **22**: 533 mg (84%) of **24**. TLC: R_f 0.23 (*C*), 0.09 (*A*). ^{1}H-NMR (CDCl$_3$): 7.44–7.40 (*m*, 2 arom. H ((MeO)$_2$Tr)); 7.32–7.16 (*m*, 7 arom. H (MeO)$_2$Tr)); 6.83–6.79 (*m*, 4 arom. H ((MeO)$_2$Tr)); 4.12–4.05 (*m*, (MeO)$_2$TrOCH$_2$CH(OH)); 3.89 (*t, J* = 5.2, CH$_2$NH$_2$); 3.79 (*s*, 2 MeO); 3.77–3.58 (overlapping *m*, 18 H); 3.26–3.01 (*m*, (MeO)$_2$TrOCH$_2$, CH$_2$CH$_2$NH$_2$).

General Method for the Preparation of dppz-Containing Amides. Equimolar amounts of **3** and the appropriate amine were co-evaporated with dry pyridine separately. To a suspension of **3** in anh. pyridine (1 ml of pyridine for 0.06 mmol of **3**) and 1,1'-carbonylbis[1*H*-imidazole] (1.5 equiv.) was added to the suspension. The mixture was stirred under N$_2$ at 85° (oil bath) until the solid particles were solubilized (*ca.* 15 min). Subsequently, a pyridine soln. of the appropriate amine (1 ml of pyridine for 0.13 mmol of amine) was added dropwise. After a further half-hour stirring under heating and a half-hour at r.t. the pyridine was evaporated. The residue was extracted with aq. sat. NaHCO$_3$ soln./CH$_2$Cl$_2$ dried (MgSO$_4$), and evaporated and the residue co-evaporated with toluene and CH$_2$Cl$_2$ twice. The product was isolated by CC (silica gel, 0–6% EtOH/ CH$_2$Cl$_2$): pale foam.

N-(2-Hydroxyethyl)-N-methyldipyrido[3,2-a:2',3'-c]phenazine-11-carboxamide (**4**). From **3** (163 mg, 0.5 mmol) and 2-(methylamino)ethanol (38 mg, 0.5 mmol): 144 mg (75%) of **4**, isolated by CC (silica gel, 0– 25% EtOH/CH$_2$Cl$_2$). ^{1}H-NMR (500 MHz, CDCl$_3$, 0°): Conformer A (33.3% of the mixture at 0°): 9.25 (*ddd, J*(3,2) = *J*(6,7) = 8.1, *J*(3,1) = *J*(6,8) = 1.7, H–C(3), H–C(6)); 9.13 (*ddd, J*(1,2) = *J*(8,7) = 4.3, *J*(1,3) = *J*(8,6) = 1.7, H–C(1), H–C(8)); 8.19 (*d, J*(10,12) = 1.8, H–C(10)); 8.18 (*d, J*(13,12) = 8.7, H–C(13)); 7.92 (*dd, J*(12,13) = 8.7, *J*(12,10) = 1.8, H–C(12)); 7.66 (*ddd, J*(2,3) = *J*(7,6) = 8.1, *J*(2,1) = *J*(7,8) = 4.3, H–C(2), H–C(7)); 4.02 (br. *t*, CH$_2$CH$_2$OH); 3.87 (br. *t*, CH$_2$CH$_2$OH); 3.19 (*s*, Me); conformer B (66.6% of the mixture at 0°): 8.93 (*dd, J*(1,2) = *J*(8,7) = 4.3, *J*(1,3) = *J*(8,6) = 1.7, H–C(1) or H–C(8)); 8.78 (*dd, J*(3,2) = *J*(6,7) = 8.0, *J*(3,1) = *J*(6,8) = 1.7, H–C(3) or H–C(6)); 8.59–8.57 (*m*, 2 H, H–C(1) or H–C(8), H–C(3) or H–C(6)); 7.76 (*d, J*(10,12) = 1.7, H–C(10)); 7.72 (*dd, J*(12,13) = 8.6, *J*(12,10) = 1.7, H–C(12)); 7.59 (*d, J*(13,12) = 8.6, H–C(13)); 7.57 (*dd, J*(2,3) = *J*(7,6) = 8.0, *J*(2,1) = *J*(7,8) = 4.3, H–C(2) or H–C(7)); 7.42 (*dd, J*(2,3) = *J*(7,6) = 8.0, *J*(2,1) = *J*(7,8) = 4.3, H–C(2) or H–C(7)); 3.96 (br. *t*, CH$_2$CH$_2$OH); 3.86 (br. *t*, CH$_2$CH$_2$OH); 3.29 (*s*, Me). ^{1}H-NMR ((D$_6$)DMSO, 90°): 9.57 (*d, J*(3,2) = *J*(6,7) = 7.3, H–C(3), H–C(6)); 9.29 (br. *s*, H–C(1), H–C(8)); 8.50 (*d, J*(13,12) = 8.9, H–C(13)); 8.45 (*s*, H–C(10)); 8.12 (*dd, J*(12,13) = 8.9, *J*(12,10) = 1.6, H–C(12)); 8.00 (*dd, J*(2,3) = *J*(7,6) = 7.3, *J*(2,1) = *J*(7,8) = 4.6, H–C(2), H–C(7)); 3.81 (br. *t*, CH$_2$CH$_2$OH); 3.69 (br. *t*, CH$_2$CH$_2$OH); 3.26 (*s*, Me).

N-[3-[(2,2-Dimethyl-1,3-dioxolan-4-yl)methoxy]propyl]dipyrido[3,2-a:2'3'-c]phenazine-11-carboxamide (**5**). From **3** (979 mg, 3 mmol) and **7** (567 mg, 3 mmol): 567 mg (38%) of **5**. ^{1}H-NMR (CDCl$_3$): 9.54–9.50 (*ddd, J* = 8.2, 4.7, 2.0, H–C(3), H–C(6)(dppz)); 9.28–9.24 (*dd, J* = 4.7, 2.0, H–C(1), H–C(8)(dppz)); 8.61 (*s*, H–C(10)(dppz)); 8.32 (*s*, H–C(12), H–C(13)(dppz)); 7.79–7.75 (*ddd, J* = 8.2, 4.7, 2.0, H–C(2), H–C(7)(dppz)); 7.49 (*t, J* = 5.0, NHC(O)); 4.42 (*quint. J* = 5.0, CH); 4.13 (*dd, J* = 8.3, 6.56, 1 H, CH$_2$); 3.86– 3.69 (*m*, 5 H, CH$_2$, OCH$_2$CH$_2$CH$_2$NH); 3.66–3.55 (*m*, CH$_2$O(CH$_2$)$_3$); 2.04 (*quint.*, *J* = 5.8, OCH$_2$CH$_2$CH$_2$NH); 1.41, 1.29 (2*s*, Me$_2$C).

N-[2-[2-[2-[3''''-[(4,4'-Dimethoxytrityl)oxy]-2''''-hydroxypropoxy]ethoxy]ethoxy]ethyl]dipyrido[3,2-a:2',3'-c]phenazine-11-carboxamide (**6**). From **3** (222 mg, 0.68 mmol) and **23** (358 mg, 0.68 mmol): 409 mg (72%) of **6**. ^{1}H-NMR (CDCl$_3$): 9.59 (2 *d, J*(3,2) = *J*(6,7) = 8.2, H–C(3), H–C(6)(dppz)); 9.31–9.24 (*m*, H–C(1), H–C(8)(dppz)); 8.86 (*d, J*(10,12) = 1.7, H–C(10)(dppz)); 8.43–8.40 (*dd, J*(12,13) = 8.9, *J*(12,10) = 1.7, H–C(12)(dppz)); 8.33 (*d, J*(13,12) = 8.9, H–C(13)(dppz)); 8.28–8.23 (br. *t*, NHC(O)); 7.84, 7.75 (2*m*, H–C(2), H–C(7)(dppz)); 7.30–7.26 (*m*, 2 arom. H ((MeO)$_2$Tr)); 7.15–7.06 (*m*, 7 arom. H ((MeO)$_2$Tr)); 6.64–6.58 (*m*, 4 arom. H ((MeO)$_2$Tr)); 4.18–4.08 (*m*, (MeO)$_2$TrOCH$_2$CH(OH)); 3.84–3.56 (overlapping *m*, 14 H); 3.66 (*s*, 2 MeO); 3.19–3.06 (*m*, (MeO)$_2$TrOCH$_2$).

N-[18'-(4,4'-Dimethoxytrityl)-17'-hydroxy-3',6',9',12',15'-pentaoxaoctadec-1'-yl]dipyrido[3,2-a:2',3'-c]phenazine-11-carboxamide (**7**). From **3** (257 mg, 0.79 mmol) and **24** (482 mg, 0.79 mmol): 324 mg (44%) of **7**. ^{1}H-NMR (CDCl$_3$): 9.61 (*td, J*(3,2) = *J*(6,7) = 8.2, *J*(3,1) = *J*(6,8) = 1.8, H–C(3), H–C(6)(dppz)); 9.28 (*dt, J*(1,2) = *J*(8,7) = 4.5, *J*(1,3) = *J*(8,6) = 1.8, H–C(1), H–C(8)(dppz)); 8.80 (*d, J*(10,12) = 1.7, H–C(10)(dppz)); 8.42–8.34 (*m*, H–C(12), H–C(13)(dppz)); 7.91 (br. *t*, NHC(O)); 7.79 (*dt, J*(2,3) = *J*(7,6) = 8.2, *J*(2,1) = *J*(7,8) = 4.5, H–C(2), H–C(7)(dppz)); 7.38–7.35 (*m*, 2 arom. H ((MeO)$_2$Tr)); 7.30– 7.10 (*m*, 7 arom. H ((MeO)$_2$Tr)); 6.76–6.71 (*m*, 4 arom. H ((MeO)$_2$Tr)); 4.02–3.94 (*m*, (MeO)$_2$TrOCH$_2$-CH(OH)); 3.78–3.48 (overlapping *m*, 22 H); 3.73 (*s*, 2 MeO); 3.19–3.07 (*m*, (MeO)$_2$TrOCH$_2$).

N-[3-(2,3-Dihydroxypropoxy)propyl]dipyrido[3,2-a:2',3'-c]phenazine-11-carboxamide (**25**). To a soln. of **5** (560 mg, 1.13 mmol) in THF (4.7 ml), 1M HCl (4.7 ml) was added, and the soln. was stirred for 1.5 h. Abs. EtOH (4.7 ml) was added, the soln. evaporated, the residue redissolved in abs. EtOH, and the soln. evaporated: 474 mg

(92%) of crude **25** which was used further without purification. ^{1}H-NMR (D$_2$O): 8.44 (*dd, J* = 7.9, 4.5, 2 H (dppz)); 7.91 (2*d, J* = 7.9, 2 H (dppz)); 7.30 (*m*, 2 H, dppz); 7.11 (*d, J*(12,13) = 8.3, H–C(12)(dppz)); 6.76 (*s*, H–C(10)(dppz)); 6.71 (*d, J*(13,12) = 8.3, H–C(13)(dppz)); 3.88 – 3.80 (*m*, CH); 3.62 – 3.42 (*m*, 6 H); 3.22 (*t, J* = 6.7, CH$_2$NH); 1.75 (*quint. J* = 6.7, OCH$_2$CH$_2$CH$_2$NH).

N-[3-[3″-[(4,4′-dimethoxytrityl)oxy]-2″-hydroxypropoxy]propyl]dipyrido[3,2-a:2′,3′-c]phenazine-11-carbox-amide (**26**). After co-evaporation from anh. pyridine, **25** (468 mg, 1.02 mmol) was suspended in dry pyridine (9.5 ml), and after cooling to 0°, (MeO)$_2$Tr–Cl (414 mg, 1.22 mmol) was added, and the mixture was stirred for 1.5 h at r.t. Then, the reaction was quenched with MeOH, the mixture poured into aq. sat. NaHCO$_3$ soln. and extracted with CH$_2$Cl$_2$, the org. phase evaporated, and the residue co-evaporated with toluene/CH$_2$Cl$_2$ and subjected to CC (silica gel, 0 – 8% EtOH/CH$_2$Cl$_2$): 707 mg (91%) of **26**. ^{1}H-NMR (CDCl$_3$): 9.52 (2 *d, J*(3,2) = *J*(6,7) = 8.2, H–C(3), H–C(6)(dppz)); 9.23 (2*d, J*(2,1) = *J*(7,8) = 4.2, H–C(1), H–C(8)(dppz)); 8.76 (*d, J*(10,12) = 1.6, H–C(10)(dppz)); 8.39 (*dd, J*(12,13) = 8.8, *J*(12,10) = 1.6, H–C(12)(dppz)); 8.39 (*dd, J*(13,12) = 8.8, H–C(13)(dppz)); 7.80 – 7.71 (*m*, H–C(2), H–C(7)(dppz)); 7.58 (2*d, J* = 4.6, NHC(O)); 7.31 – 7.28 (*m*, 2 arom. H ((MeO)$_2$Tr)); 7.18 – 7.08 (*m*, 7 arom. H ((MeO)$_2$Tr)); 6.67 – 6.63 (*m*, 4 arom. H ((MeO)$_2$Tr)); 4.21 – 4.13 (*m*, CH); 3.80 – 3.59 (overlapping *m*, 6 H); 3.68 (*s*, 2 MeO); 3.25 (*d, J* = 4.6, (MeO)$_2$TrOCH$_2$); 1.99 (*quint., J* = 5.2, OCH$_2$CH$_2$CH$_2$NH)).

General Method for the Phosphitylation of dppz-Containing Amides. The dppz-containing amide (1 equiv.) was co-evaporated with dry THF twice and dissolved in dry THF (1 ml of THF for 0.07 mmol of amide) and kept under N$_2$. Then, dry ^{i}Pr$_2$EtN (5 equiv.) was added, followed by 2-cyanoethyl diisopropylphosphoramidochlor-idite (1.5 equiv.) under vigorous stirring, which was continued for further 1.5 h. The reaction was quenched by addition of dry MeOH and the mixture stirred for 15 min. After workup with aq. sat. NaHCO$_3$ soln./CH$_2$Cl$_2$ drying (MgSO$_4$) of the org. phase, evaporation, and co-evaporation with toluene/CH$_2$Cl$_2$ (twice) the crude material was submitted to CC (silica gel, CH$_2$Cl$_2$/Et$_3$N 98 : 2).

N-[3-[3″-[(4,4′-Dimethoxytrityl)oxy]-2″-hydroxypropoxy]propyl]dipyrido[3,2-a:2′,3′-c]phenazine-11-carbox-amide 2″-(2-Cyanoethyl Diisopropylphosphoramidite) (**27**). From **26** (573 mg, 0.75 mmol): 680 mg (94%) of **27**. TLC (*B*): R_f 0.46. ^{31}P-NMR (CDCl$_3$): + 147.92.

N-[2-[2-[2-[3″″-(4,4′-Dimethoxytrityl)-2″″-hydroxypropoxy]ethoxy]ethoxy]ethyl]dipyrido[3,2-a:2′,3′-c]phen-azine-11-carboxamide 2″″-(2-cyanoethyl Diisopropylphosphoramidite) (**29**). From **6** (215 mg, 0.26 mmol): 235 mg (88%) of **29**. TLC (*B*): R_f 0.45. ^{31}P-NMR (CDCl$_3$): + 148.43.

N-[18′-(4,4′-Dimethoxytrityl)-17′-hydroxy-3′,6′,9′,12′,15′-pentaoxaoctadec-1′-yl]dipyrido[3,2-a:2′,3′-c]phen-azine-11-carboxamide 17′-(2-Cyanoethyl Diisopropylphosphoramidite) (**30**). From **7** (230 mg, 0.25 mmol): 120 mg (43%) of **30**. TLC (*B*): R_f 0.45. ^{31}P-NMR (CDCl$_3$): + 149.11.

General Method for the Preparation of the dppz-Functionalized Support. The dppz-containing amide (1 equiv). and 4-(dimethylamino)pyridine (2.1 equiv.) were dissolved in dry CH$_2$Cl$_2$ (1 ml of solvent for 0.18 mmol of amide). Then, succinic anhydride (2.03 equiv.) was added, and the soln. was stirred at r.t. for 4 h. The mixture was extracted with 0.1M citric acid and then with aq. sat. NaHCO$_3$ soln., and the org. phase dried (MgSO$_4$) and evaporated: the crude product was used directly for coupling to CPG. ^{i}Pr$_2$EtN (34 mg, 263.2 mmol) in THF (1.8 ml) was added to the succinate derivative of the dppz-containing amide (138.5 mmol), followed by isobutyl carbonochloridate (= isobutyl chloroformate; 18.9 mg, 138.5 mmol) in THF (0.5 ml). After stirring for 2 h, a soln. of ^{i}Pr$_2$EtN (0.6 ml) in dry THF (1.54 ml) and 3-aminopropyl-CPG (1 g) were added to the mixture. The suspension was shaken for 2 h and then filtered and the solid thoroughly washed with THF, CH$_2$Cl$_2$ (3 times), and Et$_2$O (4 times). (MeO)$_2$Tr release with acid and measurement at 498 nm were performed at this stage to determine the loading. The support was then suspended in dry pyridine (8 ml), 4-(dimethylamino)pyr-idine (167 mg) and Ac$_2$O (0.77 ml) were added, and the suspension was shaken for 2 h, after which the suspension was filtered and the solid thoroughly washed with pyridine, CH$_2$Cl$_2$ (4 times), and Et$_2$O (4 times) and then dried under vacuum over P$_2$O$_5$.

N-[3-[3″-[(4,4′-Dimethoxytrityl)oxy]-2″-hydroxypropoxy]propyl]dipyrido[3,2-a:2′,3′-c]phenazine-11-carbox-amide 2″-(Hydrogen Butanedioate) (**28**) *and the Corresponding 2″-[4-[(3-CPG-propyl)amino]-4-oxobuta-noate}.* From **26** (133 mg, 0.17 mmol): 122 mg (81%) of **28**. ^{1}H-NMR (CDCl$_3$): 9.47 (br. *s*, H–C(3), H–C(6)(dppz)); 9.22 (br. *s*, H–C(1), H–C(8)(dppz)); 8.73 (br. *s*, H–C(10)(dppz)); 8.39 – 8.27 (*m*, H–C(12), H–C(13)(dppz)); 7.86 (br. *s*, NHC(O)); 7.71 (br. *s*, H–C(2), H–C(7)(dppz)); 7.37 – 7.10 (*m*, 9 arom. H ((MeO)$_2$Tr)); 6.77 – 6.74 (*m*, 4 arom. H ((MeO)$_2$Tr)); 5.28 (br. *s*, (MeO)$_2$TrOCH$_2$CH); 3.80 – 3.58 (overlapping *m*, 20 H); 3.19 – 3.26 (*m*, (MeO)$_2$TrOCH$_2$); 2.68 (br. *s*, (CH$_2$)$_2$COO); 1.75 (*quint. J* = 5.2, OCH$_2$CH$_2$CH$_2$NH).

The loading of the support (1 g) was 21.9 mmol/g of CPG.

N-{2-{2-[2-[3''''-[(4,4'-Dimethoxytrityl)oxy]-2''''-hydroxypropoxy}ethoxy}ethoxy}ethyl}dipyrido[3,2-a:2',3'-c]phenazine-11-carboxamide 2''''-(Hydrogen Butanedioate) (**31**) and the Corresponding 2''''-[4-[(3-CPG-propyl)amino]-4-oxobutanoate}. From **6** (52 mg, 0.06 mmol): 47 mg (81%) of **31**. ^{1}H-NMR (CDCl$_{3}$): 9.47 (br. s, H−C(3), H−C(6)(dppz)); 9.22 (br. s, H−C(1), H−C(8)(dppz)); 8.73 (br. s, H−C(10)(dppz)); 8.39−8.27 (m, H−C(12), H−C(13)(dppz)); 7.86 (br. s, NHC(O)); 7.71 (br. s, H−C(2), H−C(7)(dppz)); 7.37−7.10 (m, 9 arom. H ((MeO)$_{2}$Tr)); 6.77−6.74 (m, 4 arom. H ((MeO)$_{2}$Tr)); 5.28 (br. s, (MeO)$_{2}$TrOCH$_{2}$,CH); 3.80−3.58 (overlapping m, 20 H); 3.19−3.17 (m, (MeO)$_{2}$TrOCH$_{2}$); 2.66 (br. s, (CH$_{2}$)$_{2}$COO).

The loading of the appropriate support (369 mg) was 22.7 mmol/g of CPG.

N-[18'-(4,4'-Dimethoxytrityl)-17'-hydroxy-3',6',9',12',15'-pentaoxaoctadec-1'-yl]dipyrido[3,2-a:2',3'-c]phenazine-11-carboxamide 17'-(Hydrogen Butanedioate) (**32**) and the Corresponding 17'-[4-[(3-CPG-propyl)amino]-4-oxobutanoate}. From **7** (100 mg, 0.107 mmol): 89 mg (80%) of **32**. ^{1}H-NMR (CDCl$_{3}$): 9.57 (br. s, H−C(3), H−C(6)(dppz)); 9.26 (br. s, H−C(1), H−C(8)(dppz)); 8.78 (br. s, H−C(10)(dppz)); 8.42−8.33 (m, H−C(12), H−C(13)(dppz)); 7.77 (br. s, NHC(O), H−C(2), H−C(7)(dppz)); 7.39−7.14 (m, 9 arom. H ((MeO)$_{2}$Tr)); 6.83−6.72 (m, 4 arom. H (MeO)$_{2}$Tr)); 5.22 (br. s, (MeO)$_{2}$TrOCH$_{2}$CH); 3.80−3.51 (overlapping m, 28 H); 3.20−3.18 (m, (MeO)$_{2}$TrOCH$_{2}$); 2.67 (br. s, (CH$_{2}$)$_{2}$COO).

The loading of the appropriate support (369 mg) was 24.3 mmol/g of CPG.

Synthesis, Deprotection, and Purification of Oligonucleotides. All ODNs were synthesized on a 1.0-μmol scale with an 8-channel *Applied-Biosystems-392*-DNA/RNA synthesizer using conventional 2-cyanoethyl phosphoramidite chemistry. Modified oligonucleotides 3'M9 (M = X, Y, or Z; **38**−**40**) and 3'M18 (M = X, Y, or Z; **51**−**53**) (*Table 1*) were synthesized on the modified supports containing the dppz chromophore tethered to the glycerol residue through linkers of 7 (M = X), 12 (M = Y), or 18 (M = Z) atoms length. Other ODNs were obtained using standard CPG-dT support. The preparation of target ODNs D11 (**60**), R11 (**61**), and D24·D24 (**62**) involved the use of standard CPG-dC or -C supports. Phosphoramidite blocks **27**, **29**, and **30** were dissolved in dry MeCN with a final concentration of 0.15M and used after filtration for solid-phase synthesis with a coupling time of 10 min (25 s for standard nucleoside phosphoramidites).

After each synthesis of the protected oligomers, the solid support was transferred directly out from the cassette to a 50-ml round-bottomed flask containing 20 ml of conc. aq. NH$_{3}$ soln. and was shaken for 2 h at 20°. After removal of CPG by filtration and evaporation of the filtrate, the residue was redissolved in conc. aq. NH$_{3}$ soln. and stirred at 55° for 17 h. The crude ODNs were purified by reversed-phase HPLC (gradient elution with A (0.1M (Et$_{3}$NH)OAc, 5% MeCN, pH 7.0) and B (0.1M (Et$_{3}$NH)OAc, 50% MeCN, pH 7.0)), and the purities of the dppz-ODN conjugates were checked by denaturing 20% polyacrylamide gel electrophoresis. The reversed-phase HPLCs of all dppz-ODN conjugates indicated the presence of several colorless fractions, and one slightly yellow fraction. The UV/VIS spectra of each colored fraction showed the characteristic bands at 369 nm and 387 nm (dppz). The detritylated oligomers were evaporated and co-evaporated with H$_{2}$O 5 times and then directly lyophilized (5 × 1 ml H$_{2}$O). All ODNs were subsequently submitted to Na^{+} exchange by passing then through a column of *Dowex-50* (Na^{+} form).

Concentrations of dppz-ODNs were determined accounting for the contribution to the absorbance at 260 nm from the dppz moiety itself. This was done by taking the ratio of the area under the UV curve for compound **25** (or products of detritylation of **6** and **7**) at 235−335 nm to that at 335−415 nm. The absorption of the dppz-labeled oligomers at 260 nm was then corrected for the dppz absorption by using the above ratio (3.78−4.06) to estimate the contribution of the dppz chromophore to the absorbance area for a given oligonucleotide at 235−335 nm. Starting from 1 μmol of thymidine residues linked to controlled-pore glass, the following amounts, measured in A_{260} units (*OD*), were obtained after purification: 3'M9: (**38**−**40**), 10.9−24.1; 5'M9 (**35**−**37**), 12.1−14.8; mM9 or mM8 (**41**−**46**), 12.1−29.5; 3'M18 (**51**−**53**), 17.1−34.8; 5'M18 (**48**−**50**), 10.4−29.5; mM18 or mM17 (**54**−**59**), 14.0−57.3.

Thermal Denaturation Experiments. UV Melting profiles were obtained by scanning A_{260} absorbance vs. time at a heating rate of 0.5°/min from 5° to 70° for triplexes and of 1.0°/min from 10° to 55° for duplexes. The melting temperature T_{m} (± 0.5°) was determined as the maximum of the first derivative of melting curves. The triplex and duplex melting experiments were performed in 20 mM Na$_{2}$HPO$_{4}$/NaH$_{2}$PO$_{4}$ and 0.1M NaCl at pH 7.3 at a hybrid concentration of *ca.* 1 mM. The approximate extinction coefficients for natural ODNs were calculated as previously described [30][31]. In cases of the dppz-tethered ODNs, the extinction coefficients were corrected for the absorbance contribution of the dppz moiety at 260 nm by subtraction. After preparation, the solns. consisting of three components (for the formation of triplexes) were heated to 70° for 5 min and then allowed to cool down to 20° for 30 min and then kept at 0° overnight. The appropriate solns. consisting of two components (for the formation of duplexes) were heated to 70° for 5 min, and then allowed to cool down to 20°

for 30 min under shaking. During the melting experiments at temp. $<$ *ca.* 15°, N_2 was continuously passed through the sample compartment to prevent moisture condensation.

REFERENCES

[1] 'Topics in Molecular and Structural Biology, Vol. 12, Oligodeoxynucleotides: Antisense Inhibitors of Gene Expression', Ed. J. S. Cohen, MacMillan, Hampshire, 1989.

[2] C. A. Stein, Y. C. Cheng, *Science (Washington, D.C.)* **1993**, *261*, 1004.

[3] N. T. Thuong, C. Hélène, *Angew. Chem., Int. Ed. Engl.* **1993**, *32*, 666.

[4] I. Radhakrishnan, D. Patel, *J. Biochem.* **1994**, *33*, 11405.

[5] C. Hélène, *Anti-Cancer Drug Res.* **1991**, *6*, 569.

[6] N. T. Thuong, U. Asseline, T. Garastier, in 'Topics in Molecular and Structural Biology; Oligodeoxynucleotides. Antisense Inhibitors of Gene Expression', Ed. J. S. Cohen, CRC Press, Inc., Boca Raton, FL 1989, pp. 25–52.

[7] J. L. Mergny, G. Duval-Valentin, C. H. Nguyen, L. Perrouault, B. Faucon, M. Rougee, T. Garastier, E. Bisagni, C. Hélène, *Science (Washington, D.C.)* **1992**, *256*, 1681; D. S. Pilch, M.-T. Martin, C. H. Nguyen, J.-S. Sun, E. Bisagni, T. Garastier, C. Hélène, *J. Am. Chem. Soc.* **1993**, *115*, 9942.

[8] R. Glaser, E. J. Gabbay, *Biopolymers* **1968**, *6*, 243; S. F. Singelton, P. B. Dervan, *Biochemistry* **1993**, *32*, 13171.

[9] D. S. Sigman, T. W. Bruice, A. Mazumder, C. L. Sutton, *Acc. Chem. Res.* **1993**, *26*, 98; G. Pratviel, J. Bernadou, B. Meunier, *Angew. Chem., Int. Ed. Engl.* **1995**, *34*, 746.

[10] B. Armitage, *Chem. Rev.* **1998**, *98*, 1171.

[11] J. E. Dickeson, L. A. Summers, *Aust. J. Chem.* **1970**, *23*, 1023.

[12] A. E. Friedman, J. C. Chambron, J. P. Sauvage, N. J. Turro, J. K. Barton, *J. Am. Chem. Soc.* **1990**, *112*, 4960.

[13] A. E. Friedman, C. V. Kumar, N. J. Turro, J. K. Barton, *Nucleic Acids Res.* **1991**, *19*, 2595.

[14] J.-C. Francois, T. Saison-Behmoaras, M. Chassignol, N. T. Thuong, C. Hélène, *J. Biol. Chem.* **1989**, *264*, 5891; C. B. Chen, A. Mazumder, J.-F. Constant, D. S. Sigman, *Bioconjugate Chem.* **1993**, *4*, 69; C. B. Chen, M. B. Gorin, D. S. Sigman, *Proc. Natl. Acad. Sci. U.S.A.* **1993**, *90*, 4206.

[15] V. W.-W. Yam, K. K.-W. Lo, K.-K. Cheung, R. Y.-C. Kong, *J. Chem. Soc., Dalton Trans.* **1997**, 2067; P. J. Carter, C.-C. Cheng, H. H. Thorp, *J. Am. Chem. Soc.* **1998**, *120*, 632.

[16] M. R. Arkin, E. D. Stemp, C. Turro, N. J. Turro, J. K. Barton, *J. Am. Chem. Soc.* **1996**, *118*, 2267.

[17] C. J. Murphy, M. R. Arkin, N. D. Ghatlia, S. H. Bossmann, N. J. Turro, J. K. Barton, *Science (Washington, D.C.)* **1993**, *262*, 1025.

[18] K. Misiura, I. Durrant, M. R. Evans, M. J. Gait, *Nucleic Acids Res.* **1990**, *18*, 4345; P. Theisen, C. McCollum, K. Upadhya, K. Jacobson, H. Vu, A. Andrus, *Tetrahedron Lett.* **1992**, *33*, 5033.

[19] E. N. Timofeev, I. P. Smirnov, L. A. Haff, E. I. Tishchenco, A. D. Mirzabekov, V. L. Florentiev, *Tetrahedron Lett.* **1996**, *47*, 8467.

[20] R. M. Hartshorn, J. K. Barton, *J. Am. Chem. Soc.* **1992**, *114*, 5919.

[21] C. R. Bertozzi, M. D. Bednarski, *J. Org. Chem.* **1991**, *56*, 4326; A. W. Schwabacher, J. W. Lane, M. W. Schiesher, K. M. Leigh, C. W. Johnson, *J. Org. Chem.* **1998**, *63*, 1727.

[22] M. C. Willis, B. Collins, T. Zhang, L. S. Green, D. P. Sebesta, C. Bell, E. Kellogg, S. C. Gill, A. Magallanez, S. Knauer, R. A. Bendele, P. S. Gill, N. Janjic, *Bioconjugate Chem.* **1998**, *9*, 573.

[23] N. Sinha, J. Biernat, H. Köster, *Tetrahedron Lett.* **1983**, *24*, 5843.

[24] 'Oligonucleotide Synthesis; A Practical Approach', Ed. M. J. Gait, IRL Press, Oxford, 1984, p. 45.

[25] L. McBride, M. Caruthers, *Tetrahedron Lett.* **1983**, *24*, 245.

[26] M. D. Matteucci, M. H. Caruthers, *J. Am. Chem. Soc.* **1981**, *103*, 3185.

[27] K. Fukui, K. Tanaka, *Nucleic Acids Res.* **1996**, *24*, 3962.

[28] D. Barawkar, K. Rajeev, V. Kumar, K. Ganesh, *Nucleic Acids Res.* **1996**, *24*, 1229.

[29] G. C. Silver, C. H. Nguyen, A. S. Boutorine, E. Bisagni, T. Garastier, C. Hélène, *Bioconjugate Chem.* **1997**, *8*, 15.

[30] C. Sund, N. Puri, J. Chattopadhyaya, *Tetrahedron* **1996**, *52*, 12275.

[31] G. Fasman, 'Handbook of Biochemistry and Molecular Biology', CRC Press, Ohio, 1975, Vol. 1, p. 589.

Synthesis and Properties of Oligothymidylates Incorporating an Artificial Bend Motif

by **Kohji Seio**, **Takeshi Wada**, and **Mitsuo Sekine***

Department of Life Science, Faculty of Bioscience and Biotechnology, Tokyo Institute of Technology, Nagatsuta, Midori-ku, Yokohama 226-8501, Japan

The uridylyl-$(3' \rightarrow 5')$-thymidine dinucleotide block **14** (cUpdU), having a cyclic structure between the 2'-hydroxy of the upstream uridine and the 5-substituent of the downstream thymidine, was synthesized (*Schemes 1* and *2*). This cyclic structure is a stable mimic of the intraresidual H-bonding found in the anticodon loop of an *E. coli* minor tRNA^{Arg}. The spectroscopic and molecular-mechanics analyses of the cyclized dinucleotides predicted two major conformers, *i.e.*, the turn and bent forms. The latter was expected to bend DNA oligomers when incorporated into them. This expectation was ascertained by incorporating the bent dimer motif into tetra-, deca-, or hexadecathymidylates by the conventional phosphoramidite method (see **18–20** in *Scheme 4*). The bending of oligonucleotides **18–20** was demonstrated by ^{31}P-NMR and CD spectra and gel-electrophoretic studies. The duplex formation of these modified oligonucleotides with oligodeoxyadenylates was also studied. The decreased thermal stability of the duplexes when compared with unmodified ones indicates distorted structures of the modified duplexes. The 3D computer model of the duplexes showed a bend of *ca.* 30° with a 'bulge-out' at the position of an adenosine residue facing the cyclized dimer. The artificially bent DNAs might become a new tool for the study of the effect of DNA bending induced in DNA/DNA-binding protein interactions.

1. Introduction. – DNA Bending induced by DNA-binding proteins (DBP) is an important event in a series of biological reactions related to the expression, recombination, and replication of genetic information [1]. The detailed tertiary structures of some bent DNAs have been unveiled to date with the aid of X-ray crystallographic or multi-dimensional NMR spectroscopic techniques [1]. Sharply bent structures have also been discovered in the active site of hammerhead ribozymes [2] as well as in the region between the anticodon and its upstream nucleoside in tRNA [3].

These bent DNAs or RNAs can apparently be stabilized by binding to proteins or their 3-dimensional structures. If such a bent structure could be created independently from these complex interactions, a variety of intrinsic studies would be developed taking advantage of its inherent properties that cannot be acquired from the usual linear DNAs or RNAs. For example, unique motifs, such as extremely stabilized loops or duplexes with a bulge-out or flip-out structure, could be generated.

Therefore, it is of great interest to create an artificially bent motif by introducing a chemically modified nucleotide unit having a rigid bent backbone structure into oligonucleotides. We previously reported our studies of chemical fixation of uridylic acid [4] and uridylyl-$(3' \rightarrow 5')$-uridine [5] in conformations similar to those of a mononucleotide unit seen in the A-type RNA duplex and a sharply bent U-turn structure existing in the tRNA anticodon loop, respectively, by introducing covalent bonding bridges into the parent structures. Preliminary molecular-mechanics calculation analysis of the cyclic uridylyl-(3′→5′)-uridine [5] having a $CH_2C(O)NHCH_2$ linker revealed a two-state equilibrium between the turn and stretch conformations. The NMR studies suggested that the latter is more stable than the former and has a bending property because of the fixed orientation of the two bases.

In this paper, we report the synthesis and conformational analysis of a uridylyl-(3′→5′)-2′-deoxyuridine derivative having the covalent-bonding linker and the synthesis of oligothymidylates containing this cyclic dimer. Conformation of the dimer unit and the modified oligothymidylates was analyzed by spectroscopic, electrophoretic, and computational methods.

2. Results and Discussion. – 2.1. *Chemical Synthesis of the 3′-Terminal Nucleoside Unit* **1.** The 3′-terminal nucleoside unit **1** was synthesized from 5-(hydroxymethyl)-2′-deoxyuridine (**2**) [6] by treatment with chlorotrimethylsilane in dioxane [7] (→**3**) and then with sodium azide in DMF *via* 5-(azidomethyl)-2′-deoxyuridine (**4**; 73% yield) (*Scheme 1*). The reaction of **4** with pivaloyl chloride (pivCl) in pyridine gave the selectively 5′-protected product **5** in 70% yield. The 3′-tetrahydropyranylation (Thp) of **5** (→**6**) followed by Pd/C-catalyzed hydrogenation furnished the 3′-terminal nucleoside unit **1** in an overall yield of 64% from **5**.

Scheme 1

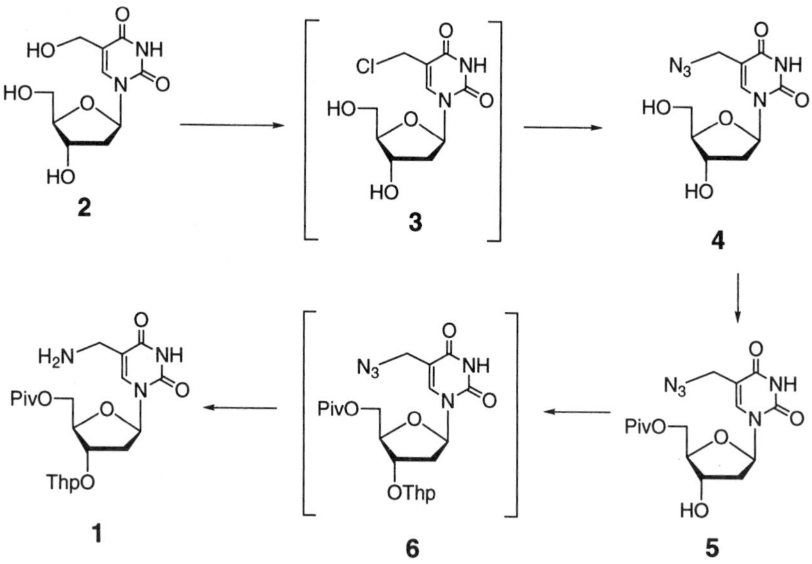

Piv = Me₃CCO, Thp = tetrahydro-2*H*-pyran-2-yl

2.2. *Chemical Synthesis of the Interresidually Cyclized Uridylyl-(3′→5′)-2′-deoxy-uridine* **14.** The interresidually cyclized uridylyl-(3′→5′)-2′-deoxyuridine **14** was synthesized from the 3′-terminal nucleoside unit **1** and the 5′-terminal nucleoside unit **7** [7] (*Scheme 2*). First, acid **7** was converted to ester **8** by treatment with *N*-

hydroxysuccinimide in the presence of *N*-[3-(dimethylamino)propyl] *N*'-ethyl-carbo-diimide hydrochloride (EDC) [8]. Condensation of the activated ester **8** with **1** in the presence of Et$_3$N gave the protected dimer **9** in 86% yield. The 'BuMe$_2$Si group of **9** was removed by treatment with tetrabutylammonium fluoride hydrate to give alcohol **10** in 71% yield, and the liberated 3'-hydroxy group of the 5'-upstream nucleoside was phosphorylated by treatment with cyclohexylammonium *S,S'*-diphenyl phosphorodi-thioate [9] in the presence of isodurenedisulfonyl dichloride (DDS) [9] and 1*H*-tetrazole [10] to give the phosphorodithioate derivative **11** in 76% yield. The conversion of **11** to the cyclized dimer **13** was accomplished in 64% yield by the intramolecular *in situ* cyclization of the phosphotriester intermediate **12**, which was obtained by alkaline hydrolysis of **11**, under the conditions using DDS as a condensing reagent and 1*H*-tetrazole as a nucleophilic catalyst. The successive deprotection of the cyclized dimer **13** by treatment with aqueous KOH solution and then with 80% AcOH gave the cyclized uridylyl-(3' → 5')-2'-deoxyuridine derivative **14** in 58% yield (*Scheme 2*). The detailed conformation of the fully deprotected cyclic dimer **14** is described in *Sect. 2.3*, and the results were used in *Sect. 2.5* to predict the 3D structure of the oligonucleotide containing the dimer unit **14**.

To incorporate the cyclized dimer unit into an oligonucleotide, the phosphoramidite unit **17** of the cyclized dimer was synthesized *via* **13** as a synthetic intermediate (*Scheme 3*). The (MeO)$_2$Tr and Thp groups of **13** were removed under acidic conditions (→ **15**), and the free 5'-hydroxy group of the 5'-upstream nucleoside unit was protected again with the (MeO)$_2$Tr group to give the 3'-hydroxy derivative **16** in 78% yield. The latter was then phosphitylated by treatment with 1.2 equiv. of 2-cyanoethyl diisopropylphosphoramidochloridite (=chloro(diisopropylamino)(2-cyano-ethoxy)phosphine) [11] in the presence of 1.2 equiv. of ethyldiisopropylamine as a base to give the desired phosphoramidite unit **17** in 71% yield.

The ^{31}P-NMR spectrum of **17** showed four signals near 150 ppm (146.99, 150.10, 150.35, and 150.44 ppm), which are derived from the phosphoramidite groups of the four diastereoisomers, and three signals around 25 ppm (24.78, 24.86, and 25.01 ppm) with the intensity ratio of 1:2:1, which are ascribed to the internucleotidic phosphoryl groups attached to the S-atom.

2.3. Conformational Analysis of the Cyclized Uridylyl-(3' → 5')-2'-deoxyuridine **14**. *2.3.1. Circular Dichroism (CD) Studies.* The CD spectra of cUpdU (**14**) and unmodified uridylyl-(3' → 5')-uridine (UpU) at 20 and 80° are shown in *Fig. 1*. The *Cotton* effect around 270 nm is the most important for the analysis of the conformational properties of these dinucleoside monophosphates because the CD pattern in this region reflects the orientation and interaction of the pyrimidine base of the 3'- and 5'-terminal nucleoside unit [12]. In the CD spectra of UpU, the intensity of this *Cotton* effect at 80° decreased to 60% at 20° indicating the conformational flexibility of UpU. In contrast, the intensity of the *Cotton* effect of cUpdU at 80° was maintained at 80% at 20°. This different temperature dependence clearly pointed to an increased conformational rigidity of cUpdU (**14**) induced by the cyclic structure. In the previous study of similarly cyclized dimers, in which the 3'-terminal nucleoside of **14** was replaced by a ribonucleoside, a similar conformational rigidity was observed in the CD spectra; the conformation was more rigid in the 3'-downstream ribonucleoside derivative than in **14** [5]. In general, the conformations of ribonucleosides are more rigid than those of

Scheme 2

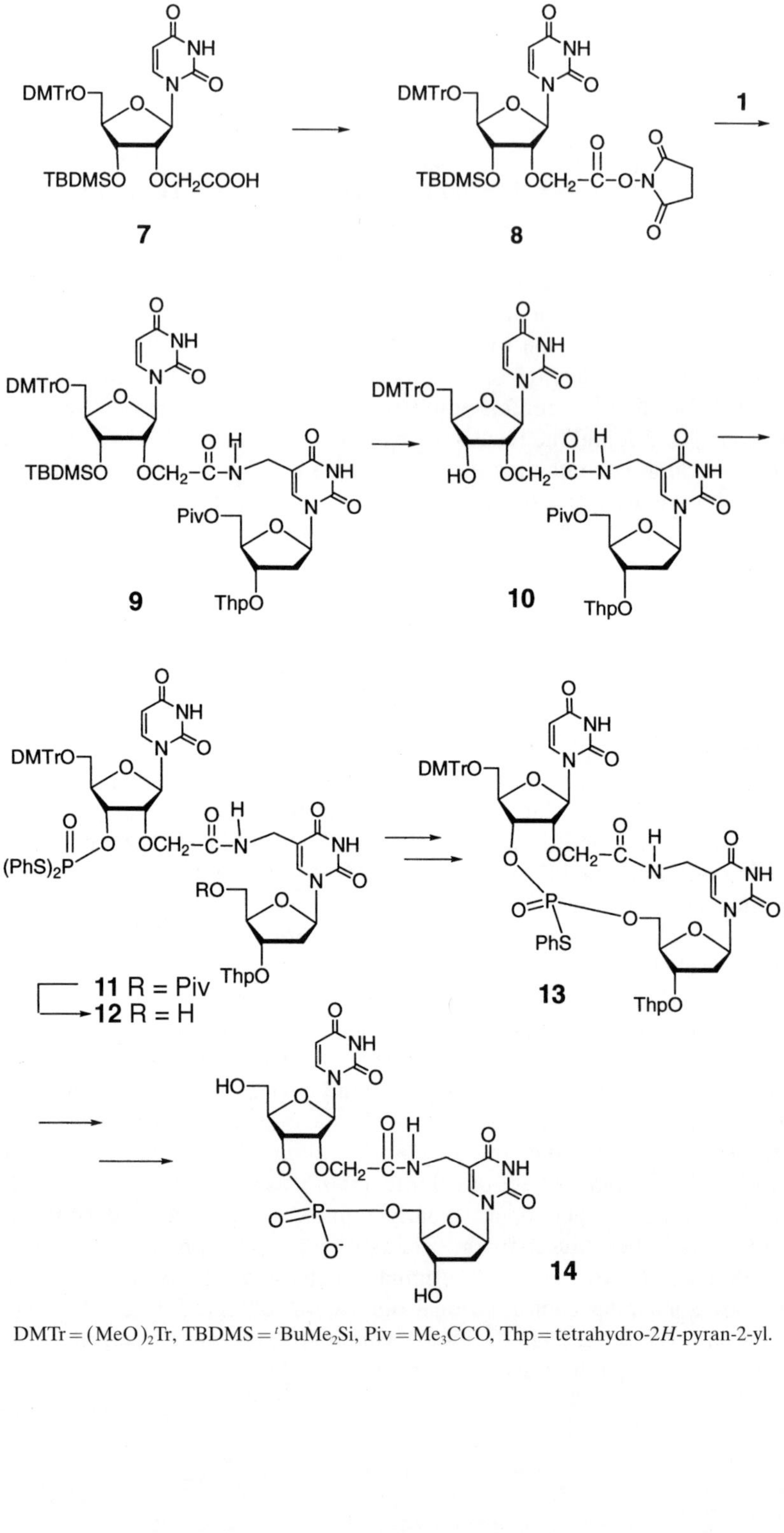

DMTr = (MeO)$_2$Tr, TBDMS = ^{t}BuMe$_2$Si, Piv = Me$_3$CCO, Thp = tetrahydro-2*H*-pyran-2-yl.

Scheme 3

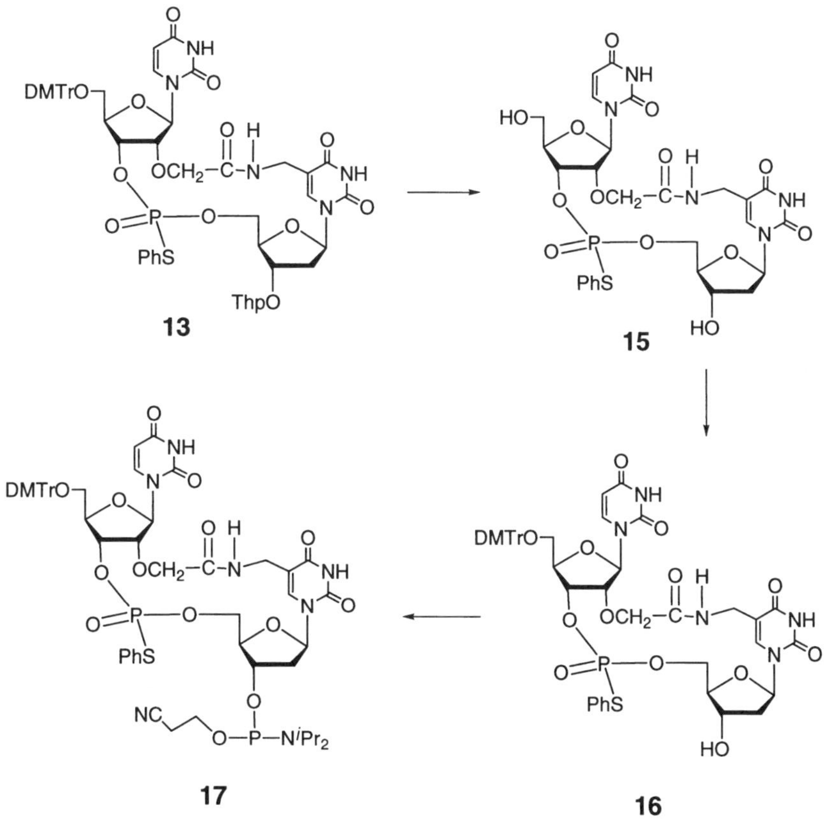

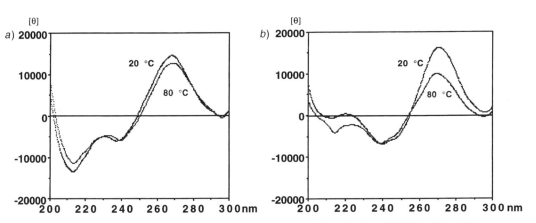

DMTr = (MeO)₂Tr, Thp = tetrahydro-2*H*-pyran-2-yl

Fig. 1. *CD Spectra of* a) *cUpdU* (**14**) *and* b) *UpU.* In 10 mM sodium phosphate (pH 7.0) at 20 and 80°.

deoxyribonucleosides because of the additional electronic and steric effect of the 2'-hydroxy group [13]. This is also true for the cyclic dimers. The conformational rigidity of cUpdU is a suitable property for the stabilization of bent structures by incorporating this cyclized dimer unit into an oligonucleotide.

2.3.2. *Analysis of the Ribose Conformation by ¹H-NMR*. The ribose moiety of the natural nucleic acids is in equilibrium between the *N*-type and *S*-type conformers [14] (*Fig. 2*). Therefore, the conformation of the ribose moiety is described by the respective population of the *N*-type conformation ($P(N$-type)) and the *S*-type conformation ($P(S$-type)). The population of each conformer is calculated from the H,H-coupling constants between the protons on the ribose or deoxyribose ring.

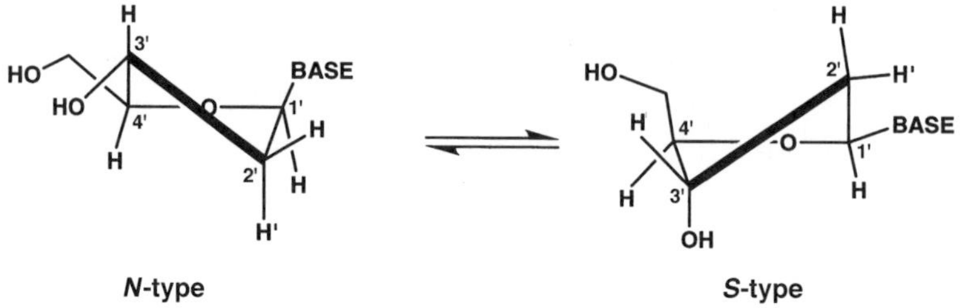

Fig. 2. *The two major conformers of deoxyribonucleosides which exist in equilibrium*

The H,H- and H,P-coupling constants of **14**, measured at 400 MHz, are shown in *Table 1*. Among the H,H-coupling constants of the ribose moiety, the $J(2',3')$ and $J(1',2')+J(3',4')$ values are of special importance. These values are nearly constant in natural or synthetic ribonucleic-acid derivatives having no structural constraints; the average values of $J(2',3')$ and $J(1',2')+J(3',4')$ are 5.2 Hz and 9.8 Hz, respectively [15]. Structurally constrained nucleic acids have values deviated from these typical ones. From this point of view, the rather smaller $J(1',2')+J(3',4')$ (7.4 Hz) of the 5'-upstream ribose of cUpdU indicates some distortion of the ribose conformation. However, the distortion might be small, because the *Karplus* equation [16] suggested that the $J(2',3')$ value of 5 Hz is realized only when the two equilibrium conformations are *N*-type and *S*-type regardless of the puckering amplitude (*Fig. 3*) and the $J(2',3')$ value of the cyclized dimer (4.8 Hz) is close to 5 Hz. Hence, the smaller $J(1',2')+J(3',4')$ observed would be explained by the *S*-type and *N*-type conformers with small distortion. In contrast, the conformation of the 3'-downstream deoxyribose moiety of **14** is not distorted because both the $J(2',3')$ (6.1 Hz) and $J(1',2')+J(3',4')$ (10.2 Hz) are close to the standard values observed for the natural 2'-deoxyribonucleosides (6.6 and 10.5 Hz, resp.). The $P(N$-type) and $P(S$-type) of both the 5'-upstream and 3'-downstream sugar moieties were calculated with the formula $P(N$-type$)=J(3',4')/(J(1',2')+J(3',4'))$ [16]. Consequently, the $P(N$-type) of the 5'-upstream ribose residue is 38%, while the $P(N$-type) of the 3'-downstream deoxyribose residue is 33%. In brief, the major conformers of both the 5'- and 3'-terminal sugars are *S*-type. The population $P(S$-type) is 60–70%.

2.3.3. *NOESY Spectrum of cUpdU*. To obtain information on the relative location of the 5'-upstream and 3'-downstream residues, the NOESY spectrum of cUpdU was

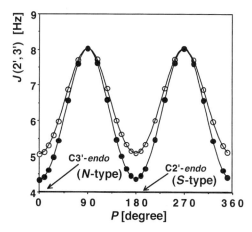

Fig. 3. *Simulation of* J*(2′,3′) by the* Karplus *equation* [16]. *P* = phase angle of pseudorotation. Plots of the values obtained from the simulations with the puckering amplitude fixed at 35° (○) and 40° (●).

Table 1. *H,H Coupling Constants* [Hz] *in the* ¹*H-NMR Spectrum* (400 MHz) *of* **14**[a])

	Coupling constant *J* [Hz]							
	J(1′,2′)	*J*(1′,2″)	*J*(2′,3′)	*J*(2″,3′)	*J*(3′,4′)	*J*(4′,5′)	*J*(4′,5″)	*J*(3′,P)[b])
5′-Terminal nucleoside	4.6	–	4.8	–	2.8	2.8	4.0	8.0
3′-Terminal nucleoside	6.7	6.7	3.7	6.1	3.5	n.d.	n.d.	

[a]) The critical values for the determination of the ribose or deoxyribose conformation are written in italics.
[b]) Measured at 109 MHz.

recorded. Although many cross-peaks are present, all of them are derived from the intraresidual H,H interactions, and no interresidual cross-peaks are observed. This result indicates that cUpdU has a conformation in which the distance between the 5′- and 3′-terminal nucleoside units is long. This distance information based on the interresidual H,H interactions was used to analyze the molecular-mechanics simulation results of the 3D structure of **14** described in *Sect. 2.3.4.*

2.3.4. *Molecular-Mechanics Studies.* Detailed three-dimensional model structures of the cyclic dimer **14** were built by molecular-mechanics calculations with the MonteCarlo conformation search and energy minimization [17]. The AMBER* force field [18][19] and GB/SA solvent model [19] were used in all simulations. Four calculation runs were carried out, in which the ribose moieties of the 5′-upstream and 3′-downstream nucleoside units were fixed in the {*N*-type and *N*-type} (**A**), {*N*-type and *S*-type} (**B**), {*S*-type and *N*-type} (**C**), and {*S*-type and *S*-type} (**D**) conformations, respectively. The ribose moieties were fixed in the standard *N*- and *S*-type conformations because the AMBER* force field did not give results compatible with the experimental data in terms of ribose conformation in our test simulations (data not shown). The structures having the lowest energies obtained from each run are shown in *Fig. 4.*

The structures **A** – **C** can be divided into two categories, namely, the 'turn' (**A** and **B** in *Fig. 4*) and 'bent' forms (**C** and **D**). The two runs in which the ribose moiety of the 5′-

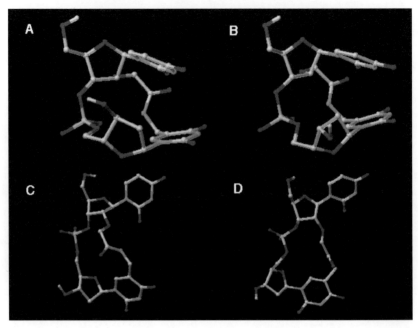

Fig. 4. *Low-energy structures of cUpdU* (**14**) *calculated by fixing the 5′-terminal ribose and 3′-terminal deoxyribose moieties in the {N-type and N-type}* (**A**), *{N-type and S-type}* (**B**), *{S-type and N-type}* (**C**), *and {S-type and S-type}* (**D**) *conformation*

upstream nucleoside unit was fixed in the *N*-type conformation gave energetically unfavorable turn conformations ($E = -882.74$ and -888.68 kJ/mol for **A** and **B**, resp.), whereas the two runs in which the ribose moiety of the 5′-upstream nucleoside units was fixed in the *S*-type conformation gave energetically more favorable bent conformers ($E = -893.02$ and -893.03 kJ/mol for **C** and **D**, resp.). In the turn conformers, the distance between the 5′-upstream and 3′-downstream residues is short enough to allow NOEs between the interresidual protons. For example, the inter-proton distances H−C(2′)(5′-upstream residue)/H−C(3′)(3′-downstream residue), H−C(2′)(5′-upstream residue)/H−C(2′)(3′-downstream residue), H−C(3′)(5′-upstream residue)/H−C(3′)(3′-downstream residue), and H−C(3′)(5′-upstream residue)/H−C(2′)(3′-downstream residue) are 4.78, 2.80, 3.18, and 2.37 Å, respectively, for conformer **A**, and 2.66, 4.42, 2.31, and 4.64 Å, respectively, for conformer **B**. In the bent conformers **C** and **D**, the interresidual distances are too long to give NOEs. Thus, the structural properties of the bent conformers are in agreement with the properties of **14** established by the H,H coupling constants (*Sect. 2.3.2*) and NOESY spectrum (*Sect. 2.3.3*), *i.e.*, the major conformer of the cyclic dimer has an *S*-type ribose at the 5′-terminus, with a long distance between the 5′- and 3′-terminal residues. Therefore, the three-dimensional structure of **14** in aqueous solution must be similar to that shown for **C** and **D** (*Fig. 4*).

From the above conformation analyses by means of CD and ¹H-NMR spectroscopy and molecular mechanics, it was concluded that the cyclic dimer **14** has a rigid conformation with the *S*-type ribose conformation and long interresidual distance. Thus, the dinucleoside monophosphate **14** having these conformational properties is

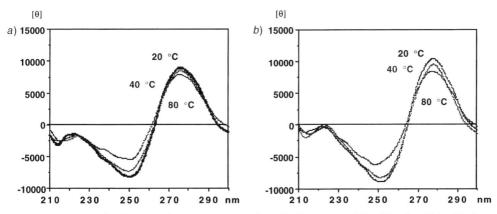

Fig. 6. *CD Spectra* a) *of decathymidylate* **19** *containing the cyclic dimer unit and* b) *of decathymidylate* T_{10}. In 10 mM sodium phosphate (pH 7.0).

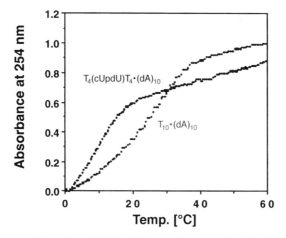

Fig. 7. *UV Melting curves of the duplexes* $T_4(cUpdU)T_4 \cdot decadeoxyadenosine$ (**19** · (dA)$_{10}$) *and of decathymidylate · decadeoxyadenosine* ($T_{10} \cdot$ (dA)$_{10}$). *Measured in 10 mM sodium phosphate and 150 mM NaCl (pH 7.0)*

Table 2. *Melting Temperature [°] of the DNA Duplexes* **19** · dA_{10}, $T_{10} \cdot dA_{10}$, *and* **20** · dA_n ($n = 14–19$). *Measured in 10 mM sodium phosphate – 150 mM NaCl buffer*

	19 · (dA)$_{10}$	$T_{10} \cdot$ (dA)$_{10}$	**20** · (dA)$_{14}$	**20** · (dA)$_{15}$	**20** · (dA)$_{16}$	**20** · (dA)$_{17}$	**20** · (dA)$_{18}$	**20** · (dA)$_{19}$
T_m [°]	8	28	26.5	30.3	32.0	33.0	33.2	33.0

conformer of **14** (*Fig. 4*, **D**) on the computer graphics (*Fig. 8*). While the decathymidylate containing the turn conformer **B** (see *Fig. 8*, **E**) does not seem suitable for duplex formation because of its extremely sharp bend, the one containing the bent conformer **D** (see *Fig. 8*, **F**) seems suitable. However, the distance between the N(1)-atoms on the 5′- and 3′-terminal pyrimidine rings of the cyclic dimer unit is *ca.* 9 Å, which is twice as long as the distance seen in the standard B-type duplex (4 Å) [22]. Therefore, the disruption of the base pairing could occur at this site.

2.5. *Conformational Properties of the Decanucleotide Having a Cyclic Dimer Unit.*
A decathymidylate **19** having the cyclic-dimer unit **14** at the center was synthesized by
the solid-phase synthesis. The decathymidylate **19** was expected to have a bent structure
induced by the cyclic structure.

It is well known that a change in three-dimensional DNA duplex structures is
reflected sensitively by the mobility in polyacrylamide-gel electrophoresis (PAGE)
[21]. The 20% PAGE of the decathymidylate **19**, T_{10}, and T_{11} are shown in *Fig. 5*. The
molecular-mass difference between **19** and T_{10} ($\Delta M_r = 57$) is smaller than that between
19 and T_{11} ($\Delta M_r = 246$). On the contrary, the difference in the gel shift between **19** and
T_{10} is larger than that between **19** and T_{11}. This result indicates a rigid structure for **19**,
which is induced by introduction of the cyclic dimer unit **14**, thereby restraining the
mobility of **19** on PAGE.

Fig. 5. *20% PAGE of $T_4(cUpdU)T_4$ (**19**), T_{11}, and T_{10}. Lanes 1 and 5, bromophenol blue; Lane 2, T_{11}; Lane 3,*
19; *Lane 4*, T_{10}.

The rigid conformation of **19** was also reflected in a smaller temperature
dependence of its CD spectra at 20, 40, and 80° compared to those of T_{10} at these
temperatures (*Fig. 6,a* and *b*, resp.). Although the cyclic structure unit of **19** reduces
the conformational flexibility, it seems only effective near the modified site and is not
likely to affect the overall conformation of the oligomer because the whole shape of the
CD spectra of **19** is similar to that of T_{10}. Oligomer **19** having a locally fixed
conformation as described above, would be suitable for the modification of a duplex
structure when hybridized to its complementary strand (see *Sect. 2.6*).

2.6. *Duplex Formation of the Deca- or Hexadecathymidylates Having a Cyclic
Dimer Unit.* The decathymidylate **19** was mixed with an equimolar amount of dA_{10}, and
the UV melting curve of the duplex formed was recorded and compared with that of
$T_{10} \cdot dA_{10}$ (*Fig. 7* and *Table 2*). The considerable decrease of the T_m value of $19 \cdot dA_{10}$
compared with that of $T_{10} \cdot dA_{10}$ ($\Delta T_m = -20°$) suggested disruption of base pairing.

The model structure of the disrupted base pairing and the whole duplex was built by
replacing the fifth and sixth thymidine units of T_{10} by the turn (*Fig. 4*, **B**) or bent

highly recommendable as the dimer unit to be introduced into the oligonucleotides as a rigid bent structure.

2.4. *Chemical Synthesis of the Oligonucleotide Having a Bent Structure.* An oligothymidylate having a bent structure was synthesized by using the phosphoramidite dimer unit **17**, derived from **14**. First, the tetrathymidylate **18** was synthesized to confirm the efficient introduction of the dimer unit under the standard conditions used in the solid-phase synthesis (*Scheme 4*). The coupling reaction, according to the general solid-phase oligonucleotide synthesis, followed by the treatments with aqueous NaOH solution and then with 80% AcOH, gave the tetramer **18**. The structure of **18** was supported by the ^{1}H- and ^{31}P-NMR spectra (D$_2$O); a low-field δ(P) indicating that the bent structure stabilized by the cyclic amide linkage was preserved even in the oligonucleotide because the low-field shift of the internucleotidic phosphate is the general indicator of the bent backbone conformation [20].

Scheme 4

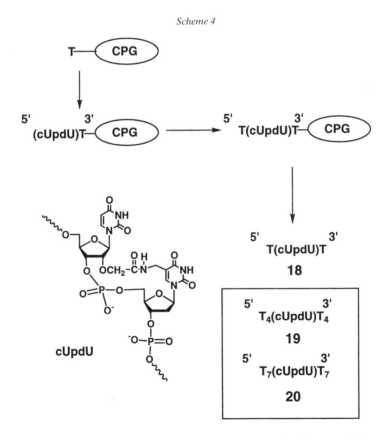

The ^{1}H-NMR spectra of tetramer **18** in D$_2$O showed the signals of the Me groups of the two thymine bases around 1.9 ppm and four signals of the 6 protons of the four pyrimidine bases around 7.5–7.9 ppm. The ^{31}P-NMR spectra gave three signals due to the three internucleotidic phosphate groups around 0 ppm. The δ(P) of the internucleotidic phosphate of the cyclic dimer unit was found in the lowest field in the ^{31}P-NMR spectrum.

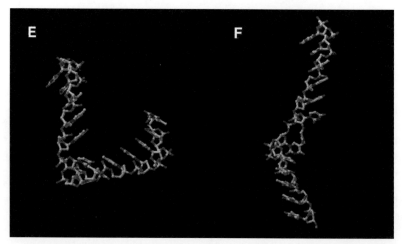

Fig. 8. *Predicted structures* **E** *and* **F** *of a dodecathymidylate* $T_5(cUpdU)T_5$ *incorporating the turn conformer* **B** *and the bent conformer* **D**, *respectively, of* Fig. 4

There are two plausible models for the disrupted duplex structure. One is the 'bulge-out' structure in which a single deoxyadenosine residue at the site opposite to the cyclic dimer unit bulges out to compensate for the widened internucleotidic distance in the oligothymidylate strand, and the other is the 'no-bulge-out' structure in which no adenosine residue bulges out but the base-pairing is distorted because of the unusual geometry. If the 'bulge-out' model structure is correct, either the 5′- or 3′-terminal deoxythymidine residue must dangle without forming any base-pairing when the lengths of the oligothymidylates containing a cyclic-dimer unit and the oligodeoxyadenylate are the same. Therefore, the validity of this model must be examined by measuring the dependence of the T_m values on the chain length of the oligodeoxyadenylate.

Because the T_m values of the oligodeoxyadenylate $\cdot$ **19** duplexes were too low to give correct T_m values when the chain length of the oligodeoxyadenylate was altered, the hexadecathymidylate **20** was synthesized, and the T_m values of the duplex formation between **20** and oligodeoxyadenylates $(dA)_{14}$ to $(dA)_{19}$ having the chain length of 14–19 units were measured (*Fig. 9* and *Table 2*). The T_m values of the duplexes formed between **20** and oligodeoxyadenylates $(dA)_{14}$ to $(dA)_{19}$ rose with the length of the oligodeoxyadenylate increasing from 14 to 17 units, and then reached a plateau (*Fig. 10*). These observations are consistent with the 'bulge-out' model described above wherein the terminal residues of the oligothymidylate strand dangle until the oligodeoxyadenylate chain is lengthened up to $(dA)_{16}$, but the number of base-pairing is constant after the chain length has reached 17 units. Shown in *Fig. 11* is the duplex model structure of $20 \cdot (dA)_{17}$ which has the bulged-out adenosine residue at the middle of the duplex. In this model, the duplex bends by *ca.* 30°, and the base-pairings are loosened at the middle of the strand. This base-pair disruption could contribute to the observed duplex instability.

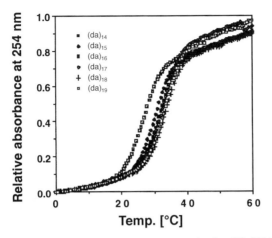

Fig. 9. *UV Melting curves of the duplexes $T_7(cUpdU)T_7 \cdot$ oligodeoxyadenylate (**20** $\cdot$ (dA)$_n$) having various chain lengths*. Measured in 10 mM sodium phosphate and 150 mM NaCl (pH 7.0). The vertical axis indicates the relative absorbance when the absorbance at 90° is normalized to unity.

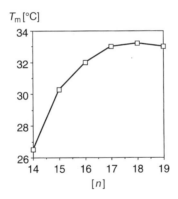

Fig. 10. *Relationship between the number* n *of nucleotide units in the oligodeoxyadenylate chain and the* T$_m$ *values* [°] *of the duplexes* $T_7(cUpdU)T_7 \cdot$ oligodeoxyadenylate (**20** $\cdot$ (dA)$_n$) *having various chain lengths (see Table 2)*

3. Conclusions. – In this study, an artificially bent DNA strand was chemically synthesized for the first time by the introduction of a conformationally rigid cyclized UpdU dimer at the center of oligothymidylates. No other report on the synthesis of such bent DNA oligomers by chemical approaches has been published to date. Since an increasing number of studies on DNA-protein and RNA-protein interactions [2] have revealed that DNA bending is a crucial step in the regulation of the function of nucleic acids [1], it is of great importance to have available an artificially bent structure of DNA or RNA that will be useful as a fixed bent motif for investigating the structure-function relationship of bent DNA or RNA.

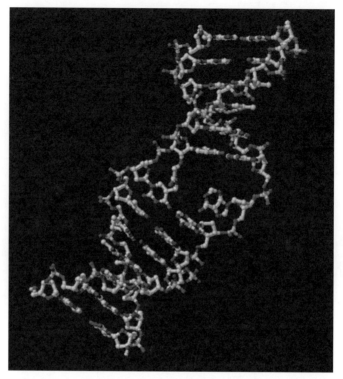

Fig. 11. *Duplex model structure of* $T_7(cUpdU)T_7 \cdot (dA)_{17}$ (**20** $\cdot$ (dA)$_{17}$) *which has a bulged-out adenosine residue at the middle of the duplex*

Experimental Part

General. Pyridine was distilled twice from TsCl and from CaH$_2$ and then stored over 4 Å molecular sieves. TLC: *Merck* silica gel *60-F-254* (0.25 mm). Column chromatography (CC): silica gal *C-200* from *Wako Co. Ltd.*; a minipump for a goldfish bowl was conveniently used to attain sufficient pressure for rapid CC separation; reversed-phase CC with μ*Bondapak-C-18* silica gel (*prep S-500, Waters*). Reversed-phase HPLC: μ*Bonda-sphere C-18* column, linear gradient starting from 0.1M NH$_4$OAc (pH 7.0) and applying 0–30% of MeCN at a flow rate of 1.0 ml/min for 30 min. UV Spectra: *Hitachi-U-2000* spectrometer; λ_{max} in nm. CD Spectra: *Jasco-J-500-C* spectrometer, 0.5-cm cell. ^{1}H-, ^{13}C-, and ^{31}P-NMR Spectra: at 270, 68, and 109 MHz, resp.; *Jeol-EX270* spectrometer; chemical shifts δ in ppm rel. to SiMe$_4$ or DSS (sodium 3-(trimethylsilyl)propane-1-sulfonate) for ^{1}H-NMR, CDCl$_3$ (= 77 ppm) or DSS (= 0 ppm) for ^{13}C-NMR, and 85% phosphoric acid (= 0 ppm) for ^{31}P-NMR, *J* in Hz; *J*(H,H) and NOESY at 400 MHz with *Varian-Unity-400* spectrometer; protons at the 5′-terminal unit are labelled with 'a', those at the 3′-terminal unit with 'b'. Elemental analyses were performed by the Microanalytical Laboratory, Tokyo Institute of Technology at Nagatsuta.

5-(Azidomethyl)-2′-deoxyuridine (**4**) [6]. To a soln. of 5-(hydroxymethyl)-2′-deoxyuridine [6] (9.0 g, 34.9 mmol) in 1,4-dioxane (250 ml) in a sealed tube, chlorotrimethylsilane (19.0 g, 175 mmol) was added. The soln. was stirred at 50° for 2.5 h and then evaporated, and the residue was dissolved in DMF (100 ml). NaN$_3$ (6.81 g, 105 mmol) was added, and the resulting soln. was stirred at 60° for 15 min. The precipitate was removed by filtration, the filtrate evaporated, and the residue submitted to CC (silica gel, CH$_2$Cl$_2$/MeOH 100:5): **4** (7.5 g, 77%). M.p. 134–136° ([6a] 133–135°). ^{1}H- and ^{13}C-NMR: identical with those of the authentic sample [6b].

5-(Azidomethyl)-2′-deoxy-5′-O-pivaloyluridine (**5**). To a soln. of **4** (200 mg, 0.69 mmol) in pyridine (5 ml), pivaloyl chloride (111 μl, 0.90 mmol) was added, and the resulting mixture was stirred for 1.5 h at r.t. The soln. was diluted with CHCl$_3$ (50 ml), washed with sat. aq. NaHCO$_3$ soln., dried (Na$_2$SO$_4$), and evaporated and the residue submitted to CC (silica gel, CHCl$_3$/MeOH 100:2): **5** (177 mg, 70%). M.p. 153–154° (hexane/AcOEt).

^{1}H-NMR (270 MHz, CDCl$_3$): 1.23 (s, 3 Me); 2.12 (m, H−C(2')); 2.50 (m, H−C(2)); 4.09−4.45 (m, H−C(3'), H−C(5'), CH$_2$N$_3$); 4.48 (m, H−C(4'), H−C(5)); 6.23 (m, H−C(1')); 7.52 (s, H−C(6)); 8.40 (br., NH). ^{13}C-NMR (68 MHz, CDCl$_3$): 27.17; 38.89; 40.74; 47.10; 63.65; 71.52; 84.75; 85.59; 109.92; 137.70. Anal. calc. for C$_{15}$H$_{21}$N$_5$O$_6$: C 49.05, H 5.76, N 19.06; found: C 49.17, H 5.86, N 18.83.

5-(Aminomethyl)-2'-deoxy-5'-O-pivaloyl-3'-O-(tetrahydro-2H-pyran-2-yl)uridine (**1**). To a soln. of **5** (5.4 g, 14.7 mmol) in CH$_2$Cl$_2$ (150 ml) 1,4-dihydro-2H-pyran (27 ml, 294 mmol) and CF$_3$COOH (570 µl, 7.4 mmol) were added, and the resulting mixture was stirred for 10 h. The mixture was washed with sat. aq. NaHCO$_3$ soln., the org. layer dried (Na$_2$SO$_4$) and evaporated, and the residue dissolved in MeOH (150 ml). To this soln., Pd/C (1 g) was added, and the mixture was stirred for 14 h under H$_2$. Pd/C was removed by filtration, the filtrate evaporated and the residue submitted to CC (silica gel, CHCl$_3$/MeOH 100:2): **1** (4.0 g, 64%). ^{1}H-NMR (270 MHz, CDCl$_3$): 1.18 (m, 3 Me); 1.82 (m, 3 CH$_2$); 2.05 (m, H'−C(2')); 2.48 (m, H−C(2')); 3.60 (s, CH$_2$N); 3.80 (m, CH$_2$O); 4.18−4.40 (m, H−C(3'), H−C(4'), 2 H−C(5')); 4.67 (s, OCHO); 6.25 (m, H−C(1')); 7.45, 7.42 (2s, H−C(6)). ^{13}C-NMR (68 MHz, CDCl$_3$): 19.21; 19.27; 25.21; 27.21; 30.55; 30.60; 37.77; 38.74; 38.83; 39.46; 62.68; 63.79; 63.90; 75.35; 76.14; 82.27; 83.04; 85.18; 85.50; 97.86; 98.40; 116.14; 135.33; 135.45; 150.01; 163.09; 178.11. Anal. calc. for C$_{20}$H$_{31}$N$_3$O$_7$ · 1/2 H$_2$O: C 55.35, H 7.34, N 9.67; found: C 55.31, H 7.56, N 8.65.

3'-O-[(tert-Butyl)dimethylsilyl]-2'-O-[2-{{{1-[2'-deoxy-5'-O-pivaloyl-3'-O-(tetrahydro-2H-pyran-2-yl)-β-D-ribofuranosyl]-1,2,3,4-tetrahydro-2,4-dioxopyrimidin-5-yl}methyl}amino]-2-oxoethyl]-5'-O-(4,4'-dimethoxytrityl)-uridine (**9**). N-[3-(Dimethylamino)propyl]-N-ethylcarbodiimide hydrochloride (58 mg, 0.3 mmol) and N-hydroxysuccimide (35 mg, 0.3 mmol) were added to a soln. of **7** (148 mg, 0.2 mmol) in DMF (2 ml). After 7.5 h, **1** (85 mg, 0.2 mmol) and Et$_3$N (28 µl, 0.2 mmol) were added, and the resulting soln. was stirred for 1 h. The mixture was partitioned between CHCl$_3$ and sat. aq. NaHCO$_3$ soln., and the org. layer was washed 5 times with sat. aq. NaHCO$_3$ soln., dried (Na$_2$CO$_3$), and evaporated. The residue was submitted to CC (silica gel, CHCl$_3$/MeOH 100:1): **9** (197 mg, 86%). ^{1}H-NMR (270 MHz, CDCl$_3$): − 0.12, 0.12 (m, 2 MeSi); 0.72, 0.83 (2s, 6 and 3 H, 3 Me); 1.18 (s, 3 Me); 1.72 (m, 3 CH$_2$); 2.22 (m, H'$_b$−C(2')); 2.48 (m, H$_b$−C(2')); 3.31 (m, H$_a$'−C(5')); 3.50 (m, H$_a$−C(5')); 3.81 (m, 2 MeO, CH$_2$O); 4.02−4.40 (m, H$_b$−C(3'), H$_b$−C(4'), 2 H$_b$−C(5'), CH$_2$N, CH$_2$CO, H$_a$−C(4')); 4.68 (br., OCHO), 5.28 (d, J(5,6) = 7.9, H$_a$−C(5)); 5.94 (s, H$_a$−C(1')); 6.25 (m, H$_b$−C(1')); 6.82−6.87 (m, 4 arom. H); 7.21−7.35 (m, 9 arom. H); 7.58 (s, H$_b$−C(6)); 8.05 (d, J(5,6) = 7.9, H$_a$−C(6)); 9.05, 8.76 (2 br. m, each 1 H, NH(3)). ^{13}C-NMR (68 MHz, CDCl$_3$): − 5.23; − 4.58; 17.70; 19.16; 25.11; 25.41; 27.10; 30.48; 35.67; 37.38; 38.35; 38.67; 55.11; 60.68; 62.43; 62.54; 63.67; 63.78; 69.53; 70.17; 75.38; 76.30; 82.14; 82.68; 82.80; 83.42; 85.21; 85.73; 86.97; 88.18; 97.66; 98.24; 102.21; 111.07; 113.14; 113.14; 127.15; 127.84; 128.23; 130.14; 134.79; 134.86; 138.35; 138.69; 139.43; 143.83; 150.10; 150.15; 150.48; 158.67; 163.54; 163.61; 169.27; 169.34; 177.95; 178.00. Anal. calc. for C$_{58}$H$_{75}$N$_5$O$_{16}$Si · H$_2$O: C 60.88, H 6.78, N 6.12; found: C 60.37, H 6.68, N 6.68.

2'-O-[2-{{{1-[2'-Deoxy-5'-O-pivaloyl-3'-O-(tetrahydro-2H-pyran-2-yl)-β-D-ribofuranosyl]-1,2,3,4-tetrahydro-2,4-dioxopyrimidin-5-yl}methyl}amino]-2-oxoethyl]-5'-O-(4,4'-dimethoxytrityl)uridine (**10**). To a soln. of **9** (547 mg, 0.48 mmol) in THF (5 ml) tetrabutylammonium fluoride monohydrate (250 mg, 0.96 mmol) was added, and the soln. was stirred for 1 h. The mixture was diluted with CHCl$_3$ (50 ml), the soln. washed with H$_2$O (3 × 50 ml), dried (Na$_2$SO$_4$), and evaporated, and the residue submitted to CC (silica gel, CHCl$_3$/MeOH 100:2): **10** (354 mg, 71%). ^{1}H-NMR (270 MHz, CDCl$_3$): 1.21 (s, 3 Me); 1.46−1.87 (m, 3 CH$_2$); 2.05 (m, H'$_b$−C(2')); 2.50 (m, H$_b$−C(2')); 3.50 (m, 3 H, CH$_2$O, 2 H$_a$−C(5')); 3.68 (m, 7 H, CH$_2$O, 2 MeO); 4.02 (m, H$_b$−C(4')); 4.16−4.39 (m, H$_a$−C(2'), H$_b$−C(3'), H$_a$−C(4'), 2 H$_b$−C(5'), CH$_2$N, CH$_2$CO); 4.49 (m, H$_a$−C(3')); 4.68 (br., OCHO); 5.33 (d, J(5,6) = 8.2, H$_a$−C(5)); 5.84 (s, H$_a$−C(1')); 6.18 (m, H$_b$−C(1')); 6.82−6.88 (m, 4 arom. H); 7.22−7.41 (m, 9 arom. H); 7.67 (s, H$_b$−C(6)); 7.97 (d, J(5,6) = 7.9, H$_a$−C(6)); 10.24, 10.90 (2 br. m, each 1 H, NH(3)). ^{13}C-NMR (68 MHz, CDCl$_3$): 19.14; 25.18; 27.21; 30.48; 36.10; 37.72; 38.80; 55.20; 61.42; 62.57; 63.65; 68.14; 69.62; 75.06; 76.17; 82.34; 83.09; 84.33; 85.43; 85.81; 86.85; 88.61; 97.75; 98.46; 102.35; 111.02; 113.12; 123.74; 127.03; 127.75; 127.92; 128.14; 129.11; 130.10; 130.17; 135.24; 135.44; 135.99; 139.10; 139.34; 139.46; 144.47; 149.70; 150.46; 151.32; 158.60; 163.23; 164.15; 170.39; 170.44; 178.17. Anal. calc. for C$_{52}$H$_{61}$N$_5$O$_{16}$ · H$_2$O: C 60.64, H 6.16, N 6.79; found: C 60.52, H 6.12, N 6.29.

2'-O-[2-{{{1-[2'-Deoxy-5'-O-pivaloyl-3'-O-(tetrahydro-2H-pyran-2-yl)-β-D-ribofuranosyl]-1,2,3,4-tetrahydro-2,4-dioxopyrimidin-5-yl}methyl}amino]-2-oxoethyl]-5'-O-(4,4'-dimethoxytrityl)uridine 3'-(S,S'-Diphenyl Phosphorodithioate) (**11**). Compound **10** (354 mg, 0.34 mmol) was rendered anhydrous by repeated co-evaporations with anh. pyridine and dissolved in anh. pyridine (5 ml). Cyclohexylammonium S,S'-diphenyl phosphorodithioate (191 mg, 0.68 mmol), isodurenedisulfonyl dichloride (= 2,4,5,6-tetramethylbenzene-1,3-disulfonyldichloride; 225 mg, 0.68 mmol), and 1H-tetrazole (95 mg, 1.36 mmol) were added, and the resulting soln. was stirred for 1 h. Then, the mixture was partitioned between CHCl$_3$ (20 ml) and sat. aq. NaHCO$_3$ soln. (20 ml). The org. phase was washed 5 times with sat. aq. NaHCO$_3$ soln., dried (Na$_2$SO$_4$), and evaporated and the residue co-evaporated repeatedly with toluene and then submitted to CC (silica gel, CHCl$_3$/MeOH 100:1): **11** (333 mg,

76%). ^{1}H-NMR (270 MHz, CDCl$_3$): 1.08 (s, 3 Me); 1.51–1.92 (m, 3 CH$_2$); 2.12 (m, H$'_b$–C(2$'$)); 2.45 (m, H$_b$–C(2$'$)); 3.38 (d, J_{gem} = 9.2, H$'_a$–C(5$'$)); 3.48 (m, CH$_2$O, H$_a$–C(5$'$)); 3.80 (m, 2 MeO, H$_a$–C(4$'$)); 3.98–4.21 (m, H$_a$–C(2$'$), 2 H$_b$–C(5$'$), CH$_2$N, CH$_2$CO); 4.30 (m, H$_b$–C(3$'$), H$_b$–C(4$'$)); 4.62 (br., OCHO); 5.12 (m, H$_a$–C(3$'$)); 5.28 (d, J(5,6) = 7.8, H$_a$–C(5)); 6.08 (m, H$_a$–C(1$'$)); 6.20 (m, H$_b$–C(1$'$)); 6.82 (m, 4 arom. H); 7.20–7.58 (m, 19 arom. H); 7.70 (d, J(5,6) = 7.9, H$_a$–C(6)); 9.34, 9.55 (2br. m, each 1 H, NH(3)). ^{13}C-NMR (68 MHz, CDCl$_3$): 19.21; 19.25; 25.20; 27.15; 27.21; 30.51; 35.76; 37.47; 38.46; 38.76; 55.20; 61.60; 62.52; 62.64; 63.76; 63.88; 70.32; 74.54; 75.44; 76.37; 82.00; 82.93; 85.36; 85.93; 86.49; 87.49; 97.74; 98.37; 102.91; 111.02; 111.05; 113.33; 125.25; 125.36; 125.75; 125.86; 127.30; 128.05; 128.14; 129.47; 129.51; 129.67; 129.72; 129.87; 129.92; 130.12; 134.65; 134.70; 135.08; 135.15; 135.51; 135.60; 138.63; 138.92; 139.05; 143.88; 149.98; 150.03; 158.78; 163.45; 163.49; 168.64; 168.73; 178.06; 178.09. ^{31}P-NMR (109 MHz, CDCl$_3$): 52.09, 52.13. Anal. calc. for C$_{64}$H$_{70}$N$_5$O$_{17}$·2 H$_2$O: C 58.58, H 5.38, N 5.33; found: C 57.94, H 5.20, N 5.48.

[5:2$'$-O-[Methyleneimino(2-oxoethane-2,1-diyl)]]-Linked 5$'$-O-(4,4$'$-Dimethoxytrityl)-P-thiouridyl-(3$' \to$ 5$'$)-2$'$-deoxy-3$'$-O-(tetrahydro-2H-pyran-2-yl)uridine S-Phenyl Ester (**13**). A soln. of **11** (750 mg, 0.58 mmol) in 0.5M KOH/pyridine 1:1 (40 ml) was stirred for 3 h. The soln. was neutralized with *Dowex 50W × 8* (pyridinium salt, 2 × 5 cm). Et$_3$N (5 ml) was added and the solvent evaporated. The residue was rendered anhydrous by repeated co-evaporations (5×) with anh. pyridine and dissolved in anh. pyridine (30 ml). Isodurenedisulfonyl dichloride (384 mg, 1.16 mmol) and 1*H*-tetrazole (326 mg, 2.32 mmol) were added, and the mixture was stirred for 20 min. The mixture was diluted with CHCl$_3$ (100 ml), the soln. washed 3 times with sat. aq. NaHCO$_3$ soln., the org. phase dried (Na$_2$SO$_4$) and evaporated, and the residue submitted to CC (silica gel, CHCl$_3$/MeOH 100:1.5): **13** (444 mg, 65%). ^{1}H-NMR (270 MHz, CDCl$_3$): 1.38–1.82 (m, 3 CH$_2$); 2.05 (m, H$'_b$–C(2$'$)); 2.52 (m, H$_b$–C(2$'$)); 3.20 (m, H$'_a$–C(5$'$)); 3.58 (m, CH$_2$O, H$_a$–C(5$'$)); 3.80 (m, 2 MeO); 4.01–4.48 (m, H$_a$–C(2$'$), H$_b$–C(3$'$), H$_b$–C(4$'$), 2 H$_b$–C(5$'$), CH$_2$N, CH$_2$CO); 4.60 (br., H$_a$–C(4$'$)); 4.61 (s, OCHO); 4.91 (m, H$_a$–C(3$'$)); 5.31 (d, J(5,6) = 8.2, H$_a$–C(5)); 6.05 (m, H$_a$–C(1$'$)); 6.21, 6.40 (2m, H$_b$–C(1$'$)); 6.83–6.99 (m, 4 arom. H); 7.18–7.58 (m, 9 arom. H); 7.79 (m, H$_a$–C(6), H$_b$–C(6)); 9.02, 9.10, 9.38, 9.44 (NH(3)). ^{13}C-NMR (68 MHz, CDCl$_3$): 19.01; 19.59; 25.18; 19.04; 29.63; 30.44; 30.64; 34.52; 38.19; 39.22; 55.22; 55.28; 62.18; 62.43; 62.61; 63.16; 68.20; 70.37; 75.29; 75.62; 75.81; 77.93; 81.38; 82.03; 82.16; 82.28; 82.43; 83.49; 85.57; 85.84; 86.02; 87.49; 87.78; 98.22; 98.40; 98.53; 102.93; 103.43; 111.52; 111.73; 112.20; 113.10; 113.32; 113.41; 124.22; 124.33; 124.44; 127.27; 127.48; 127.73; 128.12; 129.09; 129.74; 129.85; 134.14; 134.47; 134.56; 134.74; 134.91; 134.99; 138.85; 143.61; 143.90; 150.06; 150.21; 150.48; 150.57; 150.62; 158.78; 158.90; 162.17; 162.35; 162.41; 162.71; 163.04; 166.77; 168.30; 168.59. Anal. calc. for C$_{53}$H$_{56}$N$_5$O$_{16}$PS: C 58.83, H 5.22, N 6.47; found: C 58.53, H 5.78, N 6.57.

[5:2$'$-O-[Methyleneimino(2-oxoethane-2,1-diyl)]]-Linked Uridyl-(3$' \to$ 5$'$)-2$'$-deoxyuridine (**14**). A soln. of **13** (77 mg, 65 μmol) in 0.2M NaOH/pyridine 1:1 (4 ml) was stirred for 5 min. The mixture was neutralized with *Dowex 50W × 8* (pyridinium salt, 1 × 5 cm). The solvent was evaporated and the remaining pyridine was removed by repeated co-evaporations with H$_2$O. The residue was dissolved in 80% AcOH/H$_2$O and the soln. stirred for 20 h. The AcOH was removed by repeated co-evaporations with H$_2$O and the residue submitted to reversed-phase CC (*C-18*, H$_2$O/MeOH 100:8): **14** (760 A_{260}, 58%). The yield was estimated by assuming that ε_{260} of **14** was identical with that of UpU (ε_{260} 19600). ^{1}H-NMR (270 MHz, D$_2$O): 2.28 (ddd, J(1$'$,2$'$) = 6.7, J(2$'$,3$'$) = 3.7, J(2$'$,2$''$) = 13.5, H$'_b$–C(2$'$)); 2.46 (ddd, J(1$'$,2$''$) = 6.7, J(2$''$,3$'$) = 6.1, J(2$'$,2$''$) = 13.5, H$_b$–C(2$'$)); 3.90 (dd, J(5$'$,5$''$) = 12.9, J(4$'$,5$''$) = 4.0, H$'_a$–C(5$'$)); 3.98 (dd, J(5$'$,5$''$) = 12.9, J(4$'$,5$'$) = 2.8, H$_a$–C(5$'$)); 4.18 (m, 2 H$_b$–C(5$'$)); 4.24 (m, H$_b$–C(4$'$), CH$_2$N); 4.40 (m, H$_a$–C(4$'$)); 4.45 (dd, J(1$'$,2$'$) = 4.6, J(2$'$,3$'$) = 4.8, H$_a$–C(2$'$)); 4.55 (ddd, J(2$'$,3$'$) = 3.7, J(2$''$,3$'$) = 6.1, J(3$'$,4$'$) = 3.5, H$_b$–C(3$'$)); 4.82 (ddd, J(2$'$,3$'$) = 4.8, J(3$'$,4$'$) = 2.8, J(3$'$,P) = 8.0, H$_a$–C(3$'$)); 5.90 (d, J(5,6) = 8.0, H$_a$–C(5)); 6.13 (d, J(1$'$,2$'$) = 6.7, H$_a$–C(1$'$)); 6.40 (dd, J(1$'$,2$''$) = 6.7, J(1$'$,2$'$) = 6.7, H$_b$–C(1$'$)); 7.59 (s, H$_b$–C(6)); 7.90 (d, J(5,6) = 8.0, H$_a$–C(6)). ^{13}C-NMR (68 MHz, D$_2$O): 34.30; 38.95; 60.14; 65.28; 68.64; 70.56; 71.33; 80.26; 83.79; 83.88; 85.10; 85.22; 85.35; 87.42; 102.45; 111.57; 136.17; 141.54; 151.28; 151.52; 164.50; 166.09; 171.69. ^{31}P-NMR (109 MHz, D$_2$O): 0.56.

[5:2$'$-O-[Methyleneimino(2-oxoethane-2,1-diyl)]]-Linked 5$'$-O-(4,4$'$-Dimethoxytrityl)-P-thiouridyl-(3$' \to$ 5$'$)-2$'$-deoxyuridine S-Phenyl Ester (**16**). A soln. of **13** (350 mg, 0.29 mmol) in 80% AcOH (5 ml) was stirred for 5 h. The mixture was evaporated and the remaining AcOH removed by repeated co-evaporations with EtOH. The residue was rendered anhydrous by repeated co-evaporations with anh. pyridine and then dissolved in anh. pyridine (9 ml). 4,4$'$-Dimethoxytrityl chloride (268 mg, 0.78 mmol) was added, and the soln. was stirred for 20 h. The mixture was partitioned between CHCl$_3$ (20 ml) and sat. aq. NaHCO$_3$ soln. (20 ml), the org. phase washed 3 times with sat. aq. NaHCO$_3$ soln., dried (Na$_2$SO$_4$), and evaporated and the residue submitted to CC (silica gel, CHCl$_3$/MeOH 100:2): **16** (250 mg, 78%). ^{1}H-NMR (270 MHz, D$_2$O): 1.98 (m, H$'_b$–C(2$'$)); 2.38 (m, H$_b$–C(2$'$)); 3.30 (m, H$'_a$–C(5$'$)); 3.50 (m, H$_a$–C(5$'$)); 3.73 (s, 2 MeO); 3.90 (m, H$_b$–C(4$'$)); 4.00–4.10 (m, 2 H, H$_a$–C(4$'$), CH$_2$N); 4.10–4.60 (m, 7 H, 2 H$_b$–C(5$'$), H$_b$–C(3$'$), H$_a$–C(2$'$), CH$_2$N, OCH$_2$C); 5.24

(m, H$_a$–C(3')); 5.27 (d, $J(5,6) = 7.9$, H$_a$–C(5)); 6.00 (d, $J(1',2') = 5.9$, H$_a$–C(1')); 6.20 (t, $J(1',2') = J(1',2'') = 5.9$, H$_b$–C(1')); 7.62 ($d$, $J(5,6) = 7.9$, H$_a$–C(6)); 7.83 (t, $J(6,CH_2) = 3.9$, H$_b$–C(6)). ^{13}C-NMR (68 MHz, CDCl$_3$): 34.25; 34.36; 40.45; 55.01; 55.04; 61.83; 67.84; 67.96; 69.33; 69.76; 70.01; 80.94; 81.76; 81.89; 83.76; 84.17; 84.30; 85.30; 85.88; 87.31; 87.57; 103.02; 111.38; 111.73; 112.81; 113.14; 123.68; 123.76; 123.86; 127.08; 127.28; 127.85; 127.92; 127.96; 129.60; 129.63; 129.78; 129.83; 129.97; 134.09; 134.29; 134.34; 134.47; 134.67; 134.72; 134.79; 138.98; 139.26; 143.47; 143.76; 150.15; 150.24; 150.49; 158.56; 158.69; 162.77; 163.34; 168.63; 168.70; 168.98. ^{31}P-NMR (109 MHz, D$_2$O): 25.64; 24.85. Anal. calc. for C$_{48}$H$_{48}$N$_5$O$_{15}$PS · 2 H$_2$O: C 55.76, H 5.07, N 6.77; found: C 55.26, H 4.92, N 6.23.

[5 : 2'-O-[Methyleneimino(2-oxoethane-2,1-diyl)]]-Linked 5'-O-(4,4'-Dimethoxytrityl)-P-thiouridyl-(3' → 5')-2'-deoxyuridine 3'-(2-Cyanoethyl Diisopropylphosphoramidite) S-Phenyl Ester (**17**). Compound **16** (200 mg, 0.18 mmol) was rendered anhydrous by repeated co-evaporations with anh. pyridine and toluene. The residue was dissolved in anh. CH$_2$Cl$_2$ (2 ml). To this soln. were added ^{i}Pr$_2$EtN (39 μl, 0.22 mmol) and 2-cyanoethyl diisopropylphosphoramidochloridite (48 μl, 0.22 mmol), and the resulting mixture was stirred for 1 h. The mixture was diluted with CH$_2$Cl$_2$ (20 ml), the soln. washed 3 times with sat. aq. NaHCO$_3$ soln. (20 ml), dried (Na$_2$SO$_4$), and evaporated and the residue submitted to CC (silica gel, hexane/AcOEt 1 : 4 containing 1% pyridine): **17** (180 mg, 71%). ^{1}H-NMR (270 MHz, D$_2$O): major isomer: 1.30 (m, 4 Me); 2.10 (m, H$_b'$–C(2')); 2.40–2.80 (m, CH$_2$CN, H$_b$–C(2')); 3.20 (m, H$_a'$–C(5')); 3.50–3.98 (m, 2 MeO, H$_a$–C(4'), H$_a$–C(5'), CH$_2$O, NCH); 4.00–4.78 (m, H$_a$–C(2'), CH$_2$N, CH$_2$CO, H$_b$–C(4'), H$_b$–C(3'), 2 H$_b$–C(5')); 4.91 (m, H$_a$–C(2')); 5.37 (d, $J(5,6) = 8.2$, H$_a$–C(5)); 6.18 (m, H$_a$–C(1')); 6.28 (t, $J(1',2') = J(1',2'') = 6.3$, H$_b$–C(1')); 6.90 ($m$, 4 arom. H); 7.20–7.82 (m, 14 arom. H); 7.60 (m, H$_a$–C(6), H$_b$–C(6)); 9.01–9.80 (br., NH); minor isomer: 7.80 (d, H$_a$–C(6)); 6.40 (t, $J(1',2') = J(1',2'') = 5.6$, H$_b$–C(1')); 6.00 ($m$, H$_a$–C(1')). ^{13}C-NMR (68 MHz, CDCl$_3$): 20.29; 20.40; 21.39; 24.53; 34.50; 39.12; 39.93; 43.22; 43.40; 55.26; 57.92; 58.04; 58.19; 58.33; 62.12; 67.87; 69.87; 70.42; 73.17; 73.48; 74.25; 81.51; 81.98; 83.88; 84.35; 85.82; 87.46; 87.75; 102.75; 103.34; 111.64; 111.73; 112.31; 113.30; 113.37; 117.62; 117.74; 124.31; 124.42; 125.23; 127.28; 127.46; 128.10; 128.97; 129.76; 130.12; 133.96; 134.41; 134.52; 134.68; 134.74; 134.90; 134.95; 138.92; 143.59; 143.90; 150.06; 150.12; 150.57; 152.06; 150.12; 150.57; 152.06; 158.76; 158.87; 162.05; 162.26; 162.71; 163.07; 168.19; 168.50. ^{31}P-NMR (109 MHz, D$_2$O): 24.78; 24.86; 25.01; 149.69; 150.10; 150.35; 150.44. Anal. calc. for C$_{57}$H$_{65}$N$_7$O$_{16}$P$_2$S · 5/2 H$_2$O: C 55.07, H 5.68, N 7.88; found: C 54.89, H 5.34, N 7.84.

Centrally [5 : 2'-O-[Methyleneimino(2-oxoethane-2,1-diyl)]]-Linked Thymidyl-(3' → 5')-uridyl-(3' → 5')-2'-deoxyuridyl-(3' → 5')-thymidine (**18**). A commercially available DM[(MeO)$_2$Tr]T-loaded CPG support (41 mg, 1 mmol) was placed in a glass column equipped with a glass filter, a stopper, and a stopcock. The (MeO)$_2$Tr group was removed by treatment with 1% CF$_3$COOH in CH$_2$Cl$_2$ (2 ml × 2) several times. After the support was washed with CH$_2$Cl$_2$ (10 ml), the 5'-hydroxy group liberated was coupled with **17** (39 mg, 30 μmol) in the presence of 1*H*-tetrazole (21 mg, 0.3 mmol) in MeCN (200 μl) for 10 min under Ar. After the excess phosphoramidite unit and 1*H*-tetrazole were removed by washing with CH$_2$Cl$_2$ (10 ml), capping of the unreacted OH group with 0.2M Ac$_2$O in DMAP/pyridine 1 : 9 (2 ml) for 2 min, oxidation with 0.1M I$_2$ in H$_2$O/pyridine/THF 1 : 10 : 40 (2 ml) for 1 min, detritylation with 1% CF$_3$COOH in CH$_2$Cl$_2$ (2 ml × 5), and elongation of another thymidylate residue were carried out according to the standard protocol for the solid-phase oligonucleotide synthesis. After the (MeO)$_2$Tr group was removed by treatment with 1% CF$_3$COOH in CH$_2$Cl$_2$ (2 ml × 5), the removal of the alkali-labile protective groups and elution of the oligonucleotide were carried out by treatment with 0.5M NaOH/pyridine 1 : 1 (10 ml) for 30 min. The eluant was neutralized on *Dowex 50W × 8* (H$^+$ form, 3 ml), and the *Dowex 50W × 8* residue was removed by filtration. The filtrate was evaporated and the residue lyophilized. The resulting white solid was purified by anion-exchange HPLC (*Fax*; linear gradient of 50 → 770 mM NaCl in 25 mM phosphate buffer (pH 6.0) for 50 min). After evaporation, the remaining salt was removed by CC (*Sephadex G-15*, H$_2$O). The eluant was lyophilized: **18** (6 A_{260}). ^{1}H-NMR (270 MHz, D$_2$O): 7.67 (d, $J(5,6) = 8.3$, H–C(6)); 7.68 (s, H–C(6) of T); 7.62 (s, H–C(6) of T); 7.50 (s, H–C(6) of dU); 6.30 (m, 2 H, H–C(1') of T); 6.10 (d, $J(1',2') = 5.0$, 1 H, H–C(1')); 5.90 (d, $J(5,6) = 8.3$, 1 H, H–C(5) of U); 1.86 (s, 1 Me); 1.87 (s, 1 Me). ^{31}P-NMR (109 MHz, D$_2$O): 0.29; −0.29; −0.47.

General Procedure of the Synthesis of Oligothymidylates Containing the Cyclic-Dimer Unit **14**: *T-T-T-T-(cUpdU)-T-T-T-T* (**19**) *and T-T-T-T-T-T-T-(cUpdU)-T-T-T-T-T-T-T* (**20**). The fully protected oligothymidylates having the chain length of a 4-mer and 7-mer were prepared on an automated DNA synthesizer with the commercially available [(MeO)$_2$Tr]T-loaded CPG support (41 mg, 1 mmol) and the thymidine 3'-phosphoramidite unit. Then, the solid support was placed in a glass column equipped with a glass filter, a stopper, and a stopcock. The 5'-terminal (MeO)$_2$Tr group was removed by treatment with 1% CF$_3$COOH in CH$_2$Cl$_2$ (2 ml × 5), and the resin was washed with CH$_2$Cl$_2$ (10 ml). The 5'-hydroxy group liberated was coupled with the dimer unit **17** (39 mg, 30 μmol) in the presence of 1*H*-tetrazole (21 mg, 0.3 mmol) in MeCN (200 μl) for 10 min under

Ar. After the excess phosphoramidite unit and 1*H*-tetrazole were removed by washing with CH_2Cl_2 (10 ml), capping of the unreacted OH group with 0.12M AcOH in DMAP/pyridine 9:1 (2 ml) for 2 min, oxidation with 0.1M I_2 in H_2O/pyridine/THF 1:10:40 (2 ml) for 1 min was carried out. The solid support was put on an automated DNA synthesizer, and the remaining thymidylate sequences were constructed automatically. After all the couplings were achieved, the fully protected oligonucleotides were deprotected and eluted by treatment with 0.5M NaOH/pyridine 1:1 (10 ml) for 30 min. Then, the eluant was neutralized with *Dowex 50W×8* (pyridinium salt, 3 ml), and the *Dowex* resin was removed by filtration. To the filtrate was added Et_3N (5 ml), and the soln. was evaporated. The residue was placed on a *Sep-Pak C_{18}* column, and the shorter oligothymidylates were washed out with H_2O (10 ml). The $(MeO)_2Tr$ group of the desired oligothymidylates was removed by treatment with 3% CF_3COOH in CH_2Cl_2 (5 ml) on the column, and the deprotected oligothymidylates were eluted with the same solvent. The eluant was evaporated and the residue lyophilized. The resulting white solid was further purified by anion-exchange HPLC (*Fax*, linear gradient of $50 \rightarrow 770$ mM NaCl in 25 mM phosphate buffer (pH 6.0) for 50 min). After evaporation, the remaining salt was removed by CC (*Sephadex G-15*, H_2O). The eluant was lyophilized: **19** (6 A_{260}) or **20** (22 A_{260}).

19: t_R 24.3 min UV (H_2O): 262.8; min. 233.0.

20: t_R 30.5 min. UV (H_2O): 263.2; min. 233.2.

CD Spectra. Oligonucleotides were dissolved in 10 mM sodium phosphate (pH 7.0), and the absorbance at 264 nm was adjusted to 0.5. The soln. was placed in a 5-mm cell, and the spectra were accumulated 16 times at 20 nm/s scan speed. The baseline was corrected by the subtraction of the CD spectra of the solvent recorded under the same spectrometer conditions at 20°. The temp. was controlled by a *Haake-C* circulator, and the sample was equilibrated for 15 min after the cell holder had reached the desired temp. The spectra were displayed after having been smoothed by averaging. The concentration of the sample was calculated on the assumption that the ε_{260} was 19600.

Measurement of the H,H-Coupling Constants in 1D-^{1}H-NMR Spectra. The cyclic dimer **14** (100 A_{260}) was dissolved in 10 mM sodium phosphate (700 µl, pH 7.0) and then lyophilized 3 times with 99.8% D_2O. The resulting white precipitate was dissolved in 700 µl of 99.95% D_2O. The peak of H_2O was suppressed by pre-saturation. The coupling constants were measured at the 0.15 Hz of digital resolution, accomplished by the spin-simulation deconvolusion module of the spectrometer.

2D-NOESY Spectrum. Conditions: 2048×512 data points; mixing time 700 ms; data zero-filled to 2048 in the f_1 direction before *Fourier* transformation.

Polyacrylamide-Gel Electrophoresis. A soln. of 0.1 A_{260} units of **19**, T_{10}, or T_{11} in formamide (3 µl) was charged on a 20% polyacrylamide gel plate containing 7M urea (30×38 cm) containing 90 mM *Tris*-boric acid, and 3 mM EDTA for 2 h with 90 mM *Tris*-borate buffer at 1500 V. The gel was visualized by 0.2% toluidine blue in 0.4M KOAc (pH 4.8) and dried for 5 h under reduced pressure. The mobilities observed were 11.1, 11.4, and 11.1 cm for **19**, T_{10}, and T_{11}, resp.

Molecular Modeling by Molecular-Mechanics Calculation. Conformation search and energy minimization were carried out by MacroModel V. 4.5 running on a *Silicon-Graphics-Indigo-2* workstation. The initial structure was constructed from the nucleotide-unit structure extracted from either the typical A-type or B-type DNA duplex. Four calculation runs were carried out in which the ribose moieties of the 5′-terminal and 3′-terminal nucleoside unit were fixed in the {N-type and N-type}, {N-type and S-type}, {S-type and N-type}, and {S-type and S-type} conformations, resp. In the conformation search, 5000 conformers were selected in the MCMM mode, in which the conformers were generated randomly, and they were energy-minimized by the AMBER* united atom force. The low-energy structures having energy within 20 kJ/mol from the lowest energy were selected and subjected to further energy minimization by the AMBER* all-atom force. In all simulations, dielectric constant ε 1.0 and the GB/SA solvent model were used.

UV Melting Curves. Oligothymidylate (5 µmol) and oligodeoxyadenylates (5 µmol) were dissolved in 2 ml of 10 mM sodium phosphate and 150 mM NaCl (pH 7.0). The mixture was heated to 90° and then cooled to 0° within 90 min. After the mixture was equilibrated at 0° for 20 min, the UV absorbance at 260 nm was recorded as the temp. was raised at the rate of 1°/min.

REFERENCES

[1] H. M. Wu, D. M. Crothers, *Nature (London)* **1984**, *308*, 509; C. A. Frederick, J. Grable, M. Melia, C. Samudzi, L. Jen-Jacobson, B.-C. Wang, P. Green, H. W. Boyer, J. M. Rosenberg, *ibid.* **1984**, *309*, 327; A. K. Aggarwal, D. W. Rodgers, M. Drottar, M. Ptashne, S. C. Harisson, *Science (Washington, D.C.)* **1988**, *242*,

899; I. Husain, J. Griffith, A. Sancer, *Proc. Natl. Acad. Sci. U.S.A.* **1988**, *85*, 2558; R. G. Brennan, S. L. Roderick, Y. Takeda, B. W. Matthews, *ibid.* **1990**, *87*, 8165; D. B. Nicolov, H. Chen, E. D. Halay, A. A. Usheva, K. Hisataka, D. K. Lee, R. G. Roeder, S. K. Burley, *Nature (London)* **1990**, *377*, 119; C. I. Wang, J. S. Taylor, *ibid.* **1991**, *88*, 9072; S. C. Schultz, G. C. Shields, T. A. Steitz, *Science (Washington, D.C.)* **1991**, *253*, 1001; J. L. Kim, D. B. Nikolov, S. K. Burley, *Nature (London)* **1993**, *365*, 520; Y. Kim, J. H. Geiger, S. Hahn, P. B. Sigler, *ibid.* **1993**, *365*, 512; for recent studies on DNA binding, see P. A. Rice, S. Yang, K. Mizuuchi, H. A. Nash, *Cell (Cambridge, Mass.)* **1996**, *87*, 1295; F. Pio, R. Kodandapani, C. Z. Ni, W. Shepard, M. Klemsz, S. R. McKercher, R. A. Maki, K. R. Ely, *J. Biol. Chem.* **1996**, *271*, 23329; K. Swaminathan, P. Flynn, R. J. Reece, R. Marmorstein, *Nat. Struct. Biol.* **1997**, *4*, 751, F. J. Meyer-Almes, D. Porschke, *J. Mol. Biol.* **1997**, *269*, 842; N. Rajaram, T. K. Kerppola, *EMBO J.* **1997**, *16*, 2917; P. Brownlie, T. Ceska, M. Lamers, C. Romier, G. Stier, H. Teo, D. Suck, *Structure (London)* **1997**, *5*, 509; C. M. Nunn, E. German, S. Neidle, *Biochemistry* **1997**, *36*, 4792; C. R. Escalante, J. Yie, D. Thanos, A. K. Aggarwal, *Nature (London)* **1998**, *391*, 103; C. J. Squire, G. R. Clarke, W. A. Denny, *Nucleic Acids Res.* **1997**, *25*, 4072; D. A. Leonard, N. Rajaram, T. K. Kerppola, *Proc. Natl. Acad. Sci. U.S.A.* **1997**, *94*, 4913.

[2] H. W. Pley, K. M. Flaherty, D. B. McKay, *Nature (London)* **1994**, *372*, 68; W. G. Scott, J. T. Finch, A. Klug, *Nucleic Acids Symp. Ser.* **1995**, *34*, 214; W. G. Scott, J. T. Finch, A. Klug, *Cell (Cambridge, Mass.)* **1995**, *81*, 991; W. G. Scott, J. B. Murray, J. R. P. Arnold, B. L. Stoddard, A. Klug, *Science (Washington, D.C.)* **1996**, *274*, 2065; S. T. Sigurdsson, F. Eckstein, *Trends-Biotechnology* **1995**, *13*, 286.

[3] G. J. Quigley, A. Rich, *Science (Washington, D.C.)* **1976**, *194*, 796.

[4] K. Seio, T. Wada, K. Sakamoto, S. Yokoyama, M. Sekine, *J. Org. Chem.* **1996**, *61*, 1500.

[5] K. Seio, T. Wada, K. Sakamoto, S. Yokoyama, M. Sekine, *J. Org. Chem.* **1998**, *63*, 1429.

[6] a) A. Kamp, C. J. Pillar, W. J. Woodford, M. P. Mertes, *J. Med. Chem.* **1976**, *19*, 900; b) G. T. Shiau, R. F. Schinazi, M. S. Chen, W. H. Prusoff, *ibid.* **1980**, *23*, 127.

[7] M. Sekine, L. S. Peshakova, T. Hata, unpublished work.

[8] J. C. Sheehan, P. A. Cruickshank, G. L. Boshart, *J. Org. Chem.* **1961**, *26*, 2525; J. C. Sheehan, J. J. Hlavka, *ibid.* **1956**, *21*, 439.

[9] M. Sekine, J. Matsuzaki, T. Hata, *Tetrahedron* **1985**, *41*, 5279.

[10] M. D. Matteucci, M. H. Caruthers, *J. Am. Chem. Soc.* **1981**, *103*, 3185.

[11] S. L. Beaucage, *Tetrahedron Lett.* **1984**, *25*, 375; M. F. Moore, S. L. Beaucage, *J. Org. Chem.* **1985**, *50*, 2019.

[12] J. Brahms, J. C. Maurizot, A. M. Michelson, *J. Mol. Biol.* **1967**, *25*, 481; C. Cantor, M. M. Warshaw, H. Shapiro, *Biopolymers* **1970**, *9*, 1059.

[13] C. Altona, M. Sundaralingam, *J. Am. Chem. Soc.* **1972**, *94*, 8205; H. A. Gabb, S. C. Harvey, *ibid.* **1993**, *115*, 4218; M. Levitt, A. Warshal, *ibid.* **1978**, *100*, 2607.

[14] M. Sundaralingam, *Biopolymers* **1969**, *7*, 821.

[15] D. B. Davis, *Prog. Nucl. Magn. Res. Spectrosc.* **1978**, *12*, 135.

[16] D. B. Davies, S. S. Danyluk, *Biochemistry* **1974**, *13*, 4417.

[17] M. Lipton, W. C. Still, *J. Comput. Chem.* **1988**, *9*, 343.

[18] S. J. Weiner, P. A. Kollman, D. A. Case, U. C. Singh, C. Ghio, G. Alagona, S. Profeta, Jr., P. Weiner, *J. Am. Chem. Soc.* **1984**, *106*, 765; S. J. Weiner, P. A. Kollman, D. T. Nguyen, D. A. Case, *J. Comput. Chem.* **1986**, *7*, 230; D. Q. McDonald, W. C. Still, *Tetrahedron Lett.* **1992**, *33*, 7743.

[19] W. C. Still, A. Tempczyk, R. C. Hawley, T. Hendrickson, *J. Am. Chem. Soc.* **1990**, *112*, 6127.

[20] D. G. Gorenstein, *Chem. Rev.* **1994**, *94*, 1315; V. A. Roongta, C. R. Jones, D. G. Gorenstein, *Biochemistry* **1990**, *29*, 5245; C. Karslake, M. V. Botuyan, D. G. Gorenstein, *ibid.* **1992**, *31*, 1849; D. G. Gorenstein, K. Lai, D. O. Shah, *ibid.* **1984**, *23*, 6717; P. K. Kline, L. G. Marzilli, D. Live, G. Zon, *Biochem. Pharmacol.* **1990**, *40*, 97; E. M. Goldfield, B. A. Luxon, V. Bowie, D. G. Gorenstein, *Biochemistry* **1983**, *22*, 3336.

[21] J. C. Marine, S. D. Levene, D. M. Crothers, P. T. Englund, *Proc. Natl. Acad. Sci. U.S.A.* **1982**, *79*, 7664; T. E. Haran, D. M. Crothers, *Biochemistry* **1989**, *28*, 2763; M. Shatzky-Schwartz, Y. Hiller, Z. Reich, R. Ghirlando, S. Weinberger, A. Minsky, *ibid.* **1992**, *31*, 2339, and refs. cit. therein.

[22] W. Saenger, 'Principle of Nucleic Acid Structure', Springer-Verlag, New York, 1978.

Synthesis of the Anticodon Hairpin tRNA$_f^{Met}$ Containing N-{[9-(β-D-Ribofuranosyl)-9H-purin-6-yl]carbamoyl}-L-threonine (= N^6-{{[(1S,2R)-1-Carboxy-2-hydroxypropyl]amino}carbonyl}adenosine, t^{6}A)

by Valerie Boudou[a]), James Langridge[b]), Arthur Van Aerschot[a]), Chris Hendrix[a]), Alan Millar[b]), Patrick Weiss[c]), and Piet Herdewijn[a])*

[a]) Laboratory of Medicinal Chemistry, Rega Institute for Medical Research, K.U. Leuven, Minderbroederstraat 10, B-3000 Leuven (Fax. +32-16-337340; e-mail: Piet.Herdewijn@rega.kuleuven.ac.be)
[b]) *Micromass UK Ltd.* Floats Road, Wythenshawe, Manchester M23 9LZ, UK
[c]) *Xeragon AG,* Technoparkstrasse 1, CH-8005 Zürich

As part of our studies on the structure of yeast tRNA$_f^{Met}$, we investigated the incorporation of N-{[9-(β-D-ribofuranosyl)-9H-purin-6-yl]carbamoyl}-L-threonine (t^{6}A) in the loop of a RNA 17-mer hairpin. The carboxylic function of the L-threonine moiety of t^{6}A was protected with a 2-(4-nitrophenyl)ethyl group, and a (*tert*-butyl)dimethylsilyl group was used for the protection of its secondary OH group. The 2′-OH function of the standard ribonucleotide building blocks was protected with a [(triisopropylsilyl)oxy]methyl group. Removal of the base-labile protecting groups of the final RNA with 1,8-diazabicyclo[5.4.0]undec-7-ene (DBU) and then with MeNH$_2$ was done under carefully controlled conditions to prevent hydrolysis of the carbamate function, leading to loss of the L-threonine moiety.

Introduction. – tRNAs contain a whole series of modified nucleotides which influence their structure and function [1]. The exact influence of the modified nucleotide on tRNA structure and on the recognition by tRNA synthetase is largely unknown. This situation is changing slowly because of both the progress in synthetic methodologies in obtaining RNA and the availability of high-resolution NMR instruments.

As a part of our studies on the structure of yeast tRNA$_f^{Met}$, we plan to determine the solution conformation of the anticodon hairpin of this yeast tRNA. The nucleotide sequence of *Saccharomyces cerevisiae* tRNA$_f^{Met}$ in the cloverleaf form is depicted in *Fig. 1*. [2] This tRNA^{Met} was purified from baker's yeast, and it can be converted to the formylated methionyl-tRNA$_f$ by extracts of *E. coli* [2]. The same anticodon hairpin is also found in the initiator tRNA of *Torulopsis utilis* [3], a yeast belonging to *Fungi imperfecti.*

Both loop structures contain the modified-adenosine nucleoside t^{6}A (= N-{[9-(β-D-ribofuranosyl)-9H-purin-6-yl]carbamoyl}-L-threonine in the position next to the anticodon, t^{6}A representing an adenosine nucleoside substituted at the N^6-position with a L-threonine moiety *via* a carbonyl function [4]. Besides the occurrence in the above mentioned tRNA, a N-(9H-purin-6-ylcarbamoyl)-L-threonine ribonucleotide has also been discovered in several other tRNAs [5–10] and has been isolated from human urine [11][12]. The enzymatic synthesis of N-(9H-purin-6-ylcarbamoyl)-L-threonine riboside has been investigated [13][14]. The structure of N-(9H-purin-6-

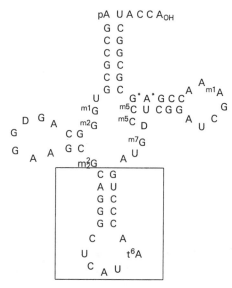

Fig. 1. *Nucleotide sequence of yeast 'RNA$_f^{Met}$ showing the hairpin structure with t^{6}A in the position next to the anticodon*

ylcarbamoyl)-L-threonine riboside was confirmed by X-ray diffraction [15][16]. It was observed that 'RNAs with codons starting with adenine mostly contain t^{6}A next to the 3′-side of the anticodon [6], and t^{6}A is thought to stabilize U·A and U·G anticodon·codon base pairs being formed adjacent to the 5′-side of the modified nucleoside [17]. It was observed that t^{6}A-deficient 'RNA is significantly less efficient in binding to ribosomes as compared to normal 'RNA, and that t^{6}A is required for proper codon·anticodon interaction [18]. Another study points to the importance of t^{6}A in the recognition of the anticodon loop of 'RNA$_1^{Ile}$ by isoleucyl-'RNA synthetase from *E. coli* [19]. NMR Studies point to the importance of t^{6}A as a binding site for magnesium ion in 'RNA [20][21].

The synthetic challenge for the incorporation of t^{6}A into RNA is situated in selecting an appropriate way to synthesize the modified nucleosides and to introduce protecting groups which are stable during RNA synthesis and can be removed easily at the end of the synthesis.

Results and Discussion. – *Synthesis.* The naturally occurring nucleoside *N*-{[(9-(*β*-D-ribofuranosyl)-9*H*-purin-6-yl]carbamoyl}-L-threonine (t^{6}A) has been synthesized before [22–26]. *Chheda* and *Hong* described the displacement of the ethoxy group of ethyl [9-(2,3,5-tri-*O*-acetyl-*β*-D-ribofuranosyl)-9*H*-purin-6-yl]carbamate with L-threonine followed by deprotection of the acetyl groups with NH$_3$/MeOH [22][25]. The same reaction scheme was used later by *Martin* and *Schlimme* [26]. The alternative of using an isocyanate intermediate of protected L-threonine was less successful [22]. As the *α*-amino group of L-threonine is more nucleophilic than the 6-amino group of the adenine base, we investigated the inverted approach and converted the protected adenosine to a *N*^{6}-isocyanate derivative.

The selection of the protecting groups for L-threonine is based on the following considerations: they should *a*) be stable during oligonucleotide synthesis, *b*) be easily removable after oligonucleotide synthesis with standard reagents, *c*) in the case of the carboxylic-acid protecting group, be removable before the ammonia-deprotection step of the oligonucleotide thus avoiding amide formation, and *d*) be removable without racemization of the L-amino acid. Therefore, we selected the (*tert*-butyl)dimethylsilyl group ('BuMe$_2$SiCl) for protecting the secondary OH function of L-threonine and the 2-(4-nitrophenyl)ethyl group as a protecting group for the carboxylic acid. The latter protecting group can be introduced and removed without causing racemizing side reactions [27]. L-Threonine (**1**) was converted into its 2-(4-nitrophenyl)ethyl ester and isolated as *p*-toluenesulfonate salt **2** (*Scheme 1*) [27]. Protection of the secondary OH group with (*tert*-butyl)dimethylsilyl chloride in pyridine catalyzed by 1*H*-imidazole yielded the protected L-threonine **3**.

Scheme 1. *Synthesis of Protected L-Threonine*

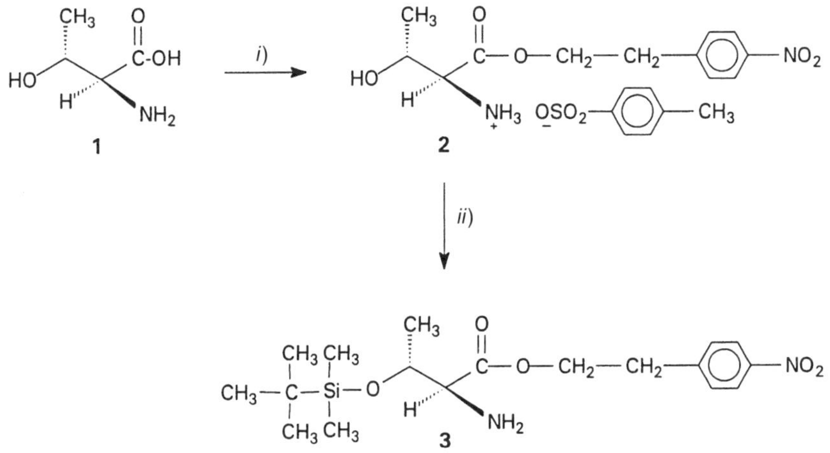

i) 4-NO$_2$C$_6$H$_4$(CH$_2$)$_2$OH, TsOH, 105°, 17 h; 63%. ii) 'BuMe$_2$SiCl, Py, 1*H*-imidazole, r.t., 17 h; 95%.

The OH groups of adenosine were first protected as acetates (→**5**; *Scheme 2*) [26]. The 6-NH$_2$ function of the protected adenosine **5** was reacted wtih triphosgene under reflux. The isocyanate **6** was not isolated but was reacted directly with the protected L-threonine **3** to give **7** in low yield (19%). The acetyl groups of **7** could selectively be removed with methanolic ammonia solution at room temperature without con-comitant conversion of the 2-(4-nitrophenyl)ethyl ester to an amide function (→**8**). Subsequent protection of the primary OH group with 4,4'-dimethoxytrityl chloride ((MeO)$_2$TrCl) in pyridine (→**9**; 94% yield), followed by reaction with 'BuMe$_2$SiCl in THF in the presence of AgNO$_3$ and pyridine gave the 2'-*O*-silylated nucleoside derivative **10** (44% yield), besides the undesired 3'-*O*-silylated compound (30% yield). The modified building block **11** for oligonucleotide synthesis was obtained by phosphitylation of the 3'-OH group of according to **10** standard procedures as described before [28].

Scheme 2. *Synthesis of the Protected t⁶A Phosphoramidites and Structure of the Protected Regular Nucleoside Phosphoramidites Used for Oligonucleotide Synthesis*

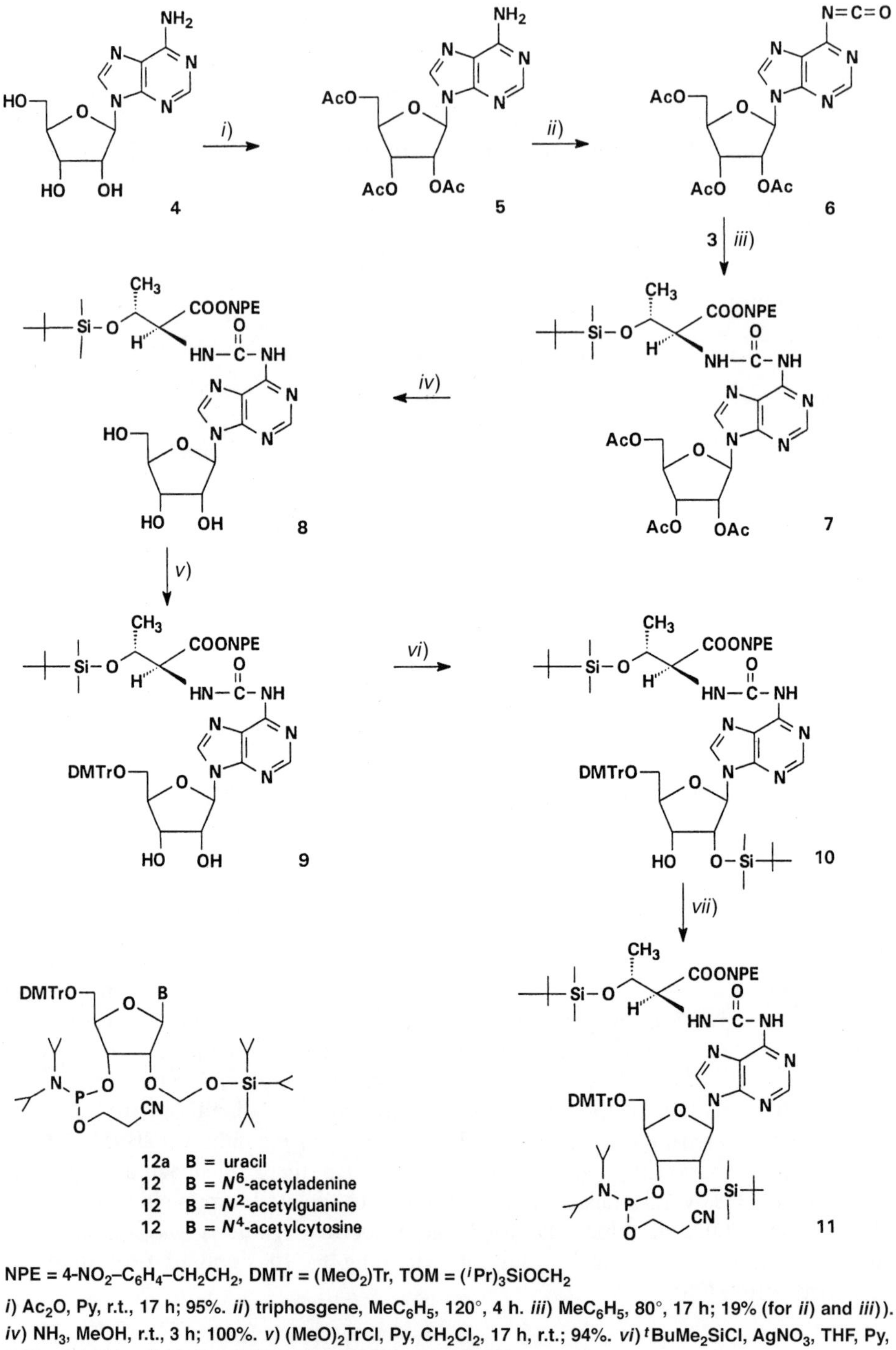

12a B = uracil
12   B = N⁶-acetyladenine
12   B = N²-acetylguanine
12   B = N⁴-acetylcytosine

NPE = 4-NO₂–C₆H₄–CH₂CH₂, DMTr = (MeO₂)Tr, TOM = (ⁱPr)₃SiOCH₂

i) Ac₂O, Py, r.t., 17 h; 95%. ii) triphosgene, MeC₆H₅, 120°, 4 h. iii) MeC₆H₅, 80°, 17 h; 19% (for ii) and iii)).
iv) NH₃, MeOH, r.t., 3 h; 100%. v) (MeO)₂TrCl, Py, CH₂Cl₂, 17 h, r.t.; 94%. vi) ᵗBuMe₂SiCl, AgNO₃, THF, Py,
17 h, r.t.; 44%. vii) ⁱPr₂EtN, CH₂Cl₂, (ⁱPr₂N)(OCH₂CH₂CN)PCl, r.t.; 72%.

Oligoribonucleotide Synthesis and Characterization. Oligoribonucleotide synthesis is currently carried out with a 2'-*O*-(*tert*-butyl)dimethylsilyl protecting group. Due to the bulkiness of this protecting group, coupling times are rather long and coupling yields are lower than in oligodeoxyribonucleotide synthesis. Therefore, we used the recently developed [(triisopropylsilyl)oxy]methyl (('Pr)$_3$SiOCH$_2$; TOM) group to protect the 2'-OH function of the standard nucleosides (see amidites **12a–d**), allowing an efficient chemical synthesis of RNA. The ('Pr)$_3$SiOCH$_2$ protecting group allowed us to carry out RNA synthesis with short coupling times (2 min) and high coupling yields (> 99%). The building blocks were used in 0.1M concentrations and were activated with 1-(benzylthio)-1*H*-tetrazole. Thus, a 17-mer corresponding to the anticodon loop of ^{t}RNA$_f^{Met}$ was synthesized in which the adenine base at the 11th position was replaced by the *N*-{[9-(β-D-ribofuranosyl)-9*H*-purin-6-yl]carbamoyl}(5'-CAGGGCU-CAU**t**^{6}**A**ACCCUG-3'). After '(MeO)$_2$Tr-off' synthesis, the solid-phase material was treated with 1,8-diazabicyclo[5.4.0]undec-7-ene (DBU) in THF to remove the 2-(4-nitrophenyl)ethyl protecting groups, and then with MeNH$_2$ in EtOH/H$_2$O. The ('Pr)$_3$SiOCH$_2$ groups were removed with Bu$_4$NF in THF. The oligoribonucleotide was desalted on a *Sephadex* (*G*10) column, purified on a *Dionex-Nucleopac-PA-100* column, and desalted again. The deprotection with DBU was necessary to obtain the correct final material; when this deprotection step was skipped and the protected oligoribonucleotide directly treated with MeNH$_2$, the methyl carboxamide of the t^{6}A insert was obtained. More vigorous deprotection conditions led to loss of the whole L-threonine moiety by hydrolysis of the carbamate function.

The modified RNA 17-mer was desalted with cation-exchange beads before it was submitted to analysis by mass spectrometry. *Fig. 2,a*, shows the ESI-MS with several peaks corresponding to multiply charged ions of the sample in the charge states [$M - 4H$]$^{4-}$ to [$M - 11H$]$^{11-}$. The Max Ent 1 processed spectrum (*Fig. 2,b*) indicates the experimental monoisotopic mass to be 5527.91 Da, confirming the identity of the RNA 17-mer (calc. 5527.80 Da).

Before starting NMR experiments, we determined the concentration-dependent T_m of the synthesized oligonucleotide. As can be seen in *Fig. 3*, the thermal stability of the oligomer did not change with increasing concentration of the oligonucleotide, indicating formation of a stable hairpin structure. Both T_m in 0.1M NaCl (64°) and the shape of the curve were independent of oligonucleotide concentration.

Conclusion. – The modified nucleoside. *N*-{[9-(β-D-ribofuranosyl)-9*H*-purin-6-yl]carbamoyl}-L-threonine (t^{6}A) was successfully incorporated into the loop of a hairpin RNA. The (*tert*-butyl)dimethylsilyl group was used for the protection of the secondary OH group of the modified nucleosides. The carboxy group was protected with a 2-(4-nitrophenyl)ethyl group. The 2'-*O*-{[(triisopropylsilyl)oxy]methyl}-protected regular nucleoside phosphoramidites were used as building blocks for RNA synthesis. The integrity of the final hairpin was established by ESI-MS. Thermal-stability studies indicated the stability of the RNA hairpin, which is now used for structural studies.

This research was supported by a grant for the K.U. Leuven (GOA 97/11). *Arthur Van Aerschot* is a research associate of the *Flemish Fund of Scientific Research*. We are grateful to Dr. *Jan Claereboudt*,

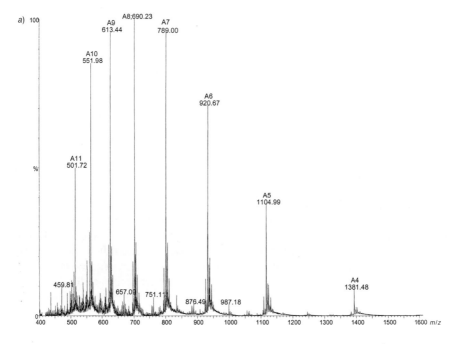

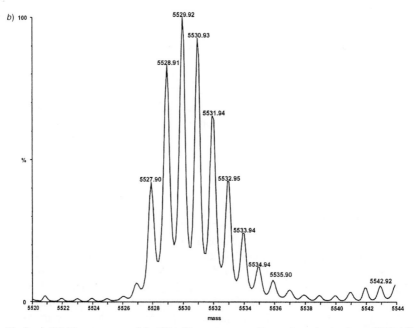

Fig. 2. a) *ESI-Mass spectrum of the RNA 17-mer corresponding to the anticodon loop of* $^tRNA_f^{Met}$ *in which the adenine base at position 11 is replaced by* N-*{[9-(β-ᴅ-ribofuranosyl)-9H-purin-6-yl]carbamoyl}-ʟ-threonine* (t⁶A) *and* b) *its Max Ent 1 processed spectrum, indicating the experimental monoisotopic mass. See Exper. Part for details.*

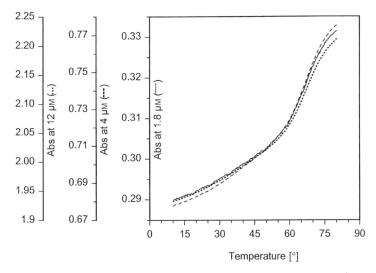

Fig. 3. *Concentration-dependent* T$_m$ *of the RNA hairpin with incorporated t⁶A*

Micromass, Belgium, for all the efforts in providing us with the Q-TOF mass spectra. We thank *Chantal Biernaux* for editorial help.

Experimental Part

General. Anh. solvents were obtained as follows: DMF was dried over molecular sieves; CH_2Cl_2 was stored over P_2O_5, refluxed, and distilled; THF was stored over Na/benzophenone, refluxed, and distilled; pyridine was stored over CaH_2 refluxed, and distilled. Column chromatography (CC): *Acros* silica gel (0.060–0.200 nm). TLC: precoated *Macherey-Nagel Alugram SIL G/UV$_{254}$* plates. ¹H- and ¹³C-NMR Spectra: *Varian-Gemini-200* spectrometer; δ in ppm rel. to $SiMe_4$ as internal standard, J in Hz; the tyrosine numbering is used for the side chain at $NH_2-C(6)$ of adenosine in **7–11**. Liquid secondary-ion mass spectra (LSI-MS): Cs^+ as primary-ion beam *Kratos-Concept-IH* (*Kratos;* Manchester, UK); solns. in NBA (= 3-nitrobenzyl alcohol), THGLY (= thioglycerol), of NPOE (= 2-nitrophenyl octyl ether); acceleration of secondary ions at 6 ZV; scans at 10 s/ decade from m/z 1500 down to m/z 50. ESI-MS. *Micromass-Q-TOF* mass spectrometer (*Whytenshawe*, Manchester, UK) fitted with a standard electrospray-ion source.

L-*Threonine 2-(4-Nitrophenyl)ethyl Ester 4-Methylbenzenesulfonate Salt* (**2**) [27]. L-Threonine (500 mg, 4.2 mmol), 2-(4-nitrophenyl)ethanol (npe-OH; 12.6 mmol, 2.11 g) and TsOH (12.6 mmol, 2.40 g) were heated in toluene (100 ml) at 105° for 17 h in a *Dean-Stark* apparatus. The soln. was cooled to r.t., and Et$_2$O (25 ml) was added. The oily residue was decanted, and the upper layer was removed to collect the oil. Precipitation of **2** was realized by adding to the oil MeOH (25 ml) and Et$_2$O (100 ml): 1.17 g (63%) of **2**, identical to the compound described in [27]. ¹H-NMR ((D$_6$)DMSO): 1.14 ($d, J = 6.6$, Me(γ)); 2.29 (s, MeC_6H_4); 3.10 ($t, J = 6.2$, CH_2CH_2O); 3.89 ($d, J = 4.0$, CH(α)); 4.0–4.1 (m, CH(β)); 4.45 ($t, J = 6.2$, CH_2CH_2O); 5.6 (br. s, OH); 7.11 ($d, J = 8.2$, 2 arom. H(Ts)); 7.48 ($d, J = 8.2$, 2 arom. H(Ts)); 7.58 ($d, J = 8.8$, 2 arom. H(npe)); 8.1–8.2 (m, NH$_3$, 2 arom. H(npe)). ¹³C-NMR ((D$_6$)DMSO): 19.96 (Me(γ)); 20.72 (MeC_6H_4); 33.83 (CH_2CHO); 57.87 (CH_2CH_2O); 64.91 (CH(α)); 65.43 (CH(β)); 123.52 (arom. C(npe)); 125.65 (arom. C(Ts)); 128.14 (arom. C(npe)); 130.38 (arom. C(Ts)); 137.73 (arom. C(Ts)); 146.32 (2 arom. C(Ts, npe)); 146.53 (arom. C(npe)); 168.23 (COO). LSI-MS (THGLY/NBA): 483 ($[M + K]^+$).

O-[(tert-Butyl)dimethylsilyl]-L-*threonine 2-(4-Nitrophenyl)ethyl Ester* (**3**). Compound **2** (500 mg, 1.15 mmol) was dissolved in dry pyridine (30 ml) and treated with one half of ^{t}BuMe$_2$SiCl (260 mg, 3.45 mmol) and 1*H*-imidazole (120 g, 3.45 mmol). After 10 min, the second half was added, and the reaction was allowed to continue for 17 h at r.t. The mixture was diluted with CH_2Cl_2 (500 ml) and washed successively with sat. NaHCO$_3$ soln. (2 × 300 ml) and H$_2$O (3 × 300 ml). The org. layer was dried, evaporated, co-evaporated with toluene and MeOH and purified by CC (silica gel, 0–6% MeOH/CH_2Cl_2): 416 mg (95%) **3**. Oil. ¹H-NMR

$((D_6)DMSO): -0.17, -0.06$ (2s, 2 MeSi); 0.75 (s, ^tBuSi); 1.11 (d, $J = 6.2$, Me(γ)); 1.6 (br. s, NH$_2$); 3.06 (t, $J = 6.2$, CH$_2$CH$_2$O); 3.17 (d, $J = 2.4$, CH(α)); 3.9–4.4 (m, CH(β), CH$_2$CH$_2$O); 7.57 (d, $J = 8.8$, 2 arom. H); 8.18 (d, $J = 8.8$, 2 arom. H). ^{13}C-NMR $((D_6)DMSO): -5.44, -4.47$ (2 MeSi); 17.59 (Me$_3$CSi); 20.42 (Me(γ)); 25.58 (Me$_3$CSi); 34.14 (CH$_2$CH$_2$O); 59.99 (CH$_2$CH$_2$O); 64.21 (CH(α)); 69.74 (CH(β)); 123.61 (2 arom. C); 130.50 (2 arom. C); 146.89 (2 arom. C); 174.39 (COO). LSI-MS (THGLY): 383 ($[M + H]^+$). Anal. calc. for C$_{18}$H$_{30}$N$_2$O$_5$Si (382.5): C 56.52, H 7.90, N 7.32; found: C 56.81, H 7.72, N 7.18.

N^6-{{{(1S,2R)-2-{[(tert-Butyl)dimethylsilyl]oxy}-1-{[2-(4-nitrophenyl)ethoxy]carbonyl}propyl}amino}carbonyl}adenosine 2',3',5'-Triacetate (**7**). A suspension of **5** [26] (2.40 g, 6.14 mmol) in dry toluene (300 ml) was treated with triphosgene (3.64 g, 12.28 mmol), and the mixture was heated under reflux (120°) for 4 h to give crude isocyanate **6**. The soln. was evaporated and then dissolved in anh. CH$_2$Cl$_2$/toluene 1 : 1 (60 ml). A soln. of **3** (1.76 g, 4.60 mmol) in dry toluene (30 ml) was added dropwise to the soln. of **6**, and the reaction was continued for 17 h at 80°. The brown mixture was evaporated and directly purified by CC (silica gel, 0–5% MeOH/CH$_2$Cl$_2$): 934 mg (19%) of pure **7**. ^{1}H-NMR $((D_6)DMSO): -0.03, 0.07$ (2s, 2 MeSi); 0.91 (s, ^tBuSi); 1.26 (d, $J = 6.2$, Me(γ)); 2.09, 2.12, 2.16 (3s, 3 MeCO); 3.03 (t, $J = 6.2$, CH$_2$CH$_2$O); 4.3–4.6 (m, CH(α), CH(β), CH$_2$CH$_2$O, H–C(4'), 2 H–C(5')); 5.68 (t, $J = 4.9$, H–C(3')); 6.06 (t, $J = 5.4$, H–C(2')); 6.24 (d, $J = 5.4$, H–C(1')); 7.30 (d, $J = 8.5$, 2 arom. H (npe)); 7.92 (d, $J = 8.5$, 2 arom. H (npe)); 8.32, 8.48 (2s, H–C(2), H–C(8)); 8.61 (br. s, NH–C(6)); 10.03 (d, $J = 9.0$, NH–C(α)). ^{13}C-NMR $((D_6)DMSO): -5.55, -4.43$ (2 MeSi); 17.69 (Me$_3$CSi); 20.31, 20.46, 20.64 (3 MeCO); 21.00 (Me(γ)); 25.44 (Me$_3$CSi); 34.66 (CH$_2$CH$_2$O); 59.58 (CH$_2$CH$_2$O); 63.04 (CH$_2$(5')); 64.37 (CH(α)); 68.59 (CH(β)); 70.66 (CH(3')); 72.93 (CH(2')); 80.43 (CH(4')); 86.68 (CH(1')); 121.04 (C(5)); 123.50 (2 arom. C (npe)); 129.72 (2 arom. C (npe)); 141.89 (CH(8)); 145.56 (2 arom. C (npe)); 150.42 (CH(2)); 151.45 (C(4), C(6)); 154.40 (NHCONH); 169.48, 169.69, 170.51 (3 MeCO); 170.99 (COOCH$_2$CH$_2$). LSI-MS (THGLY): 802 ($[M + H]^+$). Anal. calc. for C$_{35}$H$_{47}$N$_7$O$_{13}$Si (801.9): C 52.42, H 5.91, N 12.23; found: C 52.21, H 6.02, N 12.14.

N^6-{{{(1S,2R)-2-{[(tert-Butyl)dimethylsilyl]oxy}-1-{[2-(4-nitrophenyl)ethoxy]carbonyl}propyl}amino}carbonyl}adenosine (**8**). A methanolic NH$_3$ soln. (100 ml) of **7** was stirred at r.t. for 3 h. Evaporation and purification by CC (silica gel, 0–10% MeOH/CH$_2$Cl$_2$) afforded pure **8** (598 mg, quant.). ^{1}H-NMR $((D_6)DMSO): -0.13, 0.01$ (2s, 2 MeSi); 0.82 (s, ^tBuSi); 1.15 (d, $J = 6.2$, Me(γ)); 3.04 (t, $J = 6.0$, CH$_2$CH$_2$O); 3.5–3.8 (m, 2 H–C(5')); 4.0 (m, H–C(4')); 4.2 (m, H–C(3')); 4.2–4.5 (m, CH(α), CH(β), CH$_2$CH$_2$O); 4.6 (m, H–C(2')); 5.12 (t, $J = 5.5$, OH–C(5')); 5.22 (d, $J = 4.8$, OH–C(3')); 5.52 (d, $J = 5.6$, OH–C(2')); 6.01 (d, $J = 5.2$, H–C(1')); 7.48 (d, $J = 8.6$, 2 arom. H (npe)); 8.00 (d, $J = 8.6$, 2 arom. H (npe)); 8.36, 8.70 (2s, H–C(2), H–C(8)); 9.86 (d, $J = 9.2$, NH–C(α)); 9.93 (br. s, NH–C(6)). ^{13}C-NMR $((D_6)DMSO): -5.68, -4.50$ (2 MeSi); 17.44 (Me$_3$CSi); 20.90 (Me(γ)); 25.36 (Me$_3$CSi); 33.98 (CH$_2$CH$_2$O); 59.03 (CH$_2$CH$_2$O); 61.24 (CH$_2$(5')); 64.82 (CH(α)); 68.34 (CH(β)); 70.26 (CH(3')); 74.11 (CH(2')); 85.67 (CH(4')); 88.01 (CH(1')); 121.13 (C(5)); 123.37 (2 arom. C(npe)); 130.35 (2 arom. C(npe)); 142.52 (CH(8)); 146.68 (2 arom. C(npe)); 150.29 (C(4), C(6)); 154.06 (NHCONH); 170.84 (COOCH$_2$CH$_2$). LSI-MS (THGLY): 676 ($[M + H]^+$). Anal. calc. for C$_{29}$H$_{41}$N$_7$O$_{10}$Si (675.8): C 51.54, H 6.12, N 14.51; found: C 51.32, H 6.17, N 14.23.

N^6-{{{(1S,2R)-2-{[(tert-Butyl)dimethylsilyl]oxy}-1-{[2-(4-nitrophenyl)ethoxy]carbonyl}propyl}amino}carbonyl}-5'-O-(4,4'-dimethoxytrityl)adenosine (**9**). To a cooled (0°) soln. of **8** (854 mg, 1.26 mmol) in dry pyridine (15 ml), a soln. of 4,4'-dimethoxytrityl chloride (596 mg, 1.76 mmol) in dry CH$_2$Cl$_2$ (5 ml) was added dropwise. After 15 min at 0°, the mixture was allowed to warm to r.t. and the reaction continued for 17 h. MeOH (5 ml) was added, the soln. evaporated, and the residue dissolved in CH$_2$Cl$_2$ (200 ml). The org. layer was washed successively with sat. NaHCO$_3$ soln. (2 × 150 ml), brine (2 × 150 ml), and H$_2$O (2 × 150 ml), and then dried, evaporated, and co-evaporated. Purification by CC (silica gel, 0–3% MeOH/CH$_2$Cl$_2$ 1% Et$_3$N) afforded 1.16 g (94%) of **9**. ^{1}H-NMR $((D_6)DMSO): -0.13, 0.01$ (2s, 2 MeSi); 0.82 (s, ^tBuSi); 1.18 (d, $J = 5.8$, Me(γ)); 3.03 (t, $J = 6.2$, CH$_2$CH$_2$O); 3.3 (m, 2 H–C(5')); 3.70 (s, 2 MeO); 4.1 (m, H–C(4')); 4.3–4.5 (m, H–C(3'), CH(α), CH(β), CH$_2$CH$_2$O); 4.8 (m, H–C(2')); 5.26 (d, $J = 5.8$, OH–C(3')); 5.61 (d, $J = 5.6$, OH–C(2')); 6.03 (d, $J = 4.0$, H–C(1')); 6.7–7.4 (m, 13 H, (MeO)$_2$Tr); 7.48 (d, $J = 8.8$, 2 arom. H (npe)); 7.99 (d, $J = 8.8$, 2 arom. H (npe)); 8.25, 8.59 (2s, H–C(2), H–C(8)); 9.88 (d, $J = 9.2$, NH–C(α)); 9.98 (br. s, NH–C(6)). ^{13}C-NMR $((D_6)DMSO): -5.68, -4.50$ (2 MeSi); 17.44 (Me$_3$CSi); 20.93 (Me(γ)); 25.36 (Me$_3$CSi); 34.01 (CH$_2$CH$_2$O); 55.08 (2 MeO); 59.02 (CH$_2$CH$_2$O); 63.76 (CH$_2$(5')); 64.85 (CH(α)); 68.37 (CH(β)); 70.35 (CH(3')); 73.14 (CH(2')); 83.34 (CH(4')); 85.58 (CH(1')); 88.89 ((MeO)$_2$Tr); 113.23, 126.77, 127.89, 129.89, 135.75, 158.24 (arom. C(MeO)$_2$Tr)); 120.88 (C(5)); 123.40 (2 arom. C (npe)); 130.41 (2 arom. C (npe)); 143.13 (CH(8)); 145.13 (CH(2)); 146.35, 146.74 (2 arom. C (npe)); 150.50 (C(4), C(6)); 154.11 (NHCONH); 170.93 (COOCH$_2$CH$_2$). LSI-MS (THGLY) 1000 ($[M + Na]^+$). Anal. calc. for C$_{50}$H$_{59}$N$_7$O$_{12}$Si (978.1): C 61.40, H 6.08, N 10.02; found: C 61.21, H 5.81, N 9.83.

*2'-O-[(tert-Butyl)dimethylsilyl]-*N[6]*-{{{(1S,2R)-2-{{(tert-butyl)dimethylsilyl]oxy]-1-{[2-(4-nitrophenyl)eth-oxy]carbonyl]propyl]amino}carbonyl]-5'-O-(4,4'-dimethoxytrityl)adenosine* (**10**). To a soln. of **9** (1.10 g, 1.13 mmol) in dry THF (15 ml), AgNO$_3$ (230 mg, 1.35 mmol) was added. The mixture was sonicated and stirred 10 min, before 'BuMe$_2$SiCl (221 mg, 1.47 mmol) was added. After 17 h at r.t., TLC indicated incomplete reaction. Thus, more AgNO$_3$ (54 mg, 0.32 mmol), 'BuMe$_2$SiCl (51 mg, 0.34 mmol), and pyridine (100 μl) were added. After 3 h more, the mixture was filtered through *Celite* into a sat. NaHCO$_3$ soln. and extracted with CH$_2$Cl$_2$ (300 ml). The org. layer was washed with sat. NaHCO$_3$ soln. (2 × 200 ml), brine (2 × 200 ml), and H$_2$O (2 × 200 ml), dried, and evaporated. Purification by CC (silica gel, 0–2% MeOH/CHCl$_3$ +1% Et$_3$N) followed by purification with a *Chromatotron*® apparatus (4-mm layer; eluants: 35, 45, 55, 65, and 75% AcOEt/hexane) afforded pure **10** (539 mg, 44%). ^{1}H-NMR ((D$_6$)DMSO): −0.14, −0.04, −0.01 (3s, 4 MeSi); 0.74, 0.79 (2s, 2 'BuSi); 1.16 (d, J = 6.2, Me(γ)); 3.02 (t, J = 5.7, CH$_2$CH$_2$O); 3.3 (m, 2 H−C(5')); 3.70 (s, 2 MeO); 4.1 (m, H−C(4')); 4.2−4.4 (m, H−C(3'), CH(α), CH(β), CH$_2$CH$_2$O); 4.98 (t, J = 5.0, H−C(2')); 5.19 (d, J = 5.4, OH−C(3')); 6.04 (d, J = 4.8, H−C(1')); 6.7−7.5 (m, 15 arom. H, (MeO)$_2$Tr, npe); 7.98 (d, J = 8.8, 2 arom. H (npe)); 8.25, 8.61 (2s, H−C(2), H−C(8)); 9.87 (d, J = 8.8, NH−C(α)); 10.02 (br. s, NH−C(6)). ^{13}C-NMR ((D$_6$)DMSO): − 5.71, − 5.35, − 4.83, − 4.56 (4 MeSi); 17.41, 17.84 (2 Me$_3$Csi); 20.87 (Me(γ)); 25.27, 25.58, 25.82 (2 Me$_3$Csi); 33.99 (CH$_2$CH$_2$O); 55.05 (2 MeO); 58.99 (CH$_2$CH$_2$O); 63.7 (CH$_2$(5')); 64.85 (CH(α)); 68.34 (CH(β)); 70.22 (CH(3')); 74.74 (CH(2')); 83.79 (CH(4')); 85.64 (CH(1')); 88.77 ((MeO)$_2$Tr); 113.20, 126.77, 127.86, 129.89, 135.66, 158.27 (arom. C ((MeO)$_2$Tr)); 120.94 (C(5)); 123.34 (2 arom. C (npe)); 130.38 (2 arom. C (npe)); 143.16 (CH(8)); 145.19 (CH(2)); 146.32, 146.71 (2 arom. C (npe)); 150.29, 150.53 (C(4), C(6)); 154.06 (NHCONH); 170.90 (COOCH$_2$CH$_2$): LSI-MS (THGLY): 1114 ([M + Na]$^+$). Anal. calc. for C$_{56}$H$_{73}$N$_7$O$_{12}$Si$_2$ (1092.4): C 61.57, H 6.74, N 8.98; found: C 61.36, H 6.51, N 8.82.

*2'-O-[(tert-Butyl)dimethylsilyl]-*N[6]*-{{{(1S,2R)-2-{{(tert-butyl)dimethylsilyl]oxy]-1-{[2-(4-nitrophenyl)eth-oxy]carbonyl]propyl]amino}carbonyl]-5'-O-(4,4'-dimethoxytrityl)adenosine 3'-(2-Cyanoethyl Diisopropylphos-phoramidite)* (**11**). A soln. of **10** (0.40 mmol) in CH$_2$Cl$_2$ (5 ml) was phosphitylated with N,N-diisopropylethyl-amine (3 equiv.) and 2-cyanoethyl diisopropylphosphoramidochloridite (1.5 equiv.) The reaction was difficult to follow as only a slight change of R_f was seen on TLC in several systems. Therefore, after 2 h at r.t., an extra equiv. of both reagents was added, and after further 2 h, the mixture was quenched by addition of H$_2$O. The mixture was partitioned between CH$_2$Cl$_2$ (50 ml) and sat. aq. NaHCO$_3$ soln. (30 ml), the org. phase washed with brine (2 × 30 ml), dried (Na$_2$SO$_4$), and evaporated, and the residue purified by flash chromatography (silica gel (30 g), hexane/acetone/Et$_3$N 68:30:2) and precipitation in cold hexane: 0.29 mmol (72%) of **11**. LSI-MS (NPOE): 1293 ([M + H]$^+$; for C$_{65}$H$_{90}$N$_9$O$_{13}$PSi$_2$, calc. 1291.593). ^{31}P-NMR: (external ref. H$_3$PO$_4$): 149.75, 151.55.

Oligoribonucleotide Synthesis. The oligoribonucleotide containing the L-threonine-modified adenosine nucleoside was prepared on a *Pharmacia-Gene-Assembler-Special*-DNA/RNA synthesizer at 1.5-μmol scale by solid-phase 2-cyanoethyl phosphoramidite chemistry with the 2'-O-[(triisopropylsilyl)oxy]methyl (^{i}Pr$_3$SiOCH$_2$; TOM) protecting group for the standard RNA monomers **12a – d**. The synthesis was performed '(MeO)$_2$Tr-off' on LCAA CPG (500 Å) with standard RNA cycles, except for the shorter coupling times (2 min) for TOM phosphoramidite chemistry. Coupling time for the modified phosphoramidite **11** and the subsequent TOM amidite was 10 min. (Benzylthio)-1H-tetrazole (0.35M) was used as activator. Average coupling efficiency for the synthesis was 99.3% as determined by the release of the 4,4-dimethoxytrityl cation. The solid-phase material was treated with 10% DBU in THF (1 ml) for 90 min at 45° to remove the 2-(4-nitrophenyl)ethyl protecting group. After lyophilization, the solid phase was reacted with 10M MeNH$_2$ in 50% aq. EtOH (500 μl) at r.t. for 4.5 h. The oligoribonucleotide in MeNH$_2$ soln. was decanted, the solid support washed with sterile H$_2$O (2 × 200 μl) and then the soln. evaporated in a *Speedvac* concentrator. Desilylation of the 2'-O-[(triisopropylsilyl)-oxy]methyl protecting groups was achieved by dissolving the dried oligoribonucleotide in 1M Bu$_4$NF in THF (500 μl) at r.t. for 13.5 h. *Tris* buffer (pH 7.4, 500 μl) was added, and the oligoribonucleotide was desalted by CC (*Sephadex* (*G10*) sterile H$_2$O). The fully deprotected oligoribonucleotide was purified by a further CC (*Dionex Nucleopac PA-100* (4 × 250 mm), 50°, gradient 10 → 500 mM sodium perchlorate (10–35%) with in 30 min), 10 mM *Tris*, and 5M urea at pH 8.0). The desired product was eluted at 24 min at a flow rate of 1 ml/min. Then, the oligoribonucleotide was again desalted by CC (*Sephadex* (*G10*)) and lyophilized on a *Speedvac* concentrator.

For ESI-MS, the RNA sample was dissolved in H$_2$O and desalted by means of *Dowex-50W* cation-exchange beads. The supernatant was diluted by the addition of an equal volume of both MeCN and a 1% NH$_3$ soln. The *Q-TOF* mass spectrometer was operated in the negative mode with a source temp. of 80°, a counter current gas flow rate of 40 l/h and a potential of *ca.* 3200 V applied to the source needle (sample flow rate 7 μl/min). The instrument was calibrated with a two-point calibration using the singly charged ion of the monomer and dimer of raffinose.

REFERENCES

[1] A. P. Limbach, P. F. Crain, J. A. McCloskey, *Nucleic Acids Res.* **1994**, *22*, 2183.

[2] M. Simsek, U. L. RajBhandary, *Biochem. Biophys. Res. Commun.* **1972**, *49*, 508.

[3] S. Yamashiro-Matsumura, S. Takemura, *J. Biochem.* **1979**, *86*, 335.

[4] M. P. Schweizer, G. B. Chheda, L. Baczynskyj, R. H. Hall, *Biochemistry* **1969**, *8*, 3283.

[5] S. Takemura, M. Murakami, M. Miyazaki, *J. Biochem.* **1969**, *65*, 553.

[6] F. Kimura-Harada, F. Harada, S. Nishimura, *FEBS Lett.* **1972**, *21*, 71.

[7] M. Raba, K. Limburg, J. R. Katze, U. L. Rajbhandary, H. J. Gross, *Hoppe-Seyler's Z. Physiol. Chem.* **1976**, *357*, 301.

[8] R. Martin, J. M. Schneller, A. J. C. Stahl, G. Dirheimer, *Biochem. Biophys. Res. Commun.* **1976**, *70*, 997.

[9] R. Brambilla, H. Rogg, M. Staehelin, *Nature (London)* **1976**, *263*, 167.

[10] B. S. Vold, D. E. Keith Jr., M. Buck, J. A. McCloskey, H. Pang, *Nucleic Acids Res.* **1982**, *10*, 3125.

[11] G. B. Chheda, *Life Sci.* **1969**, *8*, 979.

[12] B. S. Vold, D. E. Keith, M. Slavik, *Cancer Res.* **1982**, *42*, 5265.

[13] G. B. Chheda, C. I. Hong, C. F. Piskorz, G. A. Harman, *Biochem. J.* **1972**, *127*, 515.

[14] B. N. Elkins, E. B, Keller, *Biochemistry* **1974**, *13*, 4622.

[15] R. Parthasarathy, J. M. Ohrt, G. B. Chheda, *Biochem. Biophys. Res. Commun.* **1974**, *60*, 211.

[16] R. Parthasarathy, J. M. Orth, G. B. Chheda, *Biochemistry* **1977**, *16*, 4999.

[17] J. Weissenbach, H. Grosjean, *Eur. J. Biochem.* **1981**, *116*, 207.

[18] J. P. Miller, Z. Hussain, M. P. Schweizer, *Nucleic Acids Res.* **1976**, *3*, 1185.

[19] T. Niimi, O. Nureki, T. Yokogawa, N. Hayashi, K. Nishikawa, K. Watanabe, S. Yokoyama, *Nucleosides Nucleotides* **1994**, *13*, 1231.

[20] M. P. Schweizer, W. D. Hamill, *Biochem. Biophys. Res. Commun.* **1978**, *85*, 1367.

[21] P. R. Reddy, W. D. Hamill Jr., G. B. Chheda, M. P. Schweizer, *Biochemistry* **1981**, *20*, 4979.

[22] G. B. Chheda, C. I. Hong, *J. Med. Chem.* **1971**, *14*, 748.

[23] C. I. Hong, B. G. Chheda, S. P. Dutta, A. O'Grady-Curtis, G. L. Tritsch, *J. Med. Chem.* **1973**, *16*, 139.

[24] R. W. Adamiak, M. Wiewiorowski, *Bull. Acad. Pol. Sci., Ser. Sci. Chim.* **1975**, *23*, 241.

[25] C. I. Hong, G. B. Chheda, *Nucleic Acid Chem.* **1978**, *2*, 661.

[26] D. Martin, E. Schlimme, *Z. Naturforsch., C. Biosci.* **1994**, *49*, 834.

[27] C. Buschhaus, in 'Konstanzer Dissertationen n° 239. Der *p*-Nitrophenylethyl und *p*-Nitrophenylethoxy-carbonyl-rest als neue Schutzgruppe in der Peptidsynthese', ISBN 3-89191-247-1 (W. Pfleiderer).

[28] G. Ceulemans, A. Van Aerschot, B. Wroblowski, J. Rozenski, C. Hendrix, P. Herdewijn, *Chem. Eur. J.* **1997**, *3*, 1997.

Stepwise Solid-Phase Synthesis of Peptide-Oligonucleotide Conjugates on New Solid Supports

by **Maxim Antopolsky** and **Alex Azhayev***

Department of Pharmaceutical Chemistry, University of Kuopio, P.O. Box 1627, FIN-70211 Kuopio
(tel.: + 358-17-162204; fax: + 358-17-162456; e-mail: Alex.Azhayev@uku.fi)

Several peptide-oligonucleotide and peptide-(oligonucleotide phosphorothioate) conjugates were synthesized on new solid supports. These supports are designed to link the 3′-terminus of an oligonucleotide to the C-end of a peptide *via* a phosphodiester or phosphorothioate bond in the process of stepwise solid-phase assembly.

Introduction. – Covalent oligonucleotide-peptide conjugates have received considerable attention as new antisense agents that exhibit potentially enhanced cellular uptake [1–16]. The preparation of such conjugates has been based either on postsynthetic conjugation of oligonucleotides and peptides, both bearing appropriate reactive groups [1–6][8][9][11][12][14–16], or on stepwise solid-phase assembly [7][10][13]. All the postsynthetic conjugations described so far are associated with lengthy and troublesome procedures that include several intermediate purification steps. In contrast, stepwise solid-phase procedures to prepare oligonucleotide-peptide conjugates require usually only one purification step upon the final deprotection. However, the choice of permanent protecting groups of nucleoside and amino-acid monomers, employed in the stepwise solid-phase synthesis, remains the critical factor, which explains the relatively small number of publications on this subject [7][10][13]. Moreover, these papers describe the stepwise synthesis of conjugates, in which relatively short peptides and oligonucleotides with phosphodiester bonds are incorporated. No data on conjugates of peptides and relatively long oligonucleotide phosphorothioates (15- to 30-mers), obtained by this methodology, are available to our knowledge.

Here we describe the synthesis on three different solid supports, of several peptide-oligonucleotide and peptide-(oligonucleotide phosphorothioate) conjugates designed to link the 3′-terminus of an oligonucleotide to the C-end of a peptide *via* a phosphodiester or phosphorothioate bond in the process of the stepwise solid-phase assembly.

Results and Discussion. – *Preamble.* In the present work, our objective was to synthesize certain peptide-(oligonucleotide phosphorothioate) conjugates. In these, the 3′-terminus of the 15-mer oligonucleotide phosphorothioate should be linked to the C-terminus of a 16-mer membrane-permeable motif (**MPM**) – the hydrophobic region of the signal peptide sequence of the *Kaposi* fibroblast growth factor, known to interact with lipid bilayers [17]. To achieve this goal, we had to prepare a suitable solid support, containing a (9H-fluoren-9-ylmethoxy)carbonyl(Fmoc)-protected amino function for easy standard peptide-chain elongation, a 4,4′-dimethoxytrityl((MeO)$_2$Tr)-blocked hydroxy group for standard oligonucleotide chain synthesis, and, finally, a base-labile bridge to a solid matrix, *i.e.* to a long chain alkylamino controlled pore glass (lcaa-CPG).

Solid Supports **S1** *and* **S2**. The structure of solid support **S1** seemed to fulfill the above mentioned requirements. The N-terminal L-threonine residue contained the necessary Fmoc-protected amino function and (MeO)$_2$Tr-blocked hydroxyl group. The C-terminal glycine residue served as an achiral linker between the L-threonine residue and 4-hydroxybutanoic acid, which was attached to lcaa-CPG. The ester bond of this linker could be easily cleaved under basic conditions.

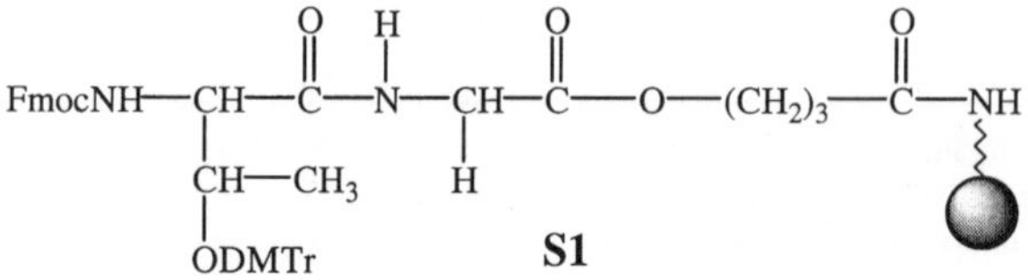

S1

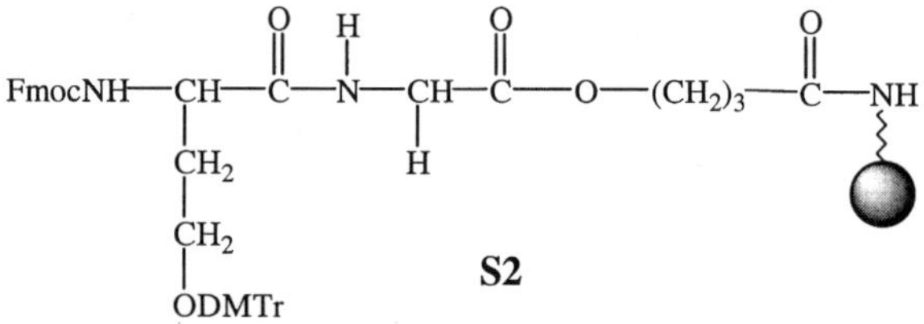

S2

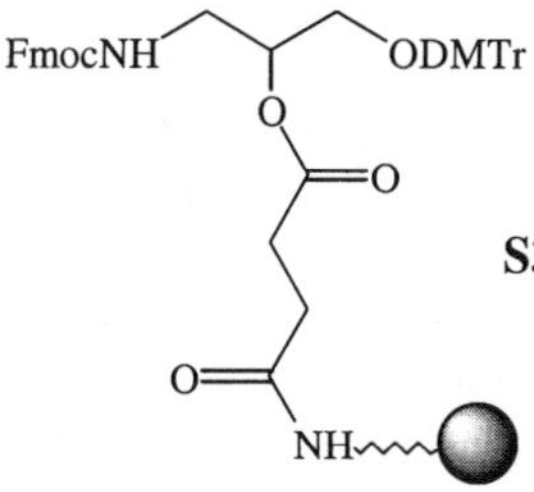

S3

DMTr=(MeO)$_2$Tr=bis(4-methoxyphenyl)phenylmethyl
Fmoc=(9H-fluoren-9-ylmethoxy)carbonyl

Of certain concern to us was the possible β-elimination of the peptide residue linked to the oligonucleotide *via* the hydroxy group of L-threonine during the final basic cleavage and deprotection step. It was obvious that the substitution of the L-threonine for a L-homoserine residue in the structure of solid matrix **S2** would ensure the absence of this possible side reaction during the synthesis of desired conjugates.

The preparation of solid supports **S1** and **S2** is outlined in *Scheme 1*. First, lcaa-CPG was derivatized with 4-hydroxybutanoic acid to give matrix **3** *via* **1** and **2**. Two dipeptides, Fmoc-Thr[(MeO)$_2$Tr]-Gly-OH (**7**) and Fmoc-Hse[(MeO)$_2$Tr]-Gly-OH

Scheme 1[a])

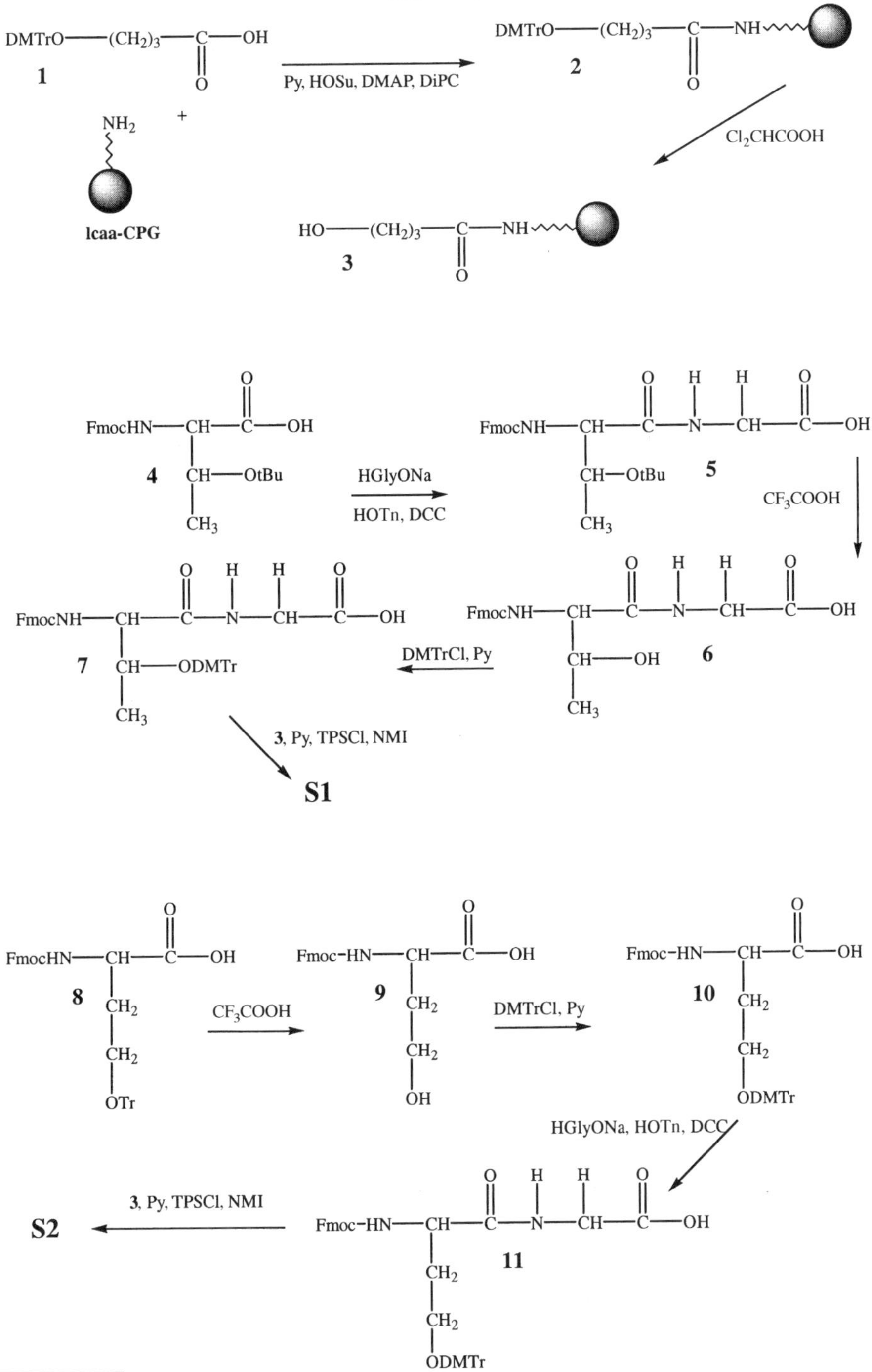

a) DMTr = (MeO)₂Tr = bis(4-methoxyphenyl)phenylmethyl; Fmoc = (9*H*-fluoren-9-ylmethoxy)carbonyl; HOSu = *N*-hydroxysuccinimide; DMAP = 4-(dimethylamino)pyridine; DiPC = diisopropylcarbodiimide; HOTn = *N*-hydroxy-8,9,10-trinorbornane-2,3-dicarboximide; DCC = dicyclohexylcarbodiimide; TPSCl = 2,4,6-triisobenzenesulfonyl chloride; NMI = 1-methyl-1*H*-imidazole.

(**11**), were synthesized in solution following routine procedures (from **4** *via* **5** and **6**, and from **8** *via* **9** and **10**, resp.), and finally linked to the matrix **3** [18] to give solid supports **S1** and **S2**. The resulting **S1** and **S2** contained 11.6 and 11.3 µmol, respectively, of (MeO)$_2$Tr groups per g of solid support [19].

Solid Support **S3**. Stepwise solid-phase assembly on supports **S1** and **S2** would lead to conjugates, incorporating the desired peptide, extended by two amino-acid residues (-Thr-Gly-, **S1**; -Hse-Gly-, **S2**) at the C-terminus. This may change the structure of some peptide moieties and, possibly, influence the cellular uptake of certain conjugates. Therefore, it seemed reasonable to design a solid support which would lead to a simple neutral linker bridging the peptide and oligonucleotide parts of conjugates.

Solid support **S3** was prepared starting from readily available (±)-3-amino-propane-1,2-diol (**12**) (*Scheme 2*). After protection of the amino function with the Fmoc group (→ **13**) and of the primary alcohol function with the (MeO)$_2$Tr group, compound **14** was linked to the lcaa-CPG by a succinate bridge between the secondary alcohol function (see **15**) and the amino group of the support [19]. The resulting **S3** contained 36.0 µmol of (MeO)$_2$Tr groups per g of solid support [19].

Scheme 2[a])

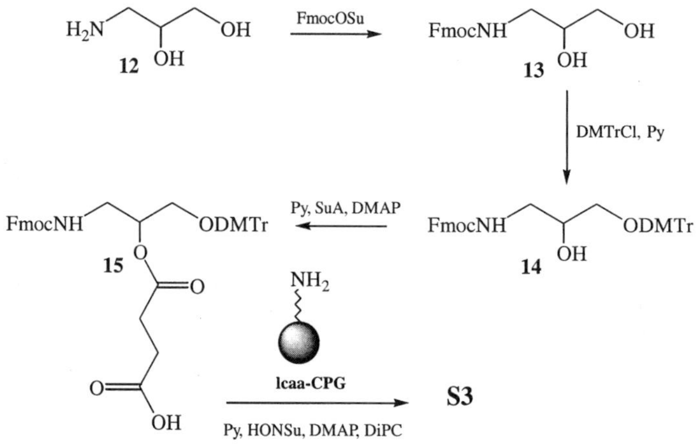

[a]) For abbreviations, see *Scheme 1*. SuA = Succinic anhydride.

Comparison of Supports **S1**, **S2**, *and* **S3** *for Synthesis of* **MPM**-*(oligo-T*$_6$*) Conjugates.* To investigate the usefulness of solid supports **S1**, **S2**, and **S3** in the stepwise solid-phase synthesis of peptide-oligonucleotide conjugates, the **MPM** sequence (-Ala-Ala-Val-Ala-Leu-Leu-Pro-Ala-Val-Leu-Leu-Ala-Leu-Leu-Ala-Pro-) was assembled on a *Milli-gen-9050* peptide synthesizer, by standard protocols based on Fmoc/2-(1*H*-benzotria-zol-1-yl)-1,1,3,3-tetramethyluronium tetrafluoroborate (TBTU)/triethylamine (TEA) chemistry (*Schemes 3* and *4*). Columns were removed from the peptide synthesizer. Portions of derivatized supports, containing 1 µmol of (MeO)$_2$Tr groups were packed into oligonucleotide-synthesizer columns and installed on a *PE-Applied Biosystems-392* DNA synthesizer. Standard phosphoramidite chemistry and recommended protocols (1.0-µmol scale) were employed to assemble the oligo-T$_6$ sequence (*Schemes 3* and *4*). The assembled conjugates were first cleaved and deprotected with

Scheme 3[a])

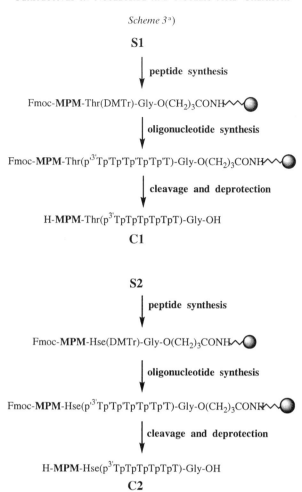

S1

↓ **peptide synthesis**

Fmoc-**MPM**-Thr(DMTr)-Gly-O(CH₂)₃CONH⁓◯

↓ **oligonucleotide synthesis**

Fmoc-**MPM**-Thr(p'³'Tp'Tp'Tp'Tp'Tp'T)-Gly-O(CH₂)₃CONH⁓●

↓ **cleavage and deprotection**

H-**MPM**-Thr(p³'TpTpTpTpT)-Gly-OH

C1

S2

↓ **peptide synthesis**

Fmoc-**MPM**-Hse(DMTr)-Gly-O(CH₂)₃CONH⁓◯

↓ **oligonucleotide synthesis**

Fmoc-**MPM**-Hse(p'³'Tp'Tp'Tp'Tp'Tp'T)-Gly-O(CH₂)₃CONH⁓●

↓ **cleavage and deprotection**

H-**MPM**-Hse(p³'TpTpTpTpT)-Gly-OH

C2

---

[a]) **MPM** = 16-mer membrane-permeable motif; p′ = 2-cyanoethyl-protected internucleosidic phosphotriester
unit; p = phosphodiester moiety; for Fmoc and DMTr, see *Scheme 1*.

conc. aqueous NH₃ solution at 25° for 2 h, and the composition of the reaction mixtures
was investigated by reversed-phase and ion-exchange HPLC.

In the case of the reaction mixture derived from support **S1**, both HPLC techniques
revealed the 21% content of conjugate **C1** (see *Table*) and *ca.* 10% of oligo-T₆. The
presence of oligo-T₆ may be explained by a β-elimination of the peptide residue linked
to the oligonucleotide *via* the hydroxy group of L-threonine during the final basic
cleavage and deprotection step. This phenomenon considerably limits the use of
support **S1** for preparation of conjugates.

In contrast, HPLC analysis of the mixtures derived from supports **S2** and **S3**
revealed higher contents of conjugate **C2** and **C3** (34 and 45%, resp.; see *Table*). As
expected, no oligo-T₆ was detected in either case. Thus, matrices **S2** and **S3** appear to

Scheme 4[a])

S3

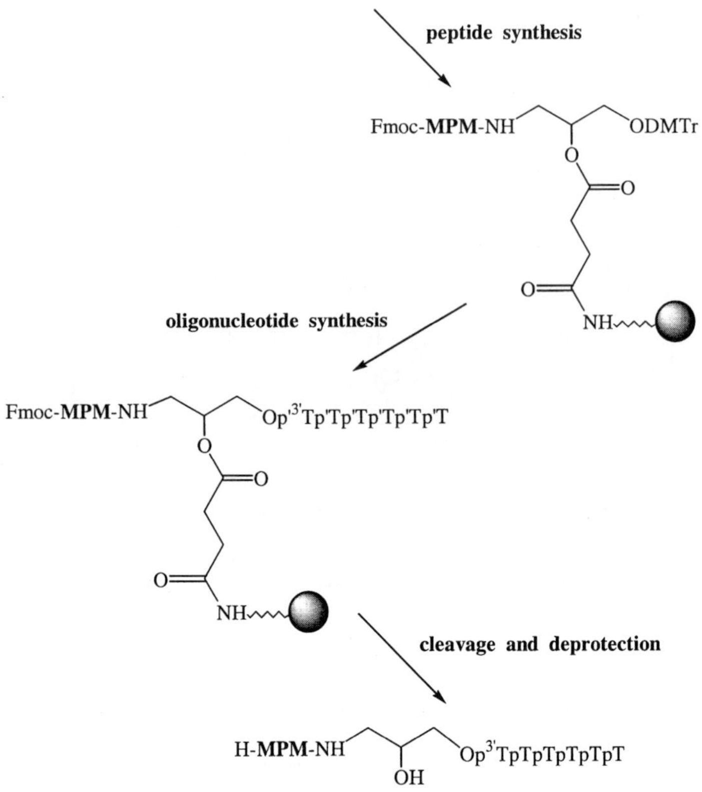

peptide synthesis

Fmoc-**MPM**-NH⌒⌒ODMTr
O
O

oligonucleotide synthesis

Fmoc-**MPM**-NH⌒⌒Op'$^{3'}$Tp'Tp'Tp'Tp'Tp'T
O
O

O
NH⌒⌒

cleavage and deprotection

H-**MPM**-NH⌒⌒Op$^{3'}$TpTpTpTpTpT
OH

C3

---

[a]) **MPM** = 16-mer membrane-permeable motif; p′ = 2-cyanoethyl-protected internucleosidic phosphotriester unit; p = phosphodiester moiety; for Fmoc and DMTr, see *Scheme 1*.

Table. *Yields, HPLC Retention Times* t_R, *and Measured (ESI-MS) and Calculated Average Molecular Masses of the Synthesized Conjugates*

	Content after cleavage and deprotection [%]	Isolated yield [%]	Reversed-phase HPLC t_R [min]	Ion-exchange HPLC t_R [min]	Calculated mass	Measured mass ± s.d. ($n = 6$)
C1	21	12.1	22.12	8.02	3499.3	3498.6 ± 0.4
C2	34	29.2	20.10	6.84	3499.3	3498.4 ± 0.7
C3	45	36.5	20.24	7.01	3414.2	3414.8 ± 0.3
C4	22	17.5	18.72	17.27	6502.0	6502.3 ± 0.5
C5	30	26.4	18.95	17.22	6416.9	6416.3 ± 1.5
O1[a])	–	–	7.12	10.28	–	–
O2[b])	–	–	12.03	22.60	–	–

[a]) **O1** = 5′-TpTpTpTpTpT-3′. [b]) **O2** = 5′-Tp$_s$Gp$_s$Gp$_s$Cp$_s$Gp$_s$Tp$_s$Cp$_s$Tp$_s$Tp$_s$Cp$_s$Cp$_s$Ap$_s$Tp$_s$Tp$_s$T-3′.

perform well in the stepwise solid-phase synthesis of peptide-oligonucleotide conjugates.

All three conjugates **C1**, **C2**, and **C3** of oligo-T_6 were isolated by ion-exchange HPLC and finally desalted (for isolated yields, see *Table*).

Peptide-(Oligonucleotide Phosphorothioate) Conjugates on Solid Supports **S2** *and* **S3**. The next question was whether our supports **S2** and **S3** are suitable for the stepwise synthesis of conjugates of **MPM** peptide and a 15-mer oligonucleotide phosphorothioate. Conjugates **C4** and **C5** were assembled analogously to the conjugates **C1**–**C3** of oligo-T_6 by the protocol recommended for phosphorothioates (*Scheme 5*). HPLC Analysis of mixtures derived from **S2** and **S3** showed reasonable contents of conjugates **C4** and **C5**, respectively (22% and 30%, resp.; see *Table*). In the ion-exchange chromatogram of the **C5** sample, after treatment with conc. aqueous NH_3 solution at 55° for 5 h, the product peak at t_R 17.22 min (see *Fig.*) corresponds to conjugate **C5**, as demonstrated by electrospray-ionization mass spectrometry (see *Table*). Compounds **C4** and **C5** were isolated by ion-exchange HPLC and finally desalted (for isolated yields, see *Table*).

Scheme 5[a])

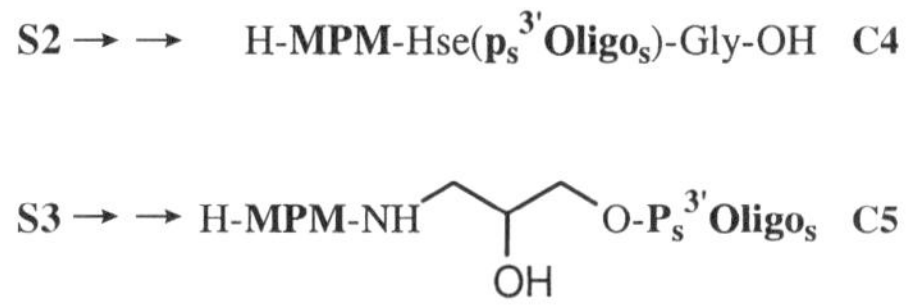

$$S2 \rightarrow \rightarrow \quad \text{H-}\textbf{MPM}\text{-Hse}(\textbf{p}_s{}^{3'}\textbf{Oligo}_s)\text{-Gly-OH} \quad \textbf{C4}$$

$$S3 \rightarrow \rightarrow \text{H-}\textbf{MPM}\text{-NH} \underset{\text{OH}}{\diagdown\diagup\diagdown} \text{O-}\textbf{P}_s{}^{3'}\textbf{Oligo}_s \quad \textbf{C5}$$

a]) **MPM** = 16-mer membrane-permeable motif; **Oligo**$_s$ = oligonucleotide phosphorothioate 5'-Tp$_s$Tp$_s$Tp$_s$Ap$_s$
Cp$_s$Cp$_s$Tp$_s$Tp$_s$Cp$_s$Tp$_s$Gp$_s$Cp$_s$Gp$_s$Gp$_s$T-3'

Hence, supports **S2** and **S3** may be successfully employed in the stepwise solid-phase synthesis of peptide-(oligonucleotide phosphorothioate) conjugates. It is noteworthy that both supports are relatively easy to prepare. While synthesis of matrix **S2** requires the use of the expensive L-homoserine monomer, preparation of **S3** employs the readily available (±)-3-aminopropane-1,2-diol (**12**). Moreover, unlike compounds derived from **S2**, conjugates synthesized on **S3** do not incorporate two additional amino-acid residues but contain only a simple neutral linker bridging the peptide and oligonucleotide moieties. Additionally, synthesis on solid support **S3** brings about higher yields of desired compounds (see *Table*). Thus, matrix **S3** appears to be the solid support of choice for the stepwise solid-phase synthesis of peptide-oligonucleotide conjugates.

Characterization of Peptide-Oligonucleotide Conjugates. Reversed-phase and ion-exchange HPLC retention times of the compounds synthesized are given in the *Table*. The purity of all conjugates, as assessed from reversed-phase HPLC, was higher than 95%. The final characterization was achieved by electrospray-ionization mass spectrometry (ESI-MS), which has proven to be a powerful method for the characterization of peptide-oligonucleotide conjugates [16]. The measured and calculated average molecular masses of the compounds reported here were in excellent agreement (see *Table*), the difference between the calculated and measured M_r being always less than 0.05%.

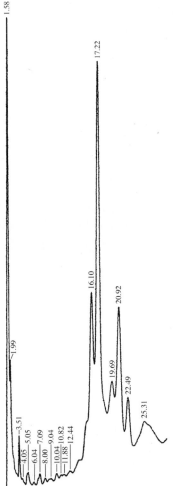

Figure. *Ion-exchange HPLC traces of the* **C5** *sample, synthesized on support* **S3**, *after treatment with conc. aqueous NH₃ solution at 55° for 5 h*

Conclusion. – We designed, synthesized, and tested new solid supports to prepare peptide-oligonucleotide and peptide-(oligonucleotide phosphorothioate) conjugates. These supports allow the linking of the 3'-terminus of an oligonucleotide to the C-end of a 16-mer peptide *via* a phosphodiester or phosphorothioate bond in the process of a stepwise solid-phase assembly. All conjugates obtained were thoroughly characterized by ESI-MS.

We thank Miss *Unni Tengvall* for ESI-MS experiments. Financial support from the *Technology Development Center of Finland* is gratefully acknowledged.

Experimental Part

General. Matrix 4-Hydroxy-N-(lcaa-CPG)butanamide. To a soln. of 4-hydroxybutanoic acid sodium salt (0.63 g, 5 mmol) in dry pyridine (50 ml), 4,4'-dimethoxytrityl chloride ((MeO)₂TrCl; 2.05 g, 6 mmol) was added.

The mixture was stirred for 4 h at 20° and the reaction finally quenched by addition of ice. The resulting soln. was evaporated and the residue dissolved in AcOEt (50 ml). The soln. was washed with 5% NaHCO₃ soln. and H₂O and evaporated, the residue dissolved in MeOH/pyridine/H₂O 1:1:1 (60 ml), and the soln. passed through a *Dowex 50* (Py⁺) column (3 × 15 cm). The resulting soln. was evaporated, co-evaporated with dry pyridine (3 × 10 ml), and finally dried: 2.1 g of *4-[bis(4-methoxyphenyl)phenylmethoxy]butanoic acid* (**1**). Pale yellow oil which was used further without any purification.

To a soln. of crude **1** (0.7 g) in dry pyridine (Py; 10 ml), *N*-hydroxysuccinimide (HOSu; 0.14 g, 1 mmol) and 4-(dimethylamino)pyridine (DMAP; 24 mg, 0.2 mmol) were added, and the soln. was treated with 2.0 g of lcaa-CPG (2.0 g). Diisopropylcarbodiimide (DiPC; 0.4 ml, 2 mmol) was added and the mixture shaken overnight. The CPG derivative was filtered off, washed twice with dry pyridine, and treated with Ac₂O/pyridine/1-methyl-1*H*-imidazole (NMI) 1:5:1 (21 ml) for 30 min. The solid matrix was filtered off, washed with pyridine, tetrahydrofuran, and Et₂O, and dried: *4-[bis(4-methoxyphenyl)phenylmethoxy]-N-(lcaa-CPG)butanamide* (**2**), containing 16.1 μmol of (MeO)₂Tr groups per g of matrix.

Matrix **2** (1 g) was treated with 2.5% Cl₂CHCOOH in CH₂Cl₂ (10 ml). After 10 min at 20°, the matrix was filtered off, washed with CH₂Cl₂, tetrahydrofuran, and Et₂O, and dried to give **3**.

Fmoc-Thr-Gly-OH (**6**). To a soln. of Fmoc-Thr(ᵗBu)-OH (**4**, 2.0 g, 4.8 mmol) and *N*-hydroxy-8,9,10-trinorborn-5-ene-2,3-dicarboximide (= *N*-hydroxybicyclo[2.2.1]hept-2-ene-2,3-dicarboximide; HOTn, 0.9 g, 5 mmol) in MeCN/DMF 4:1 (25 ml) at −20°, a soln. of dicyclohexylcarbodiimide (DCC; 1.05 g, 5 mmol) in MeCN (5 ml) was added, and the mixture was left overnight at +4°. The precipitate was filtered off, the filtrate evaporated, and the residue dissolved in MeCN/DMF (20 ml) and mixed with glycine (0.38 g, 5 mmol) in a 1M NaOH soln. in MeCN/DMF 1:2 (15 ml). The mixture was stirred overnight and then evaporated, the residue dissolved in AcOEt (50 ml), the soln. washed with 2% H₂SO₄ soln. and H₂O and evaporated, and the residue crystallized from hexane. The precipitate was washed with hexane (2 × 20 ml), dried, and treated with CF₃COOH (20 ml) for 30 min. The CF₃COOH was evaporated and the product crystallized from hexane/AcOEt 1:1. The precipitate was washed with hexane and dried: 1.65 g (86%) of (**6**). White crystals. M.p. 82–83°. ¹H-NMR (CDCl₃): 1.12 (*d*, *J* = 6.1, MeCH₂); 3.91–4.44 (*m*, CHCH₂OCO (Fmoc), CHCH₂OCO (Fmoc), CHCHCO (Thr), MeCHCH(Thr), NHCH₂COOH(Gly)); 6.31 (br. *m*, NH); 6.56 (br. *m*, NH); 7.22–7.71 (*m*, 8 arom. H).

Solid Support [O-[Bis(4-methoxyphenyl)phenylmethyl]-N-[(9H-fluoren-9-ylmethoxy)carbonyl]-L-threonyl]glycine 4-[(lcaa-CPG)amino]-4-oxobutyl Ester (**S1**). A soln. of **6** (1.6 g, 4 mmol) in 1M NaOH/pyridine 1:5 (25 ml) was evaporated and the residue dried by co-evaporation with pyridine and dissolved in dry pyridine (50 ml). (MeO)₂TrCl (1.7 g, 5 mmol) was then added, the mixture stirred for 24 h at 20°, the reaction quenched by addition of ice, the mixture evaporated, and the residue dissolved in AcOEt (70 ml). The soln. was washed with 5% NaHCO₃ soln. and H₂O and evaporated and the residue dried under vacuum. The sodium salt of **7** (0.73 g, 1 mmol) was dissolved in MeOH/pyridine/H₂O 1:1:1 (60 ml), passed through a *Dowex 50* (Py⁺) (3 × 15 cm), the soln. evaporated, and the residue dried by evaporation with pyridine and finally dissolved in dry pyridine (4 ml). Matrix **3** (1 g), 2,4,6-triisopropylbenzenesulfonyl chloride (TPSCl; 280 mg, 0.9 mmol), and 1-methyl-1*H*-imidazol (NMI; 80 μl, 0.9 mmol) were added, and the mixture was shaken overnight at 20°. Modified CPG was filtered off, washed twice with dry pyridine, and treated with Ac₂O/pyridine/NMI 1:5:1 (21 ml). After 30 min of shaking, support **S1** was filtered off, washed with pyridine, tetrahydrofuran, and Et₂O, and dried. The resulting **S1** contained 11.6 μmol of (MeO)₂Tr groups per g of solid support.

Fmoc-Hse[(MeO)₂Tr]-Gly-OH (**11**). To a soln. of Fmoc-Hse(Tr)-OH (**8**; 2.8 g, 5 mmol) and *N*-hydroxysuccinimide (0.63 g, 55 mmol) in MeCN/DMF 4:1 (25 ml) at −20°, a soln. of dicyclohexylcarbodiimide (1.05 g, 5 mmol) in MeCN (5 ml) was added, and the mixture was left overnight at +4°. The precipitate was filtered off, the filtrate evaporated, and the residue dissolved in DMF (20 ml) and mixed with a soln. of glycine (0.38 g, 5 mmol) in 1M NaOH/DMF 1:2 (15 ml). The mixture was stirred overnight and then evaporated. The residue was dissolved in AcOEt (50 ml), the soln. washed with 2% H₂SO₄ soln. and H₂O and evaporated, and the residue crystallized from hexane. The precipitate was washed with hexane (2 × 20 ml), dried, and treated with 2% CF₃COOH in CH₂Cl₂ (25 ml) for 30 min, concentrated to 5 ml, and finally diluted with ⁱPr₂O (50 ml). The precipitate was washed with hexane (2 × 20 ml) and dried. A portion of the precipitate (1.6 g) was dried by co-evaporation with pyridine (2 × 25 ml) and dissolved in dry pyridine (50 ml), and (MeO)₂TrCl (1.7 g, 5 mmol) was added. After stirring for 24 h at 20°, the reaction was quenched by addition of ice, the mixture evaporated and the residue dissolved in AcOEt (70 ml). The soln. was washed with 5% NaHCO₃ soln. and H₂O and evaporated and the residue crystallized from hexane/Et₂O: 2.34 g (87.3%) of **11**. M.p. 89–90°. ¹H-NMR (CDCl₃): 2.04 (br. *m*, CHCH₂CH₂(Hse)); 3.19 (br, *m*, CHCH₂CH₂(Hse)); 3.66 (*s*, 2 MeO); 3.71–4.27

(m, CHCH$_2$OCO (Fmoc), CHCH$_2$OCO (Fmoc), CHCHCH$_2$CH$_2$(Hse), NHCH_2COOH(Gly)); 6.28 (br. m, NH); 6.75–7.67 (m, 21 arom. H); 8.59 (br. m, NH).

Solid Support {O-[Bis(4-methoxyphenyl)phenylmethyl]-N-[(9H-fluoren-9-ylmethoxy)carbonyl]-L-homo-seryl}glycine 4-[(lcaa-CGP)amino]-4-oxobutyl Ester (**S2**). Attachment of **11** to the matrix **3** was performed as described for **S1**. The resulting **S2** contained 11.3 µmol of (MeO)$_2$Tr groups per g of solid support.

(±)-1-[Bis(4-methoxyphenyl)phenylmethoxy]-3-[[(9H-fluoren-9-ylmethoxy)carbonyl]amino]propan-2-ol (**14**). (±)-3-Aminopropane-1,2-diol (**12**, 0.77 ml, 10 mmol) was dissolved in DMF (30 ml), and *N*-{[(9*H*-fluoren-9-ylmethoxy)carbonyl]oxy}succinimide (FmocOSu; 3.37 g, 10 mmol) was added. After 1 h stirring, the mixture was evaporated, the residue dissolved in AcOEt (100 ml), the soln. washed with 2% H$_2$SO$_4$ soln. and H$_2$O and concentrated to 15 ml, and the product precipitated with hexane. The precipitate was washed with hexane (2 × 30 ml) and dried to give **13**. To a soln. of **13** in dry pyridine (25 ml), a soln. of (MeO)$_2$TrCl (3.4 g, 10 mmol) in dry dioxane (10 ml) was added dropwise, and the mixture was stirred overnight at 20°. After evaporation, the product was purified by FC (silica gel, 0→30% AcOEt/hexane +0.1% of pyridine): 4.7 g (76.5%) of **14**. White foam. ^{1}H-NMR (CDCl$_3$): 3.11–3.49 (m, NHCH$_2$CH, CHCH$_2$O); 3.76 (s, 2 MeO); 3.87 (br. m, CH$_2$CHCH$_2$); 4.19 (t, J = 7.0, CHCH$_2$OCO (Fmoc)); 4.36 (d, J = 7.0, CHCH$_2$OCO (Fmoc)); 5.01 (br. m, OH); 6.82 (d, J = 8.9, 4 arom. H); 7.30 (d, J = 8.9, 4 arom. H); 7.21–7.77 (m, 14 arom. H, NH).

Solid Support 4-[(lcaa-CGP)amino]-4-oxobutanoic Acid 2-[Bis(4-methylphenyl)phenylmethoxy]-1-[[(9H-fluoren-9-ylmethoxy)carbonyl]amino]ethyl Ester (**S3**). Compound **14** was linked to the lcaa-CPG (2.0 g) by a succinate bridge as described in [19]. The resulting **S3** contained 36.0 µmol of (MeO)$_2$Tr groups per g of solid support.

Peptide Synthesis. Fmoc-L-amino acids for peptide synthesis were purchased from *Novabiochem*. The synthesis of peptide on all solid supports was performed with a *Milligen-9050* peptide synthesizer by standard protocols based on Fmoc/TBTU/TEA chemistry.

Oligonucleotide Synthesis. The oligo-T$_6$ and oligonucleotide phosphorothioate conjugates were prepared on the **MPM**-derivatized solid supports **S1**, **S2**, or **S3** with a *PE-Applied Biosystems-392* DNA synthesizer on a 1-µM scale. 2′-Deoxyribonucleoside 3′-phosphoramidites and 'Fast sulfurizing reagent' were products of *Glen Research*.

Analysis and Isolation of Peptide-Oligonucleotide Conjugates. Cleavage and deprotection of conjugates **C1–C5** was achieved with 32% aq. NH$_3$ soln. at 25° for 2 h or at 55° for 5 h. After cleavage, all conjugates were analyzed by reversed-phase HPLC (*Genesis C4*, 300 Å, 4 µ, 4.0 × 150 mm column (*Jones*); buffer *A*, 0.05M (EtNH)OAc; buffer *B*, 0.05M (Et$_3$NH)OAc in 80% MeCN/H$_2$O; linear gradient from 0→100% *B* in 30 min, flow rate 1 ml/min) and by ion-exchange HPLC (*PolyWax LP* 300 Å, 5 µ, 4.6 × 100 mm column (*Poly LP*)) with a NH$_4$SO$_4$ gradient (buffer *A*, 0.05M KH$_2$PO$_4$ in 50% formamide/H$_2$O, pH 5.6; buffer *B*, buffer *A* + 0.6M NH$_4$SO$_4$, pH 5.6; linear gradient 0→30% *B* in 30 min, flow rate 1 ml/min) for conjugates **C1–C3** or with a NaBr gradient (buffer *A*, 0.05M KH$_2$PO$_4$ in 50% formamide/H$_2$O, pH 6.3; buffer *B*, buffer *A* + 1.5M sodium bromide, pH 6.3; linear gradient 30→100% *B* in 30 min, flow rate 1 ml/min) for conjugates **C4** and **C5**. Substances **C1–C5** were isolated by ion-exchange HPLC on prep. scale as described above and finally desalted.

Electrospray-Ionization Mass Spectrometry. Mass spectra were acquired and molecular masses were reconstructed as reported earlier [16].

REFERENCES

[1] R. Eritja, A. Pons, M. Escarceller, E. Giralt, F. Albericio, *Tetrahedron* **1991**, *47*, 4113.
[2] C.-H. Tung, M. J. Rudolph, S. Stein, *Bioconjugate Chem.* **1991**, *2*, 464.
[3] T. Zhu, S. Stein, *Bioconjugate Chem.* **1994**, *5*, 312.
[4] Z. Wei, C.-H. Tung, T. Zhu, S. Stein, *Bioconjugate Chem.* **1994**, *5*, 468.
[5] N. J. Edge, G. W. Tregear, J. Haralambidis, *Bioconjugate Chem.* **1994**, *5*, 373.
[6] J.-P. Bongartz, A.-M. Aubertin, P. G. Milhaud, B. Leblue, *Nucleic Acids Res.* **1994**, *22*, 4681.
[7] J. Robles, E. Pedroso, A. Grandas, *J. Org. Chem.* **1994**, *59*, 2482.
[8] S. Soukchareun, G. W. Tregear, J. Haralambidis, *Bioconjugate Chem.* **1995**, *6*, 43.
[9] C.-H. Tung, J. Wang, M. Leibowitz, S. Stein, *Bioconjugate Chem.* **1995**, *6*, 292.
[10] J. Robles, E. Pedroso, A. Grandas, *Nucleic Acids Res.* **1995**, *23*, 4151.
[11] J.-C. Truffert, U. Asseline, A. Brack, N. T. Thuong, *Tetrahedron* **1996**, *53*, 3005.
[12] R. K. Bruick, P. E. Dawson, S. B. H. Kent, N. Usman, G. F. Joyce, *Chem. Biol.* **1996**, *3*, 49.
[13] J. Robles, M. Maseda, M. Concernau, E. Pedroso, A. Grandas, *Bioconjugate Chem.* **1997**, *8*, 785.

[14] S. B. Rajur, C. M. Roth, J. R. Morgan, M. L. Yarmush, *Bioconjugate Chem.* **1997**, *8*, 935.

[15] L. Chaloin, P. Vidal, P. Lory, J. Mery, N. Lautredou, G. Divita, F. Heitz, *Biochem. Biophys. Res. Commun.* **1999**, *244*, 601.

[16] M. Antopolsky, E. Azhayeva, U. Tengvall, S. Auriola, I. Jääskeläinen, S. Rönkkö, P. Honkakoski, A. Urtti, H. Lönnberg, A. Azhayev, *Bioconjugate Chem.* **1999**, *10*, 598.

[17] Y.-Z. Lin, S. Yao, R. A. Veach, T. R. Torgerso, J. Hawiger, *J. Biol. Chem.* **1995**, *270*, 14255.

[18] J. Hovinen, A. Guzaev, A. Azhayev, H. Lönnberg, *Tetrahedron* **1994**, *50*, 7203.

[19] T. Atkinson, M. Smith, in 'Oligonucleotide Synthesis. A Practical Approach', Ed. M. J. Gait, IRL Press, Oxford, 1984, pp. 35–81.

Influence of the Type of Junction in DNA-3'-Peptide Nucleic Acid (PNA) Chimeras on Their Binding Affinity to DNA and RNA

by Beate Greiner, Gerhard Breipohl, and Eugen Uhlmann*[1])

Hoechst Marion Roussel Deutschland GmbH, Chemical Research G 838, D-65926 Frankfurt a.M.

The automated on-line synthesis of DNA-3'-PNA (PNA = Polyamide Nucleic Acids) chimeras 1–3 is described, in which the 3'-terminal part of the oligonucleotide is linked to the aminoterminal part of the PNA either via a N-(2-mercaptoethyl)- ($X = S$), a N-(2-hydroxyethyl)- ($X = O$), or a N-(2-aminoethyl)- ($X = NH$) N-[(thymin-1-yl)acetyl]glycine unit. Furthermore, the DNA-3'-PNA chimera 4 without a nucleobase at the linking unit was prepared. The binding affinities of all chimeras were directly compared by determining their T_m values in the duplex with complementary DNA, RNA, or DNA containing a mismatch or abasic site opposite to the linker unit. We found that all investigated chimeras with a nucleobase at the junction form more stable duplexes with complementary DNA and RNA than the corresponding unmodified DNA. The influence of X on duplex stabilization was determined to be in the order $O > S \approx NH$, rendering the phosphodiester bridge the most favored linkage at the DNA/PNA junction. The observed strong duplex-destabilizing effects, when base mismatches or non-basic sites were introduced opposite to the nucleobase at the DNA/PNA junction, suggest that the base at the linking unit contributes significantly to duplex stabilization.

Introduction. – Peptide or Polyamide Nucleic Acids (PNAs) are nucleic acid mimetics, in which the entire sugar-phosphate backbone is replaced by non-ionic N-(2-aminoethyl)glycine units [1]. They bind to complementary DNA or RNA targets with higher affinity than natural oligonucleotides following the *Watson-Crick* rules of base pairing [2], and they are resistant to enzymatic degradation [3]. These properties make them attractive candidates for the use as antisense therapeutics and diagnostics [4]. Recently, we have shown that some limitations of PNAs, such as low cellular uptake, ambiguous orientation on hybridization (parallel or antiparallel), the inability to activate RNase H, and the propensity to self-aggregation, can be overcome by using DNA-PNA chimeras [5][6]. In these chimeras (see 1), the 3'-part of the oligodeoxy-nucleotide was linked to a terminal (hydroxyethyl)glycine unit ($X = O$) of the PNA *via* a phosphodiester linkage [5][7]. For PNA-5'-DNA chimeras with the 5'-part of the DNA linked to the carboxy terminus of the PNA, a significant influence of the type of linkage at the DNA/PNA junction has been reported previously [8]. Here, we describe the synthesis and binding properties of novel DNA-3'-PNA chimeras containing all four nucleobases, which are linked by a N-(2-mercaptoethyl)glycine (2; $X = S$) or an N-(2-aminoethyl)-glycine (3; $X = NH$) [9] unit, respectively. To aid the optimal future design of DNA-3'-PNA chimeras, we directly compare the binding affinity of all three differently linked DNA-3'-PNA chimeras 1–3 in the duplex with complementary

---

[1])   Fax: + 49-69-305 5595; e-mail: eugen.uhlmann@hmrag.com

DNA and RNA. To evaluate the contribution of the linking moiety in these chimeras to their binding affinity, we have also prepared and investigated the DNA-3'-PNA chimera **4** with an abasic N-(2-hydroxyethyl)glycine unit at the DNA/PNA junction (R = H, X = O).

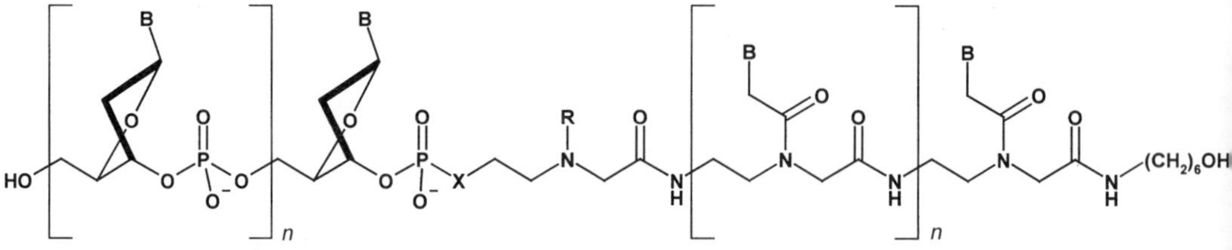

X = O, S, NH ;   R = H , R¹   R¹ =

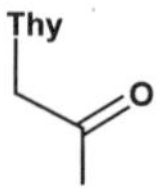

B = adenine, guanine, cytosine, or thymine

	X	R
1	O	R¹
2	S	R¹
3	NH	R¹
4	O	H

Results and Discussion. – *Synthesis of the Monomeric Linker Units.* For the incorporation of the abasic linker unit (X = O, R = H) at the DNA/PNA junction of the chimeras, we synthesized first the orthogonally protected monomeric linker unit **8** (*Scheme 1*). Reductive amination of 2-aminoethanol with glyoxylic acid in H_2O with H_2 as reducing agent and Pd/C as catalyst [10] resulted in N-(2-hydroxyethyl)glycine (**6**) in 83% yield. The amino function of **6** was then protected by reaction with (9H-fluoren-9-yl) methyl N-succinimidyl carbonate in dioxane to give the [(9H-fluoren-9-yl)methoxy]-carbonyl (Fmoc) derivative **7** in 77% yield. Finally, the abasic linker unit **8** was obtained in 70% yield by reaction of the OH group of **7** with (4-methoxyphenyl)di-phenylmethyl (Mmt) chloride in pyridine.

The monomeric linker unit **11** (*Scheme 2*), which is based on N-(2-mercapto-ethyl)glycine, was prepared from 2-mercaptoethylamine hydrochloride by reaction with bis(4-methoxyphenyl)phenylmethyl (Dmt) chloride and successive N-alkylation with ethyl 2-bromoacetate and NEt_3 in DMF to give N-[2-(Dmt-thio)ethyl]glycine ethyl ester (**10**). Intermediate **10** was coupled without further purification with thymine-3-

Scheme 1

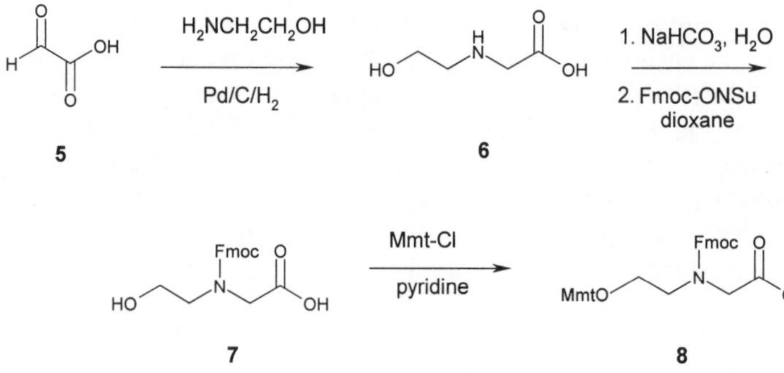

Scheme 2

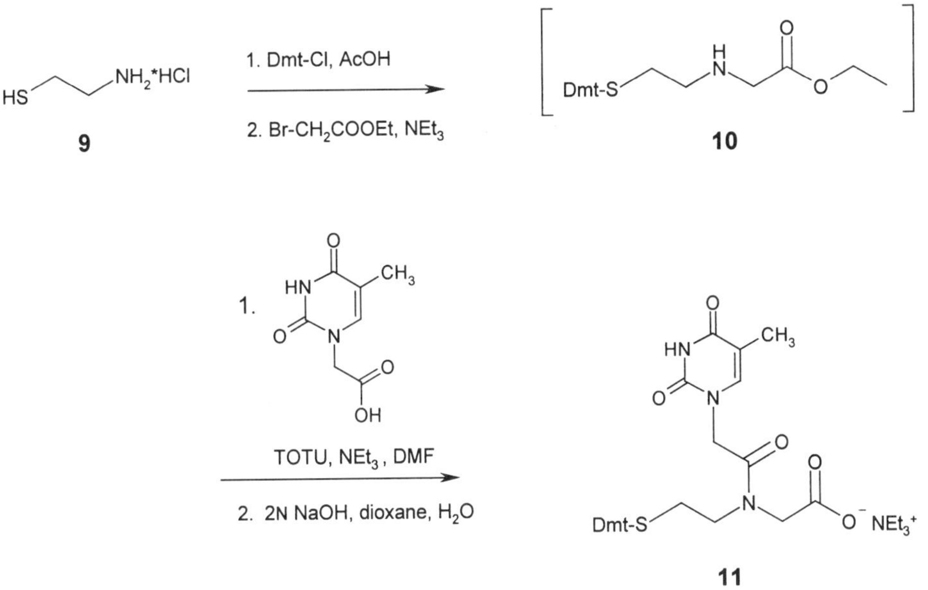

acetic acid using *O*-{[(2-cyanoethoxycarbonyl)methylidene]amino}-1,1,3,3-tetramethyl-uronium tetrafluoroborate (TOTU) and then saponified with 2N NaOH in dioxane/H$_2$O 2:1 (*v/v*). After purification of the crude product by silica-gel chromatography with 1% NEt$_3$ in CH$_2$Cl$_2$/MeOH, the (mercaptoethyl)glycine-based monomer **11** was obtained in 32% overall yield.

Solid-Phase Synthesis of DNA-3′-PNA Chimeras **15 – 18**. Mmt-protected monomers **12** based on *N*-(2-aminoethyl)glycine and a universal 6-aminohexanol-derived CPG solid support **14** [10] were used for the synthesis of the PNA part of the DNA-3′-PNA (*Scheme 3*) [5][10][11]. After synthesis of the PNA part of the chimeras, the novel linker units **8** or **11**, respectively, were coupled at the junction to allow on-line synthesis of the DNA part of the chimeras by standard phosphoramidite chemistry [12], which, after cleavage from the support and deprotection, yielded the chimeras **15** and **17**, respectively. The DNA-3′-PNA chimera **16** was synthesized using the *N*-(2-hydroxy-ethyl)glycine-derived monomer **13** [13] as linker unit, while the phosphoramidate-linked chimera **18** was synthesized with the standard PNA monomer **12** and subsequent coupling of a protected nucleoside phosphoramidite to the terminal amino group of the PNA [9]. All DNA-3′-PNA chimeras were synthesized at 2-μmol scale. Deprotection of the Mmt or Dmt group was carried out with 3% trichloroacetic acid (TCA) in CH$_2$Cl$_2$ including one intermediate washing step with MeCN and subsequent neutralization of the support with EtN(i-Pr)$_2$ (DIPEA; or *Huenig*'s base). For efficient coupling of the monomers, a pre-activation step with either *O*-(1*H*-benzotriazol-1-yl)-1,1,3,3-tetrame-thyluronium hexafluorophosphate (HBTU), *O*-(7-aza-1*H*-benzotriazol-1-yl)-1,1,3,3-tetramethyluronium hexafluorophosphate (HATU), or (7-aza-1*H*-benztriazol-1-yloxy)-tris(pyrrolidino)phosphonium hexafluorophosphate (PyAOP) in the presence of DIPEA was performed, whereby the required pre-activation time strongly depended

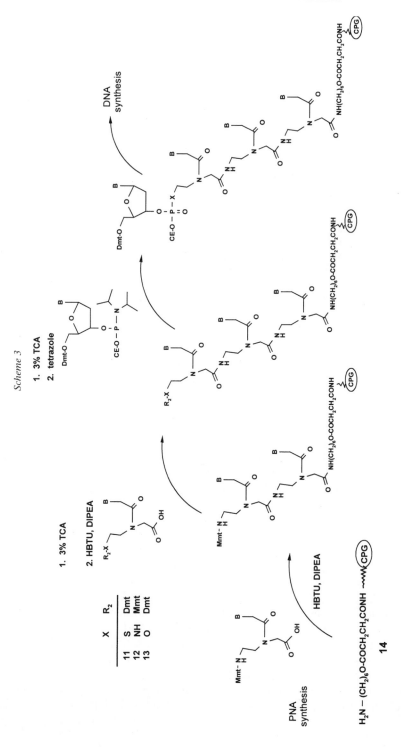

Scheme 3

on the type of coupling reagent. While HATU and PyAOP gave sufficient coupling efficiencies (90–96%) using short pre-activation times (10–60 s), HBTU required longer pre-activation (15 min) to obtain similar results. Unreacted amino functions were capped with the standard DNA-capping mixture consisting of Ac$_2$O/N-methylimidazole in THF to prevent the growth of failure sequences due to incomplete coupling reactions.

The orthogonally protected monomeric PNA building blocks with base-labile protecting groups at the heterocyclic bases and an acid-labile Mmt group for temporary protection of the terminal amino function, or a Dmt group for the OH and SH functions, respectively, are fully compatible with standard phosphoramidite DNA synthesis. Consequently, the protected chimeras could be cleaved from the solid support (2 h at 50°) and simultaneously deprotected by treatment with concentrated aqueous NH$_3$ (16 h at 50°). The crude products were purified by denaturing preparative polyacrylamide gel electrophoresis (PAGE) and desalted *via* *C-18* columns. All DNA-3′-PNA chimeras were characterized by ion-exchange HPLC using a *Gen Pack Fax* column (*Millipore-Waters*) and by negative-ion electrospray mass spectroscopy (see *Exper. Part*).

Binding Affinity of the DNA-3′-PNA Chimeras. To study the influence of the type of linking unit at the DNA/PNA junction on the binding affinity of the chimeras, the UV melting curves of the corresponding duplexes with DNA and RNA were measured at 260 nm under physiological salt conditions (140 mM KCl, 10 mM NaH$_2$PO$_4$, 0.1 mM Na-EDTA, pH 7.4) and the T_m values were calculated. All three DNA-3′-PNA chimeras **16 – 18**, which have a nucleobase at the DNA/PNA junction, form more stable duplexes with complementary DNA **20** or RNA **21** than the natural oligonucleotide **19** (*Table 1*). The duplexes with complementary DNA were stabilized by +1.7 to +4.3 K, whereas the stabilization for complementary RNA was even larger (ΔT_m +4.7 to +6 K). The influence of X in the linking unit on duplex stabilization was in the order O > S ≈ NH, rendering the phosphodiester bridge the most favored linkage at the DNA/PNA junction. If the chimera was linked with just N-(2-hydroxyethyl)glycine without a nucleobase at the junction, such as in oligomer **15**, then the duplex with complementary DNA was destabilized by −7.1 K relative to the natural duplex, and by as much as −11.4 K relative to the DNA-PNA chimera **16** with a thymine base at the junction. This result implies that the nucleobase at the PNA unit of the junction actually forms a base pair with the complementary base of the second strand.

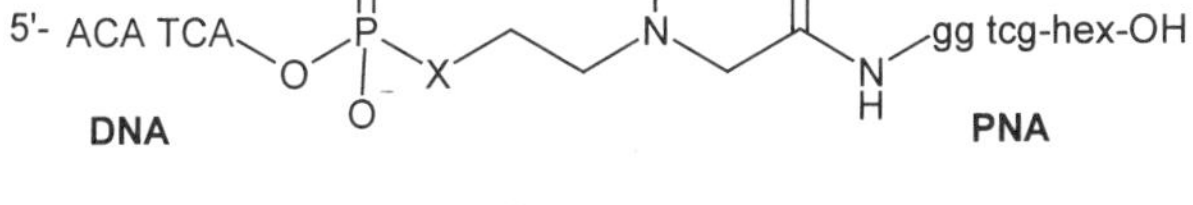

15 - 18

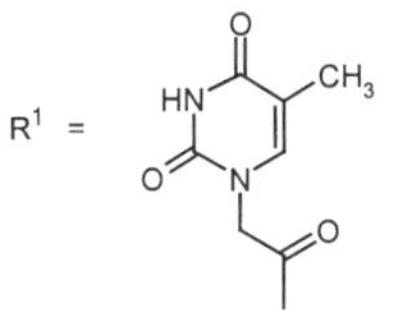

R^1 =

	X	R
15	O	H
16	O	R^1
17	S	R^1
18	NH	R^1

Table 1. T_m[a]) *Values of Duplexes of DNA-3'-PNA Chimeras* **15–18** *with Complementary DNA or RNA*

			T_m [°C]	
	X	R	DNA[b])	RNA[c])
15	O	H	42.8	43.3
16	O	R^1	54.2	54.0
17	S	R^1	52.1	52.7
18	NH	R^1	51.6	53.0
19[d])	–	–	49.9	48.0

[a]) 140 mM KCl, 10 mM NaH_2PO_4, 0.1 mM Na-EDTA, pH 7.4
[b]) Complementary DNA: 3'-T G T A G T A C C A G C-5' (**20**)
[c]) Complementary RNA: 3'-U G U A G U A C C A G C-5' (**21**)
[d]) Unmodified oligonucleotide: 5'-A C A T C A T G G T C G-3' (**19**)

To further analyze the contribution of the nucleobase-containing linker unit, we also measured the T_m values of the different chimeras **15–18** (*Table 2*) against complementary DNA having one base mismatch opposite to the base at the linking unit (mm-DNA; **22**) and against an oligonucleotide lacking the nucleobase at the junction (ab-DNA; **23**). Interestingly, introduction of a base mismatch destabilized the duplex to a similar extent (ΔT_m −7.7 to −11.5 K) as the removal of the nucleobase of the linking PNA unit (ΔT_m −11.4 K). Incorporation of an abasic unit in the complementary strand (ab-DNA; **23**) directly opposite to the linker unit confirmed the contribution of the nucleobase to binding: the T_m values of the duplexes with a base mismatch were between 40.6 to 45.4°, while the T_m values of the duplexes with an abasic unit were between 45.1 to 47.3°. Remarkably, the DNA · DNA duplex **23·24** with two abasic sites opposite to each other was significantly less stable (ΔT_m −14.5 K) than the DNA · DNA-PNA duplex **15·23**, in which the abasic site of the linker unit is opposite to the abasic site in the complementary DNA.

In conclusion, all investigated DNA-3'-PNA chimeras with a nucleobase at the junction formed more stable duplexes with complementary DNA and RNA than the corresponding unmodified DNA. The destabilizing effect of base mismatches or abasic

Table 2. *Influence of Base-Mismatches or Abasic Sites on the* T_m *Values*

			T_m [°C]		
	X	R	DNA[a])	mm-DNA[b])	ab-DNA[c])
15	O	H	42.8	45.6	46.0
16	O	R^1	54.2	45.4	47.3
17	S	R^1	52.1	40.6	45.5
18	NH	R^1	51.6	43.9	45.1
19[d])	–	–	49.9	–	31.5
24[e])	–	–	34.0	–	31.5

[a]) Complementary DNA: 3'-T G T A G T A C C A G C-5' (**20**)
[b]) Complementary mismatch-DNA: 3'-T G T A G T T C C A G C-5' (**22**)
[c]) Complementary abasic-DNA: 3'-T G T A G T D C C A G C-5' (**23**)
[d]) Unmodified sense oligonucleotide: 5'-A C A T C A T G G T C G-3' (**19**)
[e]) Abasic sense oligonucleotide: 5'-A C A T C A D G G T C G-3' (**24**)
<u>D</u>: abasic linker derived from phosphoric acid mono-[2-(hydroxymethyl)tetrahydrofuran-3-yl] ester

sites opposite to the nucleobase at the DNA/PNA junction suggests that this base very likely contributes to duplex stabilization through a *Watson-Crick* base pair. Furthermore, the phosphoramidite linkage ($X = NH$) may be an attractive alternative to the phosphodiester linkage ($X = O$) in DNA-3'-PNA chimeras, especially when hybridized to complementary RNA, since no additional building block for its incorporation is required.

We thank *H. Wenzel, S. Hein*, and *L. Hornung* for expert technical assistance, and Dr. *A. Schäfer* for measurement of mass spectra. The work was funded in part by the *Deutsche Forschungsgemeinschaft* and *Bundesministerium für Forschung und Technologie* (BMBF; project No. 0311143).

Experimental Part

1. *General.* The peptide coupling reagents HATU/1-hydroxy-7-aza-1*H*-benzotriazole (HOAt), and PyAOP were purchased from *Perspetives Biosystems*, HBTU, (9*H*-fluoren-9-yl)methyl *N*-succinimidyl carbonate (Fmoc-ONSu) and aminopropyl CPG from *Fluka* (Neu-Ulm, Germany). Standard nucleoside phosphoramidites and the reagent for introduction of the abasic linker (*dSpacer phosphoramidite*®) were purchased from *Eurogentec* (Seraing, Belgium). TLC was carried out on *Merck DC Kieselgel 60 F-254* glass plates. HPLC Analysis of PNAs: on a *Beckman System Gold* HPLC system, with a NaCl gradient (buffer *A*: 10 mM NaH$_2$PO$_4$, 100 mM NaCl in MeCN/H$_2$O 1:4 (*v/v*), pH 6.8; buffer *B*: 10 mM NaH$_2$PO$_4$, 1.5M NaCl in MeCN/H$_2$O 1:4 (*v/v*); 0 to 30% *B* in 30 min). T_m Values [°C ± 0.3] were measured under approximately physiological salt conditions (140 mM KCl, 10 mM NaH$_2$PO$_4$, 0.1 mM Na-EDTA, pH 7.4) in the cooling phase from 90° to 10° with a temp. ramp of 0.3°/min^{-1} at 260 nm on a *Varian Carey 1 Bio* UV/VIS Spectrometer at 1 μM oligomer concentration. ^{1}H-NMR Spectra: at 270 MHz in the solvents indicated; chemical shifts (δ) in ppm downfield relative to the internal standard. MS: either fast atom bombardement (FAB), electrospray (ES), or direct chemical ionization (DCI). NBA: 3-Nitrobenzyl alcohol.

2. *Monomeric Building Units.* N-*(2-Hydroxyethyl)glycine* (**6**). Glyoxylic acid monohydrate (46 g, 500 mmol) was dissolved in H$_2$O (1 l), and 2-aminoethanol (30.2 ml, 500 mmol) was added under stirring and cooling. The mixture was treated with 10% Pd/C catalyst (10 g) and hydrogenated (10 bar) in the autoclave at r.t. Then, the catalyst was filtered off, and the filtrate was concentrated *in vacuo*. The residue was co-evaporated twice with a small amount of toluene. The crude product was triturated using 250 ml of hot MeOH, filtered off, washed with MeOH, and dried: 49.56 g (83%). M.p. 178–180°(dec.). TLC (silica gel; BuOH/AcOH/H$_2$O/ AcOEt 1:1:1:1 (*v/v/v/v*)): R_f 0.36. ^{1}H-NMR (D$_2$O): 3.77 (2 H, (*m*, HOCH$_2$); 3.68 (*s*, CH$_2$COOH); 3.22 (*m*,CH$_2$NH). DCI-MS: 120 ([*M* + H]$^+$).

N-*[(9H-Fluoren-9-yl)methoxycarbonyl]-N-(2-hydroxyethyl)glycine* (**7**). NaHCO$_3$ (10.08 g, 120 mmol) was dissolved in H$_2$O (150 ml) and **6** (7.15 g, 60 mmol) was added under stirring. After a clear soln. was obtained, Fmoc-ONSu (20.22 g, 60 mmol) in dioxane (300 ml) was added dropwise. After the final addition, stirring was continued for 3 h at r.t. The soln. was filtered, and the filtrate was concentrated *in vacuo*. The residue was solved in H$_2$O (100 ml), and the pH was adjusted to 2 with a KHSO$_4$ soln. Then, the mixture was extracted with AcOEt (3 × 150 ml), the combined org. layers were washed with H$_2$O (4 × 50 ml), dried (Na$_2$SO$_4$), filtered, and evaporated *in vacuo*. Precipitation was achieved with (i-Pr)$_2$O. The precipitate was filtered off, recrystallized from AcOEt/(i-Pr)$_2$O, filtered, and dried: 15.8 g (77%) of **7**. M.p. 114–116° (dec.). TLC (silica gel; BuOH/ AcOH/H$_2$O 3:1:1 (*v/v/v*)): R_f 0.62. ^{1}H-NMR ((D$_6$)DMSO): 12.71 (br. *s*, 1 H); 7.24–7.95 (*m*, 8 H); 4.63 (br. *s*, 1 H); 4.19–4.39 (*m*, 3 H); 3.92–4.11 (*m*, 2 H); 3.11–3.56 (*m*, 4 H). ES-MS (pos. mode): 342 ([*M* + H]$^+$).

N-*[(9H-Fluoren-9-yl)methoxycarbonyl]-N-[2-[(4-methoxyphenyl)diphenylmethoxy]ethyl]glycine* (**8**). Compound **7** (14.74 g, 43 mmol) was dissolved in pyridine (160 ml), (4-methoxyphenyl)diphenylmethyl chloride (13.31 g, 43 mmol) was added, and the mixture was stirred for 2 h at r.t. After evaporation *in vacuo*, the residue was taken up in AcOEt (300 ml) and extracted with sat. aq. NaHCO$_3$ (3 × 30 ml) and H$_2$O (3 × 30 ml). The org. layer was dried (Na$_2$SO$_4$), filtered, evaporated, and precipitated with (i-Pr)$_2$O. The product was filtered and dried. 19.95 g (70%) of **8**. TLC (silica gel; AcOEt/MeOH/Et$_3$N 60:40:1 (*v/v/v*)): R_f 0.45. ^{1}H-NMR ((D$_6$)DMSO): 6.80–7.95 (*m*, 22 H); 4.09–4.31 (*m*, 3 H); 3.80, 3.91 (2*s*, 2 H); 3.76 (*s*, 3 H); 2.92–3.56 (*m*, 4 H). FAB-MS (MeOH/NBA): 658.3 ([*M* + 2Na]$^+$).

N-*(2-[[Bis(4-methoxyphenyl)phenylmethyl]thio]ethyl)-N-[(thymin-1-yl)acetyl]glycine* (**11**). To a soln. of Dmt-Cl (10.17 g, 30 mmol) in AcOH (100 ml), 2-mercaptoethylamine hydrochloride (4.43 g, 39 mmol) dissolved in H_2O (70 ml) was added. The soln. was stirred for 2 h at r.t. and then evaporated *in vacuo* to a volume of 50 ml. After addition of H_2O (200 ml), the pH was adjusted to 10 using 2N NaOH. The mixture was extracted with AcOEt (2×100 ml), and the combined org. phases were washed with sat. NaCl soln., dried, filtered, and concentrated *in vacuo* (14.75 g). Part of the resulting crude product (12.56 g) was dissolved in dry DMF (100 ml), treated with Et_3N (4.6 ml, 33 mmol) and ethyl bromoacetate (3.66 ml, 33 mmol) and stirred for 3 h. The soln. was concentrated *in vacuo*, and the residue was taken up in AcOEt (150 ml), washed with H_2O (4×40 ml). Then, the org. layer was dried (Na_2SO_4), filtered, and evaporated *in vacuo* (12.92 g). Part of the resulting crude product (**10**; 5.95 g) was dissolved in dry DMF (50 ml) and (thymin-2-yl)acetic acid (2.36 g, 12.8 mmol), Et_3N (4.45 ml, 32 mmol), and TOTU (4.2 g, 12.8 mmol) were added to the soln. The mixture was stirred for 2 h at r.t., then evaporated *in vacuo*. The residue was taken up in AcOEt (150 ml) and washed with sat. $NaHCO_3$ soln. (3×10 ml) and H_2O (2×10 ml). The org. layer was dried (Na_2SO_4), filtered, and concentrated *in vacuo*. The resulting residue was dissolved in a mixture of dioxane (80 ml) and H_2O (40 ml), and the ester was hydrolyzed by adding 2N NaOH soln. in portions (13 ml). The soln. was concentrated *in vacuo* to a volume of 50 ml and extracted with AcOEt (4×50 ml). The pH of the aq. layer was adjusted to 5 with 2N HCl, and the soln. was then extracted with AcOEt (4×50 ml). The combined org. layers were washed twice with H_2O (30 ml), dried (Na_2SO_4), filtered, and evaporated *in vacuo*. The crude product was purified by silica-gel chromatography using a step gradient of 0–5% MeOH in CH_2Cl_2 (1% Et_3N in all eluents). The fractions containing the product were pooled and dried *in vacuo*: 2.38 g (32% from Dmt-Cl) of **11**. TLC (silica gel; CH_2Cl_2/MeOH/Et_3N 100 : 10 : 1 (*v/v/v*)): R_f 0.38. FAB-MS (NBA + LiCl): 610.2 ($[M + Li]^+$), 616.2 ($[M + 2Li - H]^+$). ^1H-NMR ((D_6)DMSO): 11.4 (*s*, H–C(3) of thymine); 6.91–7.38 (*m*, 14 H, Dmt, H–C(6) of thymine); 4.16, 4.39 (2*s*, $NCOCH_2$); 3.69 (2*s*, 2 MeO); 3.56, 3.61 (2*s*, NCH_2COOH); 2.98 (*m*, CH_2N); 2.31 (*m*, CH_2S); 1.76 (*s*, Me of thymine).

2. Solid-Phase Synthesis of the DNA-3′-PNA Chimeras. The Mmt/acyl-protected monomer **12** and Dmt/acyl-protected monomer **13** were prepared as described in [13]. The Mmt-protected 6-aminohexan-1-ol/succinylamidopropyl CPG [10] with a loading 36 µmol/g was used as solid-support. Synthesis was performed on a modified *Eppendorf Biotronik Ecosyn D-300* DNA synthesizer or on an *ABI 394* DNA synthesizer at 2-µmol scale. The DNA part was synthesized according to standard procedures [12]. The following synthesis conditions were used for the synthesis of the PNA part including the *N*-(2-mercaptoethyl)glycine, *N*-(2-hydroxyethyl)-glycine, or *N*-(2-aminoethyl)glycine derived linkers: *1*) *washing step* with MeCN; *2*) *deprotection* of the Mmt group: 3% TCA in CH_2Cl_2; 110 s total treatment time interrupted by one wash with MeCN for 20 s; *3*) *washing step* with DMF/MeCN 1:1 (*v/v*); *4*) *syringe wash*; *5*) *neutralization:* washing with DMF/MeCN 1:1 (*v/v*) and EtN(i-Pr)$_2$; *6*) *coupling of monomers:* Monomers (0.2 to 0.3M solutions in DMF); etN(i-Pr)$_2$ (0.2 to 0.3M in DMF); coupling reagent (0.2–0.3M in DMF); reagents were pre-mixed, pre-activated and delivered onto the solid support. Due to the use of different coupling reagents, we had to adjust the pre-activation and reaction time; *7*) *capping.* Reagent *A:* 10% Ac_2O/10% lutidine in THF; reagent *B:* 16% *N*-methyl-1*H*-imidazole in THF (the DNA capping reagents *A* and *B* were mixed just before use).

After synthesis was complete, the chimeras were cleaved from the support (2.5 h at 50°) and deprotected (6 h at 50°) with concentrated aq. NH_3 soln. The crude product was analyzed by HPLC using a *Gen Pack Fax* column (*Millipore-Waters*), eluting with a NaCl gradient (buffer *A:* 10 mM NaH_2PO_4, 100 mM NaCl in MeCN/H_2O 1:4 (*v/v*) pH 6.8; buffer *B:* 10 mM NaH_2PO_4, 1.5M NaCl in MeCN/H_2O 1:4 (*v/v*); 0–30% *B* in 30 min). Purification was achieved by prep. PAGE (15% polyacrylamide) and desalted *via* a *C-18* column. The purified PNA/DNA chimeras were further analyzed by negative-ion ES-MS.

3. Synthesis of 5′-ACA TCA oeg(X = O, R = H)gg tcg-hex-OH (**15**). As described in *Exper. 2*, with HATU/HOAt (0.3M) as coupling reagent with a pre-activation time of 1 min, and a total coupling time of 45 min. Yield of crude product was 101 OD. Purification of 42 OD of the crude product resulted in 3.2 OD purified product. Characterization by negative-ion ES-MS: 3473.75 ± 0.13 (*M*; calc. for $C_{124}H_{162}N_{55}O_{54}P_6$: 3473.86).

4. Synthesis of 5′-ACA TCA oeg-t(X = O, R = R^1)gg tcg-hex-OH (**16**). As described in *Exper. 2*, with HBTU (0.2M) as coupling reagent with permanent pre-activation (DNA synthesizer *ABI 394*), and a total coupling time of 15 min. Yield of crude product was 26 OD. Purification of the crude product (26 OD) resulted in 13.6 OD purified product. Characterization by negative-ion ES-MS: 3596.65 ± 0.77 (*M*; calc. for $C_{129}H_{167}N_{57}O_{56}P_6$: 3597.94).

5. Synthesis of 5′-ACA TCA seg-t(X = S, R = R^1)gg tcg-hex-OH (**17**). As described in *Exper. 2*, with HBTU (0.25M) as coupling reagent with a pre-activation time of 15 min, and a total coupling time of 15 min. Yield of crude product was 60 OD. Purification of 44 OD of the crude product resulted in 9.5 OD purified product. Characterization by negative-ion ES-MS: 3613.51 ± 0.13 (*M*; calc. for $C_{129}H_{167}N_{57}O_{55}P_6S$: 3614.00).

6. *Synthesis of 5'ACA TCA aeg-t*(X = NH, R = R¹)*gg tcg-hex-OH* (**18**). As described in *Exper. 2*, with PyAOP (0.3M) as coupling reagent with a pre-activation time of 10 s, and a total coupling time of 15 min. Yield of crude product was 52 OD. Purification of 51 OD of the crude product resulted in 1.9 OD of purified product. Characterization by negative-ion ES-MS: 3596.49 ± 0.09 (M; calc. for $C_{129}H_{168}N_{58}O_{55}P_6$: 3596.98).

REFERENCES

[1] P. E. Nielsen, M. Egholm, R. H. Berg, O. Buchardt, *Science* **1991**, *254*, 1497.

[2] M. Egholm, O. Buchardt, L. Christensen, C. Behrens, S. M. Freier, D. A. Driver, R. H. Berg, S. K. Kim, B. Norden, P. E. Nielsen, *Nature* **1993**, *365*, 566.

[3] B. Hyrup, P. E. Nielsen, *Bioorg. Med. Chem.* **1996**, *4*, 5.

[4] E. Uhlmann, A. Peyman, G. Breipohl, D. W. Will, *Angew. Chem., Int. Ed.* **1998**, *37*, 2796.

[5] E. Uhlmann, D. W. Will, G. Breipohl, D. Langner, A. Ryte, *Angew. Chem., Int. Ed.* **1996**, *35*, 2632.

[6] E. Uhlmann, *Biol. Chem.* **1998**, *379*, 1045.

[7] K. H. Petersen, D. K. Jensen, M. Egholm, P. E. Nielsen, O. Buchardt, *Bioorg. Med. Chem. Lett.* **1995**, *5*, 1119.

[8] A. C. van der Laan, P. Havenaar, R. S. Oosting, E. Kuyl-Yeheskiely, E. Uhlmann, J. H. van Boom, *Bioorg. Med. Chem. Lett.* **1998**, *8*, 663.

[9] F. Bergmann, W. Bannwarth, S. Tam, *Tetrahedron Lett.* **1995**, *36*, 6823.

[10] D. W. Will, G. Breipohl, D. Langner, J. Knolle, E. Uhlmann, *Tetrahedron* **1995**, *51*, 12069.

[11] E. Uhlmann, B. Greiner, G. Breipohl, in 'Peptide Nucleic Acids: Protocols and Applications', Eds. P. E. Nielsen and M. Egholm, Horizon Scientific Press, Norfolk, UK, 1999, pp. 51–70.

[12] M. D. Matteucci, M. H. Caruthers, *J. Am. Chem. Soc.* **1981**, *103*, 3185.

[13] G. Breipohl, D. W. Will, A. Peyman, E. Uhlmann, *Tetrahedron* **1997**, *53*, 14671.

Triple Helices of Optimally Capped Duplex DNA (σ-DNA) with Homopyrimidine DNA and RNA at Neutral pH

by **Willi Bannwarth**[1])* and **Patrick Iaiza**

PRPC, *F. Hoffmann-La Roche Ltd.*, Grenzacherstrasse 124, CH-4070 Basel

An optimally capped duplex DNA (σ-DNA) was synthesized in which the cytosines in the homopyrimidine strand of σ-DNA **1** were replaced by 5-methylcytosine leading to σ-DNA **2**. Although only involving 13 base pairs, this modification resulted in very high melting temperatures above 90°. In addition, σ-DNA **2** was able to form triple helices with the corresponding homopyrimidine DNA or RNA even at neutral pH. This opens up the possibility to use σ-DNA in a triple-helix approach to modulate gene expressions on the level of the translation process.

1. Introduction. – Double-stranded DNA is able to align a third strand of DNA or RNA in its major groove to form triple-helix motifs [1][2]. In a Pyr·(Pur·Pyr) motif, the third strand is bound *via Hoogsteen* H-bonds (*Fig. 1*). Such an alignment requires the protonation of the C residues in the third strand. Although restricted to homopyrimidine or homopurine sequences, the triple-helix approach is of interest with respect to the specific control of gene expressions.

Binding of a third strand into the major groove of double-stranded target DNA can prevent the binding of transcription factors and hence, prevent the initiation of the transcription process of a gene of interest. This strategy, also dubbed the anti-gene approach, was pursued by several groups [3][4].

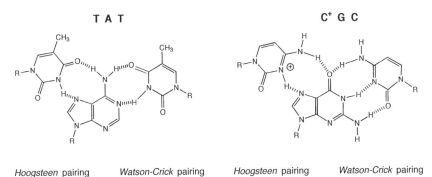

Fig. 1. Hoogsteen *pairing in the Pyr·(Pur·Pyr) mode*

---

[1]) Present address: Institute of Organic Chemistry and Biochemistry, University of Freiburg, Albertstr. 21, D-79104 Freiburg (phone: +49-761-2036073; fax: +49-761-2038705; e-mail: willi.bannwarth@organik.chemie.uni-freiburg.de).

An alternative strategy based on triple-helix formation and aiming at the inhibition or modulation of the translation process, was developed by *Kool* [5]. Application of properly designed circular homopyrimidine constructs can be used to bind single-stranded homopurine sequences. One homopyrimidine region of the circular structure is engaged in *Watson-Crick* pairing and the other one in *Hoogsteen* pairing, resulting in a tight overall binding of the circular structure to the homopurine strand, *e.g.* to an envisaged m-RNA, and thus preventing translation.

Recently, we have synthesized a variant of cyclic DNA that represents in essence an optimally capped duplex DNA [6][7]. We have dubbed this type of DNA as σ-DNA to distinguish it from 'dumbbell' DNA. We were able to demonstrate that such a σ-DNA represents a very stable mimic of long double-stranded B-DNA showing very high *Watson-Crick* melting temperatures. If one strand of the σ-DNA consisted of a homopyrimidine sequence and the other one of a homopurine sequence, we were able to align a third homopyrimidine strand in the major groove of the σ-DNA, resulting in a strong *Hoogsteen* binding. As expected, this binding was pH-dependent and decreased with increasing pH. At pH 7.0, no *Hoogsteen* binding with DNA was observed, and with RNA as third strand, the *Hoogsteen* melting temperature was only 17°.

2. Results and Discussion. – The aim of this work was to modify σ-DNA in such a way that it is able to align a homopyrimidine RNA strand in a *Hoogsteen* mode, even at neutral pH, with a binding efficiency comparable with the binding of a complementary single-stranded DNA fragment to the same target RNA strand, a strategy which one would apply to standard antisense approaches.

Our earlier results have indicated that the stability of the *Hoogsteen* binding of RNA to σ-DNA is higher than that of the corresponding DNA at all pH values investigated [7]. This is in accordance with results published by *Roberts* and *Crothers*, who demonstrated that the stability of triple helices is dependent on the nature of the individual strands engaged in the triple-helix formation (DNA or RNA) [8]. Furthermore, these experiments had indicated that the stability of the *Hoogsteen* binding was dependent on the stability of the *Watson-Crick* duplex of the σ-DNA. The more stable the σ-DNA, the more stable the *Hoogsteen* interaction. Thus, the aim was to increase the stability of this *Watson-Crick* pairing to such an extent that it was able to bind single-stranded RNA even at neutral pH with decent efficiency.

The stability of cytosine · guanine base pairs can be increased by replacing the cytosine by 5-methylcytosine [9]. The 5-methylcytosine is a naturally occurring base which plays a role in the control of transcription in eucaryotes [10]. Thus, we hoped that σ-DNA **2** in which all cytosines of σ-DNA **1** were replaced by 5-methylcytosine would be so stable as to allow for binding of the single-stranded RNA fragment **4** *via* *Hoogsteen* mode with the desired efficiency at neutral pH. At the same time, the binding to the corresponding DNA fragment **3** was also investigated (*Fig. 2*).

The strategy for the synthesis of compounds **1** and **2** is outlined in *Fig. 3*. The preparation is based on the cyclization of the linear precursors by a chemical ligation. In these linear precursors, backfolding of the overhanging ends guarantees close proximity between the phosphate group and the OH function for formation of the phosphodiester bond. The ligation can be performed starting either from the linear 5′- or the 3′-phosphorylated fragment with comparable efficiency [6]. For the synthesis of **1** and **2**,

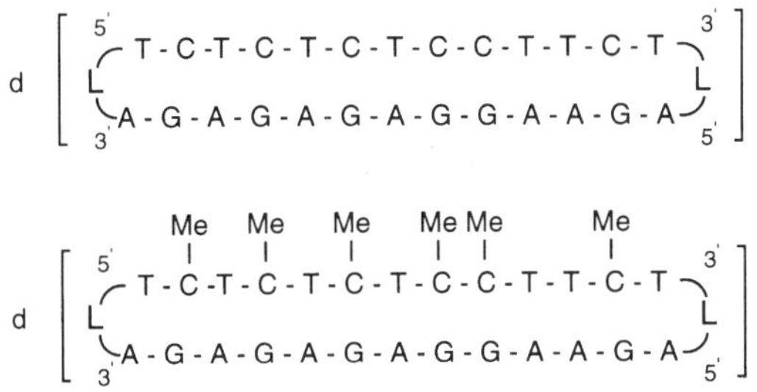

$$
d \left[\begin{array}{l} {}^{5'}\!\!\!\curvearrowright\text{T - C -T - C - T - C - T - C - C - T - T - C - T}\curvearrowright^{3'} \\ L \qquad\qquad\qquad\qquad\qquad\qquad\qquad\qquad\quad L \\ {}_{3'}\!\!\!\curvearrowleft\text{A - G - A - G - A - G - A - G - G - A - A - G - A}\curvearrowleft_{5'} \end{array} \right] \quad 1
$$

$$
d \left[\begin{array}{l} \quad\ \ \overset{\text{Me}}{|}\ \ \ \overset{\text{Me}}{|}\ \quad\ \overset{\text{Me}}{|}\ \quad\overset{\text{Me}}{|}\ \overset{\text{Me}}{|}\qquad\ \ \overset{\text{Me}}{|} \\ {}^{5'}\!\!\!\curvearrowright\text{T - C -T - C - T - C - T - C - C - T - T - C - T}\curvearrowright^{3'} \\ L \qquad\qquad\qquad\qquad\qquad\qquad\qquad\qquad\quad L \\ {}_{3'}\!\!\!\curvearrowleft\text{A - G - A - G - A - G - A - G - G - A - A - G - A}\curvearrowleft_{5'} \end{array} \right] \quad 2
$$

$$
(3'\text{-}5')\,d\,(\,T\text{-}C\text{-}T\text{-}C\text{-}T\text{-}C\text{-}T\text{-}C\text{-}C\text{-}T\text{-}T\text{-}C\text{-}T\,) \qquad 3
$$

$$
(3'\text{-}5')\,r\,(\,U\text{-}C\text{-}U\text{-}C\text{-}U\text{-}C\text{-}U\text{-}C\text{-}C\text{-}U\text{-}U\text{-}C\text{-}U\,) \qquad 4
$$

Fig. 2. *DNA and RNA sequences used in the melting experiments*

we started from the 3'-phosphates employing a disulfide linker to the support, which yields, after reduction of the disulfide bond, *via* spontaneous elimination, directly the desired 3'-phosphorylated fragment [11]. The chemical ligation was performed employing *N*-[3-(dimethylamino)propyl]-*N'*-ethylcarbodiimide hydrochloride (EDC), and a high conversion to the circular DNA **1** and **2** was achieved. The reaction was performed with the crude unpurified linear fragments.

The linker unit consists of a hexaethylene-glycol spacer containing two phosphate groups at its termini. This linker proved ideal for bridging the terminal base pairs without causing conformational strains in B-DNA. At the same time, this linker is short enough to restrict efficiently the conformational flexibility of the connected nucleotide ends [12].

The results of the melting experiments are summarized in the *Table*. Substitution of the cytosines by 5-methylcytosines resulted as expected in a greater stability of the *Watson-Crick* duplex, and this at all pH values investigated. Hence, the melting of **2** is generally above 90° although it involves only 13 base pairs. At the same time, the *Hoogsteen* melting temperatures of σ-DNA **2** with the single-stranded DNA fragment **3** and the RNA fragment **4** were higher in comparison with the corresponding ones with σ-DNA **1**. At all pH values, this increase in the *Hoogsteen* binding was more pronounced for DNA fragment **3** as the third strand as compared to the RNA fragment **4** as the third strand.

In *Fig. 4*, we have compared the melting of σ-DNA **2** with the homopyrimidine DNA **3** and the corresponding RNA sequence **4** from pH 5.0 to pH 7.5. Over this pH range, the binding of σ-DNA **2** to the RNA **4** as the third strand in the *Hoogsteen* mode was stronger than the binding of **2** to the DNA **3**. This difference became smaller with increasing pH value. Nevertheless, *Hoogsteen* binding was still observed in both experiments at neutral pH. At pH 7.0, the *Hoogsteen* melting temperature with RNA (31.6°) was comparable with the *Watson-Crick* melting of a duplex composed of the RNA sequence **4** and the complementary DNA sequence (5'-3')d(A-G-A-G-A-G-A-G-G-A-A-G-A) which showed under identical conditions a melting temperature of

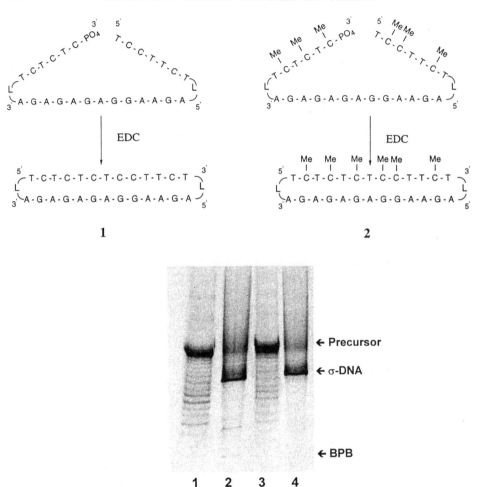

Fig. 3. *Cyclization strategy for the synthesis of the σ-DNA molecules* **1** *and* **2**, *and PAGE* (polyacrylamide-gel electrophoresis) *of the crude reaction mixtures. Lane 1:* crude linear precursor for **1**; *Lane 2: cyclization reaction leading to* **1**; *Lane 3:* crude linear precursor for **2**; *Lane 4:* cyclization reaction leading to **2**.

Table. *Results of Duplex and Triplex Melting Experiments with Sequences* **1**–**4**. Conditions as in *Figs. 4* and *5*.

σ-DNA	Hoogsteen strand	pH 5.0		pH 5.5		pH 6.0		pH 6.5		pH 7.0		pH 7.5	
		melt 1	melt 2	melt 1	melt 2	melt 1	melt 2	melt 1	melt 2	melt 1	melt 2	melt 1	melt 2
1	**3**	54.2	83.4	46.3	85.3	32.3	86.8	20.6	87.4	–	87.4	–	87.5
2	**3**	64.7	90.0	55.5	>90	43.3	>90	34.3	>90	29.1	>90	26.4	>90
1	**4**	67.6	83.8	57.8	85.8	42.4	86.6	29.2	87.7	–	87.2	–	87.3
2	**4**	74.2	>90	63.2	>90	49.0	>90	38.8	>90	31.6	>90	28.1	>90

a)
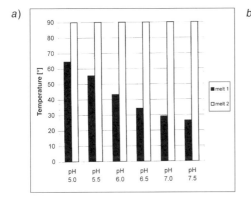

b)
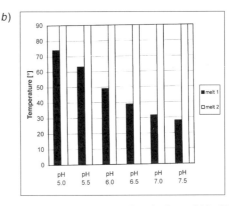

Fig. 4. a) *pH Dependency of the melting points of σ-DNA* **2** *and single-stranded DNA* **3** *as third strand.* b) *pH Dependency of the melting points of σ-DNA* **2** *and single-stranded RNA* **4** *as third strand. Melt 1: Hoogsteen melting temperature; Melt 2: Watson-Crick melting temperature. Conditions:* 1.0 μM DNA, 100 mM NaOAc/AcOH (pH 5.0–7.5); 1 mM EDTA.

30.7°. This situation is representative of the normal antisense situation with a linear unmodified DNA fragment binding to the complementary RNA.

In *Fig. 5*, we have illustrated the difference of the *Hoogsteen* melting temperatures applying either σ-DNA **1** or σ-DNA **2**. The stabilization effect obtained by substituting cytosine by 5-methylcytosine was clearly demonstrated.

a)
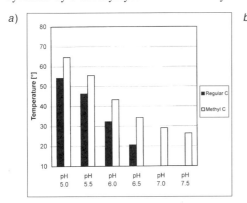

b)
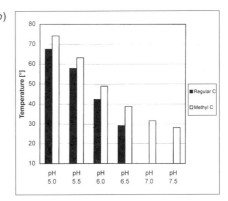

Fig. 5. a) *pH Dependency of the* Hoogsteen *melting temperatures of either σ-DNA* **1** *or σ-DNA* **2** *and DNA* **3** *as* Hoogsteen *strand.* b) *pH Dependency of the* Hoogsteen *melting temperature of either σ-DNA* **1** *or σ-DNA* **2** *and RNA* **4** *as third strand. Conditions as in Fig. 4.*

3. Conclusion and Outlook. – We were able to demonstrate that triple-helix formation of optimally capped duplex DNA (σ-DNA) is possible under neutral conditions with DNA, as well as with RNA, as the third strand. This was achieved by substitution of the cytosines in σ-DNA **1** by 5-methylcytosine leading to σ-DNA **2**. The *Hoogsteen* melting of σ-DNA **2** with the corresponding homopyrimidine RNA sequence **4** at neutral pH was in the same range as the *Watson-Crick* melting of the corresponding linear DNA · RNA hybrid. Hence, σ-DNA could be used in an alternative approach to inhibit or modulate the translation of mRNA.

Apart from the hexaethylene-glycol linker, σ-DNA consists of only naturally occurring bases and is unmodified at the phosphodiester internucleotide linkages. Thus, possible breakdown products should not lead to cell toxicity. Due to the cyclic structure of σ-DNA, such a degradation could only be initiated by double-strand specific endonucleases.

In addition, the synthesis of the σ-DNA was achieved with high efficiency by a chemical ligation of the crude phosphorylated linear precursor molecules.

Experimental Part

1. General. All solvents were of highest purity available. The 1,4-dithioerythritol (= *erythro*-1,4-dimercapto-butane-2,3-diol; DTE), *N*-[3-(dimethylamino)propyl]-*N'*-ethylcarbodiimide hydrochloride (= EDC), undecane, and *Stains-all* were from *Fluka*, $Bu_4NF \cdot 3 H_2O$ from *Aldrich*, and *NAP-10-Sephadex* columns from *Pharmacia*. All DNA and RNA syntheses were performed on an *ABI-394* synthesizer; the different solns., supports, and amidites for DNA synthesis were from *ABI*. The RNA building blocks, *i.e.*, 5'-(dimethoxytrityl-2'-[(*tert*-butyl)dimethylsilyl]-protected ribonucleoside (2-cyanoethyl diisopropylphosphoramidites), were from *Milligen*. 30% Sat. NH_3/EtOH soln. was prepared by bubbling NH_3 gas slowly through 30 ml of ice-cooled EtOH (*Aldrich*; H_2O max. 5 ppm) for 2 h. The UV-absorption/melting curves were run on a *Cary-3-UV* photometer. The hexaethylene glycol building block was synthesized as described in [7].

2. DNA Fragment **3** was prepared on controlled-pore glass (CPG; *Sigma*) as solid support [13] applying a 10-fold excess of 2-cyanoethyl phosphoramidites [14] and a 130-fold excess of 1*H*-tetrazole. The workup was done using our standard technology [15].

3. RNA Fragment **4** was synthesized on a 1-μmol scale using per coupling a 150-fold excess of 2'-*O*-[(*tert*-butyl)dimethylsilyl]-protected phosphoramidite [16] and 150-fold excess of 1*H*-tetrazole with a coupling time of 15 min and the standard RNA cycle [17]. The workup was performed under sterile conditions. The support was treated with 1 ml of sat. NH_3/EtOH soln. at 70° for 3 h. The NH_3 soln., including 1 ml of washing soln. of the support with EtOH/H_2O 6:4, was evaporated on a *Speed-vac* concentrator. After addition of 0.5 ml of fresh 1M Bu_4NF/THF, the pellet was vortexed for 16 h at r.t. until a soln. was obtained. Then 0.4 ml of 0.5M NH_4OAc soln. (pH 7.0) was added and the soln. evaporated to 300 μl and desalted by a *NAP-10* filtration. The fragment was purified by prep. 20% polyacrylamide-gel electrophoresis (PAGE) under denaturing and sterile conditions [17].

4. σ-DNA **1** was synthesized in two steps. Firstly, the 3'-phosphorylated precursors were synthesized, and in a second step, these were cyclized to yield **1**. For the synthesis of the 3'-phosphates, we used a disulfide linker as described in [11] but employing a sarcosine-modified CPG for the attachment of the linker unit. Preparation of the linear fragment was performed starting with 1 μmol of the modified support. When the linker had to be introduced, coupling with the appropriate phosphoramidite was carried out applying a standard cycle. After complete assembly, the $(MeO)_2Tr$ group was removed. Further deprotection and release as 3'-phosphate was by treatment with 1 ml of 0.2M DTE in conc. NH_3 soln. for 2 h at 70°. After evaporation of the soln. on a *Speed-vac* concentrator, the fragments were precipitated after addition of 100 μl of H_2O, 200 μl of dioxane, and 600 μl of THF. Part of the crude material was analyzed by anal. PAGE to assess the performance of the synthesis (*Fig. 3*).

The cyclization of the linear 3'-phosphorylated fragment with EDC to yield the σ-DNA **1** was performed in 1 ml of 50 mM MES (morpholine-4-ethanesulfonic acid) pH 6.0, 20 mM $MgCl_2$, and 400 mM EDC containing 200 mmol (80 *OD*) of the crude linear precursor. After 5 h, the crude mixture was desalted on a *NAP-10* column and purified by prep. PAGE. The purity was assessed by anal. PAGE after staining with *Stains-all*. MALDI-TOF-MS confirmed the correct molecular mass of the fragment.

σ-DNA **2** was prepared in the same way as **1**, except that the 2'-deoxy-5-methylcytidine phosphoramidite was used as building block instead of the corresponding 2'-deoxycytidine building block.

5. Melting Curves. The UV absorbance was measured at 260 nm at 95°, and the following extinction coefficients were employed: A 15000; C 7500; G 12500; T 8500 [18]. Each fragment (1 nmol) was dissolved in 1 ml of 1 mM EDTA and 100 mM sodium acetate (pH 7.5) adjusted to the corresponding pH with AcOH. After transfer into the cuvettes, the solns. were degassed by sonications at 90°. Then a layer of 0.1 ml of undecane was added, and the cuvettes were closed by stoppers. UV-Absorption/melting curves were measured against cuvettes containing the buffer alone. The heating program was from 10 to 95° with a 0.5° increase per min and from 95 to 10° with 0.5° per min. Each melt was performed at least twice, and the given melting temp. is the average of the two measurements. Only the melting temp. from the heating run was used for the calculation. Melting temps. were calculated from the first derivative of the raw data.

REFERENCES

[1] H. E. Moser, P. B. Dervan, *Science (Washington, D.C.)* **1987**, *238*, 645.

[2] T. Le Doan, L. Perrouault, D. Praseuth, N. Habhoub, J.-L. Decout, N. T. Thuong, J. Lhomme, C. Helene, *Nucleic Acids Res.* **1987**, *15*, 7749.

[3] C. Helene, J. J. Toulme, *Biochim. Biophys. Acta* **1990**, *1049*, 99.

[4] M. Cooney, G. Czernuszewicz, E. H. Postel, S. J. Flint, M. E. Hogan, *Science (Washington, D.C.)* **1988**, *241*, 456.

[5] G. Prakash, E. T. Kool, *J. Chem. Soc., Chem. Commun.* **1991**, 1161.

[6] W. Bannwarth, A. Dorn, P. Iaiza, X. Pannekouke, *Helv. Chim. Acta* **1994**, *77*, 182.

[7] W. Bannwarth, P. Iaiza, *Helv. Chim. Acta* **1998**, *81*, 1739.

[8] R. W. Roberts, D. M. Crothers, *Science (Washington, D.C.)* **1992**, *258*, 1463.

[9] H. H. Klump, R. Loffler, *Biol. Chem. Hoppe-Seyler* **1985**, *366*, 345.

[10] R. Z. Chen, U. Pettersson, C. Beard, L. Jackson-Grusby, R. Jaenisch, *Nature (London)* **1998**, *395*, 89.

[11] U. Asseline, N. T. Thuong, *Tetrahedron Lett.* **1989**, *30*, 2521.

[12] S. Altmann, A. M. Labhardt, D. Bur, C. Lehmann, W. Bannwarth, M. Billeter, K. Wuethrich, W. Leupin, *Nucleic Acids Res.* **1995**, *23*, 4827.

[13] S. P. Adams, K. S. Kavka, E. J. Wykes, S. B. Holder, G. R. Galluppi, *J. Am. Chem. Soc.* **1983**, *105*, 661.

[14] N. D. Sinha, J. Biernat, J. McManus, H. Koester, *Nucleic Acids Res.* **1984**, *12*, 4539.

[15] W. Bannwarth, *Chimia* **1987**, *41*, 302.

[16] S. A. Scaringe, C. Francklyn, N. Usman, *Nucleic Acids Res.* **1990**, *18*, 5433.

[17] 'RNA Synthesis, Technical Note', Milligen/Biosearch MG175, Burlington, MA.

[18] L. E. Xodo, G. Manzini, F. Quadrifoglio, D. A. van der Marel, J. H. van Boom, *Nucleic Acids Res.* **1991**, *19*, 5625.

Synthesis of 5'-C- and 2'-O-(Bromoalkyl)-Substituted Ribonucleoside Phosphoramidites for the Post-synthetic Functionalization of Oligonucleotides on Solid Support

by **Xiaolin Wu** and **Stefan Pitsch***

Laboratorium für Organische Chemie, ETH-Zentrum, Universitätstr. 16, CH-8092 Zürich

The preparation of building blocks for the incorporation of 6'-O-(5-bromopentyl)-substituted β-D-allofuranosylnucleosides and 2'-O-[(3-bromopropoxy)methyl]-substituted ribonucleosides into oligonucleotide sequences is presented (*Schemes 1* and *2*). These reactive building blocks can be modified with a variety of soft nucleophiles while the (fully protected) sequence is still attached to the solid support. As an example of this strategy, we carried out some preliminary solid-phase substitution and conjugation reactions with DNA sequences containing a 2'-O-[(3-bromopropoxy)methyl]-substituted ribonucleoside (*Scheme 3*) and determined the pairing properties of duplexes obtained therefrom.

1. Introduction. – Oligonucleotide conjugates and functionalized oligonucleotides are versatile tools for structural, biological, and biophysical studies. To access a variety of such oligonucleotide analogues within a short time, several solid-phase functionalization strategies have been developed. They are based on reactive, prefunctionalized building blocks that, after their incorporation into oligonucleotides, can be functionalized with a variety of different groups. The first examples of such 'convertible nucleosides' carried reactive nucleobases that, during deprotection with alkylamines, afforded N-alkyl-substituted nucleosides [1]. Meanwhile, a few base analogues carrying suitably protected reactive side chains have been developed; after liberation of their reactive sites, these analogues can be modified, while the sequence is still fully protected and attached to the solid support [2]. In this context, we recently reported a method for the functionalization of the sugar moiety of oligonucleotides on the solid support. Our approach was based on the incorporation of a 6'-O-(bromopentyl)-substituted allofuranosyl-cytosine phosphoramidite into sequences followed by substitution of the Br-atom with a variety of soft nucleophiles [3]. Here we describe the synthesis of the analogous 6'-O-(bromopentyl)-substituted allofuranosyl phosphoramidites containing the three other canonical nucleobases uracil, adenine, and guanosine.

Meanwhile, we have extended our concept and have prepared reactive 2'-O-[(3-bromopropoxy)methyl]-substituted ribonucleoside building blocks that allow the introduction of additional functionalities at the 2'-O-position of oligonucleotides. In this first communication, we report the synthesis of such phosphoramidites containing the four canonical nucleobases and base-pairing properties of some derivatives obtained therefrom.

2. Results and Discussion. – 2.1. *Synthesis of 6′-O-(Bromopentyl)-Substituted Allofuranosyl Phosphoramidites.* The 6′-O-(bromopentyl)-substituted β-D-allofurano-syl nucleoside building blocks containing the nucleobases adenine, uracil, and guanine were prepared from the prefunctionalized sugar building block **1** in analogy to the synthesis of the corresponding cytosine-containing phosphoramidite that has been reported in [3][4] (*Scheme 1*).

The nucleosidation of **1** was carried out under *Vorbrüggen* conditions with the *in-situ* trimethylsilylated nucleobases uracil, N^6-benzoyladenine, and N^2-acetylguanine, respectively; after aqueous workup, the products were desilylated, affording the nucleosides **2**, **3**, and **4**, respectively. Nucleoside formation with uracil proceeded smoothly in the presence of $SnCl_4$ in MeCN [5], and after treatment of the crude product with HF and HCl in H_2O/MeCN according to [4], the desilylated derivative **2** was obtained in a yield of 75%. With N^6-benzoyladenine, a mixture of products was obtained under a variety of nucleosidation and deprotection conditions. Nevertheless, the nucleoside **3** could be isolated in a moderate yield of 45% by performing the nucleosidation reaction with $SnCl_4$ in MeCN [5] and the desilylation with CF_3COOH/H_2O. The nucleosidation of **1** with N^2-acetylguanine afforded, under various conditions, always a mixture of the two isomeric N^9- and N^7-connected nucleosides, which could not be separated. The best N^9/N^7 ratio was obtained with Me_3Si-OTf in $(CH_2Cl)_2$ according to [5]. After desilylation of the crude products with HF and HCl in H_2O/MeCN [4], again an unseparable mixture of isomers was obtained in a moderate yield of 40%. The dimethoxytritylation of the intermediates **2** – **4**, respectively, was carried out with $(MeO)_2Tr$-Cl in the presence of $AgNO_3$ and *sym*-collidine (=2,4,6-trimethylpyridine) according to [4][6]. Without purification, the intermediates were then directly *O*-debenzoylated under standard conditions with NaOH in THF/MeOH/H_2O [7]. The dimethoxytritylated uracil and adenine nucleosides **5** and **6** were obtained in good yields of 87 and 92%, respectively. At this stage, the two regioisomeric N^9- and N^7-connected guanine nucleosides **7** could be separated by chromatography (silica gel) and were isolated in yields of 25 and 11%, respectively (based on **1**)[1]. Introduction of the 2′-*O*-[(triisopropylsilyloxy)methyl] (=tom) protecting group into the three nucleosides was carried out under our general conditions, by first forming a cyclic dibutyltin derivative with Bu_2SnCl_2/$^iPr_2NEt$ and then treatment with tom-Cl at 80° [4][5][9]. Under these conditions, the three 2′-*O*-alkylated nucleosides **8** – **10** were obtained as major products that could be separated by chromatography (silica gel) from the corresponding 3′-*O*-alkylated regioisomers **11** – **13**[2]). Finally, the nucleosides **8** – **10** were transformed into the phosphoramidites **14** – **16** according to standard procedures [5].

---

[1])   The connection of the nucleobase was determined by ^{13}C-NMR spectroscopy according to [8].

[2])   The correct position of the tom group was determined by ^{1}H-NMR spectroscopy according to [5]. Generally, within every pair of 5′-*O*-dimethoxytritylated, base-protected nucleosides with 2′,3′-*O*-form-aldehyde acetal-derived substituents known so far, we observed the following relationships: the 2′-*O*-alkylated nucleosides are less polar and elute faster on chromatography (silica gel) than the corresponding 3′-*O*-alkylated regioisomers; the H−C(1′) signals of the former are further downfield, and the coupling constants between H−C(1′)/H−C(2′) are smaller than the corresponding values of the latter.

Scheme 1ᵃ)

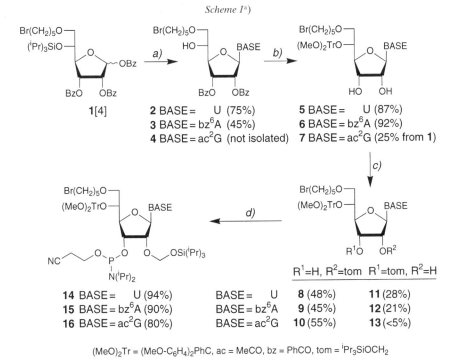

14 BASE = U (94%)   BASE = U   **8** (48%)   **11** (28%)
15 BASE = bz⁶A (90%)   BASE = bz⁶A   **9** (45%)   **12** (21%)
16 BASE = ac²G (80%)   BASE = ac²G   **10** (55%)   **13** (<5%)

(MeO)₂Tr = (MeO-C₆H₄)₂PhC, ac = MeCO, bz = PhCO, tom = ⁱPr₃SiOCH₂

a) For **2**: 1) uracil, bis(trimethylsilyl)acetamide (BSA), then SnCl₄, 2) HF, HCl in H₂O/MeCN; for **3**: *N⁶*-benzoyladenine, BSA, then SnCl₄, 2) CF₃COOH, H₂O; for **4**: *N²*-acetylguanine, BSA, then Me₃Si-OTf, 2) HF, HCl in H₂O/MeCN. *b*) (MeO)₂Tr-Cl, AgNO₃, *sym*-collidine, then NaOH in THF/MeOH/H₂O. *c*) Bu₂SnCl₂, ⁱPr₂NEt, then tom-Cl. *d*) (ⁱPr₂N)₂PCl(OCH₂CH₂CN), ⁱPr₂NEt.

─────────

a) The synthesis of the analogous *N⁴*-acetylcytosine containing phosphoramidite is reported in [3] and [4].

2.2. *Synthesis of 2'-O-[(3-Bromopropoxy)methyl]-Substituted Ribonucleosides.* The reactive, 2'-*O*-substituted ribonucleosides **25–28** were prepared by alkylation of the 5'-*O*-dimethoxytritylated, base-protected nucleosides **17–20** with 1-bromo-3-(chloromethoxy)propane (**34**) according to the conditions we developed for the introduction of related, formaldehyde acetal based 2'-*O*-protecting groups [4][5][9]. The alkylating reagent **34** was prepared in two steps. From 3-bromopropanol the *O,S*-acetal **33** was prepared with DMSO, Ac₂O, and AcOH according to [10]. This intermediate was then converted to **34** with SO₂Cl₂ according to [11]. Treatment of the nucleosides **17–20** first with Bu₂SnCl₂/ⁱPr₂NEt and then with the alkylating agent **34** afforded mixtures of the corresponding 2'-*O*- and 3'-*O*-alkylated nucleosides **21–24** and **25–28**, respectively, which could be separated easily by chromatography (silica gel)³). From the 2'-*O*-alkylated nucleosides **21–24**, the corresponding phosphorami-dites **29–32** were prepared according to standard procedures.

─────────

³) The correct position of the [(3-bromopropoxy)methyl] group was determined by ¹H-NMR spectroscopy according to [5]. See also *Footnote 2*.

Scheme 2

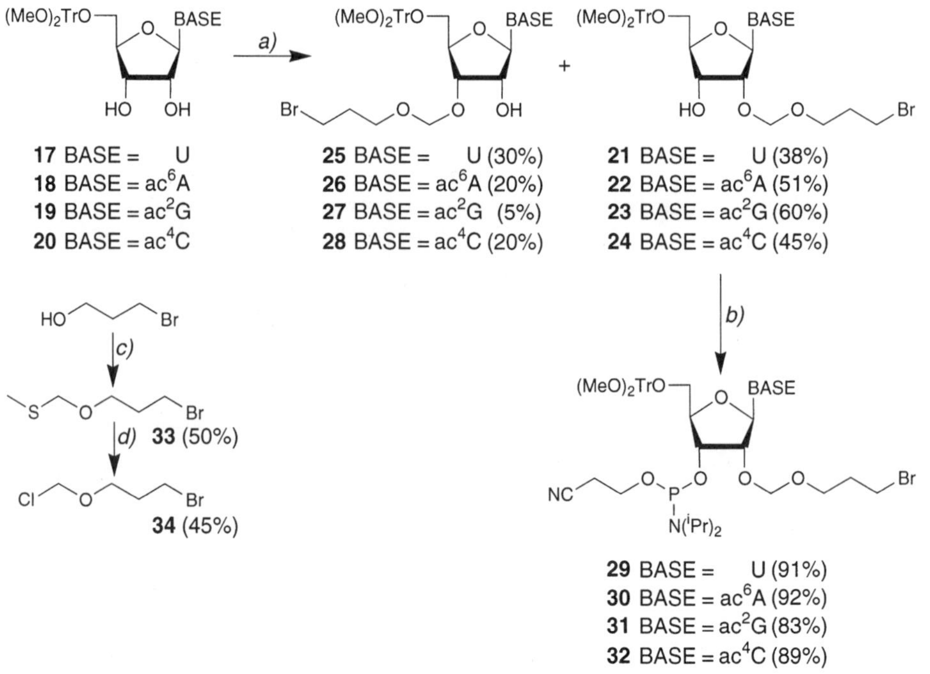

17 BASE = U
18 BASE = ac^{6}A
19 BASE = ac^{2}G
20 BASE = ac^{4}C

25 BASE = U (30%)
26 BASE = ac^{6}A (20%)
27 BASE = ac^{2}G (5%)
28 BASE = ac^{4}C (20%)

21 BASE = U (38%)
22 BASE = ac^{6}A (51%)
23 BASE = ac^{2}G (60%)
24 BASE = ac^{4}C (45%)

33 (50%)

34 (45%)

29 BASE = U (91%)
30 BASE = ac^{6}A (92%)
31 BASE = ac^{2}G (83%)
32 BASE = ac^{4}C (89%)

(MeO)$_2$Tr = (MeO-C$_6$H$_4$)$_2$PhC, ac = MeCO

a) Bu$_2$SnCl$_2$, ^{i}Pr$_2$NEt, then **34**. *b*) (^{i}Pr$_2$N)$_2$PCl(OCH$_2$CH$_2$CN), ^{i}Pr$_2$NEt. *c*) DMSO, Ac$_2$O, AcOH. *d*) SO$_2$Cl$_2$.

2.2. *Solid-Phase Substitutions.* Exploratory experiments were carried out to investigate the potential for solid-support functionalization of oligonucleotides containing 2′-*O*-[(3-bromopropoxy)methyl]substituted ribonucleosides and to collect some information about the influence of such additional substituents on the pairing behavior.

The 2′-*O*-[(3-bromopropoxy)methyl]-substituted cytidine phosphoramidite **32** was incorporated into a tetradecameric DNA sequence at two positions, once near the 3′-end (→ **35**) and once near the 5′-end (→ **36**), respectively (*Scheme 3* and *Table 1*). Under our conditions (which were developed for the assembly of RNA sequences [3][9]), the coupling yields were > 99% for the standard DNA phosphoramidites and 98% for the modified phosphoramidite **32** (*Scheme 3*). As a comparison and for comparative duplex-stability studies, the corresponding DNA sequence **37** and the complementary DNA and RNA sequences **38** and **39**, respectively, were prepared as well (*Table 1*).

Small portions of the immobilized oligonucleotides **35** and **36** were then treated with 10M MeNH$_2$ in H$_2$O/EtOH 1:1 (*Scheme 3*). In *Fig. 1*, the HPLC trace of such a crude product is presented together with the chromatogram of the analogous unmodified DNA sequence **37**. According to this analysis, from sequence **35**, a dominant main product (with essentially the same retention time as the unmodified DNA sequence **37**) was formed, together with a small by-product. The main product

Scheme 3

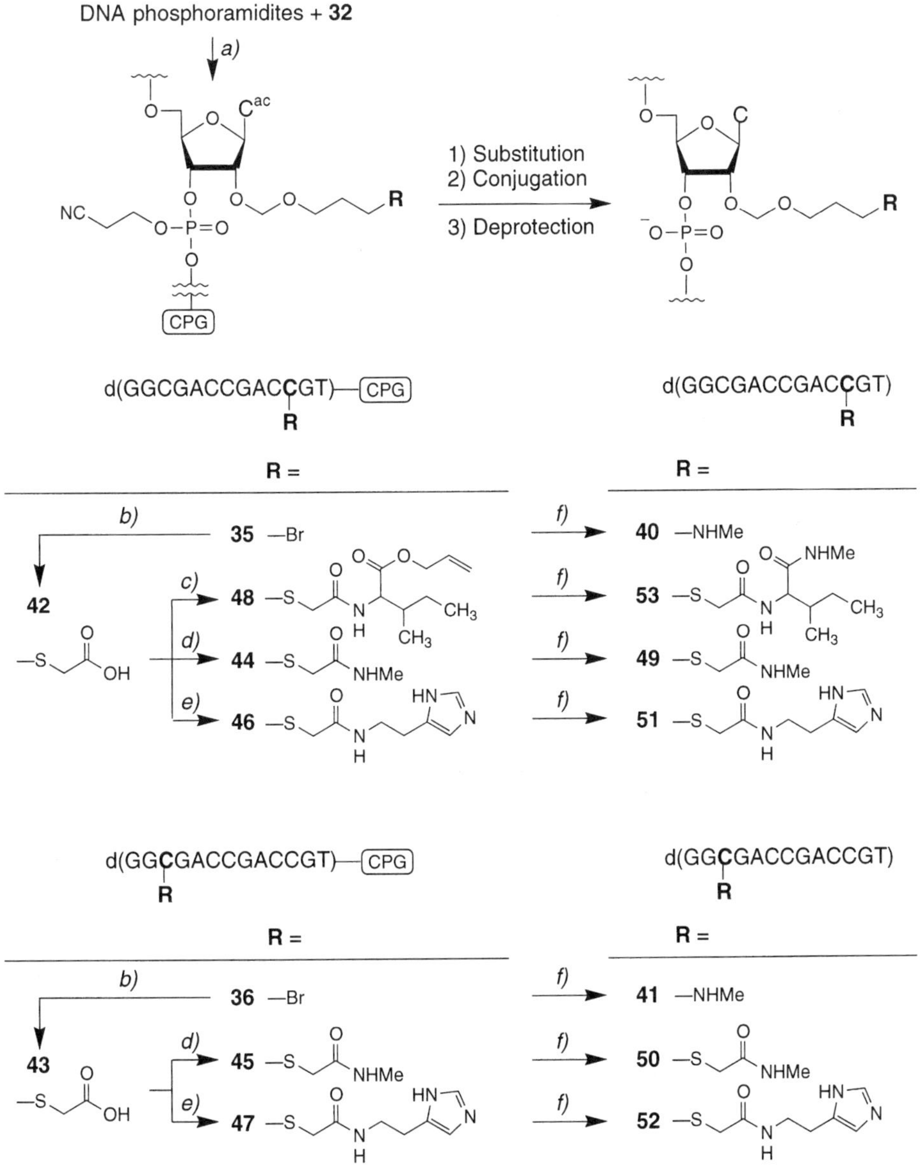

CPG = long chain alkylamino controlled pore glass

a) Oligonucleotide assembly, see *Exper. Part. b*) HSCH$_2$COOH, ^{i}Pr$_2$NEt. *c*) HOBT, TBTU, ^{i}Pr$_2$NEt, L-isoleucine allyl ester. *d*) HOBT, TBTU, ^{i}Pr$_2$NEt, MeNH$_2$. *e*) HOBT, TBTU, ^{i}Pr$_2$NEt, histamine. *f*) MeNH$_2$.

Table 1. *Characterization of Functionalized Oligonucleotides*

Sequence[a])	Product ratio[b]) [%]	Yield[c]) [%]	T_m[d]) [°]	[M − H]−[e]) calc.	obs.
40 d(GGCGACCGACWGT)	90	45	67	4078	4081
41 d(GGWGACCGACCGT)	90	40	67	4078	4080
49 d(GGCGACCGAC**X**GT)	70	40	67	4152	4153
50 d(GG**X**GACCGACCGT)	75	40	66	4152	4153
51 d(GGCGACCGAC**Y**GT)	75	35	68	4232	4234
52 d(GG**Y**GACCGACCGT)	65	30	68	4232	4235
53 d(GGCGACCGAC**Z**GT)	55	25	67	4265	4265
37 d(GGCGACCGACCGT)		60	68	3960	3959
38 d(ACGGTCGGTCGCC)		55		3952	3951
39 r(ACGGUCGGUCGCC)		30		4129	4130

[a]) For the structure of **W**, **X**, **Y**, and **Z**, see *Scheme 3*. [b]) Area-% (HPLC, capillary electrophoresis (CE)) of product signal. [c]) After purification by ion-exchange HPLC; by CE, the purity was estimated >95%. [d]) With the RNA sequence **39** as complement; conditions: c (sequences) $= 1 + 1$ μM, 150 mM NaCl, 2 mM MgCl$_2$, 10 mM *Tris* · HCl (pH 7.4). [e]) MALDI-TOF-MS; matrix, 2,4-dihydroxyacetophenone (ammonium citrate) according to [12].

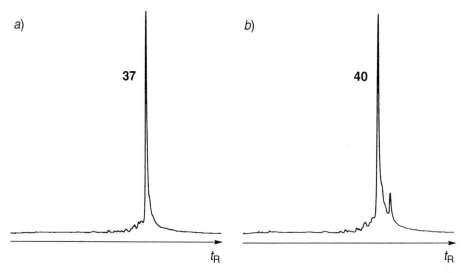

Fig. 1. *Reversed-phase HPLC traces from crude sequences:* a) *parent DNA sequence* **37**; b) *(methylamino)-substituted DNA sequence* **40**. Elution with $A \rightarrow 20\%$ B in 30 min, measured at 260 nm (see *Exper. Part*).

was isolated by prep. HPLC and identified by MALDI-TOF mass spectrometry as the (methylamino)-substituted sequence **40** (*Table 1*).

Further solid-support functionalizations were carried out according to protocols that we developed earlier [4] (*Scheme 3*). The immobilized sequences **35** and **36** were treated with thioglycolic acid (= mercaptoacetic acid) in the presence of ^{i}Pr$_2$NEt$_3$, and the resulting intermediates **42** and **43** (now containing a reactive carboxy group) were treated under peptide-coupling conditions with MeNH$_2$, histamine, and L-isoleucine allyl ester, respectively. After deprotection of the intermediates **44 – 48** with MeNH$_2$ in EtOH/H$_2$O 1:1, the main products **49 – 53** were isolated by HPLC and analyzed by

MALDI-TOF-MS (*Table 1*). According to HPLC and capillary electrophoresis (CE) of the crude products, the efficiency of the overall functionalization procedures was in the range of 55–75%. *Fig. 2* shows the CE chromatogram of the crude histamine-substituted sequence **51** (*Fig. 2,a*); this sequence was formed in a overall efficiency of *ca.* 75% and could be isolated in pure form by ion-exchange chromatography (*Fig. 2,b*).

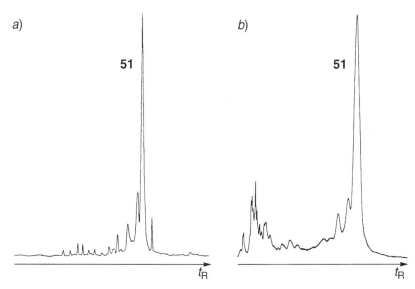

Fig. 2. a) *Capillary-electrophoresis* (CE) *trace of crude histamine-substituted sequence* **51** (conditions, see [9], measured at 260 nm); b) *ion-exchange HPLC trace of crude* **51** *obtained from a preparative run* (elution with 25% *B* → 40% *B* in 30 min, measured at 260 nm) (see *Exper. Part*).

2.4. *Thermal-Denaturation Study.* In *Table 1*, the transition temperatures (T_m) of the duplexes formed from the functionalized sequences **40**, **41**, and **49–53** and the complementary RNA sequence **39** are presented. Under physiological conditions (150 mM NaCl, 2 mM MgCl$_2$, pH 7.4), the modified duplexes had similar or slightly lower T_m values than the unmodified duplex **37·39**. A more detailed study was carried out with the two histamine-substituted sequences **51** and **52** (*Table 2*). Their duplexes with the complementary DNA sequence were significantly destabilized (as compared to the corresponding unmodified DNA duplex **37·38**). The ΔG^0 values for duplex formation of the modified DNA·RNA duplexes **51·39** and **52·39**, respectively, were more negative ($\Delta\Delta G^0 = -0.9$ and -2.8 kcal mol^{-1}, resp.) than the corresponding value of the parent DNA·RNA duplex **37·39**. In contrast to these results, the corresponding DNA·DNA duplexes **51·38** and **52·38** showed more positive ΔG^0 values than the parent DNA·DNA duplex **37·38**.

Fig. 3 illustrates the position of the additional imidazole moieties within the DNA·RNA duplexes **51·39** and **52·39**, respectively. In duplex **51·39**, the additional functional group is near the 3'-end of the modified DNA sequence and is pointing into solution, whereas in duplex **52·39**, the additional functional group is near the 5'-end and located within the minor groove of the duplex. The T_m values and the thermodynamic data reflect a higher stability of the latter duplex, indicating a positive

Table 2. *Thermodynamic Parameters of Duplex Formation*[a])

	T_m[b]) [°C]	ΔH^0 [kcal mol^{-1}]	$T\Delta S^0$ (298 K) [kcal mol^{-1}]	ΔG^0 (298 K) [kcal mol^{-1}]	$\Delta\Delta G^0$ (298 K)[c]) [kcal mol^{-1}]
37·39	68.0	−110.8	−88.0	−22.8	
51·39	68.0	−116.9	−93.2	−23.7	−0.9
52·39	67.8	−131.9	−106.3	−25.6	−2.8
37·38	65.2	−98.1	−77.6	−20.5	+0.3
51·38	63.6	−99.3	−79.1	−20.2	+0.3
52·38	62.1	−90.7	−71.7	−19.0	+1.5

[a]) In 150 mM NaCl, 2 mM MgCl$_2$, and 10 mM *Tris*·HCl (pH 7.4); thermodynamic parameters and transition temperatures were determined according to [13] (the exper. error was estimated to be ±5%). [b]) c(sequences) = 1 + 1 μM. [c]) Difference between the ΔG^0 value of the modified and the parent duplex.

interaction between the RNA strand **39** and the imidazole moiety on the modified DNA strand **52** (located in the minor groove). In the analogous duplexes with the complementary DNA strands, however, a duplex destabilization by the additional imidazole moieties was observed, and this destabilization was more important when the substituents were located in the minor groove. This difference is probably the

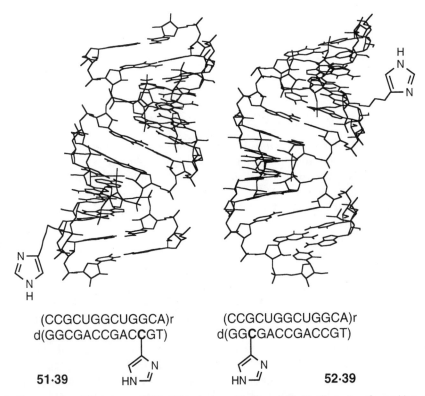

(CCGCUGGCUGGCA)r
d(GGCGACCGAC**C**GT)

51·39

(CCGCUGGCUGGCA)r
d(GG**C**GACCGACCGT)

52·39

Fig. 3. *Pictures of modified A-type DNA·RNA duplexes* **51·39** *and* **52·39**, *illustrating the position of the additional histamine substituent.* The duplex was constructed with 'MacroModel', and the substituent was added without further minimization.

consequence of a disturbance of the hydration shell within the DNA · DNA duplex **52 · 38⁴**).

In conclusion, we present novel reactive building blocks that allow the introduction of a variety of functional groups by substitution and subsequent conjugation on the solid phase according to [3]. These building blocks are fully compatible with common procedures employed for the automated assembly of DNA and RNA oligonucleotides and can be introduced by every desired position within a sequence.

This work was supported by the ETH-Zürich and the *Alfred-Werner-Stipendium*. We thank *Patrick A. Weiss* (*Xeragon AG*, Zürich) for providing us with reagents.

Experimental Part

General. Reagents and solvents from *Fluka:* $(MeO)_2$Tr-Cl, tom-Cl, nucleosides **17 – 20**, BnSTet, and tom-phosphoramidites [9] from *Xeragon AG*; standard DNA phosphoramidites (containing thymine, N^6-(phenox-yacetyl)guanine, N^2-[(4-isopropylphenoxy)acetyl]guanine, and N^4-acetylcytosine) and CPG supports from *Glen Research*; N^2-acetylguanine [15] and N^6-benzoyladenine [16] were prepared according to published procedures. Workup implies distribution of the reaction mixture between CH_2Cl_2 and sat. aq. $NaHCO_3$ soln., drying ($MgSO_4$) of the org. layer, and evaporation. Column chromatography (CC): silica gel from *Macherey & Nagel*, Al_2O_3 (act. III) from *Woelm*. TLC: precoated silica gel plates from *Macherey & Nagel*, stained by dipping into a soln. of anisaldehyde (10 ml; *Aldrich*), H_2SO_4 soln. (10 ml), and AcOH (2 ml) in EtOH (180 ml) and subsequent heating with a heat-gun. Reversed-phase HPLC: *Aquapore RP 300*, 4.6 × 220 mm (*Brownlee Labs*); eluent *A* 0.1M $(Et_3NH)OAc$ in H_2O, pH 7, and eluent *B* MeCN; flow 1 ml/min; detection at 260 nm, elution at 40°. Ion-exchange HPLC: *Mono Q HR 5/5* (*Pharmacia*); eluent *A*, 10 mM sodium phosphate in H_2O, pH 11.5, and eluent *B*, 10 mM sodium phosphate/1M NaCl in H_2O, pH 11.5; flow 1 ml/min; detection at 260 nm, elution at r.t. M.p.: uncorrected. UV Spectra: λ_{max} (ε) in nm. IR Spectra: $\tilde{v}$ in cm^{-1}. NMR Spectra: chemical shifts δ in ppm and coupling constants J in Hz. FAB-MS: positive mode; 2-nitrobenzyl alcohol (NOBA) as matrix; m/z (rel. %). MALDI-TOF-MS: according to [12]. Abbreviations: $(MeO)_2$Tr-Cl = 4,4'-dimethoxytrityl chloride, tom-Cl = (triisopropylsilyloxy)methyl chloride, BSA = bis(trimethylsilyl)acetamide, BnSTet = 1-(benzylthio)-1H-tetrazole, HOBT = 1-hydroxy-1H-benzotriazole, TBTU = O-(benzotriazole-1-yl)-N,N,N',N'-tetramethyluronium tetrafluoroborate, CPG = long-chain alkylamino controlled pore glass.

Thermal-Denaturation Studies. Absorbance *vs.* temperature profiles were recorded in fused quartz cuvettes at 260 nm on a *Cary Bio-1* spectrophotometer equipped with a *Peltier* temperature-controlling device. The samples were prepared under sterile conditions from stock solns. of the oligonucleotide, 1M *Tris* · HCl buffer (pH 7.4), 5M NaCl, and 50 mM $MgCl_2$ and subsequently degassed. A layer of silicon oil was placed on the surface of the soln. Prior to the measurements, each sample was briefly heated to 80°. The curves were obtained with both a cooling and heating ramp of 0.3°/min. The transition temperatures (T_m) were obtained after differentiation of the melting curves.

General Procedure (G.P.) for the Preparation of Phosphoramidites **14 – 16** *and* **29 – 32**. At r.t., 0.3M precursor nucleoside in CH_2Cl_2 was treated consecutively with 2.5 equiv. of $^iPr_2NEt$ and 1.2 equiv. of $(^iPr_2N)_2PCl(OCH_2CH_2CN)$. After 12 – 16 h at r.t., the mixture was subjected to CC.

1-[6'-O-(5-Bromopentyl)-2',3'-di-O-benzoyl-β-D-allofuranosyl]uracil (**2**). A suspension of **1** [4] (3.2 g, 4.0 mmol), uracil (0.5 g, 4.4 mmol), and BSA (2.5 ml, 10 mmol) in MeCN (12 ml) was stirred at 60° for 30 min. Then, $SnCl_4$ (1.7 ml, 14 mmol) was added to the clear soln., which was kept at 60° for another 20 min. After workup, the residue was dissolved in MeCN (80 ml), treated with conc. HCl soln. (0.8 ml) and 40% HF in H_2O (1.6 ml), and stirred at r.t. for 8 h. Workup and CC (silica gel, hexane/AcOEt 7:3 → 3:7) gave **2** (1.78 g, 75%). White foam. TLC (hexane/AcOEt 3:7): R_f 0.40. $[\alpha]_D^{25} = -86.0$ ($c = 1.0$, $CHCl_3$). UV (MeOH): 257 (12000), 251 (11700), 229 (29000), 216 (23300). IR ($CHCl_3$): 3619w, 3392w, 3014m, 2975w, 3939w, 1727s, 1700s, 1602m, 1453m, 1392w, 1317m, 1262s, 1222m, 1178w, 1126m, 1094m, 1070w, 1046m, 877w. ^{1}H-NMR (300 MHz, $CDCl_3$): 1.47 – 1.65 (m, $BrCH_2(CH_2)_2$); 1.82 – 1.89 (m, $Br(CH_2)_3CH_2$); 3.23 (d, $J = 1.9$, OH–C(4')); 3.39 (t, $J = 6.9$,

---

⁴) For the influence of one 2'-O-Me substituent within a DNA sequence on the minor-groove hydration, see [14].

BrCH$_2$); 3.52–3.57 (m, CH$_2$O); 3.65 (br. d, $J \approx 5.9$, 2 H–C(6')); 4.20–4.23 (m, H–C(5')); 4.37 (br. s, H–C(4')); 5.75 (dd, $J = 5.6$, 7.5, H–C(2')); 5.83 (d, $J = 8.1$, H–C(5)); 5.89 (dd, $J = 1.5$, 5.6, H–C(3')); 6.50 (d, $J = 7.5$, H–C(1')); 7.32–7.61 (m, 6 arom. H); 7.90–8.04 (m, 4 arom. H, H–C(6)); 8.18 (s, NH–C(3)). ^{13}C-NMR (75 MHz, CDCl$_3$): 24.8 (t, Br(CH$_2$)$_2$CH$_2$); 28.7 (t, BrCH$_2$CH$_2$); 32.5 (t, Br(CH$_2$)$_3$CH$_2$); 33.7 (t, BrCH$_2$); 71.0, 71.1 (2t, C(6'), CH$_2$O); 71.4, 71.8, 74.0 (3d, C(2'), C(3'), C(5')); 84.6 (d, C(4')); 86.3 (br. d, C(1')); 103.6 (d, C(5)); 123.4 (s, C(5)); 128.4, 128.5, 128.6, 129.8, 129.9 (5d, arom. C); 129.0, 133.71, 133.74 (3s, arom. C); 140.5 (d, C(6)); 150.7 (s, C(2)); 163.0 (s, C(4)); 165.3, 165.5 (2s, CO): FAB-MS: 1263 (26, [M$_2$ + H]$^+$), 633 (10, [M + H]$^+$), 631 (12, [M + H]$^+$), 521 (79), 519 (76). Anal. calc. for C$_{29}$H$_{31}$BrN$_2$O$_9$ (631.48): C 55.16, H 4.95, N 4.44; found: C 55.25, H 4.93, N 4.45.

N^6-Benzoyl-9-[2',3'-di-O-benzoyl-6'-O-(5-bromopentyl)-β-D-allofuranosyl]adenine (**3**). A suspension of **1** [4] (3.2 g, 4.0 mmol), N^6-benzoyladenine (1.14 g, 4.8 mmol), and BSA (3.5 ml, 14 mmol) in MeCN (12 ml) was stirred at 60° for 1 h. Then, SnCl$_4$ (1.9 ml, 20 mmol) was added to the clear soln., which was kept at 60° for another 20 min. After workup, the residue was dissolved in CF$_3$COOH/H$_2$O 1:1 (80 ml) and stirred at r.t. for 12 h. Workup and CC (silica gel, hexane/AcOEt 8:2 → 4:6) gave **3** (1.13 g, 45%). White foam. TLC (hexane/AcOEt 3:7): R_f 0.31. [α]$_D^{25}$ = –145.3 ($c = 1.0$, CHCl$_3$). UV (MeOH): 280 (21900), 254 (12600), 231 (32800). IR (CHCl$_3$): 3619w, 3014m, 2976w, 1730s, 1612m, 1590m, 1481w, 1457m, 1272s, 1245s, 1178w, 1123m, 1092m, 908w, 877w. ^{1}H-NMR (300 MHz, CDCl$_3$): 1.42–1.52, 1.57–1.66 (m, BrCH$_2$(CH$_2$)$_2$); 1.78–1.88 (m, Br(CH$_2$)$_3$CH$_2$); 3.33 (t, $J = 6.8$, BrCH$_2$); 3.55 (t, $J = 6.2$, CH$_2$O); 3.69–3.80 (m, 2 H–C(6')); 4.28–4.31 (m, H–C(5')); 4.67 (br. s, H–C(4')); 6.16 (dd, $J = 1.2$, 5.0, H–C(3')); 6.19 (d, $J = 2.5$, OH–C(5')); 6.35–6.43 (m, H–C(1'), H–C(2')); 7.26–7.65 (m, 6 arom. H); 7.79–8.10 (m, 4 arom. H); 8.18 (s, H–C(2)); 8.86 (s, H–C(8)); 9.10 (br. s, NH–C(6)). ^{13}C-NMR (75 MHz, CDCl$_3$): 24.9 (t, Br(CH$_2$)$_2$CH$_2$); 28.7 (t, BrCH$_2$CH$_2$); 32.6 (t, Br(CH$_2$)$_3$CH$_2$); 33.7 (t, BrCH$_2$); 70.8, 71.9, 73.7 (3d, C(2'), C(3'), C(5')); 71.1, 71.5 (2t, C(6'), CH$_2$O); 87.2, 88.2 (br. d, C(1'), C(4')); 123.4 (s, C(5)); 127.9, 128.5, 128.7, 128.9, 129.8, 133.0 (6d, arom. C); 128.3, 129.3, 133.5 (3s, arom. C); 133.7 (d, arom. C); 142.4 (d, C(8)); 150.4 (s, C(4)); 151.0 (s, C(6)); 152.6 (d, C(2)); 164.3, 164.8, 165.2 (3s, CO). FAB-MS: 761 (45, [M + H]$^+$), 760 (100, M$^+$), 759 (45, [M + H]$^+$), 758 (87, M$^+$), 519 (58), 105 (64).

1-[6'-O-(5-Bromopentyl)-5'-O-(4,4'-dimethoxytrityl)-β-D-allofuranosyl]uracil (**5**). A suspension of **2** (1.78 g, 2.8 mmol), AgNO$_3$ (476 mg, 2.8 mmol), and syn-collidine (0.93 ml, 7 mmol) in CH$_2$Cl$_2$ (10 ml) was treated with (MeO)$_2$Tr-Cl (1.42 g, 4.2 mmol) at r.t. for 1 h. After filtration and evaporation, the residue was dissolved in an ice-cold soln. of THF/MeOH/H$_2$O 5:4:1 (40 ml), treated with 10N aq. NaOH (0.8 ml), stirred at r.t. for 30 min, neutralized with AcOH (0.48 ml), and concentrated to 10 ml. Workup and CC (silica gel, CH$_2$Cl$_2$ (+2% Et$_3$N) → CH$_2$Cl$_2$/MeOH 97:3 (+2% Et$_3$N)) gave **5** (1.69 g, 87%). White foam. TLC (CH$_2$Cl$_2$/MeOH 92:8): R_f 0.45. [α]$_D^{25}$ = 31.8 ($c = 1.0$, CHCl$_3$): UV (MeOH): 269 (9500), 235 (21400), 227 (19700). IR (CHCl$_3$): 3622w, 3390w, 3029m, 1691s, 1608s, 1509m, 1461m, 1391w, 1302w, 1253s, 1226m, 1178m, 1107m, 1036s, 909w, 877w, 828w. ^{1}H-NMR (300 MHz, CDCl$_3$): 1.55–1.69 (m, BrCH$_2$(CH$_2$)$_2$); 1.90–1.94 (m, Br(CH$_2$)$_3$CH$_2$); 3.17–3.41 (m, CH$_2$O, 2 H–C(6'), OH); 3.45 (t, $J = 6.5$, BrCH$_2$); 3.60 (br. s, H–C(5')); 3.795, 3.798 (2s, MeO); 4.01 (dd, $J = 1.5$, 7.5, H–C(4')); 4.16 (dd, $J = 4.4$, 5.6, H–C(2')); 4.69 (dd, $J = 5.6$, 7.5, H–C(3')); 4.88 (d, $J = 8.1$, H–C(5)); 5.86 (d, $J = 4.4$, H–C(1')); 6.82–6.86 (m, 4 arom. H); 7.22–7.53 (m, 9 arom. H, H–C(6)). ^{13}C-NMR (75 MHz, CDCl$_3$): 27.7 (t, Br(CH$_2$)$_2$CH$_2$); 28.8 (t, BrCH$_2$CH$_2$); 32.2 (t, Br(CH$_2$)$_3$CH$_2$); 33.8 (t, BrCH$_2$); 55.3 (q, MeO); 70.7, 71.3, (2t, C(6')); 68.1, 71.5, 75.5 (3d, C(2'), C(3'), C(5')); 85.6 (d, C(4')); 87.9 (s, Ar$_2$C(Ph)); 89.2 (d, C(1')); 103.6 (d, C(5)); 113.3, 113.4 (2d, arom. C); 127.1, 127.7, 128.1, 128.4, 130.1, 130.6 (6d, arom. C); 135.6, 135.9 (2s, arom. C); 140.2 (d, C(6)); 146.3 (s, arom. C); 150.3 (s, C(2)); 158.7, 158.8 (2s, MeO–C); 163.1 (s, C(4)). FAB-MS: 747 (33), 746 (53), 727 (2, M$^+$), 725 (2, M$^+$), 303 (100).

N^6-Benzoyl-9-[6'-O-(5-bromopentyl)-5'-O-(4,4'-dimethoxytrityl)-β-D-allofuranosyl]adenine (**6**). A suspension of **3** (1.13 g, 1.5 mmol), AgNO$_3$ (255 mg, 1.5 mmol), and sym-collidine (0.5 ml, 3.8 mmol) in CH$_2$Cl$_2$ (5 ml) was treated with (MeO)$_2$Tr-Cl (0.78 g, 2.3 mmol) at r.t. for 3 h. After filtration and evaporation, the residue was dissolved in an ice-cold soln. of THF/MeOH/H$_2$O 5:4:1 (40 ml), treated with 10N aq. NaOH (0.8 ml), stirred at 4° for 15 min, neutralized with AcOH (0.48 ml), and concentrated to 10 ml. Workup and CC (silica gel, CH$_2$Cl$_2$ (+2% Et$_3$N) → CH$_2$Cl$_2$/MeOH 97:3 (+2% Et$_3$N)) gave **6** (1.13 g, 92%). Pale yellow foam. TLC (CH$_2$Cl$_2$/MeOH 95:5): R_f 0.29. [α]$_D^{25}$ = –14.0 ($c = 1.0$, CHCl$_3$). UV (MeOH): 278 (18800), 256 (11200), 232 (29100), 224 (27900). IR (CHCl$_3$): 3622w, 3113w, 2975w, 1709m, 1611s, 1585w, 1509m, 1454m, 1299w, 1251s, 1179m, 1046s, 877w, 829w. ^{1}H-NMR (300 MHz, CDCl$_3$): 1.52–1.60 (m, BrCH$_2$(CH$_2$)$_2$); 1.83–1.91 (m, Br(CH$_2$)$_3$CH$_2$); 3.09–3.13 (m, H–C(6')); 3.18–3.25 (m, CH$_2$O, H'–C(6')); 3.43 (t, $J = 6.8$, BrCH$_2$); 3.64–3.65 (m, H–C(5')); 3.75, 3.76 (2s, MeO); 4.27 (dd, $J = 3.5$, 5.0, H–C(4')); 4.67 (dd, $J = 4.6$, 5.3, H–C(2')); 4.87 (br. t, $J \approx 5.3$, H–C(3')); 5.96 (d, $J = 4.4$, H–C(1')); 6.70–6.78 (m, 4 arom. H); 7.17–7.61 (m, 12 arom. H); 7.96 (s, H–C(2)); 8.01–8.04 (m, 2 arom. H); 8.72 (s, H–C(8)); 9.08 (br. s, NH–C(6)). ^{13}C-NMR (75 MHz, CDCl$_3$): 24.8 (t, Br(CH$_2$)$_2$CH$_2$); 28.8 (t, BrCH$_2$CH$_2$); 32.5 (t, Br(CH$_2$)$_3$CH$_2$); 33.7 (t, BrCH$_2$); 55.3 (q, MeO); 70.1, 71.1 (2t, C(6')); 70.0, 72.1, 74.7

(3*d*, C(2'), C(3'), C(5)); 86.3 (*d*, C(4'), CH$_2$O); 87.2 (*s*, Ar$_2$CPh)); 89.3 (*d*, C(1')); 113.1 (*d*, arom. C); 123.0 (*s*, C(5)); 128.4 (*s*, arom. C); 126.9, 127.7, 127.9, 128.1, 128.9, 130.3, 130.4, 132.8 (8*d*, arom. C); 133.6, 136.2, 136.5 (3*s*, arom. C); 141.7 (*d*, C(8)); 145.9 (*s*, arom. C); 149.6 (*s*, C(4)); 151.2 (*s*, C(6)); 152.4 (*d*, C(2)); 158.6, 159.7 (2*s*, MeO−C); 164.3 (*s*, CO). FAB-MS: 855 (22, [*M* + H]$^+$), 854 (37, *M*$^+$), 853 (21, [*M* + H]$^+$), 852 (33, *M*$^+$), 303 (100).

N^2-*Acetyl-9-[6'-O-(5-bromopentyl)-5'-O-(4,4'-dimethoxytrityl)-β-D-allofuranosyl]guanine* (N^9-**7**) *and* N^2-*Acetyl-7-[6'-O-(5-bromopentyl)-5'-O-(4,4'-dimethoxytrityl)-β-D-allofuranosyl]guanine* (N^7-**7**). A suspension of **1** [4] (4.0 g, 5.0 mmol), N^2-acetylguanine (1.6 g, 7.5 mmol), and BSA (6 ml, 25 mmol) in (CH$_2$)Cl$_2$ (15 ml) was stirred at 80° for 1 h. Then Me$_3$Si−OSO$_2$CF$_3$ (8.1 ml, 45 mmol) was added to the clear soln., which was kept at 80° for another 4 h. After workup, the residue was dissolved in MeCN (100 ml), treated with conc. HCl soln. (1 ml) and 40% HF in H$_2$O (2 ml), and stirred at r.t. for 4 h. Workup and a filtration over silica gel gave a mixture of N^7- and N^9-nucleoside derivatives. The mixture (1.53 g, 2.0 mmol) was dissolved in CH$_2$Cl$_2$ (12 ml) and treated with *sym*-collidine (0.67 ml, 5 mmol), AgNO$_3$ (340 mg, 2 mmol), and (MeO)$_2$Tr−Cl (1.2 g, 3.6 mmol). The suspension was stirred at r.t. for 4 h. After filtration and evaporation, the residue was dissolved in an ice-cold soln. of THF/MeOH/H$_2$O 5:4:1 (50 ml), treated with 10N aq. NaOH (1 ml), stirred at 4° for 15 min, neutralized with AcOH (0.6 ml), and concentrated to 10 ml. Workup and CC (silica gel, CH$_2$Cl$_2$ (+2% Et$_3$N) → CH$_2$Cl$_2$/MeOH 97:3 (+2% Et$_3$N)) gave **42** (0.95 g, 25%) and **43** (0.56 g, 14%) as yellow foams.

Data of N^9-**7**: TLC (CH$_2$Cl$_2$/MeOH 90:10): R_f 0.32. [α]$_D^{25}$ = −26.2 (*c* = 1.0, CHCl$_3$). UV (MeOH): 275 (9300), 236 (19000), 225 (16700). IR (CHCl$_3$): 3620w, 3022w, 2937w, 1687m, 1609s, 1562m, 1509m, 1403m, 1375w, 1302m, 1252w, 1178w, 1118w, 1037s, 877w, 826w. ^{1}H-NMR (300 MHz, CDCl$_3$): 1.40−1.47 (*m*, BrCH$_2$(CH$_2$)$_2$); 1.80−1.84 (*m*, Br(CH$_2$)$_3$CH$_2$); 2.30 (*s*, Ac); 3.02−3.23 (*m*, CH$_2$O, 2 H−C(6')); 3.37 (*t*, *J* = 6.7, BrCH$_2$); 3.48−3.52 (*m*, H−C(5'), OH); 3.71, 3.73 (2*s*, MeO); 4.31−4.33 (*m*, H−C(4')); 4.51−4.54 (*m*, H−C(3')); 4.73−4.75 (*m*, H−C(2')); 5.76 (*d*, *J* = 5.6, H−C(1')); 6.70−6.76 (*m*, 4 arom. H); 7.13−7.45 (*m*, 9 arom. H); 7.58 (*s*, H−C(8)). ^{13}C-NMR (75 MHz, CDCl$_3$): 23.5 (*q*, *Me*CO); 24.8 (*t*, Br(CH$_2$)$_2$CH$_2$); 28.7 (*t*, BrCH$_2$CH$_2$); 32.5 (*t*, Br(CH$_2$)$_3$CH$_2$); 33.8 (*t*, BrCH$_2$); 55.2 (*q*, MeO); 69.7, 70.9 (2*t*, C(6'), CH$_2$O); 70.6, 72.4, 74.9 (3*d*, C(2'), C(3'), C(5')); 86.5 (*d*, C(4')); 87.2 (*s*, Ar$_2$C(Ph)); 89.1 (*d*, C(1')); 113.0 (*d*, arom. C); 120.7 (*s*, C(5)); 126.9, 127.7, 128.2, 128.3, 130.4, 130.5 (6*d*, arom. C); 136.2, 136.4 (2*s*, arom. C); 139.1 (*d*, C(8)); 145.9 (*s*, arom. C); 147.4 (*s*, C(4)); 148.2 (*s*, C(2)); 158.6, 158.7 (2*s*, MeO−C); 172.2 (*s*, CO). FAB-MS: 809 (28, [*M* + H]$^+$), 808 (63, *M*$^+$), 807 (34, [*M* + H]$^+$), 806 (54, *M*$^+$), 303 (100).

Data of N^7-**7**: TLC (CH$_2$Cl$_2$/MeOH 95:5): R_f 0.29. [α]$_D^{25}$ = 39.7 (*c* = 1.0, CHCl$_3$). UV (MeOH): 264 (14600), 255 (14300), 223 (32300). IR (CHCl$_3$): 3622w, 3152w, 3013m, 2936w, 1679s, 1609s, 1548w, 1509m, 1444w, 1373m, 1301w, 1252s, 1178m, 1116m, 1038m, 877w, 829w. ^{1}H-NMR (300 MHz, CDCl$_3$): 1.45−1.56 (*m*, BrCH$_2$(CH$_2$)$_2$); 1.83−1.87 (*m*, Br(CH$_2$)$_3$CH$_2$); 2.39 (*s*, Ac), 3.16−3.21 (*m*, CH$_2$O, H−C(6')); 3.39 (*t*, *J* = 6.7, BrCH$_2$); 3.36−3.41 (*m*, H'−C(6')); 3.64−3.69 (*m*, H−C(5')); 3.75, 3.77 (2*s*, MeO); 4.24 (br. *t*, *J* ≈ 5.0, H−C(4')); 4.31 (br. *t*, *J* ≈ 4.0, H−C(3')); 4.66 (br. *t*, *J* ≈ 5.3, H−C(2')); 6.13 (*d*, *J* = 4.3, H−C(1')); 6.76−6.81 (*m*, 4 arom. H); 7.18−7.50 (*m*, 9 arom. H); 7.85 (*s*, H−C(8)); 11.02 (br. *s*, NH−C(6)); 12.41 (br. *s*, H−N(1)). ^{13}C-NMR (75 MHz, CDCl$_3$): 24.5 (*q*, *Me*CO); 24.8 (*t*, Br(CH$_2$)$_2$CH$_2$); 28.7 (*t*, BrCH$_2$CH$_2$); 32.4 (*t*, Br(CH$_2$)$_3$CH$_2$); 33.7 (*t*, BrCH$_2$); 55.2 (*q*, MeO); 69.9, 71.0 (2*t*, C(6'), CH$_2$O); 70.1, 72.1, 77.4 (3*d*, C(2'), C(3'), C(5')); 86.2 (*d*, C(4')); 87.3 (*s*, Ar$_2$CPh); 91.5 (*d*, C(1')); 111.3 (*s*, C(5)); 113.0 (*d*, arom. C); 127.0, 127.8, 128.1, 128.3, 130.4, 130.5 (6*d*, arom. C); 136.1, 136.4 (2*s*, arom. C); 141.4 (*d*, C(8)); 145.9 (*s*, arom. C); 147.9 (*s*, C(4)); 153.4 (*s*, C(2)); 157.4 (*s*, C(6)); 158.6, 158.7 (2*s*, MeO−C); 173.5 (*s*, CO). FAB-MS: 809 (9, [*M* + H]$^+$), 808 (12, *M*$^+$), 807 (12, [*M* + H]$^+$), 806 (19, *M*$^+$), 303 (100).

1-[6'-O-(5-Bromopentyl)-5'-O-(4,4'-dimethoxytrityl)-2'-O-{[(triisopropylsilyl)oxy]methyl}-β-D-allofurano-syl]uracil (**8**) *and 1-[6'-O-(5-Bromopentyl)-5'-O-(4,4'-dimethoxytrityl)-3'-O-{[(triisopropylsilyl)oxy]methyl}-β-D-allofuranosyl]uracil* (**11**). A soln. of **5** (1.68 g, 2.4 mmol) and ^{i}Pr$_2$NEt (1.5 ml, 7 mmol) in (CH$_2$Cl$_2$) (8 ml) was treated with Bu$_2$SnCl$_2$ (729 mg, 2.4 mmol) at r.t. for 1.5 h. Then, the mixture was heated to 80°, treated with tom-Cl (700 mg, 3.1 mmol), and stirred at 75° for 15 min. Workup and CC (silica gel, hexane/AcOEt (+2% Et$_3$N) 2:8 → 7:3 (+2% Et$_3$N)) gave **8** (1.0 g, 48%) and **11** (480 mg, 23%). White foams.

Data of **8**: TLC (hexane/AcOEt 5:5): R_f 0.59. [α]$_D^{25}$ = 31.2 (*c* = 1.0, CHCl$_3$). UV (MeOH): 268 (9200), 235 (22500), 226 (20700). IR (CHCl$_3$): 3622w, 3391w, 3013m, 2945m, 1692s, 1608w, 1509m, 1462m, 1252s, 1224s, 1177w, 1038s, 996w, 881w. ^{1}H-NMR (300 MHz, CDCl$_3$): 1.05−1.11 (*m*, ^{i}Pr$_3$Si); 1.43−1.55 (*m*, BrCH$_2$(CH$_2$)$_2$); 1.77−1.87 (*m*, Br(CH$_2$)$_3$CH$_2$); 3.12−3.25 (*m*, 2 H−C(6'), CH$_2$O, OH−C(3')); 3.40 (*t*, *J* = 6.7, BrCH$_2$); 3.49−3.50 (*m*, H−C(5')); 3.80 (*s*, MeO); 4.10 (*dd*, *J* = 2.2, 4.0, H−C(4')); 4.16 (br. *t*, *J* = 5.6, H−C(3')); 4.74−4.77 (*m*, H−C(3')); 4.97, 5.17 (2*d*, *J* = 5.0, OCH$_2$O); 5.06 (*d*, *J* = 8.4, H−C(5)); 5.98 (*d*, *J* = 5.6, H−C(1')); 6.81−6.85 (*m*, 4 arom. H); 7.07−7.51 (*m*, 9 arom. H, H−C(6)); 8.05 (br. *s*, H−N(3)). ^{13}C-NMR (75 MHz, CDCl$_3$): 11.9 (*d*, Me$_2$CH)$_3$Si); 17.8 (*q*, (*Me*$_2$CH)$_3$Si); 24.8 (*t*, Br(CH$_2$)$_2$CH$_2$); 28.7 (*t*, BrCH$_2$CH$_2$); 32.5 (*t*, Br(CH$_2$)$_3$CH$_2$); 33.7

(*t*, BrCH$_2$); 55.3 (*q*, MeO); 69.7, 70.9 (2*t*, CH$_2$O, C(6')); 69.3, 72.6, 81.2 (3*d*, C(2'), C(3'), C(5')); 85.8, 86.0 (2*d*, C(1'), C(4')); 87.8 (*s*, Ar$_2$*C*(Ph)); 90.5 (*t*, OCH$_2$O); 102.3 (*d*, C(5)); 113.2, 113.3 (2*d*, arom. C); 127.1, 128.0, 130.2, 130.6 (4*d*, arom. C); 135.8, 135.9 (2*s*, arom. C); 140.2 (*d*, C(6)); 146.3 (*s*, arom. C); 150.2 (*s*, C(6)); 158.7, 158.8 (2*s*, MeO−C); 162.9 (*s*, C(4)). FAB-MS: 913 (7, *M*$^+$), 912 (4, [*M* − H]$^+$), 911 (7, *M*$^+$), 910 (4, [*M* − H]$^+$), 303 (100).

Data of **11**: TLC (hexane/AcOEt 5:5): R_f 0.48. $[\alpha]_D^{25} = -41.5$ (*c* = 1.0, CHCl$_3$). UV (MeOH): 268 (7800), 235 (21700), 227 (20200). IR (CHCl$_3$): 3620*w*, 3391*w*, 3018*m*, 2945*m*, 1716*m*, 1694*s*, 1608*w*, 1509*m*, 1462*m*, 1390*w*, 1299*w*, 1252*s*, 1226*m*, 1178*s*, 1076*m*, 1036*s*, 881*w*. ^{1}H-NMR (300 MHz, CDCl$_3$): 1.10−1.14 (*m*, ^{i}Pr$_3$Si); 1.44−1.54 (*m*, BrCH$_2$(CH$_2$)$_2$); 1.83−1.87 (*m*, Br(CH$_2$)$_3$CH$_2$); 3.10−3.20 (*m*, H−C(6'), CH$_2$O); 3.31−3.38 (*m*, H−C(5'), H'−C(6')); 3.41 (*t*, *J* = 6.8, BrCH$_2$); 3.79, 3.80 (2*s*, MeO); 3.80 (*d*, *J* = 7.5, OH−C(2')); 4.08−4.11 (*m*, H−C(4')); 4.18−4.19 (*m*, H−C(2')); 4.70 (*dd*, *J* = 2.2, 5.9, H−C(3')); 4.94, 5.20 (2*d*, *J* = 5.7, OCH$_2$O); 5.19 (*d*, *J* = 7.8, H−C(5)); 5.85 (*d*, *J* = 7.5, H−C(1')); 6.81−6.86 (*m*, 4 arom. H); 7.21−7.49 (*m*, 9 arom. H, H−C(6')); 8.44 (br. *s*, H−N(3)). ^{13}C-NMR (75 MHz, CDCl$_3$): 11.9 (*d*, (Me$_2$CH)$_3$Si); 17.8 (*q*, (*Me$_2$*CH)$_3$Si); 24.8 (*t*, Br(CH$_2$)$_2$CH$_2$); 28.9 (*t*, BrCH$_2$CH$_2$); 32.5 (*t*, Br(CH$_2$)$_3$CH$_2$); 33.6 (*t*, BrCH$_2$); 55.3 (*q*, MeO); 69.7, 70.9 (2*t*, CH$_2$O, C(6)); 71.0, 72.9, 79.2 (3*d*, C(2'), C(3'), C(5')); 84.6, 87.1 (2*d*, C(4'), C(1')); 87.8 (*s*, Ar$_2$*C*(Ph)); 90.7 (*t*, OCH$_2$O); 102.5 (*d*, C(5)); 113.2, 113.3 (2*d*, arom. C); 127.1, 128.0, 130.3, 130.5 (4*d*, arom. C); 135.8, 136.1 (2*s*, arom. C); 140.2 (*d*, C(6)); 146.2 (*s*, arom. C); 150.5 (*s*, C(6)); 158.7, 158.8 (2*s*, MeO−C); 162.8 (*s*, C(4)). FAB-MS: 913 (3, *M*$^+$), 912 (4, [*M* − H]$^+$), 911 (6, *M*$^+$), 910 (5, [*M* − H]$^+$), 303 (100).

N^6-*Benzoyl-9-[6'-O-(5-bromopentyl)-5'-O-(4,4'-dimethoxytrityl)-2'-O-{[(triisopropylsilyl)oxy]methyl}-β-D-allofuranosyl]adenine* (**9**) *and* N^6-*Benzoyl-9-[6'-O-(5-bromopentyl)-5'-O-(4,4'-dimethoxytrityl)-3'-O-{[(triisopropylsilyl)oxy]methyl}-β-D-allofuranosyl]adenine* (**12**). As described for **8/11**, with **6** (1.05 g, 1.0 mmol), ^{i}Pr$_2$NEt (0.7 ml, 4 mmol), (CH$_2$Cl)$_2$ (8 ml), Bu$_2$SnCl$_2$ (334 mg, 1.1 mmol), and tom-Cl (291 mg, 1.3 mmol). Workup and CC (silica gel, hexane/AcOEt (+2% Et$_3$N) 2:8 → 8:2 (+2% Et$_3$N)) gave **9** (0.45 g, 45%) and **12** (0.20 g, 20%) as pale yellow foams.

Data of **9**: TLC (hexane/AcOEt 4:6): R_f 0.56. $[\alpha]_D^{25} = -46.1$ (*c* = 1.0, CHCl$_3$). UV (MeOH): 278 (19400), 257 (11900), 233 (32000), 226 (31100). IR (CHCl$_3$): 3620*w*, 3016*s*, 2945*m*, 2869*w*, 1708*m*, 1611*m*, 1584*w*, 1509*m*, 1456*s*, 1391*w*, 1300*m*, 1250*s*, 1177*m*, 1048*s*, 881*w*, 862*w*. ^{1}H-NMR (300 MHz, CDCl$_3$): 0.97−1.10 (*m*, ^{i}Pr$_3$Si); 1.46−1.56 (*m*, BrCH$_2$(CH$_2$)$_2$); 1.84−1.88 (*m*, Br(CH$_2$)$_3$CH$_2$); 3.05−3.18 (*m*, H−C(6'), CH$_2$O, OH−C(3')); 3.29 (*dd*, *J* = 4.0, 10.3, H'−C(6')); 3.43 (*t*, *J* = 6.8, BrCH$_2$); 3.67 (br. *s*, H−C(5')); 3.775, 3.781 (2*s*, MeO); 4.39 (*dd*, *J* = 2.3, 5.6, H−C(4')); 4.72−4.77 (*m*, H−C(2'), H−C(3')); 4.83, 5.02 (2*d*, *J* = 4.6, OCH$_2$O); 6.07 (*d*, *J* = 6.0, H−C(1')); 6.78−6.82 (*m*, 4 arom. H); 7.21−7.61 (*m*, 12 arom. H); 7.80 (*s*, H−C(2)); 8.01−8.03 (*m*, 2 arom. H); 8.64 (*s*, H−C(8)); 9.02 (br. *s*, NH−C(6)). ^{13}C-NMR (75 MHz, CDCl$_3$): 11.8 (*d*, (Me$_2$CH)$_3$Si); 17.8 (*q*, (*Me$_2$*CH)$_3$Si); 24.9 (*t*, Br(CH$_2$)$_2$CH$_2$); 28.8 (*t*, BrCH$_2$CH$_2$); 32.6 (*t*, Br(CH$_2$)$_3$CH$_2$); 33.7 (*t*, BrCH$_2$); 55.3 (*q*, MeO); 69.1, 70.7 (2*t*, CH$_2$O, C(6)); 70.3, 72.3, 80.8 (3*d*, C(2'), C(3'), C(5')); 85.2, 86.9 (2*d*, C(1'), C(4')); 87.2 (*s*, Ar$_2$*C*(Ph)); 90.8 (*t*, OCH$_2$O); 113.1 (*d*, arom. C); 123.5 (*s*, C(5)); 126.9, 127.7, 127.8, 128.9; 130.3, 130.4, 132.8 (7*d*, arom. C); 133.7, 136.7, 136.8 (3*s*, arom. C); 142.4 (*d*, C(8)); 146.1 (*s*, arom. C); 149.5 (*s*, C(4)); 151.7 (*s*, C(6)); 152.6 (*d*, C(2)); 158.3 (*s*, MeO−C); 164.6 (*s*, CO). FAB-MS: 1040 (51, [*M* + H]$^+$), 1039 (30, *M*$^+$), 1038 (40, [*M* + H]$^+$), 1037 (35, *M*$^+$), 303 (100).

Data of **12**: TLC (hexane/AcOEt 4:6): R_f 0.40. $[\alpha]_D^{25} = -54.3$. UV (MeOH): 278 (16200), 257 (10800), 233 (30000), 226 (29400). IR (CHCl$_3$): 3621*w*, 3014*w*, 2973*w*, 2869*w*, 1710*m*, 1611*m*, 1585*w*, 1509*m*, 1456*m*, 1390*w*, 1301*m*, 1250*s*, 1178*w*, 1038*s*, 880*w*, 829*w*. ^{1}H-NMR (300 MHz, CDCl$_3$): 1.09−1.11 (*m*, ^{i}Pr$_3$Si); 1.42−1.57 (*m*, BrCH$_2$(CH$_2$)$_2$); 1.84−1.88 (*m*, BrCH$_2$)$_3$CH$_2$); 3.02−3.31 (*m*, 2 H−C(6'), CH$_2$O); 3.41 (*t*, *J* = 6.8, BrCH$_2$); 3.59−3.62 (*m*, H−C(5')); 3.77, 3.78 (2*s*, MeO); 3.99 (*d*, *J* = 6.9, OH−C(2')); 4.49−4.55 (*m*, H−C(3'), H−C(4')); 4.66−4.70 (*m*, H−C(2')); 4.95, 5.23 (2*d*, *J* = 4.6, OCH$_2$O); 5.84 (*d*, *J* = 6.9, H−C(1')); 6.78−6.84 (*m*, 4 arom. H); 7.21−7.69 (*m*, 12 arom. H); 7.82 (*s*, H−C(2)); 8.01−8.03 (*m*, 2 arom. H); 8.61 (*s*, H−C(8)); 9.02 (br. *s*, NH−C(6)). ^{13}C-NMR (75 MHz, CDCl$_3$): 11.9 (*d*, (Me$_2$CH)$_3$Si); 17.8 (*q*, (*Me$_2$*CH)$_3$Si); 24.9 (*t*, Br(CH$_2$)$_2$CH$_2$); 28.9 (*t*, BrCH$_2$CH$_2$); 32.6 (*t*, Br(CH$_2$)$_3$CH$_2$); 33.7 (*t*, BrCH$_2$); 55.3 (*q*, MeO); 68.7, 70.7 (2*t*, CH$_2$O, C(6')); 72.2, 72.6, 81.0 (3*d*, C(2'), C(3'), C(5')); 83.8 (*d*, C(4')); 87.3 (*s*, Ar$_2$*C*(Ph)); 89.0 (*d*, C(1')); 91.0 (*t*, OCH$_2$O); 113.2 (*d*, arom. C); 123.4 (*s*, C(5)); 126.9 (*s*, arom. C); 127.7, 127.8, 128.2, 128.9, 130.3, 132.7 (6*d*, arom. C); 133.7, 136.7, 136.8 (3*s*, arom. C); 142.3 (*d*, C(8)); 146.2 (*s*, arom. C); 149.4 (*s*, C(4)); 151.8 (*s*, C(6)); 152.5 (*d*, C(2)); 158.7 (*s*, MeO−C); 164.3, 170.8 (2*s*, CO). FAB-MS: 1040 (57, [*M* + H]$^+$), 1039 (39, *M*$^+$), 1038 (44, [*M* + H]$^+$), 1037 (42, *M*$^+$), 303 (100).

N^2-*Acetyl-9-[6'-O-(5-bromopentyl)-5'-O-(4,4'-dimethoxytrityl)-2'-O-{[(triisopropylsilyl)oxy]methyl}-β-D-allofuranosyl]guanine* (**10**). As described for **8/11**, with **7** (0.45 g, 0.58 mmol), ^{i}Pr$_2$NEt (0.4 ml, 2.3 mmol), (CH$_2$Cl)$_2$ (2 ml), Bu$_2$SnCl$_2$ (176 mg, 0.58 mmol), and tom-Cl (130 mg, 0.58 mmol). Workup and CC (silica gel, hexane/AcOEt (+2% Et$_3$N) 3:7 → AcOEt (+2% Et$_3$N)) gave **10** (0.31 g, 55%). Pale yellow foam. TLC

(AcOEt): R_f 0.24. $[\alpha]_D^{25} = -15.5$ ($c = 1.0$, CHCl$_3$). UV (MeOH): 282 (10900), 276 (11600), 260 (13200), 246 (11600), 231 (26100). IR (CHCl$_3$): 3215w, 3027m, 2945w, 2868m, 1704s, 1609m, 1559m, 1509m, 1464m, 1418s, 1374w, 1298m, 1251s, 1176w, 1118s, 1037m, 996w, 831w. ^{1}H-NMR (300 MHz, CDCl$_3$): 1.01–1.07 (m, ^{i}Pr$_3$Si); 1.42–1.51 (m, BrCH$_2$(CH_2)$_2$); 1.82–1.84 (m, Br(CH$_2$)$_3$$CH_2$); 2.12 ($s$, Ac); 2.95 ($dd, J = 3.1$, 10.3, H–C(6')); 3.13–3.17 (m, CH$_2$O, OH–C(3')); 3.24 ($dd, J = 4.3$, 10.3, H'–C(6')); 3.39 ($t, J = 6.8$, BrCH$_2$); 3.60–3.62 (m, H–C(5')); 3.77, 3.78 (2s, MeO); 4.23 ($dd, J = 3.1$, 3.4, H–C(4')); 4.51–4.55 (m, H–C(2')); 4.82–4.83 (m, H–C(3')); 4.89, 5.07 (2$d, J = 5.0$, OCH$_2$O); 5.85 ($d, J = 6.9$, H–C(1')); 6.74–6.79 (m, 4 arom. H); 7.20–7.54 (m, 9 arom. H, H–C(8)); 8.43 (br. s, NH–C(6)); 11.95 (s, H–N(1)). ^{13}C-NMR (75 MHz, CDCl$_3$): 11.8 (d, (Me$_2$CH)$_3$Si); 17.8 (q, (Me_2CH)$_3$Si); 24.4 (q, MeCO); 24.8 (t, Br(CH$_2$)$_2$CH$_2$); 28.7 (t, BrCH$_2$CH$_2$); 32.5 (t, Br(CH$_2$)$_3$CH$_2$); 33.8 (t, BrCH$_2$); 55.3 (q, MeO); 69.9, 70.8 (2t, CH$_2$O, C(6)); 72.5, 77.2, 81.5 (3d, C(2'), C(3'), C(5')); 85.2, 86.1 (2d, C(1'), C(4')); 87.2 (s, Ar$_2$C(Ph)); 90.9 (t, OCH$_2$O); 113.2, 113.3 (2d, arom. C); 121.6 (s, C(5)); 127.1, 127.9, 128.2, 128.4, 130.4, 130.5, (6d, arom. C); 136.3, 136.4 (2s, arom. C); 137.5 (d, C(8)); 146.1 (s, arom. C); 147.0 (s, C(4)); 148.5 (s, C(2)); 155.7 (s, C(6)); 158.7 (s, MeO–C); 171.2 (s, CO). FAB-MS: 994 (3, M^+), 992 (3, M^+), 693 (20), 691 (17), 303 (100).

1'-[6'-O-(5-Bromopentyl)-5'-O-(4,4'-dimethoxytrityl)-2'-O-[[(triisopropylsilyl)oxy]methyl]-β-D-allofurano-syl]uracil 3'-(2-Cyanoethyl Diisopropylphosphoramidite) (**14**). According to the *G.P.*, with **8** (170 mg, 0.17 mmol). CC (Al$_2$O$_3$, hexane/AcOEt 6:4 → 3:7) gave **14** (184 mg, 94%; 1:1 mixture of diastereoisomers). White foam. TLC (hexane/AcOEt 6:4): R_f 0.39. UV (MeCN): 269 (9700), 238 (23100), 226 (19100). IR (CHCl$_3$): 3408w, 3019w, 2965m, 2868w, 2360w, 1719m, 1694s, 1608w, 1509m, 1462m, 1386w, 1302w, 1252m, 1218m, 1178m, 1117w, 1036m, 980w, 882w, 828w. ^{1}H-NMR (300 MHz, CDCl$_3$): 0.95–1.05 (m, ^{i}Pr$_3$Si); 1.10–1.22 (m, (Me$_2$CH)$_2$N); 1.37–1.42 (m, BrCH$_2$(CH_2)$_2$); 1.70–1.81 (m, Br(CH$_2$)$_2$$CH_2$); 2.53 ($t, J = 6.7$, 0.5 H, OCH$_2$$CH_2$CN); 2.54 ($t, J = 6.2$, 0.5 H, OCH$_2$$CH_2$CN); 2.62 ($t, J = 6.2$, 1 H, OCH$_2$$CH_2$CN); 2.97–3.11 ($m$, CH$_2$O); 3.35–3.67 ($m$, BrCH$_2$, H–C(5'), 2 H–(6'), (Me$_2$CH)$_2$N); 3.80 ($s$, MeO); 4.30–4.36 ($m$, H–C(2'), H–C(4')); 4.68–4.81 (m, H–C(3')); 4.87, 4.92, 4.94, 4.97 (4$d, J = 5.0$, OCH$_2$O); 5.23 ($d, J = 8.1$, H–C(5)); 6.00 ($d, J = 5.8$, 0.5 H, H–C(1')); 6.01 ($d, J = 7.5$, 0.5 H, H–C(1')); 6.81–6.85 (m, 4 arom. H); 7.06–7.49 (m, 9 arom. H, H–C(6)); 8.21 (br. s, H–N(3)). ^{13}C-NMR (75 MHz, CDCl$_3$): 12.0 (d, (Me$_2$CH)$_3$Si); 17.8 (q, (Me_2CH)$_3$Si); 20.2, 20.4, (2$t, J = $(C,P) = 6.1, OCH$_2$$CH_2$CN); 23.5 ($t$, Br(CH$_2$)$_2CH_2$); 24.48, 24.50, 24.63, 24.69, 24.73, 24.74 (6q, Me$_2$CHN); 28.7, 28.8 (2t, BrCH$_2$CH$_2$); 32.4, 32.6 (2t, Br(CH$_2$)$_3$CH$_2$); 33.6, 33.7 (2t, BrCH$_2$); 43.4, 43.6 (2d, J(C,P) = 4.9, Me$_2$CHN); 55.4 (q, MeO); 57.7 (t, (C,P) = 18.3, OCH$_2$$CH_2$CN); 58.7 ($t, J$(C,P) = 15.8, OCH$_2$$CH_2$CN); 68.8, 69.2, 71.6, 71.7 (4t, CH$_2$O, C(6')); 70.7, 70.8, 70.9, 72.5 (4d, C(2'), C(3'), C(5')); 84.6, 85.6, 86.4, 87.0 (4d, C(1'), C(4')); 87.77, 87.81 (2s, Ar$_2$C(Ph)); 89.3, 89.7 (2t, OCH$_2$O); 102.5 (d, C(5)); 113.4, 113.5 (2d, arom. C); 117.8, 118.0 (2s, CN); 127.3, 128.1, 128.5, 130.5, 130.8, 130.9 (6d, arom. C); 136.28, 136.32, 136.42 (3s, arom. C); 141.0 (d, C(6)); 146.2, 146.3 (2s, arom. C); 150.3 (s, C(4)); 162.9 (2s, C(6)); 159.0, 159.1 (2s, MeO–C). ^{31}P-NMR (121 MHz, CDCl$_3$): 150.7, 151.4. FAB-MS: 1114 (6, [M+H]$^+$), 1113 (10, M^+), 1112 (7, [M+H]$^+$), 1111 (11, M^+), 303 (100).

N^6-Benzoyl-9-[6'-O-(5-bromopentyl)-5'-O-(4,4'-dimethoxytrityl)-2'-O-[[(triisopropylsilyl)oxy]methyl]-β-D-allofuranosyl]adenine 3'-(2-Cyanoethyl Diisopropylphosphoramidite) (**15**). According to the *G.P.* with **9** (560 mg, 0.17 mmol). CC (Al$_2$O$_3$, hexane/AcOEt 8:2 → 6:4) gave **15** (605 mg, 90%; a 1:1 mixture of diastereoisomers). Pale yellow foam. TLC (hexane/AcOEt 5:5): R_f 0.52. UV (MeCN): 277 (20400), 258 (14200), 234 (33300), 224 (31400). IR (CHCl$_3$): 3406w, 3067w, 2945m, 2867w, 1709m, 1611s, 1545m, 1509m, 1457s, 1365w, 1300w, 1250w, 1179m, 1120m, 1082m, 980w, 932w, 828w. ^{1}H-NMR (300 MHz, CDCl$_3$): 0.75–0.81 (m, ^{i}Pr$_3$Si); 1.21–1.27 (m, (Me$_2$CH)$_2$N); 1.45–1.50 (m, BrCH$_2$(CH_2)$_2$); 1.72–1.84 (m, Br(CH$_2$)$_3$$CH_2$); 2.53–2.57 ($m$, OCH$_2$$CH_2$CN); 2.94–2.97 ($m$, 1 H, OCH$_2$$CH_2$CN); 3.05–3.11 ($m$, CH$_2$O); 3.25–3.28 ($m$, 1 H, OCH$_2$$CH_2$CN); 3.36–3.42 ($m$, 1 H, BrCH$_2$); 3.48–3.55 ($m$, 1 H, BrCH$_2$); 3.66–3.74 ($m$, 2 H–C(6'), (Me$_2$CH)$_2$N); 3.78 ($s$, MeO); 3.81–3.87 ($m$, H–C(5')); 4.60–4.94 ($m$, H–C(2'), H–C(3'), H–C(4'), OCH$_2$O); 5.98 ($d, J = 7.4$, 0.5 H, H–C(1')); 6.02 ($d, J = 7.4$, 0.5 H, H–C(1')); 6.81–6.85 (m, 4 arom. H); 7.21–7.61 (m, 12 arom. H); 7.89, 7.91 (2s, H–C(2)); 8.00–8.02 (m, 2 arom. H); 8.58, 8.61 (2s, H–C(8)); 9.02 (br. s, NH–C(4)). ^{13}C-NMR (75 MHz, CDCl$_3$): 11.9 (d, (Me$_2$CH)$_3$Si); 17.5 (q, (Me_2CH)$_3$Si); 20.2, 20.3 (2t, J(C,P) = 6.8, CH$_2$CN); 23.6 (t, Br(CH$_2$)$_2$CH$_2$); 24.51, 24.55, 24.61, 24.70, 24.72 (5q, Me$_2$CHN); 28.90, 28.96 (2t, BrCH$_2$CH$_2$); 32.4, 32.5 (2t, Br(CH$_2$)$_3$CH$_2$); 33.8, 34.0 (2t, BrCH$_2$); 43.4 ($d, J = $(C,P) = 12.1, Me$_2$CHN); 45.1 ($d, J$(C,P) = 4.9, Me$_2$CHN); 55.3 ($q$, MeO); 57.9, 58.4 (2$t, J$(C,P) = 17.7, OCH$_2$$CH_2$CN); 67.8, 68.0, 70.5, 70.6 (4t, CH$_2$O, C(6')); 72.0, 72.3, 72.8, 73.0, 75.9, 76.4 (6d, C(2'), C(3'), C(5')); 84.0, 84.7, 87.6, 88.0 (4d, C(1'), C(4')); 87.0 (s, Ar$_2$C(Ph)); 89.1, 89.5 (2t, OCH$_2$O); 113.0 (d, arom. C); 117.7, 117.8 (2s, CN); 123.6 (d, C(5)); 127.0, 127.8, 128.6, 128.9, 130.2, 130.4, 130.5, 133.8 (8d, arom. C); 132.8, 136.8, 136.9, 137.0, 137.1 (5s, arom. C); 143.2, 143.3 (2d, C(8)); 145.8 (s, arom. C); 149.4 (s, C(4)); 151.3, 151.5 (2s, C(6)); 152.6 (d, C(2)); 158.6 (s, MeO–C); 164.4

(*s*, CO). ^{31}P-NMR (121 MHz, CDCl$_3$): 150.5, 151.4. FAB-MS: 1240 (5, $[M + H]^+$), 1239 (7, M^+), 1238 (6, $[M +$ H]$^+$), 1237 (8, M^+), 303 (100).

N²-Acetyl-9-[6'-O-(5-bromopentyl)-5'-O-(4,4'-dimethoxytrityl)-2'-O-[[(triisopropylsilyl)oxy]methyl}-α-D-allofuranosyl]guanine 3'-(2-Cyanoethyl Diisopropylphosphoramidite) (**16**). According to the *G.P.*, with **10** (170 mg, 0.17 mmol). CC (Al$_2$O$_3$, hexane/AcOEt 6:4 to 3:7) gave **16** (164 mg, 80%; 1:1 mixture of diastereoisomers). Pale yellow foam. TLC (hexane/AcOEt 2:8): R_f 0.50. UV (MeCN): 276 (13500), 270 (13000), 239 (24500), 228 (22200). IR (CHCl$_3$): 3213w, 3012m, 2945m, 2868w, 2361w, 1695s, 1609s, 1559w, 1509m, 1464m, 1403w, 1371m, 1301m, 1252s, 1179m, 1127m, 1036s, 981w, 882w, 828w. ^{1}H-NMR (300 MHz, CDCl$_3$): 0.90–1.07 (*m*, ^{i}Pr$_3$Si); 1.11–1.42 (*m*, (Me$_2$CH)$_2$N, BrCH$_2$(CH$_2$)$_2$); 1.72–1.79 (*m*, Br(CH$_2$)$_3$CH$_2$); 2.15, 2.17 (2s, MeCO); 2.74–2.78 (*m*, OCH$_2$CH$_2$CN); 2.98–3.15 (*m*, CH$_2$O); 3.35–3.61 (*m*, OCH$_2$CH$_2$CN, BrCH$_2$, 2 H–C(6'), (Me$_2$CH)$_2$N); 3.763, 3.767, 3.774, 3.778 (4s, MeO); 3.88–3.91 (*m*, H–C(5')); 4.22–4.42 (*m*, H–C(2'), H–C(4')); 4.62–4.92 (*m*, H–C(3'), OCH$_2$O); 5.80 (*d*, *J* = 5.3, 0.5 H, H–C(1')); 5.96 (*d*, *J* = 7.5, 0.5 H, H–C(1')); 6.74–6.84 (*m*, 4 arom. H); 7.21–7.53 (*m*, 9 arom. H); 9.03–9.11 (*m*, NH–C(2)); 11.96–12.07 (*m*, H–N(1)). ^{13}C-NMR (75 MHz, CDCl$_3$): 11.9 (*d*, (Me$_2$CH)$_3$Si); 17.9 (*q*, (Me$_2$CH)$_3$Si); 20.3, 20.4 (2t, *J*(C,P) = 7.2, OCH$_2$CH$_2$CN); 24.3 (*q*, MeCO); 24.8 (*t*, Br(CH$_2$)$_2$CH$_2$); 24.46, 24.51, 24.59, 24.64 (4q, Me$_2$CHN); 28.6, 28.7 (2t, BrCH$_2$CH$_2$); 32.4, 32.6 (2t, Br(CH$_2$)$_3$CH$_2$); 33.6, 33.7 (2t, BrCH$_2$); 43.4, 43.5 (2d, *J*(C,P) = 11.1, Me$_2$CHN); 55.3 (*q*, MeO); 57.8 (*t*, *J*(C,P) = 8.7, OCH$_2$CH$_2$CN); 58.7 (*t*, *J*(C,P) = 18.8, OCH$_2$CH$_2$CN); 68.7, 69.2, 70.9, 71.2 (4t, CH$_2$O, C(6')); 74.3, 74.5, 77.2, 78.1, 87.4, 87.9 (6d, C(2'), C(3'), C(5')); 84.9, 85.3 (2d, C(4')); 87.4, 87.9 (2d, C(1')); 87.3 (*s*, Ar$_2$C(Ph)); 90.4, 90.6 (2t, OCH$_2$O); 113.3 (*d*, arom. C); 117.7, 117.8 (2s, CN); 122.2 (*d*, C(5)); 127.3, 127.4, 128.1, 128.4, 128.5, 128.7, 130.4 (7d, arom. C); 135.4, 135.5, 136.1 (3s, arom. C); 137.8, 138.4 (2d, C(8)); 145.5, 146.1 (2s, arom. C); 147.7, 147.8 (2s, C(4)); 148.7, 148.9 (2s, C(2)); 155.2, 155.3 (2s, C(6)); 158.8, 158.9 (2s, MeO–C); 170.1 (s, CO). ^{31}P-NMR (121 MHz, CDCl$_3$): 150.4, 150.0. FAB-MS: 1195 (4, $[M + H]^+$), 1194 (6, M^+), 1193 (4, $[M + H]^+$), 1192 (5, M^+), 303 (100).

1-Bromo-3-[(methylthio)methoxy]propane (**33**). A soln. of 3-bromopropanol (8.8 g, 0.1 mol), DMSO (108 ml, 1.8 mol), Ac$_2$O (94 ml, 1.0 mol), and AcOH (68 ml, 1.2 mol) was kept at r.t. for 7 days. Extraction (hexane/sat. NaHCO$_3$ soln.) and distillation (75°/10 Torr) gave **33** (10 g, 50%). Yellow liquid. ^{1}H-NMR (300 MHz): 1.98–2.03 (*m*, CH$_2$); 2.14 (*s*, MeS); 3.37 (*t*, *J* = 6.5, BrCH$_2$); 3.51 (*t*, *J* = 5.4, CH$_2$O); 4.63 (*s*, OCH$_2$S). ^{13}C-NMR (75 MHz): 29.9 (*t*, CH$_2$); 31.6 (*q*, MeS); 32.2 (*t*, BrCH$_2$); 66.5 (*t*, CH$_2$O); 74.4 (*t*, OCH$_2$S).

1-Bromo-3-(chloromethoxy)propane (**34**). SO$_2$Cl$_2$ (4.3 ml, 0.1 mol) was added dropwise to a soln. of **33** (10 g, 0.05 mol) in CH$_2$Cl$_2$ (240 ml) at 0°. The soln. was stirred at r.t. for 1 h. Evaporation and distillation (53°/0.3 Torr) gave **34** (4.7 g, 42%). Yellow liquid. ^{1}H-NMR (300 MHz): 2.02–2.10 (*m*, CH$_2$); 3.43 (*t*, *J* = 6.4, BrCH$_2$); 3.61 (*t*, *J* = 5.0, CH$_2$O); 4.63 (*s*, ClCH$_2$O). ^{13}C-NMR (75 MHz): 30.1 (*t*, CH$_2$); 32.7 (*t*, BrCH$_2$); 66.5 (*t*, CH$_2$O); 82.0 (*t*, ClCH$_2$O).

2'-O-[(3-Bromopropoxy)methyl]-5'-O-(4,4'-dimethoxytrityl)uridine (**21**) *and* *3'-O-[(3-Bromopropoxy)-methyl]-5'-O-(4,4'-dimethoxytrityl)uridine* (**25**). A soln. of **17** (1.1 g, 2.0 mmol) and ^{i}Pr$_2$NEt (1.4 ml, 8 mmol) in (CH$_2$Cl)$_2$ (6 ml) was treated with Bu$_2$SnCl$_2$ (608 mg, 2.0 mmol) at r.t. for 1 h. Then the mixture was heated to 70°, treated with **34** (488 mg, 2.6 mmol), and stirred at 70° for 15 min. Workup and CC (silica gel, hexane/AcOEt (+2% Et$_3$N) 5:5 → AcOEt (+2% Et$_3$N)) gave **21** (530 mg, 38%) and **25** (418 mg, 30%) as pale yellow foams.

Data of **21**: TLC (hexane/AcOEt 7:3): R_f 0.55. $[α]_D^{25} = 17.1$ (*c* = 1.0, CHCl$_3$). UV (MeOH): 264 (9700), 256 (9500), 233 (22900), 227 (22300). IR (CHCl$_3$): 3391w, 3018m, 2959w, 1691s, 1608w, 1510m, 1461w, 1390w, 1299w, 1276w, 1253m, 1221m, 1178m, 1102w, 1036w, 909w, 830w. ^{1}H-NMR: 2.07–2.15 (*m*, CH$_2$); 2.67 (br. *s*, OH–C(3')); 3.50 (*t*, *J* = 6.4, BrCH$_2$); 3.54–3.78 (*m*, CH$_2$O, 2 H–C(5')); 3.80 (*s*, 2 MeO); 4.07–4.09 (*m*, H–C(4')); 4.26 (*dd*, *J* = 2.8, 5.3, H–C(2')); 4.47–4.49 (*m*, H–C(3')); 4.90, 5.00 (2d, *J* = 6.9, OCH$_2$O); 5.29 (*d*, *J* = 8.1, H–C(5)); 6.01 (*d*, *J* = 2.8, H–C(1')); 6.82–6.87 (*m*, 4 arom. H); 7.22–7.40 (*m*, 9 arom. H); 7.96 (*d*, *J* = 8.1, H–C(6)). ^{13}C-NMR (75 MHz, CDCl$_3$): 30.1 (*t*, CH$_2$); 32.4 (*t*, BrCH$_2$); 55.3 (*q*, MeO); 62.1, 66.1 (2t, CH$_2$O, C(5')); 74.6, 75.4, 81.9 (3d, C(2'), C(3'), C(4')); 87.1 (*s*, Ar$_2$C(Ph)); 89.6 (*d*, C(1')); 95.5 (*t*, OCH$_2$O); 102.5 (*d*, C(5)); 113.3 (*d*, arom. C); 127.3, 128.1, 128.2, 128.4, 130.1 (5d, arom. C); 135.1, 135.2 (2s, arom. C); 140.0 (*d*, C(6)); 144.2 (*s*, arom. C); 150.7 (*s*, C(2)); 158.8 (*s*, MeO–C); 163.2 (*s*, C(4)). FAB-MS: 699 (6, $[M + H]^+$), 698 (11, M^+), 697 (7, $[M + H]^+$), 696 (12, M^+), 303 (100).

Data of **25**: TLC (hexane/AcOEt 3:7): R_f 0.43. $[α]_D^{25} = 27.1$ (*c* = 1.0, CHCl$_3$). UV (MeOH): 264 (9700), 257 (9400), 234 (22100), 227 (21300). IR (CHCl$_3$): 3556w, 3390w, 2959w, 1690s, 1608w, 1510m, 1461m, 1391w, 1300w, 1253s, 1226m, 1177m, 1101m, 1036w, 909w, 830w. ^{1}H-NMR: 2.02–2.06 (*m*, CH$_2$); 3.37–3.70 (*m*, OH–C(2'), BrCH$_2$, CH$_2$O, 2 H–C(5')); 3.80 (*s*, 2 MeO); 4.25–4.32 (*m*, H–C(2'), H–C(3'), H–C(4')); 4.74, 4.78 (2d, *J* = 6.8, OCH$_2$O); 5.37 (*d*, *J* = 8.4, H–C(5)); 5.95 (*d*, *J* = 3.4, H–C(1')); 6.83–6.86 (*m*, 4 arom. H); 7.24–7.39 (*m*, 9 arom. H); 7.83 (*d*, *J* = 8.1, H–C(6)); 8.91 (br. *s*, H–N(3)). ^{13}C-NMR (75 MHz, CDCl$_3$): 30.3 (*t*, CH$_2$); 32.4 (*t*, BrCH$_2$); 55.3 (*q*, MeO); 61.7, 66.4 (2t, CH$_2$O, C(5')); 69.0, 80.0, 83.5 (3d, C(2'), C(3'), C(4')); 87.2

(s, Ar$_2$C(Ph)); 87.7 (d, C(1′)); 95.3 (t, OCH$_2$O); 102.3 (d, C(5)); 113.3 (d, arom. C), 127.2, 128.1, 128.2, 130.1, 130.2 (5d, arom. C); 128.4, 135.1, 135.2 (3s, arom. C); 140.0 (d, C(6)); 144.3 (s, arom. C); 150.2 (s, C(2)); 158.7, 158.8 (2s, MeO−C); 163.3 (s, C(4)). FAB-MS (NOBA, pos. mode): 699 (12, $[M+H]^+$), 698 (9, M^+), 697 (8, $[M+H]^+$), 696 (14, M^+), 303 (100).

N^6-*Acetyl-2′-O-[(3-bromopropoxy)methyl]-5′-O-(4,4′-dimethoxytrityl)adenosine* (**22**) *and* N^6-*Acetyl-3′-O-[(3-bromopropoxy)methyl]-5′-O-(4,4′-dimethoxytrityl)adenosine* (**26**). As described for **21/25**, with **18** (1.5 g, 2.5 mmol), ^{i}Pr$_2$NEt (1.7 ml, 10 mmol), (CH$_2$Cl)$_2$ (10 ml), Bu$_2$SnCl$_2$ (0.75 g, 2.5 mmol), and **34** (630 mg, 3.6 mmol). Workup and CC (silica gel, hexane/AcOEt (+2% Et$_3$N) 8:2 → AcOEt/EtOH 9:1 (+2% Et$_3$N)) gave **22** (941 mg, 51%) and **26** (370 mg, 20%) as pale yellow foams.

Data of **22**: TLC (AcOEt/EtOH 9:1): R_f 0.54. $[\alpha]_D^{25} = 1.3$ ($c = 1.0$, CHCl$_3$). UV (MeOH): 272 (17900), 253 (12500), 234 (18400). IR (CHCl$_3$): 3374w, 3028w, 2934w, 1728w, 1703m, 1608s, 1588m, 1509m, 1463w, 1375w, 1297m, 1251m, 1233m, 1208s, 1177m, 1098m, 1037m, 913w, 829w. ^{1}H-NMR: 1.80–1.98 (m, CH$_2$); 2.16 (s, MeCO); 2.78 (d, $J = 5.0$, OH−C(3′)); 3.37 (t, $J = 6.5$, BrCH$_2$); 3.41–3.67 (m, CH$_2$O, 2 H−C(5′)); 3.78 (s, MeO); 4.25–4.28 (m, H−C(4′)); 4.53–4.56 (m, H−C(3′)); 4.83 (s, OCH$_2$O); 4.94 (dd, $J = 5.0$, 5.4, H−C(2′)); 6.23 (d, $J = 5.4$, H−C(1′)); 6.78–6.83 (m, 4 arom. H); 7.19–7.44 (m, 9 arom. H); 8.18 (s, H−C(2)); 8.61 (s, H−C(8)); 8.66 (br. s, NH−C(6)). ^{13}C-NMR (75 MHz, CDCl$_3$): 25.6 (q, MeCO); 29.9 (t, CH$_2$); 32.2 (t, BrCH$_2$); 55.3 (q, MeO); 63.1, 66.3 (2t, CH$_2$O, C(5′)); 78.2, 80.1, 84.1 (3d, C(2′), C(3′), C(4′)); 87.1 (s, Ar$_2$C(Ph)); 89.5 (d, C(1′)); 95.9 (t, OCH$_2$O); 113.4 (d, arom. C); 127.0, 127.9, 128.1, 130.1 (4d, arom. C); 128.4, 135.5, 135.6 (3s, arom. C); 141.6 (d, C(8)); 144.4 (s, arom. C); 149.4 (s, C(4)); 151.0 (s, C(6)); 152.5 (d, C(2)); 158.6 (s, MeO−C). FAB-MS: 765 (47, $[M+H]^+$), 764 (100, M^+), 763 (59, $[M+H]^+$), 762 (94, M^+), 303 (51).

Data of **26**: TLC (AcOEt/EtOH 9:1): R_f 0.43. $[\alpha]_D^{25} = -5.7$ ($c = 1.0$, CHCl$_3$). UV (MeOH): 271 (18200), 253 (13500), 234 (24700), 229 (24300). IR (CHCl$_3$): 3373w, 3016m, 2935w, 1729w, 1704m, 1609s, 1588m, 1509m, 1462m, 1375w, 1296m, 1252m, 1226m, 1206s, 1177m, 1086w, 1036m, 912w, 833w. ^{1}H-NMR (300 MHz, CDCl$_3$): 2.04–2.08 (m, CH$_2$); 2.61 (s, MeCO); 3.17–3.71 (m, BrCH$_2$, CH$_2$O, 2 H−C(5′)); 3.77, 3.78 (2s, MeO); 4.09 (br. s, OH−C(2′)); 4.37–4.39 (m, H−C(4′)); 4.48 (dd, $J = 3.4$, 5.0, H−C(3′)); 4.80, 4.84 (2d, $J = 6.7$, OCH$_2$O); 4.82–4.88 (m, H−C(2′)); 6.04 (d, $J = 5.6$, H−C(1′)); 6.73–6.82 (m, 4 arom. H); 7.16–7.38 (m, 9 arom. H); 8.19 (s, H−C(2)); 8.62 (s, H−C(8)); 8.71 (br. s, NH−C(6)). ^{13}C-NMR (75 MHz, CDCl$_3$): 25.7 (q, MeCO); 30.1 (t, CH$_2$); 32.4 (t, BrCH$_2$); 55.2 (q, MeO); 63.0, 66.0 (2t, CH$_2$O, C(5′)); 74.4, 77.0, 83.2 (3d, C(2′), C(3′), C(4′)); 86.7 (s, Ar$_2$C(Ph)); 89.4 (d, C(1′)); 95.7 (t, OCH$_2$O); 113.2 (d, arom. C); 122.1 (s, C(5)); 127.0, 127.9, 128.1, 130.1 (4d, arom. C); 135.5, 135.6 (2s, arom. C); 141.5 (d, C(8)); 144.4 (s, arom. C); 149.3 (s, C(4)); 152.3 (s, C(6)); 158.6 (s, MeO−C). FAB-MS: 765 (34, $[M+H]^+$), 764 (67, M^+), 763 (34, $[M+H]^+$), 762 (64, M^+), 303 (100).

N^2-*Acetyl-2′-O-[(3-bromopropoxy)methyl]-5′-O-(4,4′-dimethoxytrityl)guanosine* (**23**) *and* N^2-*Acetyl-3′-O-[(3-bromopropoxy)methyl]-5′-O-(4,4′-dimethoxytrityl)guanosine* (**27**). As described for **21/25**, with **19** (1.57 g, 2.5 mmol), ^{i}Pr$_2$NEt (1.7 ml, 10 mmol), (CH$_2$Cl)$_2$ (10 ml), Bu$_2$SnCl$_2$ (760 mg, 2.5 mmol), and **34** (470 mg, 2.5 mmol). Workup and CC (silica gel, hexane/AcOEt (+2% Et$_3$N) 5:5 → AcOEt (+2% Et$_3$N)) gave **23** (1.18 g, 60%) as a pale yellow foam and 0.1 g of a mixture of by-products. From this mixture, **27** was isolated by prep. TLC (CH$_2$Cl$_2$/MeOH 9:1) as a pale yellow foam.

Data of **23**: TLC (AcOEt/EtOH 9:1): R_f 0.48. $[\alpha]_D^{25} = 1.5$ ($c = 1.0$, CHCl$_3$). UV (MeOH): 275 (13000), 271 (12900), 235 (26700), 226 (25200). IR (CHCl$_3$): 3369w, 3222w, 3022m, 2935w, 1703s, 1674s, 1563m, 1509m, 1463w, 1411m, 1301w, 1253m, 1226s, 1178w, 1095m, 1036m, 996w, 830w. ^{1}H-NMR: 1.76 (s, MeCO); 1.94–1.97 (m, CH$_2$); 3.18 (br. s, OH−C(3′)); 3.22–3.73 (m, BrCH$_2$, CH$_2$O, 2 H−C(5′)); 3.75, 3.76 (2s, MeO); 4.21 (br. s, H−C(4′)); 4.23 (br. s, H−C(3′)); 4.74, 4.80 (2d, $J = 6.8$, OCH$_2$O); 5.07–5.10 (m, H−C(2′)); 5.70 (d, $J = 6.2$, H−C(1′)); 6.75–6.84 (m, 4 arom. H); 7.15–7.50 (m, 9 arom. H); 7.87 (s, H−C(8)); 8.73 (s, NH−C(6)); 11.94 (s, H−N(1)). ^{13}C-NMR (75 MHz, CDCl$_3$): 23.7 (q, MeCO); 30.1 (t, CH$_2$); 32.2 (t, BrCH$_2$); 55.3 (q, MeO); 63.8, 66.1 (2t, CH$_2$O, C(5′)); 70.5, 78.9, 84.3 (3d, C(2′), C(3′), C(4′)); 86.5 (d, C(1′)); 87.1 (s, Ar$_2$C(Ph)); 95.7 (t, OCH$_2$O); 113.3 (d, arom. C); 122.0 (s, C(5)); 127.2, 128.1, 128.4, 130.1 (4d, arom. C); 135.5, 135.9 (2s, arom. C); 139.0 (d, C(8)); 144.9 (s, arom. C); 147.2 (s, C(4)); 148.4 (s, C(4)); 155.6 (s, C(6)); 158.8 (s, MeO−C); 171.9 (s, CO). FAB-MS: 781 (36, $[M+H]^+$), 780 (52, M^+), 779 (18, $[M+H]^+$), 778 (69, M^+), 303 (100).

Data of **27**: TLC (AcOEt/EtOH 9:1): R_f 0.40. ^{1}H-NMR: 1.64–1.67 (m, CH$_2$); 1.79 (s, MeCO); 3.11–3.16 (m, H−C(5′)); 3.26–3.58 (m, BrCH$_2$, CH$_2$O, H′−C(5′)); 3.74, 3.76 (2s, MeO); 4.10–4.12 (m, H−C(4′)); 4.30–4.33 (m, OCH$_2$O); 4.45–4.48 (m, H−C(3′)); 4.51 (br. d, $J = 5.0$, OH−C(2′)); 4.98–5.02 (m, H−C(2′)); 6.00 (d, $J = 6.5$, H−C(1′)); 6.71–6.83 (m, 4 arom. H); 7.10–7.46 (m, 9 arom. H); 7.81 (s, H−C(8)); 8.92 (br. s, NH−C(6)); 11.95 (br. s, H−N(1)). ^{13}C-NMR (75 MHz, CDCl$_3$): 23.6 (q, MeCO); 30.2 (t, CH$_2$); 31.6 (t, BrCH$_2$); 55.1 (q, MeO); 62.4, 64.8 (2t, CH$_2$O, C(5′)); 68.7, 74.9, 82.6 (3d, C(2′), C(3′), C(4′)); 86.9 (d, C(1′)); 87.9 (s, Ar$_2$C(Ph)); 94.8 (t, OCH$_2$O); 113.2 (d, arom. C); 123.8 (s, C(5)); 128.0, 128.2, 128.5, 130.1 (4d, arom. C);

135.1, 135.6 (2*s*, arom. C); 138.7 (*d*, C(8)); 144.9 (*s*, arom. C); 147.9 (*s*, C(4)); 148.4 (*s*, C(4)); 155.5 (*s*, C(6)); 158.7 (*s*, MeO−*C*); 172.1 (*s*, CO). FAB-MS: 780 (10, *M*⁺), 778 (8, *M*⁺), 653 (4), 561 (5), 303 (100).

N^4-*Acetyl-2'-O-[(3-bromopropoxy)methyl]-5'-O-(4,4'-dimethoxytrityl)cytidine* (**24**) *and* N^4-*Acetyl-3'-O-[(3-bromopropoxy)methyl]-5'-O-(4,4'-dimethoxytrityl)cytidine* (**28**). As described for **21/25**, with **20** (1.47 g, 2.5 mmol), ^{i}Pr₂NEt (1.7 ml, 10 mmol), (CH₂Cl)₂ (10 ml), Bu₂SnCl₂ (760 mg, 2.5 mmol), and **34** (610 mg, 3.3 mmol). Workup and CC (silica gel, hexane/AcOEt (+2% Et₃N) 7:3 → AcOEt/EtOH 9:1 (+2% Et₃N)) gave **24** (833 mg, 45%) and **28** (370 mg, 20%) as pale yellow foams.

Data of **24**: TLC (AcOEt/EtOH 9:1): R_f 0.63. [α]$_D^{25}$ = 40.9 (*c* = 1.0, CHCl₃). UV (MeOH): 299 (6600), 290 (6100), 283 (6500), 278 (6000), 271 (5800), 236 (26700), 226 (23300). IR (CHCl₃): 3401*w*, 3011*m*, 2961*w*, 1723*m*, 1661*s*, 1610*m*, 1554*m*, 1504*m*, 1462*s*, 1362*m*, 1306*m*, 1252*m*, 1224*s*, 1206*w*, 1177*m*, 1102*m*, 1036*m*, 909*w*, 829*w*. ¹H-NMR: 2.06−2.14 (*m*, CH₂); 2.22 (*s*, MeCO); 2.71 (*d*, *J* = 9.3, OH−C(3')); 3.48 (*t*, *J* = 6.4, BrCH₂); 3.52−3.74 (*m*, CH₂O, 2 H−C(5')); 3.81, 3.82 (2*s*, MeO); 4.11−4.13 (*m*, H−C(4')); 4.24 (br. *d*, *J* ≈ 5.0, H−C(2')); 4.43−4.46 (*m*, H−C(3')); 4.95, 5.17 (2*d*, *J* = 6.5, OCH₂O); 5.99 (*s*, H−C(1')); 6.85−6.89 (*m*, 4 arom. H); 7.08−7.44 (*m*, 9 arom. H, H−C(5)); 8.47 (*d*, *J* = 7.4, H−C(6)); 9.21 (br. *s*, NH−C(4)). ¹³C-NMR (75 MHz, CDCl₃): 24.9 (*q*, *Me*CO); 30.4 (*t*, CH₂); 32.5 (*t*, BrCH₂); 55.3 (*q*, MeO); 61.0, 66.3 (2*t*, CH₂O, C(5')); 68.0, 80.2, 83.2 (3*d*, C(2'), C(3'), C(4')); 87.1 (*s*, Ar₂*C*(Ph)); 89.5 (*d*, C(1')); 95.2 (*t*, OCH₂O); 96.7 (*d*, C(5)); 113.4 (*d*, arom. C); 127.2, 128.0, 128.1, 128.4, 130.2 (5*d*, arom. C); 135.2, 135.5, (2*s*, arom. C); 144.3 (*s*, arom. C); 144.7 (*d*, C(6)); 155.3 (*s*, C(2)); 158.7 (*s*, MeO−*C*); 162.8 (*s*, C(4)); 170.4 (*s*, CO). FAB-MS: 740 (92, [*M* + H]⁺), 739 (45, *M*⁺), 738 (74, [*M* + H]⁺), 737 (14, *M*⁺), 303 (100).

Data of **28**: TLC (AcOEt/EtOH 9:1): R_f 0.53. [α]$_D^{25}$ = 11.8 (*c* = 1.0, CHCl₃). UV (MeOH): 299 (5600), 290 (5100), 284 (5400), 271 (4700), 236 (23300), 227 (20900). IR (CHCl₃): 3400*w*, 3012*w*, 2960*w*, 1724*m*, 1658*m*, 1610*w*, 1554*m*, 1510*m*, 1482*s*, 1444*w*, 1382*m*, 1307*m*, 1252*s*, 1177*m*, 1114*m*, 1036*m*, 909*m*, 830*w*. ¹H-NMR (300 MHz, CDCl₃): 2.01−2.06 (*m*, CH₂); 2.24 (*s*, MeCO); 3.35−3.67 (*m*, BrCH₂, OCH₂, 2 H−C(5'), OH−C(2')); 3.81 (*s*, 2 MeO); 4.27−4.39 (*m*, H−C(2'), H−C(3'), H−C(4')); 4.70, 4.77 (2*d*, *J* = 6.8, OCH₂O); 5.94 (*d*, *J* = 2.8, H−C(1')); 6.84−6.88 (*m*, 4 arom. H); 7.16−7.39 (*m*, 9 arom. H, H−C(5)); 8.31 (*d*, *J* = 7.5, H−C(6)); 9.16 (br. *s*, NH−C(4)). ¹³C-NMR (75 MHz, CDCl₃): 25.0 (*q*, *Me*CO); 30.2 (*t*, CH₂); 32.4 (*t*, BrCH₂); 55.3 (*q*, MeO); 61.7, 65.9 (2*t*, CH₂O, C(5')); 74.9, 75.5, 82.4 (3*d*, C(2'), C(3'), C(4')); 87.0 (*s*, Ar₂*C*(Ph)); 92.2 (*d*, C(1')); 95.4 (*t*, OCH₂O); 96.7 (*d*, C(5)); 113.3 (*d*, arom. C); 128.0, 128.1, 128.4, 130.3 (4*d*, arom. C); 135.2, 135.3 (2*s*, arom. C); 144.1 (*d*, arom. C); 144.6 (*d*, C(6)); 155.7 (*s*, C(2)); 158.8 (*s*, MeO−*C*); 162.6 (*s*, C(4)); 170.0 (*s*, CO). FAB-MS: 740 (38, [*M* + H]⁺), 739 (25, *M*⁺), 738 (38, [*M* + H]⁺), 737 (11, *M*⁺), 303 (100).

2'-O-[(3-Bromopropoxy)methyl]-5'-O-(4,4'-dimethoxytrityl)uridine 3'-(2-Cyanoethyl Diisopropylphosphoramidite) (**29**). According to the *G.P.*, with **21** (373 mg, 0.53 mmol). CC (Al₂O₃, hexane/AcOEt 8:2 to 3:7) gave **29** (434 mg, 91%; 1:1 mixture of diastereoisomers). Pale yellow foam. TLC (hexane/AcOEt 5:5): R_f 0.42. UV (MeCN): 265 (22300), 236 (20900), 225 (18200). IR (CHCl₃): 3390*w*, 3026*s*, 2973*w*, 2360*w*, 1806*m*, 1697*s*, 1608*w*, 1509*m*, 1458*w*, 1414*w*, 1395*w*, 1251*s*, 1178*w*, 1098*w*, 1034*m*, 1004*m*, 835*w*. ¹H-NMR (300 MHz, CDCl₃): 1.03−1.29 (*m*, (*Me*₂CH₂N); 2.07−2.13 (*m*, CH₂); 2.43 (*t*, *J* = 6.4, 1 H, OCH₂*CH₂*CN); 2.66 (*t*, *J* = 6.0, 1 H, OCH₂*CH₂*CN); 3.40−3.77 (*m*, 2 H−C(5'), OCH₂CH₂CH₂Br, OCH₂CH₂CN, (Me₂*CH*)₂N); 3.80, 3.83 (2*s*, MeO); 4.18−4.19, 4.26−4.28 (2*m*, H−C(4')); 4.37−4.39, 4.42−4.45 (2*m*, H−C(2')); 4.52−4.58 (*m*, H−C(3')); 4.82−4.93 (*m*, OCH₂O); 5.24 (*d*, *J* = 8.4, 0.5 H, H−C(5)); 5.28 (*d*, *J* = 8.1, 0.5 H, H−C(5)); 6.05 (*d*, *J* = 3.7, 0.5 H, H−C(1')); 6.09 (*d*, *J* = 4.0, 0.5 H, H−C(1')); 6.82−6.86 (*m*, 4 arom. H); 7.24−7.42 (*m*, 9 arom. H); 7.90, 7.96 (2*d*, *J* = 8.1, H−C(6)). ¹³C-NMR (75 MHz, CDCl₃): 20.3 (br. *t*, CH₂CN); 24.5 (br. *q*, *Me*₂CHN); 30.4 (*t*, CH₂); 32.6 (*t*, BrCH₂); 43.2, 43.3 (2 br. *d*, Me₂*C*HN); 55.3 (*q*, MeO); 58.1, 58.2 (2*t*, *J*(C,P) = 18.5, OCH₂*CH₂*CN); 61.5, 63.9, 65.9, 66.2 (4*t*, CH₂O, C(5')); 68.0, 70.4, 73.9, 75.5, 82.7, 82.8 (6*d*, C(2'), C(3'), C(4')); 87.1, 87.2 (2*d*, C(1')); 87.7, 87.8 (2*s*, Ar₂*C*(Ph)); 95.0 (*t*, OCH₂O); 102.5, 102.6 (2*d*, C(5)); 113.3 (*s*, arom. C); 117.4, 117.6 (2*s*, CN); 127.2, 128.0, 128.2, 128.3, 130.3 (5*d*, arom. C); 134.9, 135.1, 135.2 (3*s*, arom. C); 140.0 (*d*, C(6)); 144.2, 144.3 (2*s*, arom. C); 150.06, 150.13 (2*s*, C(2)); 158.8 (*s*, MeO−*C*); 162.9 (*s*, C(4)). ³¹P-NMR (121 MHz, CDCl₃): 150.5, 151.2. FAB-MS: 900 (7, [*M* + H]⁺), 899 (16, *M*⁺), 898 (8, [*M* + H]⁺), 897 (21, *M*⁺), 595 (68), 593 (64), 303 (100).

N^6-*Acetyl-2'-O-[(3-bromopropoxy)methyl]-5'-O-(4,4'-dimethoxytrityl)adenosine 3'-(2-Cyanoethyl Diisopropylphosphoramidite)* (**30**). According to the *G.P.*, with **22** (705 mg, 0.87 mmol). CC (Al₂O₃, hexane/AcOEt 6:4 to 9:1) gave **30** (760 mg, 92%; 1:1 mixture of diastereoisomers). Pale yellow foam. TLC (hexane/AcOEt 2:8): R_f 0.35. UV (MeCN): 271 (19300), 254 (14700), 236 (22300), 229 (20800). IR (CHCl₃): 3014*m*, 2974*w*, 2337*w*, 1806*m*, 1706*w*, 1610*s*, 1590*w*, 1509*m*, 1462*m*, 1412*w*, 1373*m*, 1252*s*, 1180*m*, 1097*m*, 1035*w*, 1003*w*, 839*w*. ¹H-NMR (300 MHz, CDCl₃): 1.06−1.29 (*m*, (*Me*₂CH)₂N); 1.78−1.96 (*m*, CH₂); 2.39 (*t*, *J* = 5.6, 0.8 H, OCH₂*CH₂*CN); 2.60 (*s*, MeCO); 2.65 (*t*, *J* = 5.6, 1.2 H, OCH₂*CH₂*CN); 3.27−3.73 (*m*, 2 H−C(5'), CH₂O, BrCH₂, OCH₂CH₂CN, (Me₂*CH*)₂N); 3.77, 3.78 (2*s*, MeO); 4.32−4.35 (*m*, 0.6 H, H−C(4')); 4.40−4.42

(m, 0.4 H, H–C(4')); 4.65–4.67 (m, H–C(3')); 4.69–4.85 (m, OCH$_2$O); 5.08–5.13 (m, H–C(2')); 6.19 (d, J = 5.9, 0.4 H, H–C(1')); 6.21 (d, J = 1.5, 0.6 H, H–C(1')); 6.76–6.83 (m, 4 arom. H); 7.19–7.44 (m, 9 arom. H); 8.19, 8.20 (2s, 1 H, H–C(2)); 8.59, 8.61 (2s, 1 H, H–C(8)); 8.68 (br. s, NH–C(6)). ^{13}C-NMR (75 MHz, CDCl$_3$): 20.2, 20.4 (2t, J(C,P) = 7.1, CH$_2$CN); 24.3, 24.4, 24.5, 24.7 (4q, Me$_2$CHN); 25.7 (q, MeCO); 30.1 (t, CH$_2$); 32.3 (t, BrCH$_2$); 43.1, 43.4 (2d, J(C,P) = 6.9, Me$_2$CHN); 55.3 (q, MeO); 58.1, 58.9 (2t, J(C,P) = 13.9, OCH$_2$CH$_2$CN); 63.0, 66.0 (2t, CH$_2$O, C(5')); 71.1, 71.3, 71.7, 71.9, 83.8, 83.9 (6d, C(2'), C(3'), C(4')); 86.7, 86.8 (2d, C(1')); 87.0 (s, Ar$_2$C(Ph)); 95.1, 95.2 (2t, OCH$_2$O); 113.2 (d, arom. C); 117.4, 117.6 (2s, CN); 122.2 (s, C(5)); 127.0 (s, arom. C); 127.9, 128.2, 128.3, 130.1, 130.2 (5d, arom. C); 135.5, 135.6 (2s, arom. C); 141.9 (d, C(8)); 144.4, 144.5 (2s, arom. C); 149.2 (s, C(4)); 151.2 (s, C(6)); 152.4 (d, C(2)); 158.6, 158.7 (s, MeO–C); 170.1 (s, CO). ^{31}P-NMR (202 MHz, CDCl$_3$): 150.9, 151.1. FAB-MS: 965 (16, [M + H]$^+$), 964 (38, M^+), 963 (11, [M + H]$^+$), 962 (37, M^+), 787 (34), 785 (35), 303 (100).

N^2-*Acetyl-2'-O-[(3-bromopropoxy)methyl]-5'-O-(4,4'-dimethoxytrityl)guanosine 3'-(2-Cyanoethyl Diisopropylphosphoramidite)* (**31**). According to the *G.P.* with **23** (990 mg, 1.27 mmol). CC (silica gel, hexane/ AcOEt (+2% Et$_3$N) 6:4) → AcOEt (+2% Et$_3$N) gave **31** (1.03 mg, 83%; 1:1 mixture of diastereoisomers). Pale yellow foam. TLC (hexane/AcOEt 1:9): R_f 0.51. UV (MeCN): 281 (12900), 269 (11800), 238 (23700), 224 (19600). IR (CHCl$_3$): 3385w, 3020s, 2973m, 2241w, 1806w, 1701s, 1608m, 1560m, 1509m, 1494w, 1412m, 1301w, 1253s, 1178m, 1126w, 1094m, 1034s, 1002w, 835w. ^{1}H-NMR (300 MHz, CDCl$_3$): 1.00–1.19 (m, (Me$_2$CH)$_2$N); 1.59, 1.74 (2s, MeCO); 1.86–1.94 (m, CH$_2$); 2.31 (t, J = 6.2, 1 H, OCH$_2$CH$_2$CN); 2.72–2.78 (m, 1 H, OCH$_2$CH$_2$CN); 3.17–3.19 (m, H–C(5')); 3.26–3.29 (m, BrCH$_2$); 3.42–3.66 (m, H'–C(5'), CH$_2$O, (Me$_2$CH)$_2$N); 3.75, 3.76, 3.77 (3s, MeO); 4.23, 4.34 (2 br. s, H–C(4')); 4.52–4.61 (m, H–C(3')); 4.64–4.80 (m, OCH$_2$O); 5.13–5.22 (m, H–C(2')); 5.86 (d, J = 6.5, 0.5 H, H–C(1')); 5.98 (d, J = 8.5, 0.5 H, H–C(1')); 6.77–6.82 (m, 4 arom. H); 7.19–7.55 (m, 9 arom. H); 7.79, 7.83 (2s, H–C(8)); 8.60, 8.46 (2 br. s, NH–C(2)); 11.88 (br. s, H–N(1)). ^{13}C-NMR 75 MHz, CDCl$_3$): 20.2 (m, CH$_2$CN); 23.5, 23.6 (2q, MeCO); 24.5, 24.6, 24.7 (3q, Me$_2$CHN); 30.1 (t, CH$_2$); 32.4, 32.5 (2t, BrCH$_2$); 43.1, 45.4 (2d, J(C,P) = 12.2, Me$_2$CHN); 55.4 (q, MeO); 57.4 (t, J(C,P) = 19.5, OCH$_2$CH$_2$CN); 59.0 (t, J(C,P) = 13.4, OCH$_2$CH$_2$CN); 63.6, 63.8 (2t, C(5')); 66.0, 66.2 (2t, CH$_2$O); 70.7 (d, J(C,P) = 17.1); 71.8 (d, J(C,P) = 13.4); 76.4, 76.8 (2d); 84.5 (d, C(2'), C(3'), C(4')); 86.1, 88.0 (d, C(1')); 86.5, 86.8 (2s, Ar$_2$C(Ph)); 95.0 (t, OCH$_2$O); 113.4, 113.5 (2d, arom. C); 117.7, 118.3 (2s, CN); 122.2, 122.9 (2s, C(5)); 127.35, 127.42, 128.26, 128.33, 128.4, 130.3, 130.4 (7d, arom. C); 135.7, 135.9, 136.0, 136.4 (4s, arom. C); 138.3, 139.5 (2d, C(8)); 144.9, 145.3 (2s, arom. C); 147.1, 147.5 (2s, C(4)); 148.4, 148.7 (2s, C(2)); 155.8 (s, C(6)); 159.0 (s, MeO–C); 171.8, 171.9 (2s, CO). ^{31}P-NMR (121 MHz, CDCl$_3$): 150.4, 150.8. FAB-MS: 981 (16, [M + H]$^+$), 980 (38, M^+), 979 (10, [M + H]$^+$), 978 (36, M^+), 787 (40), 785 (38), 303 (100).

N^2-*Acetyl-2'-O-[(3-bromopropoxy)methyl]-5'-O-(4,4'-dimethoxytrityl)cytidine 3'-(2-Cyanoethyl Diisopropylphosphoramidite)* (**32**). According to the *G.P.*, with **24** (566 mg, 0.77 mmol). CC (silica gel, hexane/AcOEt 6:4 → 2:8) gave **32** (639 mg, 89%; 1:1 mixture of diastereoisomers). Pale yellow foam. TLC (hexane/AcOEt 1:9): R_f 0.40. UV (MeCN): 304 (6600), 290 (5600), 283 (6100), 237 (27900), 226 (22700), IR (CHCl$_3$): 3401w, 3011m, 2969m, 2361w, 1722m, 1664m, 1609w, 1555m, 1509m, 1482s, 1383w, 1307m, 1251m, 1179m, 1120m, 1035m, 980w, 830w. ^{1}H-NMR (300 MHz, CDCl$_3$): 0.99–1.29 (m, (Me$_2$CH)$_2$N); 2.08–2.18 (m, CH$_2$); 2.205, 2.213 (2s, MeCO); 2.40, 2.61 (2t, J = 6.2, OCH$_2$CH$_2$CN); 3.41–3.76 (m, 2 H–C(5'), CH$_2$O, BrCH$_2$, OCH$_2$CH$_2$CN, (Me$_2$CH)$_2$N); 3.81, 3.82 (2s, MeO); 4.22–4.53 (m, H–C(2'), H–C(3'), H–C(4')); 4.89–5.01 (m, OCH$_2$O); 6.06 (s, 0.5 H, H–C(1')); 6.09 (d, J = 1.5, 0.5 H, H–C(1')); 6.80–6.98 (m, 4 arom. H); 6.95 (d, J = 7.6, 0.5 H, H–C(5)); 7.00 (d, J = 7.5, 0.5 H, H–C(5)); 7.23–7.44 (m, 9 arom. H); 8.42 (d, J = 7.5, 0.5 H, H–C(6)); 8.52 (d, J = 7.8, 0.5 H, H–C(6)); 9.09, 9.13 (2 br. s, NH–C(4)). ^{13}C-NMR (75 MHz, CDCl$_3$): 20.2, 20.4 (2t, J(C,P) = 6.9, CH$_2$CN); 24.4, 24.5, 24.6, 25.0 (4q, Me$_2$CHN, MeCO); 30.7 (t, CH$_2$); 32.8 (t, BrCH$_2$); 43.1, 43.3 (2d, J(C,P) = 7.0, Me$_2$CHN); 55.2 (q, MeO); 58.2 (t, J(C,P) = 15, OCH$_2$CH$_2$CN); 60.6, 61.0, 65.2, 66.1 (4t, CH$_2$O, C(5')); 69.1, 76.6, 78.6, 79.1, 82.0, 86.4 (6d, C(2'), C(3'), C(4')); 87.08, 87.15 (2s, Ar$_2$C (Ph)); 89.8 (d, C(1')); 94.9, 95.1 (2t, OCH$_2$O); 96.4 (d, C(5)); 113.3 (d, arom. C); 117.4 (s, CN); 127.3, 128.0, 128.4, 130.1, 130.3 (5d, arom. C); 128.3, 135.1, 135.2, 135.3 (4s, arom. C); 144.1, 144.2 (2s, arom. C); 144.9 (d, C(6)); 155.2 (s, C(2)); 158.8 (s, MeO–C); 160.6 (s, C(4)); 170.1 (s, CO). ^{31}P-NMR (202 MHz, CDCl$_3$): 150.2, 151.3. FAB-MS: 941 (11, [M + H]$^+$), 940 (18, M^+), 939 (11, [M + H]$^+$), 938 (17, M^+), 303 (100).

Assembly of Oligonucleotides. Automated 1.0-μmol syntheses ('trityl-off' mode) were carried out on a *Gene Assembler (Pharmacia)* by the following protocol: 1) 1.5 min detritylation with 4% CHCl$_2$COOH in (CH$_2$Cl)$_2$; 2) 2.5 min coupling with the appropriate phosphoramidites (0.12 ml of 0.1M soln. in MeCN) and BnSTet (0.36 ml of a 0.35M soln. in MeCN); 3) 1 min capping with a 1:1 mixture of Ac$_2$O/2,6-lutidine/THF 1:1:8 and 16% 1-methyl-1H-imidazole in THF; 4) 0.5 min oxidation with I$_2$/H$_2$O/pyridine/THF 2:2:20:75. According to the trityl-assay, the average coupling yields were 99.7% for standard DNA phosphoramidites, 99.3% for tom-protected phosphoramidites [9], and *ca.* 98% for phosphoramidite **32**.

Oligonucleotides **37** *and* **38**. The assembled, immobilized sequences were treated with 12M aq. MeNH₂ (0.5 ml) and 8M MeNH₂/EtOH (0.5 ml). The suspension was shaken for 2 h at r.t. After centrifugation, the supernatant was evaporated, and the resulting crude oligonucleotides were purified by ion-exchange HPLC. After desalting of the pooled product-containing fractions, 80 OD_{260nm} (60%) of **37** and 75 OD_{260nm} (55%) of **38**, resp., were obtained (*Table 1*).

Oligonucleotide **39**. As described for **37** or **38**, but after the MeNH₂ treatment, the residue was treated with 1M Bu₄NF · 3H₂O/THF (0.5 ml); after 12 h at r.t., 1M aq. *Tris* · HCl (pH 7.4; 0.5 ml) was added, and the clear soln. was desalted on a *Sephadex G-10* column (H₂O as eluent). The product-containing fractions were collected, evaporated, and purified by ion-exchange chromatography. After desalting of the pooled product-containing fractions, 40 OD_{260nm} (30%) of **39** were obtained (*Table 1*).

Oligonucleotides **40** *and* **41**. As described for **37** and **38**, but only with 1/10 of the solid support **35** or **36** (*ca.* 0.1 µmol each), resp. After ion-exchange chromatography and desalting of the pooled product-containing fractions, 6 OD_{260nm} (45%) of **40** and 5 OD_{260nm} (40%) of **41**, resp., were obtained (*Table 1*).

Oligonucleotides **49 – 53**. Under Ar, *ca.* 80% of the solid support **35** or **36** (*ca.* 0.8 µmol each) was shaken with 1M thioglycolic acid/2M ⁱPr₂NEt/DMF (1 ml). After 14 h at r.t., the solid suppport was collected by centrifugation and washed 3 × with DMF (→ **35** and **36**, resp.). About 1/3 of the solid support (*ca.* 0.25 µmol each) was treated with a soln. of HOBT (14 mg, 0.1 mmol), TBTU (32 mg, 0.1 mmol), ⁱPr₃NEt (34 µl, 0.2 mmol), and 0.8 mmol of an amine (MeNH₂, histamine, or L-isoleucine allyl ester) in DMF. The suspension was shaken for 4 h at r.t., collected by centrifugation, and washed 2 × with DMF and EtOH, resp. Deprotection and purification were carried out as described for **37** and **38**. From **35**, 15 OD_{260nm} (40%) of **49**, 12 OD_{260nm} (35%) of **51**, and 9 OD_{260nm} (25%) of **53** were obtained with MeNH₂, histamine, and L-isoleucine allylester, resp. (*Table 1*). From **36**, 14 OD_{260nm} (40%) of **50** and 11 OD_{260nm} (30%) of **53** were obtained with MeNH₂ and histamine, resp. (*Table 1*).

REFERENCES

[1] A. E. Ferentz, G. L. Verdine, *J. Am. Chem. Soc.* **1991**, *113*, 4000; C. M. Harris, L. Zhou, E. A. Strand, T. M. Harris, *J. Am. Chem. Soc.* **1991**, *113*, 4328; A. M. Macmillan, G. L. Verdine, *Tetrahedron* **1991**, *47*, 2603; Y. Z. Xu, Q. Zheng, P. F. Swann, *J. Org. Chem.* **1992**, *57*, 3839; C. R. Allerson, S. L. Chen, G. L. Verdine, *J. Am. Chem. Soc.* **1993**, *115*, 12583; N. Schmid, J.-P. Behr, *Tetrahedron Lett.* **1995**, *36*, 1447 – 1450; Y.-Z. Xu, *Tetrahedron* **1996**, *52*, 10737.

[2] J. D. Kahl, M. M. Greenberg, *J. Am. Chem. Soc.* **1999**, *121*, 597; K. Shinozuka, S. Kohgo, H. Ozaki, H. Sawai, *Chem. Commun.* **2000**, 59.

[3] X. Wu, S. Pitsch, *Bioconjugate Chem.* **1999**, *6*, 921.

[4] X. Wu, S. Pitsch, *Nucleic Acids Res.* **1998**, *19*, 4315.

[5] S. Pitsch, *Helv. Chim. Acta* **1997**, *80*, 2286.

[6] G. H. Hakimelahi, Z. A. Proba, K. K. Ogilvie, *Can. J. Chem.* **1982**, *60*, 1106.

[7] M. J. Gait, 'Oligonucleotide Synthesis – a Practical Approach', IRL Press, Oxford, 1984.

[8] P. Garner, S. Ramakanth, *J. Org. Chem.* **1988**, *53*, 1294; M. J. Robins, R. Zhou, Z. Guo, S. F. Wuuk, *J. Org. Chem.* **1996**, *61*, 9207.

[9] S. Pitsch, P. A. Weiss, X. Wu, D. Ackermann, T. Honegger, *Helv. Chim. Acta* **1999**, *82*, 1753.

[10] J. E. Corrie, G. P. Reid, D. R. Trentham, M. B. Hursthouse, M. A. Mazid, *J. Chem. Soc., Perkin Trans. 1* **1992**, 1015; M. E. Schwartz, R. R. Breaker, G. T. Asteriadis, J. S. deBear, G. R. Gough, *Bioorg. Med. Chem. Lett.* **1992**, *2*, 1019.

[11] T. Benneche, L.-L. Gundersen, K. Undheim, *Acta Chem. Scand., Sect. B* **1988**, *42*, 384.

[12] U. Pieles, W. Zürcher, M. Schär, H. E. Moser, *Nucleic Acids Res.* **1993**, *21*, 3191.

[13] L. A. Marky, K. J. Breslauer, *Biopolymers* **1987**, *26*, 1601.

[14] P. Lubini, W. Zürcher, M. Egli, *Chem. Biol.* **1994**, *1*, 39.

[15] H. Hrebabecky, J. Farkas, in 'Nucleic Acid Chemistry', Vol. 1, J. Wiley & Sons, New York, 1978.

[16] R. K. Ness, 'Synthetic Procedures in Nucleic Acid Chemistry', Vol. 1, Eds. W. W. Zorbach and R. S. Tipson, Wiley & Sons, New York, 1968.

Synthesis and Stability of GNRA-Loop Analogs

by Karlheinz Wörner, Thorsten Strube, and Joachim W. Engels*

Institut für Organische Chemie, Johann Wolfgang Goethe-Universität, Marie Curie Strasse 11, D-60439
Frankfurt am Main (e-mail: engels@ews1.org.chemie.uni.frankfurt.de)

Nebularine, 9-(β-D-ribofuranosyl)-9H-purin-2-amine, and inosine phosphoramidites **8**, **16**, and **17**, respectively, were synthesized and incorporated into the GNRA tetraloop at different positions (see *Scheme*, *Table*, and *Fig. 4*). The oligomers were investigated by means of UV and CD spectroscopy to address the question of how the individual base-modified N-nucleosides contribute to changes in H-bonding and base-stacking interactions within the loop. Several CD spectra are given and compared with each other (*Figs. 5* and *6*). The exchange of the loop sequence in position 4 and 7 results in a distinct change in base stacking. CD-Band shifting allows us to advance the hypothesis that a transition from a GNRA-type towards a UNCG-type base stacking is observed.

1. Introduction. – RNA Structures available today are designed to meet the criteria of chemical stability and biological function. Simple structures like double helices, hairpin loops, bulges, internal loops are known. Hairpin loops result when a single strand folds back to form a double-stranded region, leaving some bases unpaired [1]. This region may vary from 2 to 10 bases at most. Those loops, found in tRNA and rRNA, are most probably present in many structures [2]. Tetraloops represent the most common class of loops in especially two families, the GNRA and UNCG types (N = any nucleotide, R = G or A) [3][4]. Atomic-resolution structures have been determined. In both tetraloop types, only two bases are free, and a non-*Watson-Crick* base pair closes the loop.

In the GNRA family, where the structure of **GCAA** has been established by NMR (*Fig. 1*), a G · A base pair has been found, which shows two H-bonds from the exocyclic NH_2 proton of the 5-positioned G to the N(7) atom of the 8-positioned A and from the exocyclic NH_2 proton of A, to the N(3) atom of G (8- and 5-positioned, resp.) [4]. In addition, phosphate-base and base-ribose H-bonds as well as base stacking stabilize the loop. The **GAAA** loop has been established by two X-ray structures. The hammerhead ribozyme and the group-I ribozyme are both virtually identical as shown by their NMR spectra, which indicate structural rigidity [5]. This unusually high stability has been explained by an internal G · A base pair, a H-bond between G and the phosphate backbone, as well as by intensive base stacking. In addition, the influence of the 2′-hydroxy group has been addressed in mixed sequences by *Serra et al.* [6]. Here, a **UUCG** tetraloop was replaced by a d(TTCG) sequence in deoxyribonucleosides, and a drop in T_m of 8° resulted.

As base substitutions, *SantaLucia et al.* [7] changed the G base for 9-(β-D-ribofuranosyl)-9H-purin-2-amine, inosine, and 2′-deoxyguanosine in the **GACC** loop.

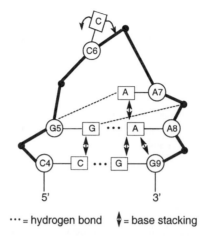

··· = hydrogen bond   ⬍ = base stacking

Fig. 1. *Schematic representation of the* **GCAA** *loop with important structural features*

In 10 mM phosphate/100 mM NaCl buffer, they observed a ΔT_m of 3° for dG, of 6° for 9-(β-D-ribofuranosyl)-9H-purin-2-amine and of 7° for inosine. This result is in accordance with the loss of H-bond contacts established by NMR data. For the overall optimization of the stacking interaction, the orientation of the closing base pair (*i.e.* whether G·C or C·G ends the stem) is very important. *Serra et al.* exchanged the sequence 5′-CG**C**-UUCG-**G**CG-3′ to 5′-CG**G**-UUCG-**C**GC and found a ΔT_m of − 10° [6].

In recent work, the influence of the ribose conformation and the 2′-*endo*/3′-*endo* equilibrium in the loop and its importance for the stability of the hairpin has been investigated [8].

Here we report the structural modification of the decamer hairpins 5′-CGC-**NAAN**-GCG-3′ and CGC-**NCAN**-GCG-3′, with N = A, G, nebularine (Ne), 9-(β-D-ribofuranosyl)-9H-purin-2-amine (A²), or inosine (I). The modifications involve individually each one of the N nucleosides to allow the mapping of the H-bond and possible base stacking.

2. Results and Discussion. – 2.1. *Synthesis of the Nucleosides.* Several phosphor-amidites with base modifications were synthesized. The monomer building blocks nebularine, 9-(β-D-ribofuranosyl)-9H-purin-2-amine, and inosine phosphoramidites **8**, **16**, and **17**, respectively (*Fig. 2*), were incorporated into RNA tetraloops.

For the synthesis of nebularine phosphoramidite **8**, commercially available inosine (**1**) was fully protected at the OH functions by reaction with benzoyl chloride (→ **2**; *Scheme*). The transformation of **2** into the thioxo derivative **3** with *Lawesson*'s reagent occurred in 90% yield, and subsequent cleavage of the benzoyl protecting groups gave **4**. To remove the S-atom, **4** was treated with *Raney*-Ni; after work-up and purification, the yield of **5** was 79%. A building block suitable for the solid-phase RNA synthesis was obtained from **5** by the phosphoramidite method. Thus, the OH functions were protected by reaction with (MeO)₂TrCl (→ **6**) and then with ʹBuMe₂SiCl (→ **7**). The latter reaction gave rise to the 3′-*O*-silyl-protected by-product in 42% yield; however,

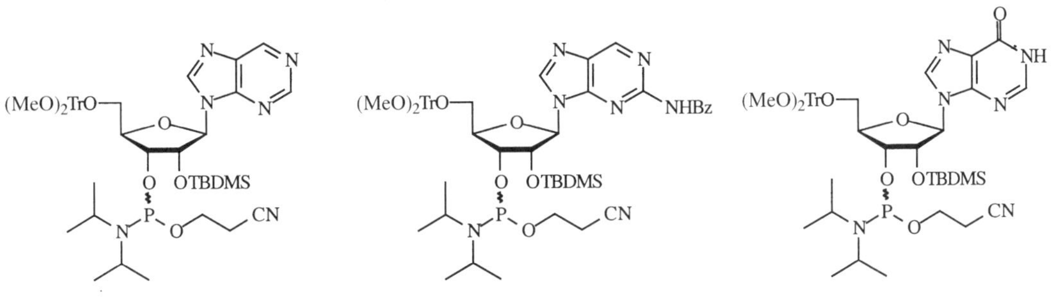

TBDMS = ^{t}BuMe$_2$Si

Fig. 2. *Monomer building blocks incorporated into RNA tetraloops by solid-phase synthesis*

the yield of the desired 2'-*O*-silylated **7** could be increased to 53% by addition of AgNO$_3$ as catalyst. The final phosphitylation reaction gave a 0.7:1 diastereoisomer mixture **8** in 79% yield; to prevent the isomerization of the silyl protecting groups, 5 equiv. of *sym*-collidine and 0.5 equiv. of 1-methyl-1*H*-imidazole were used instead of ^{i}Pr$_2$EtN. The phosphoramidite **8** was characterized by ^{1}H- and ^{31}P-NMR-spectroscopy.

The 9-(β-D-ribofuranosyl)-9*H*-purin-2-amine phosphoramidite **16** was synthesized from **9** *via* **10–15** as described above for the nebularine phosphoramidite **8** (*Scheme*). The silylation of **14** gave a mixture of the 2'-*O*- and 3'-*O*-silylated compounds in 55 and 39%, respectively, and after the final phosphitylation of the desired 2'-*O*-silylated **7**, a 1:2 diastereomer mixture **16** was obtained in 84% yield.

The inosine phosphoramidite **17** was synthesized by literature procedures starting from commercially available inosine (**1**) [9].

2.2. Oligonucleotide Synthesis. The RNA oligomers were synthesized on an *Eppendorf-D300+* synthesizer by phosphoramidite chemistry, with a coupling time for the modified monomers of 12 min [10]. The fully protected decamers were cleaved from the controlled-pore-glass (CPG) support with 32% aqueous NH$_3$ solution at 55° overnight. The 2'-*O*-silyl groups were deprotected with Et$_3$N · 3 HF within 24 h at room temperature [11]. The crude RNA oligomer was precipitated with BuOH at −20°, and the fully deprotected RNA was purified by means of anion-exchange HPLC (*NucleoPac-PA-100*). The pure oligomer was subsequently desalted (*Sephadex-G25*). All nucleotides were characterized by MALDI-TOF-MS, and the masses obtained were in good agreement with the calculated molecular masses.

2.3. Melting Curves of the Oligomers. UV/Melting profiles of the oligomers were recorded in a phosphate buffer containing NaCl (140 mmol) at a wavelength of 260 and 274 nm [12]. The temperature range was 20–85° with a heating rate of 0.5°/min, and the thermodynamic data were extracted from the melting curve by means of a two-state model for the transition from an ordered to a disordered conformation [13].

In contrast to the melting profiles of the UNCG family, the melting curves of the GNRA-family loops do not exhibit a very sharp transition between the hairpin and the molten strand (*Fig. 3*), which indicates a loss of cooperativity within the loop.

Scheme. *Synthesis of the Nebularine and 9-(β-D-Ribofuranosyl)-9H-purin-2-amine Phosphoramidites* **8** *and* **16**, *Respectively*

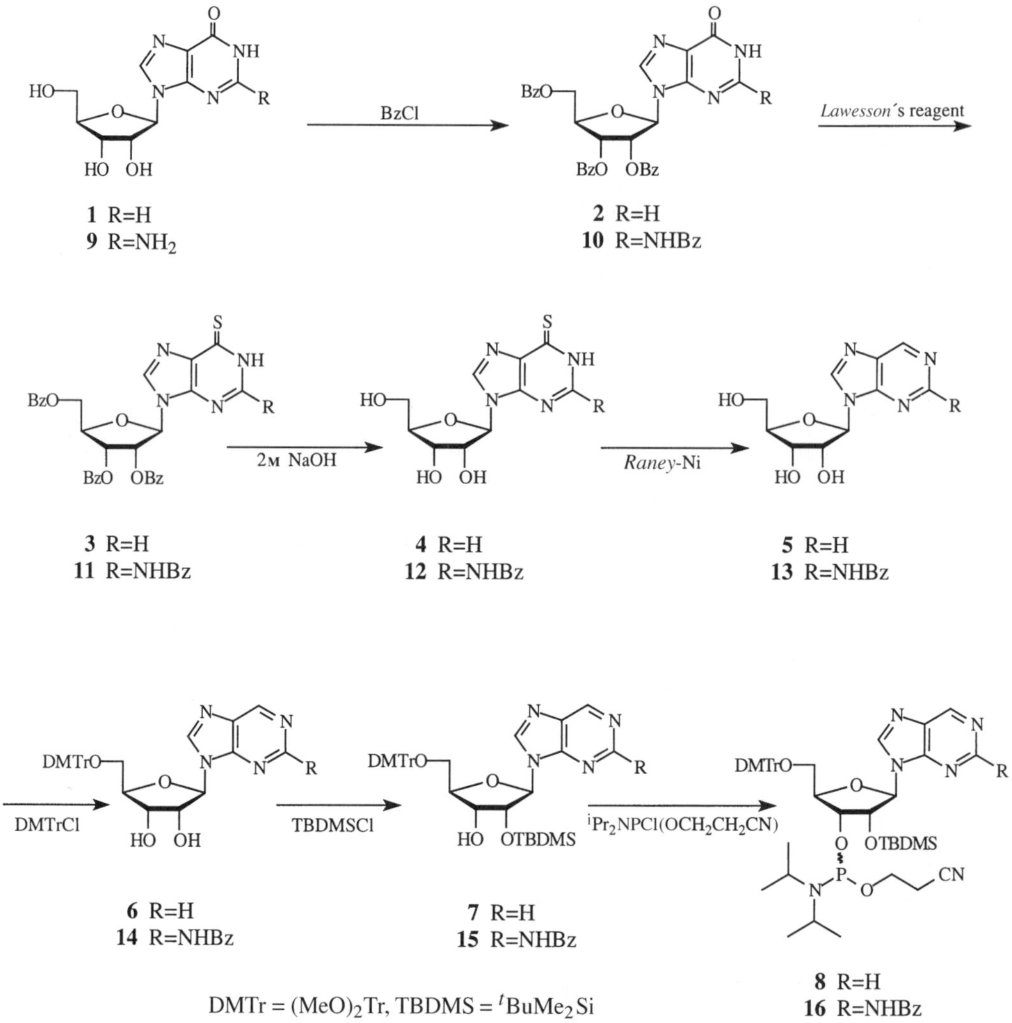
BzCl
Lawesson's reagent
1 R=H
9 R=NH_2
2 R=H
10 R=NHBz
2M NaOH
Raney-Ni
3 R=H
11 R=NHBz
4 R=H
12 R=NHBz
5 R=H
13 R=NHBz
DMTrCl
TBDMSCl
$^{i}Pr_2NPCl(OCH_2CH_2CN)$
6 R=H
14 R=NHBz
7 R=H
15 R=NHBz
8 R=H
16 R=NHBz
DMTr = $(MeO)_2Tr$, TBDMS = $^{t}BuMe_2Si$

2.4. *Thermodynamic Data.* In the CGC-**GCAA**-GCG tetraloop, the nucleoside units at position 4 and 7 (*Fig. 4*) were replaced by the analogues nebularine (Ne), inosine (I), and 9-(β-D-ribofuranosyl)-9*H*-purin-2-amine (A^2). To investigate the contribution of the C residue at position 5 to the stabilization of the loop structure, it was replaced by the nucleoside unit A.

The *Table* shows the synthesized oligonucleotides of the CGC-**NCAN**-GCG and CGC-**NAAN**-GCG type and their thermodynamic properties. In general, all hairpin loops investigated show a destabilization compared with the native **GCAA** loop (71.6°; *Entry 1*). Incorporation of the inosine or 9-(β-D-ribofuranosyl)-9*H*-purin-2-amine

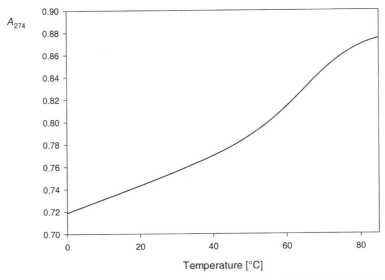

Fig. 3. *Typical melting profile of a hairpin loop of the GNRA family* (CGC-**A²CAA**-GCG)

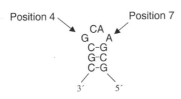

Fig. 4. *Different base analogs were incorporated at position 4 and 7*

residue at position 4 results in a 4° drop in T_m (*Entries 8* and *11*, resp.) which corresponds to missing H-bonds within the loop. The additional substitution of C by A at position 5 of the loop in the inosine sequence leads to a further 3° destabilization of the loop (*Entry 9*). As C at position 5 is looped out and does not show any H-bonding according to the structure of the **GCAA** loop established by NMR, this result suggests a change in base stacking. In comparison to the UNCG family, the contribution of position 5 to the loop formation seems to be significantly smaller.

The substitution of G by nebularine produces a large effect (56.9°; *Entry 5*) which can be explained by the missing H-bond between the NH_2 group of G at position 4 and the N(7) atom of A at position 7 in the G · A base pair. This base pair was found to be essential for the stability of the loop. Interestingly, the T_m of the sequences containing **NeCAA** (56.9°) and **NeAAA** (57.9°) is only slightly lower than that containing **AAAA** (59.2°) (*Entries 4–6*). This result is not yet fully understood, but can be explained by the strong stacking effect of the nebularine unit.

At position 7, A was replaced by the inosine, 9-(β-D-ribofuranosyl)-9*H*-purine, or nebularine moiety. The sequences with the **AAAI** (59.6°) and the **AAAA²** loop (60.3°) exhibit nearly identical T_m values (*Entries 10* and *13*, resp.). The incorporation of nebularine resulted in a T_m of 59.1° for the sequence with the **AAANe** loop (*Entry 7*)

Table. *Synthesized Hairpin Loops of the GNRA Type and Their Thermodynamic Properties*

Entry	Sequence 5′ → 3′	T_m [°] (±1°)	ΔH [kcal/mol]	ΔS [cal/mol·K]
1	CGC-**GCAA**-GCG	71.6	− 26.4	− 76.6
2	CGC-**GAAA**-GCG	67.4	− 25.9	− 76.1
3	CGC-**AAAG**-GCG	63.1	− 24.6	− 73.2
4	CGC-**AAAA**-GCG	59.4	− 18.2	− 54.7
5	CGC-**NeCAA**-GCG	56.9	− 19.5	− 59.1
6	CGC-**NeAAA**-GCG	57.9	− 19.0	− 57.4
7	CGC-**AAANe**-GCG	59.1	− 21.2	− 63.9
8	CGC-**ICAA**-GCG	67.2	− 29.7	− 87.3
9	CGC-**IAAA**-GCG	64.4	− 24.3	− 71.9
10	CGC-**AAAI**-GCG	59.6	− 20.7	− 62.2
11	CGC-**A²CAA**-GCG	67.1	− 30.2	− 88.7
12	CGC-**A²AAA**-GCG	67.5	− 29.1	− 85.4
13	CGC-**AAAA²**-GCG	60.3	− 22.5	− 67.5

which is, as for the incorporation at position 4 (*Entry 6*) quite close to that of the **AAAA**-containing sequence (59.4°; *Entry 4*).

By comparing the sequences containing **GAAA** and **AAAG**, exhibiting a T_m of 67.4° and 63.1°, respectively, the importance of the G·A base pair is stressed. The A·G base pair is less favorable, resulting in a destabilization of nearly 4°. The CD spectra reveal a change in base stacking, which may explain this drop in the T_m value. Interestingly, all the loops studied show a pronounced entropy/enthalpy compensation, which has been described for other biological systems [14].

2.5. *CD Spectra.* CD Spectra were recorded at 350–180 nm with oligonucleotide (10 μmol) solutions in sodium-phosphate buffer (pH 7) containing NaCl (140 mmol). The temperature of the measurement was 20° to ensure that only the hairpin structure of the oligonucleotides was present. *Fig. 5* shows a typical CD spectrum of a RNA hairpin loop with an A-type helix in the stem region [15]. Of the three characteristic extrema with wavelengths λ_1, λ_2, and λ_3, the strong maximum at λ_1, the small minimum at λ_2, and the weak maximum at λ_3 are always found in an A-type helix formation.

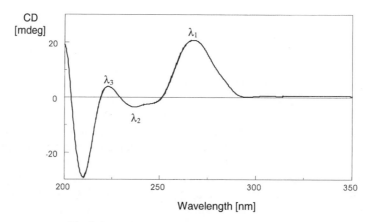

Fig. 5. *Typical CD spectrum of a decamer hairpin loop*

The intensity at λ_1 and λ_3 corresponds to the number of paired and unpaired bases as well as to the extent of base-stacking interactions within the loop and stem region. Temperature-dependent measurements of our oligonucleotides reveal a loss of intensity during the transition from hairpin to random-coil structure. In comparison to CD spectra of the UNCG-type hairpins [8], the maximum at λ_1 of the GNRA-type hairpins is shifted to shorter wavelengths and the intensity at λ_3 increases. *Fig. 6* shows different CD spectra of some of the synthesized oligonucleotides. In the case of the nebularine modification, the maximum at λ_1 has nearly the same intensity (*Fig. 6,d*). The CD spectra of the GNRA-type hairpin with the altered loop sequences **GAAA**, **GCAA**, and **AAAG** reveal a change in intensity at λ_1, while the intensity at λ_3 is nearly unchanged (*Fig. 6,a*). This result can be explained by the change of base-stacking interactions.

A striking observation is the significant difference of the position of the maximum at λ_1 between the **GAAA**- and **AAAG**-containing sequences. The change of the loop sequence leads to a shift of λ_1 of nearly 7 nm to longer wavelength, from 263.4 to 270.2 nm. A similar shift of 3 nm can be observed for the sequences with the inosine modification. The data obtained give rise to the assumption that the change of the loop sequence is accompanied by a distinct change in structure. We found a similar result when investigating the UNCG-type loop [16]. The change from **UUUG**- to **GUUU**-

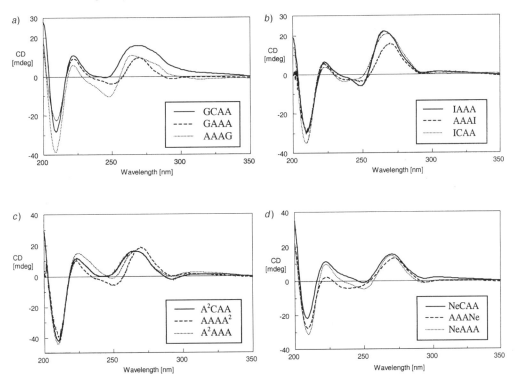

Fig. 6. *CD Spectra of the synthesized decamer hairpins of the GNRA type: a) Changes in the loop sequence without modified bases; b) incorporation of inosine; c) incorporation of 9-(β-D-ribofuranosyl)-9H-purin-2-amine; d) incorporation of nebularine*

containing sequences leads to a shift from 272.2 to 270.6 nm. In contrast to the GNRA type, we found a shift to shorter wavelengths. This opposite shift in the λ_1 CD band leads to the hypothesis that **GAAA** and **UUUG** as well as **AAAG** and **GUUU** may show similar base-stacking interactions. The CD spectra reveal that the weakening or deleting of H-bonds between the $G \cdot A$ base pair changes the base-stacking interactions significantly. The described opposite shift of the λ_1 CD band indicates a transition from the typical GNRA-type stacking between the bases in position 6 and 7 (GCAA) [17] or 5 and 6 (GAAA) [18] towards an UNCG-type stacking between position 4 and 7 [19].

We wish to thank the *Deutsche Forschungsgemeinschaft, Schwerpunktprogramm RNA-Biochemie*, for financial support.

Experimental Part

General. Inosine, guanosine, anh. pyridine, and anh. CH_2Cl_2 were obtained from *Fluka* and used without further purification. Dry MeCN (H_2O <30 ppm) for the phosphitylation reaction was purchased from *Perseptive Biosystems*. Flash column chromatography (FC): silica gel *60* (40–63 μm) from *Merck*. TLC: silica gel *60 F₂₅₄* plates from *Merck*, HPLC: anion-exchange column *NucleoPac PA-100* from *Dionex*; desalting with a *Sephadex-G25* column from *Pharmacia*. UV/Melting profiles: *Varian-Cary*-UV/VIS spectrophotometer, *Cary* temperature controller, 10-mm cuvette. CD Spectra: *Jasco-I-710* spectropolarimeter. ¹NMR: *Bruker-AM250* and *Bruker-WH270* (¹H,¹³C) and *Bruker-AMX400* (³¹P) spectrometers; δ in ppm, J in Hz. MS: *PerSeptive Biosystems* MALDI-TOF spectrometer *Voyager* DE: ESI = electron-spray ionization.

Inosine Phosphoramidite (**17**). Synthesis according to a procedure reported by *Green et al.* [9].

2′,3′,5′-Tri-O-benzoylinosine (**2**). Benzoyl chloride (6.96 ml, 60 mmol) was added dropwise to a stirred suspension of inosine (**1**; 2.68 g, 10 mmol) in dry pyridine (70 ml) at 40°. The mixture was heated for 2 h at 60° and then for 2 h at 40° and finally stirred overnight at r.t. The red soln. was treated with 5% aq. $NaHCO_3$ soln. (70 ml) and extracted with CH_2Cl_2 (3 × 40 ml). The combined org. layers were dried (Na_2SO_4), evaporated, and twice co-evaporated with toluene to remove traces of pyridine. FC (CH_2Cl_2/MeOH 95 : 5) gave **2** (5.36 g, 92.3%). Colorless foam. TLC (CH_2Cl_2/MeOH 95 : 5): R_f 0.38. ¹H-NMR (250 MHz, (D_6)DMSO): 12.49 (br. *s*, NH); 8.37 (*s*, H−C(8)); 8.01 (*s*, H−C(2)); 7.98–7.41 (*m*, 15 arom. H); 6.56 (*d*, ³$J(1′,2′)$ = 4.63, H−C(1′)); 6.40 (*t*, H−C(2′)); 6.19 (*t*, H−C(3′)); 4.85 (*m*, H−C(4′)); 4.77 (*m*, 2 H−C(5′)). ¹³C-NMR (62.9 MHz, (D_6)DMSO): 165.42, 164.79, 164.63 (C=O); 156.64 (C(6)); 148.21 (C(4)); 145.83 (C(2)); 139.75 (C(8)); 133.56, 133.52, 133.32, 129.71, 128.78, 128.71, 128.67 (arom. C); 124.71 (C(5)); 86.02 (C(1′)); 79.23 (C(4′)); 73.42 (C(2′)); 70.31 (C(3′)); 63.78 (C(5′)). ESI-MS: 581.32 ([M + H]⁺). Anal. calc. for $C_{31}H_{24}N_4O_8$ (580.55): C 64.14, H 4.17, N 9.65; found: C 64.05, H 4.31, N 9.70.

2′,3′,5′-Tri-O-benzoyl-6-thioinosine (**3**). To a stirred soln. of **2** (2.9 g, 5 mmol) in pyridine (15 ml), *Lawesson*'s reagent (4.44 g, 10 mmol) was added in one portion, and the mixture was heated at reflux temp. for 6 h. At r.t. the mixture was treated with 5% aq. $NaHCO_3$ soln. (30 ml) and extracted with CH_2Cl_2 (2 × 70 ml). The combined org. phase was dried (Na_2SO_4) and evaporated, and the residue twice co-evaporated with toluene to yield a yellow foam. FC (CH_2Cl_2/MeOH 95 : 5) gave **3** (2.7 g, 90.4%). Colorless foam. TLC (CH_2Cl_2/MeOH 95 : 5): R_f 0.52. ¹H-NMR (250 MHz, (D_6)DMSO): 8.57 (*s*, H−C(8)); 8.07 (*s*, H−C(2)); 7.99–7.40 (*m*, 15 arom. H); 6.59 (*d*, ³$J(1′,2′)$ = 4.63, H−C(1′)); 6.39 (*t*, H−C(2′)); 6.18 (*t*, H−C(3′)); 4.84 (*m*, H−C(4′)); 4.74 (*m*, 2 H−C(5′)). ¹³C-NMR (62.9, (D_6)DMSO): 176.32 (C(6)); 165.42, 164.66, 164.49 (C=O); 145.39 (C(2)); 143.59 (C(4)); 142.16 (C(8)); 135.90 (C(5)); 134.01, 133.91, 133.54, 129.39, 129.27, 128.76, 128.72, 128.56, 128.29 (arom. C); 86.55 (C(1′)); 79.46 (C(4′)); 73.36 (C(2′)); 70.69 (C(3′)); 63.24 (C(5′)). ESI-MS: 597.19 ([M + H]⁺). Anal. calc. for $C_{31}H_{24}N_4O_7S$ (596.61): C 62.41, H 4.05, N 9.36; found: C 62.16, H 4.24, N 9.16.

6-Thioinosine (**4**). To a stirred mixture of pyridine (2.5 ml), MeOH (2.5 ml), and 2M NaOH (6 ml) **3** was added (1.6 g, 2.81 mmol) at r.t. After 5 min, the deprotection was complete, and the mixture was treated with *Dowex* ion-exchange resin (*50 W × 8*) until a pH value of 6 was reached. The filtrate was evaporated and co-evaporated twice with toluene. Recrystallization from H_2O (charcoal treatment) gave **4** (682 mg, 85.4%). Colorless product. TLC (CH_2Cl_2/MeOH 8 : 2): R_f 0.25. ¹H-NMR (250 MHz, (D_6)DMSO): 8.49 (*s*, H−C(8)); 8.21 (*s*, H−C(2)); 5.87 (*d*, ³$J(1′,2′)$ = 5.55, H−C(1′)); 5.52 (br. *s*, OH−C(2′)); 5.21 (br. *s*, OH−C(3′),

OH−C(5′)); 4.48 (*m*, H−C(2′)); 4.13 (*m*, H−C(3′)); 3.94 (*m*, H−C(4′)); 3.61 (*m*, 2 H−C(5′)). ^{13}C-NMR (62.9 MHz, (D$_6$)DMSO): 176.89 (C(6)); 145.78 (C(2)); 144.02 (C(4)); 140.89 (C(8)); 135.55 (C(5)); 87.61 (C(1′)); 85.64 (C(4′)); 74.08 (C(2′)); 70.21 (C(3′)); 61.16 (C(5′)). Anal. calc. for C$_{10}$H$_{12}$N$_4$O$_4$S (284.29): C 42.25, H 4.25, N 19.71; found: C 41.98, H 4.28, N 19.47.

9-(β-D-Ribofuranosyl)-9H-purine (**5**). Compound **4** (730 mg, 2.57 mmol) was dissolved in boiling H$_2$O (50 ml), and small portions of *Raney*-Ni (4 g) were added to the stirred soln. After 6 h at reflux temp., the hot suspension was filtered and the catalyst washed with hot H$_2$O. The filtrate was evaporated, the residue dissolved in hot abs. EtOH, charcoal added, and the mixture filtered and slowly cooled to give **5** (515 mg, 79%). Colorless product. TLC (CH$_2$Cl$_2$/MeOH 8:2): R_f 0.30. ^{1}H-NMR (250 MHz, (D$_6$)DMSO): 9.21 (*s*, H−C(6)); 8.97 (*s*, H−C(2)); 8.86 (*s*, H−C(8)); 6.06 (*d*, $^3J(1′,2′) = 5.55$, H−C(1′)); 5.52 (br. *s*, OH−C(2′)); 5.25 (br. *s*, OH−C(3′)); 5.12 (br. *s*, OH−C(5′)); 4.65 (*t*, H−C(2′)); 4.21 (*t*, H−C(3′)); 3.99 (*m*, H−C(4′)); 3.61 (*m*, 2 H−C(5′)).

9-[5′-O-(4,4′-Dimethoxytriphenylmethyl)-β-D-ribofuranosyl]-9H-purine (**6**). Compound **5** (337 mg, 1.34 mmol) was dried by repeated co-evaporation with abs. pyridine and dissolved in dry pyridine (10 ml). A soln. of 4,4′-dimethoxytriphenylmethyl chloride (528 mg, 1.56 mmol) in pyridine (3 ml) was added and the mixture stirred for 7 h under Ar at r.t. The mixture was treated with 5% aq. NaHCO$_3$ soln. (20 ml) and extracted with CH$_2$Cl$_2$ (2 × 15 ml). The combined org. layer was dried (Na$_2$SO$_4$) and evaporated, and the residue twice co-evaporated with toluene to yield a yellow foam. FC (CH$_2$Cl$_2$/MeOH 95:5) gave **6** (630 mg, 84.7%). Colorless foam. TLC (CH$_2$Cl$_2$/MeOH 95:5): R_f 0.2. ^{1}H-NMR (250 MHz, (D$_6$)DMSO): 9.05 (*s*, H−C(6)); 8.82 (*s*, H−C(2)); 8.52 (*s*, H−C(8)); 7.37−6.71 (*m*, 13 arom. H); 6.11 (*d*, $^3J(1′,2′) = 4.7$, H−C(1′)); 5.46 (*d*, OH−C(2′)); 5.11 (*d*, OH−C(3′)); 4.76 (*dd*, H−C(2′)); 4.35 (*dd*, H−C(3′)); 4.16 (*m*, H−C(4′)); 3.71 (*s*, 2 MeO); 3.28 (*m*, 2 H−C(5′)). ^{13}C-NMR (62.9 MHz, (D$_6$)DMSO): 158.00 ((MeO)$_2$*Tr*); 151.87 (C(2)); 150.82 (C(4)); 147.88 (C(6)); 144.9 ((MeO)$_2$*Tr*); 144.64 (C(8)); 135.40 ((MeO)$_2$*Tr*); 134.29 (C(5)); 129.66, 129.47, 127.67, 127.45, 126.41, 112.78 ((MeO)$_2$*Tr*); 88.17 (C(1′)); 85.63 ((MeO)$_2$*Tr*); 83.50 (C(4′)); 73.37 (C(2′)); 70.37 (C(3′)); 63.41 (C(5′)); 54.78 ((MeO). ESI-MS: 555.41 ([*M* + H]$^+$).

9-[5′-O-(4,4′-Dimethoxytriphenylmethyl)-2′-O-[(tert-butyl)dimethylsilyl]-β-D-ribofuranosyl]-9H-purine (**7**). Under Ar, **6** (332 mg, 0.6 mmol) was dissolved in pyridine/THF 1:1 (6 ml). AgNO$_3$ (122 mg, 0.72 mmol) and 1M (*tert*-butyl)dimethylsilyl chloride in THF (0.72 ml) were added to the stirred soln. After 14 h, the reaction was quenched by adding 5% NaHCO$_3$ soln. (6 ml). The suspension was filtered and the filtrate extracted with CH$_2$Cl$_2$ (3 × 10 ml). The combined org. phase was dried (Na$_2$SO$_4$) and evaporated and the residue co-evaporated twice with toluene to yield a foam. FC (silica gel *60 H*, CH$_2$Cl$_2$/^{i}PrOH 98:2 → 95:5) gave **7** (214 mg, 53.2%). Colorless foam. TLC (CH$_2$Cl$_2$/^{i}PrOH 95:5): R_f 0.51. ^{1}H-NMR (250 MHz, (D$_6$)DMSO): 9.27 (*s*, H−C(6)); 8.92 (*s*, H−C(2)); 8.80 (*s*, H−C(8)); 7.45−6.86 (*m*, 13 arom. H); 6.14 (*d*, $^3J(1′,2′) = 4.88$, H−C(1′)); 5.27 (*d*, OH−C(3′)); 4.94 (*t*, H−C(2′)); 4.32 (*dd*, H−C(3′)); 4.19 (*m*, H−C(4′)); 3.77 (*s*, 2 MeO); 3.35 (*m*, 2 H−C(5′)); 0.78 (*s*, ^{t}Bu); 0.00 (*s*, MeSi); −0.11 (*s*, MeSi). ^{13}C-NMR (62.9 MHz, (D$_6$)DMSO): 158.07 ((MeO)$_2$*Tr*); 152.07 (C(2)); 150.67 (C(4)); 148.37 (C(6)); 144.83 ((MeO)$_2$*Tr*); 144.33 (C(8)); 135.45 ((MeO)$_2$*Tr*); 134.22 (C(5)); 129.68, 129.44, 127.72, 127.60, 126.63, 113.09 ((MeO)$_2$*Tr*); 88.13 (C(1′)); 85.50 ((MeO)$_2$*Tr*); 83.59 (C(4′)); 74.69 (C(2′)); 70.10 (C(3′)); 63.39 (C(5′)); 54.98 (MeO); 25.47 (*Me$_3$CSi*); 17.76 (*Me$_3$CSi*); −4.85, −5.38 (2 MeSi). ESI-MS: 669.57 ([*M* + H]$^+$). Anal. calc. for C$_{37}$H$_{44}$N$_4$O$_6$Si (668.87): C 66.44, H 6.63, N 8.38; found: C 66.46, H 6.78, N 8.13.

9-[5′-O-(4,4′-Dimethoxytriphenylmethyl)-2′-O-[(tert-butyl)dimethylsilyl]-β-D-ribofuranosyl]-9H-purine 3′-(2-Cyanoethyl Diisopropylphosphoramidite) (**8**). To a stirred soln. of **7** (200 mg, 0.3 mmol), collidine (= 2,4,6-trimethylpyridine; 200 μl, 1.5 mmol), and a catalytic amount of 1-methyl-1*H*-imidazole (12 μl, 0.15 mmol) in dry MeCN (6 ml), 2-cyanoethyl diisopropylphosphoramidochloridite (100 μl, 0.45 mmol) was added dropwise at r.t. under Ar. After 2 h, the reaction was quenched by adding 5% aq. NaHCO$_3$ soln. (10 ml), the mixture extracted with CH$_2$Cl$_2$ (3 × 10 ml), the combined org. layer dried (Na$_2$SO$_4$) and evaporated, and the residue purified by FC (silica gel *60 H*, Et$_2$O/MeCN 9:1): diastereomer mixture **8** (206 mg, 79.3%). White foam. TLC (Et$_2$O/MeCN 9:1): R_f 0.58; 2 partially overlapping spots. ^{1}H-NMR (400 MHz, CDCl$_3$, diastereomer mixture): 9.15, 9.14 (2*s*, H−C(6)); 8.88, 8.87 (2*s*, H−C(2)); 8.34, 8.30 (2*s*, H−C(8)); 7.48−6.80 (*m*, 13 arom. H); 6.12, 6.07 (2*d*, H−C(1′)); 5.08 (*m*, H−C(2′)); 4.46−4.37 (*m*, H−C(3′)); H−C(4′)); 3.96, 3.88 (2*m*, 2 Me$_2$CH); 3.78 (*s*, 2 MeO); 3.66−3.54 (*m*, CH$_2$CH$_2$O); 3.45 (*m*, 2 H−C(5′)); 2.65, 2.31 (2*m*, CH$_2$CH$_2$O); 1.21−1.05 (2 *Me$_2$CH*); 0.74, 0.71 (2*s*, ^{t}Bu); −0.03, −0.25 (2 MeSi). ^{31}P-NMR (160 MHz, CDCl$_3$): 151.7, 149.6; ratio 0.7:1.

N^2,2′-O,3′-O,5′-O-Tetrabenzoylguanosine (**10**). Benzoyl chloride (7 ml, 60.5 mmol) was added dropwise to a stirred suspension of guanosine (**9**; 2.83 g, 10 mmol) in dry pyridine (40 ml) at 40°. During the addition, the temp. rose up to 60°. After 2 h, the brown soln. was cooled down to r.t. and quenched by adding 5% aq. NaHCO$_3$ soln. (40 ml) and CH$_2$Cl$_2$ (20 ml). The mixture was extracted with CH$_2$Cl$_2$ (3 ×). The combined org. phase was dried (Na_2SO_4), evaporated, and co-evaporated twice with toluene to yield a brown foam. FC (CH_2Cl_2/MeOH 95:5) gave **10** (6.11 g, 87.3%). Colorless foam. TLC (CH_2Cl_2/MeOH 95:5): R_f 0.56. ^{1}H-NMR (250 MHz, (D_6)DMSO): 12.28 (br. s, NH); 11.72 (br. s, NHBz); 8.35 (s, H–C(8)); 8.15–7.42 (m, 20 arom. H); 6.50 (d, $^3J(1',2') = 4.95$, H–C(1')); 6.39 (t, H–C(2')); 6.21 (t, H–C(3')); 4.91 (m, H–C(4')); 4.80 (m, 2 H–C(5')). ^{13}C-NMR (62.9 MHz, (D_6)DMSO): 169.02 (NHCO); 165.47, 164.72, 164.47 (C=O); 154.98 (C(6)); 148.38 (C(2)); 148.29 (C(4)); 138.72 (C(8)); 133.97, 133.56, 133.18, 132.44, 129.39, 129.30, 129.25, 128.84, 128.80, 128.74, 128.61, 128.47, 128.24 (arom. C); 121.14 (C(5)); 86.03 (C(1')); 79.67 (C(4')); 73.42 (C(2')); 71.44 (C(3')); 63.79 (C(5')). ESI-MS: 700.15 ($[M + H]^+$). Anal. calc. for $C_{38}H_{29}N_5O_9$ (699.68): C 65.23, H 4.18, N 10.01; found: C 65.47, H 4.24, N 9.82.

N²,2'-O,3'-O,5'-O-Tetrabenzoyl-6-thioguanosine (**11**). To a soln. of **10** (1.4 g, 2.0 mmol) in abs. pyridine, (20 ml), *Lawesson's* reagent was added (1.62 g, 4.0 mmol) in one portion at r.t. The mixture was heated at reflux temp. for 6 h. At r.t. pyridine was evaporated, the residue dissolved in CH_2Cl_2 (40 ml), this soln. washed with 5% aq. $NaHCO_3$ soln. (60 ml), dried (Na_2SO_4), and evaporated, and the residue co-evaporated twice with toluene to yield a yellow foam. FC (CH_2Cl_2/MeOH 95:5) gave **11** (1.23 g, 85.6%). Yellow foam. TLC (CH_2Cl_2/MeOH 95:5): R_f 0.71. ^{1}H-NMR (250 MHz, (D_6)DMSO): 13.61 (br. s, SH); 11.97 (br. s, NHBz); 8.53 (s, H–C(8)); 8.02–7.42 (m, 20 arom. H); 6.52 (d, $^3J(1',2') = 4.95$, H–C(1')); 6.39 (t, H–C(2')); 6.20 (t, H–C(3')); 4.90 (m, H–C(4')); 4.85 (m, 2 H–C(5')). ^{13}C-NMR (62.9 MHz, (D_6)DMSO): 174.71 (C(6)); 169.25 (NHCO); 165.46, 164.68, 164.47 (C=O); 147.73 (C(2)); 144.88 (C(4)); 141.10 (C(8)); 134.03, 133.94, 133.53, 133.32 (arom. C, C(5)); 132.47, 132.20, 129.39, 129.34, 129.22, 128.82, 128.74, 128.61, 128.52, 128.22 (arom. C); 86.10 (C(1')); 79.80 (C(4')); 73.43 (C(2')); 71.41 (C(4')); 63.75 (C(5')). ESI-MS: 716.09 ($[M + H]^+$). Anal. calc. for $C_{38}H_{29}N_5O_8S$ (715.74): C 63.77, H 4.08, N 9.78; found: C 63.68, H 4.17, N 9.69.

N²-Benzoyl-6-thioguanosine (**12**). As described for **4**, with pyridine (2 ml), MeOH (1 ml), 2M NaOH (1 ml), and **11** (2.7 g, 3.77 mmol). Recrystallization from EtOH gave **12** (1.25 g, 82.4%). Colorless product. TLC (CH_2Cl_2/MeOH 8:2): R_f 0.32. ^{1}H-NMR (250 MHz, (D_6)DMSO): 13.68 (br. s, SH); 12.21 (br. s, NHBz); 8.49 (s, H–C(8)); 8.09–7.49 (m, 5 arom. H); 5.90 (d, $^3J(1',2') = 5.83$, H–C(1')); 5.50 (d, OH–C(2')); 5.19 (d, OH–C(3')); 5.06 (t, OH–C(5')); 4.48 (dd, H–C(2')); 4.16 (dd, H–C(3')); 3.93 (m, H–C(4')); 3.61 (m, 2 H–C(5')). ^{13}C-NMR (62.9 MHz, (D_6)DMSO): 174.20 (C(6)); 167.66 (NHCO); 152.49 (C(2)); 146.98 (C(4)); 139.93 (C(8)); 136.08 (C(5)); 132.43, 131.61, 131.08, 129.22, 128.47, 128.40, 128.22 (arom. C); 87.12 (C(1')); 85.69 (C(4')); 73.74 (C(2')); 70.43 (C(4')); 61.43 (C(5')). ESI-MS: 404.81 ($[M + H]^+$).

N-Benzoyl-9-(β-D-ribofuranosyl)-9H-purin-2-amine (**13**). As described for **5**, with **12** (888 mg, 2.2 mmol), boiling H_2O (50 ml), and *Raney-Ni* (4 g). After workup, the residue was dissolved in hot H_2O and the soln. treated with charcoal, filtrated, and cooled down slowly to give **13** (688 mg, 84.2%). Colorless product. TLC (CH_2Cl_2/MeOH 8:2): R_f 0.28. ^{1}H-NMR (250 MHz, (D_6)DMSO): 11.09 (s, NH); 9.10 (s, H–C(6)); 8.75 (s, H–C(8)); 8.00–7.48 (m, 5 arom. H); 6.00 (d, $^3J(1',2') = 5.73$, H–C(1')); 5.53 (d, OH–C(2')); 5.23 (d, OH–C(3')); 4.99 (t, OH–C(5')); 4.69 (dd, H–C(2')); 4.21 (m, H–C(3')); 3.96 (m, H–C(4')); 3.65 (m, 2 H–C(5')). ^{13}C-NMR (62.9 MHz, (D_6)DMSO): 165.50 (C=O); 153.02 (C(2)); 152.03 (C(4)); 148.89 (C(6)); 145.01 (C(8)); 134.30 (C(5)); 131.97, 131.22, 128.23, 128.13 (arom. C); 87.14 (C(1')); 85.86 (C(4')); 73.44 (C(2')); 70.52 (C(3')); 61.13 (C(5')). ESI-MS: 371.87 ($[M + H]^+$). Anal. calc. for $C_{17}H_{17}N_5O_9$ (371.35): C 54.98, H 4.16, N 18.86; found: C 54.91, H 4.84, N 18.74.

N-Benzoyl-9-[5'-O-(4,4'-dimethoxytriphenylmethyl)-β-D-ribofuranosyl]-9H-purin-2-amine (**14**). As described for **6**, with **13** (743 mg, 2.0 mmol), pyridine (8 ml), 4,4'-dimethoxytriphenylmethyl chloride (744 mg, 2.2 mmol), and pyridine (4 ml) (6 h). Workup with 5% aq. $NaHCO_3$ soln. (30 ml) and CH_2Cl_2 (2 × 20 ml) and FC as described for **6**: **14** (1.09 g, 80.9%). Colorless foam. TLC (CH_2Cl_2/MeOH 95:5): R_f 0.37. ^{1}H-NMR (250 MHz, (D_6)DMSO): 9.08 (s, H–C(6)); 8.62 (s, H–C(8)); 7.92–6.59 (m, 18 arom. H); 6.03 (d, $^3J(1',2') = 4.48$, H–C(1')); 5.60 (d, OH–C(2')); 5.15 (d, OH–C(3')); 4.78 (dd, H–C(2')); 4.40 (dd, H–C(3')); 4.23 (m, H–C(4')); 3.65 (s, 2 MeO); 3.22 (m, 2 H–C(5')). ^{13}C-NMR (62.9 MHz, (D_6)DMSO): 165.16 (C=O); 157.92 ((MeO)$_2$Tr); 153.06 (C(2)); 151.57 (C(4)); 148.91 (C(6)); 145.36 (C(8)); 135.63, 135.47 ((MeO)$_2$Tr); 134.42 (C(5)); 131.98, 131.23, 129.68, 128.30, 128.26, 127.64, 127.55, 126.52, 112.92 ((MeO)$_2$Tr); 88.62 (C(1')); 85.35 ((MeO)$_2$Tr); 84.04 (C(4')); 73.03 (C(2')); 70.67 (C(3')); 64.71 (C(5')); 54.92 (MeO). ESI-MS: 674.53 ($[M + H]^+$). Anal. calc. for $C_{38}H_{35}N_5O_7$ (673.73): C 67.75, H 5.24, N 10.39; found: C 67.22, H 5.29, N 10.36.

N²-Benzoyl-9-[5'-O-(4,4'-dimethoxytriphenylmethyl)-2'-O-[(tert-butyl)dimethylsilyl]-β-D-ribofuranosyl]-9H-purin-2-amine (**15**). As described for **7**, with **14** (572 mg, 0.85 mmol), pyridine/THF 1:1 (6 ml) $AgNO_3$ (172 mg, 1.015 mmol) and 1M (*tert*-butyl)dimethylsilyl chloride in THF (1.02 ml): **15** (370 mg, 55.4%). Colorless foam. TLC (CH_2Cl_2/^{i}PrOH 95:5): R_f 0.51. ^{1}H-NMR (250 MHz, (D_6)DMSO): 10.44 (s, NH); 8.99 (s, H–C(6)); 8.47 (s, H–C(8)); 7.81 (d, 2 arom. H); 7.62–7.11 (m, 13 arom. H); 6.65 (t, 3 arom. H); 6.08 (d, $^3J(1',2') = 5.15$, H–C(1')); 5.01 (t, H–C(2')); 4.61 (d, OH–C(3')); 4.36 (dd, H–C(3')); 4.16 (br. s, H–C(4')); 3.70 (s, 2 MeO);

3.38 (*m*, H−C(5′)); 0.78 (*s*, *t*Bu); 0.00 (*s*, 1 MeSi); − 0.21 (*s*, 1 MeSi). ^{13}C-NMR (62.9 MHz, (D$_6$)DMSO): 164.82 (C=O); 157.94 ((MeO)$_2$*Tr*); 153.03 (C(2)); 151.64 (C(4)); 148.61 (C(6)); 144.79 (C(8)); 144.43, 135.45, 135.40 ((MeO)$_2$*Tr*); 134.30 (C(5)); 131.49, 131.20, 129.64, 127.94, 127.86, 127.61, 127.34, 126.35, 112.67 ((MeO)$_2$*Tr*); 88.16 (C(1′)); 85.51 ((MeO)$_2$*Tr*); 84.30 (C(4′)); 74.78 (C(2′)); 70.54 (C(3′)); 64.78 (C(5′)); 54.72 (MeO); 25.41 (*Me*$_3$CSi); 17.67 (Me$_3$CSi); − 5.08 (MeSi); − 5.42 (MeSi). ESI-MS: 788.32 ([*M* + H]$^+$).

N^2-*Benzoyl-9-[5′-O-(4,4′-dimethoxytriphenylmethyl)-2′-O-[(tert-butyl)dimethylsilyl]-β-D-ribofuranosyl]-purin-9H-2-amine 3′-(2-Cyanoethyl Diisopropylphosphoramidite)* (**16**). As described for **8**, with **15** (218 mg, 0.27 mmol), collidine (179 µl, 1.35 mmol), 1-methyl-1*H*-imidazole (10.7 µl, 0.13 mmol), MeCN (10 ml), and 2-cyanoethyl diisopropylphosphoramidochloridite (90 µl, 0.41 mmol): diastereomer mixture **16** (225 mg, 84.3%). White foam. TLC (Et$_2$O/MeCN 9:1): R_f 0.59; 2 partially overlapping spots. ^{1}H-NMR (400 MHz, CDCl$_3$, diastereomer mixture): 9.19, 9.16 (2*s*, H−C(6)); 8.18, 8.15 (br. *s*, NH); 8.16, 8.13 (2*s*, H−C(8)); 7.71−6.66 (*m*, 18 arom. H); 5.99, 5.79 (2*d*, H−C(1′)); 5.41, 5.31 (2*dd*, H−C(2′)); 4.38, 4.28 (2*m*, H−C(3′)); 4.04, 3.85 (2*m*, H−C(4′)); 3.68, 3.67 (2*s*, 2 MeO); 3.65−3.48 (*m*, 2 H−C(5′), CH$_2$CH$_2$O); 2.69 (*m*, CH$_2$CH$_2$O); 1.20−0.92 (2 *Me*$_2$CH); 0.73, 0.72 (2*s*, *t*Bu); − 0.04, − 0.31 (2 MeSi). ^{31}P-NMR (160 MHz, CDCl$_3$): 152.3, 149.4; ratio 1:2. ESI-MS: 988.71 ([*M* + H]$^+$).

REFERENCES

[1] M. Delarue, D. Moras, in 'Nucleic Acids and Molecular Biology', Ed. F. Eckstein and D. M. Lilley, Springer Verlag, Berlin-Heidelberg, 1989, Vol. 3, pp. 183 – 196.

[2] C. R. Woese, S. Winkler, R. R. Gutell, *Biochemistry* **1990**, *87*, 8467.

[3] O. C. Uhlenbeck, *Nature (London)* **1990**, *346*, 613.

[4] J. A. Heus, A. Pardi, *Science (Washington, D.C.)* **1991**, *253*, 191.

[5] J. H. Cate, A. R. Gooding, E. Podell, K. Zhou, B. L. Golden, C. E. Kundrot, T. R. Cech, J. A. Doudna, *Science (London)* **1996**, *273*, 1678.

[6] M. J. Serra, T. J. Axenson, D. H. Turner, *Biochemistry* **1994**, *33*, 14289.

[7] J. SantaLucia Jr., R. Kierzek, D. H. Turner, *Science (Washington, D.C.)* **1992**, *256*, 217.

[8] N. Leulliot, V. Baumruk, M. Abdelkafi, P.-Y. Turpin, A. Namane, C. Gouyette, T. Huynh-Dinh, M. Ghomi, *Nucleic Acids Res.* **1999**, *27*, 1398.

[9] R. Green, J. W. Szostak, S. A. Benner, A. Rich, N. Usman, *Nucleic Acids Res.* **1991**, *19*, 4161.

[10] M. H. Lyttle, P. B. Wright, N. D. Sinha, J. D. Bain, A. R. Chamberlin, *J. Org. Chem.* **1991**, *56*, 4608.

[11] E. Westmann, R. Strömberg, *Nucleic Acids Res.* **1994**, *22*, 2430.

[12] M. Schweitzer, J. W. Engels, in 'Antisense – From Technology to Therapy', Ed. R. Schlingensiepen, W. Brysch, K.-H. Schlingensiepen, Blackwell Science, 1997, Vol. 6, pp. 78 – 103.

[13] L. A. Marky, K. J. Breslauer, *Biopolymers* **1987**, *26*, 1601.

[14] E. Gallicchio, M. Kubo, R. Levy, *J. Am. Chem. Soc.* **1998**, *120*, 4526.

[15] O. C. Uhlenbeck, P. N. Borer, B. Dengler, I. Tinoco Jr., *J. Mol. Biol.* **1973**, *73*, 483.

[16] K. Wörner, Dissertation 1997, Johann Wolfgang Goethe-Universität, Frankfurt am Main.

[17] J. A. Heus, A. Pardi, *Science (Washington, D.C.)* **1991**, *253*, 191.

[18] R. T. Batey, R. P. Rambo, J. A. Doudna, *Angew. Chem.* **1999**, *111*, 2472.

[19] M. Abdelkafi, N. Leulliot, V. Baumruk, L. Bednárová, P. Y. Turpin, A. Namane, C. Gouyette, T. Huynh-Dinh, M. Ghomi, *Biochemistry* **1998**, *37*, 7878.

Synthesis of 3'-Thioamido-Modified 3'-Deoxythymidine 5'-Triphosphates by Regioselective Thionation and Their Use as Chain Terminators in DNA Sequencing

by Christian Wojczewski, Karin Schwarzer, and Joachim W. Engels*

Institut für Organische Chemie, Johann Wolfgang Goethe-Universität, Marie-Curie-Str. 11,
60439 Frankfurt am Main, Germany (phone: (+49)69-798 29146; fax: (+49)69-798 29148;
e-mail: engels@ews1.org.chemie.uni-frankfurt.de)

The thioamide derivatives 3'-deoxy-5'-O-(4,4'-dimethoxytrityl)-3'-[(2-methyl-1-thioxopropyl)amino]thymidine (**4a**) and 3'-deoxy-5'-O-(4,4'-dimethoxytrityl)-3'-{{6-{[(9H-(fluoren-9-ylmethoxy)carbonyl]amino}-1-thioxohexyl}amino}thymidine (**4b**) were synthesized by regioselective thionation of the corresponding amides **3a** and **3b** with 2,4-bis(4-methoxyphenyl)-1,3,2,4-dithiadiphosphetane 2,4-disulfide (*Lawesson*'s reagent). The addition of exact amounts of pyridine to the reaction mixture proved to be essential for an efficient transformation. The thioamides were converted into the corresponding 5'-triphosphates **6a** and **6b**. Compound **6a** was chosen for DNA sequencing experiments, and **6b** was further labelled with fluorescein ($\rightarrow$ **8**).

Introduction. – Recently, the first complete sequence of a human chromosome has been published [1]. Chromosome 22 is only one step in the huge Human Genome Project, which, itself, is just one of several genome-sequencing projects that are currently being carried out [2]. The *Sanger* procedure, invented in 1977, is still the method of choice to determine the sequence of bases [3][4]. Although the development of automated methods, combined with the dye-primer or dye-terminator chemistry, has greatly accelerated the speed of sequencing, the need for high-throughput and accurate DNA-sequencing techniques is still rapidly growing.

Dye-terminator sequencing is characterized by the enzymatic incorporation of chain-terminating, fluorescent nucleotides. These terminators are attached to fluorescent dyes at their base moiety [5–8]. Apart from this labelling strategy, there are other potential positions within nucleosides that may be used for markers [9]. Different groups have demonstrated that 3'-ether- and 3'-ester-functionalized nucleotides can act as substrates for an enzyme-mediated DNA synthesis [10–12]. We have shown that 3'-amino- and 3'-amido-modified nucleoside 5'-triphosphates are excellent terminators [13][14]. However, the ability of DNA polymerases to hydrolyze ester and amide bonds at the 3'-position limits the use of these 3'-dye terminators for DNA sequencing since the label is cleaved off during the incorporation process [15].

Therefore, we decided to alter the link between dye and sugar moiety to find a linkage that is stable against enzymatic degradation. In this context, we have already reported the synthesis of 3'-thioether- [16], 3'-alkyl- [17], 3'-urea-, and 3'-thiourea-modified [18] 2',3'-dideoxynucleoside 5'-triphosphates.

Here, we describe an efficient synthesis of 3'-thioamido-modified thymidines, a new class of sugar-modified nucleosides, and the investigation of their substrate acceptance in DNA-sequencing experiments.

Results and Discussion. – The starting material 3'-azido-3'-deoxy-5'-O-(4,4'-dimethoxytrityl)thymidine (**1**) was obtained as described before [19][20] and was converted into the amino derivative **2** by hydrogenolytic reduction catalyzed by $PtO_2 \cdot H_2O$ (*Scheme 1*). This procedure allowed the isolation of **2** in high yield without further purification.

To our knowledge, no synthetic route to thioamido-tethered nucleosides has been published until now. Although a broad spectrum of reactions exists for the synthesis of thioamides [21][22], only the conversion of amides into thioamides seems to be suitable for nucleoside chemistry. Therefore, we synthesized from **2** the readily accessible amide **3a** as a model compound for the investigation of the thionation reaction. In a second synthetic approach, we chose linker **9** to introduce a free amino group for dye labelling at position 3'. This amino function had to be protected to prevent side reactions during the triphosphate synthesis. The protecting group ought to be removable under basic conditions due to the acid lability of triphosphates. Therefore, we chose the Fmoc group (*Scheme 2*), *i.e.*, the linker **10** was introduced *via* an amide coupling at the 3'-position of the sugar moiety, resulting in **3b**.

Direct action of H_2S or P_2S_5 on **3** would result in an undesired exchange of the two carbonyl O-atoms at C(2) and C(4) of the base moiety. A more sophisticated thionation reagent is 2,4-bis(4-methoxyphenyl)-1,3,2,4-dithiadiphosphetane 2,4-disulfide (*Lawesson*'s reagent), which converts carbonyl groups into thiocarbonyl groups in high yield [23–25]. Unbranched, alkyl-substituted amides can be thionated by *Lawesson*'s reagent at room temperature within minutes, whereas imides and compounds with unsaturated residues require longer reaction times and/or elevated temperatures. Thus, our aim was to utilize the different reaction conditions for a regioselective conversion at the C(3') position. *Wörner* showed that 2',3',5'-tri-O-benzoyl-N^2-benzoylguanosine and -inosine were thionated at the nucleobase C(6)=O function when refluxed with *Lawesson*'s reagent in pyridine for 6 h [26]. Therefore, **3a** was reacted with *Lawesson*'s reagent for 30 min without heating. Humidity had to be strictly eliminated to avoid formation of phosphoric acids and thus removal of the 5'-hydroxy-protecting group. Although no by-products were observed under these reaction conditions, the yield of **4a** was rather poor (22%); 33% of the educt was recovered unchanged.

The synthesis of **4b** under the same conditions resulted in even lower yields, caused mainly by 5'-hydroxy deprotection. Therefore, we tested basic solvents as scavengers for the phosphoric acids. Thionation of **3b** (*Fig. 1*) in pure anh. pyridine instead of anh. THF did not at all indicate any formation of **4b**, even when the amount of *Lawesson*'s reagent was drastically increased. Thus, the reaction of **3b** with *Lawesson*'s reagent was accomplished in mixtures of anh. THF/pyridine. We observed that lowering the amount of pyridine gradually resulted in increased yield. Finally, the addition of 1 equiv. of pyridine and 4 equiv. of *Lawesson*'s reagent in anh. THF as solvent yielded **4b** in 50%; almost 50% of the intact educt was recovered.

Both thioamides **4** were clearly identified by mass spectrometry and NMR spectroscopy. The electrospray-ionization MS (ESI-MS) showed formation of only one

Scheme 1

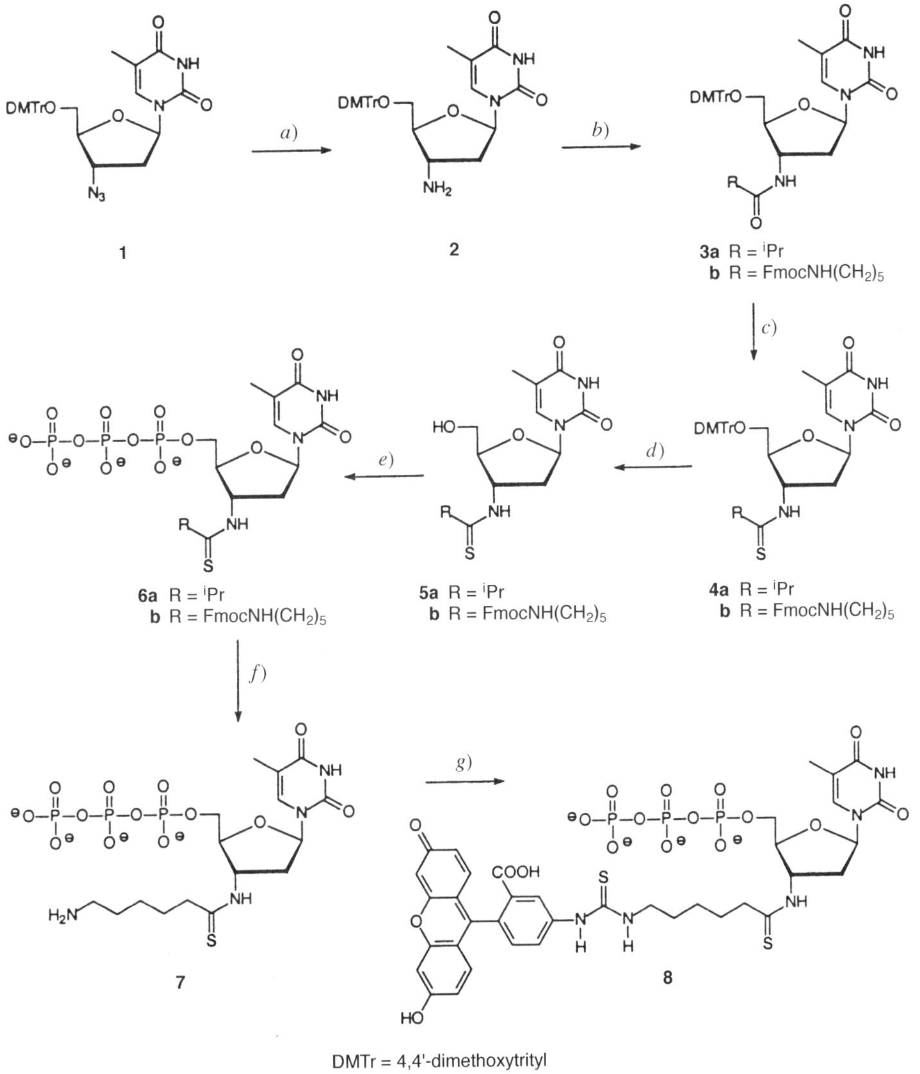

DMTr = 4,4'-dimethoxytrityl

a) PtO$_2$·H$_2$O, H$_2$, MeOH. *b*) For **3a**: (^{i}PrCO)$_2$O, pyridine; for **3b**: 6-{[(9*H*-fluoren-9-ylmethoxy)carbonyl]-amino}hexanoic acid (**10**), DMF, dioxane, H$_2$O, ^{i}Pr$_2$NEt, TSTU. *c*) For **4a**: *Lawesson*'s reagent, THF; for **4b**: *Lawesson*'s reagent, THF, pyridine. *d*) 80% aq. AcOH soln. *e*) 1. 2-chloro-4*H*-1,3,2-benzodioxaphosphorin-4-one, dioxane, pyridine, DMF; 2. (Bu$_3$NH)$_2$P$_2$O$_7$, DMF, tributylamine; 3. 1% I$_2$ in pyridine/H$_2$O 98:2; 4. 5% aq. NaHSO$_3$ soln. *f*) piperidine, pyridine, DMF. *g*) FITC, DMF, aq. NaHCO$_3$ soln. (pH 9.3).

product with an increased weight of 16 Da compared with the educt, representing the exchange of an O-atom against a S-atom. The ^{13}C-NMR spectrum revealed the identity of the carbonyl group involved. The ^{13}C-chemical shifts of C(2) and C(4) of the base moiety remained unchanged, whereas the carbonyl signal at the 3'-terminal moved downfield as expected (*Table*).

Scheme 2

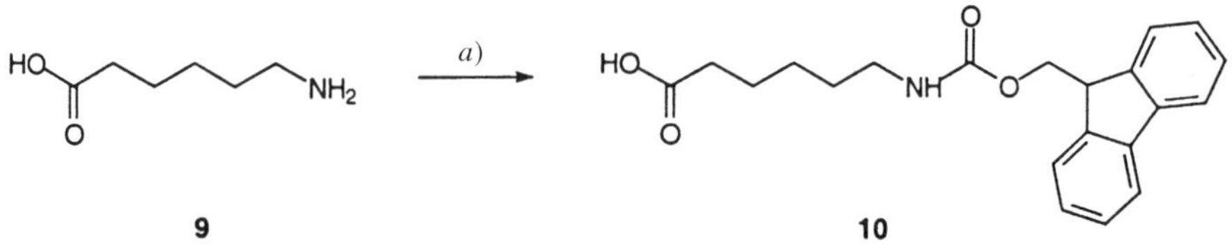

9 → **10**

a) N-[(9H-Fluoren-9-ylmethoxy)carbonyl]succinimide, NaHCO₃, H₂O, acetone.

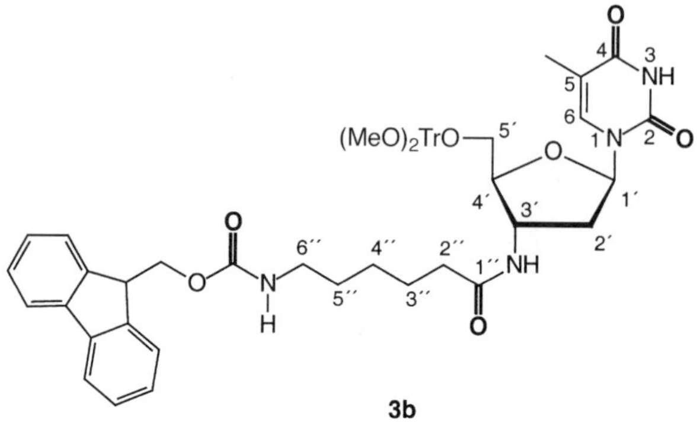

3b

Fig. 1. *5'-O-(4,4'-dimethoxytrityl)-3'-{{6-{[(9H-fluoren-9-ylmethoxy)carbonyl]amino}-1-oxohexyl}amino}thymidine* (**3b**)

Table. *Electrospray-Ionization MS (ESI-MS) and ¹³C-NMR Carbonyl Signals of Compounds **3** and **4***

	m/z	δ(C) [ppm] C(2)	C(4)	CONH–C(6″)	CXNH–C(3′)
3a (X = O)	612.7ª)	150.8	164.1	–	177.4
4a (X = S)	628.3ª)	150.7	164.2	–	213.2
3b (X = O)	896.6ᵇ)	150.8	163.9	156.5	173.2
4b (X = S)	893.6ᶜ)	150.7	164.2	156.4	206.4

ª) ESI (neg.). ᵇ) ESI (pos.) and NH₄⁺ adduct, M⁺ 879.0. ᶜ) ESI (neg.), M⁻ 895.1.

The thioamides **4a** and **4b** were deprotected in 80% aqueous AcOH solution at room temperature, and the syntheses of the triphosphates **6a** and **6b** were achieved according to the procedure of *Ludwig* and *Eckstein* [27]. The linker amino function of **6b** was deprotected with 20% piperidine in pyridine and DMF at room temperature (→ **7**) and subsequently coupled with fluorescein isothiocyanate (FITC) in aq. 0.1M NaHCO₃ (pH 9.3) and DMF at room temperature for 24 h (→ **8**).

Triphosphate **6a** was substituted for ddTTP as a terminator in DNA-sequencing experiments. No band pattern was detected with sequenase or with thermosequenase. With Taq DNA polymerase, we obtained a band pattern over a range of several hundred bases (*Fig. 2, Lanes 7–9*). This pattern did not correlate with the standard ddTTP sequence. It was, however, identical with the band pattern obtained with the 3'-(alkylthio)-3'-deoxythymidine 5'-triphosphate **11** (*Fig. 3*), an alternatively 3'-modified

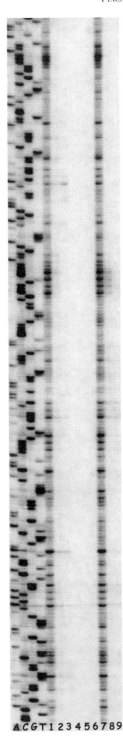

Fig. 2. *DNA Sequencing with IRD-41-labelled primer, M13 mp18 template DNA, Taq DNA Polymerase, and 2′-deoxynucleoside 5′-triphosphates* (dNTP) *as substrates.* Fluorescence detection on a *LI-COR* DNA sequencer. *Lanes A, C, G,* and *T* terminated with 2′,3′-dideoxyadenosine 5′-triphosphate (ddATP), ddCTP, ddGTP, or ddTTP, respectively; *Lanes 1–3* terminated with **11**, *lanes 7–9* terminated with **6a**. Terminator concentration: *Lanes 1* and *7*, 750 µM; *Lanes 2* and *8*, 3.75 mM; *3* and *9*, 5.25 mM.

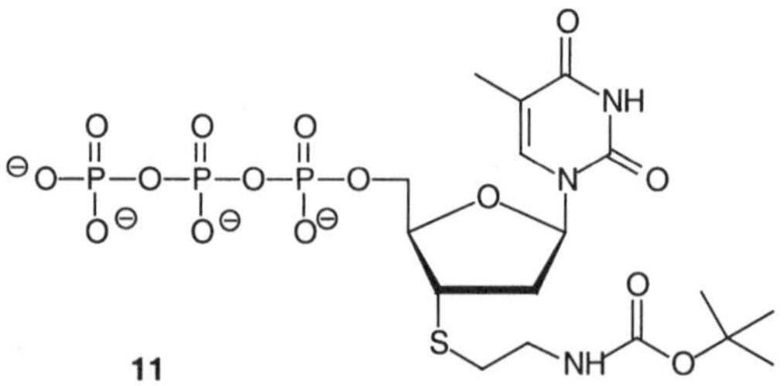

11

Fig. 3. *3'-{[2-[[(tert-Butoxy)carbonyl]amino]ethyl}thio]-3'-deoxythymidine 5'-triphosphate* (**11**)

nucleotide [16] (see *Fig. 2, Lanes 1 – 3*). Although **6a** and **11** are obviously not specific terminators for the enzymatic DNA synthesis under the chosen conditions, the result cannot be called nonspecific, since two structurally different nucleotides were incorporated at exactly the same positions within the DNA. This excludes a random process. Variation of the terminator concentration only resulted in a different fragment-length distribution. The reason for this could be a conformationally driven selection process.

Conclusion. – We demonstrated that 3'-thioamido-modified nucleosides can efficiently be synthesized from 3'-amido-tethered nucleosides by a regioselective thionation strategy applying *Lawesson*'s reagent. Addition of 1 equiv. of pyridine to the reaction mixture was necessary to prevent degradation of the nucleoside and ensure good yields at the same time. Thioamide **6a** was accepted as a substrate by the Taq DNA polymerase. However, specific incorporation did not occur. Rather the band pattern showed a distinct correlation with the 3'-thioether-tethered nucleoside triphosphate **11**.

We would like to thank the *Bundesministerium für Bildung und Forschung* (BEO0311088), the *Verband der Chemischen Industrie* and the *Deutsche Forschungsgemeinschaft (Graduiertenkolleg)* for financial support as well as *Roche Diagnostics* for cooperation and financial support. DNA-Sequencing experiments were performed by Dr. *S. Klingel* and Dr. *G. Sagner* at *Roche Diagnostics*, Penzberg, Germany.

Experimental Part

General. Solvents were of anal. grade and were dried and distilled or purchased and stored over molecular sieves. Anal. TLC: Silica gel *60 F_{254}* plates (*Merck*) for TLC. Prep. TLC: covered glass plates (gypsum silica gel *60 PF$_{254}$* from *Merck*), *Harrison Research 7924T Chromatotron*. FPLC (fast protein liquid chromatography): *Pharmacia* instrument with *LCC-500* controller, *P-500* pumps, and single-path UV monitor *UV-1*. HPLC: *Merck-Hitachi* with *L-4250-UV-VIS* detector and *Shimadzu RF-535* fluorescence monitor. DNA Sequencing: *LI-COR* automated DNA sequencer with M13 mp18 as template, Taq DNA polymerase, 5'-fluorescent-labelled primer, and **6a** or **11** (0.75 mM, 3.75 mM, 5.25 mM) as substitute for ddTTP. UV Spectra and optical density: *Varian Cary 1*; optical density measured at 270 nm (molar absorptivity in H_2O, 8700 l mol^{-1} cm^{-1}). IR Spectra: $\tilde{\nu}$ in cm^{-1}. ^{1}H-NMR Spectra: *Bruker AM250* (250 MHz) or *WH270* (270 MHz); δ in ppm; calibrating signals: δ((D_6)DMSO) = 2.50, δ(CDCl$_3$) = 7.26. ^{13}C-NMR Spectra: *Bruker AM250* (62.9 MHz) or *WH270* (67.9 MHz); calibrating signals: δ((D_6)DMSO) 39.5, δ(CDCl$_3$) 77.0. ^{31}P-NMR Spectra: *Bruker AMX 400* (162 MHz); calibrating signal: external 85% phosphoric acid. ESI-MS: *Fisons-VG-Platform-II* electrospray-ionization mass spectrometer; in *m/z*.

General Purification Procedures. a) FPLC: filtered (0.2-μm *Nalgene* syringe filter) crude product in H_2O was applied to a column filled with *Pharmacia-DEAE-Sephadex-A25* anion-exchange gel, 2.5 × 15 cm, 4°;

$A = 0.05$M (Et$_3$NH)HCO$_3$ (pH 7.5), $B = 0.5$M (Et$_3$NH)HCO$_3$ (pH 7.5); flow rate 1 ml/min, $0 \rightarrow 100\%$ *B*, 500 ml).

b) Anion-exchange HPLC: *Synchropak AX 300*, 6 μm (250 × 4.6 mm); $A = 0.05$M KH$_2$PO$_4$/50% formamide (pH 5.2), $B = 0.05$M KH$_2$PO$_4$/1M (NH$_4$)$_2$SO$_4$/50% formamide (pH 5.5); flow rate 1 ml/min, $0-30$ min $0 \rightarrow 100\%$ *B*. HPLC retention times t_R [min] refer to anion-exchange HPLC.

c) Reversed-phase HPLC: *LiChrosphere RP-18* 5 μm (3 × 125 mm); $A = 0.1$M (Et$_3$NH)OAc (pH 5.5), $B = A + 50\%$ MeCN; flow rate 1 ml/min, $0-10$ min 5% *B*.

3′-Amino-3′-deoxy-5′-O-(4,4′-dimethoxytrityl)thymidine (**2**). The 3′-azido-3′-deoxy-5′-*O*-(4,4′-dimethoxytrityl)thymidine (**1**; 2.5 g, 4.4 mmol) was filled in a heat-dried flask and dissolved in anh. MeOH (30 ml) under Ar. The soln. was degassed repeatedly by alternating evacuation and flushing with Ar. A small amount (two spatula tips) of PtO$_2 \cdot$H$_2$O was added to the stirred soln., and H$_2$ was injected for 1 h. The flask was fitted with a septum and a H$_2$-filled balloon and stirred overnight at 50°. The mixture was then filtered over *Celite*, the *Celite* washed with MeOH several times, and the combined filtrate evaporated: 2.21 g (93%) of pure **2**. No further purification was necessary. TLC (CH$_2$Cl$_2$/MeOH 95:5): R_f 0.18. IR (KBr): 2931, 1694, 1607, 1509, 1445, 1252, 1179, 1033, 829. ^{1}H-NMR ((D$_6$)DMSO): 7.58 (*d*, H$-$C(6)); 7.44$-$7.18 (*m*, 9 arom. H); 6.83 (*d*, 4 arom. H); 6.25 (*dd*, H$-$C(1′)); 3.78 (*s*, 2 MeO); 3.77$-$3.68 (*m*, H$-$C(4′), H$-$C(3′)); 3.53$-$3.32 (*m*, 2 H$-$C(5′)); 2.40$-$2.15 (*m*, 2 H$-$C(2′)); 1.52 (*s*, Me$-$C(5)). ESI-MS (neg.): 542.4 (*M*$^-$, C$_{31}$H$_{32}$N$_3$O$_6^-$; calc. 542.6).

3′-Deoxy-5′-O-(4,4′-dimethoxytrityl)-3′-[(2-methyl-1-oxopropyl)amino]thymidine (**3a**). A soln. of **2** (400 mg, 0.74 mmol) in anh. pyridine (20 ml) and isobutyryl anhydride (2 ml, 8.5 mmol) was stirred under Ar for 2 d and quenched with MeOH. The soln. was evaporated and the residue purified by prep. TLC (CH$_2$Cl$_2$/MeOH 98:2): 0.38 g (84%) of **3a**. TLC (CH$_2$Cl$_2$/MeOH 9:1): R_f 0.49. ^{1}H-NMR (CDCl$_3$): 7.65 (*d*, H$-$C(6)); 7.42$-$6.79 (*m*, 13 arom. H); 6.43 (*dd*, H$-$C(1′)); 4.70$-$4.65 (*m*, H$-$C(3′)); 4.02 (*d*, H$-$C(4′)); 3.78 (*s*, 2 MeO); 3.48 (*s*, 2 H$-$C(5′)); 2.60$-$2.31 (m, 2 H$-$C(2′), Me$_2$CHCO); 1.37 (*d*, Me$-$C(5)); 1.12 (*d*, *Me$_2$CHCO*). ^{13}C-NMR (CDCl$_3$): 177.4 (*CONH*$-$C(3′)); 164.1 (C(4)); 158.7 ((MeO)$_2$*Tr*); 150.8 (C(2)); 144.4 ((MeO)$_2$*Tr*); 135.7, 135.4 (C(6), (MeO)$_2$*Tr*); 130.1, 128.2, 127.9, 127.1, 113.3 ((MeO)$_2$*Tr*); 111.5 (C(5)); 87.0, 85.4, 84.6 (C(1′), C(4′), (MeO)$_2$*Tr* (C)); 64.1 (C(5′)); 55.2 (MeO); 51.0 (Me$_2$*CHCO*); 38.0, 35.1 (C(3′), C(2′)); 19.5, 18.8 (*Me$_2$CHCO*); 11.6 (Me$-$C(5)). ESI-MS (neg.): 612.7 (*M*$^-$, C$_{35}$H$_{38}$N$_3$O$_7^-$; calc. 612.7).

3′-Deoxy-5′-O-(4,4′-dimethoxytrityl)-3′-[(2-methyl-1-thioxopropyl)amino]thymidine (**4a**). To a soln. of **3a** (200 mg, 0.33 mmol) in anh. THF (10 ml), *Lawesson*'s reagent (66 mg, 0.16 mmol) was added. After stirring at r.t. for 30 min, sat. NaHCO$_3$ soln. (20 ml) was added. The resulting white precipitate was filtered and washed with H$_2$O. The aq. phase was extracted with CH$_2$Cl$_2$ twice, the combined org. layer dried (Na$_2$SO$_4$) and evaporated, and the residual solid purified by prep. TLC (CH$_2$Cl$_2$/MeOH 98:2): 45 mg (22%) of **4a**. TLC (CH$_2$Cl$_2$/MeOH 9:1): R_f 0.54. ^{1}H-NMR (CDCl$_3$): 7.76 (*d*, H$-$C(6)); 7.46$-$6.83 (*m*, 13 arom. H); 6.63 (*t*, H$-$C(1′)); 5.31 (*m*, H$-$C(3′)); 4.28 (*d*, H$-$C(4′)); 3.91 (*d*, 1 H$-$C(5′)); 3.79 (*s*, 2 MeO); 3.42 (*d*, 1 H$-$C(5′)); 3.05$-$2.97 (*m*, Me$_2$CHCS); 2.53$-$2.48 (*m*, 2 H$-$C(2′)); 1.42 (*d*, Me$-$C(5)); 1.26 (*d*, 3 H, *Me$_2$CHCS*); 1.23 (*d*, 3 H, *Me$_2$CHCS*). ^{13}C-NMR (CDCl$_3$): 213.2 (*CSNH*$-$C(3′)); 164.2 (C(4)); 158.7 ((MeO)$_2$*Tr*); 150.7 (C(2)); 144.6 ((MeO)$_2$*Tr*); 135.7, 135.5 (C(6), (MeO)$_2$*Tr*); 130.1, 128.2, 128.0, 127.1, 113.3 ((MeO)$_2$*Tr*); 111.7 (C(5)); 87.2, 85.0 (C(1′), C(4′)); 65.5 (C(5′)); 57.0 (Me$_2$*CHCS*); 55.3 (MeO); 42.9 (C(2′)); 37.0 (C(3′)); 23.1, 22.6 (*Me$_2$CHCS*); 11.7 (*Me*$-$C(5)). ESI-MS (neg.): 628.3 (*M*$^-$, C$_{35}$H$_{38}$N$_3$O$_6$S$^-$; calc. 628.8).

3′-Deoxy-3′-[(2-methyl-1-thioxopropyl)amino]thymidine (**5a**). A soln. of **4a** (235 mg, 0.373 mmol) in 80% AcOH/H$_2$O (20 ml) was left stirring for 2 h. About 90% of the solvent was evaporated and the resulting oil co-evaporated (3 ×) with H$_2$O to remove residual acid. The residue was purified by prep. TLC (CH$_2$Cl$_2$/MeOH 9:1): 113 mg (93%). TLC (CH$_2$Cl$_2$/MeOH 8:2): R_f 0.53. ^{1}H-NMR ((D$_6$)DMSO): 11.28 (*s*, H$-$N(3)); 10.24 (*d*, CSNH$-$C(3′)); 7.83 (*d*, H$-$C(6)); 6.30 (*t*, H$-$C(1′)); 5.15 (*t*, OH$-$C(5′)); 4.92$-$4.87 (*m*, H$-$C(3′)); 3.95 (*q*, H$-$C(4′)); 3.69 (*m*, 2 H$-$C(5′)); 2.89 (*sept.*, Me$_2$*CHCS*); 2.39$-$2.28 (*m*, 1 H$-$C(2′)); 2.20 (*ddd*, 1 H$-$C(2′)); 1.78 (*s*, Me$-$C(5)); 1.12 (*d*, *Me$_2$CHCS*). ^{13}C-NMR ((D$_6$)DMSO): 210.7 (CSNH$-$C(3′)); 163.7 (C(4)); 150.5 (C(2)); 136.0 (C(6)); 109.5 (C(5)); 85.0, 83.9 (C(1′), C(4′)); 62.2 (C(5′)); 55.6 (Me$_2$CHCS); 41.2 (C(2′)); 36.0 (C(3′)); 22.9, 22.6 (*Me$_2$CHCS*); 12.4 (Me$-$C(5)). ESI-MS (neg.): 326.3 (*M*$^-$, C$_{14}$H$_{20}$N$_3$O$_4$S$^-$; calc. 326.4).

3′-Deoxy-3′-[(2-methyl-1-thioxopropyl)amino]thymidine 5′-Triphosphate (**6a**). Compound **5a** (110 mg, 0.336 mmol) was dried by co-evaporation (3 ×) with anh. pyridine (0.7 ml)/DMF (2.75 ml). The residue was dried over P$_2$O$_5$ under vacuum overnight. The nucleoside was dissolved in anh. pyridine (0.7 ml)/DMF (2.75 ml), a freshly prepared soln. of 1M 2-chloro-4*H*-1,3,2-benzodioxaphosphorin-4-one in anh. dioxane (0.37 ml) was injected under Ar, and the mixture stirred for 20 min. Then a freshly prepared mixture (1.35 ml) of 0.5M (Bu$_3$NH)$_2$P$_2$O$_7$ in dry DMF/anh. Bu$_3$N 3:1 was instantly added. The mixture was stirred for further 30 min. Oxidation of the P-moiety and ring opening was achieved by adding 6.7 ml of 1% I$_2$ in pyridine/H$_2$O 98:2. After 20 min, remaining I$_2$ was reduced by adding dropwise a 5% aq. NaHSO$_3$ soln. until the brown color changed to light yellow. After evaporation without heating, H_2O (5 ml) was poured onto the residue, and the soln. containing the dissolved product was separated from the remaining residue by filtering through a 0.2 μm *Nalgene* syringe filter. The soln. was evaporated without heating, the residual solid dissolved in H_2O (5 ml), and this soln. filtered through a 0.2 μm *Nalgene* syringe filter and purified by FPLC (purification procedure *a*), elution concentration 0.26M ($Et_3NH)HCO_3$). The fractions containing product were lyophilized. The yield of **6a** was determined by optical-density measurement (32.9 μmol, 10%). TLC (^{i}PrOH/NH_3/H_2O): R_f 0.08. ^{31}P-NMR (D_2O): -23.3 (t, (β)); -11.8 (d, P(α)); -9.6 (d, P(γ)).

6-[[(9H-Fluoren-9-ylmethoxy)carbonyl]amino]hexanoic Acid (**10**). N-[(9H-Fluoren-9-ylmethoxy)carbonyl]succinimide (5 g, 14.82 mmol) and $NaHCO_3$ (1.24 g, 14.82 mmol) were suspended in a soln. of 6-aminohexanoic acid (**9**) (1.95 g, 14.82 mmol) in H_2O/acetone 1:1 (100 ml). The mixture was stirred for 3 h (TLC monitoring). Then the acetone was evaporated, and CH_2Cl_2 was added. The org. phase was washed with 0.1N HCl and H_2O, then dried (Na_2SO_4), and evaporated. Crystallization of the residue from CH_2Cl_2/hexane afforded **10** (5.2 g, 99%). White solid. M.p. 117–120°. TLC (CH_2Cl_2/MeOH 9:1): R_f 0.13. ^{1}H-NMR ($CDCl_3$): 7.75–7.78 (d, 2 H, Fmoc); 7.58–7.61 (d, 2 H, Fmoc); 7.26–7.43 (m, 4 H, Fmoc); 4.80 (br., NH); 4.41–4.43 (d, CH_2OCONH); 4.20–4.24 (m, H–C(9) (Fmoc)); 3.07–3.21 (m, CH_2(6)); 2.33–2.39 (t, CH_2(2)); 1.37–1.66 (m, CH_2(3), CH_2(4), CH_2(5)). ^{13}C-NMR ($CDCl_3$): 178.8 (C(1)); 156.34 (CH_2OCONH); 143.8, 141.2, 127.5, 126.9, 124.9, 119.8 (C(Fmoc)); 66.4 (CH_2OCONH); 47.1 (C(9) (Fmoc)); 40.6 (C(2)); 33.6 (C(6)); 29.4 (C(5)); 25.9 (C(3)); 24.1 (C(4)). ESI-MS (pos.): 354.1 (M^+, $C_{21}H_{24}NO_4^+$; calc. 354.3).

3′-Deoxy-5′-O-(4,4′-dimethoxytrityl)-3′-[[6-[[(9H-fluoren-9-ylmethoxy)carbonyl]amino]-1-oxohexyl]amino]-thymidine (**3b**). To a soln. of **10** (636 mg, 1.8 mmol) and **2** (1.17 g, 2.15 mmol) in DMF (5 ml), dioxane (15 ml), and H_2O (5 ml), ^{i}Pr$_2$NEt (0.4 ml, 2.4 mmol) and N,N,N′,N′-tetramethyl-O-succinimidouronium tetrafluoroborate (TSTU; 845 mg, 2.8 mmol) were added. After stirring at r.t. for 30 min (TLC monitoring), H_2O was added. The aq. phase was extracted with CH_2Cl_2 (4 ×), the combined org. layer dried ($MgSO_4$) and evaporated, and the residue purified by prep. TLC (CH_2Cl_2/MeOH 98:2): 585 mg (37%) of **3b**. TLC (CH_2Cl_2/MeOH 9:1): R_f 0.38. ^{1}H-NMR ($CDCl_3$): 9.90 (br., 1 NH); 7.99 (s, H–C(6)); 7.71–7.73 (d, 2 H, Fmoc); 6.78–7.61 (m, 19 H, (MeO)$_2$$Tr$, Fmoc); 6.36 ($t$, H–C(1′)); 5.08 (br., CH_2OCONH); 4.70–4.71 (m, H–C(3′)); 4.34–4.36 (m, CH_2OCONH); 4.20–4.22 (m, H–C(9) (Fmoc)); 4.00–4.01 (m, H–C(4′)); 3.79 (s, 2 MeO); 3.42–3.43 (m, 2 H–C(5′)); 3.12–3.14 (m, CH_2(6″)); 2.15–2.45 (m, 2 H–C(2′), CH_2(2″)); 1.25–1.62 (m, CH_2(3″), CH_2(4″), CH_2(5″)); 1.37 (s, Me–C(5)). ^{13}C-NMR ($CDCl_3$): 173.2 (CONH–C(3′)); 163.9 (C(4)); 158.6 ((MeO)$_2$CONH); 156.5 (CH_2OCONH); 150.8 (C(2)); 144.3 ((MeO)$_2$$Tr$); 143.9, 141.2 (C(Fmoc)); 135.6, 135.4 (C(6), (MeO)$_2$$Tr$); 130.1, 128.2, 127.9, 127.0 ((MeO)$_2$$Tr$); 127.6, 126.9, 124.9, 119.9 (Fmoc)); 113.2 ((MeO)$_2$$Tr$); 111.4 (C(5)); 87.0, 85.0, 84.4 (C(1′), C(4′), (MeO)$_2$$Tr$(C)); 66.5 ($CH_2$OCONH); 64.1 (C(5′)); 55.2 (MeO); 47.2 (C(9) (Fmoc)); 40.7 (C(2″)); 38.0, 35.9 (C(3′), C(2′)); 31.4, 29.5, 26.2, 25.0 (C(6″), C(5″), C(4″), C(3″)); 11.6 (Me–C(5)). ESI-MS (pos.): 896.6 (M^+, $C_{52}H_{54}N_4O_9 \cdot NH_4^+$; calc. 897.0).

3′-Deoxy-5′-O-(4,4′-dimethoxytrityl)-3′-[[6-[[(9H-fluoren-9-ylmethoxy)carbonyl]amino]-1-thioxohexyl]-amino]thymidine (**4b**). To a soln. of **3b** (1 g, 1.14 mmol) in anh. THF (70 ml) and anh. pyridine (92 μl, 1.14 mmol), *Lawesson*'s reagent (1.84 g, 4.56 ml) was added. After stirring at r.t. for 3 h, sat. $NaHCO_3$ soln. (100 ml) was added. The aq. phase was then extracted with CH_2Cl_2 (2 ×), the combined org. layer dried ($MgSO_4$) and evaporated, and the residue purified by prep. TLC (CH_2Cl_2/MeOH 98:2): 510 mg (50%) of **4b**. TLC (CH_2Cl_2/MeOH 9:1): R_f 0.44. ^{1}H-NMR ($CDCl_3$): 10.19 (br., 1 NH); 9.22 (br., 1 NH); 6.72–7.67 (m, 22 H, H–C(6), (MeO)$_2$$Tr$, Fmoc); 6.49 ($t$, H–C(1′)); 5.42 ($m$, CH_2OCONH); 4.83 (m, H–C(3′)); 4.27–4.30 (m, CH_2OCONH); 4.21 (m, H–C(9) (Fmoc)); 4.10 (m, H–C(4′)); 3.70–3.78 (m, 1 H–C(5′)); 3.68 (s, 2 MeO); 3.31–3.35 (m, 1 H–C(5′)); 3.06–3.09 (m, CH_2(6″)); 2.61–2.66 (m, 2 H–C(2′)); 2.27–2.42 (m, CH_2(2″)); 1.27–1.96 (m, CH_2(3″), CH_2(4″), CH_2(5″)); 1.35 (s, Me–C(5)). ^{13}C-NMR ($CDCl_3$): 206.4 (CSNH–C(3′)); 164.2 (C(4)); 158.4 ((MeO)$_2$$Tr$); 156.4 ($CH_2$OCONH); 150.7 (C(2)); 144.6 ((MeO)$_2$$Tr$); 143.9, 141.3 (Fmoc); 135.6, 135.4 (C(6), (MeO)$_2$$Tr$); 130.1, 128.1, 127.9, 127.05 ((MeO)$_2$$Tr$); 127.6, 126.9, 125.0, 119.9 (Fmoc); 113.3 ((MeO)$_2$$Tr$); 111.5 (C(5)); 87.1, 85.8, 85.0 (C(1′), C(4′), (MeO)$_2$$Tr$(C)); 66.5 ($CH_2$OCONH); 65.3 (C(5′)); 55.3 (MeO); 47.22 (C(9) (Fmoc)); 45.4 (C(2′)); 40.7 (C(2″)); 38.6, 37.4 (C(3′), C(6″)); 29.5, 28.9, 25.8 (C(5″), C(3″), C(4″)); 11.7 (Me–C(5)). ESI-MS (neg.): 893.6 (M^-, $C_{52}H_{53}N_4O_8S^-$; calc. 894.1).

3′-Deoxy-3′-[[6-[[(9H-fluoren-9-ylmethoxy)carbonyl]amino]-1-thioxohexyl]amino]thymidine (**5b**). To a soln. of **4b** (140 mg, 0.156 mmol) in CH_2Cl_2 (2 ml), 80% AcOH/H_2O (10 ml) was added. The mixture was stirred at r.t. for 3 h (TLC monitoring) and then evaporated. The resulting oil was co-evaporated with H_2O (3 ×) and hexane (1 ×) to remove the residual acid before purifying. The residue was purified by prep. TLC (CH_2Cl_2/MeOH 9:1): 68 mg (74%) of **5b**. TLC (CH_2Cl_2/MeOH 9:1): R_f 0.37. ^{1}H-NMR ($CDCl_3$): 10.25 (s, 1 NH); 9.17 (d, NH–C(3′)); 7.63–7.69 (m, 3 H, H–C(6), Fmoc); 7.47 (d, 2 H, Fmoc); 7.15–7.31 (m, 4 H, Fmoc); 6.23 (t, H–C(1′)); 5.15 (br., OH–C(5′)); 4.92 (m, H–C(3′)); 4.26 (m, CH_2OCONH); 4.05–4.11 (m, H–C(9) (Fmoc));

3.97 (m, H−C(4′)); 3.84 (m, 2 H−C(5′)); 3.02−3.04 (m, CH$_2$(6″)); 2.55−2.58 (m, CH$_2$(2″)); 2.18−2.38 (m, 2 H−C(2′)); 1.78 (s, Me−C(5)); 1.15−1.68 (m, CH$_2$(3″), CH$_2$(4″), CH$_2$(5″)). ^{13}C-NMR (CDCl$_3$): 206.4 (CSNH−C(3′)); 164.3 (C(4)); 156.6 (CH$_2$OCONH); 150.8 (C(2)); 143.7, 141.1 (Fmoc); 136.0 (C(6)); 127.6, 126.9, 124.8, 119.9 (Fmoc); 111.0 (C(5)); 85.6, 84.7 (C(1′), C(4′)); 66.6 (CH$_2$OCONH); 62.4 (C(5′)); 47.0 (C(9) (Fmoc)); 45.6 (C(2′)); 40.6 (C(2″)); 36.9 (C(3′)); 29.0, 25.6 (C(5″), C(3″), C(4″)); 12.5 (Me−C(5)). ESI-MS (neg.): 591.5 (M^-, C$_{31}$H$_{35}$N$_4$O$_6$S$^-$; calc. 591.6).

3′-Deoxy-3′-[[6-[[(9H-fluoren-9-ylmethoxy)carbonyl]amino]-1-thioxohexyl]amino]thymidine 5′-Triphosphate (**6b**). As described for **6a**, from **5b** (150 mg, 0.253 mmol), after co-evaporation with anh. pyridine (0.5 ml) and dissolution in anh. pyridine (0.5 ml)/DMF (2.55 ml), with 1M 2-chloro-4H-1,2,3-benzodioxaphosphorin-4-one in anh. dioxane (0.3 ml), 0.5M (Bu$_3$NH)$_2$P$_2$O$_5$ in dry DMF/Bu$_3$N (3 : 1 (1.07 ml), 5.4 ml of 1% I$_2$ in pyridine/H$_2$O 98 : 2), and FPLC (elution concentration 0.28M (Et$_3$NH) HCO$_3$): **6b** (48 μmol, 19%). TLC (^{i}PrOH/NH$_3$/H$_2$O): R_f 0.15. ^{31}P-NMR (D$_2$O): −21.8 (t, P(β)); −10.8 (d, P(α)); −5.7 (d, P(γ)). ESI-MS (neg.): 831.5 (M^-, C$_{31}$H$_{38}$N$_4$O$_{15}$P$_3$S$^-$; calc. 831.7).

3′-[(6-Amino-1-thioxohexyl)amino]-3′-deoxythymidine 5′-Triphosphate (**7**). To a mixture of pyridine (20 ml), DMF (12 ml), and piperidine (8 ml), **6b** (48 μmol) was added. This soln. was stirred at r.t. for 1 h and then evaporated without heating. The residual solid was dissolved in H$_2$O (4 ml), filtered through a 0.2-μm *Nalgene* syringe filter, and purified by FPLC (purification procedure *a*); elution concentration 0.14M (Et$_3$NH)HCO$_3$). The fractions containing product were lyophilized. The yield of **7** was determined by optical-density measurement (39 μmol, 81%). TLC (^{i}PrOH/NH$_3$/H$_2$O): R_f 0.13.

^{1}H-NMR (D$_2$O): 7.86 (s, H−C(6)); 6.26 (t, H−C(1′)); 4.95−5.00 (m, H−C(3′)); 4.46 (m, H−C(4′)); 4.18−4.28 (m, 2 H−C(5′)); 2.97−3.15 (m, CH$_2$(6″)); 2.66−2.74 (m, CH$_2$(2″), 1 H−C(2′)); 2.40−2.47 (m, 1 H−C(2′)); 1.91 (s, Me−C(5)); 1.33−1.84 (m, CH$_2$(3″), CH$_2$(4″), CH$_2$(5″)). ^{13}C-NMR (D$_2$O): 207.0 (CSNH−C(3′)); 166.5 (C(4)); 151.4 (C(2)); 137.6 (C(6)); 111.6 (C(5)); 85.3, 81.4 (C(1′), C(4′)); 66.06 (C(5′)); 42.54 (C(2′)); 39.1 (C(2″)); 35.5 (C(3′)); 28.1, 26.4, 24.3 (C(3″), C(4″), C(5″)); 11.5 (Me−C(5)). ^{31}P-NMR (D$_2$O): −22.0 (t, P(β)); −10.6 (d, P(α)); −8.2 (d, P(γ)). HPLC (purification procedures *b* and *c*): t_R 4.8. ESI-MS (neg.): 609.2 (M^-, C$_{11}$H$_{33}$N$_4$O$_{13}$P$_3$S$^-$; calc. 609.4).

Reaction of **7** *with Fluorescein Isothiocyanate: 3′-[[6-[[[[3-Carboxy-4-(6-hydroxy-3-oxo-3H-xanthen-9-yl)phenyl]amino]thioxomethyl]amino]-1-thioxohexyl]amino]-3′-deoxythymidine 5′-Triphosphate* (**8**). To a soln. of **7** (10 μmol) in aq. NaHCO$_3$ soln. (6 ml, pH 9.3), a soln. of fluorescein isothiocyanate (12 mg) in DMF (6 ml) was added. The mixture was stirred for 24 h at r.t. The purification was done by HPLC (purification procedures *b* and *c*): **8**: HPLC: t_R (**8**) 19.5; t_R (fluorescein isothiocyanate) 12.5. ESI-MS (neg.): 998.3 (M^-, C$_{37}$H$_{39}$N$_5$O$_{18}$P$_3$S$_2^-$; calc. 998.8).

REFERENCES

[1] I. Dunham, N. Shimizu, B. A. Roe, S. Chissoe, *Nature (London)* **1999**, *402*, 489.
[2] L. Rowen, G. Mahairas, L. Hood, *Science (Washington, D.C.)* **1997**, *278*, 605.
[3] F. Sanger, S. Nicklen, A. R. Coulson, *Proc. Natl. Acad. Sci. U.S.A.* **1977**, *74*, 5463.
[4] F. Sanger, *Angew. Chem.* **1981**, *93*, 937.
[5] J. M. Prober, G. L. Trainor, R. J. Dam, F. W. Hobbs, C. W. Robertson, R. J. Zagursky, A. J. Cocuzza, M. A. Jensen, K. Baumeister, *Science (Washington D.C.)* **1987**, *238*, 336.
[6] L. G. Lee, C. R. Connell, S. L. Woo., R. L. Cheng, B. F. McArdle, C. W. Fuller, N. D. Halloran, R. K. Wilson, *Nucleic Acids Res.* **1992**, *20*, 2471.
[7] B. B. Rosenblum, L. G. Lee, S. L. Spurgeon, S. H. Khan, S. M. Menchen, C. R. Heiner, S. M. Chen, *Nucleic Acids Res.* **1997**, *25*, 4500.
[8] S. R. Sarfati, A. Namane, *Tetrahedron Lett.* **1990**, *31*, 2581.
[9] C. Wojczewski, K. Stolze, J. W. Engels, *Synlett* **1999**, 1667.
[10] J. Hovinen, E. Azhayeva, A. Azhayev, A. Guzaev, H. Lönnberg, *J. Chem. Soc., Perkin Trans. 1* **1994**, 211.
[11] R. S. Sarfati, T. Berthod, C. Guerreiro, B. Canard, *J. Chem. Soc., Perkin Trans. 1* **1995**, 1163.
[12] M. L. Metzker, R. Raghavachari, S. Richards, S. E. Jacutin, A. Civitello, K. Burgess, R. A. Gibbs, *Nucleic Acids Res.* **1994**, *22*, 4259.
[13] M. K. Herrlein, R. E. Konrad, J. W. Engels, T. Holletz, D. Cech, *Helv. Chim. Acta* **1994**, *77*, 586.
[14] C. Wojczewski, K. Faulstich, J. W. Engels, *Nucleosides Nucleotides* **1997**, *16*, 751.
[15] B. Canard, B. Cardona, R. S. Sarfati, *Proc. Natl. Acad. Sci. U.S.A.* **1995**, *92*, 10859.
[16] C. Wojczewski, J. W. Engels, *Synthesis* **2000**, 149.

[17] T. Schoetzau, U. Koert, J. W. Engels, *Synthesis* **2000**, in press.
[18] K. Stolze, U. Koert, S. Klingel, G. Sagner, R. Wartbichler, J. W. Engels, *Helv. Chim. Acta* **1999**, *82*, 1311.
[19] J. P. Horwitz, J. Chua, J. A. Urbanski, M. Noel, *J. Org. Chem.* **1963**, *28*, 942.
[20] J. J. Fox, N. C. Miller, *J. Org. Chem.* **1963**, *28*, 936.
[21] W. Walter, K.-D. Bode, *Angew. Chem.* **1966**, *78*, 517.
[22] R. N. Hurd, G. DeLaMater, *Chem. Rev.* **1961**, *61*, 45.
[23] B. Yde, N. M. Yousif, U. Pedersen, I. Thomsen, S.-O. Lawesson, *Tetrahedron* **1984**, *40*, 2047.
[24] M. P. Cava, M. I. Levinson, *Tetrahedron* **1985**, *41*, 5061.
[25] R. A. Cherkasov, G. A. Kutyrev, A. N. Pudovik, *Tetrahedron* **1985**, *41*, 2567.
[26] K. Wörner, T. Strube, J. W. Engels, *Helv. Chim. Acta* **1999**, *82*, 2094.
[27] J. Ludwig, F. Eckstein, *J. Org. Chem.* **1989**, *54*, 631.

Hydrolysis of Phosphodiester Bonds within RNA Hairpin Loops in Buffer Solutions: the Effect of Secondary Structure on the Inherent Reactivity of RNA Phosphodiester Bonds

by Izabela Zagorowska[1]), Satu Mikkola*, and Harri Lönnberg

University of Turku, Department of Chemistry, FIN-20014 Turku

The hydrolysis of phosphodiester bonds of chimeric 2'-O-methyloligoribonucleotides was studied in buffer solutions. Pseudo-first-order rate constants for cleavage of phosphodiester bonds within hairpin loops were calculated and compared with those for cleavage of phosphodiester bonds within double-stranded stems and linear single-stranded oligonucleotides. No large differences in reactivity were observed: some of the hairpin structures studied were slightly less and others slightly more reactive than the linear reference. These results suggest that phosphodiester bonds within small hairpin loops are conformationally free to cleave by an in-line mechanism, but also that the secondary structure may influence the reactivity of phosphodiester bonds.

Introduction. – The cleavage of RNA phosphodiester bonds has been extensively studied during the last two decades [1][2]. Efficient catalysts for sequence-specific cleavage are being developed by several research groups [3]. It is hoped that studies of the catalysis mechanisms and factors that enhance the reactivity of phosphodiester bonds will enable rational design of such catalysts. Studies of the chemical cleavage of phosphoesters may also further our understanding of the reaction mechanisms of nucleic-acid-modifying metalloenzymes and, particularly, of ribozymes.

It may be said that the inherent reactivity of all internucleosidic phosphodiester bonds within linear oligonucleotides is approximately the same and depends only on the reaction conditions. The rate constants for cleavage of different dinucleoside monophosphates vary only slightly [4]. The differences observed can, at least in part, be attributed to the different pK_a values of the 2'- and 5'-hydroxy groups in different nucleosides. Phosphodiester bonds within poly-U oligomers also react approximately as fast as those within dinucleoside monophosphates [5]. Only under acidic conditions does poly-U appear to be more reactive, probably because protonation of phosphate groups within a polyanionic molecule is enhanced.

As opposed to the situation with linear oligonucleotides, where the reactivity of phosphodiester bonds depends only on the conditions, the secondary structure of an oligonucleotide strand may significantly influence the reactivity of phosphodiester bonds in some structures. As the most striking example, the phosphodiester bonds within RNA-RNA or RNA-DNA duplexes are relatively stable, their cleavage being very slow compared to the cleavage of phosphodiester bonds within single-stranded

---

[1])   Present address: Institute of Chemistry, Polish Academy of Sciences, Noskowskiego 12–14, PL-61-704.

regions [6]–[9]. This is true in both the absence and the presence of metal ion catalysts. It is believed that the duplex structure forces phosphodiester bonds into a conformation unfavorable for the cleavage. Phosphodiester bonds within bulges or hairpin loops, in contrast, are known to be cleaved by metal ion catalysts [6][8][9].

In the presence of metal-ion catalysts, the observed reactivity of phosphodiester bonds can be regarded as a combination of contributions of the inherent reactivity of the bond and the catalytic properties of the metal-ion species involved. The secondary structure of a substrate oligonucleotide may affect both of these factors in different ways. In addition to the restrictions brought about by the fixed structure of an RNA duplex, the strain induced by the tight bending of a RNA strand at bulged sites or in hairpin loops may destabilize phosphodiester bonds [10]. As for the metal-ion-promoted cleavage, it has been suggested that bending of an RNA strand at an RNA bulge may create strong metal-ion coordination sites that are favorable for the catalysis [11]. In contrast, our previous results show that, in the presence of metal-ion catalysts, phosphodiester bonds within hairpin loops are clearly less reactive than those within linear oligonucleotides [9]. This difference was also attributed to the coordination of metal ions.

We have previously studied the effects of the secondary structure on metal-ion catalysis by studying the cleavage of phosphodiester bonds within small RNA hairpins [9]. In this paper, we set out to study the inherent reactivity of phosphodiester bonds within bent RNA strands. The cleavage of phosphodiester bonds within hairpin loops was studied in buffer solutions, and as in the previous paper [9], the size of the hairpin and the position of the scissile bond within the loop were varied. The rate constants obtained are compared with those for cleavage of a linear single-stranded structure.

Results. – The loop structures studied are shown in the *Figure*. To study the cleavage of one particular phosphodiester bond at a time, all but one of the ribonucleoside units were converted to 2′-*O*-methylribonucleotides. 2′-*O*-Methylnucleosides were chosen, because the ring conformation of a 2′-*O*-methylribose closely resembles that of the parent ribose [12]. Compound **1** is a linear molecule that serves as a reference material. Compounds **2** and **3** constitute a pair of isomers where **2** has a 3′,5′- and **3** a 2′,5′-phosphodiester bond. In **4**, there is a scissile phosphodiester bond within the double-stranded region. In all the other cases, the reactive 3′,5′-phosphodiester bond is located within a hairpin loop.

To verify the presence of a hairpin structure under the conditions of the kinetic experiments, the melting temperatures of the molecules were determined. The hairpin structures studied appear to be very stable; in 10 mM *Tris* buffer at pH 7.0 and ionic strength of 0.1M, all exhibited concentration-independent melting temperatures (T_m) at *ca.* 90°. The T_m values were slightly lower under experimental conditions, where the T_m value of **2**, for example, was 85.5° in 0.1M CHES (= 2-(cyclohexylamino)ethanesulfonic acid) at pH 8.6, and 80.5° in 1M imidazole at pH 7.0. The structure of molecules **8**–**12** has been discussed previously [9].

Kinetic experiments were carried out at 65°, where all of the molecules can safely be assumed to exist as hairpins. Two different buffer systems were employed: 1.0M imidazole buffer at pH 7.1 and 0.1M CHES buffer, at pH 8.6. The mechanism of the cleavage probably differs in these two buffer solutions. In the imidazole buffer at high

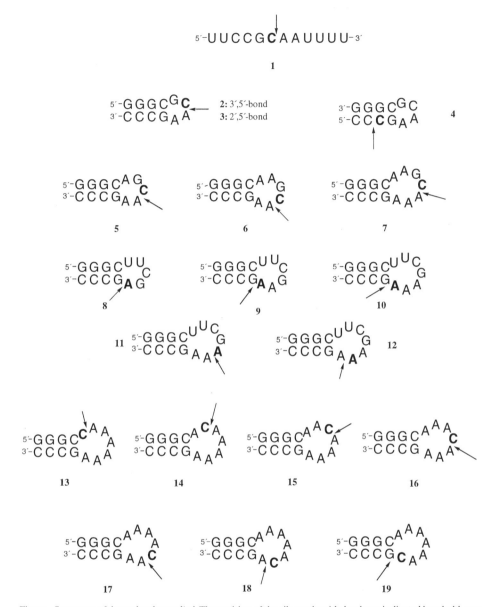

Figure. *Structures of the molecules studied.* The position of the ribonucleoside has been indicated by a bold-case letter and the scissile phosphodiester bond by an arrow.

concentration, the predominant reaction is the buffer-catalyzed cleavage of phosphodiester bonds [13][14]. Isomerization of phosphodiester bonds was insignificant under these conditions, since catalysis of phosphate migration by imidazole buffers is very modest [14]. In the CHES buffers at pH 8.6, the buffer concentration was kept low, and the predominant reaction under these conditions was the OH⁻-catalyzed cleavage of the phosphodiester bonds. The alkaline cleavage is not accompanied by the isomerization of phosphodiester bonds [4] [5].

The results of the kinetic experiments collected in the *Table* show clearly that the secondary structure of the hairpin loops under study had only a modest effect on the reactivity of phosphodiester bonds. In the CHES buffer, the reactivity of phosphodiester bonds differed only slightly between the linear molecule **1** and the hairpin structures. Similarly, the maximal difference observed between the rate constants for cleavage of 3',5'-bonds within a hairpin loop was less than 20-fold. A similar comparison between the rate constants obtained in imidazole buffer is more difficult to make, because of the unreactivity of some structures. Replacing a scissile 3',5'-phosphodiester bond with a corresponding 2',5'-bond did not appear to have any significant effect when the bond was within the loop. In keeping with the known stability of the 3',5'-bonds within double-stranded regions [6]–[9], the scissile 3',5'-phosphodiester bond of **4** placed within the stem was, in both cases, the least reactive. The differences in reactivity observed were, however, rather small: the rate constant for cleavage of **2** in the CHES buffer was only six times larger than that of **4**.

Table. *First-Order Rate Constants for Cleavage of Phosphodiester Bonds within Hairpin Loops in Buffer Solutions at 65°*

Compound	k(CHES)/10^{-6} s^{-1a})	k(imidazole)/10^{-7} s^{-1b})
1	2.3 ± 0.1	10.6 ± 0.4
2	1.8 ± 0.1	8.0 ± 0.3
3	1.0 ± 0.1	3.1 ± 0.2
4	0.3 ± 0.1	< 0.1
5	20.4 ± 0.9	9.7 ± 0.4
6	4.5 ± 0.2	8.7 ± 0.3
7	3.9 ± 0.1	10.6 ± 0.4
8	1.1 ± 0.1	5.5 ± 0.7
9	1.6 ± 0.1	< 1.0
10	1.5 ± 0.1	< 1.0
11	2.6 ± 0.1	7.9 ± 0.3
12	2.1 ± 0.1	8.2 ± 0.4
13	8.3 ± 0.3	4.6 ± 0.2
14	4.5 ± 0.3	9.0 ± 0.3
15	4.0 ± 0.1	17.1 ± 0.6
16	5.0 ± 0.3	12.8 ± 0.5
17	4.7 ± 0.3	12.2 ± 0.3
18	4.6 ± 0.3	13.0 ± 0.6
19	2.4 ± 0.2	6.3 ± 0.2

a) 0.1M CHES, pH 8.6. b) 1.0M imidazole, pH 7.1.

The most interesting results were obtained with compounds **2** and **5–7**, where the size of the loop was systematically increased by adding adenosine nucleosides. All of the molecules were slightly more reactive than the linear molecule **1**, with **5**, containing the pentaloop, being the most reactive. The rate enhancement was, however, modest; **5** was only about ten times as reactive as **1**. No obvious trend was observed, and a C-A bond in a tetraloop, as in **2**, was not found to be particularly reactive. In imidazole buffer, no such reactivity differences were observed, and **5 – 7** and **2** appear to be approximately as reactive as **1**. Compounds **8 – 10** form another series of molecules with increasing loop size, in which the scissile phosphodiester bond was placed next to the stem, and in which no large reactivity differences were observed in CHES buffer. In imidazole buffer, however, **9** and **10** were exceptionally unreactive. With all the other molecules, including **8**, the rate constants obtained in CHES buffer were two to four times larger than those obtained in imidazole buffer. The rate of cleavage of **9** and **10** in imidazole buffer was so slow that no rate constants could be obtained at all, which suggests that the rate difference is at least 15-fold.

The position of the scissile phosphodiester bond within the loop seemed to have only a very small effect, if any, on the reactivity of phosphodiester bonds. This is most clearly seen on comparing the rate constants for cleavage of **13 – 19**, where the position of the C-A bond within an adenosine-rich heptaloop is systematically changed. In both the CHES and imidazole buffers, the difference between the most and the least reactive structure was approximately fourfold. In the CHES buffer, molecules **14 – 18** were equally reactive, whereas catalysis by the imidazole buffer appears to be more sensitive to the position of the scissile phosphodiester bond, and a clear trend is observed, although the differences are very modest. In both cases, the bond next to the double-stranded stem seems to be the least reactive (**19**). Similar effects were observed with **10 – 12**: a phosphodiester bond next to the stem, as in **10**, was slightly less reactive than those of **11** or **12**.

The rate constants for cleavage of **5 – 7**, which appear to be more reactive than a linear oligonucleotide, were also determined in the presence of the $Zn^{2+}[12]aneN_3$ chelate. This catalyst was employed in our previous study on metal-ion-promoted cleavage of hairpin structures [9]. The rate constants obtained were $6.5 \times 10^{-6} \, s^{-1}$, $3.8 \times 10^{-6} \, s^{-1}$ and $3.4 \times 10^{-6} \, s^{-1}$ for structures **5 – 7**, respectively. That reported for cleavage of **1** under identical conditions was $9.2 \times 10^{-6} \, s^{-1}$. It is thus clear that, even if **5 – 7**, in the absence of metal-ion catalysts, exhibit a slightly enhanced reactivity compared to a linear molecule, no such effect is observed in the presence of a metal-ion catalyst.

Discussion. – The results discussed above show that no significant conformational restrictions are involved in the cleavage of the hairpin molecules studied; none of the phosphodiester bonds were cleaved in the CHES buffer significantly less readily than those of the linear oligonucleotide **1**. Furthermore, the 3′,5′- and 2′,5′-phosphodiester bonds within a hairpin loop appear to be equally reactive. This is in contrast to the observation where the reactivity of a 3′,5′-bond within a double helix was found to be significantly reduced compared to a 2′,5′-bond [15], and where the difference was attributed to different conformations around the phosphodiester bond to be cleaved.

There was no significant difference in the effect of secondary structure between the two different reactions studied, although slightly larger reactivity differences were observed in imidazole than in CHES buffers. The clear loss of the reactivity in the case of **9** and **10** in imidazole may well be attributed to the presence of an additional, albeit weak, A-U base pair, as was discussed before [9]. While this base pair is stabilized by the high imidazole buffer concentration at pH 7.1, it is not present, at least not to the same extent, in the CHES buffer at pH 8.6. In the imidazole buffer, the scissile phosphodiester bond lies within the double-stranded region, and could not, therefore, be cleaved efficiently. It would thus seem that the results obtained in the CHES buffer better reflect the inherent reactivity of phosphodiester bonds, because the high ionic strength of the imidazole buffer may stabilize additional interactions.

Although the reactivity differences observed in this work were small, the results obtained still suggest that secondary structure may enhance the inherent reactivity of RNA phosphodiester bonds. The tenfold rate enhancement between **1**, a linear molecule, and the hairpin structure **5**, can be attributed to the secondary structure, since in both cases a similar phosphodiester bond between C and A residues is cleaved. Since the reactivity differences are very subtle and the structure of the substrates extremely complex, the present data do not permit elucidation of which parameters might contribute to the enhanced reactivity of phosphodiester bonds in certain hairpin molecules studied. On the basis of structural information available for nucleotides within hairpin and bulge motifs, it could, however, be suggested that the conformation of the sugar ring is indicative of the reactivity. It is known that within a hairpin stem, sugar moieties mainly adopt a C(3')-*endo*-conformation, while those within a single-stranded region exist as an equilibrium mixture of C(2')- and C(3')-*endo*-conformers [16]. The proportions of the conformers vary depending on the position of the nucleoside. It is, therefore, tempting to speculate that cleavage occurs only when the nucleoside bearing the attacking nucleophile is in a C(2')-*endo*-conformation. The observed reactivity would thus be a function of the nucleoside fraction present as the C(2')-*endo*-conformer.

Experimental Part

The chimeric 2'-*O*-methyloligoribonucleotides were synthesized from commercial 2'-*O*-methylated and 2'-*O*-[1-(fluorophenyl)-4-methoxypiperidin-4-yl]-protected building blocks by means of a conventional phosphor-amidate strategy according to the standard RNA-coupling protocol of the *ABI 392* DNA/RNA synthesizer. Melting experiments were performed on a *Perkin Elmer Lambda 2* UV/VIS spectrophotometer equipped with a *PTP-6* temp. programmer. Kinetic experiments were carried out in tightly sealed glass tubes in a water bath kept at $65.0 \pm 0.1°$. The buffer systems employed were imidazole/HCl and 2-(cyclohexylamino)ethanesulfonic acid (CHES)/NaOH. Aliquots withdrawn were cooled immediately in an ice bath, and were analyzed by ion-exchange chromatography. The details of all these procedures were described before [9].

REFERENCES

[1] D. M. Perreault, E. V. Anslyn, *Angew. Chem., Int. Ed.* **1997**, *36*, 432; M. Oivanen, S. Kuusela, H. Lönnberg, *Chem. Rev.* **1998**, *98*, 961.

[2] S. Kuusela, H. Lönnberg, *Metal Ions Biol. Syst.* **1995**, *33*, 561; J. R. Morrow, *ibid.* **1996**, *33*, 561; S. Kuusela, H. Lönnberg, *Curr. Topics Sol. Chem.* **1997**, *2*, 29; B. N. Trawick, A. T. Daniher, J. K. Bashkin, *Chem. Rev.* **1998**, *98*, 939.

[3] A. De Masmaeker, R. Häner, P. Martin, H. E. Moser, *Acc. Chem. Res.* **1995**, *28*, 366; J. K. Bashkin, U. Sampath, E. Frolova, *Appl. Biochem. Biotech.* **1995**, *54*, 43; M. Komiyama, *J. Biochem.* **1995**, *118*, 665; D. Magda, S. Crofts, A. Lin, D. Miles, M. Wright, J. Sessler, *J. Am. Chem. Soc.* **1997**, *119*, 2293; D. A. Konevetz, I. E. Beck, N. G. Beloglazova, I. V. Sulimenkov, V. N. Sil'nikov, M. A. Zenkova, V. N. Shishkin, V. V. Vlassov, *Tetrahedron* **1999**, *55*, 503; J. Hovinen, A. Guzaev, E. Azhayeva, A. Azhayev, H. Lönnberg, *J. Org. Chem.* **1995**, *60*, 2205.

[4] P. Järvinen, M. Oivanen, H. Lönnberg, *J. Org. Chem.* **1991**, *56*, 5396.

[5] S. Kuusela, H. Lönnberg, *J. Chem. Soc., Perkin Trans. 2* **1994**, 2109.

[6] J. Hall, D. Hüsken, R. Häner, *Nucleic Acids Res.* **1996**, *24*, 3522.

[7] J. Ciesiolka, S. Lorenz, V. A. Erdman, *Eur. J. Biochem.* **1992**, *204*, 575; K. A. Kolasa, J. R. Morrow, A. P. Sharma, *Inorg. Chem.* **1993**, *30*, 3983.

[8] D. Hüsken, G. Goodal, M. J. J. Blommers, W. Jahnke, J. Hall, R. Häner, H. E. Moser, *Biochemistry* **1996**, *35*, 16591.

[9] I. Zagorowska, S. Kuusela, H. Lönnberg, *Nucleic Acids Res.* **1998**, *26*, 3392.

[10] A. J. Zaug, T. R. Cech, *Science* **1986**, *231*, 470.

[11] S. Portmann, S. Grimm, C. Workman, N. Usmann, M. Egli, *Chem. Biol.* **1996**, *3*, 173.

[12] M. Popenda, E. Biala, J. Milecki, R. W. Adamiak, *Nucleic Acids Res.* **1997**, *25*, 4589.

[13] R. Breslow, M. LaBelle, *J. Am. Chem. Soc.* **1986**, *108*, 2655.

[14] E. Anslyn, R. Breslow, *J. Am. Chem. Soc.* **1989**, *111*, 4473; R. Breslow, S. D. Dong, Y. Webb, R. Xu, *J. Am. Chem. Soc.* **1996**, *118*, 6588; C. Beckmann, A. K. Kirby, S. Kuusela, D. C. Tickle, *J. Chem. Soc., Perkin Trans. 2* **1998**, 573.

[15] D. D. Usher, A. H. McHale, *Proc. Natl. Acad. Sci. U.S.A.* **1976**, *73*, 1149.

[16] L. G. Laing, K. B. Hall, *Biochemistry* **1996**, *35*, 13586; M. Molinaro, I. Tinoco Jr., *Nucleic Acids Res.* **1995**, *23*, 3056; F. M. Jucker, A. Pardi, *Biochemistry* **1995**, *34*, 14416.

Duplex-Stabilization Properties of Oligodeoxynucleotides Containing N^2-Substituted Guanine Derivatives

by **Ramon Eritja**[*][a])[1]), **Antonio R. Díaz**[a]), and **Ester Saison-Behmoaras**[b])

[a]) European Molecular Biology Laboratory (EMBL), Meyerhofstrasse 1, D-69117 Heidelberg
[b]) Museum National d'Histoire Naturelle, 43 rue Cuvier, F-75231 Paris Cedex 05

Oligodeoxynucleotides **3–13** carrying different guanine derivatives with substituents at the N^2 position have been prepared from a common precursor. Duplexes containing these modified bases are more stable than unmodified duplexes. The highest stability is found in guanine derivatives carrying at N^2 an ethyl and propyl group substituted with a group that is protonated under physiological conditions, which is compatible with a possible interaction of the protonatable group with the phosphates.

1. Introduction. – The ability of oligonucleotides to form double-stranded structures with their complementary RNA and DNA sequences has been widely used for the isolation of cloned DNA sequences, the detection of specific genes, and the identification of specific mRNA sequences [1]. More recently, this ability is being used to inhibit gene expression (antisense and antigene strategy) [2] and to analyze the expression patterns of a large number of genes using oligonucleotide microarrays [3]. All these approaches rely on nucleic-acid hybridization, and they are expected to benefit from oligonucleotide analogs with better hybridization properties. A large number of nucleic-acid derivatives with enhanced binding properties have been developed [4]. Peptide nucleic acids (PNA) [5], hexitol nucleic acids (HNA) [6], 2'-fluoro-N3'-P5' phosphoramidates [7], and locked nucleic acids (LNA) [8] are examples of backbone-modified derivatives with enhanced binding properties. However, simple substitutions in the sugar moiety such as 2'-O-alkyl-RNA derivatives [9] and substitutions in the nucleobases also lead to substantial increases in duplex stability.

N^2-Substituted derivatives of guanine (*Fig. 1*) are among these base-modified analogs with interesting binding properties [10–15]. First, the interest in the preparation of oligonucleotides carrying N^2-substituted derivatives of guanine was generated by the reaction of some carcinogenic compounds with guanine to yield adducts bound to the $NH_2-C(2)$ group of guanine. These modified oligonucleotides were important intermediates in the carcinogenesis of polycyclic aromatic hydrocarbons (PAH) [16]. Later, the attachment of an imidazole moiety to the exocyclic amino group of guanine was described [11]. Oligonucleotide-imidazole conjugates were prepared to promote RNA hydrolysis [12]. Later, it was found that imidazolyl

---

[1]) Present address. *C.I.D.-C.S.I.C.*, Jordi Girona 18–26, E-08034 Barcelona, Spain (phone: +34934006145, fax: +34932045904; e-mail: recgma@cid.csic.es).

groups bound to the exocyclic amino group of G stabilized the duplex structure [12][13]. Molecular-modeling simulations suggested that the imidazole moiety stabilized the duplex by binding in the minor groove near the phosphate backbone [13]. More recently, it has been described that the incorporation of polyamine-derived groups, such as spermine- and spermidine-derived ones, and also the 3-aminopropyl group, at the N^2 position of guanine increase duplex stability [13] and strand replacement [10]. Finally, the introduction of the (naphthalen-1-ylmethyl) group at the N^2 position of guanine in G-tetrad-forming oligonucleotides increased the thrombin-inhibitory activity of these oligonucleotides [14].

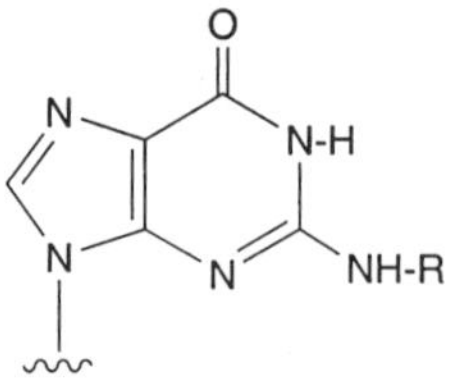

R = substituent

Fig. 1. *Structure of N^2-substituted guanine derivatives*

Two strategies have been described for the preparation of oligodeoxynucleotides carrying N^2-substituted G_d derivatives: 1) the use of a phosphoramidite carrying a protected N^2-modified G_d derivative and 2) the use of a precursor (usually a 2-fluoro-I_d derivative) [10][15] that is modified during deprotection of nucleobases by the appropriate amine (post-synthetic substitution). The first strategy is suitable for the introduction of several modifications in one oligonucleotide and the second approach is used for the introduction of fewer modifications [10b], especially when a complex amine is used, such as a PAH-substituted one [16]. A problem found when a polyamine is introduced into a guanine residue is the potential formation of different isomers [15][17]. In this communication, we describe the search for new, N^2-modified guanine derivatives with enhanced binding properties without the formation of isomers. Previous work indicated the importance of the imidazolyl group at the end of a propyl chain and the possibility that the imidazolyl group might interact with the phosphate [13]. Our results confirm this hypothesis, and we describe several new N^2-substituted derivatives of guanine with enhanced binding properties.

2. Results and Discussion. – We focused on the preparation of a dodecamer carrying a single N^2-substituted guanine. Our target oligonucleotide sequence was the dodecamer R5 (5′-d(CACCGACGGCGC-p-propanediol)-3′), complementary to a mutated Ha-*ras* oncogene with antiproliferative activity [18]. The G_d residue in bold indicates the position of N^2-substituted G_d. A propane-1,3-diol group was added at the 3′-end to prevent oligonucleotide degradation by exonucleases [19]. We decided to follow the post-synthetic substitution strategy. For this purpose, the 2-fluoro-I_d derivative protected with the 2-(4-nitrophenyl)ethyl (npe) group (see **1** in *Fig. 2*) was prepared as described [15]. The sequence was assembled on an automatic DNA synthesizer from phosphoramidites protected with the (*tert*-butyl)phenoxyacetyl group [20]. A special support carrying the 4,4′-dimethoxytrityl((MeO)$_2$Tr)-protected propanediol moiety connected by a succinyl linker to a sarcosyl-LCAA-CPG [21] (see **2** in *Fig. 2*) was prepared as described for similar supports [22]. Deprotection was performed in two steps: 1) 0.5M DBU (1,8-diazabicyclo[5.4.0]undec-7-ene) in MeCN, 2 h, at room temperature (to remove the npe and the 2-cyanoethyl phosphate protecting groups) and 2) 1M aqueous amine solution (to perform deprotection of the natural bases, modification of the 2-fluoro-I_d derivative, and cleavage of the oligonucleotide-support linkage), overnight, 55°. The resulting products were desalted over *Sephadex G-25* (*NAP-10* columns) and purified by reversed-phase HPLC using the trityl-on and trityl-off protocols. In all cases, a major peak was obtained and collected. The purified oligonucleotides were analyzed by mass spectrometry (MALDI-TOF) and exhibited the expected mass (*Table*).

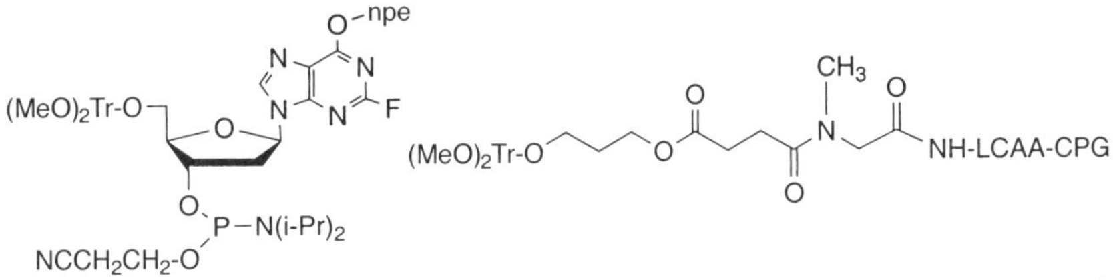

Fig. 2. *Phosphoramidite derivative **1** and solid support **2** used for the preparation of oligonucleotides **3**–**13** (see Table). LCAA-CPG = (long-chain-alkyl)amine controlled-pore glass.*

Thus, the dodecamers **3**–**13** were prepared by using appropriate amines during the deprotection-substitution step (see *Table* and *Scheme*). To avoid formation of different isomers, symmetric primary diamines (ethane-1,2-diamine (→**4**); propane-1,3-diamine (→**5**); hexane-1,6-diamine (→**8**); 2,2'-[ethane-1,2-diyl)bis(oxy)]bis(ethynamine) (→**10**)) and primary amines tethered with tertiary amines (1*H*-imidazole-1-propan-amine (→**3**); *N,N*-dimethylethane-1,2-diamine (→**6**); pyridin-2-ethanamine (→**9**) were selected. Spermidine (= *N*-(3-aminopropyl)butane-1,4-diamine (→**7**) was chosen for comparison with previous data [10][15]. The reaction of 2-fluoro-I_d with spermidine may form three isomers. Butanamine (→**11**), and 3-aminopropanol (→**12**) were selected as controls, as well as the dodecamer R5 (**13**) lacking guanine modification.

Thermal denaturation of duplexes formed by modified dodecamers and their complementary DNA sequence were studied at medium salt concentration (0.15M NaCl, 50 mM *Tris · HCl* pH 7.5). Melting temperatures are shown in the *Table*. Thermal renaturation was also studied by running a decreasing-temperature gradient. No differences were observed, indicating that strand association and dissociation is fast, as observed in unmodified oligonucleotides. All the N^2-substituted guanine derivatives prepared here stabilized the double helix. The most stabilizing substituents were the 3-(1*H*-imidazol-1-yl)propyl, the 2-aminoethyl, the 3-aminopropyl, the 2-(dimethylami-no)ethyl and the spermidine-derived groups (see **3**–**7**; 4–6° more stable than unmodified R5 (**13**), see *Table*). All these five substituents have a protonatable group (amino or imidazolyl) linked to a small chain of 2 or 3 C-atoms. A second group of substituents with moderate duplex stabilization properties was formed by the 6-

Scheme. *Preparation of Oligonucleotides Carrying N^2-Substituted Guanine Derivatives*

R, R', R": Substituents

aminohexyl, 2-(pyridin-2-yl)ethyl, and the 2-[2-(2-aminoethoxy)ethoxy]ethyl groups (see **8–10**; $2–3°$ increase). Except for the piridinylethyl group, these derivatives have an amino group tethered by a long (> 3 residues) hydrocarbon or ethyleneglycol chain. Finally, the lowest melting temperatures were obtained with guanine derivatives substituted by a butyl or 3-hydroxypropyl group devoid of a protonatable group (see **11** and **12**). These data are in agreement with previous results for oligonucleotides containing G_d derivatives with a N^2-substituent derived from $1H$-imidazole-1-propan-amine, spermine, or spermidine [10–14]. In a previous paper [15], we reported that the melting temperature of the unmodified dodecamer duplex without the propanediol moiety at the 3′ end was 51°. On discovering that the sequence of the unmodified control oligonucleotide was incorrectly introduced in the synthesizer, we repeated the melting experiments described in [15], but using the correct sequence. The melting temperature of the duplex control was found to be 63°. Considering this correction, the stabilization properties found in [15] are in agreement with the present data.

These results indicate that the most beneficial substituents at the N^2 position of guanine for duplex stabilization are ethyl and propyl groups linked to an amino or imidazolyl group. Longer linkers between the exocyclic guanine $NH_2–C(2)$ and the protonatable group reduce the duplex stability. A similar result was found on introduction of alkyl groups at the 2′-hydroxy function of RNA [23]. The O^2 position as well as N^2 position of guanine point to the minor groove where the negative phosphate groups are located, suggesting a positive interaction between the phosphate group and

Table. *Melting Temperatures and Molecular Weights of Dodecamers* 3–13 *Carrying a* N^2-*Substituted Guanine Unit*[a]

	R	T_m [°C][b]	M^+ found	calc.
3	(im)CH$_2$CH$_2$CH$_2$	71.1	3863.8	3863.1
4	NH$_2$CH$_2$CH$_2$	71.0	3798.4	3798.0
5	NH$_2$CH$_2$CH$_2$CH$_2$	70.1	3812.3	3812.1
6	Me$_2$NCH$_2$CH$_2$	69.6	3827.9	3826.1
7	NH$_2$(CH$_2$)$_4$NH(CH$_2$)$_3$[c])	69.6	3883.6	3883.2
8	NH$_2$(CH$_2$)$_6$	68.0	3854.8	3854.1
9	(py)CH$_2$CH$_2$	68.0	3865.1	3860.1
10	NH$_2$CH$_2$CH$_2$OCH$_2$CH$_2$OCH$_2$CH$_2$	67.1	3885.5	3886.1
11	MeCH$_2$CH$_2$CH$_2$	66.1	3810.7	3811.1
12	OHCH$_2$CH$_2$CH$_2$	65.6	3810.2	3813.0
13	H	65.0	3756.2	3756.0

[a]) Modified R5 dodecamer 5'-d(CAC C**G**A CGG CGC-p-propanediol)-3':

R = substituent

[b]) Complementary sequence: 3'-d(GTGGCTGCCGCG)-5', 0.15M NaCl, 50 mM *Tris* · HCl pH 7.5. Error in T_m is ±0.4°. [c]) Only one isomer of the three possible is shown.

the protonated substituent [13][23]. The results presented here are consistent with this hypothesis.

In conclusion, oligonucleotides carrying a variety of N^2-substituted guanine residues have been prepared from a common precursor. This method can be used to generate series of oligonucleotides for the screening of duplex-stabilizing substituents. Some of the modified oligonucleotides have interesting binding properties and are easy to prepare either by post-synthetic substitution or by the use of a protected phosphoramidite. The study of the antiproliferative activity of the compounds prepared is in progress.

We thank the group of Dr. *Matthias Wilm* and *Eurogentec* for obtaining mass spectra.

Experimental Part

Oligonucleotide Synthesis. The oligonucleotide sequence 5'-d(CAC C**X**A CGG CGC-p-propanediol)-3' ($\mathbf{X}_d$ = 2'-deoxy-2-fluoro-O^6-[2-(4-nitrophenyl)ethyl]inosine) was synthesized on a DNA synthesizer, model *394* (*Applied Biosystems*, USA), from (MeO)$_2$Tr- and [(*tert*-butyl)phenoxyacetyl]-protected nucleoside 3'-(2-cyanoethyl phosphoramidites) (*PerSeptive Biosystems*, USA) and 2'-deoxy-5'-*O*-4,4'-dimethoxytrityl)-2-fluoro-O^6-[2-(4-nitrophenyl)ethyl]inosine 3'-(2-cyanoethyl phosphoramidite) (**1**; *Fig. 2*). The latter was prepared as described [15]. A special support carrying the (MeO)$_2$Tr-protected propane-1,3-diol moiety connected by a succinyl linkage to sarcosyl-controlled-pore glass [21] was prepared as described [22]. Standard 1-μmol-scale synthesis cycles were used. Coupling efficiencies were higher than 98%. The oligonucleotide support was treated with 1 ml of 0.5M DBU in MeCN for 30 min at r.t. The support was washed in MeCN, 1% Et$_3$N in MeCN and MeCN and dried. The resulting support was treated with 1 ml of 1M aq. soln. of the appropriate amine (*Table*) at

60° overnight, and the soln. was filtered and evaporated. The residue was dissolved in H_2O, and the soln. was desalted on a *Sephadex G-25* (*NAP-10, Pharmacia*, Sweden) column. The oligonucleotide-containing fractions were analyzed and purified by HPLC (*PRP-1*-10 μm column, 305 × 7 mm; *Hamilton*, USA, flow rate 3 ml/min). A 20-min linear gradient from 15 to 45% MeCN over 20 mM aq. (Et₃NH)OAc was used for oligonucleotides carrying the (MeO)₂Tr group. After removal of the (MeO)₂Tr group with 80% AcOH/H_2O (30 min), the resulting oligonucleotides were purified on the same column with a 20-min linear gradient from 5 to 25% MeCN over 20 mM aq. (Et₃NH)OAc. Purified oligonucleotides were analyzed by MALDI mass spectrometry (see *Table*). On average, we obtained 25 *OD* units at 260 nm of purified dodecamer per μmol.

Melting Experiments. Melting experiments of dodecamer duplexes were performed by mixing equimolar amounts of two dodecamer strands dissolved in a soln. that contained 0.15M NaCl and 0.05M *Tris* · HCl buffer (pH 7.4). Duplexes were annealed by slow cooling from 80 to 4°. UV Absorption spectra and melting curves (absorbance *vs.* temperature) were recorded in 1-cm path-length cells using a *Cary-1/3*-UV/VIS spectrophotometer (*Varian*, Australia) with a temp. controller with a programmed temp. increase of 0.5°/min. Melting curves were determined in triplicate on duplex concentrations of 4 μM at 260 nm. Results: see *Table*.

REFERENCES

[1] K. Itakura, J. J. Rossi, R. B. Wallace, *Ann. Rev. Biochem.* **1984**, *53*, 323.

[2] E. Uhlmann, A. Peyman, *Chem. Rev.* **1990**, *90*, 543.

[3] R. J. Lipshutz, S. P. A. Fodor, T. R. Gingeras, D. J. Lockhart, *Nat. Genet. Suppl.* **1999**, *21*, 20.

[4] S. M. Frier, K. H. Altmann, *Nucleic Acids Res.* **1997**, *25*, 4429.

[5] B. Hyrup, P. Nielsen, *Bioorg. Med. Chem.* **1996**, *4*, 5.

[6] C. Hendrix, H. Rosemeyer, I. Verheggen, F. Seela, A. Van Aerschot, P. Herdewijn, *Chem.-Eur. J.* **1997**, *3*, 110.

[7] J. Wengel, *Acc. Chem. Res.* **1999**, *32*, 301.

[8] D. G. Schultz, S. M. Gryaznov, *Nucleic Acids Res.* **1996**, *24*, 2966.

[9] B. Sproat, A. I. Lamond, in 'Antisense Research and Applications', Eds. S. T. Crooke and B. LeBleu, CRC Press, Boca Raton, CA, USA, 1993, pp. 351.

[10] a) M. Schmid, J.-P. Behr, *Tetrahedron Lett.* **1995**, *36*, 1447; b) A. Adib, P. F. Potier, S. Doronina, I. Huc, J.-P. Behr, *Tetrahedron Lett.* **1997**, *38*, 2989.

[11] G. Wang, D. E. Bergstrom, *Tetrahedron Lett.* **1993**, *34*, 6725.

[12] N. V. Heeb, S. A. Benner, *Tetrahedron Lett.* **1994**, *35*, 3045.

[13] K. S. Ramasamy, M. Zounes, C. Gonzalez, S. M. Freier, E. A. Lesnik, L. Cummins, R. H. Griffey, B. P. Monia, P. D. Cook, *Tetrahedron Lett.* **1994**, *35*, 215; M. Manoharan, K. S. Ramasamy, V. Mohan, P. D. Cook, *Tetrahedron Lett.* **1996**, *37*, 7675; P. D. Cook, K. S. Ramasamy, M. Manoharan, U.S. Pat. 5,808,027, 1998, U.S. Pat. 5,587,469, 1996, and U.S. Pat. 5,459,255, 1995.

[14] G. X. He, S. H. Krawczyk, S. Swaminathan, R. G. Shea, J. P. Dougherty, T. Terhorst, V. S. Law, L. C. Griffin, S. Coutré, N. Bischofberger, *J. Med. Chem.* **1998**, *41*, 2234.

[15] A. R. Díaz, R. Eritja, R. Güimil García, *Nucleosides Nucleotides* **1997**, *16*, 2035.

[16] R. Casale, L. W. McLaughlin, *J. Am. Chem. Soc.* **1990**, *112*, 5264; H. Lee, M. Hinz, J. J. Stezowski, R. G. Harvey, *Tetrahedron Lett.* **1990**, *31*, 6773; C. M. Harris, L. Zhou, E. A. Strand, T. M. Harris, *J. Am. Chem. Soc.* **1991**, *113*, 4328; F. Johnson, I. Habus, R. G. Gentles, S. Shibutani, H. C. Lee, C. R. Iden, R. Rieger, *J. Am. Chem. Soc.* **1992**, *114*, 4923; B. Zajc, M. K. Laksham, J. M. Sayer, D. M. Jerina, *Tetrahedron Lett.* **1992**, *33*, 3409; T. Steinbrecher, C. Wameling, F. Oesch, A. Seidel, *Angew. Chem., Int. Ed.* **1993**, *32*, 404; J. Woo, S. T. Sigurdsson, P. B. Hopkins, *J. Am. Chem. Soc.* **1993**, *115*, 3407; H. Lee, E. Luna, M. Hinz, J. J. Stezowski, A. S. Kiselyov, R. G. Harvey, *J. Org. Chem.* **1995**, *60*, 5604; D. Tsarouhtsis, S. Kuchimanchi, B. L. DeCorte, C. M. Harris, T. M. Harris, *J. Am. Chem. Soc.* **1995**, *117*, 11013.

[17] W. T. Markiewicz, P. Godzina, M. Markiewicz, *Nucleosides Nucleotides* **1999**, *18*, 1449; P. Godzina, K. Adrych-Rozek, W. T. Markiewicz, *Nucleosides Nucleotides* **1999**, *18*, 2397.

[18] I. Duroux, G. Godard, M. Boidot-Forget, G. Schwab, C. Hélène, T. Saison-Behmoaras, *Nucleic Acids Res.* **1995**, *23*, 3411.

[19] A. Van Aerschot, T. Saison-Behmoaras, J. Rozenski, C. Hendrix, G. Schpers, G. Verhoeven, P. Herdewijn, *Bull. Soc. Chem. Belg.* **1995**, *104*, 717; F. Seela, K. Kaiser, *Nucleic Acids Res.* **1987**, *15*, 3113; J. Temsamani, J. Y. Tang, A. Padmapriya, M. Kubert, S. Agarwal, *Antisense Res. Dev.* **1993**, *3*, 277.

[20] N. D. Sinha, P. Davis, N. Usman, J. Pérez, R. Hodge, J. Kremsky, R. Casale, *Biochimie* **1993**, *75*, 13.

[21] T. Brown, C. E. Pritchard, G. Turner, S. A. Salisbury, *J. Chem. Soc., Chem. Commun.* **1989**, 891; K. Stengele, W. Pfleiderer, *Tetrahedron Lett.* **1990**, *31*, 2549.
[22] A. Aviñó, R. Güimil García, A. Díaz, F. Albericio, R. Eritja, *Nucleosides Nucleotides* **1996**, *15*, 1871.
[23] M. Manoharan, A. M. Kawasaki, T. P. Prakash, A. S. Fraser, *Nucleosides Nucleotides* **1999**, *18*, 1737.

Use of the Fluorescent Nucleoside Analogue Benzo[*g*]quinazoline 2′-*O*-Methyl-*β*-D-ribofuranoside to Monitor the Binding of the HIV-1 Tat Protein or of Antisense Oligonucleotides to the TAR RNA Stem-Loop

by **Andrey Arzumanov**[a]), **Frédéric Godde**[b]), **Serge Moreau**[b]), **Jean-Jacques Toulmé**[b]), **Alan Weeds**[a]), and **Michael J. Gait**[a])*

[a]) Medical Research Council, Laboratory of Molecular Biology, Hills Road, Cambridge, CB2 2QH (tel +44 1223 248011; fax +44 1223 402070; e-mail: mgait@mrc-lmb.cam.ac.uk)

[b]) INSERM U 386, IFR Pathologies Infectieuses, 146 rue Léo Saignat, Université Victor Segalen, F-33076 Bordeaux

The Tat protein is an essential *trans*-activator of HIV gene expression. It interacts with its RNA recognition sequence, the *trans*-activation responsive region TAR, as well as cellular factors. These interactions are potential targets for drug discovery against HIV infection. We have developed a new and sensitive assay for the measurement of Tat binding to TAR in solution under equilibrium conditions based on the change of fluorescence of the base analogue benzo[*g*]quinazoline-2,4(1*H*,3*H*)-dione (BgQ) incorporated into the chemically synthesized model TAR stem-loop **2** to which was added Tat-[37-72] peptide (**3**). The results show that Tat-TAR binding strength is 2–3-fold stronger than has previously been determined by mobility-shift analysis. Changes of fluorescence were used also to measure the binding of antisense 2′-*O*-methyloligonucleotides to TAR **2**.

1. Introduction. – The human immunodeficiency virus type 1 (HIV-1) Tat protein is an essential *trans*-activator required for viral replication (for recent reviews, see [1–3]). Tat interacts with an RNA sequence, the *trans*-activation responsive region TAR, a 59-residue stem-loop that occurs at the 5′-end of all viral RNA transcripts, as well as a Tat-associated kinase complex (TAK) that includes cyclin T1 and the kinase CDK9. The full mechanism of *trans*-activation by Tat is not yet established, but evidence from *in vitro* transcription techniques suggests that, in the absence of Tat, the transcription complex is unstable in the elongation phase leading to a predominance of short viral transcripts. When TAK becomes activated by Tat and TAR, the C-terminal domain of RNA polymerase II becomes hyperphosphorylated, an event that is essential for full-length Tat-dependent transcription [4][5]. The key virus-specific step in *trans*-activation, however, is Tat/TAR recognition.

Tat binds *in vitro* with high affinity near the apex of TAR to a region containing a 3-residue U-rich bulge (see below, *Figs. 1* and *2*). This interaction has been studied in great detail over many years (reviewed in [1]), including much work in our own laboratory on model TAR duplexes together with Tat protein made in *E. coli* or with synthetic Tat peptides [6][7]. More recently, synthetic TAR duplexes have been used for cross-linking studies to Tat [8–11]. Structural models of TAR either free or in the presence of a Tat peptide have also been obtained based on NMR characterization [12–16], but no complete structure of the peptide/RNA complex has been produced.

Models of the interaction are based on location of the arginine-rich region of Tat within the major groove of TAR and formation of a number of contacts with base and phosphate residues in the bulge region.

The TAR stem-loop has been a target for drug discovery and a number of small-molecule inhibitors (reviewed in [17]), and peptidomimetics [18–20] have been shown to bind to TAR and interfere with Tat action either by direct competition or by stabilizing a TAR conformation that cannot be recognized by Tat. Another interesting lead compound is the tripeptide Lys-D-Lys-Asn that was selected from a solid-phase-bound combinatorial library to interfere with Tat-TAR binding by binding the U-rich bulge [21].

An alternative approach to interfere with *trans*-activation is disruption of the TAR structure by complementary antisense oligonucleotides. The apical stem and loop region just above the bulge of TAR is well-conserved among viral strains, and disruption of this region of TAR might be doubly effective by prevention of binding both of Tat and TAK. *Vickers et al.* showed that 18–28-residue phosphodiester and phosphorothioate oligodeoxynucleotides complementary to TAR were able to bind specifically in a gel-mobility-shift assay [22]. Anti-TAR phosphorothioate oligonucleotides showed a dose-dependent activity in cellular *trans*-activation assays, but in antiviral assays, they showed no sequence specificity. Phosphodiester oligonucleotides were inactive in both assays presumably due to degradation by nucleases. We showed recently that nuclease-stabilized 16-mer 2'-*O*-methyloligoribonucleotides are effective agents in steric blocking of Tat-TAR [23][24]. The 2'-*O*-methyloligoribonucleotides targeted to TAR are also strong inhibitors of HIV reverse transcription [25]. Further, DNA aptamers and their 2'-*O*-methyl analogues have been selected to bind strongly to TAR and form kissing complexes [26]. RNA Aptamers have also been selected against the TAR element [27].

For measurement of the strength of the interaction of Tat with TAR *in vitro*, a number of standard assays have been developed involving gel-mobility shift or filter binding [28][29]. We have also used these assays to assess oligonucleotide binding to TAR [23][24]. Unfortunately, neither of these techniques is a measure of the true binding constant in solution under equilibrium conditions. We have now developed a new assay for the measurement of Tat binding to TAR in solution based on the change of fluorescence of the base analogue benzo[*g*]quinazoline-2,4(1*H*,3*H*)-dione (BgQ; see **1**) [30][31] incorporated into the chemically synthesized model TAR stem-loop **2**. Changes of fluorescence could be used also to measure the binding of antisense oligonucleotides to TAR **2**.

2. Results. – *Syntheses and Peptide Binding.* The nucleoside U^{24} in the bulge of TAR is not directly involved in binding Tat protein and may be replaced by any nucleoside [7] or by a 2'-deoxynucleoside [6]. In addition, large conformational changes in this residue are known to occur upon Tat binding [13]. Thus U^{24} is an obvious site for incorporation of a fluorescent base. BgQ is a thymine analogue that has a strong fluorescence-emission intensity which becomes quenched upon stacking within DNA duplexes and triplexes [30a][31]. The excitation spectrum of BgQ as a 2'-deoxynucleoside shows three maxima at 253 (major), 294, and 360 nm and a broad emission spectrum centered between 420 and 440 nm [30a]. The extinction coefficient at 260 nm is extremely high ($2.6 \cdot 10^5$ M^{-1} cm^{-1}), and the fluorescence quantum yield is also high (0.82) [30a]. This makes it ideal for use in studying protein interactions, since the only two amino acids with absorption in this region, tyrosine (y) and tryptophan (w), would have less than 0.5% fluorescence compared to BgQ (extinction coefficients for Y and W of $1.4 \cdot 10^3$ and $5.6 \cdot 10^3$ M^{-1} cm^{-1}, resp., and quanta yield of 0.12 and 0.20, resp., at their maxima at 270 and 280 nm, resp.). Therefore, we chemically synthesized a 27-mer RNA model, *i.e.* BgQ TAR (**2**) where U^{24} was substituted by the 2'-*O*-methylribonucleoside derivative **1** of this fluorescent base BgQ (*Fig. 1, a* and *b*), which has identical fluorescence properties to that of the corresponding 2'-deoxynucleoside. The synthesis was carried out by standard solid-phase RNA-synthesis techniques with a phosphor-

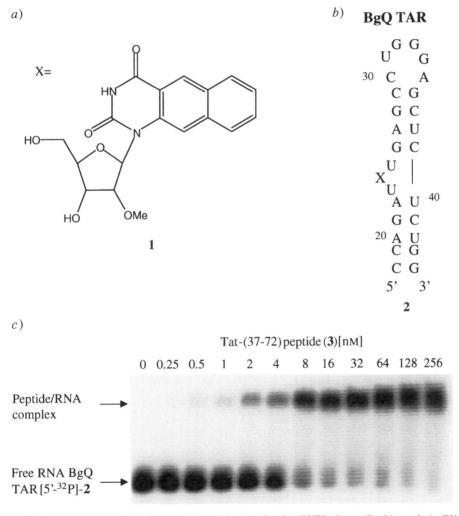

Fig. 1. a) *1-(2'-O-Methyl-β-D-ribofuranosyl)benzo[g]quinazoline-2,4(1H,3H)-dione* (**1**); b) *synthetic RNA model BgQ TAR* (**2**), *containing the modified nucleoside* (**1**) *in position X;* c) *polyacrylamide-gel-mobility-shift assay of Tat-(37–72) peptide* (**3**) *binding to* ^{32}*P-labelled RNA BgQ TAR [5'-*^{32}*P]-***2**

amidite of **1** [30b] and commercially available ribonucleoside phosphoramidite derivatives.

To check that BgQ TAR (**2**) is a good model for Tat binding, we carried out gel-mobility-shift analysis of synthetic Tat-(37-72) peptide (HIV-1 BRU; **3**) binding to **2** (*Fig. 1,c*; for the sequence of the Tat peptide **3**, see below, *Fig. 2,c*). For the interaction with **3**, an apparent K_d of 16 nM was obtained by gel scanning. In comparison, a longer TAR RNA model that is unmodified, *i.e.*, TAR BRU 39 (**4**), bound the Tat peptide **3** with an apparent K_d of 5 nM, as determined by mobility-shift analysis (*Fig. 2*). The longer TAR BRU 39 (**4**) was expected to bind about twice as strongly as a shorter 27- or 29-mer, based on previous experience [6][32]. Since Tat peptide binding to **2** is *ca.* 3-fold higher in K_d than to **4**, the replacement of U^{24} by BgQ thus results in only a very slight reduction in ability to bind Tat peptide **3**. Therefore BgQ TAR (**2**) is a good model for Tat binding.

Fluorescence Changes in BgQ TAR (**2**) *upon Tat Peptide Binding. Fig. 3* shows the excitation and emission spectra of 5 nM BgQ TAR (**2**) in 50 mM *Tris* · HCl (pH 7.5), 20 mM KCl at 20°. The intensity of both excitation and emission is quite low compared

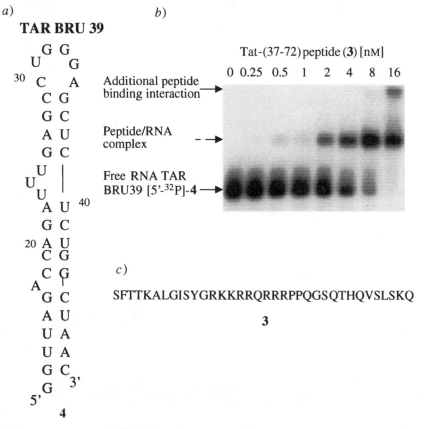

Fig. 2. a) *Synthetic RNA model TAR BRU 39* (**4**); b) *polyacrylamide-gel-mobility-shift assay of Tat-(37–72) peptide* (**3**) *binding to* ^{32}*P-labelled RNA TAR BRU 39 [5'-^{32}P]-**4**; c) *sequence of the synthetic Tat-(37–72) peptide* (**3**) *with Cys-37 replaced by Ser*

to that of the free nucleoside [30a], suggesting that the BgQ fluorescence is quenched considerably, presumably by stacking interactions within the TAR RNA. A stacking interaction of U^{24} has been established by NMR experiments within free TAR RNA [14]. In the presence of excess Tat peptide **3** (20 nM), a dramatic increase of 2–3 fold in the fluorescence was observed (*Fig. 3*). The maximum of emission was shifted upfield slightly from 424 to 429 nm.

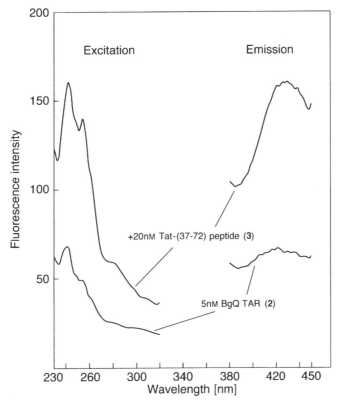

Fig. 3. *Excitation and emission spectra of BgQ TAR* (**2**) *alone and in the presence of Tat-(37-72) peptide* (**3**) *carried out in 50 mM* Tris·HCl (*pH 7.4*) *and 20 mM KCl at 20°*. The excitation fluorescence spectrum was recorded for emission wavelength set at 430 nm, and the emission spectrum recorded at 240 nm excitation wavelength

Upon titration of 20 nM BgQ TAR (**2**) with increasing concentration of Tat peptide **3** under the same buffer and relatively low salt conditions, a fluorescence-intensity curve of the emission spectra that reached a plateau at *ca.* 40 nM of Tat peptide and a *ca.* 3-fold intensity increase (*Fig. 4*) was observed. Curve fitting using a simple 1:1 binding isotherm gave an apparent K_d of 6 ± 3 nM. However, the slight deviations from this simple binding isotherm suggest a small amount of binding of a second molecule of Tat peptide **3** to BgQ TAR (**2**). Such second Tat peptide binding was not observed in the mobility-shift assay with **2** (*Fig. 1*) but can be seen at higher Tat concentrations with the longer TAR BRU 39 ((**4**); *Fig. 2*).

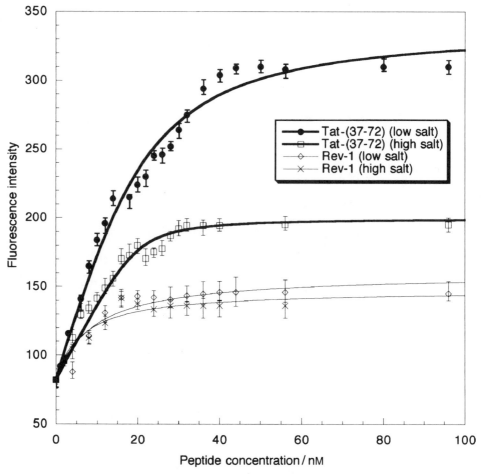

Fig. 4. *Fluorimetric emission titration curves of 20 nM BgQ TAR* (**2**) *by Tat-(37–72) peptide* (**3**) *and Rev-(34–51) peptide in low* (20 mM KCl) *and high* (80 mM KCl) *salt buffers.* Curves in bold are fitted to *Eqn. 1* to determine K_d. Note that a perfect fit to a single binding isotherm was only obtained for Tat peptide under the higher salt conditions.

Under more stringent conditions (higher salt concentration, 80 mM KCl), the fluorescence increase during titration of **2** with **3** fitted very well to a 1:1 binding isotherm (*Fig. 4*). There was a smaller fluorescence-intensity increase (2-fold), but the K_d from curve fitting showed a slightly stronger interaction with an apparent K_d of 1.5 ± 0.5 nM. We also studied the binding of a related peptide derived from the highly arginine-rich region of HIV-1 Rev, residues 34–51, which interacts with the high-affinity site of the Rev-responsive element RNA [33]. Rev-(34-51) peptide is known to bind also to TAR at low affinity and specificity [34]. Titration of BgQ TAR (**2**) with increasing Rev peptide resulted in only a very small increase in fluorescence, both at low salt and at higher salt concentration (*Fig. 4*). Further, a 1:1 binding isotherm could not be fitted to either of these curves, suggesting a complex, nonspecific binding pattern for the Rev-(34–51) peptide. This demonstrates that specific TAR binding in the bulge region by Tat peptide **3** can be distinguished from the less specific Rev-peptide binding at higher salt concentration by the higher fluorescence increase and the better fit to a 1:1 binding isotherm.

We also measured the effect of addition of Mg^{2+} ions to the binding equilibrium (*Fig. 5*). At low salt concentration (20 mM KCl), the fluorescence intensity of a mixture of 20 nM BgQ TAR (**2**) and 50 nM Tat peptide **3** was gradually quenched upon Mg^{2+} addition. The half-point of loss of fluorescence was observed at *ca.* 3 mM Mg^{2+} concentration, and full loss at 7 mM. This result suggests that the Tat-TAR interaction is not stable at moderately high Mg^{2+} concentration (10 mM), conditions quite commonly used in protein-nucleic acid interaction studies, but is relatively unaffected by concentrations of 1 mM Mg^{2+} ion or less, conditions expected to be closer to physiological. At higher salt concentration (80 mM KCl), the point of half-inhibition

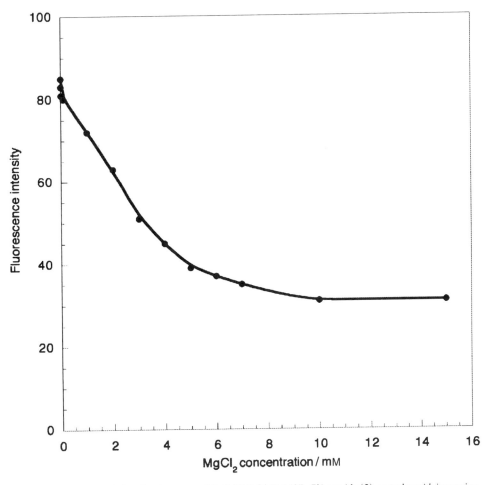

Fig. 5. *Fluorimetric emission titration curve of BgQ TAR (**2**)/Tat-(37–72) peptide (**3**) complex with increasing concentrations of MgCl₂ under lower salt conditions*

was observed at *ca.* 1 mM Mg^{2+} concentration (data not shown). This experiment shows how fluorescence changes of **2** can be used to measure the stability of Tat interaction under a variety of buffer and additive conditions.

To demonstrate further that the fluorescence-intensity changes are related to specific Tat-TAR binding, we carried out a competition assay by addition of increasing amounts of the unmodified RNA TAR BRU 39 (**4**) to a mixture of Tat peptide **3** and BgQ TAR (**2**). The fluorescence intensity decreased as the concentration of TAR **4** increased (*Fig. 6*). The concentration of **4** required to reduce fluorescence intensity by half was 14 nM, which is similar to that found by mobility-shift analysis using a competition assay (16 nM, data not shown). By curve fitting, a K_d of binding of the TAR 39-mer **4** to Tat peptide **3** was calculated to be 1.6 nM. This is a slightly stronger binding than that found for direct binding of **2** to **3** by fluorescence measurements under the same conditions (6.3 nM, *Fig. 3*) and is in line with the slightly stronger binding of **4**

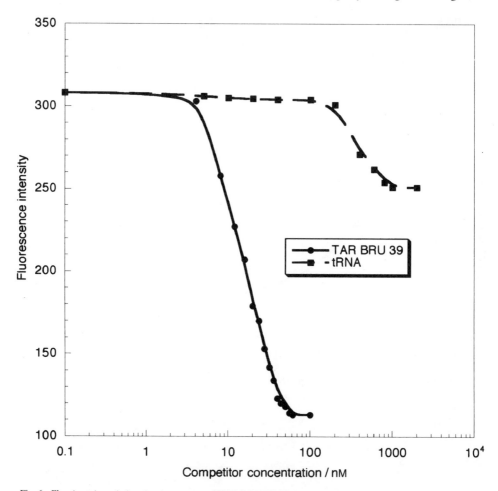

Fig. 6. *Fluorimetric emission titration curves of TAR BRU 39 (**4**) and nonspecific (tRNA) competitors with the BgQ TAR (**2**)/Tat-(37–72) peptide (**3**) complex under lower salt conditions*

than **2** as observed by mobility-shift analysis (*Figs. 1* and *2*). By contrast, addition of increasing amounts of a nonspecific competitor (tRNA) to **2/3** gave rise to a much poorer competition curve. Curve fitting showed an approximate K_d of 760 nM for tRNA binding to **2**.

Interaction of Antisense 2'-O-Methyloligonucleotides with BgQ TAR (2). A large increase of fluorescence intensity as well as a shift in the fluorescence-emission maximum was observed when a 16-mer 2'-*O*-methyloligonucleotide complementary to the RNA BgQ TAR (**2**) [23][24] was added (*Fig. 7*). A maximum value of fluorescence intensity was reached at *ca.* 300 nM of 16-mer oligonucleotide and the half-intensity at *ca.* 20 nM. This compares with *ca.* 70–80 nM for 50% complex formation by mobility-shift analysis for 16-mer binding to **2** (data not shown) and *ca.* 50–60 nM for TAR BRU 39 (**4**) [24]. At very high oligonucleotide concentration, the intensity of fluorescence was reduced again. No increase of fluorescence intensity was observed on addition of a mismatched or scrambled oligonucleotide to **2** (*Fig. 7*). Shorter 12-mer 2'-*O*-methyloligonucleotides gave different results depending on their precise alignment compared to the position of the BgQ-containing residue **1** in the bulge. The 12-mer I, which is antisense to residues 23–34 (*Fig. 1*) and would be expected to form a duplex that includes the BgQ residue **1** at position 24, showed a fluorescence increase somewhat similar to the 16-mer. By contrast, 12-mer II, which is complementary to residues 25–36 and which does not reach to the BgQ residue **1** at position 24, does not give a fluorescence shift, even though it is known to bind TAR 39 **4** with a *ca.* 50 nM apparent K_d by mobility-shift analysis [24].

3. Discussion. – The fluorescence emission or excitation spectrum of BgQ-TAR (**2**) is a sensitive measure of the binding of Tat peptide **3** to TAR RNA and gives an apparent K_d by curve fitting that is 2–3 fold lower than that determined by mobility-shift analysis. This suggests that mobility-shift analysis results in underestimation of the binding strength of the Tat-TAR interaction by a factor of 2–3. An explanation for this is that during the polyacrylamide-gel electrophoresis, there is time for some dissociation of the Tat-TAR complex to take place, whereas the fluorescence assay measures the true equilibrium binding strength. The results also show that a large fluorescence-intensity increase is observed only in binding of the cognate Tat peptide, and show that the best fit to a 1 : 1 interaction is obtained under higher salt (80 mM KCl) conditions. By contrast, titrations of BgQ TAR (**2**) with the less specific Rev-(34-51) peptide led to only a very small fluorescence increase and curves which could not be fitted to a simple binding isotherm under either salt concentration. The assay is also useful in its competition mode to assess the strength of binding of Tat peptide **3** to other unlabelled RNA constructs, by curve fitting to the fluorescence-intensity decrease as the BgQ TAR (**2**) is displaced when the competitive RNA concentration is increased.

Fluorescence increases were also used to assess the strength of binding of antisense oligonucleotides to the TAR RNA stem-loop. But this was only possible for those oligonucleotides that formed hybrids with the RNA that included the BgQ residue **1** at position 24. Again the strength of the interaction was found to be 2–3 fold stronger in the fluorescence assay as compared to mobility-shift analysis. At higher oligonucleotide concentration, a general reduction in fluorescence intensity was seen. This may be due to fluorescence quenching because of nonspecific interaction of the oligonucleotide

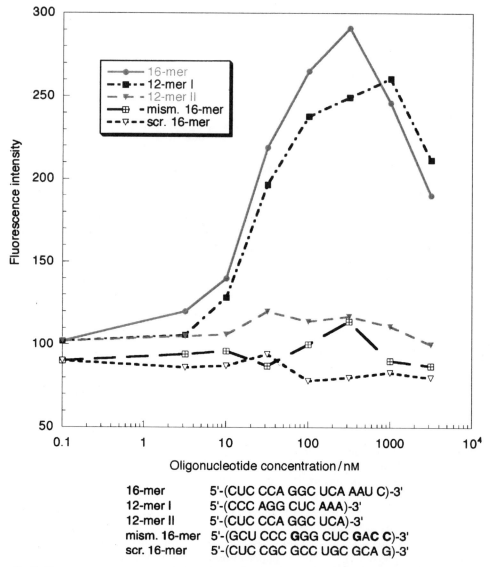

16-mer	5'-(CUC CCA GGC UCA AAU C)-3'
12-mer I	5'-(CCC AGG CUC AAA)-3'
12-mer II	5'-(CUC CCA GGC UCA)-3'
mism. 16-mer	5'-(GCU CCC GGG CUC GAC C)-3'
scr. 16-mer	5'-(CUC CGC GCC UGC GCA G)-3'

Fig. 7. *Fluorescence emission intensity of 20 nM BgQ TAR (2) at 430 nm (excitation at 240 nm) at different 2'-O-methyloligonucleotide concentrations.* The sequences of the oligonucleotides studied are shown below the graph. Nucleotides in bold represent mismatches to the TAR RNA sequence.

with the fluorophore or, more likely, to a filter effect due to the increased absorption of the oligonucleotide at the excitation wavelength (240 nm). Nevertheless, the assay could still be used up to 1 μM oligonucleotide and the binding strength measured. One way to get around a filter effect is to carry out excitation at the secondary absorbance maxima (294 or 360 nm), an approach we are currently investigating.

The fluorescence-intensity curves for BgQ TAR (**2**) are complex when both peptide and oligonucleotide are simultaneously present (data not shown), and we found that the intensity changes cannot be simply related to concentrations of the TAR binding species when arranged in this competition format. Probably, the fluorescence intensity of BgQ is dependent on to what extent it is stacked into the free TAR RNA structure, or whether it is fully or partially unstacked when the Tat peptide or an oligonucleotide is bound. Similarly, complex fluorescence changes were found when small-molecule TAR-binding molecules competed against Tat peptide (data not shown). Therefore, it seems unlikely that the assay can be used for a competition screen of potential inhibitors of the Tat-TAR interaction.

However, we have shown that changes in the fluorescence intensity can be used to assess the direct binding of molecules (peptides, oligonucleotides) to the bulge of TAR. Since the fluorescence extinction coefficient and quantum yield of BgQ is very high, small molecules, even those containing aromatic groups, should be able to be assessed for direct TAR binding in the region of the bulge, and studies of this nature are in progress. Our results suggest that the fluorescent base BgQ should be generally useful also in studying protein interactions with synthetic RNA binding sites.

We thank *Justine Michel* (INSERM) for the synthesis of BgQ TAR (**2**) and *David Owen* for the synthesis of the Tat and Rev peptides.

Experimental Part

Syntheses. The HIV-1 Tat-(37-72) peptide (**3**) with sequence SFTTKALGISYGRKKRRQRRRPPQG-SQTHQVSLSKQ (Cys[37] was replaced by Ser to avoid problems with disulfide formation) and HIV-1 Rev-(34-51) peptide (Succ-TRQARRNRRRRWRERQRK-OH) were prepared by continuous-flow Fmoc-polyamide solid-phase synthesis as previously described [35–37]. The synthesis of TAR BRU 39 (**4**) was previously reported [23]. BgQ TAR (**2**) was synthesized on a 1-µmol scale on an *Expedite 8909*-DNA/RNA synthesizer (*PE Biosystems*) using RNA monomers having (*tert*-butyl)phenoxyacetyl protecting groups at the exocyclic NH$_2$ groups and (*tert*-butyl)dimethylsilyl protection at the 2'-*O*-position. The phosphoramidite of the BgQ-containing 2'-*O*-methylribonucleoside **1** was dissolved in anh. MeCN (0.1M final concentration) and the coupling time for this monomer was increased up to 15 min. The resulting protected BgQ TAR was deprotected by standard techniques and purified by prep. polyacrylamide-gel electrophoresis visualized by UV shadowing.

The complete synthesis of the phosphoramidite of **1** will be published elsewere [30b]; the key step relies on the highly efficient procedure developed by *Ross et al.* for the synthesis of 2'-*O*-methyluridine, which is based on the opening of the 2-*O*,2'-anhydrouridine by trimethyl borate in hot MeOH [38]. We successfully extended this procedure to the benzo[*g*]quinazoline-2,4(1*H*,3*H*)-dione ribonucleoside [30a]. The ribonucleoside was obtained from the glycosylation of α-D-ribofuranose 1-acetate 2,3,5-tribenzoate by the silylated derivative of the benzo[*g*]quinazoline-2,4(1*H*,3*H*)-dione with trimethylsilyl triflate as a catalyst. The phosphoramidite was then obtained by classical dimethoxytritylation and phosphitylation procedures.

The 2'-*O*-methyloligoribonucleotides were prepared by standard solid-phase chemical synthesis from phosphoramidite reagents obtained from *Cruachem* or *Glen Research via Cambio*. The mass of each oligonucleotide was checked by MALDI-TOF mass spectrometry on a *Perspective Biosystems Voyager DE* mass spectrometer as previously described [10]. Sequences of oligonucleotides: 16-mer 5'-(CUC CCA GGC UCA AAU C)-3', 12-mer I 5'-(CCC AGG CUC AAA)-3', 12 mer II 5'-(CUC CCA GGC UCA)-3', mismatched 16-mer 5'-(GCU CCC **G**GG CUC **G**ACC)-3' (bold units show positions of mismatch), scrambled 16-mer 5'-(CUC CGC GCC UGC GCAG)-3'.

Folding of TAR RNA. For gel-mobility-shift experiments, 28.6 nM ^{32}P-labelled TAR BRU 39 [5'-^{32}P]-**4** or 35.7 nM ^{32}P-labelled BgQ TAR [5'-^{32}P]-**2** in 70 µl of 50 mM *Tris* · HCl (pH 7.4) and 20 mM KCl was incubated for 10 min at 75° and slowly cooled to 37°. Then the stock solns. (125 µl) of the corresponding TAR RNA (16 or 20 nM) in 200 mM *Tris* · HCl (pH 7.4), 80 mM KCl, 20 mM DTT, and 0.2% *Triton X-100* containing 40 units of RNase inhibitor RNasin (*Promega*) were made and kept on ice.

For fluorescence measurements, 50 or 200 nM BgQ TAR (**2**) in 50 mM *Tris* · HCl (pH 7.4) and 20 or 80 mM KCl was incubated for 10 min at 75°, slowly cooled to 37°, and then put on ice. The final portions (400 µl) of BgQ TAR **2** (5 or 20 nM) in 50 mM *Tris* · HCl (pH 7.4) and 20 or 80 mM KCl were prepared and stored on ice before use. TAR BRU 39 (**4**) (800 nM) for competitive binding assays was refolded similarly in 50 mM *Tris*·HCl (pH 7.4) and 20 mM KCl and kept on ice.

Gel-Mobility-Shift Analysis: 1) *TAR RNA with Tat-(37–72) peptide* (**3**). The 4–5 nM [5'-^{32}P]-**4** or [5'-^{32}P]-**2** was incubated with increasing amounts of Tat peptide **3** in 50 mM *Tris*·HCl (pH 7.4), 20 mM KCl, 5 mM DTT, 0.05% *Triton X-100*, and 0.08 unit/μl RNasin at 37° for 15 min. To each sample, loading buffer was added to give 0.025% bromophenol blue and 13% sucrose, and electrophoresis was carried out on 6% native polyacrylamide gel and run in 44.6 mM *Tris* borate (pH 8.3), 1 mM EDTA, 0.05% *Triton X-100*, and 0.2% glycerol at r.t. for 1 h. The gel was dried and visualized by autoradiography. The dried gels were also exposed to a phosphor storage screen (*Molecular Dynamics*) and scanned by a model *425S PhosphorImager*™ (*Molecular Dynamics*). The resulting digitized images were analyzed by *Geltrak* on a *DEC/Alpha 2100*™ (*Digital Equipment Corporation*, Maynard, MA, USA) through an X-terminal [39] and apparent K_d values estimated as the concentration required for 50% complex formation.

2) *TAR RNA with 2'-O-Methyloligonucleotides.* The 4–5 nM [5'-^{32}P]-**4** or [5'-^{32}P]-**2** was incubated with increasing amounts of 16-mer 2'-*O*-methyloligonucleotide in 50 mM *Tris*·HCl (pH 7.4), 20 mM KCl, 1 mM DTT, and 0.08 unit/μl RNasin at 37° for 15 min. To each sample, loading buffer was added to give 0.025% bromophenol blue and 13% sucrose, and electrophoresis was carried out on 8% native polyacrylamide gel and run in 20 mM *Tris* acetate (pH 8.3), 1 mM EDTA, and 0.2% glycerol at r.t. for 1 h. The gel was analyzed as described above.

Fluorescence Measurements. Fluorescence measurements were carried out on a *Perkin-Elmer LS-50B* luminescence spectrometer with thermostat control accurate to ±0.1°. The soln. of BgQ TAR (**2**) was excited at 240 nm and monitored at 430 nm. The integration time was 2 s. For calculations of K_d, the average values of three measurements were used. Measurements were carried out at 20° in buffer soln. containing 50 mM *Tris*· HCl and 20 or 80 mM KCl (pH 7.4). The solns. of **2** with increasing amounts of oligonucleotides were preincubated on ice for 1 h before recording the fluorescent spectra.

Determination of Dissociation Constants. Eqn. 1 [40] was used for the determination of the dissociation constant K_d for the interaction between BgQ TAR (**2**) and Tat-(37–72) peptide (**3**)

$$F = F_o + \Delta F\{([RNA]_o + [P]_o + K_d) - (([RNA]_o + [P]_o + K_d)^2 - 4[RNA]_o[P]_o)^{1/2}\}/2[RNA]_o \qquad (1)$$

where *F* and F_o are the fluorescence intensity of **2** in the presence and absence of **3**, resp. ΔF is the fluorescence coefficient of the **2/3** complex per nmol of complex. [RNA]$_o$ and [P]$_o$ are the initial concentrations of **2** and **3**, resp.

In the competitive binding assay, *Eqn. 2* generated from the equation for fractional saturation of labelled probe [41] was used for the calculation of the dissociation constant K_c for competitor RNA/Tat peptide complex:

$$F = F_o + \Delta F\{[RNA]_o + [P]_o + K_d + (K_d/K_c)[C]$$
$$-(([RNA]_o + [P]_o + K_d + (K_d/K_c)[C])^2 - 4[RNA]_o[P]_o)^{1/2}\}/2[RNA]_o \qquad (2)$$

where *F* is the fluorescence intensity in the presence of competitor RNA, F_o the fluorescence intensity in the absence of both peptide and competitor RNA, and ΔF, [RNA]$_o$, [P]$_o$, and K_d are the same as in *Eqn. 1*. [C] is the concentration of competitor RNA. A competition curve, $F = f([C])$, may be fitted for the best value of K_c by the nonlinear least-squares method with *KaleidaGraph* software (*Abelbeck Software*).

REFERENCES

[1]   J. Karn, *J. Mol. Biol.* **1999**, *293*, 235.
[2]   R. Taube, K. Fujinaga, J. Wimmer, M. Barboric, B. M. Peterlin, *Virology* **1999**, *264*, 245.
[3]   T. M. Rana, K.-T. Jeang, *Arch. Biochem. Biophys.* **1999**, *365*, 175.
[4]   N. J. Keen, M. J. Churcher, J. Karn, *EMBO J.* **1997**, *16*, 5260.
[5]   C. Isel, J. Karn, *J. Mol. Biol.* **1999**, *290*, 929.
[6]   F. Hamy, U. Asseline, J. A. Grasby, S. Iwai, C. E. Pritchard, G. Slim, P. J. G. Butler, J. Karn, M. J. Gait, *J. Mol. Biol.* **1993**, *230*, 111.
[7]   M. J. Churcher, C. Lamont, F. Hamy, C. Dingwall, S. M. Green, A. D. Lowe, P. J. G. Butler, M. J. Gait, J. Karn, *J. Mol. Biol.* **1993**, *230*, 90.
[8]   Z. Wang, T. M. Rana, *Biochemistry* **1996**, *35*, 6491.

[9] N. A. Naryshkin, M. A. Farrow, M. G. Ivanovskaya, T. S. Oretskaya, Z. A. Shabarova, M. J. Gait, *Biochemistry* **1997**, *36*, 3496.

[10] M. A. Farrow, F. Aboul-ela, D. Owen, A. Karpeisky, L. Beigelman, M. J. Gait, *Biochemistry* **1998**, *37*, 3096.

[11] Z. Wang, T. M. Rana, *Biochemistry* **1998**, *37*, 4235.

[12] J. D. Puglisi, R. Tan, B. J. Calnan, A. D. Frankel, J. R. Williamson, *Science (Washington, D.C.)* **1992**, *257*, 76.

[13] F. Aboul-ela, J. Karn, G. Varani, *J. Mol. Biol.* **1995**, *253*, 313.

[14] F. Aboul-ela, J. Karn, G. Varani, *Nucleic Acids Res.* **1996**, *24*, 3974.

[15] M. J. Seewald, A. U. Metzger, D. Willbold, P. Rösch, H. Sticht, *J. Biomol. Struct. Dynamics* **1998**, *16*, 683.

[16] K. S. Long, D. M. Crothers, *Biochemistry* **1999**, *38*, 10059.

[17] D. Daelemans, A.-M. Vandamme, E. De Clercq, *Antiviral Chem. Chemotherapy* **1999**, *10*, 1.

[18] F. Hamy, E. R. Felder, G. Heizmann, J. Lazdins, F. Aboul-ela, G. Varani, J. Karn, T. Klimkait, *Proc. Natl. Acad. Sci. U.S.A.* **1997**, *94*, 3548.

[19] I. Huq, X. Wang, T. M. Rana, *Nat. Struct. Biol.* **1997**, *4*, 881.

[20] N. Tamilarisu, I. Huq, T. M. Rana, *J. Am. Chem. Soc.* **1999**, *121*, 1597.

[21] S. Hwang, N. Tamilarisu, K. Ryan, I. Huq, S. Richter, W. Still, T. M. Rana, *Proc. Natl. Acad. Sci. U.S.A.* **1999**, *96*, 12997.

[22] T. Vickers, B. F. Baker, P. D. Cook, M. Zounes, R. W. J. Buckheit, J. Germany, D. J. Ecker, *Nucleic Acids Res.* **1991**, *19*, 3359.

[23] B. Mestre, A. Arzumanov, M. Singh, F. Boulmé, S. Litvak, M. J. Gait, *Biochim. Biophys. Acta* **1999**, *1445*, 86.

[24] A. Arzumanov, M. J. Gait, in 'Collection Symposium Series', Eds. A. Holy and M. Hocek, Academy of Sciences of the Czech Republic, 1999, p. 168.

[25] F. Boulmé, F. Freund, S. Moreau, P. Nielsen, S. Gryaznov, J.-J. Toulmé, S. Litvak, *Nucleic Acids Res.* **1998**, *26*, 5492.

[26] C. Boiziau, E. Dausse, L. Yurchenko, J.-J. Toulmé, *J. Biol. Chem.* **1999**, *274*, 12730.

[27] F. Ducongé, J.-J. Toulmé, *RNA* **1999**, *5*, 1605.

[28] J. Karn, M. J. Churcher, K. Rittner, A. Kelley, P. J. G. Butler, D. A. Mann, M. J. Gait, in ' HIV. A Practical Approach', Ed. J. Karn, Oxford University Press, Oxford, 1995, p. 147.

[29] M. J. Gait, D. J. Earnshaw, M. A. Farrow, J. H. Fogg, R. L. Grenfell, N. A. Naryshkin, T. V. Smith, in 'RNA-Protein Interactions: A Practical Approach', Ed. C. Smith, OUP, Oxford, UK, 1998, p. 1.

[30] a) F. Godde, S. Moreau, J.-J. Toulmé, *Antisense and Nucleic Acid Drug Develop.* **1998**, *8*, 1; b) F. Godde, J. J. Toulmé, S. Moreau, submitted to *Nucleic Acids Res.*

[31] F. Godde, J.-J. Toulmé, S. Moreau, *Biochemistry* **1998**, *37*, 13765.

[32] G. Slim, C. E. Pritchard, E. Biala, U. Asseline, M. J. Gait, *Nucleic Acids Res. Symp. Ser.* **1991**, *24*, 55.

[33] J. L. Battiste, R. Tan, A. D. Frankel, J. R. Williamson, *Biochemistry* **1994**, *33*, 2742.

[34] C. T. Rigl, D. H. Lloyd, D. S. Tsou, S. M. Gryaznov, W. D. Wilson, *Biochemistry* **1997**, *36*, 650.

[35] T. Johnson, M. Quibell, D. Owen, R. C. Sheppard, *J. Chem. Soc., Chem. Commun.* **1993**, 369.

[36] M. Quibell, W. G. Turnell, T. Johnson, *J. Org. Chem.* **1994**, *59*, 1745.

[37] M. Quibell, L. C. Packman, T. Johnson, *J. Chem. Soc., Perkin Trans. 1* **1996**, 1227.

[38] B. S. Ross, R. H. Springer, Z. Tortorici, S. Dimock, *Nucleosides Nucleotides* **1997**, *16*, 1641.

[39] J. Smith, M. Singh, *Biotechniques* **1996**, *20*, 1082.

[40] J. Cho, K. Hamasaki, R. R. Rando, *Biochemistry* **1998**, *37*, 4985.

[41] S.-Y. Linn, A. D. Riggs, *J. Mol. Biol.* **1972**, *72*, 671.

Solution Structure of a RNA Decamer Duplex, Containing 9-[2-*O*-(*β*-D-ribofuranosyl)-*β*-D-ribofuranosyl]adenine, a Special Residue in Lower Eukaryotic Initiator tRNAs

by **Ingrid Luyten**[a]), **Robert M. Esnouf**[a]), **Sergey N. Mikhailov**[b]), **Ekaterina V. Efimtseva**[b]), **Paul Michiels**[c]), **Hans A. Heus**[c]), **Cees W. Hilbers**[c]), and **Piet Herdewijn**[a])*

[a]) Rega Institute for Medical Research, K. U. Leuven, Minderbroedersstraat, 10, B-3000 Leuven

[b]) Engelhardt Institute of Molecular Biology, Russian Academy of Sciences, Vavilov Str. 32, Moscow, 117984, Russia

[c]) Nijmegen SON Research Center for Molecular Structure, Design and Synthesis, Laboratory of Biophysical Chemistry, Universiteit Nijmegen, Toernooiveld, NL-6525 ED Nijmegen

The solution structure of the self-complementary deca-ribonucleotide 5'-r(GCGA*AUUCGC)-3' containing 9-[2-*O*-(*β*-D-ribofuranosyl)-*β*-D-ribofuranosyl]adenine (A*), a modified nucleotide that occurs in lower eukaryotic methionine initiator tRNAs (tRNAs$_i^{Met}$), was determined by NMR spectroscopy. Unexpectedly, the modification has no effect on the thermal stability of the duplex. However, the extra ribose moiety is in the C(3')-*endo* conformation and takes up a well-defined position in the minor groove, which is in agreement with its position in tRNAs$_i^{Met}$ as determined by X-ray crystallography. Molecular-dynamics simulations on the RNA duplex in H$_2$O show that the position of the extra ribofuranose moiety seems to be stabilized by bridged H-bonds (mediated by two H$_2$O molecules) to the backbone of the complementary chain.

Introduction. – Oligonucleotides that hybridize sequence-specifically with mRNA can be used to control gene expression, providing that they are stable against enzymatic degradation and that they bind with high affinity to their RNA targets. Such molecules are termed antisense oligonucleotides. They can function by a variety of mechanisms, including *i*) translation arrest by blocking the progression of the ribosome and *ii*) inactivation of the mRNA by RNase H cleavage. The stability of the duplex formed between the antisense oligonucleotide and its RNA target can be influenced by introducing chemical modifications into the antisense strand. One of the simplest modifications is the introduction of an alkoxy group at the 2'-position in the deoxyribose moiety of oligodeoxyribonucleotides [1]. It has been demonstrated that 2'-*O*-alkyl-RNA · RNA hybrids where the alkyl group has the general structure ROCH(R')CH$_2$ show increased thermal stabilities when compared with duplexes containing simple alkyl groups such as propyl or allyl [1]. An alkyl group of the general structure ROCH$_2$ has no beneficial effect on the stability of the duplex when compared with the equivalent dsRNA sequence. DNA with the analogous 2'-(2-methoxyethoxy) substituent (MeOCH$_2$CH$_2$O) is currently undergoing extensive biological evaluation as an antisense construct. *Freier* and *Altmann* [2] have hypothesized that, due to the *gauche* effect between the O-atoms of the methoxyethoxy substituent, the conformation of the side chain is restricted and consistent with *A*-form duplex formation. Generally speaking, a XCH(R)CH$_2$O substituent at C(2') may stabilize the duplex when X is an electronegative group and R is any group from the series H, Me, HOCH$_2$, or MeOCH$_2$. Recently, it has been found by X-ray diffraction methods [3] that, in some structures of DNA duplexes with incorporated 2'-O-modified RNA analogues, a H₂O molecule can coordinate between the phosphate backbone and the O-atoms of the 2'-O-(methoxyethyl) substituent.

To avoid an entropic penalty during duplex formation, it could be reasoned that the 2'-O-alkoxyethyl substituent should be replaced by a conformationally more rigid substituent. A good candidate for this might be a sugar substituent such as a ribosyl unit. Furthermore, the free OH groups of the 2'-O-ribosyl substituent might influence hydration of the duplex in a similar manner as described above. On the other hand, a 2'-O-ribosyl substituent is much larger than a 2'-O-(methoxyethyl) substituent, and interaction of the ribosyl substituent with minor-groove functionalities might disrupt base pairing for steric reasons. Interestingly, this kind of modification is also present in tRNA [4]. The disaccharide nucleosides 9-[2-O-(β-D-ribofuranosyl)-β-D-ribofuranosyl]-adenine (A*) (*Fig. 1*) and 9-[2-O-(β-D-ribofuranosyl)-β-D-ribofuranosyl]guanine (G*) can bear additional phosphate esters at their 2'-O-ribosyl moieties. These modifications occur in lower eukaryotic methionine initiator tRNAs (tRNAs$_i$^Met [4]), where they are present at position 64, which is located near the junction of the T-stem and the aminoacyl stem in the tRNA tertiary structure. It was suggested that this modification might act as a discriminator for the elongation-initiator process [5] in preventing the tRNAs$_i$^Met from participating in the elongation cycle. A 3-Å crystal structure [6] of a tRNA$_i$^Met showed that this modification lies in the minor groove with the 5'*-phosphate group interacting with NH₂−C(2) of the neighboring G63. However, the exact position of the individual atoms of this 2'-O-ribosyl moiety could not be specified because of insufficient resolution. Therefore, the study of a simpler RNA structure with this disaccharide-containing nucleotide by NMR spectroscopy and restrained molecular dynamics may provide more insight into the structural features of these tRNAs$_i$^Met. We have chosen as a model a modification of the already described [7] self-complementary oligoribonucleotide sequence 5'-r(GCGAAUUCGC)-3', in which A4 is replaced by A*, i.e., 5'-r(GCGA*AUUGC)-3'. This model preserves a neighboring G on the 5' side of A*, as in the tRNA$_i$^Met structure.

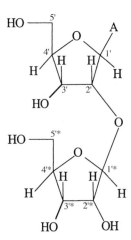

Fig. 1. *The disaccharide nucleoside 9-[2-O-(β-D-ribofuranosyl)-β-D-ribofura-nosyl]adenine (A*). To avoid confusion between the two sugar rings in A*, the extra ribose moiety has starred locants.*

Experimental. – *Sample Preparation.* The phosphoramidite of the modified nucleoside (A*) was used for solid-phase RNA synthesis, as described before [8]. The self-complementary oligoribonucleotide 5′-r(GCGA*AUUCGC)-3′ containing 9-[2-*O*-(*β*-ᴅ-ribofuranosyl)-*β*-ᴅ-ribofuranosyl]adenine (A*) was purchased from *Eurogentec*. The total yield was *ca.* 9 mg of RNA.

Melting Temperatures. Oligomers were dissolved in a buffer containing 0.1ᴍ NaCl, 0.02ᴍ potassium phosphate (pH 7.5), and 0.1 mᴍ EDTA. Concentrations were determined to be *ca.* 4 µᴍ by measuring the absorbance at 260 nm at 80°. The following extinction coefficients (ε) were used: A and A*, 15000; U, 10000; G, 12500; C, 7500. Melting curves were determined with a *Uvikon-940* spectrophotometer. Cuvette temp. was controlled by water circulation through the cuvette holder. The temp. of the soln. was measured with a thermistor directly immersed in the cuvette. Temp. control and data acquisitions were carried out automatically with an IBM-compatible computer. The samples were heated and cooled at a rate of 0.5° min⁻¹. Melting temp. were derived from the first derivative of the absorbance *vs.* temperature curves.

NMR Spectroscopy. NMR Samples were prepared by dissolving the purified RNA (*ca.* 9 mg) in D₂O and adjusting the pD to 6.8 with DCl. The sample was divided in two parts and lyophilized. One part was dissolved in D₂O (0.750 ml), and the other part was dissolved in D₂O (0.075 ml) and H₂O (0.675 ml). The samples were annealed by heating to 80° followed by slow cooling to obtain a 1.8 mᴍ concentration of the duplex.

NMR Spectra: *Varian-500 Unity* spectrometer; at 499.505 MHz unless stated otherwise; δ in ppm, *J* in Hz. Quadrature detection was achieved by the *States-Haberkorn* hypercomplex mode [9]. Spectra were processed with the programs NMRPipe [10] and XEASY [11] running on a *Silicon-Graphics-O2* workstation (IRIX version 6.3).

The 1D- and 2D-NOESY experiments from the sample dissolved in H₂O/D₂O 9:1 were recorded at 5° (*Varian Unity*(+)-*500* spectrometer, at 499.930 MHz), using a jump-return sequence as the observation pulse [12]. Sweep widths of 10000 Hz in both dimensions were used with 64 scans, 2048 data points in t_2 and 512 FIDs in t_1. The data were apodized with a shifted sine-bell square function in both dimensions and processed to a $4K \times 2K$ matrix. The 2D DQF-COSY [13], TOCSY [14], and NOESY [15] spectra from the sample in D₂O were recorded with sweep widths of 5000 Hz in both dimensions. The residual HDO peak was suppressed by presaturation. The ³¹P-decoupled DQF-COSY spectrum consisted of 4096 datapoints in t_2 and 400 increments in t_1. The data were apodized with a shifted sine-bell square function in both dimensions and processed to a $4K \times 2K$ matrix. For the TOCSY experiment, a Clean MLEV17 [16] version was used, with a low-power 90° pulse of 26.6 µs and the delay set to 69.2 µs. The total TOCSY mixing time was set to 65 ms. The spectrum was acquired with 32 scans, 4096 data points in t_2 and 256 FIDs in t_1. The data were apodized with a shifted sine-bell square function in both dimensions and processed to a $4K \times 1K$ matrix. The NOESY experiments were acquired with mixing times of 50, 100, 150, 250, and 300 ms, 64 scans, and 2048 datapoints in t_2 and 512 increments in t_1.

A ¹H,³¹P-HETCOR [17] was acquired (*Varian Unity*(+)-*500* spectrometer, at 499.930 MHz) with 32 scans, 4096 data points in the ¹H dimension, t_2, and 400 real data points in the ³¹P dimension, t_1, over sweep widths of 5000 and 2000 Hz, resp.

Restraint Generation and Refinement Procedures. Interproton-distance restraints were derived from cross-peak volumes in the 50, 100, and 150 ms NOESY spectra and were given ±20% error bounds. Cross-peaks that were only observable with longer mixing times were corrected for spin diffusion by using the H–C(5)/H–C(6) NOE (2.45 Å) as a reference for shorter distances and helical intraresidue H–C(1′)/H–C(6) or H–C(8) NOEs (3.65 Å) for longer distances [18]. All NOEs that could not be properly integrated because of overlap were assigned bounds of 1.8 to 7.0 Å. This resulted in 55 inter- and 110 intra-residue distance restraints. Conservative imino/imino (4.0 Å), imino H–C(1′) (5.0 Å) and imino H–C(2) (2.9 Å) distance restraints were obtained from the spectrum in H₂O and included in the structure calculations. H-Bond restraints were used and treated as NOE restraints to define the *Watson-Crick* base pairing. All residues (except C10) in the RNA strand gave no observable H–C(1′)/H–C(2′) cross-peaks in the DQF-COSY experiment. Therefore, dihedral restraints on H–C(1′)–C(2′)–H (99±20°) and H–C(2′)–C(3′)–H (38±20°) were used to constrain the *N*-type ribose conformation [19]. All structure calculations were performed with X-PLOR V3.851 [20]. The standard topology and parameter files were adapted for the extra 2′-*O*-ribose moiety. The torsion-angle dynamics protocol used was largely identical to that implemented for a DNA duplex [21], starting from an extended strand conformation and proceeding in four stages. *i*) A 60 ps (4000 steps of 15 fs) high-temperature torsion-angle molecular dynamics (TAMD) (simulation temp. of 20000 K) with a decreased weight on the repulsive nonbonded energy term ($\omega_{vdw} = 0.1$) to facilitate crossing rotational-energy barriers, and coefficients for the dihedral and the NOE energy terms of 5 ($\omega_{dihedral}$) and 150 (ω_{NOE}), respectively. *ii*) A 90 ps (6000 steps of 15 fs) slow cooling TAMD (from 20000 to 1000 K) with the ω_{vdw} gradually increasing to 1.0. *iii*) A 9 ps (3000 steps of

3 fs) cartesian molecular dynamics (cooling from 1000 to 300 K). *iv*) A 1000-step conjugate-gradient-energy minimization with $\omega_{dihedral} = 200$ and $\omega_{NOE} = 50$. Out of 100 calculated structures, 24 had no NOE-distance violations (>0.5 Å) and no dihedral-angle violations ($>5°$) (*Table 1*). The residual average violations were 0.091 Å and 0.7°, respectively. From the 24 structures, the 10 structures closest to the average of all 24 were used for further analysis with the program Curves 5.1. [22].

Table 1. *Structural Statistics of the Final Set of 24 Structures of the [5'-r(GCG A*AUUCGC)]₂ Duplex*

	Residues 3–5/16–18	All
Restraint Violations: NOE violations >0.5 Å	0	0
dihedral violations $>5°$	0	0
Rms Deviations: from distance restraints [Å]	0.124 ± 0.021	0.091 ± 0.012
from dihedral restraints [°]	0.7 ± 0.4	0.7 ± 0.1
from idealized geometry		
Bonds [Å]	0.0107 ± 0.0001	0.0107 ± 0.0001
Angles [°]	1.247 ± 0.022	1.169 ± 0.016
Impropers [°]	0.693 ± 0.035	0.877 ± 0.016
Pairwise rmsd for all heavy atoms [Å]	0.08	0.59

Hydration Analysis. Water molecules were included using the TIP3P force field, and molecular-dynamics simulations and energy minimizations were performed using X-PLOR V3.851 [20]. A well-equilibrated cube of H_2O molecules (125 molecules in a cube of 15.55 Å) was copied and translated many times in all three dimensions to create a cube sufficiently large to embed the RNA duplex. Thus, application of periodic boundary conditions will not give rise to spurious interactions between duplexes. The RNA duplex was placed in the center of the box, and all H_2O molecules overlapping with the RNA duplex (within 2.6 Å) were removed, which led to a total of 9000 H_2O molecules. Simulations were performed with a time step of 1 fs, and the limits for nonbonded *van der Waals* and electrostatic interactions were truncated at 12 Å. The same TIP3P force field was used for the RNA duplex supplemented with the restraints derived from the NMR data. An initial equilibration simulation of 10 ps at 300 K was performed with the RNA-duplex conformation rigidly constrained, both to lose the periodicity in the H_2O structure and to allow a reasonable H_2O structure to form around the duplex. Further simulations were performed with the RNA unconstrained (a total of 20 ps starting with a simulation temp. of 500 K and reducing it to 300 K). Individual snapshots taken during this simulation were subjected to a short energy minimization and analyzed with MidasPlus [23].

Molecular Graphics. Molecular graphics images were produced with MidasPlus [23] and Rasmol [24]. *Figs. 3* and *5* were drawn with Bobscript [25], and *Fig. 5* was produced with Raster3D [25b].

Results. – *Thermal Stability of the RNA Duplex.* Melting-temperature determinations demonstrated that the thermal stabilities of oligo $r(A)_{13} \cdot r(U)_{13}$ (T_m 15.2°) and the fully modified oligo $r(A^*)_{13} \cdot r(U)_{13}$ (T_m 15.2°) are equal within experimental error, implying that the extra sugar moieties have no effect on the stability of the RNA duplex. This unexpected result initiated the structural studies aimed at gaining insight into the position of the 2'-*O*-ribose moieties in the RNA duplex to understand why the modified nucleotide A* has no influence on the stability of a RNA duplex. Since the determination of the solution structure of a complex between oligo-U and fully modified oligo-A* (where A* represents the disaccharide nucleotide) is very complex, we have incorporated a single A* ($= 9$-[2-*O*-(β-D-ribofuranosyl)-β-D-ribofuranosyl]-adenine) unit into the well-described self-complementary oligoribonucleotide sequence [5'-r(GCGAAUUCGC)-3']₂ [7] (T_m 53°). In agreement with our previous observation, the melting point of the modified double-stranded RNA [5'-r(GCGA*AUUCGC)-3']₂ in 0.1M NaCl is identical to that found for the unmodified dsRNA duplex (T_m 53°). Determination of the solution structure of this modified duplex was done by a combination of high-resolution NMR spectroscopy and restrained molecular dynamics.

^{1}H and ^{31}P Assignments. Resonance assignment followed standard NMR methods [19]. Sequential connectivities were established following an anomeric to aromatic proton walk (*Fig. 2*). The H−C(2′) and H−C(3′) protons were assigned through H−C(2′)/H−C(3′)/aromatic H NOE connectivities. NOESY Experiments with short mixing times (50 ms) were used for H−C(2′) assignments since the internucleotide H−C(2′)/aromatic H NOEs and the intranucleotide H−C(1′)/H−C(2′) NOEs are strong. The ^{1}H,^{31}P HETCOR [17] was acquired to confirm assignment of the H−C(3′) resonances and, when possible, the H−C(4′) and H−C(5′) resonances. The rest of the sugar spin systems could be determined by a combination of DQF-COSY, TOCSY, and NOESY experiments (*Table 2*), suggesting an *A*-form helix and stacking of the A*4 and A5 bases. The COSY cross-peak pattern clearly shows that the extra ribose moiety of A* adopts a C(3′)-*endo* conformation.

The H−C(2)(A*4) and H−C(2)(A5) protons were distinguished from other base protons by their NOEs to H−C(1′) protons. The H−C(2)(A*4) exhibits NOEs to H−C(1′)(A*), H−C(1′)(A5), and H−C(1′)(C8),

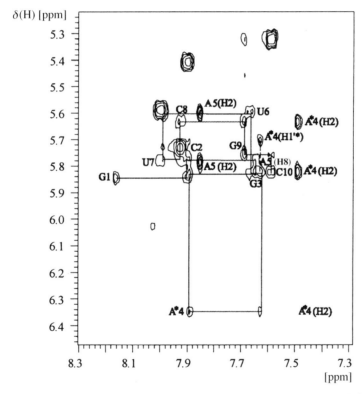

Fig. 2. *Portion of the 2D NOESY* (D$_2$O, 250 ms mixing time, 20°) *of [5′-r(GCG*AUUCGC)-3′]$_2$* (see *Fig. 1* for A*) *showing the aromatic to H−C(1′) region*. The sequential walk along the H−C(1′) and aromatic H−C(6)/ H−C(8) protons in the RNA strand is traced out by the continuous line. Assignment of the cross-peaks: H−C(8)(A5)/H−C(1*)(A*4) (=A5(H8)/A*4(H′*)) and H−C(1′)(A5)/H−C(8)(A5) (=A5(H1′)/A5(H8)) are indicated by the broken line.

Table 2. *Chemical Shift [ppm] Assignments for the Nonexchangeable ^{1}H-NMR Resonances in the [5'-r(GCGA*AUUCGC)-3']$_2$-Duplex at 20°*. Referenced to the H_2O resonance at 4.86 ppm.

	H–C(1')(= H1')	H–C(2')(= H2')	H–C(3')(= H3')	H–C(4')(= H4')	H–C(5')(= H5')	H'–C(5)(= H5'')	H–C(8)/H–C(6)(= H8/H6)	H–C(2)/H–C(5)(= H2/H5)	P^a)
G1	5.84	4.90	4.66	4.48	4.13	4.01	8.16		–0.45
C2	5.77	4.78	4.67	4.58	4.63	4.32	7.90	5.41	0.07
G3	5.82	4.70	4.70	–	4.57	4.26	7.64		–0.43
A*4	6.35	4.95	4.84	4.56	4.42	4.66	7.90	7.47	–0.60
	(5.70)^b)	(4.39)^b)	(4.55)^b)	(4.20)^b)	(4.16)^b)	(3.89)^b)			
A5	5.83	4.58	4.43	4.44^c)	4.43^c)	4.22	7.64	7.84	–0.59
U6	5.60	4.57	4.53	4.58	4.64	4.14	7.67	5.03	–0.93
U7	5.78	4.59	4.60	4.53	4.63	4.19	7.99	5.59	–0.61
C8	5.63	4.54	4.58	4.49	4.62	4.21	7.92	5.73	–0.34
G9	5.75	4.46	4.66	4.51	4.64	4.17	7.69		–0.56
C10	5.83	4.06	4.25	4.25	4.56	4.11	7.59	5.32	

^{a}) P-Assignments are to the 5'-nucleoside. ^{b}) Assignments of the equivalent sugar protons of the extra ribose moiety. ^{c}) Tentative assignments.

whereas H–C(2)(A5) exhibits NOEs to H–C(1')(A5), H–C(1')(U6), and H–C(1')(U7), as expected for an *A*-form helix. Both H–C(2)(A*4) and H–C(2)(A5) exhibit NOEs to H–C(2)(A5) suggesting that the A*4 and A5 bases are stacked. The resonance assignment of the modified nucleotide A* was based on the resonance assignment of the monomer [8]. The H–C(1')(A*4) proton resonates most downfield (6.35 ppm). The resonance of H–C(1'*) from the extra ribose moiety was assigned from the NOEs to H–C(1')(A*4) (*Table 3*, 3.5 Å) and H–C(8)(A5) (see *Fig. 2*, dashed line). The rest of the sugar-proton resonances of this extra ribose moiety could then easily be assigned on the basis the NOEs to this H–C(1'*). Furthermore, in the COSY experiment, strong cross-peaks are observed for H–C(2'*)/H–C(3'*) ($^3J(2',3') \approx 5$ Hz), H–C(3'*)/H–C(4'*) ($^3J(3'*,4'*) \approx 10$ Hz), and H–C(4'*)/H–C(5'*) and H'–C(5'*) ($^3J(4'*,5'*) \approx 8$ Hz).

Table 3. *Key Distance Restraints [Å] for Residue A* in the [5'-r(GCG A*AUUCGC)-3']$_2$ Duplex*

Proton 1	Proton 2	Distance		
		Lower	Upper	Distance^a)
H–C(1')(A*4)	H–C(1'*)(A*)	2.6	3.8	3.5
H–C(1')(A*4)	H–C(2'*)(A*4)	3.8	5.6	4.4
H–C(1')(A*4)	H–C(3'*)(A*4)	2.6	4.0	4.0
H–C(1'*)(A*4)	H–C(2'*)(A*4)	2.2	3.3	2.8
H–C(1'*)(A*4)	H–C(2'*)(A*4)	1.9	2.9	2.5
H–C(1'*)(A*4)	H–C(4'*)(A*4)	2.7	3.9	3.3
H–C(1')(A*4)	H–C(5'*)(A*4)	2.4	3.6	2.6
H–C(8)(A5)	H–C(1'*)(A*4)	3.5	5.3	4.1

^{a}) Measured distance in the RNA duplex model, which is one of the 24 selected structures as depicted in *Fig. 5,a.*

Imino-proton assignments were based on a NOESY spectrum measured at 5° in H_2O/D_2O 9:1. The H–N(3)(U6) and H–N(3)(U7) have strong NOEs with each other and with H–C(2)(A5) and H–C(2)(A*4), respectively, as expected for an *A*-form helix. These connectivities show standard *Watson-Crick* base-pairing for A*4 and A5 and normal base stacking for U6 and U7. The H–N(1)(G3) and H–N(1)(G9) protons show the normal cross-strand connectivities with the amino protons and H−C(5) of C8 and C2, respectively.

All assignments of the ^{31}P-resonances (*Table 1*) were derived from the 2D ^{1}H-detected ^{1}H,^{31}P-correlation spectrum (HETCOR).

NOE Distances, Sugar Puckers, and Backbone Torsions. With nearly complete ^{1}H- and ^{31}P-assignments, sufficient distances and torsion angles were determined to calculate a structure for the modified RNA duplex [5′-r(GCGA*AUUCGC)-3′]$_2$. The details of the refinement protocol are given in the *Exper. Part.*

Distance Restraints. Interproton distance restraints were derived from cross-peak volumes in 50, 100, and 150-ms NOESY spectra and were given ±20% error bounds. This resulted in 55 inter- and 110 intra-residue distance restraints. From the spectrum in H$_2$O, imino/imino, imino/H−C(1′) and imino/H−C(2) distance restraints were included in the structure calculations. H-Bond restraints were used and treated as NOE restraints.

Conformation of Ribose Rings in RNA. The conformation of the furanose ring can be described by the pseudorotation angle (*P*) and the puckering amplitude (Ψ_m). These are related to the endocyclic torsion angles that, in turn, are correlated to the vicinal ^{1}H,^{1}H-coupling constants *via* the *Karplus* equation. The combination of the small coupling constants $J(1′,2′)$ (extremely sharp anomeric signals and no visible cross-peaks in *J*-correlated spectra) and the large $J(3′,4′)$ (*ca.* 10 Hz) in the ribose moieties, is very characteristic of *N*-type sugars [19].

Backbone Dihedral-Angle Restraints. In an RNA strand, the backbone dihedral angle δ is correlated with the sugar puckering. In the *N*-type sugars, this dihedral angle is restrained to 80 ± 20°. The small passive couplings in the H−C(5′)/H′−C(5′) cross-peaks of the DQF-COSY experiment allowed the backbone dihedral angles γ to be restrained in the RNA strand to 60 ± 35°. The ε backbone dihedral angles were restrained to − 130 ± 40°, based on the large $J(P,3′)$ (*ca.* 11 Hz) measured in the ^{1}H,^{31}P-HETCOR spectrum. The lack of P/H−C(5′) and P/H′−C(5′) cross-peaks in this spectrum is an indication of small $J(P,5′)$ and $J(P,5″)$ coupling constants which correspond to the β backbone dihedral angles having a *trans* conformation (180 ± 30°). The ^{31}P-chemical shift was used to constrain the α and ζ backbone dihedral angles within the region 0 ± 120°.

Structure Determination and Analysis. Starting from two extended strands, a set of 100 structures was generated by torsion-angle molecular dynamics (*cf. Exper. Part*). Out of 100 calculated structures, 24 were selected for analysis that had no NOE-distance violations >0.5 Å and no dihedral-angle violations >5°. The residual average violations were 0.091 Å and 0.7°, respectively. Superimpositions were made for all heavy atoms of base pairs 3−5/16−18 on one side of the duplex for the 24 structures. *Fig. 3* shows the models after superimposition of the ten selected structures closest to the average. The part of the duplex around the modified adenosine (*Fig. 3,c*) is well-defined because of the high number of restraints per residue as well as for the extra ribose ring, but the precision is lost progressively going away from this part of the molecule as a result of the lack of long-range distance restraints in the structure determination of RNA helices by NMR [26]. The local superimposition of the domain 3−5/16−18 containing the modified adenosine yields a root mean square deviation (rmsd) of 0.08 Å, whereas the global rmsd gives a value of 0.59 Å (*Table 1*).

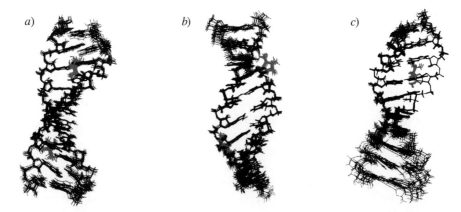

Fig. 3. *Views of the three-dimensional structure of the RNA structure modified with 9-[2-O-(β-D-ribofuranosyl)-β-D-ribofuranosyl]adenine* (A*). a) *All heavy atom superimposition of the ten structures closest to the average of the 24 final structures.* b) *The same superimposition rotated by 90 degrees.* c) *Overlays of the same ten structures, but the superimposition is limited to the heavy atoms in the residues 3–5/16–18.*

The precision in the definition of these structures is further evidenced by the small variation in the torsion angles (*Fig. 4*), shown for the 10 selected structures closest to the average. The values also correspond to a helix with an *A*-type conformation.

Orientation and Conformation of the Extra Ribose Ring. The extra ribose ring has a fixed orientation in the minor groove (*Fig. 5* shows a part of the structure around the modification of one of the 24 selected structures) and adopts a C(3')-*endo* conformation. The latter can be deduced from the lack of an observable H−C(1'*)/H−C(2'*) cross-peak in the DQF-COSY and because $J(3'*,4'*)$ is *ca.* 10 Hz. We have used the three-bond ^{1}H,^{1}H coupling constants, *i.e.*, $^3J(2'*,3'*) \approx 5$ Hz and $^3J(3'*,4'*) \approx$ 10 Hz to restrain the δ dihedral angle to $80 \pm 20°$, as found in a C(3')-*endo* conformation. In the free disaccharide nucleoside (*i.e.*, monomeric A*) [8], the extra ribose moiety also adopts a C(3')-*endo* conformation. The observed $J(4'*, 5'*)$ of *ca.* 8 Hz corresponds to mixed populations of *gauche* and *trans* rotamers of the C(4'*)−C(5'*) bond. Therefore, we have not restrained this γ dihedral angle.

The key NOEs, shown in *Table 3*, lead to an orientation of the 2'-O-ribose moiety as depicted in *Fig. 5*. The O(2')−C(1'*)−C(2'*)−O(2') torsion angle is roughly −134°, and the extra ribose moiety is oriented more or less perpendicular to the plane of the A* · U6 base pair and more or less perpendicular to the A* ribose such that H−C(5'*) and H−C(1'*) are only 2.6 Å apart, whereas the H−C(3'*) and the H−C(1*) are separated by roughly 4 Å. H−C(1'*) points in the direction of A5 and is only 4.1 Å away from H−C(8)(A5).

Discussion. – This NMR study reveals that the duplex RNA substituted with one 9-[2-O-(β-D-ribofuranosyl)-β-D-ribofuranosyl]adenine (A*) [8] maintains an *A*-type helical geometry and that the modified adenosine has no profound effect on the RNA structure when paired opposite to a uridine residue and stacked between two other purine nucleotides. Nevertheless, the extra sugar ring does take

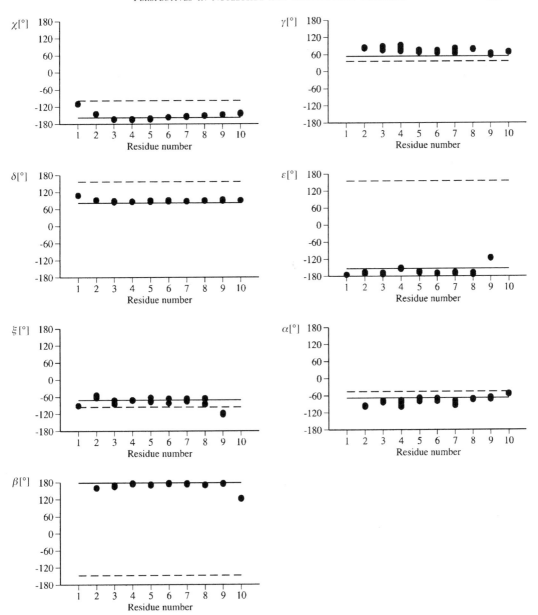

Fig. 4. *Variation of the torsion angles for the individual residues of the 10 structures. A-* and *B*-form values are given by dashed and dotted lines, respectively.

up a well-defined position in the minor groove, as illustrated in *Fig. 5*, which is in agreement with the X-ray data [6] of tRNAs$_i^{Met}$ where the extra sugar moiety is placed in the minor groove with its 5'-phosphate group pointing to $NH_2-C(2)$ of the neighboring 5'-G residue. *Fig. 5,b* shows that the distances between $NH_2-C(2)$ of G3 and $O(5'*)$ of A*4 and between $O(2'*)$ of A*4 and $O(4')$ of A5 are within H-bond

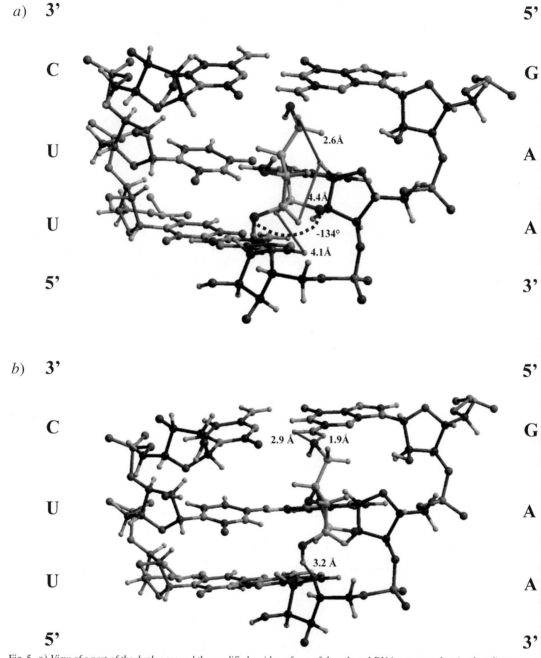

Fig. 5. a) *View of a part of the duplex around the modified residue of one of the selected RNA structure showing key distances as green lines.* b) *View a) slightly rotated to point out the possible H-bonds* ((A*4)O(5′*) ··· H(21)−N(2)(G3) or (A*4)(O5′*) ··· H(22)−N(2)(G3) and/or (A*4)O(2′*)−H(2′*) ··· O(4′)(A5)). *H−C(5′*)(A*) and H−C(1′)(A*) are only 2.6 Å apart, whereas the H−C(3′*)(A*) and H−C(1′)(A*) are separated by 4.4 Å. The H−C(1′*)(A*) points to the next A5 and is only 4.1 Å away from H−C(8)(A5). The O(2′)−C(1′*)−C(2′*)−O(2′*) torsion angle is* −134° *(dashed red arc).*

range. These H-bonds probably drive the extra sugar ring into a well-defined orientation.

As this study was being completed, X-ray crystallography studies of related modified RNA duplexes containing 2'-O-(methoxyethyl) and 2'-O-(ethoxymethyl) modifications appeared, giving rationales for the observed increase in T_m for these molecules [3]. Increase in T_m can be achieved only if the two O-atoms in the alkoxyalkyl side chain are separated by an ethylene spacer, giving rise to a *gauche* conformation between these two O-atoms. In this conformation, a new hydration site is formed whereby a H_2O molecule interacts with both O-atoms of the side chain and also with $O-C(3')$ of the backbone sugar moiety. With the ethoxymethyl group, the spacing is wrong for an additional hydration. These results are supported by recent molecular-dynamics simulations [27] which suggest that H_2O molecules may help stabilize the *gauche* conformation, thus generating an electrostatic groove in the duplex that does not disturb the H_2O shell. In this study with the ribofuranosyl substituent, the $O(2')-C(1'^*)-C(2'^*)-O(2'^*)$ torsion angle ($-134°$) does not have the *gauche* conformation (*Fig. 5,a*), responsible for the enhanced stability of the 2'-O-(methoxyethyl) derivatives. This result may partially explain the lack of an effect on stability due to the extra ribofuranosyl group, and hence the lack of a change in the observed melting temperature.

Besides the possible H-bonding as shown in *Fig. 5,b*, the H_2O structure may also play a crucial role in positioning the extra ribofuranosyl group and, therefore, we have performed molecular-dynamics simulations on the RNA duplex in H_2O. Although limited in their scope, these simulations showed that it was not possible to create a H_2O-mediated interaction between $O(2')$ and $O(2'^*)$ of the extra ribofuranose moiety and $O-C(3')$ of the backbone ribose moiety. The position of the extra ribofuranosyl substituent appeared to be stabilized by bridged H-bonds (mediated by two H_2O molecules) to the backbone of the complementary chain. These interactions involved one or both of the atoms $O(3'^*)$ and $O(5'^*)$ of the extra ribofuranosyl substituent, but the exact geometry of the interactions and the particular atoms involved in the complementary chain varied during the course of the simulations.

The extra ribose moiety disrupts the H-bonding between $O(2')$ of that residue and $O(4')$ of the next residue. However, our H_2O simulations also showed that it was possible to compensate for the loss of this H-bond by formation of a H-bond between $O(2'^*)$ and a H_2O molecule.

Conclusions. – The NMR data, represented by the measured NOEs and coupling constants, are consistent with the values found for an *A*-form double helix with all ribose residues (including the ribosyl substituent of the disaccharide nucleotide) adopting a C(3')-*endo* (*N*-type) conformation. The extra ribofuranosyl unit of the disaccharide nucleotide occupies a well-defined position in the minor groove (*Fig. 5*), probably driven by H-bonding as shown in *Fig. 5,b*. Simulations incorporating H_2O suggested that bridged H-bond interactions with the complementary strand may be additionally responsible for defining the orientation of the extra ribofuranosyl unit. The conformation of the extra ribose unit is characterized by an $O(2')-C(1'^*)-C(2'^*)-O(2'^*)$ torsion angle of $-134°$ (*Fig. 5,a*). This means that the atoms $O(2')$ and $O(2'^*)$ are not positioned in the *gauche* relationship that was necessary for the enhanced stability of RNA duplexes containing 2'-O-(methoxyethyl) modifications. Consequently, the typical hydration site found in the 2'-O-(methoxyethyl)-RNA is not present in the 2'-O-ribosyl congener. The fixed orientation of the extra ribose unit is such that it does not influence duplex stability in either a positive or negative way. Our NMR data are in agreement with the X-ray data [6], but we could also define the exact position of the individual atoms of this 2'-O-ribose moiety. This study will be followed by an analysis of the T-stem of a tRNA$_i^{Met}$ containing the 5'*-O-phosphorylated analogue of A*.

This work was supported by a grant from the *Onderzoeksfonds K. U. Leuven* (GOA 97/11). *Ingrid Luyten* thanks the *F.W.O.* for a fellowship. We thank *R. Busson* and *E. Lescrinier* for the NMR technical support and *Guy Schepers* for determining the melting temperatures.

REFERENCES

[1] P. Martin, *Helv. Chim. Acta* **1995**, *78*, 486.

[2] S. M. Freier, K. H. Altmann, *Nucl. Acids Res.* **1997**, *25*, 4429.

[3] V. Tereshko, S. Portmann, E. C. Tay, P. Martin, F. Natt, K. H. Altmann, M. Egli, *Biochemistry* **1998**, *37*, 10626.

[4] J. Desgrés, G. Keith, K. C. Kuo, C. W. Gehrke, *Nucl. Acids Res.* **1989**, *17*, 865.

[5] S. Kiesewetter, G. Ott, M. Sprinzl, *Nucl. Acids Res.* **1990**, *18*, 4677.

[6] R. Basavappa, P. B. Sigler, *EMBO J.* **1991**, *10*, 3105.

[7] P. V. Sahasrabudhe, R. T. Pon, W. H. Gmeiner, *Biochemistry* **1996**, *35*, 13597.

[8] E. V. Efimtseva, B. S. Ermolinsky, M. V. Fomitcheva, S. V. Meshkov, N. S. Padyukova, S. N. Mikhailov, J. Rozenski, A. Van Aerschot, P. Herdewijn, *Coll. Czech. Chem. Commun.* **1996**, *61*, S206.

[9] D. J. States, R. A. Haberkorn, D. J. Ruben, *J. Magn. Res.* **1982**, *48*, 286.

[10] F. Delaglio, S. Grzesiek, G. Vuister, G. Zhu, J. Pfeifer, A. Bax, *J. Biomol. NMR* **1995**, *6*, 277.

[11] T. Xia, C. Bartels, 'XEASY 1994', Institute of Molecular Biology and Biophysics, Zürich, Switzerland.

[12] P. Plateau, M. Guéron, *J. Am. Chem. Soc.* **1982**, *104*, 7310.

[13] M. Rance, O. W. Sørensen, G. Bodenhausen, G. Wagner, R. R. Ernst, K. Wüthrich, *Biochem. Biophys. Res. Commun.* **1983**, *117*, 479.

[14] A. Bax, D. G. Davies, *J. Magn. Res.* **1985**, *65*, 355.

[15] J. Jeener, B. H. Meier, P. Bachman, R. R. Ernst, *J. Chem. Phys.* **1979**, *71*, 4546.

[16] C. Griesinger, G. Otting, K. Wüthrich, *J. Am. Chem. Soc.* **1988**, *110*, 7870.

[17] V. Sklenar, H. Miyashiro, G. Zon, H. T. Miles, A. Bax, *FEBS Lett.* **1986**, *208*, 94.

[18] I. L. Barsukov, L.-Y. Lian, in 'NMR of Macromolecules. A Practical Approach', Ed. G. Roberts, Oxford University Press, New York, 1993, p. 315.

[19] S. S. Wijmenga, M. W. Mooren, C. W. Hilbers, in 'NMR of Macromolecules. A Practical Approach', Ed. G. Roberts, Oxford University Press, New York, 1993, p. 217.

[20] A. T. Brügner, 'X-PLOR. A System for X-Ray Crystallography and NMR', Yale University Press, New Haven, CT, 1992.

[21] E. G. Stein, L. M. Rice, A. T. Brügner, *J. Magn. Reson.* **1997**, *14*, 51.

[22] a) R. Lavery, H. Sklenar, *J. Biomol. Struct. Dynam.* **1988**, *6*, 63; b) R. Lavery, H. Sklenar, *J. Biomol. Struct. Dynam.* **1989**, *6*, 655.

[23] T. E. Ferrin, C. C. Huang, L. E. Jarvis, R. Landgridge, *J. Mol. Graph.* **1988**, *6*, 13.

[24] R. Sayle, E. James Milner-White, *Trends. Biochem. Sci.* **1995**, *20*, 374.

[25] R. M. Esnouf, *J. Mol. Graph. Modell.* **1997**, *15*, 132; D. J. Bacon, W. F. Anderson, *J. Mol. Graph.* **1988**, *6*, 219; E. A. Merritt, M. E. P. Murphy, *Acta Crystallogr., Sect. D* **1994**, *50*, 869.

[26] F. H.-T. Allain, G. Varani, *J. Mol. Biol.* **1995**, *250*, 333; C. W. Hilbers, S. S. Wijmenga, H. Hoppe, H. A. Heus, in 'NMR of Biological Systems', Ed. O. Jardetzky, Wiley, New York, 1996, pp. 193–207.

[27] K. E. Lind, V. Mohan, M. Manoharan, D. M. Ferguson, *Nucl. Acids Res.* **1998**, *26*, 3694.

Solution Structure of a Hexitol Nucleic Acid Duplex with Four Consecutive T·T Base Pairs

by **Eveline Lescrinier**[a]), **Robert M. Esnouf**[b]), **Jan Schraml**[c]), **Roger Busson**[a]), and **Piet Herdewijn**[* a])[1])

[a]) Laboratory of Medicinal Chemistry, Rega Institute for Medical Research and Faculty of Pharmacy, Katholieke Universiteit Leuven, Minderbroedersstraat 10, B-3000 Leuven
[b]) Division of Structural Biology, Wellcome Trust Centre for Human Genetics, Roosevelt Drive, Headington, Oxford, OX3 7BN
[c]) Institute of Chemical Process Fundamentals, Rozvojova 135, 165 02 CZ-Prague 6

The structure of the hexitol nucleic acid (HNA) h(GCGCTTTTGCGC) was determined by NMR spectroscopy. This unnatural nucleic acid was developed as a mimic for *A*-RNA. In solution, the studied sequence is forming a symmetric double-stranded structure with four central consecutive T·T wobble pairs flanked by G·C *Watson-Crick* base pairs. The stem regions adopt an *A*-type helical structure. Discrete changes in backbone angles are altering the course of the helix axis in the internal loop region. Two H-bonds are formed in each wobble pair, and base stacking is preserved in the duplex, explaining the stability of the duplex. This structure elucidation provides information about the influence of a (T)$_4$ fragment on local helix geometries as well as on the nature of the T·T mismatch base pairing in a TTTT tract.

1. Introduction. – The sequence of RNA has an influence on its final three-dimensional structure. Due to frequently occurring base-pair mismatches, RNA is allowed to adopt secondary-structure motifs that are different from the typical *A*-form geometry of double-stranded RNA (dsRNA). Moreover, several RNA sequences have been shown to occur as equilibria between different structures. Among them, the hairpin-duplex equilibrium is most characteristic. Some of these conformations might be more biologically relevant than others, but when they represent only a minor part of the total RNA in solution, it is difficult to study their structure in detail. The fraction of noncanonical base pairs is estimated to 6% of all base pairs in ribosomal RNA, with 50% of those being G·G and 17% U·U [1]. The U·U base pair is very flexible, and it occurs frequently in RNA [2–4]. An analogous T·T mismatch is observed in DNA in the solution structure of d(m^{5}CCTCC$_2$)[5] and of other i-motifs [6] where T·T wobble pairs are formed between strands that are oriented parallel to each other. Therefore, mismatches can be considered as structural elements of nucleic acids rather than errors in sequences.

Both NMR-based solution structures and crystallographic structures of U·U-containing duplexes have been studied intensively. Self-complementary ribonucleic acids with U·U tandem mismatches often occur as duplexes. For example, the r(CGCUUGCG) sequence forms a stable duplex in 1M NaCl [7][8]. The X-ray structure of two noncanonical U·U base pairs flanked by C·G (CUUG) has been

---

[1])   Tel.: + 32-16-33 73 87, Fax: + 32-16-33 73 40, e-mail: Piet.Herdewijn@rega.kuleuven.ac.be

studied in r(GGCGCUUGCGUC). This dodecamer forms a hairpin loop at low salt concentration and a duplex structure under crystallization conditions [9]. The formation of an A-type duplex structure containing U · U wobble pairs in an internal loop causes only small changes in backbone angles and preserves base stacking in the helix. For both U · U base pairs, O(2) and N(3) of one uracil are H-bonded to N(3) and O(4) of the uracil base in the opposite strand. Interestingly, this base pairing alters the course of the helix axis by $11 - 12°$.

The structure of a double-stranded helical RNA structure with an internal loop of two U · U mismatches flanked by U · G standard wobble pairs was determined by X-ray diffraction [10]. Superimposition of the crystallographic model onto a canonical A-form RNA structure showed that the central mismatches cause no major distortion of this helix. Both U · U mismatches form wobble pairs, but one of them is so highly twisted that only one H-bond can be formed. The loss of one H-bond is compensated by a potential intraresidue H-bond and by a bound H_2O molecule that bridges the two O(4) atoms of this wobble pair with two H-bonds.

Thermodynamic studies on loop formation in RNA tetraloops demonstrated biphasic melting points for the r(GGACUUUUGUCC) sequence, indicating the presence of significant amounts of two species – most likely hairpin and duplex structures. Other reported DNA [11] and RNA [12] sequences with four central T (DNA) or U (RNA) residues only adopt hairpin structures. So far, the structure of a UUUU tract in RNA has only been determined as a hairpin [13]. The formation of this hairpin structure implies conformational changes in backbone angles in the loop region.

The behavior of RNA is similar to that described for pyranosyl-RNA (pRNA) [14], a nucleoside analogue in which the five-membered ribose is replaced by a six-membered pyranose. The pRNA sequence pr(GCGTTCGC) forms a duplex. Increasing the number of thymine nucleotides at the center of the oligomer shifts the equilibrium toward the hairpin conformation. The pyranosyl sequence pr(GCGTTTTCGC) adopts an exclusive 'hairpin structure (at 0.15M NaCl) with a stability comparable to that of the corresponding RNA hairpin. These conclusions are derived from concentration-dependent melting temperature (T_m) measurements. Thus, neither flexible furanose nucleotides (*i.e.* ribose or deoxyribose) nor pyranosyl RNA can be used to study self-complementary duplexes with an internal loop formed by a UUUU or TTTT tract.

Folding a TTTT or UUUU sequence into a tetraloop hairpin requires an S-type sugar conformation to bridge the ends of an A-type stem with N-type sugars [15]. The formation of a hairpin structure might be avoided by using conformationally restricted nucleotides, where the carbohydrate moiety is not able to adopt an S-type conformation. Recently, we have synthesized hexitol nucleic acids (HNAs) as mimics of A-type RNA [16]. A HNA is a modified nucleic acid consisting of a phosphorylated 1,5-anhydrohexitol backbone and natural nucleobases (Fig. 1). HNAs are distinguished from pRNAs by having strong and selective hybridization properties with complementary natural nucleic acids. The building block of HNA (a 1,5-anhydrohexitol nucleotide) can be considered a conformationally restricted mimic of a natural nucleoside in its N-conformation [17]. The stiffness of the six-membered anhydrohexitol rings prevents HNAs from undergoing drastic conformational alterations. Therefore, HNA can be used as a model nucleic acid in which folding into a TTTT (or UUUU) tetraloop is avoided.

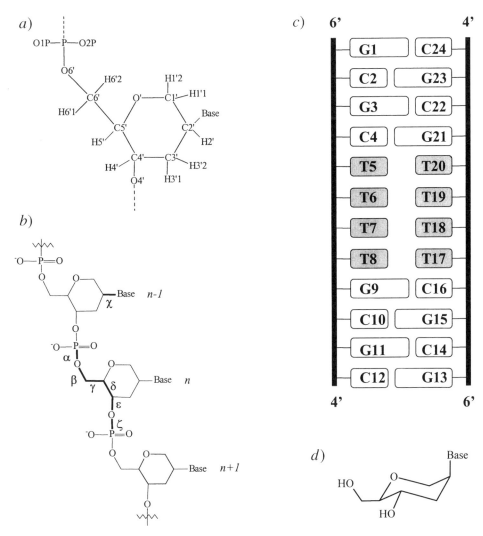

Fig. 1. a) *Chemical structure and atomic numbering in a HNA nucleotide* (base = guanine, cytosine, or thymine). b) *Schematic representation of a HNA oligonucleotide* (main torsion angles are indicated and labeled in boldface). c) *Diagram of dsHNA* (thymine bases forming the internal loop of non-standard base pairing are shaded). d) *Preferred conformation of HNA monomers as determined by X-ray crystallography and NMR spectroscopy.*

During studies of the hybridization properties of HNA, we observed that HNA oligothymidylate (oligo(hT)) self-associates at high NaCl concentrations (1M) in the presence of MgCl$_2$, but not at low NaCl concentrations (0.1M) [18–20]. Although neither the characterization of this hybridization nor the orientation of the oligo(hT) strands were studied, this demonstrates the occurrence of T·T base pairing in HNA. Therefore, we studied the structure of the HNA oligomer h(GCGCTTTTGCGC) that adopts an *A*-type helix structure in solution with an internal loop formed by a TTTT tract. This structure elucidation could provide information about the influence of a $(T)_4$ fragment on local helix geometries as well as the nature of the T·T mismatch base pairing in a TTTT tract.

The NMR-derived structure we report here shows remarkable similarities to the crystal structure of r(GGCGUUGCGUC) [9]. This may indicate that the structure presented here could be considered a model for the duplex structure adopted by r(GGACUUUUGUCC), which cannot be determined due to its instability.

2. Results and Discussion. – 2.1. *Preamble*. The constitution of HNAs, their atom numbering, and the definitions of the main torsion angles are depicted in *Fig. 1, a* and *b*. Individual residues in the sequence studied (*Fig. 1, c*) are composed of six-membered 1,5-anhydrohexitol rings, substituted at C(2') with nucleobases that can be either guanine (G), thymine (T), or cytosine (C). Adjacent residues are connected by a 4'(*n*) to 6' (*n*+1) phosphodiester linkage. Due to the C(2') positioning of the base moiety, these nucleoside analogues lack an anomeric center. Replacement of the furanose ring by an anhydrohexitol ring results in a reduced conformational flexibility of the carbohydrate moiety of HNA compared to natural nucleic acids. The most stable conformation of HNA monomers, as determined by X-ray crystallography [21] and ¹H-NMR spectroscopy [22] is depicted in *Fig. 1, d*.

2.2. *Formation of the Duplex Structure*. A natural nucleic acid with sequence 5'-GCGCTTTTGCGC-3' would be expected to adopt a hairpin structure. However, as explained above, HNA should not be able to fold similarly.

The one-dimensional (1D) ¹H-NMR spectrum of h(GCGCTTTTGCGC) in D_2O showed a mixture of two different conformations of the oligomer (*Fig. 2, middle*). Increasing temperature caused a decrease of one set of signals, while the second set simultaneously increased. At 70°, only one structural form could be observed (*Fig. 2, top*). This is probably the single-stranded HNA (ssHNA), since the T_m of h(GCGCTTTTGCGC) as a 4 μM solution in H_2O with 0.1M NaCl is 52°. When NaCl was gradually added to the sample at 20°, signals from ssHNA decreased, while those belonging to the second structural form increased. At 0.3M NaCl, only signals of the second structural form were visible in the 1D ¹H-NMR spectrum (*Fig. 2, bottom*). This sample was used for further studies, in which this structure was shown to be double-stranded HNA (dsHNA).

2.3. *Determination of the Duplex Structure*. The assignment of non-exchangeable protons was performed in two stages. First, ¹H-NMR signals of the HNA dodecamer were assigned by TOCSY, NOESY, and COSY. This assignment allowed the determination of the conformation for the twelve hexitol rings in the molecule. Second, the sequential assignment of residues based on interresidue NOE contacts was performed.

Expansions of a NOESY plot with mixing time 50 ms (*Fig. 3*) show the intraresidue NOE effects used to link the aromatic proton signals to a hexitol proton spin system (*Fig. 3, b*) and the observed interresidue H^2–C(3') (=H3'2)(*n*) to H–C(8)/ H–C(6)(= H8/H6)(*n*+1) cross-peaks (*Fig. 3, a*) needed for sequential assignment in the dodecamer.

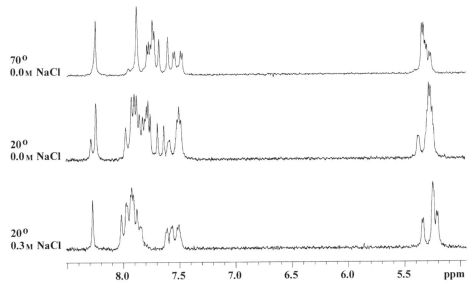

Fig. 2. *Section of the one-dimensional ^{1}H-NMR spectrum of a 2 mM solution of h(GCGCTTTTGCGC) in D$_{2}$O, measured under different experimental conditions*

The temperature dependence of imino-proton signals is depicted in *Fig. 4*. The four signals between 10 and 12 ppm are characteristic of two different T·T base pairs. This base-pairing evidence is corroborated by strong pairwise NOEs of these signals in a Watergate NOESY spectrum at 0°, used for sequential assignment of the imino signals (*Fig. 5*). The occurrence of T·T base pairing implied that, under the experimental conditions, h(GCGCTTTTGCGC) is forming a symmetrical duplex with four central T·T mismatches flanked by four G·C base pairs on each side. Due to the symmetry of the duplex, signals from both strands degenerate. Duplex formation is confirmed by the absence of a 'turning phosphate', which would be expected in a hairpin structure and which would show a downfield shift compared to phosphate signals in the duplex region of the hairpin stem. In this sample, all ^{31}P nuclei resonated within a 1 ppm range, suggesting that no large backbone distortion had occurred [23].

In one of the mismatch T·T pairs, a much larger spectral separation exists for the imino signals between 10 and 12 ppm compared to the separation for the second pair. In the Watergate NOESY spectrum, these separated signals were assigned to the hT5·hT20 and hT8·hT17 mismatches. This is in agreement with the description of tandem U·U mismatches in the sequence track 5'-YUUR-3'/5'-YUUR-3' (R = G, Y = C or U) [7] that have one downfield shifted line at *ca.* 11.3 ppm and an upfield shifted imino signal at *ca.* 10.5 ppm. In a thorough NMR analysis of ^{15}N- and ^{13}C-labeled RNA in combination with multidimensional homo- and heteronuclear experiments [24], the downfield signal was assigned to the U-3' neighbor of Y preceding the U·U wobble pairs. Difference in stacking and shielding effects in a more-or-less regular *A*-type RNA helix were considered possible reasons for the large spectral separation of both imino protons of the U·U mismatch. The similarity to this tandem U·U mismatch was used to assign the signal at 11.5 ppm to H−C(3) of hT5 and hT17 (the 4' neighbors of hC4

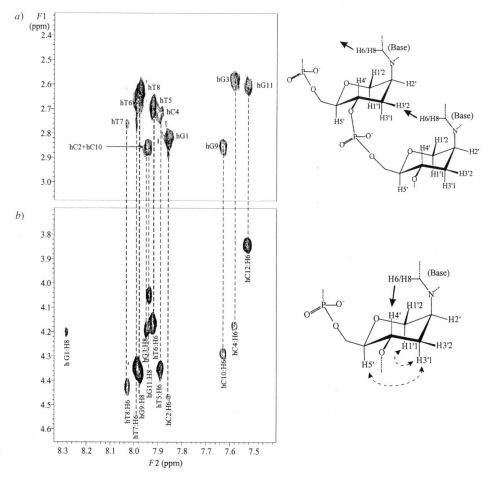

Fig. 3. *Expansions of a 50-ms mixing time NOESY spectrum of a 2 m*M *solution of h(GCGCTTTTGCGC) in* D_2O *at 20° and 0.3*M *NaCl. a) Region showing the interresidue* $H-C(6)/H-C(8)$ (= H6/H8) (n + 1) *to* $H^2-C(3')$ (= H3'2)(n) *cross-peaks needed for sequential assignment in the oligomer* (indicated by solid arrows on the right), *and* b) *section with intraresidue* $H-C(6)/H-C(8)$ (= H6/H8) *to* $H-C(4')$ (= H4') *NOE interactions used to link the hexitol ring proton spin system to the nucleobase at its C(2') position* (indicated by solid arrows on the right). Extra NOE interactions, not visible in the depicted spectra but used to determine the hexitol ring conformation, are indicated by dashed arrows on the right.

and hC16) and the upfield shifted signal at 10.6 ppm to $H-C(3)$ of hT8 and hT20. This assignment is in agreement with the strong intrastrand NOE we observed between $H-C(3)$ of hT5 and Me of hT6 (and $H-C(3)$ of hT17 and Me of hT18). Other imino signals were assigned by intra- and interstrand NOE cross-peaks in the Watergate NOESY spectrum schematically represented in *Fig. 5*.

All ^{31}P-NMR signals were assigned in a ^{1}H-detected ^{1}H-^{31}P HETCOR. A full listing of all ^{1}H- and ^{31}P-resonances is given in *Table 1*.

2.4. *Conformation of the Hexitol Rings.* The hexitol ring can adopt a conformation with either an axially (chair) or an equatorially (inverse chair) oriented base moiety.

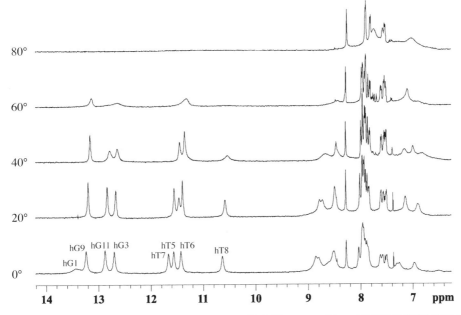

Fig. 4. *Imino- and aromatic-proton region of 'jump-return' ¹H-NMR spectra of 2 mM h(GCGCTTTTGCGC) in H₂O/D₂O 9:1 with 0.3M NaCl at different temperatures* (0, 20, 40, 60, and 80°)

Table 1. *Chemical Shifts* [ppm] *of* ¹H- *and* ³¹P-*Resonances at 20°. The* $\delta(^1H)$ *and* $\delta(^{31}P)$ *are referenced to acetate* (1.92 ppm) *and TMP, resp.* (n.a. = not applicable).

	H−C(8)/ H−C(6)	H−C(2)/ H−C(5)	H−C(1)/ H−C(3)	H¹− C(1')	H²− C(1')	H− C(2')	H¹− C(3')	H²− C(3')	H− C(4')	H− C(5')	H¹−C(6')/ H²−C(6')	P
hG1/hG13	8.27	–	13.41	4.11	4.36	4.68	2.18	2.82	4.20	3.64	3.99/3.83	–
hC2/hC14	7.85	5.28	–	4.11	4.54	4.73	2.02	2.86	4.46	3.61	4.18/4.01	− 3.43
hG3/hG15	7.94	–	12.67	4.03	4.39	4.63	2.05	2.59	4.19	3.62	n.a	− 3.31
hC4/hC16	7.58	5.24	–	4.00	4.28	4.65	1.92	2.74	4.17	3.52	n.a.	− 3.61
hT5/hT17	7.88	1.69	11.46	3.92	4.43	4.52	2.00	2.70	4.35	3.55	4.18/4.02	− 3.48
hT6/hT18	7.91	1.64	11.40	3.86	4.17	4.43	1.98	2.67	4.16	3.53	n.a.	− 3.29
hT7/hT19	7.98	1.63	11.55	3.93	4.33	4.60	2.02	2.77	4.33	3.57	4.16/4.07	− 3.53
hT8/hT20	8.02	1.76	10.58	3.94	4.33	4.52	2.09	2.63	4.42	3.60	4.10/4.04	− 3.56
hG9/hG21	7.96	–	13.20	4.04	4.23	4.76	2.03	2.85	4.37	3.66	n.a./4.13	− 3.51
hC10/hC22	7.62	5.37	–	4.00	4.39	4.72	1.93	2.85	4.29	3.59	n.a.	− 3.43
hG11/hG23	7.92	–	12.67	3.97	4.51	4.57	1.99	2.61	4.05	3.61	n.a.	− 3.05
hC12/hG24	7.52	5.27	–	3.93	4.16	4.69	1.83	2.06	3.84	3.36	4.05	− 3.74

The orientation has a profound influence on duplex geometry. We have previously demonstrated that an axial orientation of the nucleobase is needed to allow HNA to hybridize strongly and selectively with RNA [17]. A HNA·RNA heteroduplex has a helical structure, very similar to the *A*-type helices of natural double-stranded nucleic acids. If the base moiety is in an equatorial position, HNA is unable to hybridize with DNA or RNA. The reason for this is the higher stabilization of the helical structure due to stacking interactions when base moieties are axially oriented, compared to the stabilization of linear structures with equatorially oriented nucleobases. However, this

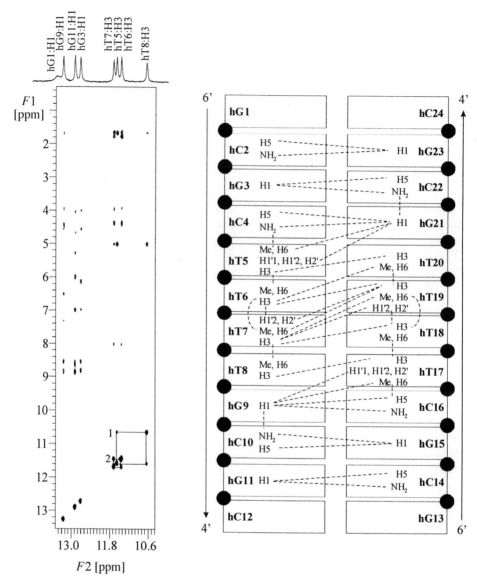

Fig. 5. *Portion of the Watergate NOESY plot measured at 0° in H₂O/D₂O 9 : 1 with 0.3M NaCl showing the cross-peaks from the imino signals of h(GCGCTTTTGCGC).* Boxes 1 and 2 indicate the NOE interactions of hT5 to hT20 (hT20 to hT5) and hT6 to hT19 (hT18 to hT7), respectively. Imino signals were assigned by intra- and interstrand NOE cross-peaks in the Watergate NOESY spectrum as represented schematically on the right.

does not mean that dsHNA may not adopt structures other than the typical *A*-type helix.

Before studying the overall structure, the conformation of the hexitol rings has to be verified. The strong H−C(4′)/H¹−C(3′) and H−C(4′)/H−C(5′) cross-peaks in the DQF-COSY, indicative of large active $J(4′,3′1)$ and $J(4′,5′)$ coupling constants, correspond to axial positions for the three involved protons in the anhydrohexitol ring. Furthermore, very weak or non-observable cross-peaks from H−C(2′) to the axially positioned H¹−C(3′), correspond to an equatorial position for the H−C(2′) protons in the HNA strand. In the NOESY plot, there are strong intraresidue cross-peaks from the base protons H−C(8)/H−C(6) to H−C(4′) and close contacts between H¹−C(1′), H¹−C(3′), and H−C(5′). All these data correspond to anhydrohexitol rings in chair conformations with axially positioned bases (*Fig. 1, d*). This is in agreement with the previously determined conformation of monomeric [21] and dimeric [16] HNA.

2.5. *Base-Pairing in T·T Mismatches.* The observed imino signals between 10 and 12 ppm are characteristic of T·T base pairing where both H−N(3) imino protons are involved in H-bonding and in slow exchange with the surrounding H_2O. However, two different arrangements are possible for T·T wobble pairs (*Fig. 6*). In one pairing mode, the O(4) of the thymine in strand 1 extends into the major groove, and the O(4) of the thymine in strand 2 is involved in base pairing (*Fig. 6, a*), while the other pairing mode is just the opposite. Both pairing modes are present in the h(GCGCTTTTGCGC) structure (*Fig. 7*). When the O(4) of thymine is extended into the major groove, the λ angle formed by N(1) and C(2′) of this residue and C(2′) of the residue in the other strand is decreased, while λ is increased if O(4) is H-bonded (*Fig. 6* and *Table 2*). Since both pairing modes are possible, we did not include any H-bonding restraints for the four central T·T base pairs in the duplex during the structure determination. The first calculation stage of torsion-angle molecular dynamics started from two extended strands and yielded 42 accepted structures out of 100 calculated. The 25 lowest-energy structures were subjected to further refinement. Twelve refined structures showed nice base stacking for the four central wobble pairs in the duplex, despite the absence of H-bonding restraints in this part of the molecule. The mismatch pairs in the structure closest to the average of the twelve final structures are depicted in *Fig. 7, b*. Average H-bonding distances in the two outer mismatches (hT5·hT20 and hT8·hT17) are larger than those in the two central T·T wobble pairs. These increased distances correspond to the increased line broadening at 20° for the imino protons in the two outer mismatches

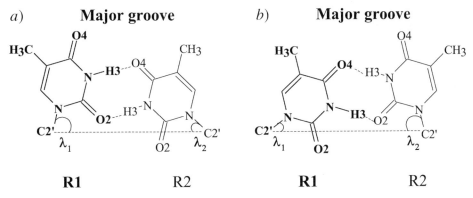

Fig. 6. *The two possible T·T wobble pairs: a) with H-bonds between HN(3) and O(2) of the thymine in strand 1 (R1) and O(4) and HN(3) of the nucleobase in strand 2 (R2), respectively, or b) with H-bonds between HN(3) and O(4) of the thymine in strand 1 (R1) and O(2) and HN(3) of the nucleobase in strand 2 (R2), respectively*

a)

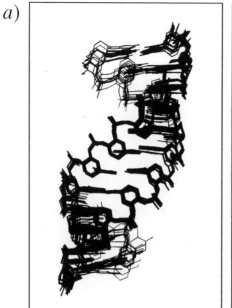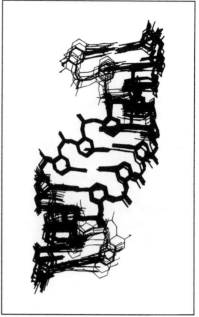

b)

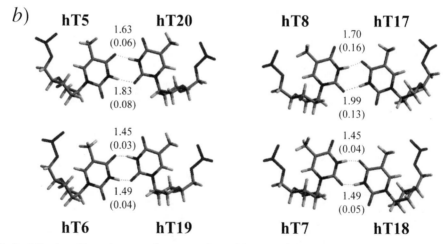

Fig. 7. a) *Overlay of the twelve accepted structures obtained from two extended strands by applying torsion-angle molecular dynamics followed by conjugate-gradient minimization without H-bonding restraints in the internal loop region* (stereo view). b) *Mismatch T·T wobble pairs in the central part of the duplex, closest to the average of the twelve refined structures that were accepted.* Average H-bonding distances are given with their standard deviations.

Table 2. *Average Values and Standard Deviations for Base-Pair Parameters in the Twelve Final Structures.* The angles λ_1 and λ_2 are defined as depicted in *Fig. 6*.

Base pair	X-displacement [Å]	C(2') to C(2') [Å]	λ_1 [°]	λ_2 [°]
hG1·hC24	−4.5 (0.3)	10.33 (0.04)	53.2 (0.3)	56.5 (1.0)
hC2·hG23	−4.6 (0.3)	10.34 (0.02)	55.3 (0.8)	53.3 (0.6)
hG3·hC22	−4.7 (0.2)	10.32 (0.02)	53.2 (0.7)	55.1 (1.1)
hC4·hG21	−4.6 (0.2)	10.20 (0.03)	58.3 (1.0)	53.3 (0.4)
hT5·hT20	−4.2 (0.3)	8.77 (0.09)	72.4 (1.1)	45.2 (1.0)
hT6·hT19	−4.3 (0.4)	8.20 (0.07)	80.6 (0.9)	43.3 (0.6)
hT7·hT18	−4.3 (0.4)	8.22 (0.08)	43.2 (0.7)	80.4 (1.2)
hT8·hT17	−4.3 (0.3)	8.81 (0.14)	45.6 (2.2)	72.1 (1.9)
hG9·hC16	−4.6 (0.2)	10.22 (0.04)	53.2 (0.3)	57.8 (0.8)
hC10·hG15	−4.7 (0.2)	10.32 (0.02)	55.5 (0.8)	53.4 (0.6)
hG11·hC14	−4.6 (0.2)	10.36 (0.02)	53.0 (0.5)	55.2 (0.5)
hC12·hG13	−4.6 (0.2)	10.35 (0.05)	56.5 (0.9)	53.2 (0.4)

Table 3. *Structure Statistics for the Final Set of Twelve Structures after Refinement with All Restraints*

NOE Violations (> 0.5 Å)	0
Dihedral violations ($> 5°$)	1 ± 1
R.m.s.d. from distance restraints [Å]	0.025 ± 0.001
R.m.s.d. from dihedral restraints [°]	0.571 ± 0.110
R.m.s.d. from average structure for all heavy atoms [Å]	0.75

compared to line-widths for other imino signals (*Fig. 4*). This line broadening is an indication of weaker H-bonding, allowing more possibility for the exchange of imino protons with the surrounding H_2O. A summary of structure statistics for accepted structures is listed in *Table 3*.

Our data confirm that four successive pyrimidine·pyrimidine base pairs can be formed within a double-stranded structure. The stability of the double-stranded structure is due to H-bonding in the mismatch region, as well as to increased base stacking for the hT7 and hT19 bases in the central part of the molecule. Therefore, the backbone of both oligonucleotide strands should come closer in the mismatch region. Due to local helix adjustments in the internal loop region, hT7 is located above hT19. This results in an extra interstrand stacking interaction compared to the central base-pair steps in the reported structures with two central U·U tandem mismatches [9][10] (*Fig. 8*). The stacking diagrams of the latter show more similarity to the hT5·hT20 to hT6·hT19 (or hT17·hT8 to hT18:·hT7) base-pair step that is flanked by the hC4·hG21 (or hC16·hG9) *Watson-Crick* base pair. Stacking interactions at the base-pair steps from hC4·hG21 to hT5·hT20 are comparable to the corresponding rC5/rU5·rG20 to rU6·rU19 wobble pairs (or rC17/rU17·rG8 to rU18·rU7) in the reported r(GGCGCUUGCGUC) [9] and r(GGACUUUGGUCC) [10] duplex structures determined by X-ray crystallography.

2.6. *Structure of the Double Helix.* The HNA dodecamer forms a double-stranded helical structure with an internal loop of four T·T wobble pairs. The overall structure is comparable to duplexes of the *A*-type family formed by natural DNA and RNA (*Fig. 9*). Superimposition of P-atoms of the structure determined by NMR onto

a) h(GCG**CTTTTG**CGC)   *b*) r(GGCG**CUUG**CGUC)   *c*) r(GGAC**UUUG**GUCC)

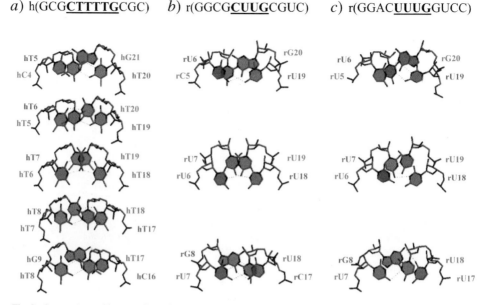

Fig. 8. *Comparison of base-stacking diagrams in the mismatch region viewed perpendicular to the mean value plane through both base pairs in each step for* a) *our HNA duplex and* b), c) *two reported RNA structures* [9] [10]. The base pairs at the top of each step have green-colored C-atoms and base planes, while these are gray-colored in the lower base pairs.

corresponding P-atoms of a canonical *A*-form RNA duplex with the same sequence (*Fig. 10*) demonstrates the influence of the central mismatches on the overall structure of h(GCGCTTTTGCGC). When all P-atoms are used in the superposition, a root mean square deviation (r.m.s.d.) of 3.6 Å is obtained. The large deviation from canonical *A*-form RNA is mainly caused by the internal loop. Superimposing the P-atoms in the G·C and T·T regions separately, yields r.m.s.d.s of 0.5 Å for regions with *Watson-Crick* pairing (green *vs.* red and blue *vs.* yellow) and 1.5 Å for the mismatch region (cyan *vs.* orange) (*Fig. 10,b* and *c*). The substitution of the five-membered ribose ring in RNA by a six-membered anhydrohexitol ring in HNA does not influence the overall *A*-type geometry of the helix in the G·C parts of the duplex. In both regions, the phosphate atoms of HNA (green and blue) and RNA (red and yellow) superimpose well, *Watson-Crick* base pairing occurs, and base stacking interactions are very similar. However, the structure of the central internal loop region of HNA (cyan) deviates substantially from that of canonical *A*-type RNA (orange). In the mismatch region, the ribbons through P-atoms in both structures deviate substantially, and compared to canonical *A*-type RNA, the T·T bases from opposite HNA strands come closer to each other, and the axial rise is increased with RNA.

The helix adjustments in the internal loop cause bending of the local helix axis in the central part of the HNA duplex, while in G·C regions, the local helix axes are rather straight and parallel to each other (*Fig. 9*). The helical axis calculated for the internal loop (base pairs hT5·hT20 to hT8·hT17) makes an angle of 25° (θ_1) with the axes calculated for each of the G·C regions. The 'swinging motion' of the helix axis in

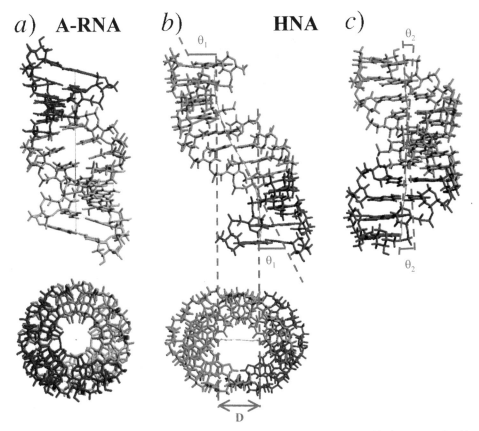

Fig. 9. *Comparison of the HNA duplex structure to a canonical dsRNA structure with the same nucleoside sequence.* a), b) *Top view and side view into the major groove for each duplex* (the helical axis calculated for the internal loop in HNA (cyan) makes an angle of 25° (θ_1) with the axis calculated for in both G · C regions (green and blue); the 'swinging motion' of the helix axis in the mismatch region causes an 8.9-Å dislocation (D) of the axis for both stems of the duplex. c) *The 10° (θ_2) twist of both stem axes relative to each other is visible in 90° rotated view of the dsHNA helix shown in b.*

the mismatch region introduces a 8.9 Å dislocation (D) between the axes for each stem of the duplex and a twist of 10° (θ_2) for the stem axes relative to each other (*Fig. 9*).

2.7. *Geometric Features of the Duplex.* 2.7.1. *Backbone Torsion Angles.* Since HNA was designed as a conformationally restricted mimic of RNA in its *A*-type conformation, it is obviously necessary to compare the structure of the HNA duplex to that of the dsRNA *A*-helical form. Deviation from the regular structure in the mismatch region is shown by comparing the present structure to that of a previously determined HNA · RNA hybrid containing only canonical base pairs and to that of the G · C base pairing regions of the dsHNA structure.

Average values for the χ, α, β, γ, δ, ε, and ζ torsion angles (*Fig. 1, b*) are listed in *Table 4*. The χ torsion angles are slightly increased compared to the average value in *A*-helical structures. The δ torsion angles in the six-membered hexitol rings center

a) *b)* *c)*

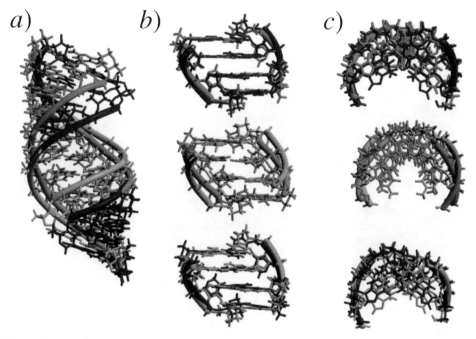

Fig. 10. *Overlay of the P-atoms of the HNA duplex to those of a canonical dsRNA structure with the same nucleoside sequence* (see *Fig. 9* for coloring scheme): a) *overlay of all P-atoms,* b) *view into the major groove, and* c) *top view when P-atoms in the top stem* (green *vs.* red), *the internal loop* (cyan *vs.* orange), *and the bottom stem* (blue *vs.* yellow) *are overlaid*

around the value of $62°$ compared to $82°$ in a five-membered $C(2')$-*endo* ribose ring in A-type RNA. In the regions where $G \cdot C$ base pairing occurs, the backbone angle β deviates significantly ($>10°$, except for hG3 and hG15) from the average A-type dsRNA value, but remains close to the value measured in the HNA strand of the HNA $\cdot$ RNA heteroduplex (HNA*). In the mismatch region, several dihedral angles are altered to allow for local helix adjustments. The largest variations occur in the ε and ζ dihedral angles. There is a significant decrease in both torsion angles for hT6 (hT18) compared to the corresponding angles in the GCGC part of the duplex and HNA*. In the same base pair step (hT6 $\cdot$ hT19 to hT7 $\cdot$ hT18), the γ torsion angle in hT7 (hT19) is slightly increased relative to the γ torsion angles in the rest of the duplex. There is also a slight increase of the ε and ζ dihedral angles for the hC4 (hC16) residue. For the same hC4 $\cdot$ hG21 to hT5 $\cdot$ hT20 (hC16 $\cdot$ hG9 to hT17 $\cdot$ hT8) base-pair step but in the opposite strand, these angles are slightly decreased for the hT8 (hT20) residue. The γ torsion angle in hG9 (hG21) is smaller compared to the average values in HNA* and the measured values in the stem parts of the presented duplex. An enlarged ε dihedral angle occurs for hT7 (hT19), while ζ for this residue is decreased. In the same phospho-diester bridge, linking hT7 to hT8 (hT19 to hT20), the α torsion angle is significantly increased, while β is decreased relative to other torsion angles in the duplex.

It can be concluded that local geometry adjustment in the mismatch region is due to the cooperative effect of all backbone torsion angles. The smallest variation is found for

Table 4. *Average Backbone Torsion Angles [°] for Both Strands in the Twelve Final Structures Refined with All Experimental Restraints.* Standard deviations from average values are indicated in parentheses.

	χ	α	β	γ	δ	ε	ζ
hG1	220.0 (1.8)	–	–	–	60.9 (1.3)	213.4 (5.0)	289.7 (4.2)
hG13	219.3 (1.4)	–	–	–	60.9 (0.9)	213.9 (2.9)	288.9 (2.8)
hC2	222.5 (1.2)	299.6 (6.8)	163.6 (6.6)	64.3 (2.0)	64.9 (1.3)	215.1 (3.9)	289.1 (2.1)
hC14	222.6 (1.1)	301.7 (3.3)	162.3 (3.4)	64.7 (0.9)	65.3 (1.1)	215.6 (2.6)	290.0 (1.5)
hG3	224.5 (1.3)	289.7 (8.5)	174.8 (12.0)	62.0 (0.7)	64.0 (3.6)	217.6 (1.3)	289.5 (4.5)
hG15	224.6 (2.0)	291.7 (3.8)	172.0 (6.0)	62.2 (1.4)	63.2 (2.1)	217.4 (1.7)	288.5 (2.5)
hC4	220.7 (1.0)	301.3 (6.8)	157.4 (3.6)	64.9 (1.1)	65.0 (0.6)	221.1 (1.1)	293.7 (2.3)
hC16	221.1 (1.1)	303.3 (2.6)	156.7 (2.1)	64.5 (0.9)	64.6 (0.8)	221.6 (0.9)	292.6 (1.7)
hT5	221.9 (1.5)	296.5 (4.5)	168.4 (2.8)	62.7 (1.0)	64.4 (0.3)	216.1 (1.2)	288.6 (1.8)
hT17	221.1 (1.3)	298.8 (2.3)	167.1 (2.2)	62.3 (1.1)	64.3 (0.6)	216.4 (1.1)	287.9 (1.9)
hT6	222.8 (1.6)	298.6 (2.4)	158.4 (1.8)	63.6 (0.6)	65.5 (0.4)	204.9 (1.2)	278.9 (1.1)
hT18	221.9 (1.9)	300.0 (3.1)	158.0 (2.0)	63.4 (0.8)	65.5 (0.6)	204.9 (1.5)	279.3 (1.3)
hT7	215.7 (1.1)	298.3 (1.3)	163.2 (1.1)	66.9 (0.5)	62.2 (0.7)	234.9 (0.6)	274.7 (1.7)
hT19	215.5 (1.0)	298.8 (1.4)	163.3 (1.6)	66.8 (0.7)	62.1 (0.5)	234.8 (1.1)	274.9 (1.4)
hT8	215.4 (0.9)	316.9 (2.0)	144.4 (1.3)	64.4 (0.9)	60.0 (1.8)	212.2 (4.3)	282.9 (3.0)
hT20	215.3 (0.7)	316.8 (2.3)	144.6 (1.3)	64.4 (0.7)	60.0 (1.6)	212.2 (5.8)	283.8 (4.4)
hG9	215.0 (2.5)	200.9 (4.5)	168.5 (5.0)	58.3 (0.8)	67.2 (1.3)	218.1 (1.8)	286.5 (2.6)
hG21	214.9 (1.6)	299.3 (8.1)	170.2 (10.1)	58.0 (1.6)	67.2 (1.9)	216.5 (2.7)	287.9 (3.4)
hC10	220.4 (1.2)	306.1 (3.9)	157.8 (3.0)	64.1 (1.3)	64.8 (1.2)	214.6 (2.6)	289.5 (3.3)
hC22	221.4 (1.8)	304.2 (5.2)	160.2 (5.8)	64.4 (1.0)	65.2 (0.9)	216.0 (3.3)	290.3 (2.2)
hG11	223.4 (1.8)	298.3 (4.1)	165.8 (4.2)	62.4 (1.3)	62.9 (1.6)	220.8 (3.7)	285.6 (3.1)
hG23	224.1 (2.0)	297.0 (3.0)	165.6 (2.3)	62.4 (0.9)	63.4 (2.6)	219.8 (3.3)	288.1 (4.5)
hC12	229.4 (1.6)	305.2 (3.3)	158.5 (5.5)	62.6 (0.9)	64.9 (0.7)	–	–
hc24	229.0 (1.3)	302.6 (5.1)	160.5 (5.0)	62.5 (1.0)	64.2 (0.8)	–	–
A-RNA[a])	202	292	178	54	82	207	289
B-DNA[a])	262	314	213	36	157	155	264
HNA*[b])	222.0 (2.0)	292.8 (13.5)	163.9 (6.4)	65.3 (8.8)	67.4 (1.8)	215.6 (3.6)	289.4 (3.2)

[a]) Data for *A*-RNA and *B*-DNA are taken from *Saenger* [45]. [b]) Values listed for HNA* are average torsion angles in the HNA strand of a HNA · RNA duplex determined by NMR spectroscopy [46].

dihedral angles γ and δ, while large changes are found for the β and ε torsion angles in hT8 (hT20) and hT6 (hT18), respectively.

2.7.2. *Helical Parameters and Stacking Interactions.* To enable H-bonding between two pyrimidines in opposite strands, the distance between the C(2′) atoms of the hexitol rings to which the nucleobases are attached is reduced (*Table 2*). The minor-groove width drops from 9.9 to 8.6 Å in the hT5 · hT8 (hT17 · hT20) wobble pair and to 7.6 Å in the two central hT6 · hT7 (hT18 · hT19) mismatches (*Fig. 11*). To close the gap, the C(2′) atoms in each wobble pair move towards each other and towards the helix axis causing a reduction of the X-displacement from -4.7 Å for the C · G *Watson-Crick* pairs to -4.3 Å in the central T · T wobble pairs (*Table 2*). Similar features are reported for the crystal structure of an RNA dodecamer with tandem U · U mismatches flanked by C · G base pairs [9]. The base-pair steps in the mismatch region are different from those in regular *A*-type RNA. Axial twist and the related axial rise show an obvious change in the mismatch region of the final structures (*Fig. 11,a*). Due to the large shear in T · T wobble pairs, measurement of twist is strongly influenced by the choice of base-pair reference frame [25], which can be located either at the center of a base pair (CompDNA [26], Curves [31], RNA [27]) or either based on C(6)(Y)/C(8)(R)

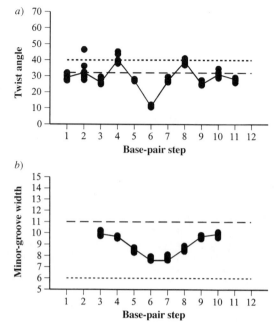

Fig. 11. *Variation of* a) *the twist angles and* b) *the minor-groove width in the twelve final structures as determined by* Curves 5.0 [31]. Average *A-* and *B*-form values are given by dashed and dotted lines, respectively. Measured averages for the twelve structures are connected by solid lines.

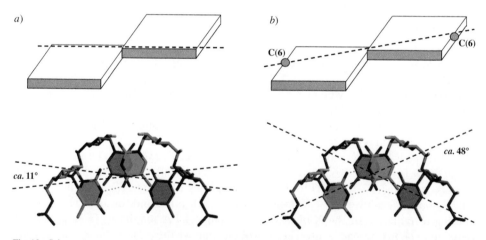

Fig. 12. *Schematic representation of the influence of a large shear on the Y axis in* a) *a reference frame located at the center of the base pair and* b) *a reference frame based on C(6) atoms at the outer edge of the hT·hT wobble pair.* The central hT6·hT19 to hT7·hT18 base-pair step is shown as an example of how the choice of a reference frame influences the measured twist (for coloring scheme, see *Fig. 8*).

atoms at the outer edge of a base pair (CEHS [28], FreeHelix [29], NUPARM [30]) (*Fig. 12* and *Table 5*). As with a $^{5'}_{3'}\!\!\begin{smallmatrix}CUUG\\GUUC\end{smallmatrix}\!^{3'}_{5'}$ tandem mismatch in RNA [9], Curves5.0 [31] measures an over-winding of the helix at the $^{6''}_{4'}\!\!\begin{smallmatrix}CT\\GT\end{smallmatrix}\!^{4'}_{6'}$ and $^{6''}_{4'}\!\!\begin{smallmatrix}TG\\TC\end{smallmatrix}\!^{4'}_{6'}$ steps (average twist angles of

39.8 and 39.1° in HNA, 38.8 and 40.7° in RNA, PDB id 280D). This over-winding is followed by an unwinding in the next base-pair steps between two neighboring wobble pairs (average twist angles of 27.4 and 25.1° in HNA, 21.8 and 21.5° in RNA, PDB id 280D). Most striking is the increase in unwinding between the two central T·T wobble pairs in the HNA duplex (average twist angles of 9.1°).

Table 5. *Twist Angles in Each Base-Pair Step in the Structure Closest to the Average of All Accepted Structures Measured with Two Programs Using Different Reference Frames.* Values for base-pair steps in which hT·hT wobble pairs are involved are listed in italics.

	Curves 5.0	CEHS
hG1·hC24 to hC2·hG23	29.9	27.3
hC2·hG23 to hG3·hC22	32.1	32.3
hG3·hC22 to hC4·hG21	25.6	24.0
hC4·hG21 to hT5·hT20	*40.6*	*27.1*
hT5·hT20 to hT6·hT19	26.7	*23.5*
hT6·hT19 to hT7·hT18	*11.4*	*48.2*
hT7·hT18 to hT8·hT17	*26.8*	*22.9*
hT8·hT17 to hG9·hC16	*40.6*	*26.8*
hG9·hC16 to hC10·hG15	24.8	23.4
hC10·hG15 to hG11·hC14	30.1	30.9
hG11·hC14 to hC12·hG13	27.3	27.0

3. Conclusion. – The HNA duplex structure presented here demonstrates that an internal loop of four successive anti-parallel T·T wobble pairs can exist in an *A*-type helical structure where there is a conformationally restricted 'sugar-phosphate' backbone that favors duplex formation. The stability of the internal TTTT loop is due to the strong H-bonding and stacking interactions within the helix that remains relatively undistorted. A combination of small changes in backbone torsion angles reduces the minor-groove width, allowing T residues from opposite strands to approach each other and enabling H-bonding in T·T wobble pairs. The strongest interaction occurs in the two central T·T base pairs, as reflected in the sharp NMR signals from H-bonded imino protons in these wobble pairs and the short H-bond distances in the calculated structures. The T·T mismatches flanked by G·C *Watson-Crick* pairs have slightly longer H-bonds characterized by broader imino signals that disappear first when the temperature is increased, two spectral features that are indicative of weaker H-bonding within these wobble pairs. The central part of the h(GCGCTTTTGCGC) duplex shows striking differences in geometry. Local helix adjustments in the mismatch region give rise to a 'swinging motion' of the helix axis over the course of the HNA duplex, and an 8.9 Å displacement of the helix axes of the stems relative to each other. Conformational changes at the interface between the internal loop and both stems introduce an overall bend of 10° in the duplex.

Similar H-bonding in wobble pairs and helix adjustments have been reported for an RNA dodecamer with an internal loop of two U·U mismatches [9]. The displacement is not as pronounced since the internal loop in this structure contains only two non-standard base pairs. The strong resemblance of this RNA structure to our dsHNA structure is striking and suggests that the HNA duplex can be considered a model for the structure of dsRNA with an internal loop of four successive U · U mismatches. The occurrence of such an RNA duplex has been described in the sequence r(GGA-CUUUUGUCC), but its structure has not been determined yet. HNA, with its conformationally rigid N-type building blocks, may be used as a tool to stabilize metastable RNA structures. The synthesis and study of chimeric RNA · HNA oligonucleotides may thus shed light on RNA structures of low stability and allow an analysis of the dynamics of conformational changes in RNA.

We are grateful to the K.U. Leuven for financial support (GOA 97/11). *E. Lescrinier* is a research associate of the *Fund of Scientific Research*. The authors thank Dr. *B. De Bouvere* for synthesizing the HNA monomers, Mr. *G. Schepers* for synthesizing the HNA oligonucleotide, and Dr. *A. Van Aerschot* for supervising the oligonucleotide synthesis.

Experimental Part

1. *Sample Preparation.* The HNA molecule h(GCGCTTTTGCGC) was synthesized by phosphoramidite chemistry as described previously [16].

About 5 mg of HNA dodecamer was dissolved in D_2O (0.5 ml), and the pD was adjusted to 7 by adding small amounts of 0.01N HCl in D_2O. Small amounts of NaCl were added to obtain a single structural form of the HNA molecule. After adding each aliquot of NaCl, the mixture was briefly heated to 80° and slowly cooled to r.t. to promote duplex formation. Finally, the sample was lyophilized and redissolved in D_2O (0.650 ml), yielding a soln. of 2 mm oligomer and 0.3m NaCl (pD 7). To measure imino signals, the sample was relyophilized and dissolved in H_2O/D_2O (9 : 1) (0.650 ml).

2. *NMR Measurements.* NMR Spectra. *Varian-500-Unity* spectrometer operating at 499.505 MHz. Quadrature detection was achieved by the *States-Haberkorn* hypercomplex mode [32]. Spectra were processed using the FELIX 97.00 software package (*Biosym Technologies*, San Diego, CA) running on a *Silicon-Graphics-Indigo2-R10000* workstation (IRIX version 6.2). The one-dimensional (1D) spectra in H_2O were recorded with a jump-return observation pulse [33]. The two-dimensional (2D) NOESY in H_2O were recorded at 0° and 20° with 50-ms mixing time using the Watergate sequence [34]. A sweep width of 11000 Hz was used in both dimensions with 64 scans, 4096 data points in t_2 and 512 FIDs in t_1. Data were apodized with a shifted sine bell square function in both dimensions and processed to a $4K \times 2K$ matrix. The 2D DQF-COSY [35], TOCSY [36], and NOESY [37] experiments (D_2O) were recorded with a sweep width of 4200 Hz in both dimensions. The residual HDO peak was suppressed by presaturation. The DQF-COSY consisted of 4096 datapoints in t_2 and 360 increments in t_1. These data were apodized with a shifted sine bell square function in both dimensions and processed to a $4K \times 1K$ matrix. Both ^{31}P-decoupled and ^{31}P-coupled spectra were recorded under the same conditions. For the TOCSY experiment, a Clean MLEV17 [38] version was used, with a low-power 90° pulse of 26.6 μs and the delay set to 69.2 μs. The total TOCSY mixing time was set to 65 ms. The spectrum was acquired with 32 scans, 2048 data points in t_2 and 512 FIDs in t_1. The data were apodized with a shifted sine bell square function in both dimensions and processed to a $4K \times 2K$ matrix. The NOESY experiments were acquired with mixing times 50, 100, 150, 200, 250, and 300 ms, 64 scans, 2048 datapoints in t_2 and 512 increments in t_1. The $^1H,^{31}P$ HETCOR [39] was recorded with 64 scans, 4096 datapoints in the proton dimension, t_2, and 400 real datapoints in the phosphorus dimension, t_1, over sweep widths of 5000 and 3000 Hz, resp. The data were apodized with a shifted sine bell square function in both dimensions and processed to a $4K \times 1K$ matrix.

3. *Structural Restraints. Distance Restraints.* The build-up curves calculated from NOESY spectra in D_2O with mixing times 50, 100, 150, and 200 ms were used to calculate 202 inter-proton distances for nonexchangeable protons. The distances H^1−C(3')(hG1)/H^2−C(3')(hG1) (1.75 Å), H^1−C(3')(hG1)/H−C(5')(hG1) and H^1−C(3')(hC12)/H−C(5')(hC12) (2.72 Å), H−C(5)(hC2)/H−C(6)(hC2), and H−C(5)(hC10)/H−C(6)(hC10) (2.46 Å) were used as internal references. From the Watergate NOESY spectrum at 20°, 33 NOE distance restraints involving imino protons were defined according to the same internal reference distances as mentioned above. All distance restraints were calculated with the Felix97.0 software package. Upper and lower bounds for the measured distances were set to ±20% of the calculated value. H-Bonding restraints on G · C base pairs were considered as NOE distance restraints, and no H-bonding restraints were applied in the T · T mismatch region of the molecule.

Dihedral Angle Restraints. The conformation of the anhydrohexitol rings was restrained by dihedral restraints on O'−C(1')−C(2')−C(3') (60 ± 20°), C(4')−C(2')−C(3')−H(3'1) (−60 ± 20°), C(2')−C(3')−

C(4')–C(5') (60° ± 20°), C(1')–O'–C(5')–C(4') (60° ± 20°), and C(3')–C(4')–C(5')–O' (−60° ± 20°). The backbone dihedral angle ε was restrained to $−130 ± 40°$ based on the large $J(P,4')$ (*ca.* 11 Hz) measured in the ^{1}H,^{31}P HETCOR spectrum. Absence of P/H^1–C(6') and P/H^2–C(6') crosspeaks in this spectrum, indicative for small $J(P,6'1)$ and $J(P,6'2)$ coupling constants, were used to restrain β to the *trans* conformation (180 ± 30°). The small passive couplings observed in the H^1–C(6')/H^2–C(6') cross-peaks in the ^{31}P-decoupled DQF-COSY were used to restrain the γ angles to 60 ± 35°.

4. *Structure Calculation.* All structure calculations were performed with X-PLOR V3.851 [40]. The topallhdg.dna and parallhdg.dna were modified to include HNA oligonucleotides. In the topology file, the 6'-phosphorylated HNA monomers were introduced as individual residues that can be patched to each other to form an HNA oligonucleotide, comparable to the treatment of RNA and DNA in the standard X-PLOR package. Typical parameters for the hexitol ring were added to the standard parallhdg.dna parameter file.

The torsion-angle-dynamics protocol used was largely identical to that proposed for a DNA duplex [41]. A set of 100 structures was generated by torsion-angle dynamics starting from two extended strands. The torsion-angle dynamics procedure consisted of four different stages, and only in the final refinement stage, no planarity or symmetry restraints were applied. In the first stage, there was an initial search (60 ps in steps of 0.015 ps) of torsion-angle space at high temp. (2000 K) with a decreased weight on the repulsive energy term ($\omega_{vdw} = 0.1$) to facilitate rotational-barrier crossing. The coefficient for the dihedral-angle energy term ($\omega_{dihedral}$) and the NOE energy term (ω_{NOE}) were 5 and 150, resp. During the second stage (90 ps in steps of 0.015 ps), the temp. of the system was gradually reduced from 20000 to 1000 K, while ω_{vdw} was gradually increased to 1.0. The third stage was a slow cooling cartesian molecular dynamics simulation of 6 ps (with 0.003 ps timesteps). Finally, a 500-step conjugate-gradient minimization was performed with $\omega_{dihedral} = 400$ and $\omega_{NOE} = 50$.

Backbone angles and helix parameters were calculated with Curves 5.0 [31], which had been modified to handle HNA. For comparison, twist angles were recalculated with CEHS [28].

Molecule figures were generated with Bobscript2.4 [42], a modified version of Molscript1.4 [43]. *Fig. 7,b, 8 – 10,* and *12* were produced with Raster3D [44].

The refined coordinates of the ten structures closest to the average are deposited in the Brookhaven Protein Data Bank (PDB ID 1EC4).

REFERENCES

[1] R. R. Gutell, 'Ribosomal RNA Structure, Evolution and Function in Protein Synthesis', CRC Press, NY, USA, 1996.
[2] R. R. Gutell, M. W. Gray, M. N. Schnare, *Nucleic Acids Res.* **1993**, *21*, 3055.
[3] R. R. Gutell, *Nucleic Acids Res.* **1994**, *22*, 3502.
[4] T. R. Cech, S. H. Damberger, R. R. Gutell, *Nat. Struct. Biol.* **1994**, *1*, 273.
[5] S. Nonin, J. L. Leroy, *J. Mol. Biol.* **1996**, *261*, 399.
[6] X. Han, J. L. Leroy, M. Gueron, *J. Mol. Biol.* **1998**, *278*, 949.
[7] J. SantaLucia, R. Kierzek, D. H. Turner, *Biochemistry* **1991**, *30*, 8242.
[8] M. Wu, J. A. McDowell, D. H. Turner, *Biochemistry* **1995**, *34*, 3204.
[9] S. E. Lietzke, C. L. Barnes, J. A. Berglund, C. E. Kundrot, *Structure* **1996**, *4*, 917.
[10] K. Baeyens, H. De Bondt, S. Holbrook, *Nature Struct. Biol.* **1995**, *2*, 56.
[11] D. R. Hare, B. Reid, *Biochemistry* **1986**, *25*, 5341.
[12] V. P. Antao, I. Tinoco Jr., *Nucl. Acids Res.* **1992**, *20*, 819.
[13] C. Richter, B. Reif, K. Wörner, S. Quant, J. P. Marino, J. Engels, C. Griesinger, H. Schwalbe, *J. Biomol. NMR* **1998**, *12*, 223.
[14] R. Micura, M. Bolli, N. Windhab, A. Eschenmoser, *Angew. Chem.* **1997**, *36*, 870.
[15] H. A. Heus, A. Pardi, *Science* **1991**, *253*, 191.
[16] H. De Winter, E. Lescrinier, A. Van Aerschot, P. Herdewijn, *J. Am. Chem. Soc.* **1998**, *120*, 5381.
[17] Y. Maurinsh, J. Schraml, H. De Winter, N. Blaton, O. Peeters, E. Lescrinier, J. Rozenski, A. Van Aerschot, E. De Clercq, R. Busson, P. Herdewijn, *J. Org. Chem.* **1997**, *62*, 2861.
[18] C. Hendrix, H. Rosemeyer, I. Verheggen, F. Seela, A. Van Aerschot, P. Herdewijn, *Chem.–Eur. J.* **1997**, *3*, 110.
[19] P. Herdewijn, H. De Winter, B. Doboszewski, I. Verheggen, K. Augustijns, C. Hendrix, T. Saison-Behmoaras, C. De Ranter, A. Van Aerschot, 'Hexopyranosyl-Like Oligonucleotides', in 'Carbohydrate Modifications in Antisense Research', Eds. Y. S. Sanghvi and P. D. Cook, American Chemical Society Symposium Series 580, Washington D.C.; 1994, p. 80.

[20] C. Hendrix, H. Rosemeyer, B. De Bouvere, A. Van Aerschot, F. Seela, P. Herdewijn, *Chem.–Eur. J.* **1997**, *3*, 1513.

[21] I. Verheggen, A. Van Aerschot, S. Toppet, R. Snoeck, G. Janssen, J. Balzarini, E. De Clercq, P. Herdewijn, *J. Med. Chem.* **1993**, *36*, 2033.

[22] I. Verheggen, A. Van Aerschot, L. Van Meervelt, J. Rozenski, L. Wiebe, R. Snoeck, G. Andrei, J. Balzarini, P. Claes, E. De Clercq, P. Herdewijn, *J. Med. Chem.* **1995**, *38*, 826.

[23] D. Gorenstein, '^{31}P-NMR, Principles and Applications', Academic Press, New York, 1984.

[24] E. P. Nikonowicz, A. Pardi, *J. Mol. Biol.* **1993**, *232*, 1141.

[25] X.-J. Lu, W. K. Olson, 'On Comparative Conformational Analysis of Nucleic Acid Structures', poster present. 11th Conversation at Albany, June 15–19, **1999** (http://rutchem.rutgers.edu/ ~ xiangjun).

[26] A. A. Gorin, V. B. Zhurkin, W. K. Olson, *J. Mol. Biol.* **1995**, *247*, 34.

[27] M. S. Babcock, E. P. D. Pednault, W. K. Olson, *J. Mol. Biol.* **1994**, *237*, 125.

[28] M. A. el Hassan, C. R. Calladine, *J. Mol. Biol.* **1995**, *251*, 648.

[29] R. E. Dickerson, *Nucleic Acids Res.* **1998**, *26*, 1906.

[30] M. Bansal, D. Bhattacharyya, B. Ravi, *Comput. Appl. Biosci.* **1995**, *11*, 281.

[31] R. Lavery, H. Sklenar, *J. Biomol. Struct. Dynam.* **1988**, *6*, 63.

[32] D. J. States, R. A. Haberkorn, D. J. Ruben, *J. Magn. Reson.* **1982**, *48*, 286.

[33] P. Plateau, M. Guéron, *J. Am. Chem. Soc.* **1982**, *104*, 7310.

[34] M. Piotto, V. Saudek, V. Sklenar, *J. Biomol. NMR* **1992**, *2*, 661.

[35] M. Rance, O. W. Sørensen, G. Bodenhausen, G. Wagner, R. R. Ernst, K. Wüthrich, *Biochem. Biophys. Res. Commun.* **1983**, *117*, 479.

[36] A. Bax, D. G. Davis, *J. Magn. Reson.* **1985**, *65*, 355.

[37] J. Jeener, B. H. Meier, P. Bachmann, R. R. Ernst, *J. Chem. Phys.* **1979**, *71*, 4546.

[38] C. Griesinger, G. Otting, K. Wüthrich, R. R. Ernst, *J. Am. Chem. Soc.* **1988**, *110*, 7870.

[39] V. Sklenar, H. Miyashiro, G. Zon, H. T. Miles, A. Bax, *FEBS Letters* **1986**, *208*, 94.

[40] A. T. Brünger, 'X-PLOR, in A System for X-Ray Crystallography and NMR', Yale University Press, New Haven, CT, 1992.

[41] E. G. Stein, L. M. Rice, A. T. Brünger, *J. Magn. Reson.* **1997**, *124*, 154.

[42] R. M. Esnouf, *J. Mol. Graphics* **1997**, *15*, 132.

[43] P. J. Kraulis, *J. Appl. Crystallogr.* **1991**, *24*, 946.

[44] E. A. Merritt, D. J. Bacon, *Meth. Enzymol.* **1997**, *277*, 505.

[45] W. Saenger, 'Principles of Nucleic Acid Structure', Springer-Verlag, New York, 1984.

[46] E. Lescrinier, J. Schraml, R. Busson, R. Esnouf, H. Heus, C. Hilbers, P. Herdewijn, 'Structure Determination of a HNA · RNA Hybrid by NMR Spectroscopy', in preparation.

Nonenzymatic Oligomerization Reactions on Templates Containing Inosinic Acid or Diaminopurine Nucleotide Residues

by Igor A. Kozlov and Leslie E. Orgel*[1])

The Salk Institute for Biological Studies, P.O. Box 85800, San Diego, CA, 92186, USA

The template-directed oligomerization of nucleoside-5′-phosphoro-2-methyl imidazolides on standard oligonucleotide templates has been studied extensively. Here, we describe experiments with templates in which inosinic acid (I) is substituted for guanylic acid, or 2,6-diaminopurine nucleotide (D) for adenylic acid. We find that the substitution of I for G in a template is strongly inhibitory and prevents any incorporation of C into internal positions in the oligomeric products of the reaction. The substitution of D for A, on the contrary, leads to increased incorporation of U into the products. We found no evidence for the template-directed facilitation of oligomerization of A or I through A – I base pairing. The significance of these results for prebiotic chemistry is discussed.

Introduction. – Because the nucleic acids occupy a central position in biochemistry, studies of template-directed synthesis have concentrated on the oligomerization of activated derivatives of the standard nucleotides, U, A, C, and G [1–7], although a few experiments with 2,6-diaminopurine nucleotide [8][9] and other nucleotide analogues [10] have been reported. Oligomerization reactions involving inosinic acid (I) or 2,6-diaminopurine nucleotide (D), although they have limited application to biochemistry, are of considerable interest for prebiotic chemistry, since potentially prebiotic syntheses that yield adenine and guanine typically also lead to the formation of hypoxanthine and 2,6-diaminopurine [11]. Studies of template-directed chemistry might help to explain why A and G were chosen as components of RNA (DNA) while I and D were excluded. It has been suggested that a primitive genetic system might have been based on A – I pairing [12]. Template-directed reactions of activated A-derivative on I-containing templates and *vice versa* might throw light on this hypothesis. Here, we investigate template-directed synthesis in a variety of relevant systems involving I and D, and compare our results to those obtained using the standard bases.

Results. – *Oligomerization of Activated Mononucleotides on C_4XC_4 (X = G, I, A, or D) DNA Templates (Fig. 1,b).* The presence of a C_4XC_4 (X = G, I, A, or D) DNA template leads to oligomerization of guanosine-5′-(2-methylimidazol-1-yl phosphate) (2-MeImpG) (*Fig. 1,a*; X′ = G) alone to give G_3 and small amounts of G_4 (data not shown). An equimolar mixture of 2-MeImpG and 2-MeImpC in the presence of a C_4GC_4 template leads to generation of G_4CG_n products up to at least octamers (*Fig. 2,b*) but, in the presence of a C_4IC_4 template, gives no products longer than tri-

---

[1])   Tel.: (858)453-4100; fax: (858)558-7359; e-mail: orgel@salk.edu.

and tetramers (*Fig. 2,a*). An equimolar mixture of 2-MeImpG and 2-MeImpU in the presence of a C_4AC_4 template leads to formation of G_4UG_n products up to nonamers (*Fig. 2,c*). This reaction becomes about three times more efficient when C_4DC_4 is substituted for C_4AC_4 (*Fig. 2,d*).

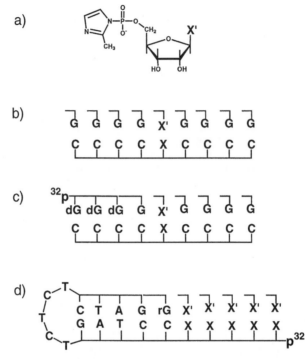

Fig. 1. a) *Structure of activated nucleoside 5′-phosphates* 2-MeImpX′ (X′ = G, C, A, I, or U). b) *Schematic representation of oligomerization of activated monomers* 2-MeImpG *and* 2-MeImpX′ (X′ = C or U) *on a* C_4XC_4 *template* (X = G, I, A, or D). c) *Schematic representation of the primer* $^{32}p(dG)_3G$ *extension reaction with* 2-MeImpG *and* 2-MeImpX′ (X′ = C or U) *on a* C_4XC_4 *template* (X = G, I, A, or D). d) *Schematic representation of the oligomerization of* 2-MeImpX′ (X′ = C, U, A, or I) *on a deoxyribonucleotide hairpin template* 5′-*XXXXXCCTAGTCTCTCTAGrG-3′* (X = D, I, or A).

Extension of a $^{32}p(dG)_3G$ *Primer on* C_4XC_4 (X = G, I, A, or D) *DNA Templates* (*Fig. 2,c*). The product distributions in the reactions of ^{32}P-labeled p(dG)$_3$G with 2-MeImpG, or an equimolar mixture of 2-MeImpG with 2-MeImpC or 2-MeImpU on C_4XC_4 (X = G, I, A, or D) DNA templates are shown in *Fig. 3*. Significant extension of the primer p(dG)$_3$G with 2-MeImpG alone does not take place on any of the DNA templates (*Fig. 3, Lanes 1, 3, 5,* and *7*). The extension of the primer p(dG)$_3$G with an equimolar mixture of 2-MeImpG and 2-MeImpC in the presence of a C_4GC_4 DNA template leads to conversion of more than 80% of the primer to p(dG)$_3$GCG$_n$ ($n = 1$–4) products (*Fig. 3, Lane 2*). However, the same reaction on a C_4IC_4 DNA template leads to conversion of only *ca.* 30% of the primer to a p(dG)$_3$GC product, and does not yield any longer products (*Fig. 3, Lane 4*). The extension of the primer p(dG)$_3$G with an equimolar mixture of 2-MeImpG and 2-MeImpU on a C_4AC_4 DNA template leads to conversion of less than 5% of the primer mainly to a p(dG)$_3$GUG$_3$ product (*Fig. 3,*

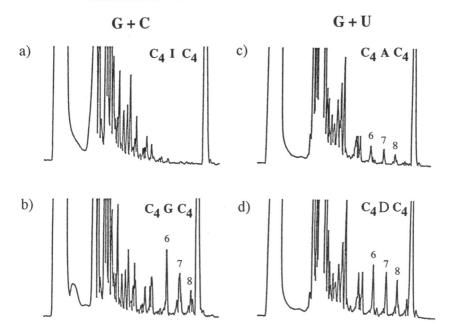

Fig. 2. *Elution profiles from an RPC5 column of the products from the oligomerization of an equimolar mixture of 2-MeImpG and 2-MeImpC* (G + C), *or an equimolar mixture of 2-MeImpG and 2-MeImpU* (G + U) a) *on a C_4IC_4 template,* b) *on a C_4GC_4 template,* c) *on a C_4AC_4 template,* d) *and on a C_4DC_4 template.* The reaction time was 14 days. The numbers above the peaks indicate the length of the all 3′-5′-linked oligoribonucleotide products.

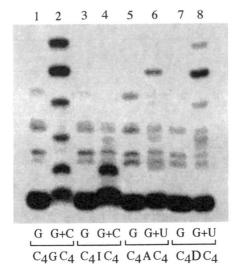

Fig. 3. *Extension of a $^{32}p(dG)_3G$ primer on C_4XC_4 DNA templates* (X = G, I, A, or D) *after 5 days.* G represents 2-MeImpG, G + C represents an equimolar mixture of 2-MeImpG and 2-MeImpC, G + U represents an equimolar mixture of 2-MeImpG and 2-MeImpU. The fastest-moving band in the diagram corresponds to the $^{32}p(dG)_3G$ primer.

Lane 6). The same reaction on a C_4DC_4 DNA template leads to conversion of more than 10% of the primer to $p(dG)_3GUG_3$ and $p(dG)_3GUG_4$ products in a ratio of *ca.* 10:1 (*Fig. 3, Lane 8*).

Oligomerization Reactions of 2-MeImpU and 2-MeImpC on Templates Containing Several D or I Residues. We studied oligomerization reactions of 2-MeImpU and 2-MeImpC on oligodeoxynucleotide D_{10} and I_{10} templates, respectively. We also studied 'primer extension' reactions in oligodeoxynucleotide hairpin templates, 5'-XXXXXCCTAGTCTCTCTAGrG-3' (*Fig. 1,d*; X = D or I, X' = U or C) that are 3'-terminated with a single ribonucleotide [6][7]. No template-directed facilitation of oligomerization was detected.

Oligomerization Reactions of 2-MeImpA and 2-MeImpI on Templates Containing Several I or A Residues. We studied in some detail the oligomerization reactions of 2-MeImpA and 2-MeImpI on oligodeoxynucleotide I_{10} and A_{10} templates, respectively. We also used oligodeoxynucleotide hairpin templates (*Fig. 1,d*, X = I or A, X' = A or I) and oligodeoxynucleotide hairpin templates containing A–I base pairs (*Fig. 4*) to explore template-directed primer extension. In no case were we able to detect an effect of the template on oligomerization or primer extension.

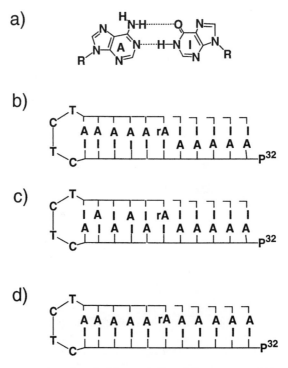

Fig. 4. a) *Schematic representation of an A–I base-pair.* Dotted lines represent H-bonds. b)–d) Schematic representation of attempted oligomerizations of 2-MeImpA or 2-MeImpI on deoxyribonucleotide hairpin templates containing A–I base pairs.

Discussion. – The results obtained in this study imply that replacement of the A – U base pair by the D – U base pair improves the efficiency of nonenzymatic template-directed oligomerization reactions. In some cases, D – U base pairs support reactions that are comparable in efficiency to those involving G – C base pairs (compare *Fig. 2,b* and *d*). The difference between A – U and D – U pairs is probably attributable to the presence of three H-bonds in the D – U pair compared with only two H-bonds in the A – U pair (*Fig. 5*). In this respect, the D – U pair resembles a G – C pair, both pairs having three H-bonds.

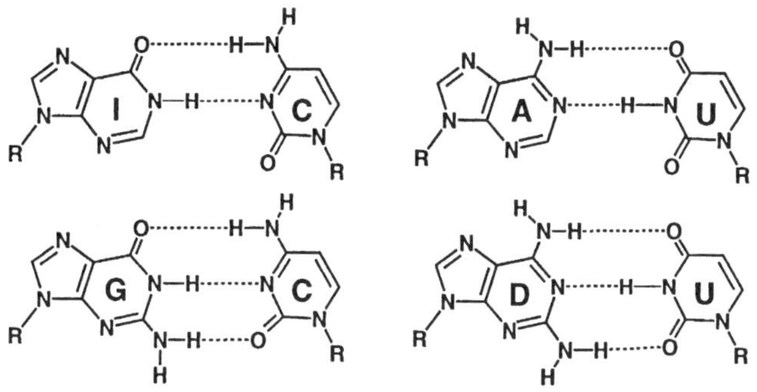

Fig. 5. *Schematic representation of base-pairings I – C, G – C, A – U, and D – U.* Dotted lines represent H-bonds.

The substitution of a G – C base pair by an I – C pair leads to a very large decrease in oligomerization efficiency. The I – C pair is much less efficient than the A – U pair in facilitating oligomerization (*cf. Fig. 2,a,* and *c; Fig. 3, Lanes 4* and *6*), although both are held together by two H-bonds. Our results show that the I – C base pair must adopt a conformation that inhibits template-directed synthesis strongly, especially at the stage when a primer is terminated by a C residue opposite I and needs to be extended by the addition of a G residue. We do not understand the structural basis of this inhibition. Our attempts to oligomerize 2-MeImpU and 2-MeImpC on D_{10} and I_{10} templates were unsuccessful, as were analogous attempts at primer-extension with oligodeoxyribonucleotide hairpin templates (*Fig. 1,d,* X = D or I, X' = U or C).

Our results confirm that the G – C base pair is exceptional in providing the necessary conformation for efficient nonenzymatic RNA synthesis using 2-methyl imidazolides of nucleoside 5'-phosphates as substrates. Substitution of I for G in this reaction leads to poor incorporation of C and negligible extension of the resulting primer terminated by C. It is possible that a different activated derivative of C might polymerize efficiently, but our results suggest that the absence of I from replicating nucleic acids may have a basis in the conformation of double helices containing I – C base pairs.

A similar explanation to that given above cannot account for the exclusion of D from nucleic acids. Our results suggest that the replacement of A by D would lead to more efficient synthesis (*cf. Fig. 2,c* and *d; Fig 3, Lanes 6* and *8*), so the choice of A rather than D is likely to reflect factors other than efficiency of replication. Availability in the prebiotic environment is one possibility. Alternatively, optimization rather than maximization of the stability of double-helical RNA may have led to the selection of A rather than D.

Purine nucleosides can be synthesized more easily than pyrimidine nucleosides under prebiotic conditions [11]. Consequently, it has been suggested that purine–purine, A–I, pairing may have made possible the development of the first nucleic acid genetic system [12]. Our extensive efforts to demonstrate facilitation of adenosine nucleotide oligomerization on templates containing I, and *vice versa*, like several less complete earlier studies in our laboratory, have failed. We cannot exclude the possibility that experiments using a different form of activation would succeed, but our results do not provide any evidence supporting A–I pairing as a mechanism of complementary replication. They argue, although not conclusively, against the hypothesis of a genetic system based on A–I pairing.

Experimental Part

Unless otherwise noted, all chemicals were reagent grade, were purchased from commercial sources and used without further purification. Nucleoside 5'-(2-methylimidazol-1-yl phosphates) (2-MeImpX', X' = G, C, A, U, I) were synthesized by a published method in at least 95% yield [13]. The oligodeoxyribonucleotides were synthesized and purified as previously described [14]. 2-Amino-2-deoxyadenosine-β-cyanoethyl phosphoramidite and deoxyinosine-β-cyanoethyl phosphoramidite (*Glen Research*) were used under standard conditions to introduce D or I residues into oligodeoxyribonucleotides.

Reaction conditions for the oligomerization of 2-MeImpG (or its mixture with an equal amount of 2-MeImpC or 2-MeImpU) on DNA C_4XC_4 (X = G, I, A or D) templates were chosen to permit comparison with earlier published work [5][15]. Reactions were run at 0° for 14 days in 0.2M 2,6-lutidine-HCl buffer (pH 7.9 at 25°) containing 1.2M NaCl, 0.2M $MgCl_2$, and 0.5 mM of a template. In one set of reactions, the soln. also contained 0.1M 2-MeImpG. In another set, the soln. contained 0.1M 2-MeImpC and 2-MeImpG with the C_4GC_4 or C_4IC_4 template, or 0.1M 2-ImpU and 0.1M 2-MeImpG with the C_4AC_4 or C_4DC_4 templates. The same conditions were used for reactions on oligodeoxynucleotide I_{10}, D_{10}, and A_{10} templates; the concentration of activated substrate was always 0.1M. The reaction mixtures were analyzed by HPLC on an *RPC5* column as previously described [15].

Reaction conditions for p(dG)$_3$G primer extension reactions on DNA C_4XC_4 (X = G, I, A, or D) templates were again chosen to permit comparison with earlier published work [5–7][15]. The reactions were incubated for 5 days at 0° in 0.2M 2,6-lutidine buffer (pH 7.9 at 25°) containing 1.2M NaCl, 0.2M $MgCl_2$, 20 μM of a template, and 20 nM of the primer. In one set of reactions, the solns. also contained 50 mM 2-MeImpG. In a second set of reactions, the soln. contained not only 50 mM 2-MeImpG but also 50 mM of 2-MeImpC with C_4GC_4 and C_4IC_4 templates or 50 mM of 2-MeImpU with C_4AC_4 and C_4DC_4 templates. The same conditions were used in experiments with DNA hairpin templates (*Fig. 1,d*; X = A, D, I; *Fig. 4*). In these experiments, the concentration of template was 20 μM and the concentration of the corresponding activated monomer, 50 mM. The reaction mixtures were analyzed by electrophoresis in 20% PAG containing 8M urea as previously described [14].

We thank *Aubrey R. Hill, Jr.* for technical assistance and *Bernice Walker* for manuscript preparation. This work was supported by grants from *NASA* NAG5-4118 and NAG5-4546 NSCORT/Exobiology.

REFERENCES

[1] A. Kanavarioti, *J. Org. Chem.* **1998**, *63*, 6830.
[2] A. R. Hill Jr., L. E. Orgel, T. Wu, *Origins Life Evol. Biosphere* **1993**, *23*, 285.
[3] M. Kurz, K. Göbel, C. Hartel, M. W. Göbel, *Angew. Chem., Int. Ed.* **1997**, *36*, 842.
[4] M. Kurz, K. Göbel, C. Hartel, M. W. Göbel, *Helv. Chim. Acta* **1998**, *81*, 1156.
[5] J. G. Schmidt, P. E. Nielsen, L. E. Orgel, *Nucl. Acid Res.* **1997**, *25*, 4797.
[6] T. Wu, L. E. Orgel, *J. Am. Chem. Soc.* **1992**, *114*, 7964.
[7] T. Wu, L. E. Orgel, *J. Am. Chem. Soc.* **1992**, *114*, 5496.
[8] T. R. Webb, L. E. Orgel, *Nucl. Acids Res.* **1982**, *10*, 4413.
[9] K. Grzeskowiak, T. R. Webb, L. E. Orgel, *J. Mol. Evol.* **1984**, *21*, 81.
[10] H. Rembold, R. K. Robins, F. Seela, L. E. Orgel, *J. Mol. Evol.* **1994**, *38*, 211.

[11] S. L. Miller, L. E. Orgel, 'The Origins of Life on the Earth', Prentice-Hall, Inc., Englewood Cliffs, New Jersey 1974.

[12] F. H. C. Crick, *J. Mol. Biol.* **1968**, *38*, 367.

[13] G. F. Joyce, T. Inoue, L. E. Orgel, *J. Mol. Biol.* **1984**, *176*, 279.

[14] I. A. Kozlov, S. Pitsch, L. E. Orgel, *Proc. Natl. Acad. Sci. U.S.A.* **1998**, *95*, 13448.

[15] I. A. Kozlov, B. De Bouvere, A. Van Aerschot, P. Herdewijn, L. E. Orgel, *J. Am. Chem. Soc.* **1999**, *21*, 5856.

Index

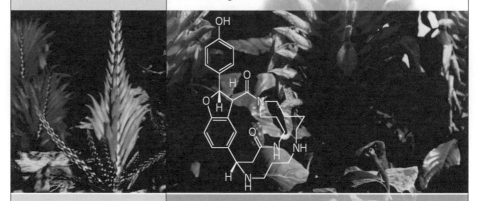
 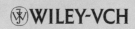